FRONTIERS OF MOLECULAR SPECTROSCOPY

T0348832

FRONTIERS OF MOLECULAR SPECTROSCOPY

Edited by

JAAN LAANE

Professor of Chemistry and Professor of Physics at
Texas A & M University,
College Station, Texas 77843-3255 USA

ELSEVIER

Amsterdam • Boston • Heidelberg • London • New York • Oxford
Paris • San Diego • San Francisco • Singapore • Sydney • Tokyo

Elsevier
Radarweg 29, PO Box 211, 1000 AE Amsterdam, The Netherlands
Linacre House, Jordan Hill, Oxford OX2 8DP, UK

First edition 2009

British Library Cataloguing in Publication Data
A catalogue record for this book is available from the British Library

Library of Congress Cataloging-in-Publication Data
A catalog record for this book is available from the Library of Congress

ISBN: 978-0-444-53175-9

For information on all Elsevier publications
visit our web site at books.elsevier.com

Printed and bound by CPI Group (UK) Ltd, Croydon, CR0 4YY
Transferred to Digital Print 2011

Contents

4 Vibrational Potential Energy Surfaces in Electronic Excited States 63
Jaan Laane

5 Raman Spectroscopy in Art and Archaeology: A New Light on Historical Mysteries 133
Howell G.M. Edwards

6 Real-Time Vibrational Spectroscopy and Ultrafast Structural Relaxation 175
Takayoshi Kobayashi

16 Raman Spectroscopy of Viruses and Viral Proteins 553

Daniel Němeček and George J. Thomas, Jr.

17 Vibrational Spectroscopy via Inelastic Neutron Scattering 597

Bruce S. Hudson

PREFACE

Although Aristotle recognized that light was necessary for color to exist and Aristophanes made use of lenses before 400 BC, the origins of spectroscopy are perhaps best traced back to Newton's observation of a spectrum obtained from a prism in 1666. A dozen years later, Huygens proposed the wave theory of light. Other developments in optics, physics, and chemistry ensued over the next century, but it was not until 1800 that F. W. Herschel recognized the existence of infrared radiation and 1801 that J. W. Ritter discovered the presence of ultraviolet light. In the next quarter of the century, during which John Dalton proposed the atomic theory, W. H. Wallaston and Josef Fraunhofer observed specific absorption lines, and W. H. F. Talbot recognized that different salts yield different colors in flames. In the middle of the 19th century, many researchers including Bequerel, Draper, Kirchoff, Bunsen, and Angstrom contributed extensively to experimental data and were able to correlate spectral properties to the nature of matter, and in 1873, J. C. Maxwell's equations unifying electricity and magnetism were revealed. In 1885, a Swiss high-school teacher, J. J. Balmer, finally achieved a major spectroscopic breakthrough by showing that the wavelengths of the visible spectral lines of atomic hydrogen could be predicted by a mathematical formula. Rydberg and Ritz, among others, soon developed a more universal expression for explaining the atomic spectrum of hydrogen ranging from the infrared region to the ultraviolet region. It was clearly demonstrated that each observed spectral frequency was determined from the difference of two fixed quantities. In 1913, Niels Bohr devised a model of the hydrogen atom based on quantum theory, and this laid the groundwork for understanding atomic spectroscopy.

The pioneering work of Planck, Einstein, de Broglie, Heisenberg, and others finally led to Schrödinger's discovery of the wave equation in 1926, which today serves not only as the basis for understanding atomic and molecular structure but also as the basis of molecular spectroscopy. Work of R. W. Wood in the United States and A. Smekal in Germany helped pave the way for C. V. Raman's discovery of the effect bearing his name in 1928. A few years later in 1933, C. E. Gleeton and N. H. Williams carried out the first microwave absorption experiments. Meanwhile, infrared absorption had been evolving through the work of William Coblentz, H. M. Randall, Gerhard Herzberg, and others, and by 1950, it had become a major tool for qualitative and quantitative chemical analysis, thanks in part to commercial instruments developed by the Beckman and Perkin-Elmer companies.

During the last half of the 20th century, the development of both lasers and computers immensely helped to advance the field of molecular spectroscopy. C. Townes, A. L. Schalaw, and T. H. Maiman developed the laser in 1958, and this has not only led to the field of Raman spectroscopy flourishing but also to innovative work in fluorescence spectroscopy and a variety of nonlinear techniques. The commercial availability of Fourier transform infrared (FT-IR) spectrometers from about 1970, based on the Michelson interferometer (1880) and Peter Fellgett's theory (1949), rapidly advanced the field of infrared spectroscopy, including near-infrared. Commercial instruments, by Bomem for example, could cover the range from 3 to 50,000 cm^{-1}, well into the ultraviolet region.

During the 20th century, 22 scientists received Nobel Prizes that can be directly or indirectly related to spectroscopy. Most of them are well-known names, and many have had effects or theories named after them. These include A. A. Michelson, H. A. Lorentz, P. Zeeman (magnetic splitting), J. Stark (electric splitting), C. V. Raman, N. E. Lamb, R. S. Mulliken, G. Herzberg, C. H. Townes, A. L. Schawlow, N. Blombergen (nonlinear optics), and A. Zewail (chemical dynamics using femtosecond lasers).

As is evident from all of this, spectroscopy is arguably the one most important tool which has taught us the most about the nature of atoms and molecules, much of this coming in the 20th century. As we now proceed through the 21st century, it is appropriate to examine where we are today in the development of spectroscopic tools that can provide us further insight in a way that only spectroscopy can do. This is the purpose of this book; we have 19 chapters, written individually by eminent scientists in selected areas of spectroscopy.

The book starts off with a chapter by Sir Harry Kroto, Nobel Laureate in 1996 for his work on buckyballs. Although his research has at times extended well into the ultraviolet region, much of his research (and his chapter) has been devoted to microwave spectroscopy. In later chapters, Walther Caminati and Jens-Uwe Grabow describe the methods and results from microwave spectroscopy in the 21st century. Several chapters are devoted to recent work using Raman spectroscopy. Hiro-o Hamaguchi describes several applications including the study of living and dying yeast cells. George Thomas and Daniel Nemecek discuss Raman studies of viruses and viral proteins, and H. G. M. Edwards contributes a colorful presentation of the use of Raman to investigate art and archeology. Surface-enhanced Raman spectroscopy (SERS) is covered in two chapters. T. Itoh, A. Sujith, and Y. Ozaki cover a range of studies, whereas S. Schlücker and W. Kiefer focus on the detection of biomolecules. In other Raman investigations, D. Pestov, A. Sokolov, and M. Scully discuss applications to the detection of bacterial spores. Jaan Laane's chapter describes how several spectroscopic techniques, including fluorescence excitation and ultraviolet absorption spectroscopies, are utilized to determine potential energy surfaces and molecular structures in electronic excited states. Cavity ring-down spectroscopy is described in depth by Kevin Lehmann, a pioneer in developing this high-sensitivity technique, and by H. Huang. The cutting edge research on high Rydberg states, much of it at the ETH in Zürich, is covered nicely by Martin Schäfer and Frederic Merkt. The rapidly developing area of two-dimensional correlation spectroscopy is described by Isao Noda, a leading investigator in that area. There have also been many new developments in terahertz spectroscopy, and S. Schlemmer, T. Giesen, and T. Lewen describe astrophysical applications of this technique. In the time-resolved area, Takayoshi Kobayashi elaborates on the use of sub-femtosecond spectroscopy for molecular systems. Other work of a dynamical nature is discussed by C. Y. Ng, whose research has concentrated on photoionization and photoelectron methods in the infrared and vacuum ultraviolet regions. The two leading theorists in the area of symmetry, Philip Bunker and Per Jensen, present an insightful journey into spectroscopy and broken symmetry. Bruce Hudson's chapter on vibrational studies using neutron scattering demonstrates that photons are not necessarily required for these studies, while physicists A. M. Burzo and A. V. Sokolov show how broadband modulation of light can be achieved by coherent molecular oscillations.

It is clear that the 20th century was a period of unprecedented advances in molecular spectroscopy. However, there is much yet left to be learned. Hopefully, this book helps to present a broad perspective of the many avenues of work underway in the 21st century. From these and other spectroscopic studies, we will continue to understand our universe at the molecular level even better.

As the editor of this volume, I thank all the authors for their excellent contributions and also acknowledge with gratitude the able assistance of Linda Redd in producing the final copy.

Jaan Laane
College Station, TX
January 2008

Acknowledgments

First and foremost, I wish to thank the 30 authors of the 19 individual chapters for their excellent contributions. It is a delight to see so much spectroscopy gathered together in one volume. I also wish to thank my own students, post docs, and visiting professors, past and present, for all of their achievements in our laboratories over the years. Last but not least, I would like to thank Linda Redd for her efficient and outstanding assistance in working with all of the authors and helping to get the material into its final form.

Jaan Laane
July 2008

CHAPTER 1

Old Spectroscopists Forget a Lot but They do Remember Their Lines

Harold Kroto

Contents

Abstract

In this chapter, a personal perspective is presented on the fascinatingly wide impact of molecular spectroscopy – in particular microwave (rotational) spectroscopy – on an amazingly wide range of key areas in the sciences. In addition to its fundamental role in revealing a wealth of detailed intrinsic information about molecular structure from bond lengths to bond angles, it has also revealed intimate details of molecular dynamics, including internal rotation parameters and the effects of quantum mechanical tunnelling. No less important has been its more general contributions, where it has led to the development of new areas of mainline chemistry as well as the birth of the field of interstellar chemistry and our understanding of how stars start to form. The impetus for the experiments that uncovered the existence of the fullerenes can be traced back to microwave spectroscopy studies, which had initiated radio astronomy detection of carbonaceous molecules in space. It is also interesting to note that the inventor of the maser/laser and one of the proponents of the leading theory of superconductivity were pioneers in molecular microwave spectroscopy.

Keywords: molecular spectroscopy; rotational spectroscopy; molecular structure; microwave spectroscopy; molecular dynamics; astrophysical chemistry

1. Introduction

Spectroscopy is arguably (and I would argue it) the most fundamental of the experimental physical sciences. After all we obtain most of our knowledge through our eyes and it is via the quest for an in-depth understanding of what light is, and what it can tell us, that all our true understanding of the universe has been obtained.

Answers to questions about light have led to many of our greatest discoveries, not least our present description of the way almost everything works both on a macroscopic and a microscopic scale. In the deceptively simple question of why objects possess colour at all – such an everyday experience that few think it odd – lies the seed for the development of arguably our most profound and far-reaching theory – quantum mechanics. The realisation that colour is not explainable classically was one catalyst of quantum mechanics and quantum electrodynamics. The Theory of Relativity also was born out of observations on the subtle behaviour of light with very deep consequences. Few scientific quests have been of more fundamental significance than those involving the corpuscular and wave nature of electromagnetic radiation that Newton and Huygens discussed which, together with the work on electricity and magnetism of Ampère and Faraday, led to the unification that Maxwell uncovered when he showed that electromagnetic interactions travelled at the velocity of light. A beautiful set of four lectures by Richard Feynman on this subject was recorded at the University of Auckland in New Zealand (1979) and can be viewed streaming from the Vega Science Trust website (www.vega.org.uk) [1]. The breakthrough that, more than any other, signified the birth of not only astrophysics but also spectroscopy in general lies in the work on the dark sharp absorption lines in the solar spectrum that Joseph von Fraunhofer realised could be identified with atoms of elements known on Earth, such as sodium. It often amuses me to think that the NaD doublet was labelled "D" only because – presumably – it was the fourth feature to be labelled by Fraunhofer. Amazingly, spectroscopy allows us to carry out chemical analysis on objects not just millions of miles away but billions of light years away.

The original quantum mechanical work of Heisenberg indicates that he and others such as Max Born were attempting, among other things, to understand the quantum harmonic oscillator – the fundamental mechanical system in engineering as well as classical and quantum physics [2]. Once these and other fundamental quantum issues had been resolved, almost the whole of the modern description of world followed – at least in the more-or-less non-relativistic regime. Few concepts capture the imaginations of non-scientists, no less than scientists, than the Big Bang Theory of the origin of our universe which finds its major support in spectroscopic observations such at the (Red) frequency shifts observed for atomic lines in stellar spectra and of course the blackbody contour of the microwave background.

2. MICROWAVE SPECTROSCOPY: A FUNDAMENTAL ANALYSIS

This chapter concentrates primarily on various aspects of spectroscopy – mainly microwave – and astrophysical spectroscopy from a personal perspective of the way it has impacted on other areas of the science in ways that I find interesting. This perspective developed over a long period, at one stage encapsulated in a monograph in 1974 [3] and then in 1981 in a review [4]. The astrophysical perspective was also encapsulated in a review some time ago [5] but more recently in a set of seminars streaming from the website www.vega.org.uk [6]. This chapter is not by any means meant to be a complete survey and I apologise if many very important advances have been omitted – partly because my perspective is limited and partly because I am not as close to the field as I once was. It is somewhat ironic for me that the one field that I vaguely understood required that the molecules to be studied possess dipole moments and I was dragged away from the field screaming by a spheroidal molecule that has almost no useful moments at all – let alone a dipole moment. I had first become fascinated by molecular spectroscopy while studying the visible spectrum of a Bunsen burner flame through a spectroscope. I still have my practical book containing my Birge–Sponer plot for the C_2 radical. I became completely enamoured in general during an undergraduate lecture course on molecular spectroscopy by Richard Dixon at Sheffield University. I was introduced to elegant rotational branch structures which indicated that molecules could count. From there I was led on to carry out research on the spectra of small free radicals produced, detected and studied by flash photolysis and electronic spectroscopy, the technique pioneered by George Porter, who was then the Professor of Physical Chemistry at Sheffield University. Dixon had built a large Eagle grating spectrometer to detect the spectra at what was then "very" high resolution just as the first rumblings of the possibilities of using something called an "optical maser" to carry out spectroscopy could be heard. In 1964 I went on to the National Research Council (NRC) in Ottawa, where Gerhard Herzberg and Alec Douglas had created the legendary "Mecca" of spectroscopy. At first I worked with Don Ramsay and then I discovered microwave spectroscopy and transferred to Cec Costain's group; from that moment on the future direction of my career as a researcher was sealed. I gained a high degree of satisfaction from measurements made at high resolution on the rotational spectra of molecules and in particular from the ability to fit the frequency patterns with theory to the high degree of accuracy that this form of spectroscopy offered. At that time tunable lasers were not available and UV and IR techniques just did not compare in attainable resolution. Great intellectual satisfaction comes from knowing that the parameters deduced – such as bond lengths, dipole moments, quadrupole and centrifugal distortion parameters – are well-determined quantities both numerically and also in a physically descriptive sense. Some sort of odd schizoid understanding seems to develop as one gains more and more facility (rather than familiarity?) with the abstract quantum mechanical (mathematical) approaches to spectroscopic analyses. The quantitative perspective adds to the (subliminal?) classical physical descriptions needed to convince oneself that one really does know what is going on. One thing "fundamentalist" spectroscopists (microwavers?) can say is, "We may not know much, but what we know we really do know – a lot about very little – things". I was to learn later that such levels of satisfying certainty of knowledge are a rarity in many other branches of science and in almost all areas of life in general. It gives one a very clear view of how the scientific mindset develops and what makes science different from all other professions and within the sciences a clear vision of what it means to really "know" something.

The equations of Kraitchman [7] and the further development of their application by Costain [8] to isotopic substitution data in the 1960s resulted in a wealth of structural information on small- to moderate-sized molecules from microwave measurements on rotational structure. At Sussex in 1974, my colleague David Walton and I put together a project for an undergraduate researcher, Andrew Alexander, to synthesise some long(ish) linear carbon chain species starting with HC_5N and study

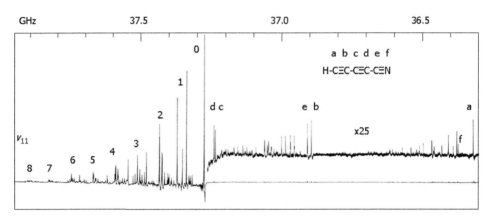

Figure 1   The microwave spectrum of HC_5N. To the RHS of the ground vibrational state line can be seen the weak spectra, in natural abundance, of the ^{13}C ($\sim$1%) and ^{15}N ($\sim$0.3%) isotopic modifications. These satellites are also depicted enhanced in relative intensity by a factor of $\sim \times 25$.

their spectra – IR and NMR as well as microwave [9]. Of key interest for me was a study of the low-frequency bending vibration. One of Alex's rotational bands is shown in Figure 1 [3,9], where a particularly neat phenomenon can be semi-quantitatively observed. Kraitchman and Costain [6,7] showed that for a linear molecule, with moment of inertia I, on substitution (new moment of inertia I^*) the distance of a substituted atom r^* from the centre of mass of the unsubstituted molecule is governed by the relation $I^* - I = \mu r^{*2}$, where $\mu = \Delta m M/(\Delta m + M)$, with M being the total mass and Δm the increase in mass on substitution. As the rotational constant $B \propto I^{-1}$ and BB^* is almost constant especially for ^{13}C and ^{15}N substitutions, where the increase is only 1 AU, the change in B value on substitution follows the relation $\Delta B \propto r^{*2}$ quite closely. The centre of mass of

$$H-C_a{\equiv}C_b-C_c{\equiv}C_d-C_e{\equiv}N$$

lies near the midpoint of the $C_c{\equiv}C_d$ triple bond and so substitution of ^{13}C at C_c and C_d gives rise to the smallest shift; substitutions at C_b and C_e yield shifts roughly $3^2 \times$ larger, and ^{13}C and ^{15}N substitutions at the C_a and the N atom positions respectively give shifts roughly $5^2 \times$ larger. The spectra of the ^{13}C- and ^{15}N-substituted species can be observed at natural abundance in Figure 1. In such simply appreciated effects lie some of the greatest delights in science [3,4,9].

The most personally satisfying research breakthrough was our detection of $CH_2{=}PH$, the first (smallest and simplest) phosphaalkene [4,10,11]. It was this study and the allied study on $MeC{\equiv}P$ [12] that essentially led to the birth of the fields of phosphaalkene and phosphaalkyne chemistry. The fact that these fields were born out of a series of microwave studies (carried out originally with Nigel Simmons and John Nixon), which started in 1974, has been lost in the mists of time. The massive catalogue of fundamental structural data on small- to medium-sized molecules is used with scarcely a thought about how hard it has been to assemble. Alas it seems it is the lot of the microwave community to be so little appreciated! It is also worth noting that our studies [4] were facilitated by the synergistic effects of combining microwave with photoelectron experiments in tandem. One or other of these techniques often enabled detection of a given species more readily than the other and this facilitated study by the technique presenting more difficulties.

Microwave measurements can, however, yield other important molecular quantities. Wilson, Herschbach and co-workers [13,14], Nielsen [15,16], and Dennison and co-workers [17] showed how one could extract internal rotational barriers [3]. Centrifugal distortion parameters can be analysed to extract vibrational force field data, and quadrupole moment splittings can yield bond electron density properties [3]. Early on in my career I had wondered about the spectrum of acetylene studied by Ingold and King [18] and the way in which shape changes might affect the spectrum. I wondered about how one applied group theory in this case where molecular structure changed from linear to trans-bent on electronic excitation. Later I started to learn about quasi-linearity and quasi-planarity. It was at about this time that Richard Dixon, still then my PhD supervisor, realised that puzzling patterns in the electronic spectra of species such as NH_2 and PH_2 could be explained if there were a hump in the large amplitude bending potential [19]. As a research student, I did attempt to prepare carbon suboxide, which was known to have an exceptionally low vibrational frequency in an aborted attempt to study its quasi-linearity. With my first PhD student at Sussex, Terry Morgan, I studied the interesting case of C_3, which has a very low 63 cm^{-1} bending frequency in the $^1\Sigma$ ground electronic state and a $^1\Pi$ Renner-split excited state with "standard" roughly harmonic linear molecule bending frequency. I had taken some plates of the extensive C_3 electronic spectrum while a postdoctoral fellow at NRC and was attracted by the unique possibility of using the Frank Condon factors for the copious number of vibronic transitions to probe the potential surface to high energies. Morgan's ground-breaking vibronic treatment [20] – for some reason unpublished in the standard literature – was superceded by the elegant and more complete ro-vibronic study, some 10 years later, by Jungen and Merer [21].

At Sussex much later we obtained a microwave spectrum that afforded great delight and a unique visual insight into quasi-linearity. This was in the microwave spectrum of NCNCS shown in Figure 2 which Mike King and Barry Landsberg studied [22,23]. The fact that this spectrum was discovered accidentally when $S(CN)_2$ was simply thermolysed by passage through a hot quartz tube added significantly to the pleasure. Its spectrum reveals a fascinating degree of insight into the "meaning" of quasi-linearity. When the bending vibrational assignments of the lines have been disentangled, as indicated by the branch markers above the spectrum, one sees that the A-type R branch rotational patterns for the ground vibrational state exhibits a pattern fairly characteristic of a bent system. The pattern is that of an asymmetric top in which the $K_A = 1$ lines are split apart by roughly $(B - C)(J + 1)$ and the $K = 0$ and $K_A = 2$ lines lie in a bunch, roughly midway between them. As the bending vibration becomes more and more excited, the pattern changes quite quickly such that when four quanta are excited, the pattern has metamorphosed into one much more characteristic of a linear molecule. The rotational spectrum at high v is now more usefully described in terms of a vibrational angular momentum quantum number l (rather than K_A). The $l = 0$ line lies to low frequency and $l \neq 0$ lines march out in a sequence to high frequency, with the $l = 1$ lines split apart

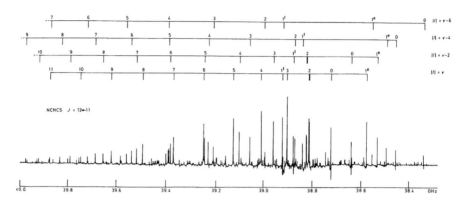

Figure 2 The microwave spectrum of NCNCS depicted together with lead markers that identify the rotational lines belonging to the states with $v = 0, 1, 2$ and 4 in the low-frequency bending vibration.

significantly by l-type doubling. For the lowest vibrational states, the centrifugal effects that result from increasing rotation about the A axis also leads to more linear character in the spectrum as K_A increases and these lines march out to high frequency. Basically the spectroscopic patterns are governed by a Mexican sombrero hat potential in which the maximum has less and less effect as the vibrational excitation increases and the levels rise above the tip of the hat. Brenda and Manfred Winnewisser and their co-workers [24] have, in their beautifully detailed study of the fascinating quasi-linear dynamics in this molecule, found evidence for quantum monodromy. The top of the barrier to linearity in the two-dimensional anharmonic sombrero potential for the bending mode forms the critical monodromy point at which there is a discontinuity in the mathematical description of the eigenstates that lie above and below this point. The study has thus taken the description of the associated motion on to a new level of elegance and understanding.

Microwave measurements also enabled researchers such as Sunny Chan, Bill Gwinn and co-workers [25] and Jaan Laane [26] and others to develop our in-depth understanding of some beautiful effects of quasi-planarity in cyclic ring molecules such as trimethylene oxide [3]. These studies enabled us [27] to recognise that the reason for some puzzling features in the electronic spectrum of pyridine which I studied with Don Ramsay when I first arrived at NRC, which had perplexed early spectroscopists such as Sponer and Teller many years earlier, were due to non-planarity of the electronically excited state.

The major contributor to the plethora of fundamental observational spectroscopic advances in molecular spectroscopy was the laboratory of Gerhard Herzberg [28] and Alec Douglas at NRC in Canada. It catalysed several brilliant theoretical breakthroughs in understanding the basic theory of molecular spectroscopy. Taking off from where Van Vleck [29] and others such as Bright Wilson, Nielsen and Dennison had left off, Jon Hougen had refined the theoretical treatment of molecular electronic spectroscopy – much encapsulated in Gerhard Herzberg's [28] fourth book. Improved understanding of flexible systems was achieved and the associated Group Theory developed by Hougen [30], Longuet-Higgins [31],

Jim Watson [32] and Phil Bunker [33]. Hougen and Bunker with John Johns [34] developed the theoretical treatment of quasi-linear small molecules further. Watson had, in a series of remarkable papers [35–37], clarified some complex issues involving the way centrifugal distortion governs the energy levels of an asymmetric top molecule. He went on to make an even more ingenious breakthrough when he developed the Watson Hamiltonian for vibration rotation, which re-cast the complicated and somewhat ungainly Wilson and Howard formalism [38] into an elegantly simplified and useful form [39].

During the decade from 1968 onwards another revolution took place. It has its origins in the discoveries by Townes' group using radio astronomy to detect the interstellar spectra of ammonia and water in a molecular cloud just behind the famous Orion emission nebula [40]. In the book of Townes and Schawlow [41] on microwave spectroscopy, there is a couple of succinctly prescient pages that detail the ideas that led to the birth of interstellar molecular radio astronomy [5]. The role that the discovery of the existence of copious numbers of molecules in molecular clouds played in a true revolution in our understanding of interstellar space should not be underestimated. The first rumblings of the impending revolution are perhaps to be found in the 1963 paper on the detection of OH by Weinreb et al. [42]. It is interesting to note how important the ammonia inversion microwave spectrum has been for science in general. As well as for the interstellar breakthrough [40], the first microwave spectrum observed was of that of ammonia by Cleeton and Williams [43], and the theoretical treatment of quantum mechanical tunnelling developed by Dennison and Uhlenbeck [44] was a major step to understanding this fundamental quantum mechanical phenomenon. Furthermore, Townes and co-workers [45] chose the ammonia spectrum in his invention of the maser.

At this point it is worth pointing out that, quite curiously, just after the discovery of the three-degree microwave background by Penzias and Wilson in 1967 [46], it was suddenly realised that the measurement had already been made in 1941 in a study of the interstellar optical spectrum of three lines of the CN radical observed by Adams [47]. The $R(0)$ line originates in the lowest rotational level of the ground electronic state and the $P(1)$ and $R(1)$ in the first excited rotational level; thus from the ratio of the intensities, one can deduce in a fairly straight-forward manner (assuming "equilibrium" conditions of course) that the temperature of the CN molecules in the interstellar medium (ISM) is $\sim$2.7 K [5,6]. With hindsight, the most odd aspect of this digression lies in what Gerhard Herzberg [48, p. 496] in the fascinating chapter of The Spectra of Diatomic Molecule devoted to astrophysics says of this measurement – "...that it has of course only very restricted meaning." In retrospect it did have a really revolutionary "restricted" meaning – especially had the meaning been recognised in 1937 [6].

A key discovery was made about this time and that was that CO was a major component of the ISM [49]. The interstellar CO rotational spectrum is collisionally excited by H_2, which was then, and still is, effectively invisible and so the mass of the clouds is deduced from CO measurements. It then became clear that the thermal energy of the clouds leaked out by radiative emission from the collisionally excited rotational levels of CO and this was crucial in the initial cooling stages enabling the collapse of interstellar clouds. The thermal energy must leak out (at the speed of light), otherwise Boyle's volume/pressure gas law would prevent collapse. So we owe the birth of stars, planets, the biosphere as well as our very existence to the CO molecule. Bill Klemperer has pointed out that if thermodynamics were applicable to the chemistry of the ISM, there shouldn't be one CO molecule in the whole universe and we should all be dead (cf. www.vega.org.uk [50]). So, thank heavens, thermodynamics is not applicable – to us!

A key breakthrough was made when Klemperer [51] imaginatively suggested that the carrier of a hitherto unidentified very strong interstellar radio line might be protonated carbon monoxide HCO^+. Subsequently, work by Claude Woods, Rich Saykally and co-workers [52] in plasma microwave spectroscopy experiments proved this conjecture to be correct. Thus was born the field of ion–molecule chemistry. Klemperer together with Solomon [53] and Herbst [54] laid the foundations of a satisfyingly robust theoretical picture of how ion–molecule chemistry governs the molecular composition of the interstellar clouds that we now recognise are the birthplaces of stars and planets.

As all types of lasers, in particular tunable ones, evolved enabling laser-induced fluorescence (LIF), multiple resonance and other techniques to be developed, they invaded and revolutionised high-resolution spectroscopy – an area that originally had been the preserve of (microwave) rotational spectroscopy. The major breakthrough of Takeshi Oka [55] in developing laser IR techniques for detecting ions such as H_3^+ and Rich Saykally's [56] further imaginative development of the velocity modulation refinement were beautiful developments that enabled the study of ions, which had hitherto been well-nigh impossible, to be carried out. I cannot do justice to all the numerous elegant breakthroughs, but certainly the development by the late Bill Flygare [57] of the Fourier-transform pulsed-nozzle microwave spectrometer must be highlighted as it led to another breakthrough, the field of weakly bound molecular cluster rotational spectroscopy. Tony Legon and others [58] have used this technique to rewrite the book on weakly bound molecular complexes involving hydrogen bonding and van der Waals forces. In parallel with these developments was that of Rick Smalley, who with Don Levy and Lennard Wharton [59], combined the supersonic nozzle technique with LIF, which led to a revolution in the analysis of complex spectra. Smalley then went on to develop the laser vapourisation cluster beam technique [60], which led to another revolution – in this case in the study of refractory and other clusters. Many other high-resolution approaches have been developed but one that really seems, in the hands of researchers such as John Maier in Basel [61] and Rich Saykally at Berkeley [62], to be taking spectroscopy on to new exciting heights is cavity ring-down spectroscopy.

About 1974 I received a preprint of a paper by my friend and former laboratory colleague (at NRC) Takeshi Oka which described radio observations on interstellar H_2CS. This was an old nemesis molecule of mine – but that is another story. I wrote to Takeshi to see if he might be interested in a collaboration to detect the HC_5N molecule in the ISM by radio astronomy. His response (Figure 3) – as exemplified in the fourth line – is a prime example of customary Japanese understated enthusiasm. In the event, we detected HC_5N in Sagittarius [63]. This led on to a project to synthesise HC_7N, specifically for radio detection. David Walton devised a synthesis and Colin Kirby took time off from some at the time highly exciting work on boron suphides to prepare it [4,64]. Our series of detections of HC_nN ($n = 5$, 7 and 9) together with Canadian astronomers Lorne Avery, Norm

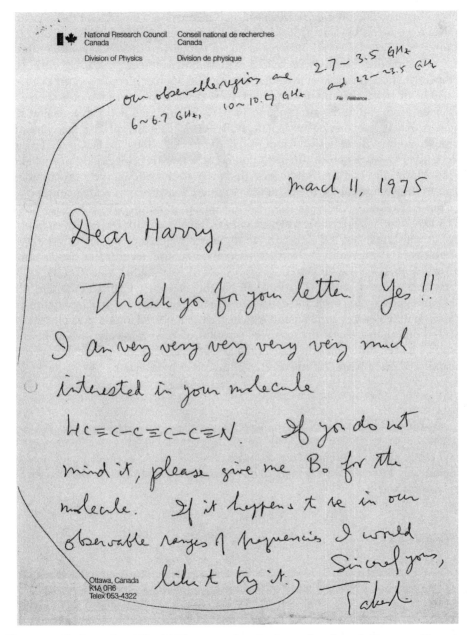

Figure 3   The letter from Takeshi Oka suggesting that he was "moderately" interested in trying to detect HC_5N – the first long-chain polyyne in the interstellar medium (ISM).

Broten and John McLeod, between 1975 and 1978 [63,65,66], led me to the opinion, rightly or wrongly, that ion-molecule reactions were probably too slow to create all these chains in the abundance that we had found in the ISM. I also suspected that grain-surface catalysis was also an unlikely formation process as it was likely to be very difficult in general to get such molecules to desorb – at the very low temperatures in the ISM. It had indeed been pretty difficult to get HC_7N into the microwave cell to detect its spectrum at room temperature. Whichever way, those were my thoughts in the early 1980s which I occasionally discussed with Bill Klemperer, who felt that his ion-molecule reaction approach could indeed produce them. My view was that the molecules we were detecting were much more likely to have been produced in the atmospheres of stars. What certainly became clear at the time was that at least one old red giant, $ICR + 10,217$, was expelling the chains in copious quantities. Thus, it was that during Easter in 1984 Bob Curl had invited me to Rice University after one of the great conferences that Jim Boggs held every other year in Austin. Jim had been most kind in almost always finding me a few hundred dollars to help me scrounge the extra travel money I needed from other sources.

Thus began the fateful year 1985 for me. It started on a cold very snowy New Year's Day with a 12-day observing session at Greenbank with Les Little, an astronomer at the University of Kent, in an unsuccessful attempt to detect NCCP, a molecule whose microwave spectrum Keichi Ohno had recorded at Sussex some time earlier [67]. I always pictured this molecule as two mighty trans-Canada steam locomotives – one Canadian National (CN) and the other Canadian Pacific (CP) pulling in opposite directions and getting nowhere. This epitomised perfectly the 12-day session in which Les and I sat and watched what appeared to be a possible feature, observed on the first day at the expected position with poor signal to noise, take 12 days to be beaten down into the baseline noise. The year ended with the discovery of C_{60} and, with its appearance, my spectroscopy career evaporated.

ACKNOWLEDGEMENTS

I thank all the numerous friends and colleagues, many of them mentioned by name in this review, with whom I have worked or who have clarified many problems for me.

REFERENCES

[1] R.P. Feynman,"Today's Answers to Queries about Light"; Sir Douglas Robb Lectures; University of Auckland – four lectures (4 × ca. 1.5 h) 1979. Donated by Auckland University to the Vega Science Trust and streaming free at www.vega.org.uk.
[2] B.L. Van der Waerden, Sources of Quantum Mechanics, Dover, New York, 1967.
[3] H. W. Kroto, Molecular Rotation Spectra, originally published 1974 by Wiley. Then republished by Dover in 1992 as a paperback, with an extra preface including many spectra. Now republished in Phoenix editions 2003.
[4] H.W. Kroto, Chem. Soc. Rev., 11: 435, 1981.
[5] H.W. Kroto, Int. Rev. Phys. Chem., 1: 309, 1981.
[6] H.W. Kroto, The Vega Science Trust, Vega Science Lectures, Series of 8 lectures entitled "Astrophysical Chemistry" streaming at www.vega.org.uk/video/subseries/16/
[7] J. Kraitchman, Am. J. Phys., 21: 17–24, 1953.
[8] C.C. Costain, Phys. Rev., 82: 108, 1951.

[9] A.J. Alexander, H.W. Kroto, and D.R.M. Walton, J. Mol. Spectrosc., 62: 175–180, 1976.
[10] M.J. Hopkinson, H.W. Kroto, J.F. Nixon, and N.P.C. Simmons, Chem. Phys. Lett., 42: 460–461, 1976.
[11] H.W. Kroto, J.F. Nixon, and K. Ohno, J. Mol. Spectrosc., 90: 367–373, 1981.
[12] H.W. Kroto, J.F. Nixon, and N.P.C. Simmons, J. Mol. Spectroc., 77: 270–285, 1979.
[13] R.W. Kilb, C.C. Lin, and E.B. Wilson, J. Chem. Phys., 26: 1695–1703, 1957.
[14] D.R. Herschbach, J. Chem. Phys., 31: 91–108, 1959.
[15] H.H. Nielsen, Phys. Rev., 40: 445–456, 1932.
[16] H.H. Nielsen, Phys. Rev., 77: 130–135, 1950.
[17] D.G. Burckhard and D.M. Dennison, Phys. Rev., 84: 408–417, 1951.
[18] C.K. Ingold, and G.W. King, J. Chem. Soc., 2702–2704, 1953.
[19] R.N. Dixon, Trans. Faraday Soc., 60: 1363–1368, 1964.
[20] T.F Morgan, Ph.D. Thesis, Electronic Spectroscopy, University of Sussex, 1970.
[21] Ch. Jungen and A.J. Merer, Mol. Phys., 40: 95–114, 1980.
[22] M.A. King and H.W. Kroto, J. Chem. Soc. Chem. Commun., 13: 606, 1980.
[23] M.A. King, H.W. Kroto, and B.M. Landsberg, J. Mol. Spectrosc., 113: 1–20, 1985.
[24] B. Winnewisser, M. Winnewisser, I.R. Medvedev, M. Behnke, F.C. De Lucia, S.C. Ross, and J. Koput, Phys. Rev. Lett., 95: 243002/1–4, 2005.
[25] S.I. Chan, J. Zilm, J. Fernandez, and W.D. Gwinn, J. Chem. Phys., 33: 1643–1655, 1960.
[26] J. Laane, Annu. Rev. Phys. Chem., 45: 179–211, 1994.
[27] J.P. Jesson, H.W. Kroto, and D.A. Ramsay, J. Chem. Phys., 56: 6257–6258, 1972.
[28] G. Herzberg, Molecular Spectroscopy Molecular Structure, Vol. 3: Electronic Spectra of Polyatomic Molecules, Van Nostrand, New York, 1966.
[29] J.H. Van Vleck, Rev. Mod. Phys., 23: 213–227, 1951.
[30] J.T. Hougen, Pure App. Chem., 11: 481–495, 1965.
[31] H.C. Longuet-Higgins, Mol. Phys., 6: 445–460, 1963.
[32] J.K.G. Watson, Can. J. Phys., 43: 1996–2007, 1965.
[33] P.R. Bunker, Ann. Rev. Phys. Chem., 34: 59–75, 1983.
[34] J.T. Hougen, P.R. Bunker, and J.W.C. Johns, J. Mol. Spectrosc., 34: 136–172, 1970.
[35] J.K.G. Watson, J. Chem. Phys., 45: 1360–1361, 1966.
[36] J.K.G. Watson, J. Chem. Phys., 46: 1935–1949, 1967.
[37] J.K.G. Watson, J. Chem. Phys., 48: 181–185, 1968.
[38] E.B. Wilson and J.B. Howard, J. Chem. Phys., 4: 260–268, 1936.
[39] J.K.G. Watson, Mol. Phys., 15: 479–490, 1968.
[40] A.C. Cheung, D.M. Rank, C.H. Townes, D.D. Thornton, and W.J. Welch, Phys. Rev. Lett., 21: 1701–1705, 1968.
[41] C.H. Townes and A. Schawlow, Microwave Spectroscopy, McGraw Hill, New York, 1955.
[42] S. Weinreb, A.H. Barrett, M.L. Meeks, and J.C. Henry, Nature, 200: 829–831, 1963.
[43] C.E. Cleeton and N.H. Williams, Phys. Rev., 45: 234–237, 1934.
[44] D.M. Dennison and G.E. Uhlenbeck, Phys. Rev., 41: 313–321, 1932.
[45] J.P. Gordon, H.J. Zeiger, and C.H. Townes, Phys. Rev., 99: 1264–1274, 1955.
[46] A.A. Penzias and R.W. Wilson, Astrophys. J., 142: 419–425, 1965.
[47] W.S. Adams, Astrophys. J., 93: 11, 1941.
[48] G. Herzberg, Molecular Spectroscopy Molecular Structure, Vol. 1: The Spectra of Diatomic Molecules, 2nd edition, Van Nostrand, New York, 1950.
[49] W.J. Wilson, K.B. Jefferts, and A.A. Penzias, Astrophys. J., 161: L43–L44, 1970.
[50] W. Klemperer, The Vega Science Trust, http://www.vega.org.uk/video/programme/64, Royal Institution Discourse "The Chemistry of Interstellar Space."
[51] W. Klemperer, Nature, 227: 1230, 1970.
[52] R.C. Woods, T.A. Dixon, R.J. Saykally, and P.G. Szanto, Phys. Rev. Lett., 35: 1269–1272, 1975.
[53] P.M. Solomon and W. Klemperer, Astrophys. J., 178: 389–421, 1972.
[54] E. Herbst and W. Klemperer, Astrophys. J., 185: 505–533, 1973.
[55] T. Oka, Rev. Mod. Phys., 64: 1141–1149, 1992.
[56] C.S. Gudeman and R.J. Saykally, Annu. Rev. Phys. Chem., 35: 387–418, 1984.

[57] T.J. Balle and W.H. Flygare, Rev. Sci. Instrum., 52: 33–45, 1981.

[58] A.C. Legon, Angew. Chem. Int. Ed. Engl., 38: 2687–2714, 1999.

[59] R.E. Smalley, L. Wharton, and D.H. Levy, J. Chem. Phys., 63: 4977–4989, 1975.

[60] T.G. Dietz, M.A. Duncan, D.E. Powers, and R.E. Smalley, J. Chem. Phys., 74: 6511–6512, 1981.

[61] J.P. Maier, A.E. Boguslavskiy, H. Ding, G.A.H. Walker, and D.A. Bohlender, Astrophys. J., 640: 369–372, 2006.

[62] J.B. Paul and R.J. Saykally, Anal. Chem., 69: 287A–292A, 1997.

[63] L.W. Avery, N.W. Broten, J.M. MacLeod, T. Oka, and H.W. Kroto, Astrophys. J., 205: L173–L175, 1976.

[64] C. Kirby, H.W. Kroto, and D.R.M. Walton, J. Mol. Spectrosc., 83: 261–265, 1980.

[65] H.W. Kroto, C. Kirby, D.R.M. Walton, L.W. Avery, N.W. Broten, J.M. MacLeod, and T. Oka, Astrophys. J., 219: L133–L137, 1978.

[66] N.W. Broten, T. Oka, L.W. Avery, J.M. MacLeod, and H.W. Kroto, Astrophys. J., 223: L105–L107, 1978.

[67] H.W. Kroto, J.F. Nixon, and K. Ohno, J. Mol. Spectrosc., 90: 512–516, 1981.

CHAPTER 2

FRONTIERS OF LINEAR AND NON-LINEAR RAMAN SPECTROSCOPY

FROM A MOLECULE TO A LIVING CELL

Hiro-o Hamaguchi

Contents

Abstract

Recent advances in our laboratory of applications of linear and non-linear Raman spectroscopy are reviewed. We first discuss ultrafast ion association/dissociation dynamics of the sulfate ion in water and protonation/deprotonation dynamics of N,N-dimethylacetamide (DMAA) in hydrochloric acid on the basis of band-shaped analysis. Second, the discovery of the "Raman spectroscopic signature of life" and its application to monitor the life and the death of a single yeast cell are described. Then, we introduce a new non-linear Raman method, the coherent anti-Stokes Raman scattering (CARS) signal spatial distribution measurements, to probe local structures and their temporal evolution in an ethanol/water mixture. Finally, we describe our recent finding of the "molecular near-field effect" in which solvent vibrations intervene in the process of resonance hyper-Raman (HR) scattering. This effect has a potential for selectively detecting the solvent molecules in the close vicinity of the probe β-carotene. It may lead to single molecule spectroscopy of ensemble of molecules.

Keywords: Raman spectroscopy; non-linear Raman; CARS; hyper-Raman; protonation; deprotonation; Raman spectroscopic signature of life; living cell; single cell; cell death; molecular near-field effect

1. INTRODUCTION

Raman spectroscopy, born in 1928 and continuously growing up since then [1,2], seems to have reached a new horizon. Its application now extends over all basic sciences including physics, chemistry, and biology as well as various research fields in industrial, pharmaceutical, medical, and agricultural science and technology. It is true that we owe this explosive development of Raman spectroscopy to recent advances in lasers, optical instruments, detectors, and computers. However, these advances should also have benefited other spectroscopies. Why are Raman applications so extensive? What is special with Raman?

Raman spectroscopy has its intrinsic width and depth that are not associated with other spectroscopies (Figure 1). The width comes from the fact that it can measure a sample as it is or under any extreme conditions. We can use an optical fiber probe to detect Raman scattering directly from a human body for clinical applications or from a high-pressure and high-temperature cell for basic physical studies. Raman spectroscopy can measure all kinds of materials from a molecule to a living cell. They all give the same Raman spectra that are written in the same language of molecular fingerprint. The depth comes from the fact that any kind of light–matter interaction that involves a Raman resonance gives rise to the corresponding Raman spectroscopy. Spontaneous Raman spectroscopy is related linearly to the electric field E of the incident light, hyper-Raman (HR) spectroscopy to E^2, and coherent

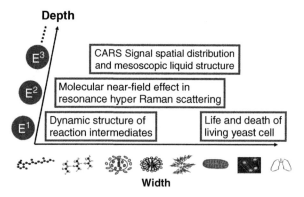

Figure 1 Width and depth of Raman spectroscopy and the four topics discussed this article (See color plate 1).

anti–Stokes Raman scattering (CARS) spectroscopy to E^3. Other non–linear Raman spectroscopies including stimulated Raman gain/loss, and coherent Stokes Raman scattering spectroscopy have also been demonstrated [3,4].

In this chapter, we briefly overview the recent advances in our laboratory of applications of linear and non–linear Raman spectroscopy. We first describe ultrafast chemical exchange dynamics of reaction intermediates as investigated by the band shape analysis of linear Raman spectra. Ion association/dissociation dynamics of the sulfate ion in water and protonation/deprotonation dynamics of N,N-dimethylacetamide (DMAA) in hydrochloric acid are discussed. Second, we show results of time- and space-resolved Raman spectroscopy of single living yeast cells, again using linear Raman spectroscopy. The discovery of the "Raman spectroscopic signature of life" is described followed by its application to monitor the life and the death of a single yeast cell at the molecular level. Then we introduce a new non–linear Raman method, the CARS signal spatial distribution measurements, to probe local structures in liquids and solutions. Local structures and their temporal evolution in an ethanol/water mixture are discussed. Finally, we describe our recent discovery of the "molecular near–field effect" in resonance HR scattering. Resonance HR spectra of β–carotene in a few organic solutions show extra band attributable to the solvent. We interpret this finding as due to a new photomolecular effect in which solvent vibrations intervene in the process of resonance HR scattering. This effect, named by us the "molecular near–field effect," has a potential for selectively detecting the solvent molecules in the close vicinity of the probe β–carotene. It may lead to single molecule spectroscopy of ensemble of molecules.

2. ULTRAFAST CHEMICAL EXCHANGE DYNAMICS OF REACTION INTERMEDIATES ELUCIDATED BY RAMAN BAND SHAPE ANALYSIS

There are many important chemical reactions that proceed thermally at room temperature. Biochemical reactions in our body and environmental reactions on the Earth are good examples. However, our understanding of these thermal reactions is much less advanced compared with that of photochemical reactions. For photochemical reactions, various kinds of ultrafast time-resolved spectroscopies provide direct information on how they proceed [5]. This is not the case with thermal reactions, which take place statistically and which cannot be triggered by light. As an alternative to the time domain approach using time-resolved spectroscopies, we have been developing a frequency-domain approach based on the Raman band shape analysis. In particular, we have been using the two-state dynamic exchange model to derive quantitative dynamical information on short-lived reaction intermediates in thermal reactions. The approach was first used for *trans*-stilbene in the first-excited singlet (S_1) state [6–10] and then for studying the ion association dynamics of the sulfate ions in aqueous solutions [11] and the acid-catalyzed dehydration reaction of *tert*-butanol [12]. In the following, we first discuss the prototypical case of the sulfate ion and then show a more general case of the protonation reaction of *N,N*-DMAA in hydrochloric acid [13].

2.1.   Ion association dynamics in aqueous solutions of magnesium sulfate

The ion association of the sulfate ion has been studied extensively as one of the simplest chemical reactions in water. Based on ultrasonic experiments, Eigen and Tamm proposed a model that assumes four species, $[Mg^{2+} + SO_4^{2-}]$, $[Mg^{2+}(H_2O)_2SO_4^{2-}]$, $[Mg^{2+}(H_2O)SO_4^{2-}]$, and $[Mg^{2+}SO_4^{2-}]$, co-existing in aqueous solutions of magnesium sulfate [14]. Raman spectroscopy has also been used for studying the aqueous solutions of magnesium sulfate. Davis and Oliver [15] found that the shape of the a_1 band (totally symmetric SO stretch) of the sulfate ion changed with the salt concentration. They argued that the observed band consisted of two components: one component corresponding to the contact ion pair ($[Mg^{2+}SO_4^{2-}]$) and the other to the remaining three species ($[Mg^{2+} + SO_4^{2-}]$, $[Mg^{2+}(H_2O)_2SO_4^{2-}]$, and $[Mg^{2+}(H_2O)SO_4^{2-}]$). According to this interpretation, the two components of the a_1 Raman band should have different depolarization ratios. The unassociated sulfate ion has a T_d symmetry, whose depolarization ratio must be equal to zero. The associated sulfate ion, however, is expected to have non-zero depolarization ratio because of the lowered symmetry. We used polarization-resolved CARS spectroscopy, which can measure the depolarization ratio very accurately (±0.002) [16] to resolve the a_1 band. We found, contrary to what was expected, that the band was not separated into two components with different depolarization ratios [16]. The a_1 band of magnesium sulfate must be considered as composed of a single component as long as it is measured by Raman spectroscopy. To explain this unexpected result, it is necessary to consider the association/dissociation dynamics of ions in a timescale faster than the vibrational dephasing time.

We adopt an approach based on the probability theory. This theory was first developed by Anderson [17] to treat NMR band shapes and later used by us to study the Raman band shapes in connection with the photoisomerization dynamics of *trans*-stilbene [6–10]. We use the two-state exchange model for the sulfate ion in aqueous solutions (Scheme 1); the sulfate ion exists either in the free or in the associated state.

The frequency of the a_1 mode takes the value ω_1 in the free state and ω_2 in the associated state. The sulfate ion transforms between the two states with the association rate W_1 and the dissociation rate W_2. The vibrational correlation

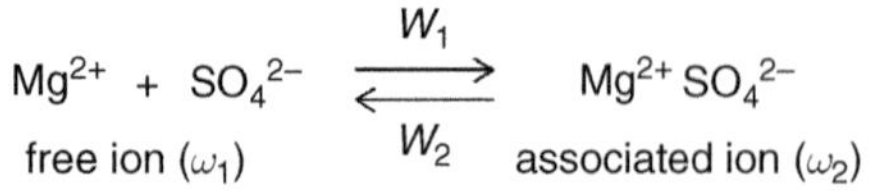

$$Mg^{2+} + SO_4^{2-} \underset{W_2}{\overset{W_1}{\rightleftarrows}} Mg^{2+}SO_4^{2-}$$

free ion (ω_1)   associated ion (ω_2)

Scheme 1   The chemical exchange of magnesium sulfate in water.

function ϕ_{ex} under this exchange scheme is given as follows by the theory of probability [17].

$$\phi_{ex}(t) = \mathbf{1} \cdot \exp[i\omega_L t - \Pi |t|] \mathbf{W}_0 \tag{1}$$

Here, $\mathbf{1}$, ω_L, Π, and $\mathbf{W}_0$ are the vectors or the matrices defined as

$$\mathbf{1} = \begin{pmatrix} 1 \\ 1 \end{pmatrix}; \omega_L = 2\pi c \begin{pmatrix} \omega_1 & 0 \\ 0 & \omega_2 \end{pmatrix}; \Pi = 2\pi c \begin{pmatrix} W_1 & -W_2 \\ -W_1 & W_2 \end{pmatrix}; \mathbf{W}_0 = \frac{1}{W_1 + W_2} \begin{pmatrix} W_2 \\ W_1 \end{pmatrix}.$$

Note that all the parameters we use, ω_1, ω_2, W_1, and W_2, have the unit of wavenumber for simplicity of expression, and c is the light speed.

If we assume that the Raman amplitude of the sulfate ion, or the intensity of Raman scattering, is the same for each of the two states, the band shape, $I_{ex}(\omega)$, is given as the Fourier transform of Equation (1), which can be calculated analytically to be

$$I_{ex}(\omega) = \frac{2W_1 W_2 (\omega_2 - \omega_1)^2 / (W_1 + W_2)}{(\omega - \omega_1)^2 (\omega - \omega_2)^2 + [W_2(\omega - \omega_1) + W_1(\omega - \omega_2)]^2}. \tag{2}$$

If the dephasing by another origin needs to be considered, an exponential function that decays with the time constant $2\pi c \Gamma_0$ is multiplied to the correlation function (1). The band shape in this case is the convolution of $I_{ex}(\omega)$ with a Lorentzian whose half width is Γ_0.

The band shape (2) is given as a function of four parameters that characterize the exchange dynamics: ω_1, ω_2, W_1, and W_2. To make this formula further tractable, we introduce the asymmetric exchange limit approximation and reduce the number of parameters while keeping the physical soundness of Equation (2). This approximation holds for systems in which one of the two exchanging states is very predominant over the other. In the case of the sulfate system, this limit corresponds to diluted solutions in which the sulfate ions stay most of the time as free ions. Under this condition, the band shape (2) becomes a Lorentzian having the peak $\omega_1 + \Delta\Omega$ and the width $\Gamma_0 + \Delta\Gamma$. The peak shift and the band width increase $\Delta\Omega$ are given as follows, respectively, as functions of W_1 and τ

$$\Delta\Omega = \frac{\tau}{1 + \tau^2} W_1, \tag{3}$$

$$\Delta\Gamma = \frac{\tau^2}{1 + \tau^2} W_1. \tag{4}$$

Here, τ is a parameter defined as follows using the frequency difference $\delta\omega = \omega_2 - \omega_1$ and the dissociation rate W_2.

$$\tau = \frac{\delta\omega}{W_2} \tag{5}$$

This parameter τ corresponds to the mean phase shift accompanying one ion encounter that results in the association. From Equations (3) and (4), the following relation is derived.

$$\Delta\Gamma = \tau\Delta\Omega \tag{6}$$

As shown in the following, the observed ratio $\Delta\Gamma/\Delta\Omega$ for dilute solutions of magnesium sulfate is independent of the salt concentration. This fact indicates that Equation (6) based on the asymmetric exchange limit approximation holds well for dilute solutions and that τ is determined experimentally. Then, we can determine the association rate W_1 directly from Equations (3) and (4). The dissociation rate W_2 is obtained with the least-squares fitting based on Equation (5).

The concentration dependence of the a_1 Raman band shape of the sulfate ion in aqueous magnesium sulfate solutions is shown in Figure 2. The depolarization ratio of this band is very small (<0.003) [16], indicating that the anisotropic part of the Raman scattering tensor is negligibly small. Therefore, the observed Raman spectra can be regarded as due to the isotropic part of the Raman scattering tensor that is free from the rotational broadening. Figure 2 shows that the peak position moves to higher frequencies and the bandwidth increases with increasing ion concentration.

By extrapolating the plot of the peak position and the bandwidth against concentration, we obtain the limiting frequency $\omega_1 = 981.25 \pm 0.1$ cm^{-1} and bandwidth $\Gamma_0 = 2.7 \pm 0.1$ cm^{-1} that correspond to infinite dilution. Using these values, the peak shift $\Delta\Omega$ and the width increase $\Delta\Gamma$ can be calculated. The $\Delta\Gamma$ values thus obtained are plotted against $\Delta\Omega$ in Figure 3. A clear linear relation is seen between and $\Delta\Omega$ and $\Delta\Gamma$. From this result, we learn that the two-frequency exchange model in the asymmetric limit is appropriate for the ion association dynamics in aqueous solutions of magnesium sulfate. Note that the peak position and the bandwidth have no intrinsic relation with each other and the observed linear relationship cannot be explained with any conventional static model.

Using Equation (6), the value $\tau = 0.78 \pm 0.03$ is obtained from the slope of the plot in Figure 3. Now that three parameters, ω_1, Γ_0, and $\tau = \delta\omega/W_2$, out of the four are determined, it is possible to fit the observed spectra with the theoretical

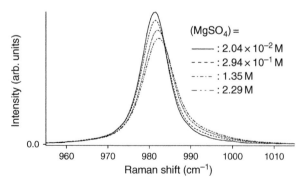

Figure 2 Concentration dependence of the a_1 Raman band shape of the sulfate ion in aqueous solutions of magnesium sulfate.

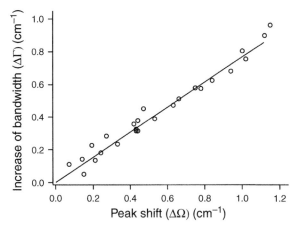

Figure 3 Plot of $\Delta\Gamma$ versus $\Delta\Omega$ for the a_1 Raman band of the sulfate ion in aqueous solutions of magnesium sulfate.

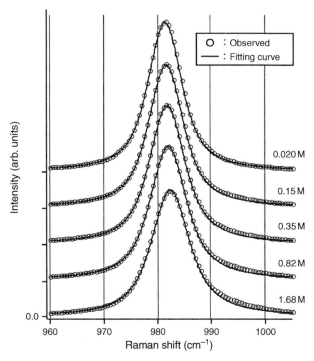

Figure 4 The observed (circles) and theoretical (full lines) band shapes of the a_1 band of the sulfate ion.

formula and determine the remaining one parameter using Equation (2). The fitting results are shown in Figure 4, where the observed and theoretical band shapes agree excellently with each other. The tendency of the peak shift and the band broadening with increasing the ion concentration is reproduced very well.

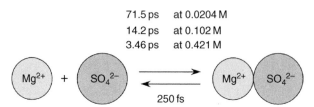

Figure 5   Chemical exchange dynamics of the sulfate ion in aqueous solutions of magnesium sulfate.

From the fitting, the optimized value of w_2 is obtained as $w_2 = 998 \pm 2$ cm^{-1}. We can then calculate $W_2 = 21 \pm 3$ cm^{-1} using Equation (5). The reciprocal of W_2 gives the mean lifetime of the associated ions as $T_2 = 0.25 \pm 0.04$ ps. Thus, all the four parameters, w_1, w_2, W_1, and W_2, are obtained. These values indicate that the ionic association is a very fast process as shown in Figure 5. The association time is in the order of a few picoseconds to tens of picosecond depending on the salt concentration. The mean lifetime of the associated ion, $T_2 \sim 250$ fs, is much shorter than the vibrational dephasing time, $T_0 \sim 2.0$ ps. This result is fully consistent with the fact that the observed Raman band is broadened homogeneously and that the associated ion cannot be detected independently by vibrational spectroscopy, as had been proved by our polarization-resolved CARS experiments [16].

2.2.   Protonation/deprotonation dynamics of *N,N*-dimethylacetamide in hydrochloric acid

Protonation is one of the most fundamental chemical reactions involved in acid/base-catalyzed reactions. Here, we study the protonation/deprotonation dynamics of *N,N*-DMAA using a more general two-state exchange model (Figure 6). In this model, in addition to the four parameters w_1, w_2, W_1, and W_2 in the previous section, we introduce the Raman amplitudes α_1 and α_2 and the intrinsic dephasing rates Γ_1 and Γ_2. The amplitude corresponds to the norm of the isotropic

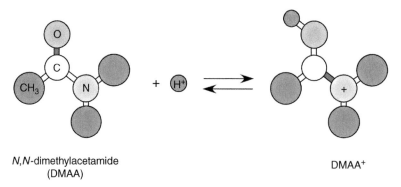

Figure 6   The protonation/deprotonation dynamics of *N,N*-dimethylacetamide.

component of the Raman scattering tensor and the intrinsic dephasing to the bandwidth that is observed in the absence of the exchange dynamics.

The vibrational correlation function under this model is given as follows [13]:

$$\phi(t) = \frac{1}{W_1 + W_2} \begin{pmatrix} a_1 & a_2 \end{pmatrix} \exp\left[\begin{pmatrix} i\omega_1 - W_1 - \Gamma_1 & W_2 \\ W_1 & i\omega_2 - W_2 - \Gamma_2 \end{pmatrix} t\right]$$
$$\times \begin{pmatrix} W_2 & 0 \\ 0 & W_1 \end{pmatrix} \begin{pmatrix} a_1 \\ a_2 \end{pmatrix}. \tag{7}$$

The band shape is given by the Fourier transform of (7) as

$$I(\omega) = \mathrm{Re}\left[\frac{1}{W_1 + W_2} \begin{pmatrix} a_1 & a_2 \end{pmatrix} \begin{pmatrix} i(\omega - \omega_1) + W_1 + \Gamma_1 & -W_2 \\ -W_1 & i(\omega - \omega_2) + W_2 + \Gamma_2 \end{pmatrix}^{-1}\right.$$
$$\left. \times \begin{pmatrix} W_2 & 0 \\ 0 & W_1 \end{pmatrix} \begin{pmatrix} a_1 \\ a_2 \end{pmatrix}\right]. \tag{8}$$

Figure 7 shows the acid concentration dependence of the CC/CN anti-symmetric stretch Raman band of the DMAA/DMAAH$^+$ system. The observed band shapes are given in circles, and two theoretical band shapes are given in full and dotted lines. In the neutral solution without acid ([HCl] $= 0$ M), the effect of protonation can be neglected. Therefore, the values of ω_1, a_1, and Γ_1 can be determined straightforwardly from the observed peak position, the amplitude, and the bandwidth. The following values have been obtained: $\omega_1 = 965.7 \pm 0.1 \mathrm{cm}^{-1}$, $a_1 = 3.30 \pm 0.05$, and

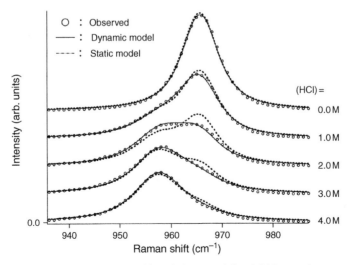

Figure 7 Experimental and theoretical band shapes of the CC/CN anti-symmetric stretch band of the DMAA/DMAAH$^+$ system in hydrochloric acid.

$\Gamma_1 = 3.5 \pm 0.1 \text{cm}^{-1}$. The same logic is applicable for the opposite limit. From the spectral changes of three other Raman bands with the acid concentration (data not shown), we know that the equilibrium of the reaction shifts to the protonated side almost completely at $[\text{HCl}] = 4.0 \text{ M}$ and that a great majority of molecules exist as DMAAH^+ at this concentration. Therefore, the spectrum at $[\text{HCl}] = 4.0 \text{ M}$ must be very close to that of DMAAH^+, providing us with highly reliable initial values of ω_2, a_2, and Γ_2. With these initial values of ω_2, a_2, and Γ_2 together with the prefixed values of ω_1, a_1, and Γ_1, we can easily optimize the rate parameters W_1 and W_2 to reproduce the band shape changes in Figure 7. As shown by full lines, the two-state exchange model reproduces the observed band shape changes very well. The optimized values are $\omega_2 = 957.4 \pm 0.2 \text{cm}^{-1}$, $a_2 = 2.82 \pm 0.05$, and $\Gamma_2 = 5.1 \pm 0.3 \text{cm}^{-1}$.

At this point, we need to check whether the band shape changes in Figure 7 cannot be reproduced by linear combinations of the two independent spectral components for DMAA and DMAAH^+. The spectrum of DMAA is obtained from the neutral solution. The other spectral component can be taken as the spectrum at $[\text{H}^+] = 4.0 \text{ M}$, which is very close to the spectrum of DMAAH^+ as already discussed. We have tried to fit the observed band shape changes as linear combinations of these two components. The fitting results are shown as dotted lines in Figure 7. The static approach obviously fails to reproduce the experimental band shape changes.

The concentration dependence of the optimized rate parameters is shown in Figure 8 (left). The protonation rate W_1 increases linearly with increasing the acid concentration, indicating that the protonation reaction obeys the first-order reaction kinetics. Assuming the complete dissociation of hydrochloric acid, the rate constant k_1 is obtained as $k_1 = 10.1 \pm 0.2 \times 10^{10} \text{ s}^{-1} \text{M}^{-1}$. On the contrary, W_2 decreases with increasing the acid concentration. This result indicates that the deprotonation process is not controlled by spontaneous proton release but that it requires a proton-accepting water molecule. Assuming that the concentration of

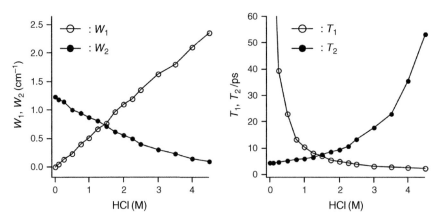

Figure 8 Concentration dependence of the rate parameters (left) and characteristic times (right) of the protonation/deprotonation dynamics of the $\text{DMAA}/\text{DMAAH}^+$ system.

free water molecule is 50 M in the neutral DMAA solution, the rate constant k_2 is calculated to be $k_2 = 4.6 \times 10^9\,\mathrm{s}^{-1}\,\mathrm{M}^{-1}$. This value is less than a half of that of the protonation reaction. Nevertheless, deprotonation occurs faster than protonation because of the high concentration of water. From the two rate parameters, the characteristic times of the reaction, T_1 and T_2, are calculated as $T_1 = (2\pi c W_1)^{-1}$ and $T_2 = (2\pi c W_2)^{-1}$, where c is the light speed. The calculated values are plotted in Figure 8 (right). From the figure, we can see that the protonation occurs in a few tens of picosecond time in dilute acids and that it occurs much more frequently in higher acid concentrations. The lifetime of the protonated form is as short as several picoseconds in dilute acids but it is prolonged to a few tens of picosecond in concentrated acid solutions.

3. TIME- AND SPACE-RESOLVED RAMAN SPECTROSCOPY OF SINGLE LIVING YEAST CELLS

One of the ultimate goals of science is to elucidate life at the molecular level. Experimental approach to this goal definitely needs the help of molecular spectroscopy. Confocal Raman microspectroscopy provides a powerful non-destructive means with high time, space, and molecular specificity, which are very essential for in vivo molecular–level studies of living cells. Typical measurement time needed for obtaining a cell spectrum under microscope is 100 s. Thus, it is possible to trace the cellular dynamics that proceeds in a much slower time scale. As shown below, we are able to trace the mitosis process of a fission yeast (*Schizosaccharomyces pombe*) that takes a few hours for one cycle. If visible light is used for excitation, we can easily achieve sub-micrometer spatial resolution. We use the 632.8 nm line of a He–Ne laser for excitation to achieve a spatial resolution of 250 nm for the xy and 1.8 mm for z (depth) directions. This spatial resolution is good enough to measure separately the Raman spectra of organelles like nucleus, a group of mitochondria, septum, and vacuole in a yeast cell. Raman spectra are often called "molecular fingerprints." Once a Raman spectrum is obtained, we can identify the molecule, obtain direct information on its structure and dynamics, and also specify the environment in which it is placed. Thus, time- and space-resolved Raman spectroscopy using a confocal microscope provides an ideal tool for studying the structure and dynamics of living cells at the molecular level. In fact, since the pioneering work by Puppels and co-workers [18], there have been many reports on Raman spectroscopic studies of "living" cells. However, there has been no direct evidence that these studies were actually looking at a cell in the living state.

3.1. Time- and space-resolved Raman spectra of a dividing fission yeast cell [19]

Yeast is the simplest eukaryote that serves as a prototype model for studying a variety of biological phenomena at the cellular level. It is well established that what is true for yeast is likely to be also true for higher organisms including human cells. Yeast is also expected to play a major role in cell engineering in which useful properties and functions are provided to a cell through gene manipulation. Fission yeast (*Schizosaccharomyces pombe*) is a kind of yeast that repeats cell division according to the cycle G2→M→G1/S→G2. Although a number of detailed microscopic and electron-microscopic observations have been reported, molecular-level tracing of yeast mitosis in vivo was not possible till the present time. We have succeeded in recording the time- and space-resolved Raman spectra (time resolution: 100 s, spatial resolution: 250 nm) of a dividing *S. pombe* cell on a confocal Raman microspectrometer. We used an *S. pombe* cell whose nuclei were labeled by Green Fluorescent Protein so that the position and the number of nuclei are directly seen as fluorescence images. Figure 9 shows the Raman spectra of the central part (the area surrounded by small red circles) of an *S. pombe* cell in the M–G2 period of the mitosis [19]. The Raman spectrum of the dividing nucleus at the M period (0 min) is dominated by known protein bands. As the mitosis proceeds, the spectrum changes to that of mitochondria, which contain many strong bands of phospholipids, then it further changes to that of septum and finally to that of the cell wall. This is the first in vivo observation of the dynamic behavior of biomolecules during the cell division. The change indicates clearly that the cell is actually "living" during the Raman measurement.

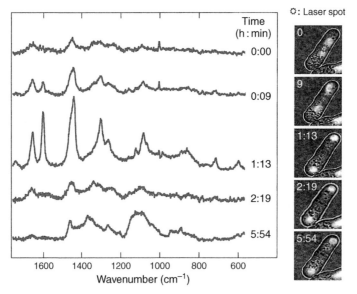

Figure 9   Time- and space-resolved Raman spectra of a dividing *Schizosaccharomyces pombe* cell (See color plate 2).

3.2. Discovery of the "Raman spectroscopic signature of life" in the mitochondria of a living fission yeast cell [20,21]

An intense Raman band is observed at $1602\,\mathrm{cm}^{-1}$ in the mitochondria of a living fission yeast cell (spectrum at 1 h and 13 min in Figure 9). From a Raman mapping experiment of a cell whose mitochondria are GFP labeled, we proved that the $1602\,\mathrm{cm}^{-1}$ band came exclusively from mitochondria. Figure 10 shows the (i) microscopic optical image, (ii) GFP image, (iii) Raman images at $1602\,\mathrm{cm}^{-1}$, and (iv) that at the phospholipid band at $1446\,\mathrm{cm}^{-1}$. The GFP image shows the location of mitochondria, which coincides very well with the Raman image at $1602\,\mathrm{cm}^{-1}$. This coincidence clearly indicates that the molecular species giving rise to the $1602\,\mathrm{cm}^{-1}$ band are localized within mitochondria. The Raman image at $1446\,\mathrm{cm}^{-1}$ produces the shape of the cell with several strong areas that overlaps very well with the mitochondria distribution. As mitochondria are highly rich in phospholipids, because of its double membrane structure, the mapping with the phosphlipid band gives a Raman image of mitochondria distribution.

We also found that the intensity of this band was correlated strongly with the living activity of the cell [20]. This relationship of the $1602\,\mathrm{cm}^{-1}$ band with the cell activity was examined by a respiration inhibiting experiment using KCN (Figure 11). In this experiment, we first measured the space–resolved Raman spectra of a fission yeast cell in water ($-5\,\mathrm{min}$). Then we added KCN to make the final concentration 0.5 mM. Soon after adding KCN (3 min), the $1602\,\mathrm{cm}^{-1}$ band became weaker while the other Raman bands ascribed to phospholipids remained unchanged. Then, the $1602\,\mathrm{cm}^{-1}$ band became further weaker (11 and 19 min) and disappeared eventually (36 min). The phospholipid bands change their shapes gradually after 19 min. We consider that the spectral changes in Figure11 reflect the effect of KCN in two steps. First, the respiration is inhibited by KCN and the metabolic activities in mitochondria are subsequently lowered. The intensity of the $1602\,\mathrm{cm}^{-1}$ band is most likely to be the measure of the metabolic activity in mitochondria. Then, with the lowering of metabolic activities, the double membrane structures of mitochondria are degraded. The gradual changes of the phospholipid bands reflect this structural degradation of mitochondria. Thus, we

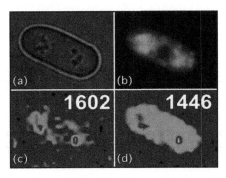

Figure 10 Optical image (a), GFP image, (b) and Raman images (c and d) of an *Schizosaccharomyces pombe* cell whose mitochondria are GFP labeled (See color plate 3).

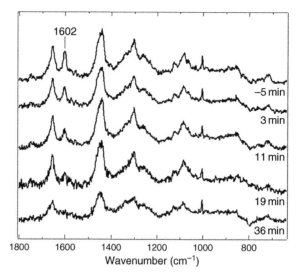

Figure 11 The effect of KCN on the Raman spectra of mitochondria of a living *Schizosaccharomyces pombe* cell.

are looking at the early dying process of a living yeast cell, which is induced by respiration inhibition with KCN, by Raman spectroscopy. In this regard, we call the $1602\,cm^{-1}$ band the "Raman spectroscopic signature of life" [20,21]. The discovery of this signature has opened up a new possibility of molecular level elucidation and quantification of life. In fact, we use this signature to observe the life and death of a starving budding yeast cell as shown in the next section.

3.3. Molecular level pursuit of a spontaneous death process of a starving budding yeast cell by time-resolved Raman mapping [22]

It is known that a particle called dancing body (DB) occasionally appears and moves vigorously in a vacuole of a budding yeast (*Schizosaccharomyces cerevisiae*) cell under a starving condition. We have found that once a DB is formed, the vacuole was eventually lost and the cell dyes subsequently. This spontaneous death process of a starving budding yeast cell was studied by time-resolved Raman mapping experiments (Figure 12). In this experiment, Raman images were obtained at 1602, 1440, 1160, and $1002\,cm^{-1}$. The $1602\,cm^{-1}$ band, the "Raman spectroscopic signature of life," indicates the distribution of active mitochondria. The $1440\,cm^{-1}$ band shows the distribution of phospholipids, in particular, the location of mitochondria that contain high concentration of phospholipids. The band at $1160\,cm^{-1}$ is due to the symmetrical stretching vibration of the PO_2^{-} moiety of polyphosphates. Thus, this band indicates the polyphosphates distribution in the cell. The $1002\,cm^{-1}$ band, which is assigned to the breathing mode of phenylalanine residues, shows the protein distribution in the cell.

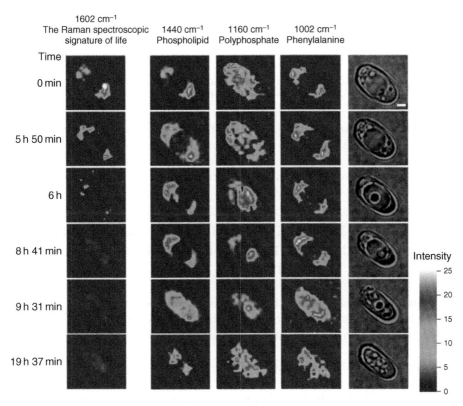

Figure 12 Time–resolved Raman mapping of a spontaneous cell death process of a staving *S. cerevisiae* cell (See color plate 4).

At 0 min, the $1602\,\text{cm}^{-1}$ image shows that there exist active mitochondria in the cell. The cell is in a normal condition. The 1440 and $1002\,\text{cm}^{-1}$ images indicate that phospholipids and proteins are located only outside the vacuole. At 5 h 50 min, the cell seems to be still under normal conditions with a clear pattern of the $1602\,\text{cm}^{-1}$ image. Between 5 h 50 min and 6 h, a DB appears and, concomitantly, the activity of mitochondria is significantly lowered. The Raman image at $1440\,\text{cm}^{-1}$ indicates that the phospholipids distribution does not change much and the distribution of mitochondria remains unchanged at 6 h. The protein distribution ($1002\,\text{cm}^{-1}$) does not change either. Polyphosphates appear to exist all over the vacuole in the $1160\,\text{cm}^{-1}$ image because of the trapping of the DB by the laser field. At 8 h 41 min, the mitochondrial activity completely stops according to the $1602\,\text{cm}^{-1}$ image. There is no active mitochondrion in the cell at this time. The DB stops moving so that the $1160\,\text{cm}^{-1}$ band gives a distribution localized at one end of the vacuole. It is evident that phospholipids and proteins are still localized outside of the vacuole and keep normal distribution. Between 8 h 41 min and 9 h 31 min, the vacuole disappears. The optical image at 9 h 31 min indicates that the cell structure is lost and the cell does not seem to be alive. The molecular distributions of phospholipids, polyphosphates and proteins become completely disordered at 19 h 37 min. The structure in the cell is totally lost at this time.

The metabolic activity in mitochondria completely stops at 8 h 41 min. However, the phospholipid and protein distributions do not change much between 6 h and 8 h 41 min. It is not easy to say when the cell is dead at the molecular level. Note that the cell death is conventionally detected by the dye permeability of the cell wall, which occurs much later than the process we look at in Figure 12 at the molecular level. In this way, the spontaneous cell death process following the DB formation was traced successfully by time-resolved Raman mapping. This result is of great interest regarding the definition of cell death at the molecular level.

4. LOCAL STRUCTURES AND THEIR TEMPORAL EVOLUTION IN AN ETHANOL/WATER MIXTURE PROBED WITH CARS SIGNAL SPATIAL DISTRIBUTION

Aqueous alcohol solutions, basic binary systems containing water, play major roles in chemistry as solvents for organic syntheses, liquid chromatography, solvent extraction, and so on. Ethanol/water mixtures are undoubtedly one of the most familiar solutions in our daily life as beverages. It is well known that the physico-chemical properties of alcohol/water mixtures are rather unusual [23]. For instance, the entropy increase upon mixing alcohol and water is much smaller than that expected for an ideal solution. Such unique behavior of alcohol/water mixtures has motivated many structural studies over the past decades. Evidence for incomplete mixing of alcohol and water at the molecular level has emerged from recent neutron scattering [24], X–ray emission [25], and low–frequency Raman scattering [26] experiments to explain the anomalous properties of alcohol/water mixtures. However, a decisive description of their microscopic structure is yet to be established.

We have recently demonstrated [27,28] that the measurement of the spatial distribution of the CARS signal quantitatively probes the local structures that are formed in liquids/solutions with high molecular specificity. The molecular specificity arises from the vibrational resonance in the CARS process. The sensitivity to the spatial size of local structures originates from the fact that the phase-matching condition of CARS cannot be perfectly satisfied and is relaxed to some extent when the sample medium is optically inhomogeneous due to the local structure formation. Figure 13 shows the CARS signal spatial distributions for polystyrene beads with different sizes (100 and 350 nm in diameter) dispersed in water [26]. In this figure, the observed CARS intensities are plotted against the angle θ between the phased matched direction and the direction of observation. The theoretical values of CARS intensities calculated on a simple model are also given in full lines showing a good agreement between experiment and theory. This result indicates that local structures with larger size tend to give narrower spatial distribution of the CARS signal.

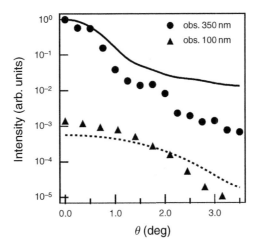

Figure 13 Coherent anti-Stokes Raman scattering (CARS) signal spatial distributions of polystyrene beads dispersed in water. Beads with 100 nm size (triangles) and those with 350 nm size (circles). Theoretical curves are given in dotted and full lines, respectively.

Figure 14 shows the spatial distribution patterns of the CARS signal of the C—C—O symmetric stretch mode of ethanol ($883\,cm^{-1}$ in pure ethanol and $880\,cm^{-1}$ in the ethanol/water mixture). Data for pure ethanol and ethanol/water mixtures (molar fraction of ethanol $x_E = 0.17$ or 40 vol %) are shown. The sample was prepared by mixing high-performance liquid chromatography (HPLC)-grade ethanol and distilled water. It was sonicated for 1 min and then sealed in a glass cuvette (path length, 5 mm) to prevent possible changes in the sample caused by volatilization, moisture absorption, and so on. All experiments were conducted at room temperature.

First, we compare the spatial distribution patterns for pure ethanol (broken line) and the ethanol/water mixture ($x_E = 0.17$, dotted line) prepared just before the

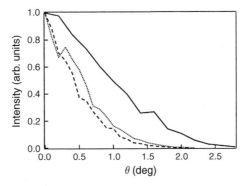

Figure 14 Coherent anti-Stokes Raman scattering (CARS) spatial distribution curves for neat ethanol (broken line), a 40% ethanol/water mixture just after mixing (dotted line), and the same mixture after 2 weeks (full line).

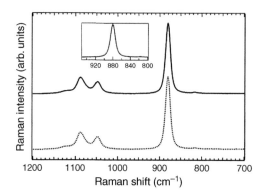

Figure 15   Raman spectra of the same 40% ethanol/water mixtures just after (full line) and 2 weeks after preparation (dotted line).

measurement. The mixture shows a slightly but meaningfully broadened pattern compared with pure ethanol. Even more intriguing is the fact that the distribution pattern exhibits *temporal evolution*. As can be seen from Figure 14, the pattern measured 2 weeks after mixing (full line) is apparently broader than that taken immediately after mixing (dotted line). Note that these samples were originally identical with each other, and attention was paid so as not to degrade the sample while being kept for the 2 weeks.

We also measured spontaneous Raman spectra of the two mixtures to check the possibility of sample degradation. The results are given in Figure 15, where all observed Raman bands are assignable to ethanol vibrations. From the inset of Figure 15, it can be clearly seen that peak position, band width, and relative intensity of the C—C—O symmetric stretch band are the same in these two spectra within experimental uncertainties. Similar agreement is found for other Raman bands as well. This fact suggests that the variation of the CARS signal spatial distribution does not result from the sample degradation. Rather, it reflects the formation and temporal evolution of local structures in the mixture that are not detectable by ordinary Raman spectroscopy.

As mentioned earlier, the smaller the local structure is, the broader the CARS signal spatial distribution becomes. Note that the coherent superposition of CARS polarization waves is more hampered by a larger number of smaller local structures. Hence the following scenario can be drawn for interpreting the present observation. In the ethanol/water mixture just after mixing, ethanol and water each form local structures with a size similar to that of bulk, engendering incomplete mixing at the microscopic level. Such bulk–like structures of ethanol and water seem to be qualitatively consistent with a previous result [29], in which the local structure of water in a methanol/water mixture resembles the corresponding structure in pure water. As more homogeneous mixing state is realized due probably to the re-formation of hydrogen bonding, the size of local structures becomes small (Figure 16). Consequently, the CARS signal spatial distribution becomes broadened. If this is the case, the formation and temporal evolution of the local structures in alcohol/water mixtures have been demonstrated for the first time.

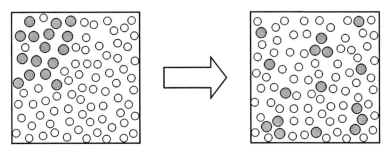

Figure 16 Possible formation and temporal evolution of the local structures in a 40% ethanol/ water mixture.

Combination of the present experiment with molecular dynamics simulation will shed more quantitative light on this interesting problem of ethanol/water mixture (which is known as whisky).

5. MOLECULAR NEAR-FIELD EFFECT IN RESONANCE HYPER-RAMAN SCATTERING

HR scattering bears vibrational information that is distinct either from infrared absorption or from Raman scattering [30]. In the last two decades, considerable research efforts have been directed toward HR scattering under two–photon resonance (resonance HR scattering). The mechanism of resonance HR scattering is now well understood in terms of a vibronic theory [31]. Recently, we have developed HR microspectroscopy that provides vibrational information on Raman inactive but infrared active modes of molecules under a microscope [32]. The spatial resolution of HR microspectroscopy is more than an order of magnitude better than that of the infrared counterpart. During the course of our HR microspectroscopic studies of all-*trans*-β-carotene, we found a new interesting phenomenon, enhancement of HR signals of solvents by a solute [33].

HR spectra of all-*trans*-β-carotene in cyclohexane, carbon tetrachloride, benzene, and CS_2 (all 2 mM concentration) are shown in Figure 17, together with that of a microcrystal. A strong band at $1574 \, \text{cm}^{-1}$ and two weaker bands at 1370 and $1322 \, \text{cm}^{-1}$ are observed in common to all the five HR spectra. They are unequivocally assigned to the a_u (infrared active and Raman inactive) vibrations of all-*trans*-β-carotene with reference to a normal coordinate analysis [34]. On the contrary, the spectrum in cyclohexane shows three extra bands at 1542, 2857, and $2933 \, \text{cm}^{-1}$, which are not present in the microcrystalline spectrum. Similarly, one extra band appears strongly at $794 \, \text{cm}^{-1}$ in the spectrum in carbon tetrachloride. The spectra in benzene and in CS_2 also show extra bands that are solvent specific. It has turned out that these extra HR bands agree very well with those of the infrared bands of the solvents. No HR signal is detected from neat solvents without all-*trans*-β-carotene. These results indicate that the extra HR bands originate from the solvents and that they are enhanced by an unknown mechanism associated with resonance HR scattering.

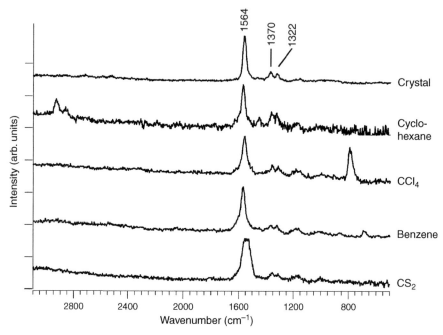

Figure 17 Hyper-Raman spectrum of crystalline all-*trans*-β-carotene and those of all-*trans*-β-carotene in four different solvents.

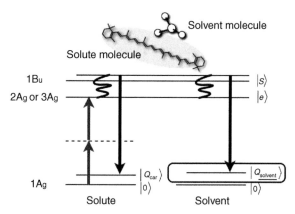

Figure 18 Intra- and intermolecular vibronic coupling and hyper-Raman (HR) process of all-*trans*-β-carotene including a solvent vibration.

In resonance HR scattering of all-*trans*-β-carotene, a two-photon allowed A_g state, and a one-photon allowed $1A_u$ ($1B_u$ under C_{2h} symmetry) state must be coupled via an a_u vibration (the so-called B-term [31]). The mixing of an A_g and an A_u states may also be induced by the solvent field if the energy gap between the two states is small (Figure 18). Through such intermolecular vibronic coupling, if exists, solvent vibrations can intervene (molecular near-field effect) the resonance HR process and give rise to the extra HR bands of the solvents. Only the solvent molecules in the close vicinity of all-*trans*-β-carotene can cause this effect. We expect that the present finding will lead to a new vibrational spectroscopy, "Molecular Near-field Hyper-Raman Spectroscopy," which selectively detects the molecules existing in the vicinity of a HR probe.

ACKNOWLEDGMENTS

The work described in this chapter has been done in collaboration with Dr. Daisuke Watanabe (Raman band shape analysis); Drs Yu-San Huang, Yasuaki Naito, Takeshi Karashima and Professors Y. Yamamoto and A. Toh-e (living cell Raman); Dr. Shinsuke Shigeto (CARS spatial distribution); and Dr Hideaki Kano and Mr. Rintaro Shimada (molecular near-field effect in resonance HR scattering). The author is grateful to these collaborators and all the other members of his laboratory at the University of Tokyo who contributed greatly to deepen the science involved in this work.

REFERENCES

[1] D.A. Long, "Raman Spectroscopy," McGraw-Hill (Tx) (1977).
[2] H. Hamaguchi and A. Hirakawa, "Raman Spectroscopy (in Japanese and Korean)," Gakkai Shuppan Center (1988).
[3] Y.R. Shen, "The principle of Noninear Optics," John Wiley and Sons, (1984).
[4] H. Hamaguchi, in "Nonlinear Specrtroscopy for Molecular Structure Determination" R.W. Field, E. Hirota, J.P. Maier, and S. Tsuchiya eds., Chapter 8 203–222, Blackwell Science, (1998).
[5] H. Hamaguchi, and T.L. Gustafson, Annu. Rev. Phys. Chem., **45** (1994) 593 and references therein.
[6] H. Hamaguchi and K. Iwata, Chem. Phys. Lett., **208**, 465 (1993).
[7] V. Deckert, K. Iwata, and H. Hamaguchi, J. Photochem. Photobiol., **102**, 35 (1996).
[8] H. Hamaguchi, Mol. Phys., **89**, 463 (1996).
[9] K. Iwata, R. Ozawa, and H. Hamaguchi, J. Phys. Chem. A, **106**, 3614 (2002).
[10] H. Hamaguchi and K. Iwata, Bull. Chem. Soc. Jpn., **75**, 883–897 (2002).
[11] D. Watanabe, and H. Hamaguchi, J. Chem. Phys., **123**, 034508 (2005).
[12] D. Watanabe and H. Hamaguchi, Chem. Phys., (in press).
[13] D. Watanabe, Chemical Reaction Dynamics in Solutions as Studied by Vibrational Band Shape Analyses, PhD thesis, The University of Tokyo (2007).
[14] Von M. Eigen and K. Tamm, Z. Elektrochem., **66**, 107 (1962).
[15] A.R. Davis and B.G. Oliver, J. Phys. Chem., **77**, 1315 (1973).
[16] Y. Saito and H. Hamaguchi, Chem. Phys. Lett., **339**, 351 (2001).
[17] P.W. Anderson, J. Phys. Soc. Jpn., **9**, 316 (1954).
[18] G.J. Puppels, F.F. de Mul, C. Otto, J. Greve, M. Robert-Nicoud, D.J. Arndt-Jovin, and T.M. Jovin, Nature, **347**, 301–303 (1990).
[19] Y.-S. Huang, T. Karashima, M. Yamamoto, and H. Hamaguchi, J. Raman Spectrosc., **34**, 1 (2003).
[20] Y.-S. Huang, T. Karashima, M. Yamamoto, T. Ogura, and H. Hamaguchi, J. Raman Spectrosc., **35**, 525 (2004).

[21] Y.-S. Huang, T. Karashima, M. Yamamoto, and H. Hamaguchi, Biochemistry, **44**, 10009 (2005).
[22] Y. Naito, A. Toh-e, and H. Hamaguchi, J. Raman Spectrosc., 36 837 (2005).
[23] F. Franks and D.J.G. Ives, Quart. Rev. Chem. Soc., **20**, 1 (1966).
[24] S. Dixit, J. Crain, W.C.K. Poon, J.L. Finney, and A.K. Soper, Nature, **416**, 829 (2002).
[25] J.H. Guo, Y. Luo, A. Augustsson, S. Kashtanov, J.E. Rubensson, D.K. Shuh, H. Ågren, and J. Nordgren, Phys. Rev. Lett., **91**, 157401 (2003).
[26] K. Egashira and N. Nishi, J. Phys. Chem. B, **102**, 4054 (1998).
[27] S. Shigeto and H. Hamaguchi, Chem. Phys. Lett., **417**, 149 (2006).
[28] S. Shigeto and H. Hamaguchi, Chem. Phys. Lett., **427**, 329–332 (2006).
[29] S. Dixit, J. Crain, W.C.K. Poon, J.L. Finney, and A.K. Soper, Nature, **416**, 829 (2002).
[30] D.A. Long and L. Stanton, Proc. R. Soc. London Ser. A, **318**, 441 (1970).
[31] Y.C. Chung and L.D. Ziegler, J. Chem. Phys, **88**, 7287 (1988).
[32] R. Shimada, H. Kano, and H. Hamaguchi, Opt. Lett., **31**, 320–322 (2006).
[33] R. Shimada, H. Kano, and H. Hamaguchi, J. Raman spectrosc., **37**, 469–471 (2006).
[34] S. Saito and M. Tasumi, J. Raman. Spectrosc., **14**: 310 (1983).

CHAPTER 3

Structure and Dynamics of High Rydberg States Studied by High-Resolution Spectroscopy and Multichannel Quantum Defect Theory

Martin Schäfer *and* Frédéric Merkt

Contents

Abstract

The properties of high atomic and molecular Rydberg states and their dependence on the principal quantum number n are presented together with the spectroscopic methods used to study these states at high resolution. Such studies require multiphoton excitation schemes or single-photon excitation with vacuum ultraviolet (VUV) radiation. Narrow-bandwidth laser systems are used in pulsed-field-ionization zero-kinetic-energy photoelectron spectroscopy (PFI-ZEKE-PES) and Rydberg-state-resolved threshold-ionization spectroscopy (RSR-TIS) experiments, which yield precise values of ionization energies and rovibrational energy structures of cations. The main limitation in resolution originates from the Doppler effect and can be overcome by VUV-millimeter wave double-resonance experiments. In such experiments, even the hyperfine structure of Rydberg states can be resolved as is demonstrated for the rare gases krypton and xenon and for ortho-hydrogen. Multichannel quantum defect theory (MQDT) is the most appropriate and reliable tool for the analysis of the spectra of high atomic and molecular Rydberg states. The theory has recently been extended to treat the hyperfine structure of high Rydberg states.

Keywords: Rydberg states; molecular cations; autoionization; hyperfine structure; high-resolution spectroscopy; vacuum ultraviolet lasers; pulsed-field-ionization zero-kinetic-energy photoelectron spectroscopy; Rydberg-state-resolved threshold-ionization spectroscopy; VUV-millimeter wave double-resonance spectroscopy; multichannel quantum defect theory.

1.   INTRODUCTION

Thanks to recent progress in experimental techniques, the measurement of the finest details of the energy level structure of highly excited Rydberg states of atoms and molecules has become possible. From such measurements, a better understanding of atomic and molecular photoionization can be reached and the role of nuclear spin(s) in atomic and molecular photoionization can be studied for the first time. This chapter summarizes this experimental progress and explains how the combination of high-resolution spectroscopy and multichannel quantum defect theory (MQDT) can be used to obtain new information on atomic and molecular photoionization, on the energy level structure of highly excited electronic states, and on atomic and molecular cations.

Rydberg states are a common feature of atoms and molecules [1–4]. These electronically excited states form series that converge to each quantum state of a cation (Figure 1). These series, if they are not perturbed by other series or electronic states, can be described by the well-known Rydberg formula:

$$E(n, \ell; \Xi^+) = E_{\text{ion}}(\Xi^+) - \frac{R_M}{(n - \delta_\ell)^2}, \tag{1}$$

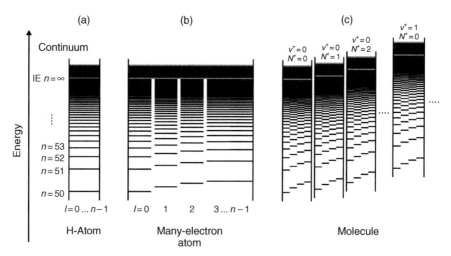

Figure 1 Schematic representations of Rydberg series in (a) the hydrogen atom, (b) a nonhydrogenic atom, and (c) a diatomic molecule. In a molecule, the Rydberg structure of a nonhydrogenic atom in panel (b) is repeated for each rovibrational state of the molecular cation.

where n is the principal quantum number, ℓ the orbital angular momentum quantum number of the Rydberg electron, R_M the mass-dependent Rydberg constant, δ_ℓ the quantum defect, and Ξ^+ represents the ensemble of quantum numbers of the ion. R_M is defined as $R_M = R_\infty \left(\frac{M}{M+m_e} \right)$, where M is the mass of the ionic core, m_e the electron mass, and $R_\infty = 109\,737.3157 \text{ cm}^{-1}$ is the Rydberg constant. The quantum defect δ_ℓ reflects the interaction between the Rydberg electron and the ionic core; for hydrogen-like systems or high-ℓ ($\ell > 3$) nonpenetrating Rydberg states, δ_ℓ is zero or close to zero, respectively. For low-ℓ Rydberg states, the electron penetrates the inner electron shells and is exposed to an increasing electronic charge for $r \rightarrow 0$. The deviation from the Coulomb potential at short range and the interaction between the Rydberg electron and the inner shell electrons are responsible for a nonvanishing quantum defect $\delta_\ell > 0$, which is generally weakly energy dependent. Extrapolation of Rydberg series to $n \rightarrow \infty$ yields accurate values of $E_{ion}(\Xi^+)$, i.e., the ionization energy and the energy level structure of the ion, which might be difficult to obtain by other spectroscopic methods. Rydberg series extrapolation has so far been used to determine the ionization potentials and the cationic electronic or rovibronic energy level structures of atoms and mostly small molecules such as H_2 [5], NO [6], N_2 [7], NH_3 [8], C_6H_6 [9], and C_6H_6–Rg (where Rg symbolize a rare gas atom) [10,11].

The simple Rydberg formula [Equation (1)] is no longer valid if there are series converging to different ionic states and interacting with each other. A more elaborate treatment in the form of MQDT [12–15] is needed in such cases. MQDT treats the Rydberg states from the perspective of a collision between an ion core and a Rydberg electron with an attractive electrostatic potential. The word "channel" indicates a set of states that consist of an electron of arbitrary energy and of a target ion in a specific quantum state; specification of the angular momenta of the electron and cationic core, and of their coupling, completes the identification of a channel [13,16]. If the energy of the electron lies below the ionization limit, the state belongs to the series of discrete bound states (Rydberg series) and the channel is said to be "closed"; if the energy of the electron is higher than the limit, the state belongs to the adjoining continuum and the channel is said to be "open". Perturbations between Rydberg series and configuration interaction (resulting from correlations between Rydberg electron and core electrons) are treated as channel mixing. Autoionization is described in MQDT as scattering of an electron in a closed channel into an open channel by collision with the ionic core [16]. A remarkable feature of MQDT is its ability to represent the effects of all interactions by means of a small set of physically meaningful parameters: eigen-quantum defects μ_α representing short-range electron–core interactions, dipole matrix elements D_α, the energy levels of the ion $E_i = E_{ion}(\Xi^+)$, and the frame transformation $U_{i\alpha}$ between the angular momenta coupling schemes of the close-coupling eigenchannels α and of the dissociation or ionization channels i [13,16,17]; μ_α, D_α, and $U_{i\alpha}$ generally depend weakly on the energy and, in molecules, on the spatial arrangement of the nuclei (i.e., the internuclear distances). Explicit formulae for special cases are given below in Section 3.

Table 1 Properties of Rydberg atoms[a]

Property	n dependence	Na(10d)[b]	H(100d)
Binding energy	n^{-2}	$1100 \, \text{cm}^{-1}$	$11 \, \text{cm}^{-1}$
Energy between adjacent n states	n^{-3}	$190 \, \text{cm}^{-1}$	$0.22 \, \text{cm}^{-1}$
Threshold ionization field	n^{-4}	$33 \, \text{kV cm}^{-1}$	$3.3 \, \text{V cm}^{-1}$
Orbital radius	n^2	$147 \, a_0$	$1.50 \times 10^4 a_0$
Geometric cross section	n^4	$6.8 \times 10^4 \, a_0^2$	$7.1 \times 10^8 a_0^2$
Dipole moment $\langle n\ell \mid er \mid n(\ell+1)\rangle$	n^2	$143 \, ea_0$	$1.50 \times 10^4 ea_0$
Polarizability	n^7	$0.21 \, \text{MHz cm}^2 \, \text{V}^{-2}$	$2 \times 10^6 \, \text{MHz cm}^2 \, \text{V}^{-2}$
Radiative lifetime	n^{3c}	$1.0 \, \mu\text{s}$	$0.53 \, \text{ms}^d$

[a] Atomic units: Bohr radius $a_0 = 0.5292 \times 10^{-10}$ m; dipole moment $ea_0 = 8.478 \times 10^{-30}$ Cm $= 2.542$ D.
[b] From Ref. [1].
[c] The lifetime of Rydberg molecules under typical PFI-ZEKE-PES or MATI conditions scales with n^5 (see text).
[d] Extrapolated from the values in Ref. [120].

An interesting aspect of Rydberg states is that their physical properties scale with integer powers of the principal quantum number n (see Table 1). These properties determine the behavior of Rydberg states and must be considered when planning spectroscopic experiments. The spectral density of Rydberg states scales with n^3, and thus a high spectral resolution is required if the high members of the Rydberg series are to be resolved. The long lifetimes of these states imply narrow linewidths. The low binding energy of the Rydberg electron (the threshold field for field ionization is proportional to n^{-4}) and the very high polarizability of these states ($\propto n^7$) render the elimination of stray electric fields imperative [18]. These fields may result from rest or patch surface potentials on the walls surrounding the measurement region or from ions or dipolar particles in the measurement volume. Therefore, a careful design of the experimental chamber is necessary. With a residual electric field of about $45 \, \mu\text{V cm}^{-1}$, Neukammer et al. [19] were able to resolve $6snd \, ^1D_2$ Rydberg states of barium up to $n = 520$. In general, the power of the radiation source used to observe transitions to or from Rydberg states must be kept low, particularly if low-frequency radiation is used. Rydberg atoms and molecules are strongly affected by blackbody radiation at room temperature [2]. The unusually strong effect of thermal radiation originates from two sources. First, the energy spacing ΔE between high-n Rydberg levels is similar to the thermal energy kT at 300 K. Second, the dipole matrix elements for transitions between Rydberg states are large, so that blackbody radiation induces transitions to energetically nearby states or can also photoionize Rydberg atoms or molecules. In addition, blackbody radiation shifts the energy levels by a second-order AC Stark effect [2].

The n-scaling of the physical properties of Rydberg atoms and molecules (see Table 1) can be derived from the radial part of the Rydberg electron wavefunctions in the classically allowed region between the inner and the outer classical turning point [1,2]

$$r_{\text{o,i}} = \left(n^2 \pm n\sqrt{n^2 - \ell(\ell+1)} \right) a_0. \tag{2}$$

The expectation values $\langle r^\sigma \rangle$ with positive powers $\sigma > 0$ of r are primarily determined by the location of the outer classical turning point $r_o \approx 2n^2 a_0$, i.e., the "size" of the Rydberg orbit, from which we can estimate

$$\langle r^\sigma \rangle \propto n^{2\sigma}. \tag{3}$$

The cross sections for collisions or interaction with millimeter wave radiation are therefore proportional to n^4. The dipole moment matrix element $\langle n\ell | er | n(\ell + 1) \rangle$ is proportional to n^2 and the properties proportional to the dipole moment scale accordingly, e.g., the intensity of a transition between adjacent Rydberg states is proportional to n^4. The dipole moment induced by an electric field and, as a consequence, the energy shift for Stark states of a hydrogenic Rydberg atom is proportional to kn (where k is the difference between the parabolic quantum numbers n_1 and n_2 and $|k| < n$) or, for the outermost Stark states, proportional to n^2.

The expectation values with negative powers of r depend on the behavior of the wavefunction at small values of r, i.e., in the core region $r < r_c$, and on ℓ, which determines the inner classical turning point $r_i \approx \ell(\ell + 1)a_0$. The amplitude of the wavefunction in the core region is given by the normalization ($\psi(r < r_c) \propto n^{-3/2}$), thus, at small r and $n \gg \ell$,

$$\langle r^{-|\sigma|} \rangle \propto |\psi(r < r_c)|^2 \propto n^{-3}. \tag{4}$$

The radiative decay rate of high-n Rydberg states depends on the oscillator strength for transitions to the lowest states and therefore on the square of the corresponding transition dipole matrix element, which is proportional to the spatial overlap of the radial wavefunctions. Therefore, the oscillator strength and the decay rate scale with n^{-3} and the radiative lifetime of Rydberg states scales with n^3 [2]. Molecular Rydberg states are subject to decay processes such as predissociation and rotational and vibrational autoionization. These processes involve interactions between the Rydberg electron and the ion core and are important in penetrating Rydberg states only. The decay rate of low-ℓ Rydberg states is expected to be proportional to the Rydberg electron density in the core region scaling with n^{-3}. Thus, the lifetime of low-ℓ molecular Rydberg states also scales with n^3. The lifetime of Rydberg molecules under typical zero-kinetic-energy photoelectron spectroscopy (ZEKE-PES) conditions, however, scales as n^α with α between 4 and 5, because high-n Rydberg states are ℓ-mixed by residual electric fields so that the penetrating character of the ℓ-mixed states is reduced and the lifetimes prolonged by a factor of about n. Inhomogeneous stray fields such as those originating from ions in the vicinity of high Rydberg states may further induce m-mixing which can prolong the lifetime by an additional factor of n [4,20–22]. The n^3-scaling of the lifetime of Rydberg states can also be obtained from the period of a classically orbiting Rydberg electron. According to Kepler's third law, the square of the period of the orbiting electron is proportional to the cube of the semi-major axis of its orbit, $T \propto a^{3/2}$. Because the latter scales with n^2, the period scales with n^3 and the lifetime is proportional to the orbiting period if decay processes are thought to take place only in the region of the ion core.

Perturbation theory can be used to derive the n-scaling laws of other properties such as the polarizability, which is proportional to $\Sigma_m \dfrac{|\langle m|\mu|n\rangle|^2}{E_m - E_n}$ and scales with $n^2 \cdot n^2/n^{-3} = n^7$. The n-scaling laws of the coefficients for the long-range dispersion forces between Rydberg atoms can be evaluated in the same way; for example, the C_6 coefficient of the van der Waals interaction $V(R) = -C_6/R^6$ is obtained from a similar expression with four dipole matrix elements in the numerator and scales with n^{11} [23,24].

High-n Rydberg states can be efficiently and state-selectively ionized with a pulsed electric field and the accelerated charged species (ions or electrons) subsequently detected with high efficiency (>50 percent) using a microchannel plate (MCP) detector. Pulsed-field ionization is at the heart of the techniques of pulsed-field-ionization ZEKE-PES (PFI-ZEKE-PES) [25,26] and mass-analyzed threshold ionization (MATI) spectroscopy [27]. In these techniques, molecules are excited to long-lived high-n Rydberg states and the electrons or ions are detected after application of a small electric field. To obtain the correct values of the ionic energy levels from ZEKE and MATI experiments, it is necessary to understand the field-ionization dynamics of high Rydberg states and to apply the appropriate field correction (see Section 2.2).

For most atoms and molecules, the direct (single-photon) access of Rydberg states from the ground state requires light in the UV or VUV range. The term vacuum ultraviolet (VUV) refers to radiation with wavelength $\lambda < 200$ nm – light which is strongly absorbed by the oxygen and nitrogen in the air. The high-energy range of the VUV is often referred to as extreme ultraviolet (XUV) and lies beyond the so-called lithium fluoride (LiF) cutoff ($\lambda < 105$ nm), where no materials are known to transmit VUV light. The development of narrow-bandwidth VUV laser systems is described in Section 2.1. High Rydberg states can be accessed from the ground state by resonant or nonresonant multiphoton excitation. The resolution can be significantly enhanced by combination of high-resolution VUV lasers with low-frequency radiation such as microwave or millimeter wave radiation [2, 28–33]. UV/VIS or VUV-millimeter wave double-resonance experiments with selective field ionization were used to measure the properties of Rydberg states of sodium [30,34,35], rubidium [36], argon [31], krypton [18,33,37,38], xenon [39], benzene [40], and H_2 [41,42] and are described in Section 2.3.

In Section 3 applications of high-resolution spectroscopy to the study of the hyperfine structures of Rydberg states of exemplary atomic and molecular systems are described, and the evolution of the hyperfine pattern and the angular momentum coupling hierarchy with increasing n values is discussed. MQDT represents a particularly powerful way to describe this evolution. MQDT has recently been extended to treat the hyperfine structure of high Rydberg states and was used to derive the hyperfine structures of $^{83}Kr^+$ [38,43], $^{129/131}Xe^+$ [39,44], and ortho-H_2^+ [42,45]. The presence of nuclear spins and their coupling with electron spins lead to a wide range of dynamical processes, which are just beginning to be unravelled.

2. EXPERIMENTAL TECHNIQUES

2.1. VUV lasers

Early studies of Rydberg states by photoabsorption and photoionization spectroscopy used gas discharge lamps (e.g., helium, $\lambda = 60$–100 nm, or argon, 110–140 nm) as sources for VUV radiation [46–48]. With the development of the synchrotron, a broadly tunable and intense source of VUV radiation became available [48–51]. For both types of sources, the resolution is limited by the monochromators used to disperse the radiation to about 0.5 cm^{-1} ($\Delta\nu/\nu \approx 5 \times 10^{-6}$) [7,49,51]. A much higher spectral resolution, however, is obtained with VUV laser systems.

Tunable narrow-band VUV/XUV laser radiation is conveniently generated by nonlinear optical frequency conversion of visible or ultraviolet radiation in gases [52–54]. Because there are no transparent media for the XUV range, the frequency conversion (nonresonant frequency tripling or sum-frequency mixing) has to occur in a free gas jet [55,56]. VUV/XUV radiation up to 20 eV [57] is produced efficiently by resonant four-wave mixing in a rare gas: $\nu_{\mathrm{VUV}} = 2\nu_1 \pm \nu_2$, where $2\nu_1$ corresponds to a two-photon transition of the rare gas. Examples of two-photon resonances with which high VUV intensities can be reached and the wavenumber ranges of the VUV radiation that can be produced with these transitions are summarized in Table 2. The generated VUV/XUV radiation is separated from the fundamental laser radiation (ν_1 and ν_2) by using a toroidal dispersion grating, which can be used to collimate the diverging VUV radiation or even focus it into the photoexcitation region [58]. Using pulsed dye lasers with an intracavity étalon, VUV/XUV radiation with a spectral bandwidth of 0.1 cm^{-1} (12 μeV) and maximal XUV intensities of about 10^9–10^{10} photons/pulse after the monochromator can be produced, sufficient to resolve adjacent Rydberg states up to $n \approx 120$ [58].

Replacing the pulsed dye lasers by continuous-wave (cw) single-mode ring dye lasers, which have spectral bandwidths of less than 1 MHz, and amplifying the laser radiation in dye cells pumped by an injection-seeded Nd:YAG laser, near-Fourier-transform-limited VUV radiation with a spectral bandwidth of 250 MHz or

Table 2 Two-photon transitions in rare gases and VUV/XUV production ranges[a]

	Transition[b]	$2\tilde{\nu}_1$ (cm^{-1})	$2\tilde{\nu}_1 - \tilde{\nu}_2$ (cm^{-1})	$2\tilde{\nu}_1 + \tilde{\nu}_2$ (cm^{-1})
Xe	$5p^5 6p[1/2]_0 \leftarrow 5p^6\ (^1S_0)$	$80{,}118.962(3)^c$	$\leq 76{,}800$	$92{,}200$–$142{,}000$
	$5p^5 6p'[1/2]_0 \leftarrow 5p^6\ (^1S_0)$	$89{,}860.015(3)^c$		
Kr	$4p^5 5p[1/2]_0 \leftarrow 4p^6\ (^1S_0)$	$94{,}092.8632(14)^d$	$\leq 86{,}700$	$107{,}000$–$151{,}000$
	$4p^5 5p'[1/2]_0 \leftarrow 4p^6\ (^1S_0)$	$98{,}855.0703(14)^d$		
Ar	$3p^5 4p[1/2]_0 \leftarrow 3p^6\ (^1S_0)$	$107{,}054.2773(30)^{e,f}$	$55{,}000$–$96{,}000$	$121{,}500$–$161{,}000$
	$3p^5 4p'[1/2]_0 \leftarrow 3p^6\ (^1S_0)$	$108{,}722.6247(30)^{e,f}$		

VUV, vacuum ultraviolet; XUV, extreme ultraviolet.
[a] Tuning range of variable frequency laser $\tilde{\nu}_2$: $13{,}000$–$52{,}000$ cm^{-1} (770–190 nm).
[b] The notation of the Rydberg levels of rare gases is explained in Section 3.1.
[c] Value for the isotopic center of gravity of natural Xe [121].
[d] Value for the most abundant isotope ^{84}Kr [121].
[e] Value for the most abundant isotope ^{40}Ar [121].
[f] $\tilde{\nu}_1 + \tilde{\nu}_1'$ where $\tilde{\nu}_1' = 63{,}439.322(40)$ cm^{-1} (157.6 nm F$_2$ excimer line) [57].

$0.008 \, \mathrm{cm}^{-1}$ can be obtained, with which Rydberg series can be resolved up to $n \approx 200$ [59–63]. A smaller bandwidth requires longer pulses, which can be generated by using Ti^{3+}-doped sapphire (Ti:Sa) crystals instead of dye cells. The much longer lifetime of the population inversion in the crystal compared to the dye solution permits the generation of longer laser pulses, but the amplification factor is by many orders of magnitude smaller. Therefore, many more amplification steps are required; this can be achieved by guiding the laser beam to be amplified many times through the Ti:Sa crystals [64]. A solid–state VUV laser system with a 55 MHz ($0.0018 \, \mathrm{cm}^{-1}$) bandwidth delivering about 10^8 photons/pulse (at a repetition rate of 25 Hz) is described in References [65,66] and depicted schematically in Figure 2. This bandwidth enables the resolution of adjacent members of a series up to $n > 300$.

The highest resolution can be obtained using cw radiation sources. Such sources (in the UV/VIS range) can be used to study high Rydberg states of alkali and alkaline earth metal atoms [67] or of the rare gas atoms from the lowest $^3P_0$ and $^3P_2$

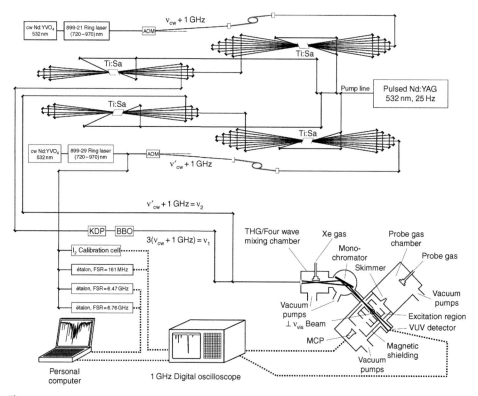

Figure 2 Schematic representation of the solid-state vacuum ultraviolet (VUV) laser system and the experimental set-up described in Ref. [66]. The upper part displays the generation of near infrared (NIR) pulses from cw ring lasers using pulsed accusto-optic modulators (AOMs) and multipass arrangements for the pulse amplification in Nd:YAG-pumped Ti^{3+}-doped sapphire (Ti:Sa) crystals. The amplified output from one ring laser is tripled in a potassium dihydrogen phosphate (KDP) and a beta–barium borate (BBO) crystal to produce the radiation (with frequency ν_1) used in the four-wave mixing process $\nu_{\mathrm{VUV}} = 2\nu_1 \pm \nu_2$ in a rare gas (Xe). The lower part shows the vacuum chambers, the calibration set-up, and the data acquisition systems.

metastable levels [68]. However, cw radiation in the VUV range is very difficult to generate and to scan [69].

2.2. PFI-ZEKE photoelectron spectroscopy and Rydberg-state-resolved threshold-ionization spectroscopy

In classical photoelectron spectroscopy, the sample is irradiated by light of a fixed frequency ν and the kinetic energy of the ejected photoelectron E_{kin} is measured; the ionization potential is obtained from the energy difference $h\nu - E_{kin}$ [70,71]. The resolution of this technique is limited to about $20\,\mathrm{cm}^{-1}$ by technical difficulties associated with the measurement of the kinetic energy of the electrons. In threshold photoelectron spectroscopy (TPES), the wavelength of the light is scanned and a signal is observed for photoelectrons with a given (low) kinetic energy; this occurs at the ionization threshold if photoelectrons with quasi-zero kinetic energy are observed [72–76].

In the 1980s, Müller-Dethlefs, Schlag and coworkers [26] developed ZEKE-PES. The original idea was to photoexcite the sample to the ionization potential and to extract the photoelectrons with zero kinetic energy after a certain delay time that allows electrons with nonzero kinetic energy to escape the extraction region. It was later realized that these "zero-kinetic-energy" electrons are actually electrons produced by field-ionization of high Rydberg states and that the ionization potentials have to be corrected by a field-induced shift $\Delta E/hc \approx 4\,\mathrm{cm}^{-1}\sqrt{F/(\mathrm{V\,cm}^{-1})}$ [20,25,77]. Therefore, the method is nowadays referred to as PFI-ZEKE-PES. Application of a suitable sequence of pulsed electric fields with a first discrimination field of opposite polarity enabled a higher selectivity in the pulsed-field ionization of high Rydberg states with $n > 100$ and a resolution of $0.055\,\mathrm{cm}^{-1}$ [63] (Figure 3).

The precision and accuracy of the determination of ionization potentials can be improved with Rydberg-state-resolved threshold-ionization spectroscopy (RSR–TIS) [78–80], where the resolution is limited by the bandwidth of the tunable photoexcitation source (see Section 2.1). A sequence of pulsed electric fields is used to field-ionize first the highest Rydberg states that cannot be resolved and subsequently those with $n = 100 - 200$ (Figure 4). Extrapolation of the observed Rydberg series leads to accurate values of the ionization thresholds. This technique has been applied to determine the ionization potentials of Ar [78], Kr [79], N_2 [78,80], CO [80], NH_3 [8], and the spin–rotation splitting in some levels of N_2^+, CO^+ [80], and NH_3^+ [8] (see Figure 4). An advantage of this technique is that it can simplify the analysis of overlapping Rydberg series, which lead to a congested spectrum in the standard photoionization spectrum, by a large reduction of the number of peaks in the spectrum [78]. This is an experimental alternative to the automatic computer analysis of dense Rydberg spectra using the crosscorrelation ionization energy spectroscopy (CRIES) method [9].

2.3. VUV-millimeter wave double-resonance spectroscopy

In Section 2.1 we have discussed the development of narrow-bandwidth VUV laser systems. When carrying out spectroscopic measurements with these lasers, the resolution is limited by the Doppler width caused by residual velocity components in the transverse direction of the skimmed supersonic beam [66]. Because the Doppler

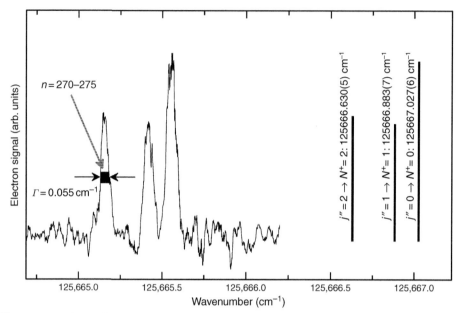

Figure 3 High-resolution pulsed-field-ionization zero-kinetic-energy (PFI-ZEKE) photo-electron spectrum of the $N^+ - J'' = 0$ branch of the N_2^+ X $^2\Sigma_g^+(v^+ = 0) \leftarrow N_2$ X $^1\Sigma_g^+(v = 0)$ transition obtained by selective field-ionization of Rydberg states with n between 270 and 275. The vertical lines mark the field-free ionization thresholds. (Adapted from Reference [63].)

broadening is proportional to the frequency of the radiation, a higher resolution can be attained by using low-frequency radiation to probe the Rydberg states. Low-frequency radiation such as microwave or millimeter wave radiation has, in rotational spectroscopy, the disadvantage of small absorption coefficients, which are proportional to the frequency and the square of the dipole moment, and therefore low sensitivities [81]. For Rydberg–Rydberg transitions, however, the transition dipole moments are very large (see Table 1) and the transitions can be efficiently detected by selective pulsed-field ionization of the final Rydberg states; therefore, microwave and millimeter wave spectroscopy turns out to be a very sensitive technique.

In our VUV-millimeter wave double-resonance experiments, a narrow-bandwidth VUV laser, which excites the atoms or molecules of interest to high Rydberg states, is combined with a continuous millimeter wave source. Millimeter wave radiation is produced by phase- and frequency-stabilized backward wave oscillators (BWOs) operating in the frequency ranges 118–180 GHz and 240–380 GHz with spectral bandwidths of less than 1 kHz [31,33]. The spectral bandwidth of the latter system has been found to be limited by the phase noise of the reference microwave synthesizer and is much smaller than the error in the determination of transition frequencies because of broadening effects and frequency shifts as discussed below. The output powers of our BWOs are of the order of 1–3 mW for the low and up to 20 mW for the high-frequency source; the output power of the latter, however, varies significantly with frequency, and about 10–20 percent of the radiation is used for the frequency stabilization. These output powers are much higher than what is

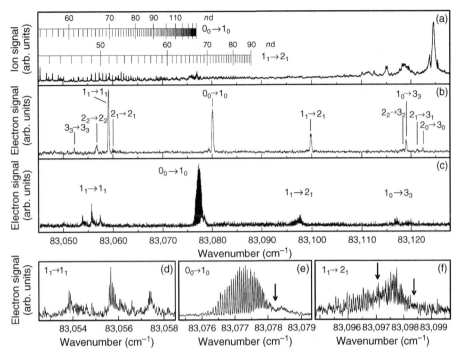

Figure 4 Photoionization spectrum (a), pulsed-field-ionization zero-kinetic-energy (PFI-ZEKE) photoelectron spectrum (b), and Rydberg-state-resolved threshold-ionization (RSR-TI) spectrum (c) of NH_3 in the region of the 2_0^1 band of the $\tilde{X} \to \tilde{X}^+$ photoionization transition. The transitions are labeled with the rotational quantum numbers J_K and $N_{K_+}^+$ of the ground and ionic state, respectively. Panels (d)–(f) show the three main lines of the RSR-TI spectrum on an expanded scale. The Rydberg series in panels (e) and (f) consist of states with n between 140 and 180. The nodal structures (marked with arrows) arise from the superposition of two Rydberg series converging to closely separated ionization limits corresponding to the two spin–rotational components of the ionic state with $J^+ = N^+ \pm 1/2$. (Adapted from Reference [8].)

actually needed for the double-resonance experiments because the transition dipole moments between Rydberg states with neighboring principal quantum numbers are very large. Experiments with the low-frequency source have shown that power densities of the order of $100\,\text{nW cm}^{-2}$ are sufficient to notably broaden transitions between neighboring Rydberg states so that the radiation has to be attenuated by 50–60 dB to about 0.5–$5\,\text{nW cm}^{-2}$. Even for transitions with larger Δn such as transitions between $n = 115$ and $n = 200$ Rydberg states, it was necessary to attenuate the power density to about $10\ \mu\text{W cm}^{-2}$ in order to avoid power-broadening [31,32].

Laser and millimeter wave radiation cross the skimmed gas beam at right angles in order to minimize the Doppler shift and broadening. The millimeter wave transitions are detected by selective field ionization of the final or initial Rydberg states [32]. The applied electric field accelerates the produced ions or electrons to a MCP detector. The detection of the ions offers the possibility to obtain mass-selective spectra based on the different times of flight of ions of different masses. This feature is particularly useful when studying the hyperfine structure of selected isotopes or isotopomers in a natural sample.

As already mentioned in the introduction, the polarizability of Rydberg states is very high, so that even small electric fields may induce large electric dipole moments and transition frequencies are shifted due to the Stark effect. Effects of stray electric fields are illustrated in Figure 5, which also shows how stray electric fields can be measured by millimeter wave spectroscopy of Rydberg states and how they can be compensated in the longitudinal direction (i.e., in the direction toward the detector), so that the transition frequencies can be determined with an accuracy of better than 1 MHz [18]. The increased resolution of the millimeter wave spectrum with respect to the optical laser spectrum is illustrated in Figure 6: the laser spectrum in panel (a) covers a frequency region 7.5 times larger than that of the millimeter wave overview spectrum in panel (b).

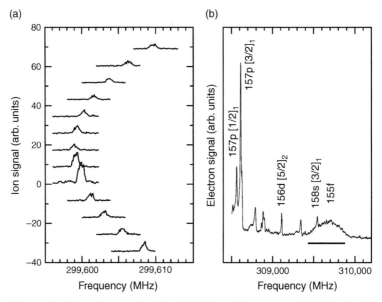

Figure 5 Determination of stray electric fields from the spectra of high-n Rydberg states. Panel (a) shows the millimeter wave spectra of the 70 d$[1/2]_1 \leftarrow$60p $[1/2]_1$ transition of ^{84}Kr in the presence of different electric fields. The spectra have been shifted along the vertical axis by an offset corresponding to the value of the applied field in mVcm^{-1} (top spectrum 69.0 mVcm^{-1}, bottom spectrum -34.5 mVcm^{-1}). This transition exhibits a quadratic Stark effect. The line positions form a parabola, whose apex corresponds to the field necessary to compensate the stray field (about 15 mVcm^{-1}). Panel (b) shows the millimeter wave spectrum of transitions from the 87d$[3/2]_1$ Rydberg state to high-n Rydberg states of krypton. Because of the presence of a weak stray field, mixing between different ℓ states occurs and transitions to ns, nd, and high-ℓ Rydberg states are observable. The broad unresolved structure at the position of the 155f Rydberg levels marked with a horizontal line corresponds to Stark states of the high-ℓ manifold. These nearly degenerate states undergo a linear Stark effect, and the magnitude of the stray field (here about 5 mVcm^{-1}) can be deduced from the width of this Stark manifold given by the relation $\Delta v/\mathrm{MHz} = 3.84 n^2 (F/\mathrm{Vcm}^{-1})$ ($3n^2 F$ in atomic units).

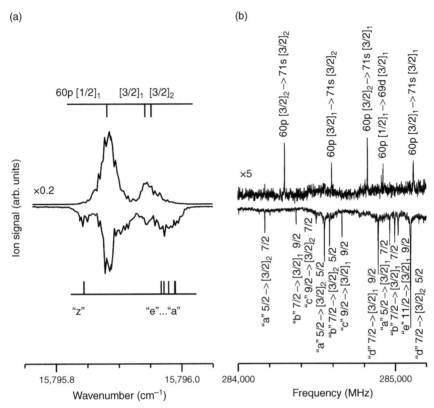

Figure 6 Comparison of the pulsed-field-ionization laser spectrum of the 60p ← 4d[1/2]₁ transitions (panel a) and a section of the millimeter wave spectrum of the 71s ← 60p transitions of krypton (panel b). The upper traces are the signals observed in the ^{84}Kr mass channel, and the lower inverted traces are those in the ^{83}Kr mass channel; note that in panel (a) the ^{84}Kr signal was slightly leaking into the ^{83}Kr mass channel. Assignment bars in panel (a) indicate the positions of the levels probed by millimeter wave spectroscopy. The millimeter wave spectra in panel (b) were recorded when the tunable dye laser was set to excite the high-frequency end of the 60p state (in the range of the levels "a"–"c" of ^{83}Kr). Note that the labels of the 71s $[K]_J F$ ← 60p $[K]_J F$ transitions of ^{83}Kr in panel (b) are given in abbreviated form, where the 60p hyperfine level is designated by a specific letter and the F value. (Adapted from Reference [38].)

The line widths of the millimeter wave transitions can be reduced by attenuating the laser and the millimeter wave intensities and increasing the interaction time, i.e., the time between the laser and field ionization pulses [18,31,32]. In most cases, the linewidths are limited by the interaction time (typically a few μs) or the lifetime of the Rydberg state: lines as narrow as 60 kHz have been observed for krypton with an interaction time of 18 μs [18]. For sufficiently long-lived Rydberg states, the interaction time is ultimately limited by the transit time of the gas particles through the interaction region. In a supersonic expansion, the typical velocity is about 2900, 550, and 305 ms^{-1} for an expansion of pure H$_2$, Ar, and Xe, respectively, from a room temperature reservoir [82,83]. The velocity of a light molecule can be reduced if it is seeded in a supersonic expansion of a heavy rare gas (so called carrier gas); the velocity becomes nearly equal to that of the carrier gas. A further reduction of the velocity can be achieved by exploiting the high polarizability of Rydberg states: electric fields induce large dipole moments, which can be used to decelerate Rydberg atoms or molecules (see Outlook below).

3. HYPERFINE STRUCTURE OF RYDBERG STATES

3.1. Rare gas atoms: krypton and xenon

The study of Rydberg states of rare gases have played an important role in the understanding of photoionization [84,85] and in the development of MQDT [12,86–89]. Rare gases other than helium have two ion core levels in the $(np)^5$ ground state configuration ($^2P_{3/2}$ and $^2P_{1/2}$) well separated from the next higher electronic terms (see Figure 7). Because there are only two ion core levels, a graphical method that displays, in a so-called Lu–Fano plot, the (near-)periodicity

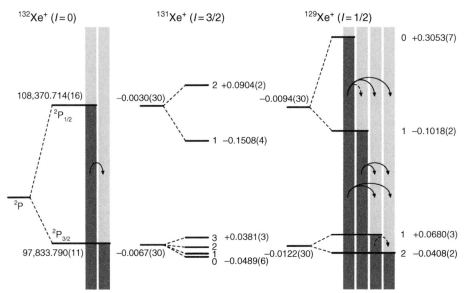

Figure 7 Energy level diagram of the ^2P ground state of $^{132}Xe^+$, $^{131}Xe^+$, and $^{129}Xe^+$ as determined from high-resolution spectra [39,44]. For each spin–orbit component of the ion ($^2P_{3/2}$ and $^2P_{1/2}$), the wavenumbers (cm^{-1}) of the ionization energies of ^{132}Xe from the $^1S_0$ ground state and the isotope shifts (cm^{-1}) for ^{131}Xe and ^{129}Xe are given. The hyperfine levels for $^{131}Xe^+$ and $^{129}Xe^+$ are labeled with the quantum number F^+ and their positions (cm^{-1}) with respect to the center of gravity of the hyperfine structure are given. For ^{132}Xe and ^{129}Xe, the different autoionization processes are indicated schematically. Dark gray regions are closed channels representing series of discrete Rydberg states and light gray regions are open channels, i.e., the adjoining continua. Spin–orbit autoionization processes are indicated by full arrows and hyperfine autoionization processes by dashed arrows.

of the MQDT equations and the perturbation between different Rydberg series can be easily applied to extract MQDT parameters [86]. Beutler [84] observed autoionizing Rydberg series between the two ionization limits; these series exhibit narrow and broad lines with asymmetric shapes, which have been analyzed by Fano [85,90] and have subsequently been called Beutler–Fano profiles. Transitions between Rydberg states of krypton have been proposed as wavelength standards; it may be specially noted that the 6d[1/2]$_1$–5p[1/2]$_1$ transition of ^{86}Kr ($\lambda_{vac} = 605.780210$ nm) was used from 1960 until 1983 to define the length standard of the meter [91,92]. A vast amount of research has been carried out on the study of spectra of rare gases. However, only few studies have observed the hyperfine structure in the spectra of rare gases; studies of the hyperfine structure of high-n Rydberg states are particularly scarce (References [38,43,44] and references therein). Of the heavier rare gases, argon has no naturally occurring isotope with nuclear spin $I \neq 0$ and neon only one (^{21}Ne) with very low natural abundance (0.27 percent). In contrary, krypton has one (^{83}Kr with $I = 9/2$, 11.5 percent) and xenon two (^{129}Xe with $I = 1/2$, 26.4 percent, and ^{131}Xe with $I = 3/2$, 21.2 percent) isotopes with appreciable natural abundance, and these two gases represent ideal systems to study the role of the nuclear spin in photoionization dynamics.

In atomic and molecular physics, the process of autoionization is classified as electronic, vibrational, spin–orbit, or rotational according to the type of energy that is transferred from the ion core to the Rydberg electron. In rare gas isotopes with $I = 0$, autoionization results in a change of the spin–orbit state of the ion core, whereas in $I \neq 0$ isotopes, the autoionization may also involve a change in the hyperfine state of the ion core either with or without a change of spin–orbit state (see Figure 7). This offers the possibility to produce ions in a selected hyperfine state. To study these dynamical processes, MQDT was extended to include nuclear spins in order to derive partial photoionization cross sections to selected hyperfine states of the ion [44].

The Rydberg states of the rare gases are usually labeled using the $n\ell[K]_J(F)$ notation based on the pair coupling ($j\ell$ or jK coupling) scheme [93]:
$$\vec{L}^+ + \vec{S}^+ = \vec{J}^+, \; \vec{J}^+ + \vec{\ell} = \vec{K}, \; \vec{K} + \vec{s} = \vec{J}, \; \vec{J} + \vec{I} = \vec{F}, \text{ where } \vec{L}^+ \text{ and } \vec{S}^+$$
represent the orbital and electron spin angular momenta of the ionic core, $\vec{\ell}$ and $\vec{s}$ the corresponding angular momenta of the Rydberg electron, and $\vec{I}$ the nuclear spin. In the pair coupling scheme, the electrostatic quadrupole energy for given J^+ and ℓ values is diagonal with respect to K if the short-range Rydberg electron–core electron interactions (e.g., the spin–orbit interaction, which splits the K states into doublets with $J = K \pm 1/2$) are small compared to the fine-structure splitting of the ionic core; this is especially the case for the nonpenetrating high-ℓ Rydberg states [12,94–96]. Rydberg series converging to the first ionization threshold ($^2P_{3/2}$) are designated ns, np, nd, etc., those converging to the second one ($^2P_{1/2}$) are designated ns', np', nd', etc.

By single-photon excitation from the $(np)^6{}^1S_0$ ground state, only ns and nd Rydberg states with $J = 1$ are accessible in isotopes with $I = 0$ as a consequence of the standard selection rules for electric dipole transitions. For isotopes with nuclear spin $I \neq 0$, the ΔJ selection rule has to be replaced by the corresponding rule for ΔF. Therefore, it is possible to access hyperfine levels of ns and nd Rydberg states of [83]Kr or [129/131]Xe with $J = 2$ (and $J = 3$) from the [1]S_0 ground state (see Figures 8 and 9) [43,44].

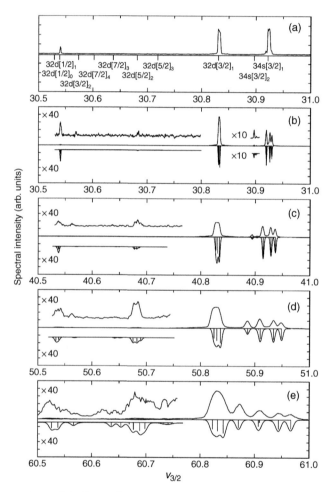

Figure 8 Comparison of the hyperfine structure in experimental spectra and multichannel quantum defect theory (MQDT) simulations (inverted traces and stick spectra) of ns and nd Rydberg states of [83]Kr corresponding to single-photon excitation from the [1]S_0 ground state [panels (b)–(c)]. For direct comparison of the hyperfine structure at different n values, the spectra are displayed as a function of the effective principal quantum number $v_{3/2}$ defined with respect to the center of gravity of the hyperfine structure of the [2]$P_{3/2}$ ground state of [83]Kr[+]. In the uppermost panel (a), the spectrum of [84]Kr is shown for comparison, and the positions of the different ns and nd Rydberg levels as obtained in Reference [38] are indicated. This spectrum was taken under conditions where the strong transitions were saturated, so that the weak transition to the 32d[1/2]$_1$ level became clearly visible. (Adapted from Reference [43].)

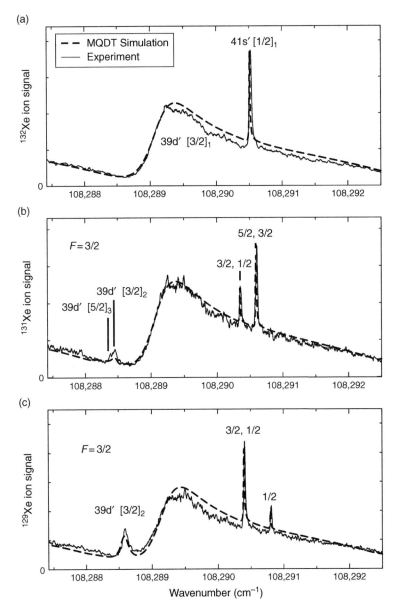

Figure 9 Experimental relative photoionization cross section of xenon in the region of the autoionizing 39d′[3/2]₁ state (solid line) and its multichannel quantum defect theory (MQDT) simulation (dotted line) for (a) ^{132}Xe, (b) ^{131}Xe, and (c) ^{129}Xe. (Adapted from Reference [44].)

The hyperfine structure of the ns and nd Rydberg series converging to the ^{2}P$_{3/2}$ ground state of ^{83}Kr$^+$ has been studied by Wörner *et al.* using a high–resolution VUV laser system (see Section 2.1) and mass–selective detection of the ions after pulsed–field ionization [43]. The mass resolution made it possible to obtain precise values of the ionization thresholds for the different isotopes [79]. For the ^{83}Kr isotope, it was found that hyperfine levels of ns and nd series start to mix around $n = 70$, which enables the study of the interaction between s and d channels, which results from the nonspherical symmetry of the ion core. In order to obtain quantitative values for this interaction and more accurate values of the hyperfine structure of the ion, a millimeter wave spectroscopy study of these states was undertaken [38]. Krypton atoms were first excited via the 4d $[1/2]_1$ state to np Rydberg states with $n = 58-62$ with a two-photon laser excitation scheme and subsequently with millimeter waves to ns and nd Rydberg states with $n = 68-76$ (see Figure 6). This excitation scheme allows access to significantly more levels than direct excitation from the ground state: for 69d/71s, 8 (out of 10) fine structure levels of ^{84}Kr and 40 (out of 48) hyperfine levels of 70d/72s of ^{83}Kr have been observed [38]. Such an almost complete set of experimental data on hyperfine levels represents an excellent basis to test the extension of the multichannel quantum defect method to treat nuclear spins [43] and to derive accurate values for the hyperfine structure of the ^{83}Kr$^+$ ^{2}P$_{3/2}$ ground state.

The hyperfine structure of Rydberg series converging to the lower (^{2}P$_{3/2}$) and upper (^{2}P$_{1/2}$) spin–orbit component of the 5p^5 ground state of $^{129/131}$Xe$^+$ (see Figure 7) has been studied using millimeter wave [39] and high-resolution VUV laser [44] spectroscopies, respectively. The autoionizing Rydberg series of even isotopes ($I = 0$) exhibit the typical Beutler–Fano line shape pattern (with narrow ns' and broad nd' lines), whereas for ^{129}Xe and ^{131}Xe, the ns' series exhibit an obvious splitting resulting from ns' series converging to the two hyperfine levels of the ^{2}P$_{1/2}$ state. Moreover, nd' series with $J = 2$ appear with increasing intensities (see Figure 9).

3.2. MQDT calculations for rare gases

The MQDT treatment of the Rydberg states of rare gases (without nuclear spins) was developed by Lu, Lee, Fano, and others [86–89, 97–99]. In the close-coupling region, the eigenchannels α are nearly LS coupled ($\vec{L}^+ + \vec{\ell} = \vec{L}$, $\vec{S}^+ + \vec{s} = \vec{S}$, $\vec{L} + \vec{S} = \vec{J}$, $\vec{J} + \vec{I} = \vec{F}$), whereas the jj angular momentum coupling scheme ($\vec{L}^+ + \vec{S}^+ = \vec{J}^+$, $\vec{J}^+ + \vec{I} = \vec{F}^+$, $\vec{\ell} + \vec{s} = \vec{j}$, $\vec{F}^+ + \vec{j} = \vec{F}$) is used to describe the dissociation channels i (at large distance). In the discrete part of the spectrum, the equation

$$\sum_{\alpha_F} U_{i_F \alpha_F} \sin\left[\pi\left(\mu_{\alpha_F} + \nu_{i_F}\right)\right] A_{\alpha_F} = 0, \tag{5}$$

which requires the wavefunction of the bound levels to vanish at infinity, is used to determine the positions of the bound Rydberg states. ν_{i_F} is an effective principal quantum number $\nu_{J^+F^+}$ defined by

$$E = E_{\text{ion}}\left(J^+F^+\right) - \frac{R_M}{\nu_{J^+F^+}^2} \tag{6}$$

with the mass-dependent Rydberg constant R_M and the ion energy level $E_{ion}(J^+F^+)$ associated with the dissociation channel i. The transformation matrix $U_{i_F\alpha_F}$ differs slightly from the jj–LS frame transformation matrix $\langle LSJF|J^+F^+jF\rangle$ because of the spin–orbit interaction and the deviation of the electrostatic potential from a pure Coulomb potential. The coefficients A_{α_F} enable the expansion of the dissociation channels in the basis of the close-coupling eigenchannels. Equation (5) has nontrivial solutions when

$$\det\left|U_{i_F\alpha_F}\sin\left[\pi\left(\mu_{\alpha_F}+\nu_{i_F}\right)\right]\right|=0, \tag{7}$$

and the values of ν_{i_F} satisfying this relation correspond to the bound Rydberg levels. The intensities $\mathcal{I}$ of the transitions from the ground state to the bound Rydberg states can be calculated from the dipole amplitudes D_{α_F} and the expansion coefficients A_{α_F} according to

$$\mathcal{I}\propto W_F\left(\sum_{\alpha_F}D_{\alpha_F}A_{\alpha_F}\right)^2, \tag{8}$$

where W_F is a weighting factor that accounts for the multiplicity of the Rydberg states $W_F=(2F+1)/(2I+1)$ [43,44].

If the total energy lies between the lowest and the highest ionic level included in the MQDT model, some dissociation channels are closed (forming an ensemble denoted Q) and some are open (forming an ensemble labeled P). In addition to the boundary condition represented by Equation (5) for the closed channels, the open-channel wavefunctions should behave at large r as collision eigenfunctions of the open channels, labeled ρ, with a phase shift τ_ρ; this boundary condition is represented by the following equation:

$$\sum_{\alpha_F}U_{i_F\alpha_F}\sin\left[\pi\left(-\tau_{\rho_F}+\nu_{i_F}\right)\right]A_{\alpha_F}=0. \tag{9}$$

For each value of the total energy in the autoionizing region, there are as many solutions τ_{ρ_F} and associated vectors of expansion coefficients $\mathbf{A}^{\rho_F}$ as open channels. These coefficients are obtained in a single step by solving the equation [13]

$$\Gamma\mathbf{A}^{\rho_F}=\tan\left(\pi\tau_{\rho_F}\right)\Lambda\mathbf{A}^{\rho_F}, \tag{10}$$

where

$$\Gamma_{i_F\alpha_F}=\begin{cases} U_{i_F\alpha_F}\sin\left[\pi\left(\mu_{\alpha_F}+\nu_{i_F}\right)\right] & \text{for } i\in Q, \\ U_{i_F\alpha_F}\sin\left(\pi\mu_{\alpha_F}\right) & \text{for } i\in P, \end{cases} \tag{11}$$

$$\Lambda_{i_F\alpha_F}=\begin{cases} 0 & \text{for } i\in Q, \\ U_{i_F\alpha_F}\cos\left(\pi\mu_{\alpha_F}\right) & \text{for } i\in P. \end{cases} \tag{12}$$

Because the electrostatic interaction in the close-coupling region is much bigger than the hyperfine interaction, the same sets of eigenquantum defects μ_α and dipole transition amplitudes D_α can be used for all isotopes and all F values. In this approximation, the parameter set for $^{A=2n+1}\mathrm{Rg}$ $(I \neq 0)$ differs from that of $^{A=2n}\mathrm{Rg}$ $(I=0)$ only by the additional parameters describing the hyperfine structure of the ion $E_{\mathrm{ion}}(J^+F^+)$ and a small isotope shift of the ion energy level with respect to the ground state of the neutral atom.

This MQDT model is capable of reproducing the experimentally determined fine and hyperfine structure of high-n Rydberg states of $^{84}\mathrm{Kr}$ and $^{83}\mathrm{Kr}$ within the experimental accuracy of $<1\,\mathrm{MHz}$ reached in millimeter wave spectroscopy experiments [38] (see Figure 10). The MQDT analysis enables one to derive the hyperfine structure of the $^{83}\mathrm{Kr}^+$ $^2P_{3/2}$ ground state with MHz accuracy (see Figures 10 and 11). Figure 11 shows the evolution of the hyperfine structure from low-n to high-n states. At low n, the hyperfine splitting is small compared to the spacing between the Rydberg states, and the quantum defect does not differ significantly from that of the $I=0$ isotope. A typical hyperfine interval of about $1\,\mathrm{GHz}$ $(0.03\,\mathrm{cm}^{-1})$ is indeed negligible compared to the interval between adjacent members of the same Rydberg series at low n $(\approx 10\,000\,\mathrm{cm}^{-1})$. Consequently, the

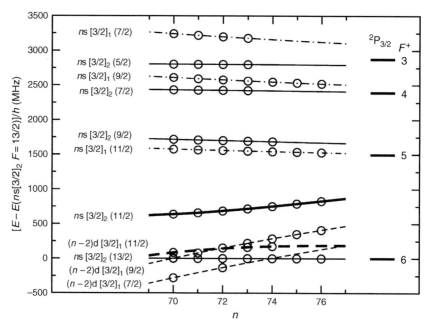

Figure 10 Hyperfine structure of high ns Rydberg states of $^{83}\mathrm{Kr}$: experimental values are indicated with circles and lines connect the calculated level positions. The avoided crossing between the $F=11/2$ levels of $ns\,[3/2]_2$ and $(n-2)\mathrm{d}[3/2]_1$ is marked with bold lines. The size of the circles does not represent the uncertainty of the experimental measurements, which is better than $1\,\mathrm{MHz}$. On the right the hyperfine structure of the $^2P_{3/2}$ ground state of $^{83}\mathrm{Kr}^+$ as derived from the MQDT analysis is given. The intervals are 484(15) MHz, 894(7) MHz, and 1496(15) MHz, respectively [38].

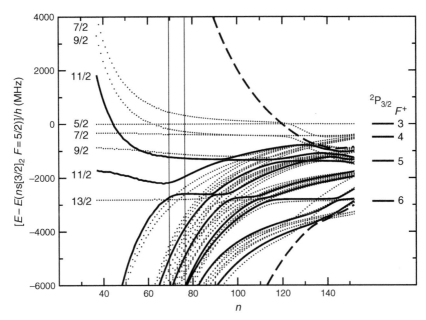

Figure 11 Hyperfine structure of the ns ($37 \leq n \leq 152$) Rydberg states of ^{83}Kr and of the $^2P_{3/2}$ state of ^{83}Kr$^+$ as derived from the multichannel quantum defect theory (MQDT) analysis. The $F = 11/2$ levels are highlighted by solid black curves. The limits of the region studied in millimeter wave experiments (see Figure 10) are indicated by thin vertical lines. For $n > 60$, avoided crossings due to interactions between ns and $(n-2)$d hyperfine levels can be observed. For $n > 120$, the hyperfine levels mix with those of the next higher or lower n, indicated by the broad dashed lines representing the lowest $F = 11/2$ level of $(n-1)$d/ $(n+1)$s and the highest $F = 11/2$ level of $(n-3)$d/$(n-1)$s, respectively. (Adapted from Reference [38].)

relative error in the quantum defect that results from neglecting the hyperfine structure is of the order of 10^{-5}. At higher n (above 60 for ^{83}Kr), hyperfine levels of nd and $(n+2)$s states approach each other and typical hyperfine intervals become comparable with intervals between adjacent Rydberg states of the same series. Moreover, avoided crossings occur as a result of interactions between different channels (see also Figure 10). For $I = 0$ isotopes, such s–d interactions can be studied only at low n between series converging to different ionization levels. As soon as the spacing between adjacent Rydberg levels of the same series becomes smaller than the hyperfine splitting of the ion, hyperfine levels of different n states start to mix and the hyperfine structure gets complicated.

For the low-n states, the use of μ eigenquantum defects occasionally poses problems because the determinantal equation [Equation (7)] yields unphysical solutions for $\ell > 0$. This can be avoided when η quantum defects [12,100] are used instead. They have the additional advantage that the description of their energy dependence (for $\ell < 3$) requires less parameters to obtain the same precision as can be achieved when using μ defects [101].

3.3.   Ortho-hydrogen

Hydrogen H_2 and its cation H_2^+ are the simplest molecular systems. They have played a crucial role in the development of molecular physics and served as a theoretical and experimental proving ground for quantum mechanics; the Rydberg states of H_2 have also been the testing ground for molecular MQDT for many decades [5,13,17,42,45,102–110]. The normal isotopomer of H_2 consists of two nuclear spin isomers: ortho-H_2 with total nuclear spin $I=1$ and para-H_2 with $I=0$. For symmetry reasons, ortho-H_2 occurs in the electronic ground state only with odd rotational quanta ($N=1,3,\ldots$) and para-H_2 only with even rotational quanta ($N=0,2,\ldots$). The nonzero nuclear spin of ortho-H_2 makes it a good model system to study the role of nuclear spins in the photoionization of molecular systems and to investigate how it influences, for example, the rotational or vibrational autoionization dynamics. The same procedure as outlined above for the rare gas atoms can also be used to extract the hyperfine structure of selected rotational levels of ortho-H_2^+ at sub–MHz precision and accuracy (see Figure 12) [42].

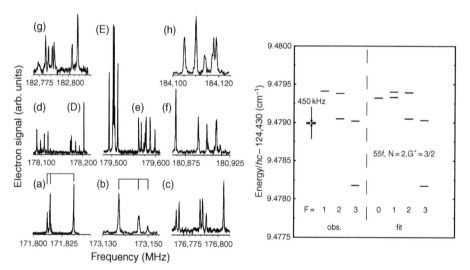

Figure 12   Hyperfine structure of high Rydberg states of ortho-H_2 converging to the $X\,^2\Sigma_g^+$, $v^+=0$, $N^+=1$ level of H_2^+. The left side displays millimeter wave spectra of transitions from the hyperfine structure components of the $51d1_1$ Rydberg state to the hyperfine structure components of the $55p1_0$, $55p1_1$, and $55f1_2$ Rydberg states (Hund's case (d) labels $n\ell N^+{}_N$). (a) $55p1_0$ ($S=0$) ← $51d1_1$ ($G^+=3/2$), (b) $55p1_0$ ($S=0$) ← $51d1_1$ ($G^+=1/2$), (c) $55p1_1$ ($S=1$) ← $51d1_1$ ($G^+=3/2$), (d) $55p1_1$ ($S=1$) ← $51d1_1$ ($G^+=1/2$), (D) $55f1_2$ ($G^+=1/2$) ← $51d1_1$ ($G^+=3/2$), (E) $55f1_2$ ($G^+=1/2$) ← $51d1_1$ ($G^+=1/2$), (e) $55f1_2$ ($G^+=3/2$) ← $51d1_1$ ($G^+=3/2$), (f) $55f1_2$ ($G^+=3/2$) ← $51d1_1$ ($G^+=1/2$), (g) $55p1_1$ ($S=0$) ← $51d1_1$ ($G^+=3/2$), (h) $55p1_1$ ($S=0$) ← $51d1_1$ ($G^+=1/2$). Because the $55p1_0$ ($S=0$) Rydberg state is not split by the hyperfine interaction, the (inverted) hyperfine structure of the two G^+ components of the initial $51d1_1$ state is directly observable in traces (a) and (b). On the right-hand side, the experimentally determined hyperfine structure of the $55f1_2$ ($G^+=3/2$) Rydberg state is compared with the calculated positions. The experimental accuracy of about 450 kHz in this range is delimited by the two vertical arrows. (Adapted from Reference [42].)

The angular momenta in diatomic Rydberg molecules allow a large number of possible different angular momenta coupling schemes. For sufficiently high Rydberg states, Hund's case (d) limit is approached, in which the angular momentum of the Rydberg electron ℓ is coupled to the core rotation $\vec{N}^+$ to form $\vec{N} = \vec{N}^+ + \vec{\ell}$, and the Rydberg states are labeled $n\ell N^+_N$. The coupling schemes for the spins, however, depend on n and ℓ. In the range $n = 50-65$, two situations have been observed. For the penetrating s and p series, the exchange interaction between core and Rydberg electron is sufficiently large so that the core electron spin $\vec{S}^+$ couples to the Rydberg electron spin $\vec{s}$ and not the nuclear spin $\vec{I}$. As a result, S remains an approximately good quantum number. For the nonpenetrating f Rydberg series, and to a lesser extent the d series, the total core spin quantum number G^+ (but not S) is almost a good quantum number, and the f (and also the d) states have mixed singlet and triplet characters [42].

In Figure 13, high-resolution laser spectra of transitions between the $H\,^1\Sigma_g^+ (v=0, J=3)$ state of ortho-H$_2$ and $np3_N$ ($N=2,3,4$) Rydberg states converging to the $X\,^2\Sigma_g^+$, $v^+ = 0$, $N^+ = 3$ level of H$_2^+$ are displayed. Because the $N=2$ Rydberg levels can rotationally autoionize even in the absence of nuclear spins (interaction with the $N=2$ continuum of the $v^+ = 0$, $N^+ = 1$ level of ortho-H$_2^+$),

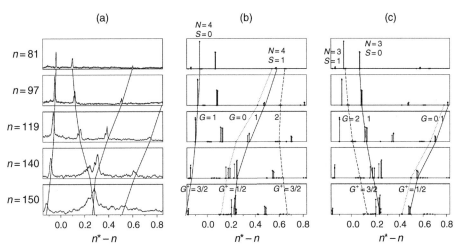

Figure 13 Comparison of (a) experimental photoionization spectra and (b) and (c) MQDT calculations of the hyperfine structure of ortho-H$_2$ rotationally auto-ionizing $np3_4$ and $np3_3$ Rydberg series. The spectra are represented on the scale of an effective quantum number n^* defined with respect to the center of gravity of the hyperfine components of the $X\,^2\Sigma_g^+$, $v^+ = 0$, $N^+ = 3$ level of H$_2^+$. The integer value of n^* for each spectrum is given at the left margin. The $G=0$, 1, and 2 levels are connected by dotted, full, and dashed lines, respectively; in the experimental spectra (panel a), only transitions to Rydberg levels with $G=1$ bear significant intensities. For $n < 80$, the fine and hyperfine structures are dominated by the singlet–triplet splitting, whereas for $n > 140$, the G^+ structure of the ion is dominating the hyperfine structure. (Adapted from Reference [45].)

they appear only as broad features at low n values ($n < 80$). The $N = 3$ and 4 levels cannot autoionize (if interactions with the nuclear spins can be neglected) and appear as sharp lines. The Rydberg series show a transition from the coupling case called (d$_{\beta S}$) $\left[\vec{S} = \vec{S}^+ + \vec{s}, \vec{G} = \vec{I} + \vec{S}, \vec{F} = \vec{N} + \vec{G}\right]$ for low-n states to the coupling case (d$_{\beta S+}$) $\left[\vec{G}^+ = \vec{I} + \vec{S}^+, \vec{G} = \vec{G}^+ + \vec{s}, \vec{F} = \vec{N} + \vec{G}\right]$ for high-n states. In both cases, an intermediate angular momentum quantum number G corresponding to the total (electronic and nuclear) spin represents a good quantum number to very good approximation [45]. At $n \approx 80$, the interval between the singlet and triplet levels is comparable to the hyperfine splitting of the ion; thus, neither the total electronic spin quantum number S nor the total core spin quantum number G^+ are good quantum numbers. At $n \approx 140$, the Hund's case (d) fine-structure splittings are smaller than the G^+ splitting of the ion, and the spectrum is dominated by the G^+ ionic structure. Because the Rydberg levels are excited from a nearly pure singlet level of ortho-H_2 with $G = 1$ and only Rydberg levels with $G = 1$ are observed, it can be concluded that the total spin quantum number G remains a good quantum number up to $n \approx 170$ even though the total electron spin is not conserved. From this conservation rule, one can conclude that rotational autoionization connecting channels of different N values induced by the nuclear spins only becomes significant beyond $n = 170$, when the interaction between the core rotation N^+ and the core spin S^+ starts to efficiently mix channels of different N values. The diversity and complexity of physical phenomena in high Rydberg states become remarkable when the effects of nuclear spins are also considered. The systematic investigation of these phenomena is just beginning and promises to provide a better understanding of photoionization in atoms and molecules, to uncover so far unknown processes, and to lead to new methods of preparing ions in selected hyperfine levels.

4. CONCLUSION AND OUTLOOK

The systematic improvement of spectral resolution and sensitivity in spectroscopic measurements of high Rydberg states over the past decades enables one today to study the finest details of the energy level structure of atoms and molecules, including their fine and hyperfine structures. The combination of high-resolution spectroscopy and MQDT provides a powerful method to determine the hyperfine structure of ions and to study the photoionization dynamics at an unprecedented degree of detail.

Further improvements of the resolution of spectroscopic measurements of high Rydberg states will necessitate the preparation of translationally cold samples of Rydberg atoms and molecules at rest in the laboratory frame. Because of the large induced dipole moments of Rydberg states, Rydberg atoms and molecules in supersonic beams can be decelerated with inhomogeneous electric fields so that the time of interaction of the atoms or molecules with the electromagnetic radiation and thus the spectral resolution of spectroscopic measurements can be increased. Static electric fields have been used to decelerate or deflect argon [111], krypton [112], and hydrogen molecules [113,114]; using time-dependent electric fields, argon atoms could be more efficiently decelerated [115,116], and hydrogen atoms could be decelerated, reflected, and trapped [117–119]. The deceleration of Rydberg atoms and molecules in supersonic beams offers the prospect of a new generation of high-resolution spectroscopic experiments.

REFERENCES

[1]   T. F. Gallagher, Rep. Prog. Phys., 51: 143–188, 1988.
[2]   T. F. Gallagher, Rydberg Atoms, Cambridge University Press, Cambridge, 1994.
[3]   G. Herzberg, Annu. Rev. Phys. Chem., 38: 27–56, 1987.
[4]   F. Merkt, Annu. Rev. Phys. Chem., 48: 675–709, 1997.
[5]   G. Herzberg and Ch. Jungen, J. Mol. Spectrosc., 41: 425–486, 1972.
[6]   E. Miescher, Can. J. Phys., 54: 2074–2092, 1976.
[7]   K. P. Huber and Ch. Jungen, J. Chem. Phys., 92: 850–861, 1990.
[8]   R. Seiler, U. Hollenstein, T. P. Softley and F. Merkt, J. Chem. Phys., 118: 10024–10033, 2003.
[9]   R. G. Neuhauser, K. Siglow and H. J. Neusser, J. Chem. Phys., 106: 896–907, 1997.
[10] K. Siglow, R. Neuhauser and H. J. Neusser, J. Chem. Phys., 110: 5589–5599, 1999.
[11] K. Siglow and H. J. Neusser, Faraday Discuss., 115: 245–257, 2000.
[12] M. J. Seaton, Rep. Prog. Phys., 46: 167–257, 1983.
[13] C. H. Greene and Ch. Jungen, Adv. At. Mol. Phys., 21: 51–121, 1985.
[14] Ch. Jungen (Ed.), Molecular Applications of Quantum Defect Theory, Institute of Physics Publishing, Bristol and Philadelphia, 1996.
[15] M. Aymar, C. H. Greene and E. Luc-Koenig, Rev. Mod. Phys., 68: 1015–1123, 1996.
[16] U. Fano, J. Opt. Soc. Am., 65: 979–987, 1975.
[17] U. Fano, Phys. Rev. A, 2: 353–365, 1970; Phys. Rev. A, 15: 817, 1977.
[18] A. Osterwalder and F. Merkt, Phys. Rev. Lett., 82: 1831–1834, 1999.
[19] J. Neukammer, H. Rinneberg, K. Vietzke, A. König, H. Hieronymus, M. Kohl, H. J. Grabka and G. Wunner, Phys. Rev. Lett., 59: 2947–2950, 1987.
[20] W. A. Chupka, J. Chem. Phys., 98: 4520–4530, 1993.
[21] W. A. Chupka, J. Chem. Phys., 99: 5800–5806, 1993.
[22] F. Merkt, J. Chem. Phys., 100: 2623–2628, 1994.
[23] M. Marinescu and A. Dalgarno, Phys. Rev. A, 52: 311–328, 1995.
[24] C. Boisseau, I. Simbotin and R. Côté, Phys. Rev. Lett., 88: 133004, 2002.
[25] G. Reiser, W. Habenicht, K. Müller-Dethlefs and E. W. Schlag, Chem. Phys. Lett., 152: 119–123, 1988.
[26] K. Müller-Dethlefs and E. W. Schlag, Annu. Rev. Phys. Chem., 42: 109–136, 1991.
[27] L. Zhu and P. Johnson, J. Chem. Phys., 94: 5769–5771, 1991.
[28] T. F. Gallagher, R. M. Hill and S. A. Edelstein, Phys. Rev. A, 13: 1448–1450, 1976.
[29] K. A. Safinya, T. F. Gallagher and W. Sandner, Phys. Rev. A, 22: 2672–2678, 1980.
[30] C. Fabre, P. Goy and S. Haroche, J. Phys. B: At. Mol. Phys., 10: L183–L189, 1977.
[31] F. Merkt and H. Schmutz, J. Chem. Phys., 108: 10033–10045, 1998.
[32] F. Merkt and A. Osterwalder, Int. Rev. Phys. Chem., 21: 385–403, 2002.
[33] M. Schäfer, M. Andrist, H. Schmutz, F. Lewen, G. Winnewisser and F. Merkt, J. Phys. B: At. Mol. Opt. Phys., 39: 831–845, 2006.
[34] C. Fabre, S. Haroche and P. Goy, Phys. Rev. A, 18: 229–237, 1978.
[35] P. Goy, C. Fabre, M. Gross and S. Haroche, J. Phys. B: At. Mol. Phys., 13: L83–L91, 1980.
[36] W. Li, I. Mourachko, M. W. Noel and T. F. Gallagher, Phys. Rev. A, 67: 052502, 2003.
[37] F. Merkt, R. Signorell, H. Palm, A. Osterwalder and M. Sommavilla, Mol. Phys., 95: 1045–1054, 1998.
[38] M. Schäfer and F. Merkt, Phys. Rev. A, 74: 062506, 2006.

[39] M. Raunhardt, M. Schäfer and F. Merkt, unpublished results.
[40] A. Osterwalder, S. Willitsch and F. Merkt, J. Mol. Struct., 599: 163–176, 2001.
[41] A. Osterwalder, R. Seiler and F. Merkt, J. Chem. Phys., 113: 7939–7944, 2000.
[42] A. Osterwalder, A. Wüest, F. Merkt and Ch. Jungen, J. Chem. Phys., 121: 11810–11838, 2004.
[43] H. J. Wörner, U. Hollenstein and F. Merkt, Phys. Rev. A, 68: 032510, 2003.
[44] H. J. Wörner, M. Grütter, E. Vliegen and F. Merkt, Phys. Rev. A, 71: 052504, 2005; Phys. Rev. A, 73: 059904(E), 2006.
[45] H. J. Wörner, S. Mollet, Ch. Jungen and F. Merkt, Phys. Rev. A, 75: 062511, 2007.
[46] R. E. Huffman, J. C. Larrabee and D. Chambers, Appl. Opt., 4: 1145–1150, 1965.
[47] R. E. Huffman, J. C. Larrabee and Y. Tanaka, Appl. Opt., 4: 1581–1588, 1965.
[48] J. Berkowitz, Photoabsorption, Photoionization and Photoelectron Spectroscopy, Academic Press, New York, 1979.
[49] K. Ito, T. Namioka, Y. Morioka, T. Sasaki, H. Noda, K. Goto, T. Katayama and M. Koike, Appl. Opt., 25: 837–847, 1986.
[50] P. A. Heimann, M. Koike, C. W. Hsu, D. Blank, X. M. Yang, A. G. Suits, Y. T. Lee, M. Evans, C. Y. Ng, C. Flaim and H. A. Padmore, Rev. Sci. Instrum., 68: 1945–1951, 1997.
[51] L. Nahon, C. Alcaraz, J.-L. Marlats, B. Lagarde, F. Polack, R. Thissen, D. Lepère and K. Ito, Rev. Sci. Instrum., 72: 1320–1329, 2001.
[52] C. R. Vidal, Appl. Opt., 19: 3897–3903, 1980.
[53] B. P. Stoicheff, P. R. Herman, P. F. LaRocque and R. H. Lipson, in: Laser Spectroscopy VII (T. W. Hänsch and Y. R. Shen, ed.), Springer-Verlag, Berlin, 174–178, 1985.
[54] R. Hilbig, G. Hilber, A. Lago, B. Wolff and R. Wallenstein, Comments At. Mol. Phys., 18: 157–180, 1986.
[55] A. H. Kung, Opt. Lett., 8: 24–26, 1983.
[56] C. T. Rettner, E. E. Marinero, R. N. Zare and A. H. Kung, J. Phys. Chem., 88: 4459–4465, 1984.
[57] P. Rupper and F. Merkt, Rev. Sci. Instrum., 75: 613–622, 2004.
[58] F. Merkt, A. Osterwalder, R. Seiler, R. Signorell, H. Palm, H. Schmutz and R. Gunzinger, J. Phys. B: At. Mol. Opt. Phys., 31: 1705–1724, 1998.
[59] E. Cromwell, T. Trickl, Y. T. Lee and A. H. Kung, Rev. Sci. Instrum., 60: 2888–2892, 1989.
[60] W. Ubachs, K. S. E. Eikema, W. Hogervorst and P. C. Cacciani, J. Opt. Soc. Am. B, 14: 2469–2476, 1997.
[61] K. S. E. Eikema, W. Ubachs, W. Vassen and W. Hogervorst, Phys. Rev. A, 55: 1866–1884, 1997.
[62] U. Hollenstein, H. Palm and F. Merkt, Rev. Sci. Instrum., 71: 4023–4028, 2000.
[63] U. Hollenstein, R. Seiler, H. Schmutz, M. Andrist and F. Merkt, J. Chem. Phys., 115: 5461–5469, 2001.
[64] F. Brandi, I. Velchev, D. Neshev, W. Hogervorst and W. Ubachs, Rev. Sci. Instrum., 74: 32–37, 2003.
[65] R. Seiler, Th. Paul, M. Andrist and F. Merkt, Rev. Sci. Instrum., 76: 103103, 2005.
[66] Th. A. Paul and F. Merkt, J. Phys. B: At. Mol. Opt. Phys., 38: 4145–4154, 2005.
[67] K. Singer, M. Reetz-Lamour, T. Amthor, S. Fölling, M. Tscherneck and M. Weidemüller, J. Phys. B: At. Mol. Opt. Phys., 38: S321–S332, 2005.
[68] D. Klar, M. Aslam, M. A. Baig, K. Ueda, M.-W. Ruf and H. Hotop, J. Phys. B: At. Mol. Opt. Phys., 34: 1549–1568, 2001.
[69] K. S. E. Eikema, J. Walz and T. W. Hänsch, Phys. Rev. Lett., 86: 5679–5682, 2001.
[70] D. W. Turner and M. I. Al Jobory, J. Chem. Phys., 37: 3007–3008, 1962.
[71] D. W. Turner, C. Baker, A. D. Baker and C. R. Brundle, Molecular Photoelectron Spectroscopy: A Handbook of He 584 Å Spectra, John Wiley & Sons, New York, 1970.
[72] D. Villarejo, R. R. Herm and M. G. Inghram, J. Chem. Phys., 46: 4995–4996, 1967.
[73] W. B. Peatman, T. B. Borne and E. W. Schlag, Chem. Phys. Lett., 3: 492–497, 1969.
[74] T. Baer, W. B. Peatman and E. W. Schlag, Chem. Phys. Lett., 4: 243–247, 1969.
[75] R. Spohr, P. M. Guyon, W. A. Chupka and J. Berkowitz, Rev. Sci. Instrum., 42: 1872–1879, 1971.
[76] P. M. Guyon, J. Chim. Phys., 90: 1313–1324, 1993.
[77] R. Lindner, H.-J. Dietrich and K. Müller-Dethlefs, Chem. Phys. Lett., 228: 417–425, 1994.
[78] R. Seiler, U. Hollenstein, G. M. Greetham and F. Merkt, Chem. Phys. Lett., 346: 201–208, 2001.

[79] U. Hollenstein, R. Seiler and F. Merkt, J. Phys. B: At. Mol. Opt. Phys., 36: 893–903, 2003.

[80] R. Seiler, Rydbergzustandsaufgelöste Schwellenionisationsspektroskopie, PhD thesis, ETH Zürich, Switzerland, 2004.

[81] H. W. Kroto, Molecular Rotation Spectra, Wiley & Sons, New York, 1975.

[82] E. J. Campbell, L. W. Buxton, T. J. Balle, M. R. Keenan and W. H. Flygare, J. Chem. Phys., 74: 829–840, 1981.

[83] H. S. Zivi, A. Bauder and Hs. H. Günthard, Chem. Phys., 83: 1–18, 1984.

[84] H. Beutler, Z. Phys., 93: 177–196, 1935.

[85] U. Fano, Phys. Rev., 124: 1866–1878, 1961.

[86] K. T. Lu and U. Fano, Phys. Rev. A, 2: 81–86, 1970.

[87] K. T. Lu, Phys. Rev. A, 4: 579–596, 1971.

[88] C.-M. Lee and K. T. Lu, Phys. Rev. A, 8: 1241–1257, 1973.

[89] C. M. Lee, Phys. Rev. A, 10: 584–600, 1974.

[90] U. Fano, G. Pupillo, A. Zannoni and C. W. Clark, J. Res. Natl. Inst. Stand. Technol., 110: 583–587, 2005.

[91] V. Kaufman and C. J. Humphreys, J. Opt. Soc. Am., 59: 1614–1628, 1969.

[92] BIPM, Bureau international des poids et mesures / Organisation intergouvernementale de la Convention du Mètre (Ed.), Le Système international d'unités, STEDI Media, Paris, 8th edition, 2006.

[93] G. Racah, Phys. Rev., 61: 537, 1942.

[94] E. S. Chang and H. Sakai, J. Phys. B: At. Mol. Phys., 15: L649–L653, 1982.

[95] D. E. Kelleher and E. B. Saloman, Phys. Rev. A, 35: 3327–3338, 1987.

[96] D. E. Kelleher, Phys. Rev. A, 42: 1151–1154, 1990.

[97] J. Geiger, Z. Phys. A, 282: 129–141, 1977.

[98] W. R. Johnson, K. T. Cheng, K.-N. Huang and M. Le Dourneuf, Phys. Rev. A, 22: 989–997, 1980.

[99] M. Aymar, O. Robaux and C. Thomas, J. Phys. B: At. Mol. Phys., 14: 4255–4270, 1981.

[100] F. S. Ham, Solid State Phys., 1: 127–192, 1955.

[101] M. Schäfer, F. Merkt and Ch. Jungen, unpublished results.

[102] Ch. Jungen and O. Atabek, J. Chem. Phys., 66: 5584–5609, 1977.

[103] Ch. Jungen and D. Dill, J. Chem. Phys., 73: 3338–3345, 1980.

[104] Ch. Jungen and M. Raoult, Faraday Discuss., 71: 253–271, 1981.

[105] Ch. Jungen, Phys. Rev. Lett., 53: 2394–2397, 1984.

[106] N. Y. Du and C. H. Greene, J. Chem. Phys., 85: 5430–5436, 1986.

[107] S. C. Ross and Ch. Jungen, Phys. Rev. A, 49: 4364–4377, 1994.

[108] Ch. Jungen and S. C. Ross, Phys. Rev. A, 55: R2503–R2506, 1997.

[109] M. Telmini and Ch. Jungen, Phys. Rev. A, 68: 062704, 2003.

[110] Th. A. Paul, H. A. Cruse, H. J. Wörner and F. Merkt, Mol. Phys., 105: 871–883, 2007.

[111] E. Vliegen, H. J. Wörner, T. P. Softley and F. Merkt, Phys. Rev. Lett., 92: 033005, 2004.

[112] D. Townsend, A. L. Goodgame, S. R. Procter, S. R. Mackenzie and T. P. Softley, J. Phys. B: At. Mol. Opt. Phys., 34: 439–450, 2001.

[113] S. R. Procter, Y. Yamakita, F. Merkt and T. P. Softley, Chem. Phys. Lett., 374: 667–675, 2003.

[114] Y. Yamakita, S. R. Procter, A. L. Goodgame, T. P. Softley and F. Merkt, J. Chem. Phys., 121: 1419–1431, 2004.

[115] E. Vliegen and F. Merkt, J. Phys. B: At. Mol. Opt. Phys., 38: 1623–1636, 2005.

[116] E. Vliegen, P. Limacher and F. Merkt, Eur. Phys. J. D, 40: 73–80, 2006.

[117] E. Vliegen and F. Merkt, J. Phys. B: At. Mol. Opt. Phys., 39: L241–L247, 2006.

[118] E. Vliegen and F. Merkt, Phys. Rev. Lett., 97: 033002, 2006.

[119] E. Vliegen, S. D. Hogan, H. Schmutz and F. Merkt, Phys. Rev. A, 76: 023405, 2007.

[120] A. Lindgård and S. E. Nielsen, At. Data Nucl. Data Tables, 19: 533–633, 1977.

[121] J. E. Sansonetti and W. C. Martin, J. Phys. Chem. Ref. Data, 34: 1559–2259, 2005.

CHAPTER 4

VIBRATIONAL POTENTIAL ENERGY SURFACES IN ELECTRONIC EXCITED STATES

Jaan Laane

Contents

Abstract

Several spectroscopic methods, including infrared and ultraviolet absorption, Raman, jet-cooled laser-induced fluorescence (LIF), and cavity ringdown, have been utilized to map out the vibrational quantum states of molecules in their ground and excited electronic states. Data on the higher excited vibrational levels for large-amplitude vibrations such as ring-puckering, ring-twisting, ring-flapping, and internal rotation allow one- or two-dimensional potential energy surfaces (PESs) to be accurately determined. In many cases, ab initio and/or DFT computations are utilized to complement the experimental work. Following a discussion of theory, experimental methods, and computational methods, the spectroscopic results and PESs for several types of molecules are presented. First, the PESs for the carbonyl wagging vibration of seven cyclic ketones in their $S_1(n,\pi^*)$ excited states are reviewed. Except for 2-cyclopentenone (2-CP), which is conjugated and planar, the PESs have a barrier to planarity which increases with angle strain. PESs for the ring-bending and ring-twisting vibrations were also determined for these ketones in both their ground and excited states. The LIF study of trans-stilbene and two substituted stilbenes allowed two-dimensional PESs for the internal rotations of the phenyl groups to be calculated for both ground and $S_1(\pi,\pi^*)$ states. The torsion about the C=C double bond in the $S_1(\pi,\pi^*)$ state was also investigated to understand the photoisomerization. The ring-puckering and ring-flapping of four molecules in the indan (IND) family were investigated in both ground and $S_1(\pi,\pi^*)$ states, and the PESs, which differ substantially between the electronic states, were determined. Results for 1,3-benzodioxole (13BZD) were particularly interesting in that it possesses the anomeric effect that results in a non-planar five-membered ring. In its ground state, the barrier to planarity is relatively small due to suppression of the anomeric effect due to interactions with the benzene ring. However, the suppression is considerably reduced in the $S_1(\pi,\pi^*)$ state and the barrier almost doubles. Spectroscopic results and PESs for two dihydro-naphthalenes, tetralin (TET), and two benzodioxoles are also reviewed as is the work on 2-indanol. The latter molecule possesses intramolecular hydrogen bonding that was analyzed in both ground and excited electronic states. Finally, the cavity ringdown spectra (CRDS) of 2-CP and 2-cyclohexenone (2-CHO) are presented along with their ring-puckering and ring-twisting PESs. This highly sensitive technique allows the $T_1(n,\pi^*)$ triplet levels and the corresponding PESs to be determined for 2-CP. The flipped spin for the triplet state results in a barrier to planarity, whereas the ground and $S_1(n,\pi^*)$ states are planar.

Keywords: molecular vibrations; potential energy surfaces; molecular structure; infrared spectroscopy; Raman spectroscopy; laser-induced fluorescence; large-amplitude vibrations; electronic excited states; fluorescence spectra

1.   INTRODUCTION

Many discussions of potential energy functions concentrate on the use of the harmonic oscillator. This is the case, for example, for representing the vibration of a diatomic molecule. Whenever greater accuracy is desired, perturbations to the harmonic oscillator are used and this is known as the anharmonicity. Similarly, the vibrations of polyatomic molecules can be considered to be nearly harmonic, and quadratic force fields have often been utilized. Although these kinds of investigations can be very informative about bonding forces, direct structural information can generally not be obtained. Potential functions that are not harmonic, however, can be much more informative, especially if their energy minima do not correspond to the coordinate origins. The much studied inversion vibration of ammonia provides an illustrative example. Dennison [1,2] first recognized in 1932 that there would be inversion doubling for ammonia due to the fact that the inversion vibration has a double minimum potential energy function. Since then, this vibration has been studied many dozens of times [3–5]. Figure 1 presents our own representation of the potential energy function for the ammonia inversion vibration along with the experimentally observed energy spacings. This is assumed to be a one–dimensional system where the only coordinate is that for the inversion motion. Although ammonia has six vibrations, the inversion can be considered independently as all of the other vibrations are either of different symmetry

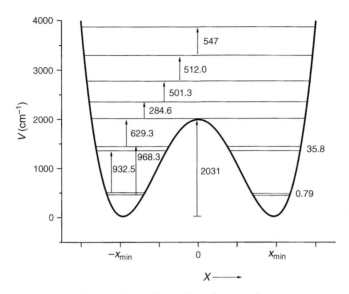

Figure 1   Potential energy function for the inversion of ammonia.

(E symmetry species instead of A_1 for the inversion) or are much higher in frequency (ν_1, the symmetric stretching mode). The function in Figure 1 is well represented by a harmonic potential energy function with a Gaussian barrier

$$V = Ax^2 + Be^{-Cx^2}, \tag{1}$$

where x represents the inversion coordinate and where $x = 0$ represents a planar ammonia molecule. A, B, and C are potential energy parameters. The function in Figure 1 has a barrier of 2031 cm^{-1} (5.80 kcal/mol), and this value is the energy difference between the planar form of ammonia and its equilibrium structure. The energy minima correspond to the equilibrium values of the out-of-plane distance x. When used in the Schrödinger equation, this three-parameter function reproduces the experimental data very accurately. As we showed many years ago [6], a simpler function also reproduces the observed frequencies almost as well. This has the form

$$V = Ax^4 + Bx^2, \tag{2}$$

where a negative B value is utilized to produce the barrier. As we shall see, this form of the potential energy function has proven to be extremely valuable for our work on the ring systems to be discussed below.

Another type of vibration that has a potential function very different from a harmonic oscillator was recognized by Bell in 1945 [7]. He postulated that a four-membered ring molecule such as cyclobutane should have a ring-puckering vibration governed by a quartic potential energy function

$$V = Ax^4. \tag{3}$$

As it turns out, and as will be discussed below, the function in Equation (2) more appropriately represents these types of ring-puckering vibrations, which are low in frequency and have large amplitudes of motion. The first far-infrared observation of the ring-puckering transitions was the study of trimethylene oxide (TMO) reported from the R. C. Lord laboratory at MIT nearly half a century ago [8,9]. Later studies there and at Berkeley established that TMO has a tiny barrier to planarity and a small negative B coefficient for Equation (2)[10–13]. Studies of cyclobutanone (CB) [13] and trimethylene sulfide [14] at Berkeley by the Gwinn and Strauss research groups and of silacyclobutane (SCB) at MIT by Laane and Lord [15] further paved the way for the dozens of far-infrared investigations to follow. Laane and Lord further recognized that molecules such as cyclopentene [16] and 1,4-cyclohexadiene [17] could be considered to be "pseudo-four-membered rings" and their ring-puckering spectra could be fit well with the one-dimensional function in Equation (2). As a result, the ring-puckering of many five- and six-membered rings were studied in the following decades. Much of the work on four-, five-, and six-membered rings through 1979 was reviewed by Carreira et al. [18].

The situation with cyclopentane and other unsaturated five-membered rings, however, is more complicated. Pitzer and co-workers [19,20] initially showed that the two out-of-plane ring motions of cyclopentane could be described by a nearly free pseudorotation and a radial mode. They postulated that the pseudorotation could be represented by a one-dimensional potential energy model with $V = 0$ and that this would give rise to energy levels $2B$ apart, where B is the pseudorotational constant. In 1968, Durig and Wertz [21] observed a number of the pseudorotational transitions as combinations with a CH_2 bending mode, and in 1988, Bauman and Laane [22] presented improved spectra, a comprehensive analysis, and a two-dimensional function for this molecule. One- and two-dimensional energy calculations based on experimental data have followed for many other five-membered ring molecules, and reviews have discussed these [23,24]. Two-dimensional studies were also extended to six-membered ring molecules such as cyclohexene and to bicyclic molecules, and these will be discussed later.

In the 1970s, vapor-phase Raman spectroscopy also became available to complement the far-infrared studies. TMO was investigated by Kiefer and Bernstein [25] in 1970 and cyclopentene by Chao and Laane [26] in 1972 in early work. Many studies since then, including the study of cyclobutane [27], utilized vapor-phase Raman as an additional tool. In addition to out-of-plane ring vibrations and inversions, internal rotations (or torsions) have also been investigated by far-infrared and Raman spectroscopy. One- and two-dimensional periodic potential energy functions have been determined for many molecules, and a number of reviews are available [28,29]. The energy level calculations have generally been based on methods developed in the Laane laboratories [30].

These types of potential energy studies took a major step forward in the 1980s with the development of supersonic jet cooling and with the availability of tunable laser sources in the ultraviolet region. This allowed the vibronic energy states of electronically excited states to be determined directly and therefore made it possible to determine the potential energy functions and structures for these states. Much of the discussion here will focus on these excited state investigations.

2. THEORY

2.1. The quartic potential energy

Bell theorized [7] that cyclobutane should have a quartic potential energy function for its ring-puckering vibration, but he never detailed his derivation. This was done in 1991 by Laane [31]. Figure 2 shows a four-membered ring, such as the four carbon atoms of cyclobutane, with four equal bond angles ϕ. The degree of puckering is measured by the puckering coordinate x, which reflects the distance that each carbon atom has moved away from a planar structure. Hence, the difference between the two ring diagonals is $2x$. The structure of the ring is fully defined by x and the carbon—carbon bond distance R, which is assumed to be fixed as bond stretching force constants are much larger than angle bending constants. The angle strain at each angle ϕ is assumed to be the sole contributor to the potential energy, and the angle bending forces are assumed to be harmonic.

The derivative of the resulting potential energy function is based on some geometrical relationships:

$$\sin(\phi/2) = A/R \tag{4}$$

$$A^2 + 4x^2 = B^2 \tag{5}$$

and

$$A^2 + B^2 = R^2. \tag{6}$$

From Equations (5) and (6)

$$A = \left(\frac{1}{2}R^2 - 2x^2\right)^{1/2} \tag{7}$$

and then

$$\sin(\phi/2) = (1/2 - 2x^2/R^2)^{1/2}. \tag{8}$$

Differentiation gives

$$\cos(\phi/2)d\phi = (-4x/R^2)(1/2 - 2x^2/R^2)^{-1/2}dx. \tag{9}$$

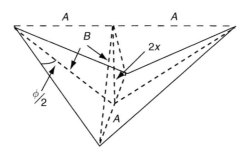

Figure 2 The ring-puckering coordinate x and the geometrical parameters for the cyclobutane ring.

From Equations (6) and (7),

$$B = \left(\frac{1}{2}R^2 + 2x^2 \right)^{1/2} \tag{10}$$

and since

$$\cos(\phi/2) = B/R, \tag{11}$$

we have

$$\cos(\phi/2) = (1/2 + 2x^2/R^2)^{1/2}. \tag{12}$$

Then from Equations (9) and (12)

$$d\phi = -8xdx \, (R^4 - 16x^4)^{-1/2}. \tag{13}$$

As $x = 0$ for the planar structure, $dx = x$. Then $\Delta\phi = d\phi$ is given by

$$\Delta\phi = -8x^2 \, (R^4 - 16x^4)^{-1/2}. \tag{14}$$

The bond distance R is typically an order of magnitude greater than the puckering coordinate x at the equilibrium position. Thus, $R^4 \rangle\rangle 16x^4$ and

$$\Delta\phi \cong -8x^2/R^2. \tag{15}$$

Bell's conclusion that the potential energy is quartic (Equation (3)) must clearly be based on the assumption that the potential energy function for the ring-puckering vibration arises from a harmonic dependence on each angle change $\Delta\phi$ for each of the four angles:

$$V = 4[1/2k_\phi (\Delta\phi)^2] = 128k_\phi x^4/R^4. \tag{16}$$

Thus, a in Equation (3) is given by

$$a = 128k_\phi/R^4 \tag{17}$$

However, Equation (16) neglects initial strain present at each of the already strained C—C—C angles in the four-membered ring. Assuming that each of the ring angles prefers to have an ideal value of ϕ_{ideal} (probably tetrahedral), instead of Equation (16) we then have

$$V_{ang} = 2k_\phi (\phi_{ideal} - \phi)^2 \tag{18}$$

and

$$\phi = \phi_0 + \Delta\phi = \phi_0 - 8x^2 \, (R^4 - 16x^4)^{-1/2} \cong \phi_0 - 8x^2/R^2, \tag{19}$$

where ϕ_0 is the internal ring angle for the planar conformation (90°). Defining the initial strain by

$$S_0 = \phi_{ideal} - \phi_0 \tag{20}$$

then yields the potential energy

$$V_{ang} \cong 2k_\phi (S_0 + 8x^2/R^2)^2 = 2k_\phi S_0^2 + (32k_\phi S_0/R^2)x^2 + (128k_\phi/R^4)x^4 \quad (21)$$

or

$$V_{ang} = a_{ang}x^4 + b_{ang}x^2 + V_0 \quad (22)$$

where

$$a_{ang} = 128k_\phi/R^4, \ b_{ang} = 32k_\phi S_0/R^2, \ V_0 = 2k_\phi S_0^2. \quad (23)$$

In quantum mechanical calculations for the energy levels, the value of V_0 can be set equal to zero so that $V(0) = 0$. It should also be noted that if $\Delta\phi$ is given by the exact expression in Equation (14) rather than Equation (15), then each R^2 in Equation (23) should be replaced by $(R^4 - 16x^4)^{1/2}$.

The effect of torsional interactions is not included in Equation (23). For cyclobutane the planar conformation has each of the four CH_2 groups eclipsed by two adjacent groups, and the torsional contribution to the potential function is at a maximum at $x = 0$. A three-fold potential function can be used to approximate the internal rotation about each bond. When this is converted to a dependence on x, we have a power series in x^2 to represent this contribution

$$V_{tors} = b_{tors}x^2 + a_{tors}x^4 + \cdots \quad (24)$$

Thus, when Equation (24) is added to Equation (22) to determine the overall potential function, the function will be of the double-minimum type when b_{tors} is negative and $|b_{tors}| > b_{ang}$. It will have a single minimum when $b_{ang} + b_{tors} > 0$. It can also be seen that a_{tors} will contribute somewhat to the quartic coefficient in the total potential function, but a_{ang} in Equation (22) will dominate.

The derivation described above only applies directly for cyclobutane, but the same principles apply for other four-membered ring molecules. Moreover, they can also be applied to "pseudo-four-membered rings" such as cyclopentene, 1,4-cyclohexadiene, and their analogs. Figure 3 shows how the ring-puckering coordinate is defined for these larger ring molecules.

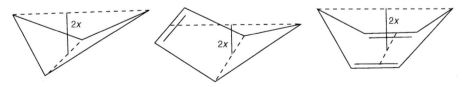

Figure 3   Definition of the ring-puckering coordinate for three types of ring molecules.

2.2.   Calculation of energy levels

Initial studies of ring-puckering potential energy functions assumed a fixed reduced mass μ for the computations. In this case the one-dimensional wave equation is

$$(-\hbar^2/2\mu)\,d^2\psi/dx^2 + V\psi = E\psi. \tag{25}$$

This can be transformed to the reduced form [6] using

$$Z = (2\mu/\hbar^2)^{1/6} a^{1/6} x \tag{26}$$

$$E = A\lambda \tag{27}$$

$$A = (\hbar^2/2\mu)^{2/3} a^{1/3} \tag{28}$$

and

$$B = (2\mu/\hbar^2)^{1/3} a^{-2/3} b. \tag{29}$$

This results in

$$-d^2\psi/dZ^2 + (Z^4 + BZ^2)\psi = \lambda\psi, \tag{30}$$

where Z is the dimensionless coordinate. B defines the quadratic contribution that may be positive or negative, and the λ are the eigenvalues that are proportional to the vibrational quantum states E. Each value of B defines a set of eigenvalues that may be computed and then scaled by the parameter A to best fit the experimental data. Hence, the effective potential energy function is

$$V = A(Z^4 + BZ^2). \tag{31}$$

Figure 4 shows the first 17 eigenvalues calculated for this function for different values of B (with $A = 1$). This data were presented in tabular form in 1970 when the computation of eigenvalues was slow and tedious and became known as Laane's tables [6]. Laane utilized numerical methods based on Milne to calculate the eigenvalues. His laboratory and others nowadays utilize matrix diagonalization methods based on harmonic oscillator functions to calculate energy levels. For $B=0$, the function is that of a pure quartic oscillator. For positive B, a mixed quartic/quadratic function exists, and a pure harmonic oscillator is approached as $B \to \infty$. When B is negative, the function represents a double minimum potential with a barrier of $B^2/4$, and pairs of energy levels begin to merge below the barrier. When $B \to -\infty$, the pairs of energy levels become equally spaced and also approach those of a harmonic oscillator. Figure 5 shows the experimentally determined potential energy functions for the ring-puckering vibrations of several different molecules with values of B ranging from $B=+4.68$ to -9.33. 2,5-Dihydrothiophene [32] (DHT) has a mixed quartic/quadratic potential ($B=4.68$) while 3-silacyclopent-1-ene [33] (SCP) is a nearly perfect quartic oscillator with $B=-0.17$. Both molecules are planar. TMO [8–13], 1,3-disilacyclobutane [34] (DSCB), cyclopentene [16], and SCB [15] have increasingly negative B values (−1.47, −4.65, −6.17 and −9.33, respectively) and increasing barriers to planarity. Their potential energy functions show how the energy levels merge to produce (inversion

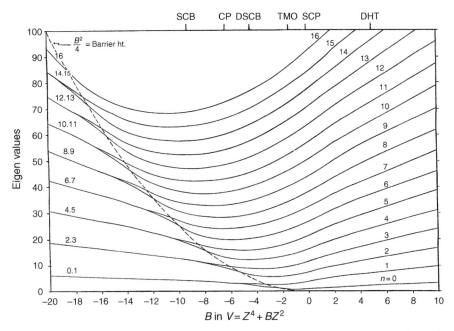

Figure 4 Eigenvalues λ as a function of B for the potential energy of the form $V = z^4 + Bz^2$.

doubling.) As the barrier increases, more and more levels can be seen to become nearly doubly degenerate. This is also clear in Figure 4 where the lower λ values merge as B becomes more and more negative. The eigenvalues at the specific B values shown in Figure 5 correlate to the energy levels on the potential energy curves when the appropriate scaling factor A is used.

2.3. Kinetic energy functions

Many of the original ring-puckering studies utilized the wave equation (Equation (25)) with the mixed quartic/quadratic potential energy (Equation (2)) and a fixed reduced mass to fit the experimental data. However, as this vibration has a large amplitude and the reduced mass changes with the vibrational coordinate, the calculations can be improved by utilizing a computed reduced mass function that depends on the coordinate. In one dimension, the Hamiltonian becomes

$$\mathcal{H}(x) = \left(-\hbar^2/2\right)\partial/\partial x(g_{44}(x))\partial/\partial x + V(x), \tag{32}$$

where g_{44} is the coordinate-dependent reciprocal-reduced mass function that can be represented by a polynomial

$$g_{44} = g_{44}^{(0)} + g_{44}^{(2)}x^2 + g_{44}^{(4)}x^4 + g_{44}^{(6)}x^6, \tag{33}$$

where the $g_{44}^{(i)}$ are the parameters determined computationally to best fit the coordinate dependence. The odd-powered terms in this polynomial representation

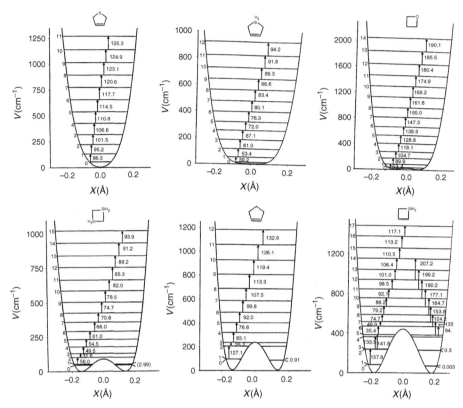

Figure 5   Experimentally determined potential energy functions for various different B values.

are zero if puckering up and down are equivalent. Typically, only the terms up through the sixth power are included. The subscripts on g_{44} reflect the fact the values 1 to 3 are reserved for the molecular rotations. As shown by Malloy [35], the reduced mass for the puckering motion can be readily calculated at different coordinate values using vector methods. The difficult part is to correctly model the vibrational motions as derivatives of the type $\partial \bar{r}_i / \partial Q_j$ must be computed. These are then utilized to set up a G matrix from which the g_{44} terms as a function of the coordinate Q_j can be determined. Laane and co-workers [36–41] have described the computation of the kinetic energy expressions for several different types of vibrations using vector methods for both one- and two-dimensional cases. As an example, Figure 6 shows the two-dimensional kinetic energy (reciprocal–reduced mass) function calculated for the ring-puckering of phthalan (PHT) [42] in terms of its ring-puckering (x_1) and ring-flapping (x_2) coordinates. The coordinate dependence is clearly evident. In two dimensions, the Hamiltonian becomes

$$\mathcal{H}(x_1, x_2) = -\hbar^2/2[\partial/\partial x_1 (g_{44}(x_1, x_2))\partial/\partial x_1 + \partial/\partial x_2 (g_{55}(x_1, x_2))\partial/\partial x_2$$
$$+ \partial/\partial x_1 (g_{45}(x_1, x_2))\partial/\partial x_2)\partial/\partial x_2 (g_{45}(x_1, x_2))\partial/\partial x_1]$$
$$+ V(x_1, x_2), \tag{34}$$

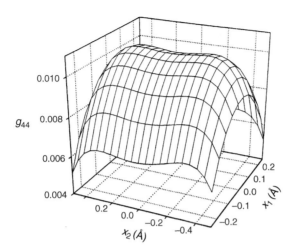

Figure 6 Kinetic energy function for the ring-bending (g_{44}) of phthalan in terms of the bending (x_1) and twisting (x_2) coordinates.

where $V(x_1,x_2)$ typically has the two-dimensional form to be discussed later. The $g_{ij}(x_1,x_2)$ for $i,j = 4,5$ is calculated as a function of both coordinates, again using vector methods. Utilization of these kinetic energy functions rather than a fixed reduced mass tends to affect the potential energy parameters by less than 5% but does reduce the deviation between observed and calculated frequencies by about a factor of 2 (typically from an average 1–2 cm^{-1} deviation for the fixed reduced mass model to half of that).

3. EXPERIMENTAL METHODS

Figure 7 shows the types of spectroscopic transitions used in the experimental work for determining the vibrational energy levels for both the ground (S$_0$) and the first excited S$_1$ electronic states of the molecules to be discussed. The diagram shows far-infrared absorption and Raman transitions that are used to determine the vibrational quantum states for the electronic ground state. These generally have quantum number changes of $\Delta v = 1$ and 2 for the principal transitions. Laser-induced fluorescence (LIF) of the jet-cooled molecules results from transitions which for the most part originate from the vibrational ground state in S$_0$. However, if the molecular jet is warmed somewhat, or if the ground state is nearly degenerate, weak transitions from the first excited vibrational level of low-lying states can also be seen. For LIF, the laser system is tuned, and when the energy of the vibronic level is reached, the fluorescence signal is detected as fluorescence excitation spectra (FES). For molecules excited from the $v = 1$ level, fluorescence results when the frequency matches the separation between $v = 1$ and the vibronic level. Figure 8 shows a schematic diagram of the LIF system in our laboratories for FES and single vibronic level fluorescence (SVLF) investigations. Sensitized phosphorescence excitation spectroscopy (SPES) can also be carried out. A Nd : YAG laser is used to

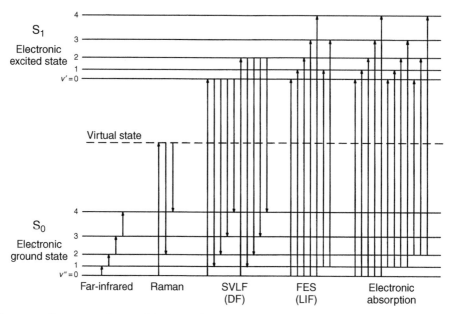

Figure 7 Spectroscopic transitions involving the ground and excited electronic states of molecules.

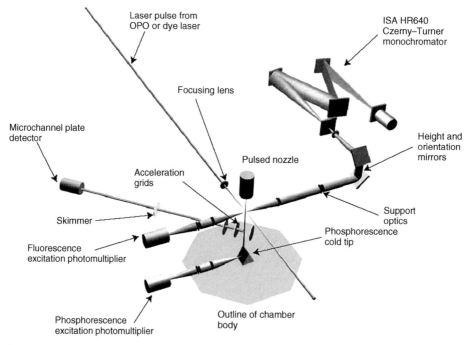

Figure 8 Laser-induced fluorescence (LIF) system in the Laane laboratories. (See color plate 5).

drive an optical parametric oscillator (OPO). The molecules are cooled through a supersonic jet expansion using a pulsed valve. Detection is with a photomultiplier tube (PMT) or by a time-of-flight mass spectrometer. Dispersed fluorescence or SVLF spectra can also be recorded using a single 0.64-m monochromator and PMT or CCD detection. More details can be found elsewhere [43–45].

The ultraviolet absorption spectra are recorded at room temperature and hence transitions can originate from more than a dozen of the low-energy vibrational levels in the S_0 state. The symmetry of the molecules restricts which transitions are allowed, and generally only $v =$ even to even or odd to odd transitions are observed. The jet-cooled LIF spectra are most valuable in that they clearly show which transitions originate from the vibrational ground state. However, the Franck–Condon factor limits the number of S_1 vibronic states that can be accessed. In these cases, the ultraviolet absorption spectra are very helpful in that many of the higher vibronic states can be reached only when the transitions originate from upper vibrational levels in S_0. The ultraviolet absorption spectra in our laboratory are recorded on a Bomem 8.02 Fourier transform spectrometer, which is capable of $0.02\ cm^{-1}$ resolution and operates up to $50{,}000\ cm^{-1}$. In some cases, as many as 100,000 scans were averaged to achieve high signal to noise ratios.

The far-infrared (3–400 cm^{-1}) and mid-infrared (400–4000 cm^{-1}) spectra are also recorded on the Bomem instrument. Heatable long path multiple-reflection cells up to 20 m are used to investigate weak signals. Before FT-IR spectroscopy became common in the 1970s, the best far-infrared instrument was constructed by Jarrell Ash for the R. C. Lord laboratory [46] at M.I.T. at a cost of about one quarter million 1964 dollars. This instrument is shown in Figure 9. It produced many excellent spectra including those of SCB [15] and cyclopentene [16]. Its monochromator had a 5-m path length and its long path cell, sticking out to the left, had a volume of more than 200 L, so it required about a gram of sample to produce one torr of vapor pressure. Many grating and filter changes

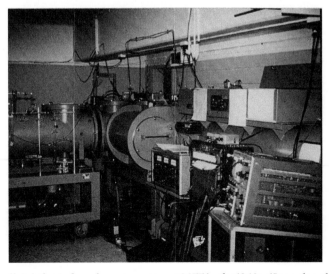

Figure 9   Jarrell Ash far-infrared spectrometer at MIT in the 1960s. (See color plate 6).

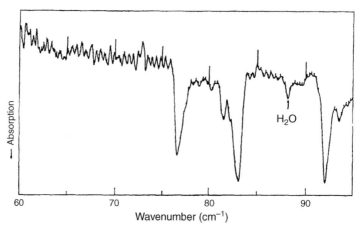

Figure 10 A portion of the far-infrared spectrum of cyclopentene recorded on the MIT Jarrell Ash instrument.

were required to investigate the 25–400 cm^{-1} region, and several days of scanning were required to do this. Figure 10 shows a small portion of the far-infrared spectrum of CP [16] recorded with this instrument. This shows the rotational fine structure of the nearly symmetric top molecule as well as several ring-puckering bands.

Raman spectra are recorded on a JY U-1000 monochromator equipped with a CCD detector. Either a Coherent Innova 20 argon ion laser or a Coherent Verdi 10 operating at 532 nm has been used as the excitation source. As many of the samples have high boiling points, a special cell had to be constructed [47] for heating samples as high as 350° C to obtain substantial vapor pressures. This is shown in Figure 11.

Figure 11 High-temperature Raman cell for vapors. (See color plate 7).

4. ELECTRONIC GROUND STATE

4.1. One-dimensional potential functions

Figure 5 already showed several experimentally determined one-dimensional potential energy functions. Here additional detail will be provided for other examples. Figure 12 shows the far-infrared spectrum of 2,3-dihydrofuran [48] and Figure 13 shows its potential energy function determined from the data for its ring-puckering vibration. The experimental data were fit very well using the simple two-parameter potential

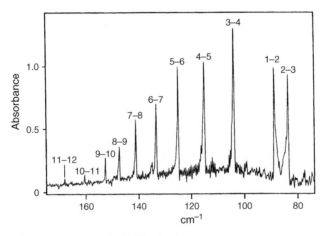

Figure 12   Far-infrared spectrum of 2,3-dihydrofuran.

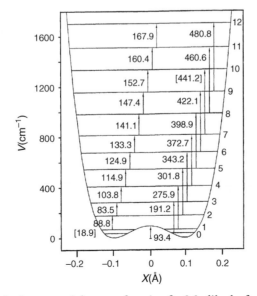

Figure 13   Ring-puckering potential energy function for 2,3-dihydrofuran.

function of Equation (2). This molecule has only a small barrier to planarity of 93 cm^{-1}, but this is sufficient to pucker it with a dihedral angle of 22°. The barrier to planarity arises from the torsional interaction between the two CH$_2$ groups.

Another especially interesting molecule is 1,3-dioxole in that it might be expected to be planar as it has no CH$_2$—CH$_2$ torsional interactions. However, it shows the anomeric effect that results in a puckered molecule. Figures 14 and 15 show its far-infrared and Raman spectra of the vapor, and Figure 16 presents the potential energy function that has a substantial barrier of 325 cm^{-1} [49]. The double quantum jumps

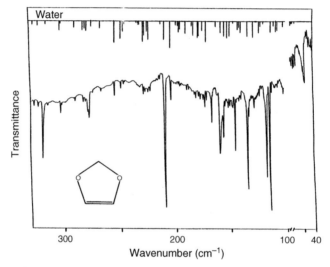

Figure 14   Far-infrared spectrum of 1,3-dioxole vapor.

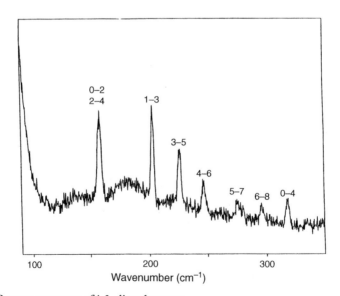

Figure 15   Raman spectrum of 1,3-dioxole vapor.

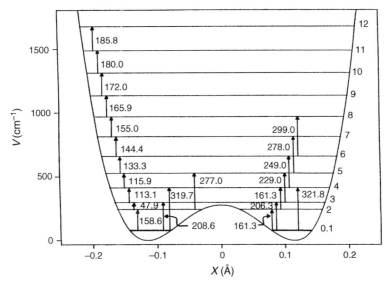

Figure 16 Ring-puckering potential energy function of 1,3-dioxole.

of the puckering observed in the Raman spectra were very important for complementing the far-infrared data and helping to make the assignments for the spectral bands. The anomeric effect, which is present in molecules with the —O—CH$_2$—O— configuration, is responsible for twisting the molecule out-of-plane. This will be examined in more detail when the 13BZD molecule is discussed later.

Another study worth noting is that of 1,3-cyclohexadiene as this was achieved using high-temperature vapor-phase Raman spectroscopy [50]. Figures 17 and 18 show the Raman spectra and potential energy function, respectively. The number of Raman transitions observed is remarkable because they extend above the high barrier to planarity (1132 cm^{-1}) for this molecule.

The experimentally determined potential energy functions and barriers of the four-membered ring and pseudo-four-membered ring molecules for which Equation (2) was utilized have been presented elsewhere [18] and in many of the references herein. As discussed above, the a and b potential energy parameters for the most part reflect angle strain and torsional forces, respectively.

4.2. Two-dimensional potential energy functions

The one-dimensional approximation described above can only be applied when a single vibration such as the ring–puckering can be investigated independently from all the other vibrations. When a second low-frequency vibration is present, two-dimensional potential energy surfaces (PESs) can be utilized. These generally have the form

$$V = ax_1^4 + bx_1^2 + cx_2^4 + dx_2^2 + ex_1^2 x_2^2. \tag{35}$$

Two-dimensional studies have been carried out for molecules in the cyclopentane and cyclohexene families, both of which have ring-bending and ring-twisting

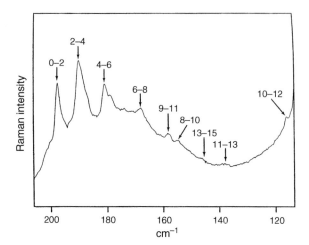

Figure 17 Vapor-phase Raman spectra of 1,3-cyclohexadiene at 150° C.

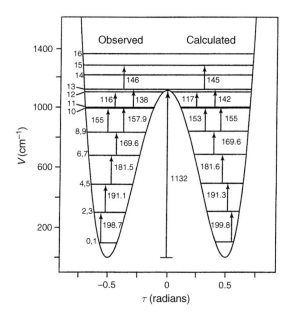

Figure 18 Potential energy function for the ring-twisting vibration of 1,3-cyclohexadiene.

vibrations, and for bicyclic molecules similar to IND, which has ring–bending and ring–flapping vibrations.

The cyclopentane molecule is a particularly interesting case as it has 10 equivalent bent forms and 10 equivalent twisted forms [19,20]. Moreover, the energy difference between bent and twisted forms in less than a few cm^{-1} [51]. Figure 19 shows how these 20 structures readily interconvert into one another resulting in the pseudorotational process. Figure 20 shows the combination bands of the

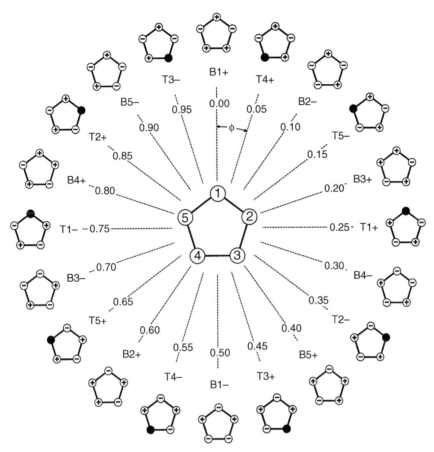

Figure 19 Pseudorotation scheme for cyclopentane. B and T refer to bent and twist structures, respectively.

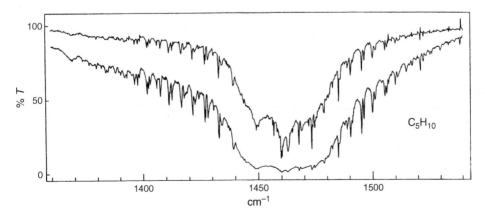

Figure 20 Cyclopentane pseudorotational bands in combination with a CH_2 bending mode.

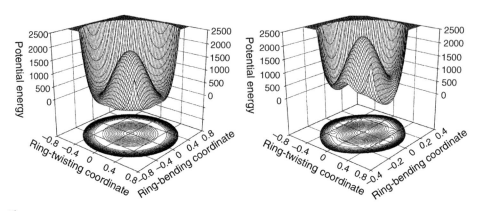

Figure 21 Two-dimensional potential energy surface (PES) for the ring-bending (x_1) and ring-twisting (x_2) of cyclopentane (left) and 1,3-oxathiolane (right).

pseudorotation with a CH_2 bending mode [52]. These are characteristic of nearly free pseudorotation where the energy levels are separated by $2\overline{B}_{pseudo}$, the pseudorotational constant. In this case, $\overline{B}_{pseudo} \cong 2.7\ cm^{-1}$. Figure 21 (left) shows the two-dimensional PES determined for cyclopentane by Bauman and Laane [52], which also fit the data for the radial mode near $272\ cm^{-1}$. A similar result was first presented by Carreira, Mills, and Person [53]. This function has a barrier to planarity of $1808\ cm^{-1}$, and the bent and twisted structures all have the same energy. What this PES shows is that all 20 conformations of cyclopentane can interconvert essentially without encountering any energy barriers, and the molecule can totally invert in this manner without passing through the planar structure. When the cyclopentane ring is substituted with a heteroatom, a barrier to pseudorotation results. This is the case for 1,3-oxathiole, and its two-dimensional PES is also presented in Figure 21 (right) [54,55]. As can be seen, the molecule is twisted, and the bent structures correspond to saddle points that lie $570\ cm^{-1}$ above the energy minima. The anomeric effect helps to reduce the energy of the bent form. The planar structure is $2289\ cm^{-1}$ higher in energy. Although there is a barrier to pseudorotation at the bent conformation, the molecule can still invert more readily along this pathway than by passing through the high energy planar structure.

Cyclohexene [56,57] and similar molecules containing heteroatoms have also been investigated using far-infrared spectroscopy and two-dimensional PESs [58,59]. These molecules in general are twisted with high barriers to planarities. The lowest frequency spectra arise from the ring-bending motion, but twisting bands and twisting-bending combinations were also utilized to determine the PESs.

We have also reported results for numerous molecules in the indan family where these bicyclic systems have interacting low-frequency ring-bending and ring-flapping modes. The definitions of these two coordinates are shown in Figure 22. As an example, Figure 23 shows the far-infrared spectrum of phthalan [60]. Figure 24 shows its PES in terms of these two coordinates [42]. The molecule is quasi-planar with only a tiny barrier to planarity of $35\ cm^{-1}$. A one-dimensional potential function was not able to fit the experimental data well, but the two-dimensional PES together

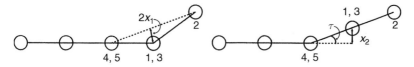

Figure 22 Ring-bending (x_1) and ring-flapping $(x_2$ or $\tau)$ coordinates for indan and related molecules.

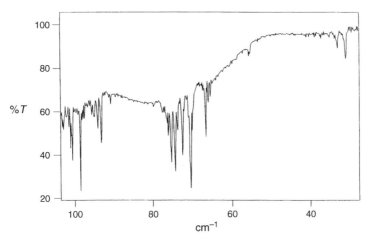

Figure 23 Far-infrared spectrum of phthalan in the 25–105 cm^{-1} region.

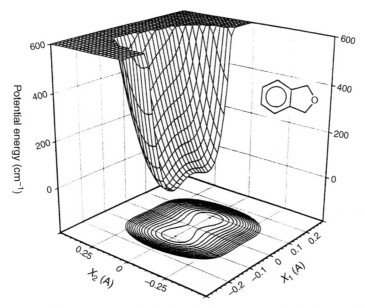

Figure 24 Potential energy surface (PES) for phthalan in terms of its ring-bending (x_1) and ring-flapping (x_2) modes.

with the calculated kinetic energy expansion similar to Equation (34) very nicely fit the data. The energy level pattern was highly unusual due to the kinetic energy interactions. Far-infrared data and PESs for the related molecules coumaran (COU) [61,62], 13BZD [63], and IND [64] have also been published, and these will be discussed later.

5. ELECTRONIC EXCITED STATES

5.1. Cyclic ketones

When a ketone molecule undergoes an $n \rightarrow \pi^*$ transition, the oxygen atom of the carbonyl group typically bends out of the plane of the adjoining carbon atoms. Similarly, the formaldehyde molecule has long been known to be non-planar in its $S_1(n,\pi^*)$ excited state [65]. In the 1990s, we carried out a number of investigations using FES of cyclic ketones including 2-cyclopenten-1-one (2CP) [66], 3-cyclopenten-1-one (3CP) [67], CP [68,69], CB [70], bicyclo[3.1.0]hexan-3-one [71] (BCH), tetrahydrofuran-3-one (THFO), and tetrahydrothiophen-3-one (THTP) [72].

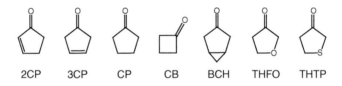

2CP   3CP   CP   CB   BCH   THFO   THTP

In each case, we analyzed the vibronic bands resulting from the carbonyl wagging motion in the $S_1(n,\pi^*)$ state and determined the one-dimensional potential energy function governing this vibration. The bending and twisting vibrations associated with the conformational changes of the rings were also examined in detail. Except for 2CP, where sufficient conjugation is retained after the $n \rightarrow \pi^*$ transition to keep the carbonyl oxygen in the plane of the ring, the other cyclic ketones each have double-minimum carbonyl wagging potential energy functions for the $S_1(n,\pi^*)$ state demonstrating that the carbonyl groups are bent out of the ring planes. In the following sections, the results on these molecules will be presented.

5.1.1. 2-Cyclopenten-1-one (2CP)
Figure 25 shows the survey FES of 2CP [66]. The spectra were recorded in a region from below the electronic origin (27,210 cm^{-1}) to about 1800 cm^{-1} beyond it. In comparison to 3CP and CP, which will be considered later, the 2CP electronic band origin is considerably lower. This is the expected result from the conjugation between the C=O and the C=C groups, which results in a lower energy π^* orbital. The electronic origin for 2CP is also extremely intense in contrast to the $S_1(n,\pi^*)$ origins of similar molecules. This is the result of a planar excited state structure and a high Franck–Condon factor. The frequencies of some of the fundamental vibrations determined for the electronic ground state and S_1 excited state are given in Table 1.

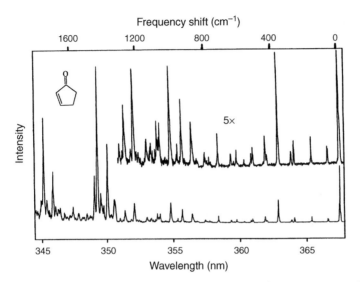

Figure 25 Fluorescence excitation spectrum of jet-cooled 2-cyclopentenone (2CP).

Table 1 Vibrational frequencies (cm^{-1}) for the ground and excited $S_1(n, \pi^*)$ states of 2CP

Approx. description	Ground	Excited
ν_5 C=O stretch	1748	1357
ν_6 C=C stretch	1599	1418
ν_{13} Ring mode	1094	1037
ν_{14} Ring mode	999	974
ν_{15} Ring mode	912	906
ν_{16} Ring mode	822	849
ν_{17} Ring mode	753	746
ν_{18} Ring mode	630	587
ν_{19} C=O def (ll)	464	348
ν_{26} α-CH bend	750	768
ν_{28} C=O def ($\perp$)	537	422
ν_{29} C=C twist	287	274
ν_{30} Ring-puckering	94	67

In addition to these, more than 50 other bands were seen in the FES, with 34 of these involving ν_{30}, the ring-puckering, and 20 involving ν_{29}, the ring-twisting (several are associated with both). Combinations with the C=O in-plane (ν_{19}) and out-of-plane (ν_{28}) wags, which occur at 348 and 422 cm^{-1}, are also common. These modes are both considerably lower in frequency than in the electronic ground state reflecting the decrease in π character of the C=O bond. The intense bands at 1357 and 1418 cm^{-1} for 2CP are due to the C=O and C=C stretches, respectively. These two vibrations are Franck–Condon active due to the increased bond lengths for both bonds. Intense combination bands for each of these stretches were observed with the carbonyl in-plane wag, 19_0^1. All of the other ring mode transitions were also observed, and many of them were also found to be associated with combination bands.

The ring-puckering potential functions for the S_0 and $S_1(n,\pi^*)$ states, which will be compared later to that in the $T_1(n,\pi^*)$ state, were determined based on the far-infrared [73] and FES [66] data, both of which show series of band progressions. Equation (2) fits the data very well for both the ground and the excited states. The molecule is planar in both states but becomes much less rigid in the excited state due to the decreased conjugation resulting from the transition to the antibonding orbital.

5.1.2. 3-Cyclopenten-1-one (3CP)

Figure 26 shows the jet-cooled fluorescence excitation spectrum of 3CP [67]. The band origin is observed at $30{,}238\,\text{cm}^{-1}$. For the molecule lying in the xz plane, each $v=0$ (in the electronic ground state) to $v=n$ (in the A_2 electronic excited state) transition of the C=O wag has B_2 vibrational symmetry for $n=$ odd but has A_1 vibrational symmetry for $n=$ even. Only transitions to the $n=$ odd states can be observed. These show up as intense Type B bands arising from $A_2 \; x \; B_2 = B_1$ symmetry. The first five of these transitions are shown in Figure 26. Because the $v=0$ and $v=1$ levels in the $S_1(n,\pi^*)$ state are near-degenerate, the band origin lies very close to the $0 \rightarrow 1$ frequency. The other bands in the spectrum include many combinations of the C=O wag with the ring-puckering vibration and also combinations of these with other fundamentals including the C=O stretch.

For 3CP, the reduced mass and the carbonyl wagging potential energy parameters that best fit the observed frequency separations are listed in Table 2. The experimentally determined potential energy function is shown in Figure 27 along with both the observed and the calculated frequency separations. The minimum energy corresponds to wagging angles of $\pm 26°$, and the barrier to inversion is $926\,\text{cm}^{-1}$ (2.65 kcal/mol). For 3CP, the ring-puckering frequency of $127\,\text{cm}^{-1}$ in

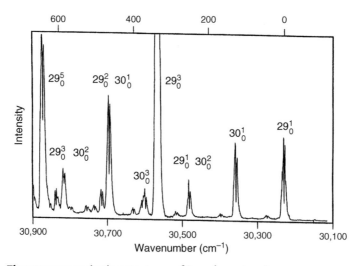

Figure 26 Fluorescence excitation spectrum of 3–cyclopenten-1-one.

Table 2 Potential energy parameters and reduced masses for C=O wagging vibrations in the $S_1(n,\pi^*)$ electronic state

Molecule	μ (au)	$V = ax^4 + bx^2$		Barrier (cm^{-1})	ϕ_{min}
		a(cm^{-1}/Å^4)	b(cm^{-1}/Å^2)		
THTP	3.572[a]	13.4 × 10³	−5.94 × 10³	659	20°
CP	5.569[b]	10.49 × 10³	−5.34 × 10³	680	22°
BCH	6.49[c]	7.84 × 10³	−5.18 × 10³	873	23°
3CP	5.260[d]	8.11 × 10³	−5.48 × 10³	926	26°
THFO	5.203[e]	8.51 × 10³	−6.26 × 10³	1152	26°
CB	4.244[f]	2.47 × 10³	−4.38 × 10³	1940	41°

[a] $g_{44} = 0.2799 - 0.1872x^2 + 0.0887x^4 - 0.0165x^6$.
[b] $g_{44} = 0.17957 - 0.049144x^2 + 0.014227x^4 - 0.002181x^6$.
[c] $g_{44} = 01541 + 0.1483x - 0.0428x^2 + 0.0143x^3 - 0.04199x^4 + 0.0586x^5 + 0.0239x^6$; V includes $-0.097 \times 10^3 x^3$.
[d] $g_{44} = 0.19012 - 0.054853x^2 + 0.016335x^4 - 0.002554x^6$.
[e] $g_{44} = 0.1920 - 0.0546x^2 + 0.0152x^4 - 0.0021x^6$.
[f] $g_{44} = 0.23565 - 0.076454x^2 + 0.024096x^4 - 0.003922x^6$.

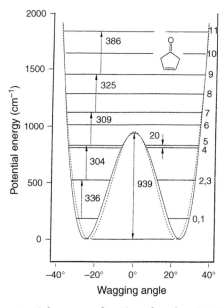

Figure 27 Vibrational potential energy function for the C=O out-of-plane wagging vibration of 3-cyclopenten-1-one.

the S_1 state is considerably higher than the value of 83 cm^{-1} in the ground state indicating a stiffer and asymmetric puckering potential function.

5.1.3. Cyclopentenone (CP)

The FES spectrum of cyclopentenone [68,69] is shown in Figure 28. The band origin is at 30,276 cm^{-1}. In the ground state, the molecule is twisted, and in the C_{2v} approximation, the vibrational ground state is nearly doubly degenerate

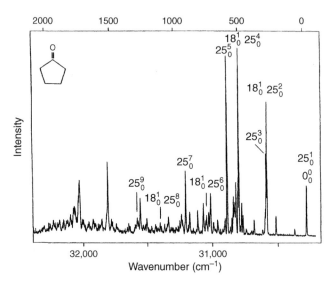

Figure 28   Fluorescence excitation spectrum of cyclopentanone.

with symmetry species A_1 and A_2. The twisting conformation (and degeneracy) carries through to the electronic excited state. The purely electronic transition is $^1A_2 \leftarrow\, ^1A_1$, which is forbidden in the C_{2v} approximation. However, combinations with odd quantum transitions of the C=O wagging with B_2 symmetry results in Type B bands from B_1 symmetry. If either the ground or excited electronic state is also in combination with the near-degenerate A_2 twisting state, the even quanta C=O wagging transitions can also be observed as Type A (A_1) bands $[A_2 \times A_2 \times (B_2)^n = A_1$ for $n =$ even$]$. As can be seen in Figure 29, the transitions for both even and odd quantum states of the C=O wag in the S_1 state are readily observed.

Figure 30 shows the C=O wagging potential energy function for cyclopentanone. The barrier is $680\ \mathrm{cm}^{-1}$ and the energy minima are at $\pm 22°$. The kinetic and potential energy terms are given in Table 2. The comparison between the S_0 and S_1 states for the fundamental vibrational frequencies of several other modes is given in Table 3.

The observed data for the ring-twisting and ring-bending motions of CP in the S_0 and S_1 states have also been analyzed [69]. The fundamental frequencies for these two modes are changed little in the two states as the ring has a similar twisted conformation for each state. However, the two-dimensional PESs for these two modes have been determined for both states and they are quite different. In S_0, the barrier to planarity is $1408\ \mathrm{cm}^{-1}$ and that for pseudorotation is $1358\ \mathrm{cm}^{-1}$. The twist angle is $29°$. These barrier values become 1445 and $596\ \mathrm{cm}^{-1}$, respectively, for the excited state. Figure 31 shows the PES for the $S_1(n,\pi^*)$ state. The lower pseudorotation barrier shows that the bent conformation has become considerably lower in energy.

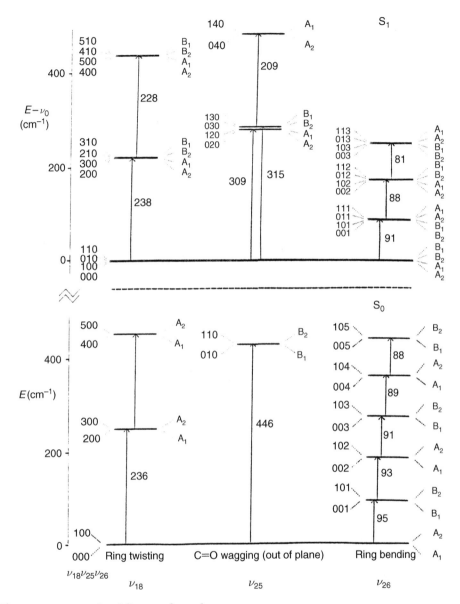

Figure 29   Energy level diagram for cyclopentanone.

5.1.4.   Cyclobutanone (CB)

The FES spectrum of CB is shown in Figure 32 [70]. The ν_{26} carbonyl bending bands are labeled, and transitions up to the 11th quantum state can be seen. In addition to the carbonyl wagging, puckering data for the $S_1(n,\pi^*)$ state can also be deduced from the low-frequency data shown in Figure 33. The potential energy function for the carbonyl wagging is shown in Figure 34, and this highly strained

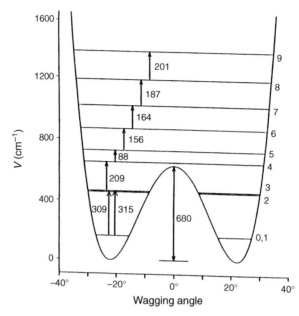

Figure 30 C=O wagging potential energy function and observed energy spacings for cyclopentanone.

four-membered ring can be seen to have a wagging barrier ($2149\,\mathrm{cm}^{-1}$) much higher than the other ketones. At its energy minimum, the wagging angle is a large 39°. The presence of the wagging barrier also complicates the ring-puckering vibration. As shown in Figure 35, both the puckering and the carbonyl inversion can change the conformation of the molecule and result in two puckered forms of different energy. Hence, the puckering potential function is asymmetric in the $S_1(n,\pi^*)$ state. Figure 36 shows several similar one-dimensional ring-puckering potential energy functions which reproduce the experimental data very well. Figure 37 shows the two-dimensional PES in terms of the carbonyl wagging and ring-puckering modes.

5.1.5.   Bicyclo[3.1.0]hexan-3-one

The bicyclic BCHO molecule is similar to CP except that it has an attached three-membered ring that induces asymmetry for both the carbonyl wagging and the ring-puckering modes. The FES spectrum shows about three dozen bands within $1000\,\mathrm{cm}^{-1}$ of the 0_0^0 band at $30{,}262\,\mathrm{cm}^{-1}$ [71]. Many of these are associated with the ring-puckering (ν_{22}) and/or the carbonyl wagging (ν_{21}), and these were analyzed to determine the one-dimensional ring-puckering potential energy function in Figure 38 and the function for the wagging in Figure 39. The molecule is not puckered but shows the expected asymmetry. The barrier to carbonyl inversion is $873\,\mathrm{cm}^{-1}$, and the asymmetry resulting from the energy difference between the two inversion directions is approximately very small, $36\,\mathrm{cm}^{-1}$.

Table 3 Comparison of frequencies of several vibrations of cyclic ketones in the ground and S1(n,π^*) electronic excited states

Vibration	2CP		THTP		CP		BHO		3CP		THFO		CB	
	S_0	S_1	S_0	S_1	S_0	S_1	S_0	S_1	S_0	S_1	S_0	S_1	S_0	S_1
Ring–puckering	94	67	67	58	95	91	86	134	83	127	59	82	36	106
Ring–twisting	287	274	170	—	38	238	—	—	378	377	228	224	—	—
C=O wag o.p.	537	422	427	326	446	309	—	315	450	336	463	344	395	355
C=O wag i.p.	464	348	483	329	467	342	—	—	458	339	463	365	454	392
C=O stretch	1748	135	1760	1240	1770	1230	—	—	1773	1227	1755	1232	1816	1251

i.p., in-plane; o.p., out-of-plane.

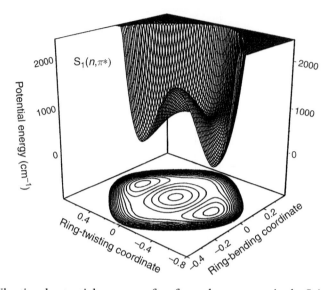

Figure 31 Vibrational potential energy surface for cyclopentanone in the $S_1(n,\pi^*)$ electronic excited state. The contour lines are 150 cm^{-1} apart.

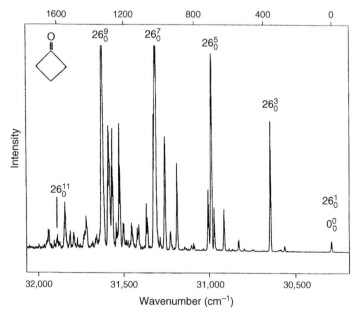

Figure 32 Fluorescence excitation spectrum of cyclobutanone.

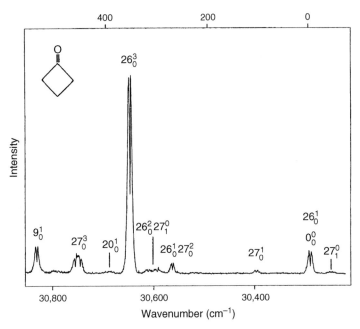

Figure 33 Low-frequency region of the cyclobutanone fluorescence excitation spectrum.

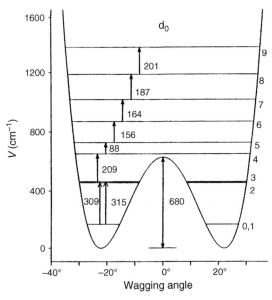

Figure 34 Carbonyl wagging potential energy functions and observed energy spacings for cyclobutanone.

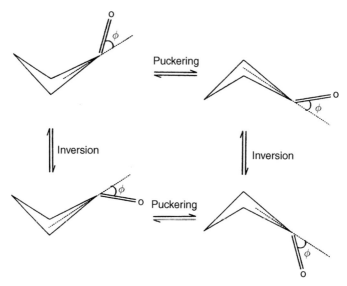

Figure 35 Conformations of cyclobutanone in the $S_1(n,\pi^*)$ state resulting from the ring-puckering or C=O wagging (inversion) vibrations. For the (x_1,x_2) notation, the $(+,+)$ and $(-,-)$ conformations are equivalent as are the $(+,-)$ and $(-,+)$ conformations.

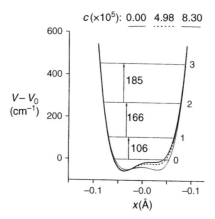

Figure 36 One-dimensional potential energy curves of the form $V = ax_2^4 + bx_2^2 + cx_2^3$ for ring-puckering of cyclobutanone in the $S_1(n,\pi^*)$ state.

5.1.6.   Tetrahydrofuran-3-one (THFO) and tetrahydrothiophen-3-one (THTP)

The FES spectra of THFO and THTP were investigated to determine their carbonyl wagging potential energy functions in their $S_1(n,\pi^*)$ states [72]. Both spectra are unusual in that the band intensities die out by about 1500 cm^{-1}. The THFO spectrum is shown in Figure 40. The carbonyl wagging bands (ν_{28}) are

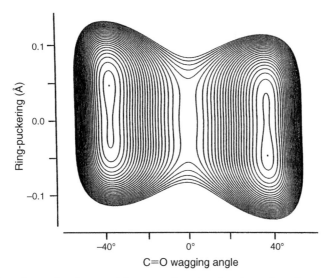

Figure 37   Two-dimensional potential energy function for the carbonyl wagging and ring-puckering coordinates of cyclobutanone. The dots mark the energy minima. The first contour line lies 50 cm^{-1} above each minimum point. The others are 100 cm^{-1} apart.

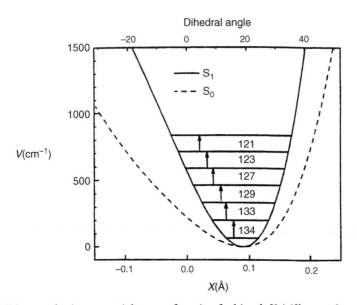

Figure 38   Ring-puckering potential energy function for bicyclo[3.1.0]hexan-3-one (BCHO) in the $S_1(n,\pi^*)$ electronic state. The function has been translated so that $x = 0$ represents the planar structure for the five-membered ring. This is compared to the function in the S_0 ground state (dashed curve).

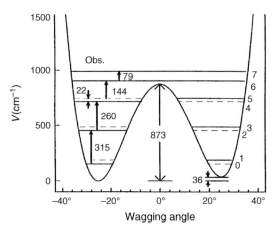

Figure 39 Carbonyl wagging potential energy function for bicyclo[3.1.0]hexan-3-one (BCHO).

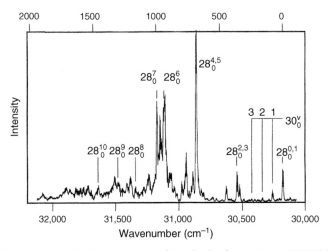

Figure 40 Fluorescence excitation spectrum of tetrahydrofuran-3-one (THFO).

labeled, and these were used to determine the potential function of Figure 41. The bending modes (ν_{30}) for the hindered pseudorotation are also labeled. The THTP spectra, which can be found elsewhere [72], were also used to determine its wagging potential function which is shown in Figure 42. As can be seen, the THTP barrier of 659 cm^{-1} is considerably less than that for THFO (1152 cm^{-1}).

5.1.7. Summary

The studies of these cyclic ketones have demonstrated a number of interesting properties. One of these is that the inversion barrier for the carbonyl group increases with the angle strain. Molecular mechanics (MM3) calculations were used to estimate the CCC angle at the carbonyl atom, and Figure 43 plots the

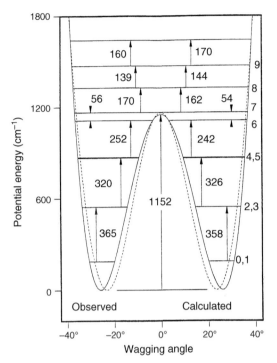

Figure 41 Two similar potential energy functions for the carbonyl wagging of tetrahydrofuran-3-one (THFO).

barrier height versus this angle [72]. As this angle becomes more strained, the barrier increases. The cyclopentanone molecule appears to have too low a barrier, but this may result from its lower CCC angle resulting from the effect of torsional forces.

Table 2 summarizes the results for the carbonyl inversion potential energy functions, and Table 3 summarizes data for several of the other relevant vibrations of these cyclic ketones.

5.2. Stilbenes

Both *trans-* and *cis*-stilbene are stable molecules with the conformation of the *trans* form slightly lower in energy by about 4.6 kcal/mol and with the barrier to internal rotation (isomerization) of 48.3 kcal/mol [74]. This precludes any isomerization about the double bond occurring in the ground state, but the *trans* to *cis* photoisomerization in the $S_1(\pi,\pi^*)$ state has been often investigated [74,75]. In this excited state, the lowest energy form of stilbene has a twisted configuration and only a small barrier to internal rotation exists between the *trans* and the twist forms. A dynamical study in 1992 concluded that this barrier is about 1200 cm^{-1} [76].

In 1995, Chiang and Laane [75] studied the LIF spectra of *trans*-stilbene. The FES is shown in Figure 44 and the SVLF in Figure 45. The analysis of the spectra

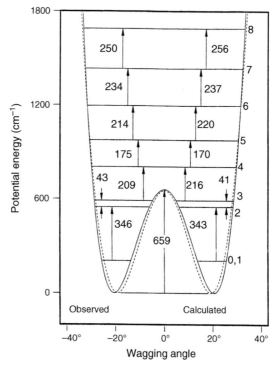

Figure 42 Potential energy function for the carbonyl wagging of tetrahydrothiophen-3-one (THTP). The solid curve corresponds to V_1 and the dashed curve to V_2.

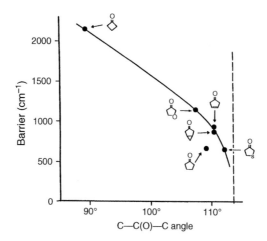

Figure 43 Correlation of inversion barrier with θ (CCC angle at the carbonyl carbon atom). The dashed line indicates the "strain-free" angle according to the MM3 calculation.

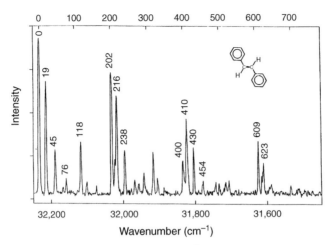

Figure 44   Dispersed fluorescence spectra of the 0_0^0 band of *trans*-stilbene.

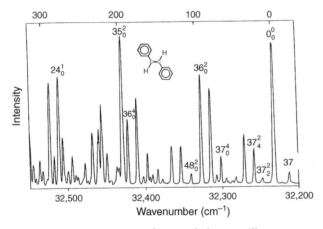

Figure 45   Fluorescence excitation spectra of jet-cooled *trans*-stilbene.

is greatly complicated by the fact that *trans*-stilbene has eight low-frequency vibrations and these needed to be assigned before the internal rotation vibrations could be analyzed. This assignment was greatly aided by the vapor-phase Raman spectra of this molecule [77], which clearly identified the Raman active modes. From the Raman and the LIF spectra, the energy diagram in Figure 46 was determined, and the analysis of the three torsional (internal rotational) modes became feasible.

   Figure 47 defines the ϕ_1 and ϕ_2 coordinates that represent the phenyl torsions and θ that represents the internal rotation about the C=C bond. The phenyl torsions were analyzed two dimensionally, and a computer program was written to

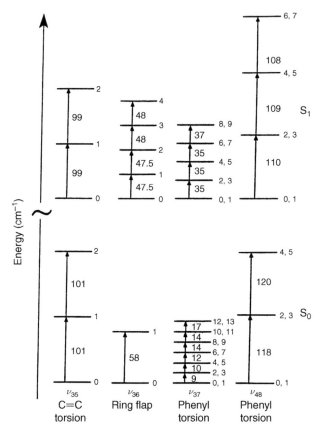

Figure 46 Vibrational energy spacings for four low-frequency vibrations of *trans*-stilbene in its ground and excited states.

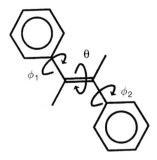

Figure 47 Torsional coordinates for *trans*-stilbene.

calculate the coordinate-dependent kinetic energy function in terms of ϕ_1 and ϕ_2. This was then used in the potential energy computation with

$$V(\phi_1, \phi_2) = \tfrac{1}{2} V_2(2 + \cos 2\phi_2) + V_{12}\cos 2\phi_1 \cos 2\phi_2 + V'_{12} \sin 2\phi_1 \sin 2\phi_2. \quad (36)$$

A separate computer program was written to determine the energy levels for this PES and to optimize the potential energy parameters V_2, V_{12}, and V'_{12}. Figure 48 shows the PES for the $S_1(\pi,\pi^*)$ excited state. The coordinates are defined so that at $\phi_1 = \phi_2 = 90°$, both phenyl groups are in the —C—C=C—C— plane. The potential energy has its maxima of $3000\,\mathrm{cm}^{-1}$ ($2V_2$) at $\phi_1 = \phi_2 = 0°$. There are four equivalent energy minima at $\phi_1 = \pm90°$ and $\phi_2 = \pm90°$, where the entire skeleton of the molecule lies in a plane. The barrier to rotating a single phenyl ring is $1670\,\mathrm{cm}^{-1}$. The S_0 PES is qualitatively similar with a barrier of $3100\,\mathrm{cm}^{-1}$. The barrier to rotating a single phenyl group is only $875\,\mathrm{cm}^{-1}$. The S_0 and S_1 surfaces each require only three potential energy parameters, and these do an excellent job of fitting all of the observed energy spacings.

For the internal rotation about the C=C double bond, a computer program was written to calculate the coordinate-dependent kinetic energy function. For the S_0 ground state, energy level calculation was used together with the one-dimensional potential function

$$V(\theta) = {}^{1}/{}_{2}\,V_1(1-\cos\theta) + {}^{1}/{}_{2}\,V_2(1-\cos 2\theta) + {}^{1}/{}_{2}\,V_4(1-\cos 4\theta). \qquad (37)$$

Here, $\theta = 0°$ and $180°$ correspond to the *trans* and *cis* isomers, respectively, V_1 is the energy difference between the *trans* and the *cis* forms, V_2 almost entirely determines the barrier, and V_4 is a shaping parameter. When the literature values [66] $V_1 = 1605\,\mathrm{cm}^{-1}$ (4.6 kcal/mol) and $V_2 = 16{,}892\,\mathrm{cm}^{-1}$ (48.3 kcal/mol) are used, excellent frequency agreement between the experimental and the calculated data is obtained using $V_4 = -900\,\mathrm{cm}^{-1}$. The data is confined to the bottom region of a potential energy function that has a very high barrier, so the calculation of V_2 by extrapolation does not give a very accurate value ($15{,}000 \pm 3000\,\mathrm{cm}^{-1}$), but this is consistent with the literature values. The lower half of Figure 49 shows the potential energy curve for the S_0 state.

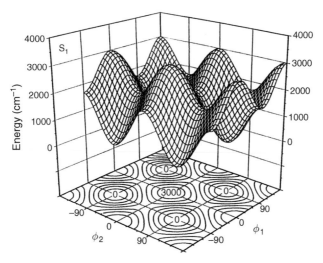

Figure 48   Potential energy surface for the phenyl torsions of *trans*-stilbene in its $S_1(\pi,\pi^*)$ state.

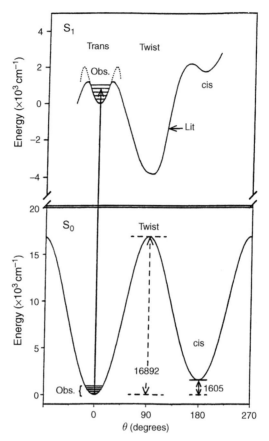

Figure 49   Potential energy function for the internal rotation of *trans*-stilbene about the C=C bond. $\theta = 0°$ corresponds to the *trans* isomer.

The C=C torsion data for the $S_1(\pi,\pi^*)$ state were only observed for the *trans* conformation due to the Franck–Condon principle. Figure 49 (top section) shows the qualitative potential energy curve estimated in the literature. The figure also shows the vibronic energy levels observed within the *trans* potential well. No FES could be observed going to levels in the twist well. In addition, the *cis* well is apparently too shallow to allow transitions originating from the S_0 state of *cis*-stilbene. Consequently, the fluorescence data allow only the shape of the *trans* well to be calculated, and this is shown as the dotted line in Figure 49. This can be represented by [75]

$$V(\theta) = {}^{1}\!/_{2}\, V_1(1-\cos\theta) + {}^{1}\!/_{2}\, V_2(1-\cos 2\theta) \qquad (38)$$
$$+ {}^{1}\!/_{2}\, V_4(1-\cos 4\theta) + {}^{1}\!/_{2}\, V_8(1-\cos 8\theta).$$

V_1 represents the *trans/cis* energy difference, and V_2 primarily determines the depth of the twist well. V_4 primarily determines the *trans* $\rightarrow$ twist barrier while V_8 is a minor shaping term. Analysis of the experimental data, which extend 1230 cm^{-1} above the *trans* well minimum, shows that the *trans* $\rightarrow$ twist barrier is somewhat higher than the 1200 cm^{-1} value estimated from the dynamics data [76].

Investigations have also been carried out for 4,4'-dimethoxy- and 4,4-dimethyl-*trans*-stilbene [78,79], and similar results have been obtained for the internal rotations. In addition, the data for the methyl torsions of the dimethyl compound were also analyzed and the quantum states for these resembled those of *m*-xylene rather than *p*-xylene. Figure 50 shows the energy diagram for the methyl internal rotations for both S_0 and S_1 states, and these were fit with the one-dimensional periodic functions shown in Figure 51. The S_0 state has a tiny six-fold barrier, whereas the $S_1(\pi,\pi^*)$ state has a small but higher three-fold barrier.

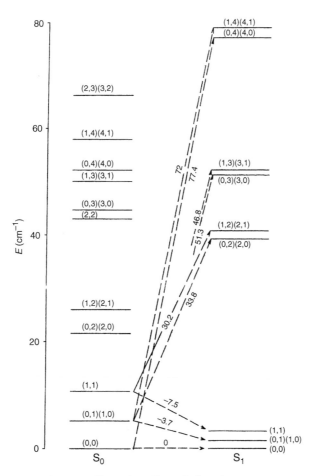

Figure 50 4,4-dimethyl-*trans*-stilbene (DMS) methyl torsion quantum states and transitions for S_0 and S_1 electronic states. The transitions to the (0,3) and (1,3) are not completely shown.

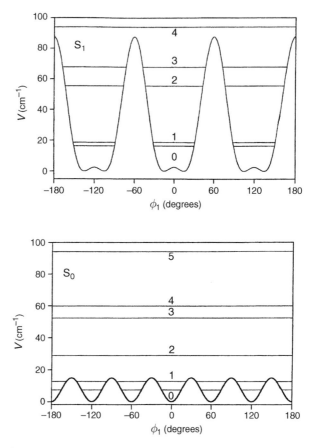

Figure 51 One-dimensional potential energy functions for a single methyl torsion in DMS in its S_0 and S_1 electronic states.

5.3. Bicyclic aromatics

5.3.1. Indan and related molecules

Indan (IND), phthalan (PHT), coumaran (COU) and 1,3-benzodioxole (13BZD), shown below, are "pseudo-four-membered-ring" molecules because their ring-puckering vibrations resemble those of four-membered rings as the two atoms of the five-membered ring joined to the benzene ring tend to move together as a single unit.

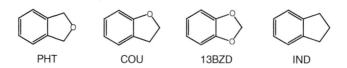

PHT COU 13BZD IND

These molecules also have low-frequency ring-flapping vibrations (also called butterfly motions) of the same symmetry species as the ring-puckering, and these

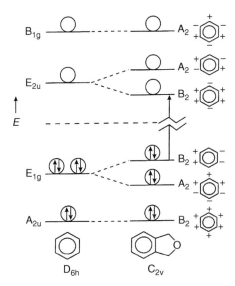

Figure 52 Molecular orbital diagram correlating to π orbitals of benzene to those of phthalan. The skeletal atoms of phthalan are assumed to be in the xz plane.

can interact strongly with the puckering motions. Figure 22 showed the definition of these two low-frequency vibrations. The puckering (x_1 or τ) is basically the out-of-plane motion of the apex CH_2 group or oxygen atom relative to all the other atoms in the two rings. The flapping (x_2) is the motion of the entire five-membered ring relative to the benzene ring. As with the other molecules discussed, vector methods have been developed to represent the motion of all the atoms relative to the two coordinates so that the kinetic energy (reciprocal-reduced mass) expansions for the molecules can be calculated as a function of the coordinates. These expansions have been presented elsewhere [59–64].

These bicyclic molecules have been studied in both their ground and excited electronic states. The discussion here will concentrate on the $S_1(\pi,\pi^*)$ excited states. Figure 52 shows a molecular orbital diagram correlating the benzene π orbitals to those of PHT (and the other three molecules in this group). The reduction of the benzene D_{6h} symmetry to C_{2v} in PHT produces the splitting of the degenerate E_{1g} and E_{2u} orbitals into A_2, B_2 pairs. The transition resulting in the $S_1(\pi,\pi^*)$ state is also shown in the figure.

5.3.1.1. Phthalan
The far-infrared spectrum of PHT in Figure 23 shows not only single quantum transitions but also weaker double and triple quantum jumps [59,60]. The primary series is in the 30–105 cm^{-1} region, and this also shows side bands arising from transitions in the flapping excited state. Figure 53 shows the energy map and observed transitions for the S_0 ground state for the ring-puckering (ν_P) and ring-flapping (ν_F) states. The primary puckering sequence is irregular and cannot be fit well with a one-dimensional function [59]. However, when the ring-flapping

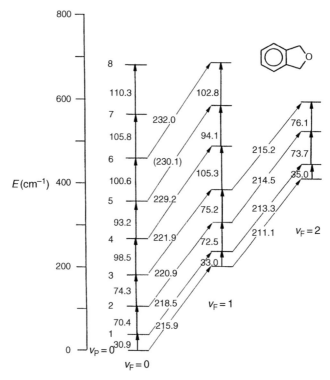

Figure 53 Energy level diagram for the ring-puckering vibration (ν_P) in different ring-flapping (ν_F) states of phthalan.

coordinate is included for a two-dimensional calculation that includes a cross-kinetic-energy term, an excellent agreement with the experimental data is obtained using a PES with a tiny barrier to planarity of 35 cm^{-1} [60].

The LIF of jet-cooled PHT and its ultraviolet absorption spectrum were used to determine the puckering and flapping quantum states for the $S_1(\pi,\pi^*)$ excited state [80]. Figure 54 shows the region of the UV spectrum near the band origin at 37,034.2 cm^{-1}. The puckering transitions to the vibronic levels in S_1 are labeled. A two-dimensional PES was again required to reproduce the experimental data. However, in this case, the PES has no barrier and the surface is considerably stiffer along the puckering coordinate. Figure 55 shows a cut of the PES along the ring-puckering coordinate. The puckering component is nearly pure quartic while the flapping is quadratic. The excited state energy level separations are also shown. Ab initio calculations also predict a stiffer excited state PES for PHT.

5.3.1.2. Coumaran

The original study of the far-infrared spectrum of COU [61] identified an inversion splitting of 3.1 cm^{-1}, and this was assigned to the $v=2-3$ separation. However, Ottavani and Caminati [81] later utilized millimeter wave spectroscopy and showed that the $v=0-1$ splitting was 3.12 cm^{-1}, and this assignment was then utilized to

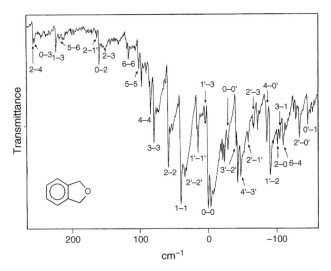

Figure 54 Ultraviolet absorption spectra of phthalan vapor near the electronic band origin. The ring-puckering transitions are labeled. Primes refer to the $\nu_F = 1$ state.

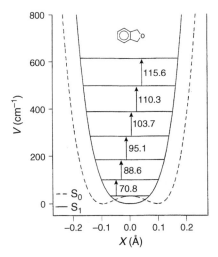

Figure 55 Ring-puckering potential energy curves for the phthalan $S_1(\pi, \pi^*)$ excited state compared to the S_0 ground state. The flapping coordinate x_2 is set equal to zero.

determine the one-dimensional ring-puckering potential energy function for COU [62]. The inversion barrier was found to be $154\,\mathrm{cm}^{-1}$, and the puckering angle at the energy minima is $25°$. A triple zeta ab initio calculation predicted a barrier of $238\,\mathrm{cm}^{-1}$ and a dihedral angle of $26.5°$.

The LIF spectra of COU were first reported by Watkins and co-workers [82] but were not correctly assigned due in part to the incorrect far-infrared assignments [61]. Yang and co-workers [83] in the Laane laboratory later reported the

FES, SVLF, and UV absorption spectra that not only confirmed the electronic ground state assignments but also provided the data for determining the energy map for both S_0 and $S_1(\pi,\pi^*)$ states. Figures 56 and 57 show the FES and SVLF spectra of COU, and Figure 58 shows the energy diagram. As was the case for the electronic ground state, the ring–puckering data can be fit very nicely with a one-dimensional potential energy function for S_1. In the excited state, however,

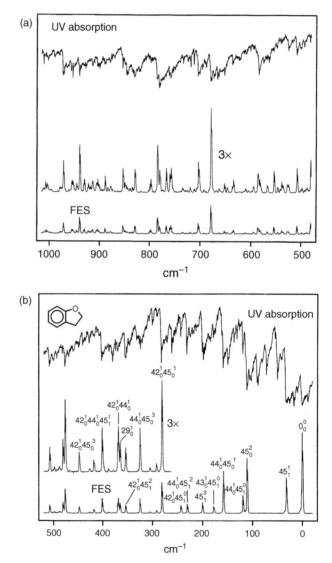

Figure 56 Fluorescence excitation spectra of jet-cooled coumaran and the corresponding ultraviolet absorption spectra at ambient temperatures.

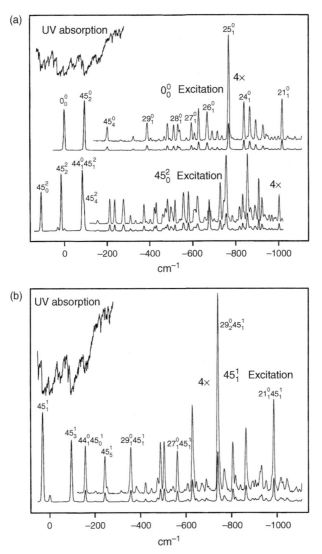

Figure 57 Single vibronic level fluorescence (SVLF) spectra of coumaran from the 0_0^0 band at $34{,}965.9\,\text{cm}^{-1}$ and the 45_0^2 and 45_1^1 bands which are 110.8 and $31.9\,\text{cm}^{-1}$ higher, respectively.

the barrier drops to $34\,\text{cm}^{-1}$ and the puckering dihedral angle drops to $14°$. Figure 59 compares the two ring-puckering potential energy functions to each other. The barrier in each case is due to the single CH_2—CH_2 torsional interaction. For the excited electronic state, the barrier is reduced because of the increased angle strain in the five-membered ring arising from the reduced π character of the benzene ring.

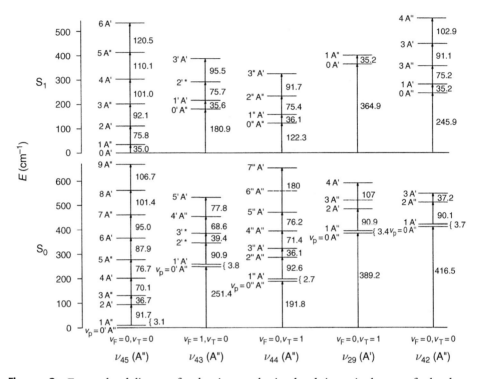

Figure 58 Energy level diagram for the ring-puckering levels in excited states of other low-frequency modes of coumaran in its S_0 and $S_1(\pi,\pi^*)$ states.

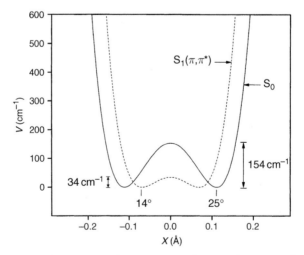

Figure 59 Comparison of the ring-puckering potential energy functions of coumaran in its S_0 and $S_1(\pi,\pi^*)$ states.

5.3.1.3. 1,3-Benzodioxole

13BZD is one of the most interesting molecules we have studied in our laboratory. Because the molecule has no CH_2—CH_2 interactions, which would be expected to pucker the five-membered ring, it would seem likely that the molecule would be planar. However, like 1,3-dioxole discussed earlier, 13BZD possesses the anomeric effect due to the —O—CH_2—O— configuration, and this provides an impetus for the five-membered ring to pucker. Duckett et al. [84] initially reported the far-infrared spectra of 13BZD but incorrectly assigned the spectra based on a planar structure. Later, far-infrared and Raman spectra by Sakurai and co-workers [85], however, showed that the molecule has a barrier to planarity of $164\,cm^{-1}$ and puckering angles of $\pm 24°$. A one-dimensional potential energy function was able to fit the data moderately well, but all of the experimental ring-puckering and ring-flapping data were much better fit with a two-dimensional PES. The $164\,cm^{-1}$ barrier provided evidence for the anomeric effect, but its magnitude was only about half of what had been found for 1,3-dioxole. This indicated that the anomeric effect was suppressed by the presence of the benzene ring.

The FES and UV absorption spectra of 13BZD were also investigated by Laane and co-workers [86] to study the $S_1(\pi,\pi^*)$ excited state. Figure 60 shows the jet-cooled FES and UV absorption spectra for this molecule. This is a beautiful example of how the two types of spectra complement each other extremely well. The FES shows only transitions from the vibrational ground state plus weaker transitions from the nearly degenerate $v=1$ state, which is $9.6\,cm^{-1}$ higher. The UV spectrum, on the contrary, shows a very large number of bands arising from the many populated S_0 vibrational levels at room temperature. Figure 61 shows the

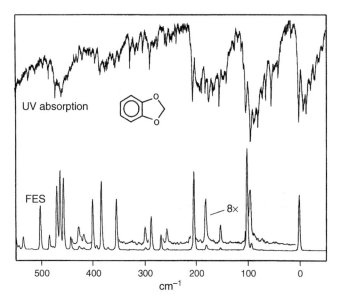

Figure 60 Fluorescence excitation spectrum (FES) and ultraviolet absorption spectrum of 1,3-benzodioxole. The band origin is at 34,789.8 cm^{-1}.

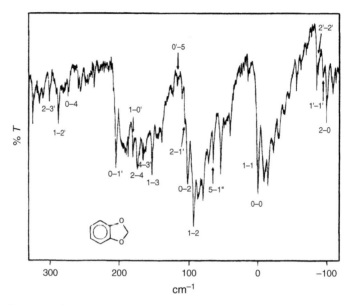

Figure 61 Electronic absorption spectra and assignments for 1,3-benzodioxole near the band origin.

kind of very valuable detail available from the UV absorption spectra. Figure 62 shows the ring-puckering and ring-flapping energy map generated from the spectral data. As can be seen from the significant differences between the puckering levels in the flapping $v_F = 0$ and $v_F = 1$ states, there is a large amount of interaction between these two modes. This is not unexpected as they are both of low frequency and have the same symmetry species (B_2 for C_{2v} symmetry). The $S_1(\pi,\pi^*)$ data were also utilized for a two-dimensional PES calculation, and Figure 63 shows the resulting surface that has a barrier of 264 cm^{-1}. To compare the inversion barriers for the S_0 and $S_1(\pi,\pi^*)$ states, Figure 64 shows the one-dimensional cut of the two PESs along the puckering coordinate. As is obvious, the barrier for the $S_1(\pi,\pi^*)$ state nearly doubles.

The experimental result that the 13BZD has a higher barrier and hence an increased anomeric effect in the electronic excited state supports the view that this effect is suppressed in the S_0 ground state. Apparently, as shown in Figure 65, there is competition between the anomeric effect and the interaction of the oxygen non-bonded p orbital with the benzene ring. The anomeric effect also utilizes this p orbital as it interacts with the σ^* orbital of the C—O bond involving the other oxygen atom. In the S_0 ground state, this $p(O)$-benzene interaction is higher and competes with the anomeric effect. However, in the $S_1(\pi,\pi^*)$ state, the π system of the benzene ring has been disturbed and thus the $p(O)$-benzene interaction is diminished allowing the anomeric effect to dominate. The $S_1(\pi,\pi^*)$ barrier height of 264 cm^{-1} is similar to that of 1,3-dioxole [48], where there is no suppression of this effect.

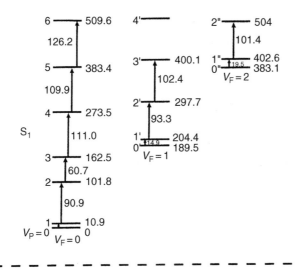

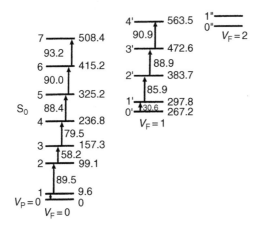

Figure 62 Energy diagram for the ring-puckering (ν_P), ring-flapping (ν_F), and ν_{37} vibrations of 1,3-benzodioxole in its ground (S_0) and excited (S_1) electronic states. Single and double primes on the ring-puckering (ν_P) quantum numbers indicate the $\nu_P = 1$ and 2 states, respectively.

5.3.1.4. *Indan*

The far-infrared spectra of IND were first reported by Smithson et al. [87] in 1984 and these are of high quality showing detail for both the ring-puckering and ring-flapping levels. They analyzed the puckering spectra with a one-dimensional potential energy function and reported a barrier of 1900 cm^{-1} for the S_0 ground state. As this value appeared to be inconsistent with the 232 cm^{-1} barrier of CP [16], Arp and co-workers [64] reinvestigated the far-infrared work in 2002 using a two-dimensional PES. This PES had a barrier of 488 cm^{-1}, minima at puckering angles of $\pm 30°$, and reproduced the experimental data

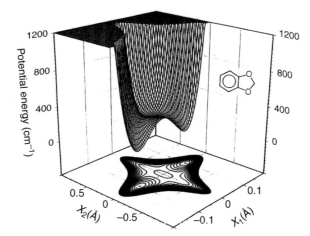

Figure 63 Two-dimensional potential energy surface (cm^{-1}) of 1,3-benzodioxole in its S$_1$(π,π^*) state: x_1, ring-puckering; x_2, ring-flapping.

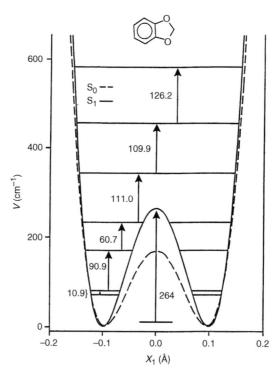

Figure 64 Comparison of the S$_0$ and S$_1$(π,π^*) vibrational potential energy surfaces along the ring-puckering coordinate (x_1). The flapping coordinate $x_2 = 0$.

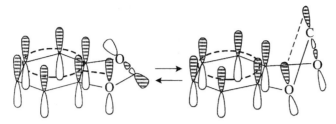

Figure 65 Depiction of the interaction between the benzene π system and the oxygen atoms. Only one oxygen non-bonded p orbital is shown along with a * (C—O) orbital with which the anomeric effect is achieved.

very well. In this case, the barrier arises from the two —CH$_2$—CH$_2$— torsional interactions, which tend to pucker the molecule so that the methylene groups do not eclipse each other.

Hollas et al. [88] reported the first UV absorption spectra of IND in 1977 and then the LIF in 1991 [89]. They relied on the Smithson and co-workers [87] assignments and reported barriers of 1979 and 1800 cm^{-1} for the S$_0$ and S$_1(\pi,\pi^*)$ states, respectively. The Laane laboratory reinvestigated the jet-cooled LIF of IND and reported new data and a two-dimensional PES analysis in 2002 [64]. Figure 66 shows the FES and the UV absorption spectra, and Figure 67 shows the vibronic levels for the ring-puckering, ring-flapping, and ring-twisting states in S$_1(\pi,\pi^*)$. The two-dimensional PES analysis showed the barrier to be 441 cm^{-1}, slightly less than that in the ground state and also somewhat less than the ab initio value of 528 cm^{-1}.

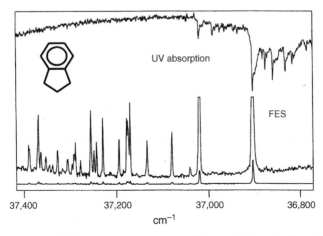

Figure 66 Laser-induced fluorescence spectrum (jet-cooled) and ultraviolet absorption spectrum of indan (25° C).

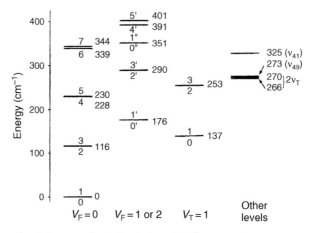

Figure 67   Energy level diagram for indan in its $S_1(\pi,\pi^*)$ excited state.

5.3.1.5.   Summary

The investigation of these four members of the IND family in both their ground and excited electronic states has provided some valuable insights into these molecules. Some of the quantitative results for the experimental and calculated barriers to planarity are shown in Table 4. The PHT molecule has no CH_2—CH_2 torsional interactions, so angle strain is expected to keep the molecule planar in both electronic states. This is almost true, but the electronic ground state does in fact have a small $35\ cm^{-1}$ barrier, and the ab initio calculation also predicts a small barrier. This may be because of a weak interaction involving the oxygen non-bonded orbital. The barrier disappears in the $S_1(\pi,\pi^*)$ state. COU does have a single CH_2—CH_2 interaction, which is responsible for the puckering of the five-membered ring in both electronic states. In the S_0 state, the barrier is $154\ cm^{-1}$, a magnitude that is fairly typical of a single methylene—methylene interaction. For 2,3-dihydrofuran, the barrier is $93\ cm^{-1}$ [47]. In its $S_1(\pi,\pi^*)$ excited state, the barrier almost disappears, and this reflects increased angle strain in the five-membered ring. IND has two methylene—methylene torsional interactions, and, as expected, it has higher and similar barriers to planarity in both S_0 and $S_1(\pi,\pi^*)$ states. There is little effect on the barrier as a result of the electronic transition. The most interesting case is that of 13BZD, which possesses the anomeric effect because of its two oxygen atoms in a 1,3-configuration. As was the case for 1,3-dioxole [48], the five-membered ring puckers to increase the overlap between the non-bonded oxygen p orbital and the σ^* orbital of the adjacent C—O bond. In the electronic ground state, this effect is suppressed as the benzene π system competes for interaction with the oxygen p orbital. Hence, the barrier is only about half of that determined for 1,3-dioxole. When the benzene π system is perturbed by the $\pi \rightarrow \pi^*$ excitation, however, its suppression of the anomeric effect is substantially diminished, and the barrier to planarity is almost doubled.

Table 4 Comparison of experimental and ab initio results for molecules in the indan family

Molecule	Ground state				S1(π,π^*) excited state					
	Barrier		Dihedral angles (cm^{-1})[a]		Barrier		Dihedral angles		ν_0^0(cm^{-1})	
	Exp.	Ab initio	Exp.	Ab initio	Exp.	Ab initio	Exp.	Ab initio	Exp.	Ab initio
Phthalan	35	91	0	23°	0	11	0	0	37,034	46,635
1,3-Benzodioxole	164	171	24°, 3°	25°, 3°	264	369	24°	29°	34,790	36,986
Coumaran	154	258	25°	27°	34	21	14°	—	34,870	38,965
Indan	488	662	30°	32°	441	528	39°	—	36,904	—

[a] Puckering angle except for 1,3-benzodioxole for which the flapping angle is also given.

5.3.2. Other bicyclic aromatics
5.3.2.1. Dihydronaphthalenes

The ground and $S_1(\pi,\pi^*)$ excited states of both 1,2-dihydronaphthalene (12DHN) and 14DHN have been studied using LIF. Autrey and co-workers [90]

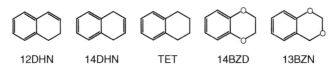

12DHN   14DHN   TET   14BZD   13BZN

reported the LIF and UV absorption spectra along with ab initio calculations of 12DHN in 2003. In addition to the determination of the PESs, vibrational assignments were reported for both states showing the effect of the electronic excitation. 12DHN is analogous to 1,3-cyclohexadiene [50] discussed earlier, but it possesses one extra out-of-plane ring mode, the ring-flapping, as shown in Figure 68. Figure 69 shows the FES and UV absorption spectra of 12DHN. As can be seen, the FES is very rich, but the UV shows much less detail than what is the case for many of the other systems discussed in this work. Figure 70 shows the energy diagram for both the ground and the excited electronic states. Vibrations 51–54 are the out-of-plane ring modes while 35 is an in-plane ring-bending mode. Numerous combinations between these vibrations were observed in the FES. Analysis of the data show the molecule to be twisted, and a one-dimensional potential energy function showed the barrier to inversion in the ground state to be $1363 \pm 100 \text{ cm}^{-1}$, with the uncertainty coming from the need to extrapolate the function above the observed experimental data. This agrees well with the triple-zeta ab initio value of 1524 cm^{-1}. In the $S_1(\pi,\pi^*)$ state, the experimental data show that the barrier increases substantially, probably to about 3000 cm^{-1}, but an accurate value could not be determined. The increased barrier in the excited state reflects the decreased conjugation in the π system, and this leads to lower angle strain.

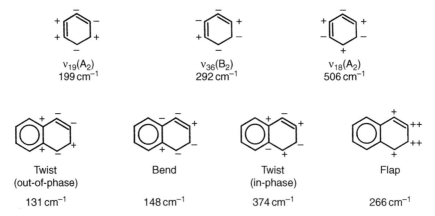

$\nu_{19}(A_2)$
199 cm^{-1}

$\nu_{36}(B_2)$
292 cm^{-1}

$\nu_{18}(A_2)$
506 cm^{-1}

Twist
(out-of-phase)
131 cm^{-1}

Bend
148 cm^{-1}

Twist
(in-phase)
374 cm^{-1}

Flap
266 cm^{-1}

Figure 68 Low-frequency out-of-plane vibrations of 1,3-cyclohexadiene and 1,2-dihydronaphthalene.

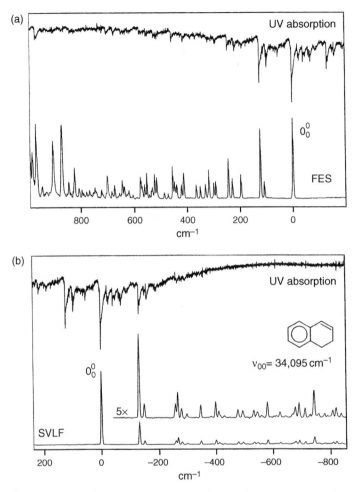

Figure 69 Fluorescence excitation spectrum (FES) and ultraviolet absorption spectrum of 1,2-dihydronaphthalene (12DHN).

14DHN can be classified as a "pseudo-four-membered ring." Its LIF was originally studied by Chakraborty and co-workers [91], but improved FES and UV data [92] in the Laane laboratories have led to a more reliable analysis. Figure 71 shows its FES and UV absorption spectra, which were used to generate the quantum energy map for the low-frequency modes, including ν_{54}, the ring-puckering. In the S_0 state, a regular sequence of energy separations beginning at 33.8 cm^{-1} correspond to a very non-rigid ring system. In the $S_1(\pi,\pi^*)$ state, the same type of sequence beginning at 77.4 cm^{-1} arises from a more rigid ring system. Figure 72 compares these one-dimensional ring-puckering potential energy functions to each other and also to 1,4-cyclohexadiene [17]. The stiffer potential energy function for the excited state again reflects increased angle strain [92].

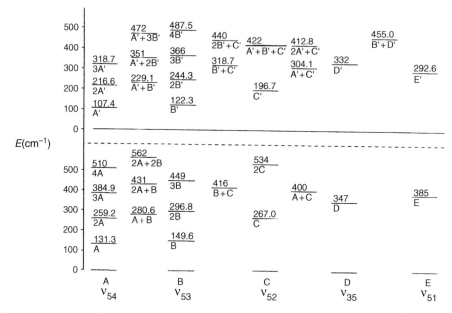

Figure 70 Energy diagram for the low-frequency vibrations in their S_0 and $S_1(\pi,\pi^*)$ electronic states of 1,2-dihydronaphthalene (12DHN).

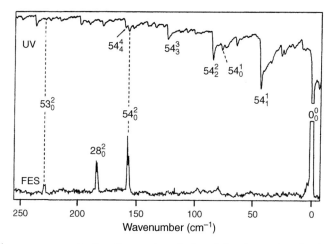

Figure 71 Fluorescence excitation (bottom) and ultraviolet absorption (top) spectra of 1,4-dihydronaphthalene (14DHN). The wavenumbers are relative to the 0_0^0 band at 36,788.6 cm^{-1}.

5.3.2.2. Tetralin and benzodioxans

Tetralin (TET), 1,4-benzodioxan (14BZD), and 1,3-benzodioxan (13BZN) are all analogous to the cyclohexene family of molecules and hence have twisted structures with high barriers to planarity. Because of their low vapor pressures, they have not been studied by far-infrared spectroscopy, but their S_0 vibrational data have been obtained using SVLF spectra of the jet-cooled molecules and

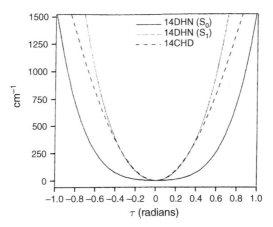

Figure 72   Comparison of the 1,4-dihydronaphthalene (14DHN) S_0 and S_1 potential energy functions with the 1,4-cyclohexadiene (14CHD) function.

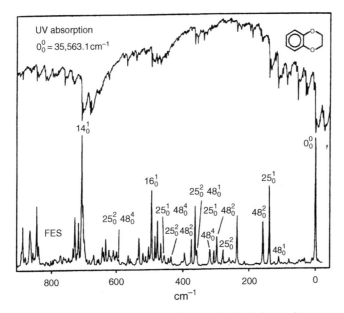

Figure 73   Fluorescence excitation spectra of jet-cooled 1,4-benzodioxan and ultraviolet absorption spectra at ambient temperature in the 0–900 cm^{-1} region. The band origin 0_0^0 is at 35,563.1 cm^{-1}.

high-temperature vapor-phase Raman spectra. The FES data were also recorded. Figures 73 and 74 show the FES and SVLF spectra of 14BZD, respectively, and this data were used to produce the quantum energy map of the low-frequency modes shown in Figure 75 [93]. Because the inversion barrier is so high and the experimental data extend to only about 700 cm^{-1} above the energy minima, an accurate PES could not be reliably obtained. However, Figure 76 shows the calculated relative energies of the twisted, bent, and planar conformations, and it

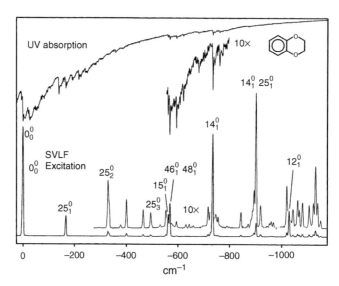

Figure 74   Single vibronic level fluorescence spectra of jet-cooled 1,4-benzodioxan with excitation of the 0_0^0 band at 35,563.1 cm^{-1} and the ultraviolet absorption spectra at ambient temperature.

also shows the direction and frequency values for the bending and twisting modes. A two-dimensional calculation produced a PES with a barrier of 3906 cm^{-1} in reasonable agreement with the ab initio value of 4095 cm^{-1}. For the S$_1$ excited state, the bending and twisting frequencies drop considerably to 79.8 and 139.6 cm^{-1} reflecting a lower barrier to planarity. The two-dimensional PES computed from the data yields a barrier of 1744 cm^{-1}.

The FES and SVLF data for 13BZN have also been recorded and analyzed [94]. This molecule undergoes the anomeric effect due to the presence of the O—CH$_2$—O configuration. 13BZN is also twisted with a high barrier to planarity. Its bending and twisting frequencies are 108.4 and 158.4 cm^{-1} for the ground state and drop to 96.3 and 101.7 cm^{-1} in S$_1$, again reflecting a decrease in barrier in the excited state.

The complete vibrational spectra of TET have been assigned for the S$_0$ state, and 17 of its 60 vibrations were determined for the S$_1(\pi,\pi^*)$ state including the out-of-plane ring modes [95]. Figure 77 shows its FES spectrum along with some of the assignments. As was done for the other molecules, a quantum energy map was generated and utilized to analyze the ring-bending and ring-twisting modes in particular. In the S$_0$ ground state, the bending and twisting frequencies are 94.3 and 141.7 cm^{-1}, respectively, and these decrease to 85.1 and 94.5 in the excited state. The barriers to planarity are close to 5000 cm^{-1} but cannot be determined with great accuracy.

5.3.2.3.   2-Indanol

2-Indanol is a fascinating molecule in that it can exist in four different conformations, one of which has intramolecular hydrogen bonding between the OH group and the benzene ring. Both LIF and detailed ab initio calculations were carried out for this molecule [96]. The four structures are shown in Figure 78. Conformation A possesses the intramolecular hydrogen bond and is calculated to be about 450 cm^{-1} lower in energy than the other three conformations, all of which have similar

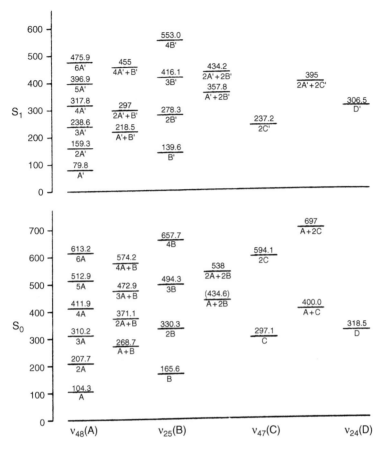

Figure 75   Energy level diagram for the low-frequency modes of 1,4-benzodioxan in its S_0 and $S_1(\pi,\pi^*)$ states. A is the ring-bending, B and D are ring-twisting modes, and C is ring-flapping.

energies. The planar ring structure is calculated to be $1308\ \mathrm{cm}^{-1}$ higher in energy. The four conformers can interconvert into one another either through the ring-puckering or the OH internal rotation vibrations. These motions encounter barriers from about 400 to $1700\ \mathrm{cm}^{-1}$. Figure 79 shows the computed theoretical PES for this molecule in terms of the ring-puckering angle and the OH internal rotation angle. The lowest energy form with the hydrogen bond corresponds to a puckering angle of $35.4°$ and an internal rotation angle of $180°$. The other forms are at angles of $-35.1°$ and $180°$ (b), $31.1°$ and $66.9°$ (c), and $-35.0°$ and $54.5°$ (d). Figure 80 shows the jet-cooled FES spectrum of 2-indanol, which strongly supports the theoretical calculations. Bands from each of the four conformers are present with the 0_0^0 band of A, the hydrogen bonded structure, by far the strongest. The band intensities indicate a distribution of 82% A, 3% B, 13% C, and 9% D, and these values are similar to those computed from the relative energies of the conformers. As is evident in Figure 80, the ring-puckering (ν_{29}) bands can also be seen for each conformer.

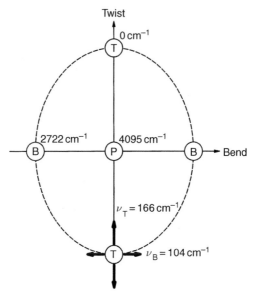

Figure 76 Representation of the two-dimensional potential energy for the ring-twisting and ring-bending vibrations of 1,4-benzodioxan in the S_0 state. P, planar; T, twisted; B, bent structure.

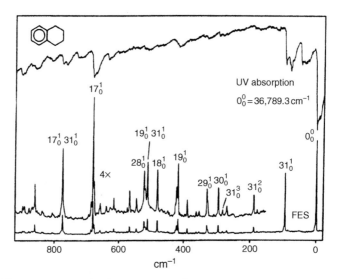

Figure 77 Fluorescence excitation spectra of jet-cooled tetralin and the ultraviolet absorption spectra at ambient temperature in the 0–900 cm^{-1} region.

5.4. Cavity ringdown spectroscopy of enones

In a collaboration between the Drucker and Laane laboratories, the cavity ringdown spectra (CRDS) of 2-cyclopenenone (2CP) and 2CH have been recorded and analyzed. The CRDS system in the Drucker laboratory is capable of very high

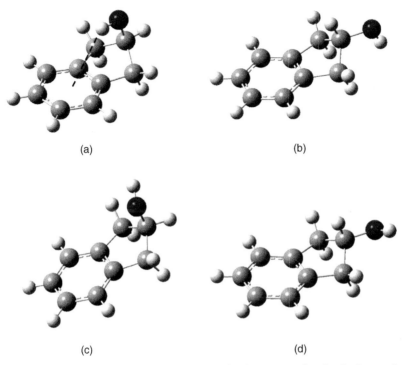

(a)   (b)

(c)   (d)

Figure 78 Four stable conformations of 2-indanol. The intramolecular hydrogen bonding present in the most stable conformer (structure A) is represented by a dotted line.

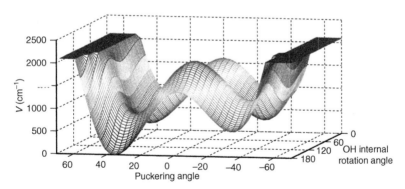

Figure 79 Calculated potential energy surface of 2-indanol in terms of its ring-puckering angle (degrees) on OH internal rotation (degrees relative to $180°$ at the A conformation). (See color plate 8).

sensitivity, and its details have been described elsewhere [97]. The high sensitivity, which arises from thousands of reflections between two highly reflecting mirrors, has allowed very weak absorption signals to be recorded in the UV region for these two molecules. Both 2CP and 2CH are conjugated possessing adjoining C=C and C=O double bonds. As already discussed, 2-CPO is planar due to this conjugation

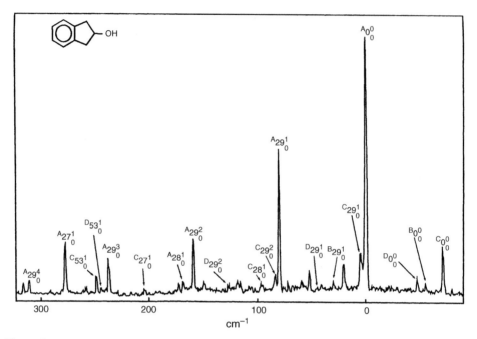

Figure 80 Laser-induced fluorescence excitation spectrum of 2-indanol.

in the electronic ground state. The lowest frequency electronic transition for these molecules is the $n \rightarrow \pi$ transition, and this results in weaker conjugation in the electronic excited state. As already shown in Table 1 for 2CP, the C=O and C=C frequencies are decreased in the $S_1(n,\pi^*)$ state. However, the single bond between these two double bonds is strengthened, and one of the other ring-stretching modes increases in frequency. Figure 81 presents the survey CRDS spectrum of 2CP showing the absorption by both transitions to the singlet $S_1(n,\pi^*)$ and the triplet $T_1(n,\pi^*)$ states [98]. Transitions to triplet states are spin forbidden, so observation of such transitions is quite unusual. The $S_1 \leftarrow S_0$ transitions confirm those previously discussed in the FES spectrum [66], but the $T_1 \leftarrow S_0$ had not been previously seen. The weaker absorption bands near the triplet band origin arise from transitions between the ring-puckering (ν_{30}) levels of the S_0 and T_1 states. Namely, 30_0^1, 30_0^2, 30_0^3, 30_4^3, 30_1^1, 30_1^2, 30_1^3, 30_2^2, 30_3^3 and many other transitions were observed. Figure 82 shows the energy level spacings for three isotopomers of 2CP for both the singlet ground state and the triplet excited state. Figure 83 shows the one-dimensional potential energy function with the form of Equation (2) for the triplet state. This is the first determination of a ring-puckering potential function for a triplet state. The function has a barrier of 42 cm^{-1}, which is totally produced by the flipped electron spin as the S_0 and S_1 states are both planar. Figure 84 compares the S_0, S_1, and T_1 ring-puckering functions. In S_0, 2CP is planar and rigid, in the S_1 state it becomes floppy because of the loss of conjugation, and in the triplet state the molecule remains floppy, but also takes on a non-planar structure. This structural change and the presence of the barrier produced by the spin flip is highly unusual.

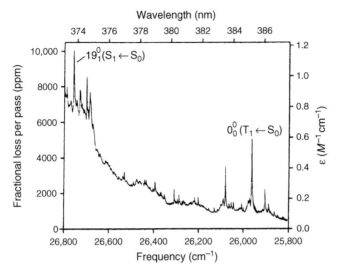

Figure 81 Room-temperature CRD spectrum of 2-cyclopentenone (2CP) vapor showing the $T_1 \leftarrow S_0$ origin band near 385 nm, as well as the onset of $S_1 \leftarrow S_0$ transitions near 375 nm.

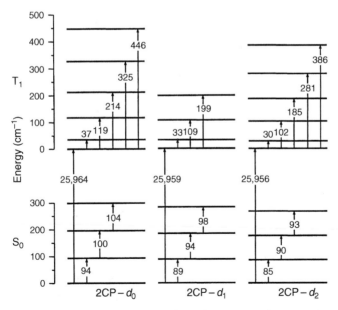

Figure 82 Energy-level diagram, determined from the present vibronic assignments (triplet state) and previous far-infrared results (ground state), for the ring-bending vibrational mode of 2CP-d_0, 2CP-d_1, and 2CP-d_2.

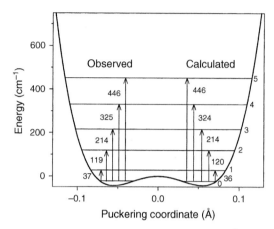

Figure 83 Ring-bending potential-energy function for the $T_1(n,\pi^*)$ state of 2-cyclopentenone (2CP).

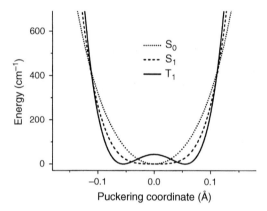

Figure 84 Comparison of experimentally determined ring-bending potential energy functions for the S_0, $T_1(n,\pi^*)$, and $S_1(n,\pi^*)$ states of 2-cyclopentenone (2CP).

To more fully understand the changes in 2CP, theoretical DFT and ab initio calculations were carried out for the molecule in its S_0, $S_1(n,\pi^*)$, $T_1(n,\pi^*)$, and $T_2(\pi,\pi^*)$ states [99]. These supported the experimental results and confirmed that the T_1 state arises from the $n \rightarrow \pi^*$ transition. In addition, the calculations provided data for many of the other vibronic levels. The ab initio calculations predicted a barrier of $8\ \mathrm{cm}^{-1}$ for the $T_1(n,\pi^*)$ state compared to the experimental value of $43\ \mathrm{cm}^{-1}$. The barrier for the $T_2(\pi,\pi^*)$ state was calculated to be much higher at $999\ \mathrm{cm}^{-1}$.

The CRDS of 2CHO have also been analyzed [100,101] and the potential energy functions for the inversion vibration have been determined for both the ground and $S_1(n,\pi^*)$ excited states. Earlier Raman [102] and infrared work [103] were used to postulate potential energy functions with barriers of 935 and $3379\ \mathrm{cm}^{-1}$, respectively, but both of these proved to be erroneous. Figure 85 shows the CRDS of 2CHO which has the $S_1(n,\pi^*)$ band origin at $26{,}081.3\ \mathrm{cm}^{-1}$. As can be seen, bands in both the

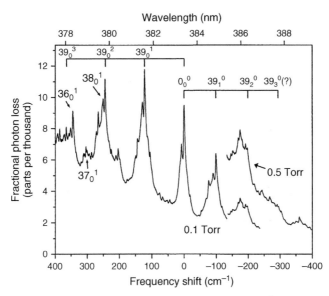

Figure 85   Cavity ringdown spectra of 2CHO relative to the $S_1\ 0_0^0$ band at 26,089.1 cm^{-1} in the difference band region.

39_n^0 and 39_0^m series were readily observed where n is the S_0 state quantum number for the inversion vibration (ν_{39}) and m is that for the $S_1(n,\pi^*)$ state. The 39_n^0 series together with the far-infrared spectrum [103] allowed the inversion and twisting quantum states for the S_0 electronic ground state to be correctly assigned, whereas the 39_0^m series provided the data for the $S_1(n,\pi^*)$ state. A one-dimensional potential function for the ground state determined the barrier to be 1900 ± 300 cm^{-1}, in good agreement with the DFT value of 2090 cm^{-1} [100]. The excited state data extended to about 800 cm^{-1} above the energy minima and were used to determine a one-dimensional potential function with a barrier of 3550 ± 500 cm^{-1} for the S_1 state. Ab initio calculations predicted a value of only 2265 cm^{-1}. The increased barrier height for the electronic excited state results from the loss of conjugation and decreased angle strain allowing the molecule to distort more from a planar structure.

For both 2CPO and 2CHO the absorption spectra are very weak and have not been observed by conventional methods. These two studies demonstrate the power of CRDS to get these results. Direct observation of the triplet state spectra for 2CPO was especially gratifying.

6.   SUMMARY AND CONCLUSIONS

As demonstrated by the many studies discussed here, one- and two-dimensional PESs can be most invaluable for determining the structures of molecules in both their ground and excited electronic states. They also help to elucidate the forces responsible for the structures and also provide quantitative detail on the energy differences between various molecular structures. The results for electronic excited states, including triplet states, are particularly valuable in that photochemical processes proceed through such states after the molecules have been structurally distorted because of electronic transitions. It should also be emphasized that these investigations validate the use of the Schrödinger equation along with the approximation that the vibrations of interest can be studied independently from all the other vibrations in these molecules. This is based on the fact that the low-frequency vibrations do not interact with the higher frequency modes or with those of different symmetry. While tables comparing observed and calculated frequencies for the PESs, often based on Equation (2), have not been shown here, it should be noted that in virtually all of these cases the agreement is fantastic. For SCB [15], for example, 34 frequencies are fit with the two-parameter potential function of Equation (2) with an average accuracy of better than $1 \, cm^{-1}$. Examples such as this should be added to physical chemistry texts, which typically only discuss the rather boring harmonic oscillator.

ACKNOWLEDGMENTS

The author expresses his sincere gratitude to the more than 40 students, postdocs, and visiting faculty members who have contributed to this work. He also thanks the National Science Foundation and the Robert A. Welch Foundation for many years of financial support. He is also grateful to Ms. Linda Redd for assistance with the preparation of this manuscript.

REFERENCES

[1] D. M. Dennison and G. E. Uhlenbeck, Phys. Rev., 41: 313–321, 1932.
[2] D. M. Dennison, Rev. Mod. Phys., 12: 175–321, 1940.
[3] L. Coudert, A. Valentin, and L. Henry, J. Mol. Spectrosc., 120: 185–204, 1986.
[4] V. Spirko and W. P. Kraemer, J. Mol. Spectrosc., 133: 331–344, 1989.
[5] D. J. Rush and K. B. Wiberg, J. Phys. Chem. A, 101: 3143–3151, 1997.
[6] J. Laane, Appl. Spectrosc., 24: 73–80, 1970.
[7] R. P. Bell, Proc. R. Soc., A183: 328–337, 1945.
[8] A. Danti and R. C. Lord, J. Chem. Phys., 30: 1310–1313, 1959.
[9] A. Danti, W. J. Lafferty, and R. C. Lord, J. Chem. Phys., 33: 294–295, 1960.
[10] S. I. Chan, J. Zinn, and W. D. Gwinn, J. Chem. Phys., 33: 295–296, 1960.
[11] S. I. Chan, J. Zinn, and W. D. Gwinn, J. Chem. Phys., 33: 1643–1655, 1960.
[12] S. I. Chan, J. Zinn, and W. D. Gwinn, J. Chem. Phys., 34: 1319–1329, 1961.
[13] S. I. Chan, T. R. Borgers, J. W. Russell, H. L. Strauss, and W. D. Gwinn, J. Chem. Phys., 44: 1103–1111, 1966.
[14] T. R. Borgers and H. L. Strauss, J. Chem. Phys., 45: 947–955, 1966.
[15] J. Laane and R. C. Lord, J. Chem. Phys., 48: 1508–1513, 1968.
[16] J. Laane and R. C. Lord, J. Chem. Phys., 47: 4941–4945, 1967.
[17] J. Laane and R. C. Lord, J. Mol. Spectrosc., 39: 340–344, 1971.
[18] L. A. Carreira, R. C. Lord, and T. B. Malloy, Top. Curr. Chem., 82: 1–95, 1979.
[19] J. E. Kilpatrick, K. S. Pitzer, and R. Spitzer, J. Am. Chem. Soc., 69: 2483–2488, 1947.
[20] K. S. Pitzer and W. E. Donath, J. Am. Chem. Soc., 81: 3213–3218, 1959.
[21] J. R. Durig and D. W. Wertz, J. Chem. Phys., 49: 2118–2121, 1968.

[22] L. E. Bauman and J. Laane, J. Phys. Chem., 92: 1040–1051, 1988.

[23] J. Laane, Vib. Spectra Struct., 1: 25–50, 1978.

[24] H. Strauss, Ann. Rev. Phys. Chem., 34: 301–328, 1983.

[25] W. Kiefer, H. J. Bernstein, M. Danyluk, and H. Wieser, Chem. Phys. Lett., 12: 605–609, 1972.

[26] T. H. Chao and J. Laane, Chem. Phys. Lett., 14: 595–597, 1972.

[27] F. A. Miller and R. J. Capwell, Spectrochim. Acta, 28A: 603–618, 1972.

[28] Internal Rotation and Inversion(Lister, MacDonald, and Owen, eds), Academic Press, London.

[29] P. Groner, J. F. Sullivan, and J. R. Durig, Vib. Spectra Struct., 9: 405–496, 1981.

[30] J. D. Lewis, T. B. Malloy, Jr., T. H. Chao, and J. Laane, J. Mol. Struct., 12: 427–449, 1972.

[31] J. Laane, J. Phys. Chem., 95: 9246–9249, 1991.

[32] T. Klots, S. N. Lee, and J. Laane, J. Phys. Chem., 103: 833–837, 1999.

[33] J. Laane, J. Chem. Phys., 50: 776–782, 1969.

[34] R. M. Irwin, J. M. Cooke, and J. Laane, J. Am. Chem. Soc., 99: 3273–3278, 1977.

[35] T. B. Malloy, J. Mol. Spectrosc., 44: 504–535, 1972.

[36] J. Laane, M. A. Harthcock, P. M. Killough, L. E. Bauman, and J. M. Cooke, J. Mol. Spectrosc., 91: 286–299, 1982.

[37] M. A. Harthcock and J. Laane, J. Mol. Spectrosc., 91: 300–324, 1982.

[38] R. W. Schmude, M. A. Harthcock, M. B. Kelly, and J. Laane, J. Mol. Spectrosc., 124: 369–378, 1987.

[39] M. M. Tecklenburg and J. Laane, J. Mol. Spectrosc., 137: 65–81, 1989.

[40] M. M. Strube and J. Laane, J. Mol. Spectrosc., 129: 126–139, 1988.

[41] J. Yang and J. Laane, J. Mol. Struct., 798: 27–33, 2006.

[42] S. Sakurai, N. Meinander, and J. Laane, J. Chem. Phys., 108: 3537–3542, 1998.

[43] C. M. Cheatham, M. Huang, N. Meinander, M. B. Kelly, K. Haller, W.-Y. Chiang, and J. Laane, J. Mol. Struct., 377: 81–92, 1996.

[44] C. M. Cheatham, M.-H. Huang, and J. Laane, J. Mol. Struct., 377: 93–99, 1996.

[45] K. Morris, Instrumentation for the Study of Jet-Cooled Molecules and Spectroscopic Investigation of Low-Frequency Vibrations, Ph.D. Thesis, Texas A&M University, 1998.

[46] T. M. Hard and R. C. Lord, Appl. Opt., 7: 589–598, 1968.

[47] J. Laane, K. Haller, S. Sakurai, K. Morris, D. Autrey, Z. Arp, W.-Y. Chiang, and A. Combs, J. Mol. Struct., 650: 57–68, 2003.

[48] D. Autrey and J. Laane, J. Phys. Chem., 105: 6894–6899, 2001.

[49] E. Cortez, R. Verastegui, J. R. Villarreal, and J. Laane, J. Am. Chem. Soc., 115: 12132–12136, 1993.

[50] D. Autrey, J. Choo, and J. Laane, J. Phys. Chem., 105: 10230–10236, 2001.

[51] T. H. Chao and J. Laane, J. Mol. Spectrosc., 70: 357–360, 1978.

[52] L. E. Bauman and J. Laane, J. Phys. Chem., 92: 1040–1051, 1988.

[53] L. A. Carreira, I. M. Mills, W. B. Person, J. Chem. Phys., 56: 1444–1448, 1972.

[54] S. J. Leibowitz, J. R. Villarreal, and J. Laane, J. Chem. Phys., 96: 7298–7305, 1992.

[55] S. J. Leibowitz and J. Laane, J. Chem. Phys., 101: 2740–2745, 1994.

[56] V. E. Gaines, S. J. Leibowitz, and J. Laane, J. Am. Chem. Soc., 113: 9735–9742, 1991.

[57] J. Laane and J. Choo, J. Am. Chem. Soc., 116: 3889–3891, 1994.

[58] M. M. Tecklenburg and J. Laane, J. Am. Chem. Soc., 111: 6920–6926, 1989.

[59] M. M. Tecklenburg, J. Villarreal, and J. Laane, J. Chem. Phys., 91: 2771–2775, 1989.

[60] T. Klots, S. Sakurai, and J. Laane, J. Chem. Phys., 108: 3531–3536, 1998.

[61] T. Klots, E. Bondoc, and J. Laane, J. Phys. Chem., 104: 275–279, 2000.

[62] J. Yang, K. Okuyama, K. Morris, Z. Arp, and J. Laane, J. Phys. Chem. A, 109: 8290–8292, 2005.

[63] S. Sakurai, N. Meinander, K. Morris, and J. Laane, J. Am. Chem. Soc., 121: 5056–5062, 1999.

[64] Z. Arp, N. Meinander, J. Choo, and J. Laane, J. Chem. Phys., 116: 6648–6655, 2002.

[65] J. C. D. Brand, J. Chem. Soc. 858–872, 1956..

[66] C. M. Cheatham and J. Laane, J. Chem. Phys., 94: 7734–7743, 1991.

[67] P. Sagear and J. Laane, J. Chem. Phys., 102: 7789–7797, 1995.

[68] J. Zhang, W.-Y. Chiang, and J. Laane, J. Chem. Phys., 98: 6129–6137, 1993.

[69] J. Choo and J. Laane, J. Chem. Phys., 101: 2772–2778, 1994.
[70] J. Zhang, W.-Y. Chiang, and J. Laane, J. Chem. Phys., 100: 3455–3462, 1994.
[71] W.-Y. Chiang and J. Laane, J. Phys. Chem., 99: 11640–11643, 1995.
[72] P. Sagear, S. N. Lee, and J. Laane, J. Chem. Phys., 106: 3876–3883, 1997.
[73] C. M. Cheatham and J. Laane, J. Chem. Phys., 94: 5394–5401, 1991.
[74] D. H. Waldeck, Chem. Rev., 91: 415–436, and references therein, 1991.
[75] W.-Y. Chiang and J. Laane, J. Chem. Phys., 101: 8755–8767, and references therein, 1994.
[76] L. Banares, A. A. Heikal, and A. H. Zewail, J. Chem. Phys., 96: 4127–4130, 1992.
[77] K. Haller, W.-Y. Chiang, A. del Rosario, and J. Laane, J. Mol. Struct., 379: 19–23, 1996.
[78] W.-Y. Chiang and J. Laane, J. Phys. Chem., 99: 11823–11829, 1995.
[79] Z. Arp, J. Laane, A. Sakamoto, and M. Tasumi, J. Phys. Chem., 106: 3479–3484, 2002.
[80] E. Bondoc, S. Sakurai, K. Morris, W.-Y. Chiang, and J. Laane, J. Chem. Phys., 112: 6700–6706, 2000.
[81] P. Ottavani and W. Caminati, Chem. Phys. Lett., 405: 68–72, 2005.
[82] M. J. Watkins, D. E. Belcher, and M. C. R. Cockett, J. Chem. Phys., 116: 7855–7867, 2002.
[83] J. Yang, M. Wagner, K. Okuyama, K. Morris, Z. Arp, J. Choo, N. Meinander, and J. Laane, J. Chem. Phys., 125: 034308/1–034308/9, 2006.
[84] J. A. Duckett, T. L. Smithson, and H. Wieser, Chem. Phys. Lett., 64: 261–265, 1979.
[85] S. Sakurai, N. Meinander, K. Morris, and J. Laane, J. Am. Chem. Soc., 121: 5056–5062, 1999.
[86] S. Sakurai, E. Bondoc, J. Laane, K. Morris, N. Meinander, and J. Choo, J. Am. Chem. Soc., 122: 2628–2634, 2000.
[87] T. L. Smithson, J. A. Duckett, and H. Wieser, J. Phys. Chem., 88: 1102–1109, 1984.
[88] J. M. Hollas, J. Mol. Spectrosc., 66: 452–464, 1977.
[89] K. M. Hassan and J. M. Hollas, J. Mol. Spectrosc., 147: 100–113, 1991.
[90] D. Autrey, Z. Arp, J. Choo, and J. Laane, J. Chem. Phys., 119: 2557–2568, 2003.
[91] T. Chakraborty, J. E. Del Bene, and E. C. Lim, J. Chem. Phys., 98: 8–13, 1993.
[92] M. Z. M. Rishard, M. Wagner, J. Yang, and J. Laane, Chem. Phys. Lett., 442: 182–186, 2007.
[93] J. Yang, M. Wagner, and J. Laane, J. Phys. Chem. A, 110: 9805–9815, 2006.
[94] K. McCann, M. Wagner, and J. Laane, to be published.
[95] J. Yang, M. Wagner, and J. Laane, J. Phys. Chem. A, 111: 8429–8438, 2007.
[96] A. A. Al-Saadi, M. Wagner, and J. Laane, J. Phys. Chem. A., 110: 12292–12297, 2006.
[97] S. Drucker, J. L. Van Zanten, N. D. Gagnon, E. J. Gilles, and N. R. Pillsbury, J. Mol. Struct., 692: 1–16, 2004.
[98] N. Pillsbury, J. Choo, J. Laane, and S. Drucker, J. Phys. Chem. A, 107: 10648–10654, 2003.
[99] J. Choo, S. Kim, S. Drucker, and J. Laane, J. Phys. Chem. A, 107: 10655–10659, 2003.
[100] E. J. Gilles, J. Choo, D. Autrey, M. Rishard, S. Drucker, and J. Laane, Can. J. Chem., 82: 867–872, 2004.
[101] M. Z. M. Rishard, E. A. Brown, L. A. Ausman, S. Drucker, J. Choo, and J. Laane, J. Phys. Chem. A, 112, 38–44, (2007).
[102] L. A. Carreira, T. G. Towns, and T. B. Malloy, Jr., J. Chem. Phys., 70: 2273–2275, 1979.
[103] T. L. Smithson and H. Wieser, J. Chem. Phys., 72: 2340–2346, 1980.

CHAPTER 5

RAMAN SPECTROSCOPY IN ART AND ARCHAEOLOGY: A NEW LIGHT ON HISTORICAL MYSTERIES

Howell G.M. Edwards

Contents

Abstract

The major advantages of the adoption of Raman spectroscopy for the analysis of artefacts and works of art are twofold: the technique is nondestructive and requires little or no chemical and mechanical pretreatments of the specimen, and the molecular signatures from both the organic and inorganic components are obtained in the same spectrum, hence affording the opportunity for assessing the interactions and relative stabilities to chemical, biological and environmental changes operating on the specimen. In this chapter, the results of several selected case studies that have been undertaken in collaboration with archaeologists, art historians and conservation scientists will be presented to illustrate the use of the analytical data derived from Raman molecular spectroscopic applications in the following scenarios: the biodegradation of wall paintings, the restoration of mediaeval and Renaissance frescoes, the restoration of a badly damaged, historic marine textile (the foretopsail of *HMS Victory* from the Battle of Trafalgar), the identification of localised biological attack on human mummified and skeletal remains from diverse burial environments (Egyptian mummies, ice cave mummies and the 6th/7th century monastic Towyn-y-Capel grave site) and the identification of ivories and resins in museum specimens. The examination of such a diverse range of artefacts and specimens using nondestructive laser Raman spectroscopic analysis can provide the acquisition of novel historical information about their origins and degradation suffered in a new light.

Keywords: archaeology; Raman spectroscopy; pigments; mummies; art history; linen textiles; ivories; resins; biodeterioration; restoration; conservation

1. INTRODUCTION

The importance of chemical analysis in the characterisation of ancient objects and materials first appeared in a paper to the Royal Society of London authored by Sir Humphry Davy [1] entitled "Some experiments and observations on the colours used in painting by the Ancients", which reported the study of pigments in wall-painting fragments from the ruins of Pompeii and several Imperial Roman palaces in Rome; Davy was concerned even then with conservation aspects that have much resonance today and hinted at the need for minimal sampling consistent with the provision of chemical compositional data.

The major problem associated with the wet chemical analysis of museum artefacts and excavated relics using traditional methods is the destruction of the object being analysed, or at least a significant part of it; hence, in the chemical analyses of Eccles and Rackham [2] of a range of porcelains in 1922 for the Victoria and Albert Museum in London, following an earlier study of Church [3], whole pieces (cups, plates and saucers) were sacrificed from antique china services. What could not be analysed before the advent of microchemical techniques were paintings and large art objects in which the removal of substantial amounts of material would have created severe problems for either the integrity or the aesthetic appearance of the art work. The advent of spectroscopic analysis enabled analytical information to be obtained without the despoiling of the art object, although severe limitations were still apparent in the selection of problems that could be addressed [4,5].

The viability of the Raman spectroscopic technique for the analysis of art works and archaeological specimens was demonstrated comprehensively by Guineau [6] for manuscript pigments in the 1980s. In this, the illumination of small areas of specimen by a focussed laser beam and the collection of scattered radiation by a microscope lens provided a powerful combination of one of the oldest and the newest scientific instruments for the analysis of art work. These early years saw major advances made in the Raman spectroscopic characterisation of mineral pigments and their mixtures, which incidentally gave a parallel insight into ancient technologies of production and provided the first chemical evidence of specific deterioration that provided the impetus for its adoption in a conservation context and in the provision of novel information for the restoration of damaged art works. Several examples of the power of the emerging Raman spectroscopic analytical technique in art applications are the identification of adulteration of pigments and the substitution of cheaper alternatives for expensive semiprecious minerals [7], the admixture of coloured pigments to achieve the desired tonal qualities [8], the detection of unrecorded restoration using atypical synthetic materials for the original natural minerals [9], the recognition of degraded mineral pigments in deteriorated art work [10] and the exposure of fraudulent works of art through the use of inappropriate materials [11].

Additional information was quickly forthcoming from Raman spectroscopic microanalyses of paint layers and their interactions with substrates, which was previously not suspected when separation processes had to be undertaken, the use of additives such as siccatives, gums and resins to achieve the adhesion of pigments to manuscripts and canvas, and the detection of mixtures of organic dyes and inorganic pigments in combination on art work. An example of the latter, namely the admixture of two red pigments, one mineral (*cinnabar*) and the other an organic extract (*kermes*, carmine), to create a special effect is shown in Figure 1; here the Fourier-transform Raman spectrum of a specimen of red pigment from an ancient Indian statue group, *Kali distressing Siva*, which is now in the Horniman Museum, London, indicates clear signatures of the low wavenumber modes in the cinnabar component and other modes at higher wavenumbers characteristic of the carmine component. The key factor in all this work was the use of molecular fingerprinting for different minerals and organic components, which could now be

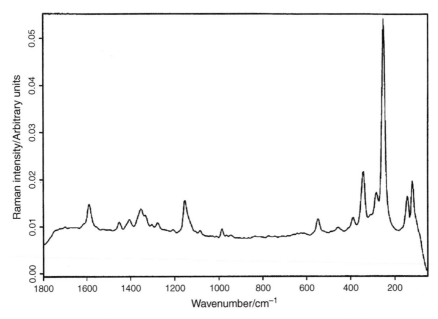

Figure 1 Fourier-transform Raman spectrum of a red pigment from the Indian statue group of *Kali distressing Siva*, The Horniman Museum, London; acquired in 1896 but of indeterminate age; 1064 nm excitation; wavenumber range 100–1800 cm^{-1}. The spectral signatures of mercury (II) sulphide, *cinnabar*/vermilion, and of *kermes*/carmine demonstrate the viability of Raman spectroscopy to detect organic and inorganic components in admixture in an art work.

related to the solution of art historical problems from vibrational spectroscopic studies that had hitherto been the realm of preparative inorganic and organic chemists. In this respect, the construction of relevant databases that had been undertaken for several laser wavelengths provided an integral part of the spectroscopic interpretation needed [12,13,14]. Further opportunities for an extension of these spectroscopic applications were made in the analytical studies of art works that were exposed to environmental effects, such as wall paintings and prehistoric cave art [15,16]; the information that could now be accessed from these sources was often the only analytical data that could be obtained and therefore was extremely useful in that respect, thereby opening up the field of study to archaeological casework. For example, the application of Raman spectroscopy to deteriorated wall paintings and frescoes was a triumph for the technique, which upon extension to mediaeval, Renaissance and Romanesque examples gave novel information about the technology of preparation of the pigments and the methodology of their application to the substrates [17,18,19]. A recent example of this theme is provided by Raman spectroscopic data from badly damaged, late Renaissance ceiling frescoes in the church of Sant Joan del Mercat in Valencia, Spain, which resulted from deliberate attempts to destroy the art work by gunfire and conflagration during the Spanish Civil War in the 20th century (Figure 2) [20]; here, the

Figure 2 Severely damaged late 17th century frescoes in the vaulted nave of the church of Sant Joan del Mercat, Valencia, Spain; only some 20% of the frescoes now remain after a series of seven conflagrations and gunfire during the Spanish Civil War in 1936. In places, the ground substrate of the frescoes has been damaged through the *intonaco, arricio* and *arenato* zones to the bare brick of the church ceiling.

temperatures reached in different parts of the nave of the church could be evaluated from the Raman measurements of the surviving mineral pigments and knowledge of their transition and conversion temperatures. This information is essential for restorers to create the original colours in the damaged frescoes and could not be derived from other historical means.

A rather different area of application of Raman spectroscopy in art and archaeology is the characterisation of biomaterials and their degradation; often, the deterioration of sensitive organic materials such as textiles, ivories, skin, hair, paper and parchment is far advanced in their burial environments and their conservation problems are consequently urgent and immense. Differential deterioration occurs, dependent upon the localised conditions, and this can be especially troublesome where biological colonisation is concerned. Recent examples to be found in the literature are the variable preservation of the skin of Egyptian mummies [21], the variable degradation of human skeletal remains from a stone cist burial [22], the survival of hair from a waterlogged archaeological excavation [23], the identification of ivory fragments from a Viking settlement [24], the restoration of an important relic battle-damaged textile [25] and the discrimination between bone fragments and stained mineral deposits in a Roman grave site [26].

There follows a series of case studies that have been selected to illustrate many of the applications of analytical Raman spectroscopy described above, from which novel knowledge has been contributed to art history and archaeology. Most of these studies have involved the collaboration of museum scientists, art historians, archaeologists and conservation scientists at what has been termed the interface of "hard" and "soft" science, and both spectroscopists and non-spectroscopists have thereby been made aware of the importance of spectroscopic data at this arts/science interface. Finally, there is an obvious further application too in the area of forensic science, strictly, "science in the public forum", and in the identification of forgeries and fakes where the legal implications emerge through fraud and the intent to deceive through an enhancement of the object's value. However, it must be accepted that Raman spectroscopy can never prove that an art object is authentic even though the vibrational spectroscopic data can still expose a fake. The succinct statement of Orna [27] that

> "*Scientific methods can only be used to unmask or de-authenticate a given artefact; they can never prove an artefact is genuine*"

is one that will be addressed several times in this chapter. In one example, an antique sperm whale ivory tooth that has been carved or inscribed in recent times to increase its value will still have the Raman spectroscopic signatures indicative of the genuine article [28] (Figure 3) (*scrimshaw*), unlike the quill pen holder from the same collection and comprising a hollowed-out section of sperm whale tooth shown in Figure 4, where the spectroscopic signatures clearly indicate a resin composition comprising polymethylmethacrylate and polystyrene in comparison with the genuine ivory of the sperm whale tooth. Likewise, Figure 5 shows an insect inclusion inside a piece of Baltic amber [29], which forms a very commercially desirable piece of jewellery. The Raman spectra of the insect and of the

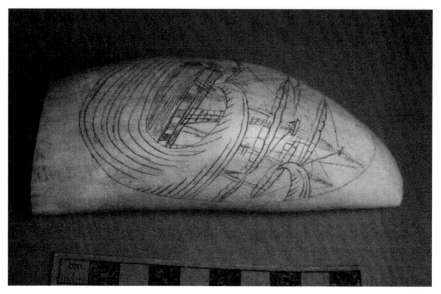

Figure 3 Sperm whale tooth *scrimshaw*; the carving artwork was very popular during the 18th and 19th centuries by whalers operating in the South Seas in the preparation of keepsakes for loved ones at home. These items are now highly valuable and collectable, but are subject to much faking. This is a genuine item from the Merseyside Maritime Museums, Liverpool, and forms part of the Lord Lever Collection.

Figure 4 Quill pen holder, carved *scrimshaw*, appears genuine but the Raman spectral signatures indicate that it is a modern copy composed of polymethylmethacrylate and polystyrene, stained with carbon-based ink.

Figure 5   Stinging insect inclusion in a Baltic amber matrix, ~30 My old. (See color plate 9).

amber matrix indisputably confirmed that both were genuine in the specimen in that they have the correct spectral signatures for both components. However, it is not possible spectroscopically to distinguish a genuine specimen of a prehistoric insect in an amber matrix from a specimen that has been deliberately "faked" by the modern insertion of an insect into the molten resin matrix to increase its apparent value, unless the thermal treatment of the matrix has left an indicator, which is not evident in the specimen under consideration here. The conclusion must, therefore, be that even when the spectral markers indicate an authenticity of composition, there could well be certain situations arising in which the object may have been faked for commercial reasons and for which spectroscopic analysis alone cannot provide a clear and definitive answer.

An important corollary of this tenet is that Raman spectroscopic data derived from the constituent components can nevertheless place an art work in the correct period, which is a critical factor when coupled with stylistic diagnostic opinion and historical issues of provenance for the correct attribution of a work of art. Two very famous cases of this aspect of Raman spectroscopic attribution of art works are provided in the recent literature, both involving investigations of oil paintings, one of which was supposedly painted by Vermeer in the 17th century and the other attributed to the hand of Raphael in the 16th century [30,31].

2.   RAMAN SPECTROSCOPY

Raman spectroscopic techniques have some special advantages for application in art and archaeological analysis; the provision of spectral data from microscopic specimens typically in the nanogram to picogram range, the ability to examine samples with minimal or no chemical and mechanical pretreatment, the versatility of selection of a range of laser excitation wavelengths to interrogate coloured specimens, the acceptance of specimens of a size range from milligram to kilogram and, because of the low Raman scattering cross-section for water, the accessibility of data from hydrated archaeological specimens all contribute to the unique niche position that is occupied by Raman spectroscopy in this field today [4]. Despite these advantages, however, there are several important factors underlying spectroscopic analysis generally and Raman spectroscopy in particular for art and archaeology applications, which have been addressed in the first book dedicated to this subject based on a conference held at the British Museum and published in 2005 [32]:

- In situ analysis using portable instrumentation or the laboratory analysis of excised specimens by microsampling of art work is highly favoured among exponents of Raman spectroscopic methods of analysis for artwork provenancing: although this is in itself appreciated in terms of the preservation of the art work essentially intact, it does pose a fundamental question that lies at the heart of most microanalytical experiments and procedures: Is the analysis of perhaps picograms of material in a laser focal cylinder truly representative of an object that may have a mass of several kilogram and is possibly of a heterogeneous composition? This point is usually addressed by using as many replicate samples or specimens as possible, consistent with the maintenance of the integrity of the sample; various methods have been employed in the spectroscopic literature to circumvent possible damage caused by excised specimen microsampling of paintings and manuscripts, comprising the taking of specimens from the edge of the painting usually covered by the frame and from the spallated pigment detritus that collects between the leaves of parchment or paper manuscript pages from use with time.
- Spectroscopic analysis provides qualitative and possibly quantitative information about the molecular composition of the materials comprising the artwork; this "hard" information can then be compared with what may have been expected from an artwork of the expected historical period. It is the latter "soft" science comparison that provides potential errors in the conclusion that could then be drawn, because the historical opinion about materials used by an artist at a particular time can only be based on known recipes, treatises or manuscripts. There is strong evidence that even in well-respected published treatises on their paintings, artists could be deliberately misleading about their favoured pigments, for example, as found from the Raman spectroscopic analysis of the 16th century "Armada Jewel" of the Court limnologist Nicholas Hilliard [33] made for Queen Elizabeth I after the defeat of the Spanish Armada in 1588, where Raman spectroscopy found pigments present that were positively not recommended for this type of work by the painter. In the case of historical manuscripts, the situation can be even more tenuous; *mosaic gold*, tin (IV) sulphide, was found during our own Raman spectroscopic studies [34] of a 13th century polychrome statue of Santa Ana in the monastery of Santa Maria la Real, Silos, Spain (Figure 6). The problem here centred on a manuscript in a Naples scriptorium dating from the 14th century, which was the first to mention the use of *mosaic gold* as a yellow pigment in Europe – but this manuscript was dated some 100 years *after* the Santa Ana statue was made and decorated. Does this mean that *per se* the Santa Ana statue is therefore a fake? However, in this context, there

are Chinese manuscripts that report the manufacture of *mosaic gold* which predate the Neapolitan manuscripts by 2000 years, so clearly, there was a preparatory knowledge in existence at the time that the Santa Ana statue was made and an earlier nonsurviving European manuscript could possibly have made reference to this. During the course of the Raman spectroscopic investigation of the Santa Ana statue, it was found that the burnishing of the applied *mosaic gold* pigment in the simulation of genuine gold leaf is related very closely to the stability and composition of the underlying ground pigment composition, yet this process has hitherto not been reported historically – this example illustrates very well the gap in our historical knowledge of technological achievements in antiquity [35]. In this

(a)

Figure 6   Statue of Santa Ana, monastery of Real Santo Domingo de Silos, Castille y Leon, Spain; pigments on marble, dating from the late 13th century. The Raman spectrum of the golden yellow hem of the cape of this statue indicated that it was not gold leaf as expected (which does not possess a Raman spectrum) but rather *mosaic gold*, a pigment that apparently did not exist in Europe in the 13th century, but was well known in China at that time. (See color plate 10).

(b)

Figure 6   (Continued)

particular case, the spectroscopic analysis has suggested the presence of an ancient, long-lost processing and materials technology hitherto unsuspected by archaeologists and art historians, which we were able to model in the laboratory to the specification recognised for the 600-year-old statue decoration. The stackplotted Raman spectra shown in Figure 7 for some sulphide minerals of arsenic, tin and mercury show the key vibrational features used in the Raman spectroscopic discrimination between some important yellow, orange and red pigments, all of which were used in antiquity. In these spectra, an important vibrational spectroscopic criterion is that all the Raman bands that characterise these pigments occur at low wavenumber shifts and these would not be accessible in the normal way using infrared spectroscopy.

- Another example of a spectroscopic/historical discrepancy is provided by the comparison of the palette actually adopted by Acislo Palomino [36] at the end of the 17th century in his frescoes in the Sant Joan del Mercat church in Valencia, Spain, and those recommended in his well-respected three-volume paintings treatise, which was highly regarded in Europe during the 18th century. Although Palomino expressly discouraged the use of certain pigments in fresco work and recommended others specifically for this purpose, his chosen palette for his frescoes in Valencia was actually quite different; for example, the presence of lead white and Naples yellow, a lead antimonate, in the frescoes was clearly identified from Raman spectroscopy and confirmed from the presence of elemental indicators in associated analytical techniques such as SEM/EDAXS. Nevertheless, Palomino had strongly advised against the use of these lead-based pigments for frescoes in his own treatise on the subject. It should not be assumed from this discovery, however, that Palomino did not execute the art work – the historical evidence is unambiguous that he did create the art work himself, but for some reason he decided to be secretive about his pigment usage, although from confocal Raman spectroscopic studies he prepared his *intonaco*, *arricio* and

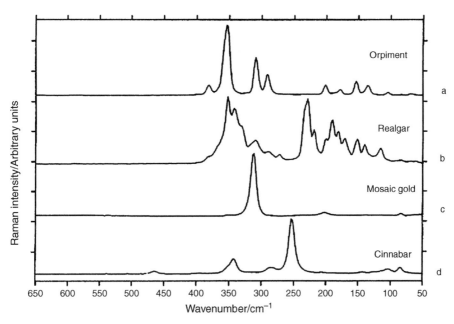

Figure 7 Raman spectral stackplot of some key heavy metal sulphide pigments that were used in antiquity, indicating the discrimination that is possible from the observation of these spectral signatures.

arenato ground substrates precisely according to his recommended methods in his treatise.

- The effects of deterioration with time and storage or of degradation in changing environments are often extremely difficult to quantify, yet this information is critical for conservators and restorers. The discovery that the *Angels with Black Faces* manuscript was not a mediaeval work of art with a unique theme but rather one of several that had now become distinguished from a blackening of the *lead white*, basic lead (II) carbonate pigment, by sulphides to form *galena*, a black lead (II) sulphide, is illustrative of the important role of Raman spectroscopy in pigment research [37]. The blackening, whilst indisputable from its presence, has a questionable origin, which has been ascribed variously to atmospheric pollution and the use of sulphur-containing materials such as egg-tempera in the pigment bodies and possibly to the proximal location of other sulphide pigments to the lead white pigment [38]; this last suggestion has already been recognised in antiquity by Roman wall-painters, who were warned by the chroniclers Pliny and Vitruvius about the instability of sulphide pigments, especially cinnabar and orpiment, in association with lead compounds, especially in the presence of light. Nevertheless, it is still quite common that cinnabar and lead white are found in close proximity to each other in ancient art – sometimes resulting in the colour deterioration described here.
- In a recent study of Egyptian pigments, the presence of *dragon's blood*, a highly-prized resin obtained from *Dracaena* botanical sources in ancient times [39], was

questioned by historians, despite spectroscopic evidence for its presence on the artefact, because it had never been seen before archaeologically in this context. It has since been identified as a component in a funerary sarcophagus from the Graeco-Roman period from a totally different suite of specimens in another museum [40] which had hitherto not been subjected to analysis. In another recent investigation on a similar theme, the yellow wash on a Graeco-Roman period Egyptian sarcophagus (Figure 8) was identified from Raman spectroscopy as a degraded terpenoid resin and confirmed from localised destructive sampling to be a *Pistacia* resin from GCMS (gas chromatography mass spectrometry) studies [41]. This study provides an example of the use of Raman spectroscopy to illustrate the important areas where strictly limited, but local sampling could be undertaken for more destructive analysis if the result was deemed to be significant in an historical context; in the case described here, the museum authorities judged that the presence of this resin clearly was indicative of a trade route with Egypt from geographical sources that were only hypothetical previously.

- Perhaps, the most troublesome effect that is noticed in the applications of Raman spectroscopy to materials analysis in art and archaeology is that of fluorescence emission [32]; this can easily swamp the Raman spectral bands and is particularly problematic for degraded biomaterials. Several techniques can be adopted to try to combat fluorescence, including long wavelength laser excitation, which practically now means at 785, 830 or 1064 nm, and special spectral processing

(a)

Figure 8 (a) Fragment of Egyptian sarcophagus, Graeco-Roman period, showing several pigments and a human figure with hieroglyphics; the Raman spectrum of the yellow wash analysed as a degraded terpenoid resin of the *Pinus/Pistacia* variety, confirmed by GCMS analysis to be a degraded *Pistacia* resin, not native to Egypt, which has some implications of possible novel trade routes of interest to art historians. (b) Fragment of plaster from a Graeco-Roman Sarcophagus containing dragon's blood resin pigment. (See color plate 11).

(b)

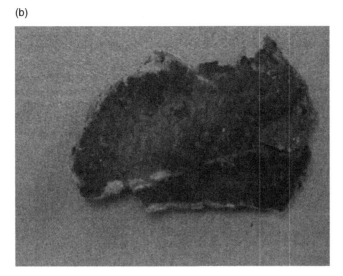

Figure 8   (Continued)

methods such as wavenumber-shifted spectroscopy. The analysis of biomaterials nondestructively is extremely valuable for the characterisation of organic dyes, resins, waxes, gums, tissues and textiles in artefacts in museum collections. Of special importance in this respect is the early-warning identification of biological colonisation of artefacts, which can often be achieved before visual deterioration can be detected; this is a vital piece of information for conservation scientists as it affords a means of assessing the regions of the art work that are most gravely at risk of serious deterioration if left untreated.

3.   CASE STUDIES

3.1.   Biodeteriorated wall paintings

3.1.1.   Rock art

The cave art in the rock shelters (Figure 9) of the palaeoIndian settlements of the Pecos culture in the Rio Grande–Devil River region of Texas/New Mexico provides an early prehistoric record of human occupation dating back more than 5000 years. Raman spectroscopic studies [16,42] have shown that the spallated fragments collected from the Seminole Canyon and Big Bend sites contain organic signatures that, in the case of the black paint used for animal hunting scenes, has been identified with ancient DNA from a bison bone marrow additive and, in the case of the red paint with signatures of calcium oxalate monohydrate, a by–product of the lichen hyphal metabolism reaction of oxalic acid with calcareous ions in the rock substratum.

Figure 9 PalaeoIndian rock art, Pecos culture, Seminole Canyon, Rio Grande–Devil River region, Texas/New Mexico; ca. 3000 BC. (See color plate 12).

3.1.2. Wall paintings/frescoes

The damage caused by lichen hyphal penetration of mediaeval and Renaissance artwork can be illustrated by the wall paintings in ancient churches in Spain – where damage to 14th century artwork [43,44] has been attributed to lichen colonisation, despite high concentrations of lead, antimony and mercury in the mineral pigments of the wall paintings. Lichens have successfully invaded the artwork and caused problems for archaeologists responsible for their preservation. Although the integrity of the substrate has been compromised, it is clear that Raman spectroscopic analysis have identified areas most "at risk" from the biosignatures of lichen chemicals present – hence the analyses represent an *early warning* method for conservators. An interesting point relates to the hierarchical use of expensive pigments in mediaeval artwork; in the fresco of *Christ in Majesty* at Basconcillos del Tozo [45] (Figure 10), *cinnabar* was used for the cloak of the Christ figure and *lapis lazuli* for the Virgin Mary's cloak (both very expensive pigments), whereas St. Peter's cloak was painted with an adulterated mixture of *haematite* and *minium*. The Raman spectrum of the red pigment used in St. Peter's cloak is shown in Figure 11.

In the Palazzo Farnese, a 16th century palace at Caprarola (Figure 12) some 60 miles North of Rome, the frescoes painted in 1560 by Zuccari have been very significantly damaged by the invasion of aggressive lichen colonies of *Dirina massiliensis* forma *sorediata*, an organism that can produce up to 50% of its biomass as hydrated calcium oxalate. A typical Raman spectrum of the lichen encrustation is shown in Figure 13, where the characteristic calcium oxalate monohydrate bands are seen at 1492, 1460, 903 and 505 cm^{-1}, the stretching and bending vibrations of the oxalate anion. With some 80% of the paintings covered by lichen, Raman analyses have indicated that approximately 1 kg of calcite substrate per square metre has been converted into fragile calcium oxalate, so destroying the platform on which the artwork is based [46,47]. Raman analytical studies on cored samples have also shown that lichen

Figure 10 The 14th century fresco of *Christ in Majesty* in central nave of the church of SS Damien and Cosmo at Basconcillos del Tozo, Castille y Leon, Spain. Source of religious hierarchical use of pigments detected by Raman spectroscopy. (See color plate 13).

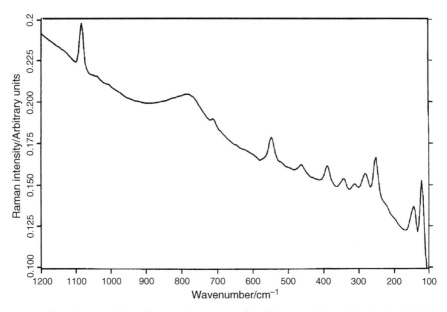

Figure 11 Fourier-transform Raman spectrum of red pigment from the cloak of St. Peter, Basconcillos det Tozo frescoes, indicating the use of lower grade adulterated *haematite* and *minium* pigments in place of the more expensive *cinnabar* as used for the cloak of Christ in Figure 10.

Figure 12 Palazzo Farnese, Caprarola, Italy, 16th century, built by Vignale in 1548 and containing frescoes painted by Zuccari in an inner courtyard in 1560 that have been significantly biodeteriorated by the invasive lichen, *Dirina massiliensis* forma *sorediata*.

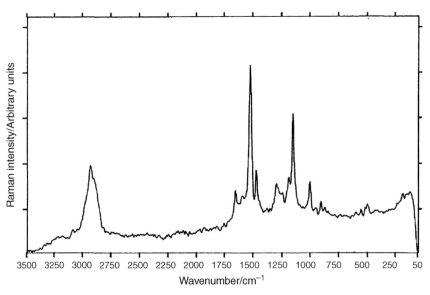

Figure 13 Raman spectrum, 1064 nm excitation, of the *Dirina* lichen encrustation on the biodeteriorated Palazzo Farnese frescoes; spectral signatures of calcium oxalate and carotenoids can be identified, along with substrate calcite inclusions.

hyphae are present up to 10 mm inside the basal calcite – hence, the mere removal of surface growths for restorative procedures will not cure the problem.

In the Convento del Peregrina at Sahagun, on the Camino del Santiago in Spain, which was closed in the 14th century because of an outbreak of pestilence and reopened in the 1960s, severe damage has been caused to some interesting mediaeval pigmented fresco reliefs (Figure 14), which bear a resemblance to the Moorish geometric designs from the decoration of the Alhambra palace in Granada. However, the Camino del Santiago at Sahagun was never occupied by the Moors, and the origin of Islamic decoration inside this Christian church is a matter for conjecture. The source of the deterioration of these frescoes can be assigned to the presence of unglazed windows, through which strong light and lichen spores of *Diploschistes scruposus*, identified from its Raman spectrum, have invaded and destroyed much of the art work [44]; the Raman spectrum that has facilitated the identification of this lichen inside the Sahagun church is shown in Figure 15.

3.2. Roman wall paintings

Raman spectroscopic analyses of Roman wall paintings have been carried out to determine the pigments used in several locations. In the north of Spain, near Burgos in Castille y Leon, a wide range of mineral pigments have been identified [48]. Of special significance, confocal Raman microscopy of a small crystal of cinnabar on a plaster fragment (Figure 16a) from a Roman villa at Banos de Valcabado enabled the sourcing of the mineral to the local mine at Tarna, rather than to the main cinnabar mines of Almaden, on account of the calcite inclusion within the crystal. In addition, no less than three different methods of producing a green pigment were identified from the Raman spectra in Figure 16b, involving *terre verte*, verdigris and a mixture of limonite and azurite. A very different situation was seen in a similar study of Romano-British villas near Peterborough and Colchester from the 1st century AD where a simpler range of pigments was identified [49]; however, at the latter site, which had been attacked during the Boudiccan revolt of 60–61 AD, an intact terracotta paint pot (Figure 17) was recovered from a Roman villa and provided evidence of the presence of anatase in the pigment mixture [50, 51]. *Anatase*, a rare mineral form of titanium (IV) oxide, was found underlying the black ink on the Vinland Map attributed to early navigators who defined the north-east coast of North America some 70 years before Columbus. Micro-Raman spectroscopic studies [52] have shown the spectral signature of anatase in small, spherical particulate form, which was not expected from natural sources – and anatase was synthesised commercially only in 1920 – from these studies the Vinland map was pronounced to be a fake. However, the presence of anatase in ancient art works is now being recognised in a wider regime [53] and the debate on the authenticity or otherwise of the Vinland map is continuing [54].

3.3. Egyptian mummies

Raman spectroscopy has provided some classic examples of the characterisation of degraded biodeteriorated tissue from archaeological environments. The height of mummification in Ancient Egypt occurred in the Middle Kingdom and the

Figure 14 The 14th century Convento de la Peregrina, Sahagun, Castille y Leon, Spain; pigmented Moorish geometric designs exhibiting significant biodeterioration since closure of the chapel following an outbreak of pestilence in the 15th century.

mummy of Nekht-Ankh (Figure 18) (12th dynasty; ca. 2000 BC) has been analysed by Raman spectroscopy – this mummy, from the "Tomb of the Two Brothers" excavated by Flinders Petrie in 1906, is now in the Manchester Museum. This mummy is of particular significance for analytical science as it was the first to be

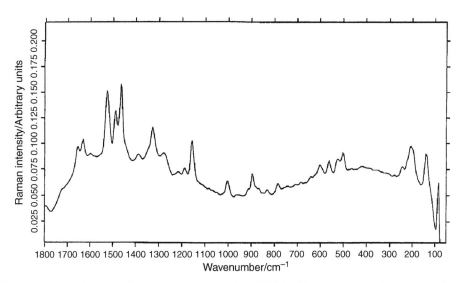

Figure 15 Fourier-transform Raman spectrum of *Diploschistes scruposus*, the source of the biodeterioration of the pigments and the substrate in the Sahagun frescoes.

subjected to a scientific unwrapping by Margaret Murray in the University of Manchester before a packed lecture theatre in 1906 (Figure 19). The skin of the mummy is shown from Raman spectroscopy to be in a variable state of preservation [55] – spectral stackplots show that in some specimens the skin is well preserved as evidenced by the clearly defined amide I and associated modes of skin proteins, whereas in others the degradation of the proteins is shown by the broad, diffuse spectra. It is interesting that in the region where residual bands of the mummification chemicals still remain, namely those of sodium sulphate from the natron used, the skin protein bands indicate a very badly preserved skin structure (Figure 20). Recent studies of the Raman spectra of natron samples from the Wadi Natrun, Egypt, have shown the presence of a bacterial contamination, *Natromonas pharaonis*, a halotrophic extremophile that can adapt to survival in the strongly alkaline *natron* salt matrix comprising sodium carbonate, sulphate and bicarbonate; the presence of biological contamination could explain some of the highly variable states of preservation of surviving mummies from this period.

A unique specimen of an eye-bead from an Egyptian 18th dynasty cat mummy (Figure 21) was believed to be either amber or brown glass by archaeologists; it was neither, as the Raman spectrum shows in Figure 22 – the eye-bead spectrum is characteristic of keratin and is closely matched with that of a claw or horn, which suggested a hitherto unrecognised funerary practice was perhaps being operated [56].

3.4. Tissues and resins

Hair provides an example of the use of Raman spectroscopy for assessment of biodeterioration in an archaeological or museological context; hair consists of

(a)

(b)

Figure 16 (a) Crystal of cinnabar in the red pigmented region of a wall painting fragment from a 1st century Roman villa at Banos de Valcabado, Castile y Leon, Spain. The detection of an inclusion of calcite inside the cinnabar crystal indicates the possible local sourcing of the mineral from the mines at Tarna. (b) Fragment of a 1st century Roman villa wall painting near Burgos, Spain, showing the Raman spectroscopic detection of different combinations of coloured pigments for the production of green colours on the same fragment, namely the green earth, *terre verte*, *verdigris* and a mixture of azurite and limonite.

keratinous proteins and can survive for considerable times in adverse burial conditions, which is currently exciting much forensic interest. The hair from a waterlogged skeletal burial [57] from the late 18th century (Figure 23) shows evidence of broad protein bands and significant deterioration, despite its apparent survival when all other soft tissue had been degraded. In comparison [58], the historical hair specimens belonging to Robert Stephenson, locomotive engineer and bridge builder, who built the first passenger carrying locomotive called *Locomotion No. 1* in 1825 and who died in 1859, and of Sir Isaac Newton, the "grandfather of spectroscopy", show an excellent state of preservation as they had been kept in archival storage. The difference in the quality of the Raman spectra from the ancient historical and archaeological hair specimens is significant and the presence

Figure 17 Intact terracotta paint pot from the archaeological excavation of a 1st century Romano-British villa near Castor, Northants, UK, which was destroyed in the Boudiccan revolt of 60–61 AD; the paint pot still contains residues of pigment, including anatase and haematite.

of sharper protein features in the historical specimens is striking; perhaps, the most significant spectral feature possessed by both Stephenson's and Newton's hair, however, is that of the ν (S-S) stretching mode near $500\,\mathrm{cm}^{-1}$, which has disappeared from the archaeological specimen but which can still be recognised in the historical specimens, indicating that the latter are well-preserved.

Other human and animal tissues from archaeological excavations investigated by Raman spectroscopy include teeth, bone and nail; for all of these, evidence of ancient funerary practices and preservation procedures are forthcoming. An ancient Brazilian *tembeta* or lip-plug made of organic material is believed unique, and Raman spectroscopy has assisted in the identification of the resin in its reconstruction for restoration purposes [59]. The presence of haematite on 3000-year-old skeletal remains [60] from an ancient hunter-gatherer culture giving rise to the so-called "ochred bone" has also indicated the adoption of an ancient funerary technology from the Raman microscopic identification of key spectral biomarker features assignable to a "limewash" formed by the calcination of seashells (aragonite). From this discovery, an archaeological reassessment of structures in the excavations previously attributed to cooking hearths to primitive lime-kilns has now resulted.

In 1545, following an engagement with a French naval force in the English Channel off Portsmouth, the *Mary Rose*, flagship of King Henry VIII's navy and shown here in Figure 24 from an ancient manuscript, the Antony Roll in the Pepys Library at Magdalene College, Cambridge, sank with the loss of 345 lives.

Figure 18 Sarcophagi from the 12th dynasty *Tomb of the Two Brothers*, Der Rifeh, Egypt, ca. 2100 BC, excavated by Sir William Flinders Petrie in 1905; the tomb with the mummies of Nekht-ankh and Khnum-nakht was intact on discovery and contained items of funerary furniture, some of which are seen in the photograph. The sarcophagi and mummies are now in the Manchester Museum. (See color plate 14).

Archaeological excavations from the 1980s have recovered many interesting arte-facts of Tudor naval life, including a chest from the Barber-Surgeon's cabin. The chest and cabin yielded ceramic and wooden vessels that contained residues of *materia medica*, used in the treatment of wounds and disease. We have subjected 10 specimens from these containers to Raman spectroscopic analysis [61] and have identified inorganic and organic materials that are recognised as staples of 16th century medicine, including sulphur, carbon, frankincense, myrrh and beeswax. Examples of the Raman spectra of some of the organic components of the medicine chest are shown in Figure 25. The carbon is believed to arise from

Figure 19 Archival photograph of the first forensic examination of an Egyptian mummy, Nekht-ankh, from the *Tomb of the Two Brothers*, carried out by Dr Margaret Murray in 1906, in the University of Manchester Chemistry Lecture Theatre.

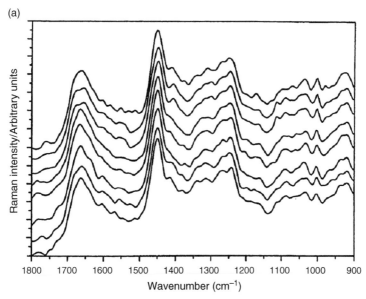

Figure 20 Fourier-transform Raman spectra of skin samples from the mummy of Nekht-ankh, showing the variability of preservation following mummification and funerary procedures. The stackplotted spectra above show evidence of well-preserved protein structures, with essentially alpha-helical amide I bands, whilst that below in contrast shows deteriorated protein and the presence of degraded amide bands, of beta-sheet and random coil structures. A significant conclusion form the series of spectra on the right is the presence of inorganic residues from the *natron* desiccant applied during the mummification process 4000 years ago.

(b)

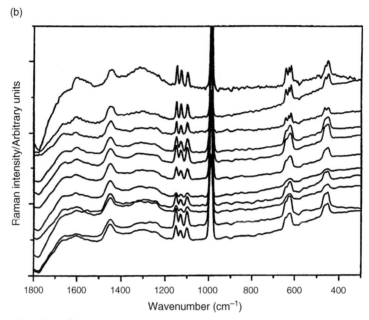

Figure 20 (Continued)

Figure 21 Mummified cat, 18th dynasty Egyptian, ca. 1350 BC, from the cat necropolis at Beni Hassan; on unwrapping the mummy, the head of the cat was seen to have an unusual brownish-yellow eye-bead (amber, glass?).

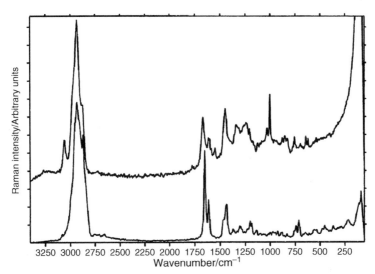

Figure 22 Raman spectrum of cat's eye-bead from the cat mummy shown in Figure 21; this has the vibrational modes characteristic of a keratotic material (upper spectrum), similar to modern claw or horn, unlike amber resin shown in the lower spectrum.

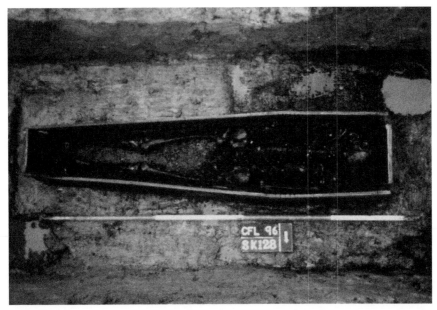

Figure 23 Archaeological excavation of a waterlogged burial site during an extension for the Newcastle Royal Infirmary, UK. The soft tissue has decomposed completely, but the hair remains along with the bones that have been stained black in the depositional environment.

Figure 24 The *Mary Rose*, flagship of King Henry VIII's navy, sunk during an engagement with a French naval force in the English Channel in 1545. From the Antony Roll in the Pepys Library, Magdalene College, University of Cambridge; reproduced with the permission of the Master and Fellows of Magdalene College, Cambridge. (See color plate 15).

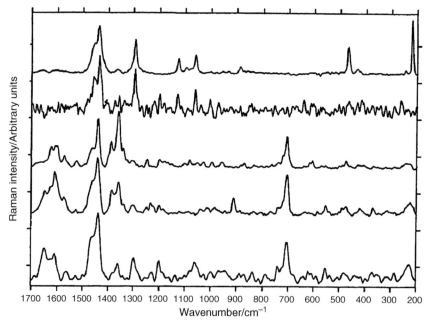

Figure 25 Fourier-transform Raman stackplotted spectra of organic components of the contents of the Surgeon-Barber's medicine chest excavated from the *Mary Rose*; signatures of frankincense, myrrh and amber are noted.

Figure 26 Mediaeval woodcut of the alchemical requirements for the use of *dragon's blood* resins; many of these are still necessary today for the interpretation of the vibrational spectra of materials relevant to art and archaeology.

magdaleones, bandages used in the treatment of burns. Most specimens were contaminated with sand and aragonite, which was consistent with their burial in the marine environment.

The geographical sourcing of resins used in antiquity using nondestructive Raman spectroscopy has proved useful in relation to the spectroscopic signatures obtained from generic materials obtained from different countries. A good example of this application is *dragon's blood* resin, which can be sourced form several botanical species ranging from the Mediterranean Sea, through North Africa to the East Indies. *Dragon's blood* has acquired much mystique over the centuries and has been used in different cultures; it was a staple of mediaeval alchemy and a print of a mediaeval woodcut shown in Figure 26 indicated the qualities required of the alchemical practitioner who handled *dragon's blood*. These qualities are still applicable to analytical science today! *Daemonorops draco* and *Dracaena cinnibari* are two such examples of the different botanical sources of *dragon's blood* resins, the former coming from the East Indies and the latter from Socotra, East Africa. The discrimination of Raman spectroscopy [62] for these dragon's blood resins (Figure 27) is a powerful analytical protocol for the identification of the source of materials via established trade routes; by this means, the type of dragon's blood resin used on an ancient art work (the Singer Dish in the National Museum of Scotland, Edinburgh) could be classified so that contemporary restoration could be undertaken with the original materials used in its construction.

3.5. Textiles

This case study provides an example of a *forensic conservation* exercise on a unique historical relic that was necessary not only because of its bad state of repair upon its re-discovery but also for the need to ascertain what the damage was scientifically so

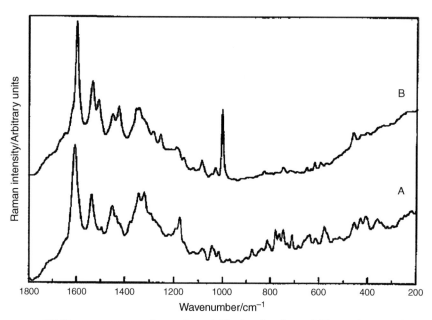

Figure 27 FT Raman spectra of two *dragon's blood* resins from different botanical sources; wavenumber range 200–1800 cm^{-1}, excitation 1064 nm. The lower spectrum is that of a resin from *Dracaena cinnabari* from Socotra, East Africa, believed to be the true *dragon's blood* of ancient usage, whilst the upper spectrum shown is that of *Daemonorops draco*, from the East Indies, which came into use in Europe from late mediaeval times. The Raman spectroscopic signatures are different and offer a means of sourcing *dragon's blood* resins from different botanical sources and the possible identification of trade routes.

that steps could be taken to undertake its conservation using appropriately aged modern materials for the sympathetic restoration of the artefact for its eventual display. This particular course of action had arisen because of the rather complex nature of the deterioration suffered by the artefact through both natural environmental exposure and from the additional and very obvious deliberate damage caused by human agency; the requirement was for a restorative repair to be undertaken with modern simulates of the ancient materials that had been aged, so that the stress of the repair work on the restored artefact would be minimal. Clearly, the first objective in this exercise would be the analytical spectroscopic characterisation of the relict material followed by the spectroscopic monitoring of the effect of several degradative processes conducted in the laboratory on a selected modern replicate to simulate the aged material for the eventual sympathetic restoration of the historical artefact.

The relict artefact under study was the foretopsail of *HMS Victory*, which was carried into battle on 21 October 1805, when Admiral Lord Nelson led a numerically inferior force against the combined might of the Franco-Spanish naval alliance off Cape Trafalgar in what is now appreciated as the last great naval engagement [63] between "fighting sail". Sixty battleships in the opposing navies fought an intense action on that day and the *Victory sail*, as it is now known, was the only sail to survive the ferocity of the battle. Several stratagems were adopted in that engagement, the most notable being that of Nelson, who divided his force into a Weather Column led by himself in the *Victory* and a Lee Column under the command of Admiral Lord Collingwood in the *Royal Sovereign*; the Franco-Spanish fleet were under the command of Admiral Ville-neuve in the *Bucentaure* and Admiral Gravina in the *Santissima Trinidad*, which was the most powerful ship on either side, being a four-decked first-rater with 144 guns. Nelson and Collingwood attacked the enemy line orthogonally at the two points represented by the command battleships of the French and Spanish Admirals, but in doing so they were forced to receive intense bombardment from the Franco-Spanish fleet for about 30 min before they were able to respond with broadsides into the unprotected flanks of the *Bucentaure* and *Santissima Trinidad*. Franco-Spanish naval strategy was to fire their broadsides "on the roll", which caused a disproportionate damage to the masts and rigging of the enemy ships; hence, most of the British battleships that survived the action were completely dismasted.

That the *Victory sail* survived as the sole example of its kind from the 37 sails that she carried into the battle is ascribed to the foretopmast being brought down by gunfire from the *Neptune* very early in the action as the *Victory* penetrated the enemy line of battleships. This is depicted in the painting by J.M.W. Turner of the Battle of Trafalgar commissioned by King George IV in 1825 and shown in Figure 28. This can be best considered as a montage of events that occurred that day, including the breaking of the line by the *Victory*, the sinking of the *Redoubtable*, the capture of the *Bucentaure*, the burning of the *Achilles* and, very appropriately for the subject under study here, the bringing down of the foretopmast and sail of the *Victory*, whose sailors were able to cut free and put to one side early in the engagement – thereby directly achieving its future preservation and facilitating its Raman spectroscopic study today.

We have used Raman spectroscopy [64] to study the condition of the *Victory sail*, which has suffered both from its exposure to the elements during Nelson's blockade of Cadiz in the early 19th century battle damage and from unsympathetic storage in the intervening 200 years. A conservation scientist's picture of the foretopsail of the *Victory* as removed from storage some 3 years ago is shown in Figure 29. The sail actually contains over 90 holes and tears, most of which can be directly attributed to the battle damage; however, another problem was identified as the significant decrease in tensility of the linen fibres of the sail, which is now only some 30% of the strength of new sail linen. Repair of the sail with modern sailcloth linen to facilitate its re-hanging as part of the bi-centenary commemoration of the Battle of Trafalgar in 2005 in the Portsmouth Historic Dockyard could therefore cause severe internal stresses in the relic that could promote further damage and thus be detrimental to the survival of the artefact. Molecular spectroscopic analysis of the sail fibres was needed to assess the damage and a monitoring of the process of artificial ageing of modern linen sailcloth would be needed to simulate the aged, degraded material, thereby effecting a better compatibility between the old

Figure 28 Painting by J.M.W. Turner of *HMS Victory* breaking the line and attacking the Franco-Spanish fleet at Trafalgar, 21 October 1805; commissioned by King George IV in 1825 for St. James' Palace and now in the National Maritime Museum at Greenwich. The foretopmast with foretopsail attached is shown being brought down early in the engagement by gunfire from the *Neptune* as the *Victory* is engaging the *Bucentaure* and *Redoubtable*. Reproduction by courtesy of the National Maritime Museum, Greenwich. (See color plate 16).

and replacement materials that would assist in the preservation of this ancient, historic marine textile.

In cyclical ageing experiments, specimens of modern linen sailcloth (Banks) were subjected to the following sequential procedures:

1. Soaking in brine for 1h, followed by drying at 70°C for a further 1h, repeated cyclically for 1 week;
2. Heating to 70°C for a period of 7 weeks at 100% relative humidity;
3. Exposure to UV-radiation for 1 week with an irradiance of 200 W m^{-2}
4. Multiple sequencing of the above operations, representing about seven complete cycles.

The Raman spectrum of an ancient textile linen fibre from the *Victory sail* can be compared with that of a modern linen Banks sailcloth; the dominant features in the vibrational spectrum are those of the COC glycosidic stretching and ring stretching bands in the poly-D-glucose structure of the linen. The major differences between the Raman spectra can be attributed to the degradation of the linen structure, which can occur through temperature changes, radiation fission of the beta-1, 4-glycosidic linkages, oxidative fission, hydrolysis,

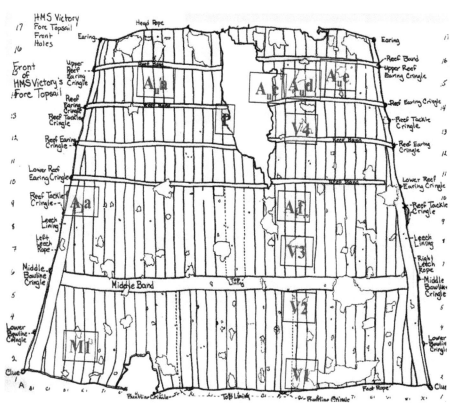

Figure 29 Conservator's working sketch of the *Victory* foretopsail from the Battle of Trafalgar recovered from naval storage in Portsmouth, showing the extent of the battle damage in which some 90 holes and tears can be discerned. Samples taken for spectroscopic and rheological measurements are indicated.

bacterial action resulting in ring-opening and corrosion effects on the integrity of the natural polymer [65]. The resulting embrittlement of the fibre and detritus deposited can be observed in the formation of coloured mucilage and acidic debris. The average tensility of the historically aged linen fibres of the *Victory sail* is now only some 30% of that measured for the new sailcloth, and this implies that the degradation of the ancient marine textile is far advanced.

We have previously identified vibrational spectroscopic features in the Raman spectra of linen specimens which can be assigned to the major functionalities associated with the cellulosic structure and assignments and several bands can be ascribed to the addition of jute, which was combined with linen to strengthen the weave; a good example that can be seen in the Raman spectrum of the modern linen is the aromatic quadrant stretching band at 1605 cm^{-1} which is assignable to the phenolic lignin component of jute. Although this feature is not observed in the comparative spectra of some ancient archaeological linens, such as those analysed from ancient Egyptian Dynastic burials [66], it is seen here at much reduced

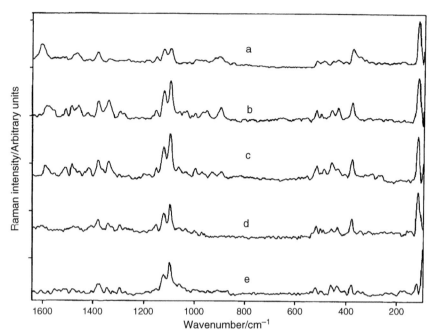

Figure 30 Spectral stackplot of specimens of modern sailcloth (Banks) that have been subjected to a series of cyclical artificial ageing stresses, indicated here by a, b, c, and d, for comparison with the *Victory* foretopsail specimen, labelled e. For details of the cyclical experiments, see text.

intensity in the stackplotted spectra of the historic sail which could reflect either a smaller jute content or probably the destruction of the lignin component in the ancient sail consequent upon its oxidative and radiation-induced degradation over two centuries of use and storage.

Figure 30 gives a stackplotted example of the effect of the cyclical ageing experiments upon the observed Raman spectra of modern Banks sail linen fibres. Of particular interest is the lower spectrum (e), which represents the accumulation of the extreme conditions imposed by the multiple cyclical simulated ageing processes carried out in the laboratory and which is a considered to be very close match for the spectra of the specimens of historical linen taken form the *Victory sail*. The laboratory extreme specimen, resulting from the combination of ageing cycles involving heat, dehydration, brine saturation and radiation exposure on modern linen fibres also afforded specimens which exhibited similar decreased tensility to those observed for the historical specimens and gave similar Raman spectra to that shown in Figure 30e. Some significant relative intensity changes can be seen in this spectral stack plot which clearly reflect the progressive degradation of the cellulose biopolymer; in particular, attention can be drawn to the relative intensities in the components of the doublet at 1122 and 1097 cm^{-1} and the ratio of the intensities of this doublet and the band at 120 cm^{-1}.

3.6.   Ivories

Ivory is the generic name for the exoskeletal tusks of terrestrial mammals such as the African and Asian elephant, wart hog, hippopotamus and the extinct mammoth, and marine mammals such as the narwhal and walrus; sperm whale teeth also come into this category. It has been a decorative material that has attracted the attention of many cultures for thousands of years, and ivory artefacts are to be found in many museums and from archaeological excavations. In addition to the preservation of carved historical ivories, which can be damaged by degradation of the keratotic protein component in the hydroxyapatite inorganic matrix, the recovery of ivories from an archaeological depositional environment is fraught with difficulty in stabilising their structures. Ivory was much traded and the origin of ivory artefacts is of importance for the definition of trade routes in ancient cultures. For example, the discovery of an ivory die (Figure 31) in a 1st century Roman villa excavation in Frocester, England, revealed from Raman spectroscopic analysis that it was from a hippopotamus tusk and therefore came from Central Africa, rather than a marine origin that would have indicated a possible source from Northern Europe (Figure 32).

The characterisation of the mammalian species of ivory artefacts using Raman spectroscopy has occupied several groups over many years, and a high discrimination can be achieved particularly when good quality spectra can be analysed using chemometric methods [67, 68, 69]. The impetus for these studies has been generated from a forensic context and the need to acquire materials sourcing data for the identification of illicit importations and contraband, especially because there are some extremely good simulates and fakes now in circulation [70]. Raman spectroscopy has played an effective role in the exposure of fake ivories with the example of scrimshaw as indicated

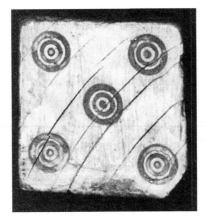

Figure 31   Ivory Roman die from 1st century Romano–British villa excavation at Frocester Court, Gloucestershire, UK.

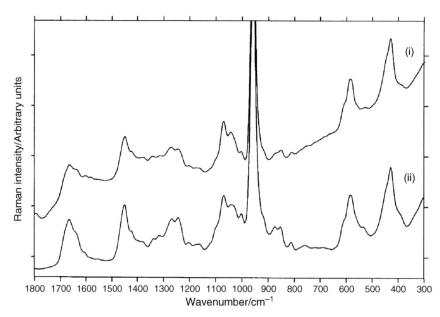

Figure 32 Fourier-transform Raman spectra of the Roman die shown in Figure 31 with that of a standard, hippopotamus tusk, from which it is possible to conclude that the die originated from an African source. Wavenumber range 300–1800 cm^{-1}, excitation 1064 nm.

earlier in this chapter. Experts can normally distinguish real ivory from the fake variety, but in some cases, particularly where the object has been heavily carved or hollowed out so that surface morphological features have been removed and for ivory fragments from burial sites, it is a matter for conjecture and informed opinion. On occasion, the Raman spectra of an assumed ivory object have given a surprising result and some examples are provided here. Figure 33 shows a spectral stackplot of two specimens of mammoth ivory from archaeological excavations dating from about 120 Ky to 15 Ky ago in upper and lower spectra, with that of a modern African elephant in the middle; here, the different composition of keratotic and hydroxyapatite content in the specimens can be seen from the relative band intensities of the amide III protein band at 1650 cm^{-1} and the phosphate stretching band at 960 cm^{-1}, which are not solely dependent upon the time of burial but also upon their depositional chemical environment. In a case of real versus fake ivory identification, the two spectra shown in Figure 34 show the spectrum of genuine African elephant ivory from an 18th century tobacco jar at the bottom and that of a cat that was believed to be of a similar age in the upper spectrum. Clearly, the Raman spectrum of the cat is not characteristic of ivory and can be assigned to a modern composition of polymethylmethacrylate and polystyrene, with aromatic CH stretching bands at 3060 and 1600 cm^{-1} and carbonyl stretching at 1720 cm^{-1}, to which the manufacturer has added powdered calcite evidenced by the strong band at 1086 cm^{-1}; the resultant composite has the same specific

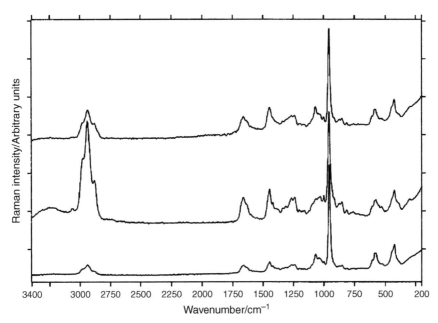

Figure 33 Full range Raman spectra of ivory specimens, 200–3400 cm^{-1}, 1064 nm excitation; the top and bottom spectra are from excavated mammoth ivory tusks aged 120 ky and 15 ky, respectively, whilst the middle spectrum is that of modern African elephant ivory. The common spectral signatures are to be noted in all three spectra but the diminution of the organic component in the archaeological ivories as evidenced by the intensity of the CH band near 3000 cm^{-1} is notable; minor differences in the amide I and phosphate bending modes can also be seen near 1660 and 600–450 cm^{-1}, respectively, as can the influence of the assimilation of carbonate into the inorganic matrix in eth band envelope near 1070 cm^{-1}.

gravity as genuine ivory, which with the carving and textural treatments makes it difficult to identify as fake ivory. Finally, a collection of bangles from the Victorian period, mid- to late- 19th century (three of which are shown in Figure 35), has been analysed by Raman spectroscopy and the results by comparison with genuine ivory in Figure 34 indicate that none of the specimens is ivory, but all are simulates made of cellulose nitrate, polymethylmethacrylate and polystyrene and a polymethylmethacrylate, polystyrene and calcite mixture, as shown in Figure 36.

Following the public and museum interest in early plastics, Raman spectroscopy is now finding application also in the evaluation of artefacts that are "at risk" in the storage environment through the effect of residual monomers and initiators from incomplete polymer processing and from degradation through changes in humidity and exposure to light, especially where junctions with metal components are involved, such as moving parts and steel pins in figures and household objects that have been corroded by acidic residues and which catalyse the further decomposition of the polymer bodies. The survival of early

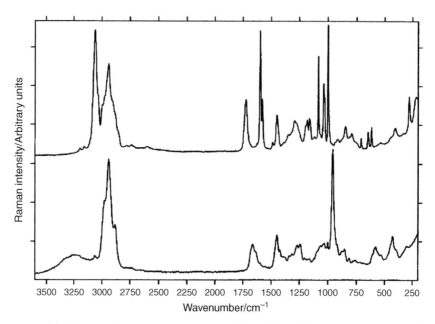

Figure 34 Fourier-transform Raman spectra of two "ivory" museum specimens; lower spectrum, 18th century tobacco jar is characterised and confirmed to be genuine African elephant ivory whereas the upper spectrum of a cat, appearing alongside some other genuine ivory artefacts, is assignable to a modern fake, being a composition of polymethylmethacrylate and polystyrene to which calcium carbonate has been added in simulation of the specific gravity of genuine ivory.

Figure 35 Three bangles from a collection of Victorian "ivory" adornment dating from the last quarter of the 19th century.

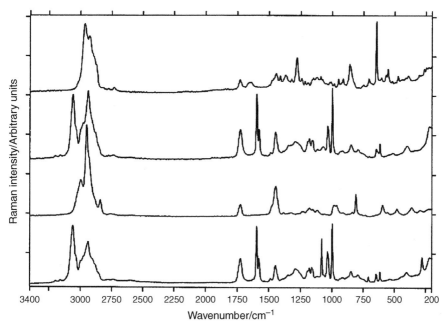

Figure 36   Stackplotted Raman spectra of four specimens of "ivory" bangles dating from the 1880s; none is genuine ivory and all can be classified as composed of cellulose nitrate, polymethylmethacrylate and polystyrene, and a similar resin with the addition of calcite.

cellulose acetate cinematographic films stored in metal canisters in museum archives is already compromised by similar processes.

4.   CONCLUSIONS

The applications of Raman spectroscopic techniques to the nondestructive analysis of art works and historical artefacts have given rise to some 1200 papers in the last decade, as cited in the most recent review of this subject area published this year [71]; in 2005 alone, almost 200 papers on Raman spectroscopy in the theme of art and archaeology were published as cited in the Web of Science. Several meetings are now devoted to this topic, including the Raman in Art and Archaeology series, of which number IV is scheduled in September 2007 in Modena, Italy. The Infrared and Raman Users Group is active in art and archaeological applications of Raman spectroscopy and held an IRUG VII meeting in New York in 2006. The GeoRaman series has been devoting special sessions to art and archaeological applications since 1999 and GeoRaman VI, with a strong emphasis in the art and archaeological applications of Raman spectroscopy, was held in Almunecar, Spain, in 2006. The first book dedicated to Raman spectroscopy in archaeology and art history was published in 2005, following a successful meeting at the British Museum in 2001 attended by 140 delegates, and contains many examples of the novel applications of the technique [32], including some problems that have not been addressed in the current article. The prognosis for the successful growth of application of Raman spectroscopy to art and archaeology is therefore excellent, especially with the advent of new portable Raman spectroscopic instrumentation which facilitates access to large objects of art, to objects where even strictly limited sampling cannot be undertaken and to those items which cannot be removed from their museum environments; the increasing drive in this direction has been surveyed in a very recent publication [72].

ACKNOWLEDGEMENTS

The author would like to record his gratitude to Dr Paul Wyeth of the Textile Conservation Institute, University of Southampton at Winchester, Lt. Cdr. Frank Nowosoielski, Commanding Officer of *HMS Victory*, Peter Goodwin, Keeper and Curator of *HMS Victory*, and Mark Jones, Head of Collections, The Mary Rose Trust, for the provision of the historic samples of the foretopsail of *HMS Victory*. Also, the author acknowledges the invaluable contribution of doctoral research students Nik Nikhassan, Rachel Brody, Emma Newton, Suzanne Petersen, Hassan Ali and to Paul Garside, Research Fellow at the AHRC Research Centre for Textile Conservation and Textile Studies. The collaboration of the following colleagues was invaluable: Dr Anita Quye and Dr Kathy Eremin (The National Museum of Scotland, Edinburgh), Professor Rosalie David (University of Manchester), Tracy Seddon (Merseyside Maritime Museums), Chris Brooke (University of Nottingham), Hugh Prendergast (Royal Botanic Gardens, Kew), Professor Ole Nielsen and Dr Monika Gniadecka (University of Copenhagen), Paul Middleton (Peterborough Regional College), Professor Mark Seaward, Dr Andy Wilson, Dr Ben Stern, Dr Sonia O'Connor, Dr Mike Hargreaves and Dr Rob Janaway (University of Bradford), Professor Susana Jorge Villar (University of Burgos), Professor Fernando Rull (University of Valladolid, Spain), Dr Dalva de Faria (University of Sao Paulo), Dr Marisa Afonso (Museum of Ethnography, Sao Paulo, Brazil), Dr Luiz de Oliveira (Federal University of Juiz de Fora, Brazil), Professor Jose-Maria Alia (University of Castilla la Mancha, Spain), Professors Maria Teresa and Antonio Domenech-Carbo (Polytechnic University of Valencia and University of Valencia, respectively), Tim Benoy (The de Brecy Trust) and Professor Peter Vandenabeele (University of Ghent).

Finally, the author thanks Dennis Farwell, a valued colleague for many years who has shared in much of the excitement generated by the discoveries of Raman spectroscopy in the emerging area of art and archaeology from the very start, as they actually happened!

The permission of The National Maritime Museum, Greenwich, for the reproduction of J.M.W. Turner's painting of *HMS Victory* at the Battle of Trafalgar is also especially acknowledged.

REFERENCES

[1] H. Davy, Philos. Trans. R. Soc. London, 105: 97–110, 1815.
[2] H. Eccles and B. Rackham, Analysed Specimens of English Porcelain, V & A Museum, London, 1922.
[3] A.H. Church, English Porcelain, South Kensington Museum Handbook, London, 1886.
[4] H.G.M. Edwards, "Art Works Studied using IR and Raman Spectroscopy" in Encyclopaedia of Spectroscopy and Spectrometry, J.C. Lindon, G.E. Tranter and J.L. Holmes (Eds.), Academic Press, London, pp. 2–17, 1999.
[5] E. Ciliberto and G.Spoto (Eds.), Modern Analytical Methods in Art and Archaeology, Chemical Analysis Series, Volume 155, J. Wiley & Sons, Chichester, 2000.
[6] B. Guineau, Stud. Conserv., 34: 38–44, 1989.
[7] G.D. Smith and R.J.H. Clark, J. Arch. Sci., 31: 1137–1160, 2004.
[8] S.P. Best, R.J.H. Clark and R. Withnall, Endeavour, 16: 66–73, 1992.
[9] C. Coupry, Analusis, 28: 39–46, 2000.
[10] L. Burgio, R.J.H. Clark, and S. Firth, Analyst, 126: 222–227, 2001.
[11] L. Burgio and R.J.H. Clark, J. Raman Spectrosc., 31: 395–401, 2001.
[12] M. Bouchard and D.C. Smith, Spectrochim. Acta Part A, 59: 2247–2266, 2003.
[13] I.M. Bell, R.J.H. Clark and P.J. Gibbs, Spectrochim. Acta Part A, 53: 2159–2179, 1997.
[14] L. Burgio and R.J.H. Clark, Spectrochim. Acta Part A, 57: 1491–1521, 2001.
[15] H.G.M. Edwards, Spectroscopy, 17: 16–40, 2002.
[16] H.G.M. Edwards, L. Drummond and J. Russ, J. Raman Spectrosc., 30: 421–428, 1999.
[17] D.C. Smith and A. Barbet, J. Raman Spectrosc., 30: 319–324, 1999.
[18] D. Bikiaris, S. Daniilia, S. Sotiropoulou, O. Katsimbiri, E. Pavlidou, A.P. Moutsatsou and Y. Chryssoulakis, Spectrochim. Acta Part A, 56: 3–18, 1999.
[19] L.F.C. de Oliveira, H.G.M. Edwards, R.L. Frost, J.T. Kloprogge and P.S. Middleton, Analyst, 127: 536–541, 2002.
[20] H.G.M. Edwards, M.T. Domenech-Carbo, M. Hargreaves and A. Domenech-Carbo, J. Raman Spectrosc., 39: 444–452, 2008.
[21] S. Petersen, O.F. Nielsen, D.H. Christensen, H.G.M. Edwards, D.W. Farwell, A.R. David, P. Lambert, M. Gniadiecka, J.P. Hart Hansen, and H.C. Wulf, J. Raman Spectrosc., 34: 375–382, 2003.
[22] H.G.M. Edwards, A.S. Wilson, N.F. Nikhassan, A. Davidson and A.D. Burnett, Anal. Bioanal. Chem., 387: 821–828, 2007.
[23] A.S. Wilson, H.G.M. Edwards, D.W. Farwell and R.C. Janaway, J. Raman Spectrosc., 30: 367–373, 1999.
[24] H.G.M. Edwards, R.H. Brody, N.F. Nikhassan, D.W. Farwell and S. O'Connor, Anal. Chim. Acta, 559: 64–72, 2006.
[25] H.G.M. Edwards, N.F. Nikhassan, D.W. Farwell, P. Garside and P. Wyeth, J. Raman Spectrosc., 37: 1193–1200, 2006.
[26] D.A. Long, H.G.M. Edwards and D. W. Farwell, J. Raman Spectrosc., 39: 322–330, 2008.
[27] M.V. Orna, Chem. Australia, 470–472, 1966.
[28] H.G.M. Edwards, D.W. Farwell, T. Seddon and J.K.F. Tait, J. Raman Spectrosc., 26: 623–628, 1995.
[29] H.G.M. Edwards, D.W. Farwell and S.E. Jorge Villar, Spectrochim. Acta Part A, 68: 1089–1095, 2007.
[30] L. Burgio, R.J.H. Clark, L. Sheldon, and G.D. Smith, Anal. Chem., 77: 1261–1267, 2005.
[31] H.G.M. Edwards and T.J. Benoy, Anal. Bioanal. Chem., 387: 837–846, 2007.
[32] H.G.M. Edwards and J.M. Chalmers (Eds.), Raman Spectroscopy in Archaeology and Art History, Royal Society of Chemistry Publishing, Cambridge, 2005.
[33] A. Derbyshire and R. Withnall, J. Raman Spectrosc., 30: 185–188, 1999.
[34] H.G.M. Edwards, D.W. Farwell, E.M. Newton, F. Rull Perez and S.E. Jorge Villar, J. Raman Spectrosc., 31: 407–413, 2000.

[35] H.G.M. Edwards, E.L. Dixon, I.J. Scowen and F.R. Perez, Spectrochim. Acta Part A, 59: 2291–2297, 2003.

[36] A. Palomino de Castro y Velasco El Museo Pictorio y Escala Optica:Practica de la Pintura. Madrid, 1714.

[37] R.J.H. Clark and P.J. Gibbs, Anal. Chem., 70: 99A–104A, 1998.

[38] G.D. Smith and R.J.H. Clark, J. Cult. Heritage, 3: 101–105, 2002.

[39] K. Eremin, A. Quye, H.G.M. Edwards, S.E. Jorge Villar and W. Manley, "Colours in Ancient Egyptian Funerary Artefacts in the National Museums of Scotland", in Colours in the Ancient Mediterranean World, L. Cleland, K. Stears and G. Davies (Eds.), British Archaeological Reports, International Series, Hadrian Books, Oxford, 1267: 1–8, 2004.

[40] H.G.M. Edwards, S.E. Jorge Villar, A.R. David and D.L.A. de Faria, Anal. Chim. Acta, 503: 223–229, 2004.

[41] H.G.M. Edwards, B. Stern, S.E. Jorge Villar and A.R. David, Anal. Bioanal. Chem., 387: 829–836, 2007.

[42] J. Russ, R.L. Palma, D.H. Loyd, D.W. Farwell and H.G.M. Edwards, Geoarchaeology, 10: 43–52, 1995.

[43] F. Rull Perez, H.G.M. Edwards, A. Rivas and L. Drummond, J. Raman Spectrosc., 30: 301–307, 1999.

[44] H.G.M. Edwards and F. Rull Perez, Biospectroscopy, 5: 47–53, 1999.

[45] H.G.M. Edwards, D.W. Farwell, F. Rull Perez and S.E. Jorge Villar, J. Raman Spectrosc., 30: 307–312, 1999.

[46] H.G.M. Edwards, E.R. Gwyer and J.K.F. Tait, J. Raman Spectrosc., 28: 677–682, 1997.

[47] M.R.D. Seaward and H.G.M. Edwards, J. Raman Spectrosc., 28: 691–697, 1997.

[48] S.E. Jorge Villar and H.G.M. Edwards, Anal. Bioanal. Chem., 382: 283–289, 2005.

[49] H.G.M. Edwards, P.S. Middleton, S.E. Jorge Villar and D.L.A. de Faria, Analytica Chim. Acta, 484: 211–221, 2003.

[50] A.P. Middleton, H.G.M. Edwards, P.S. Middleton and J. Ambers, J. Raman Spectrosc., 36: 984–987, 2005.

[51] H.G.M. Edwards, P.S. Middleton and J.P. Wild, J. Roman Pottery Stud., 12: 204–208, 2006.

[52] K.L. Brown and R.J.H. Clark, Anal. Chem., 74: 3658–3702, 2002.

[53] H.G.M. Edwards, N.F. Nikhassan and P.S. Middleton, Anal. Bioanal. Chem., 384: 1356–1365, 2006.

[54] G. Harbottle, Archaeometry, 50: 177–189, 2008.

[55] S. Petersen, O.F. Nielsen, D.H. Christensen, H.G.M. Edwards, D.W. Farwell, A.R. David, P. Lambert, M. Gniadiecka, J.P. Hart Hansen, and H.C. Wulf, J. Raman Spectrosc., 34: 375–380, 2003.

[56] H.G.M. Edwards, D.W. Farwell, C.A. Heron, H. Croft and A.R. David, J. Raman Spectrosc., 30: 139–144, 1999.

[57] A.S. Wilson, H.G.M. Edwards, D.W. Farwell and R.C. Janaway, J. Raman Spectrosc., 30: 378–382, 1999.

[58] H.G.M. Edwards, N.F. Nikhassan and A.S. Wilson, Analyst, 129: 870–879, 2004.

[59] D.L.A. de Faria, H.G.M. Edwards, M.C. de Afonso, R.H. Brody and J.L. Morais, Spectrochimica Acta Part A, 60: 1505–1513, 2004.

[60] H.G.M. Edwards, D.W. Farwell, D.L.A. de Faria, A.M.F. Monteiro, M.C. Afonso, P. De Blasis and S. Eggers, J. Raman Spectrosc., 32: 17–23, 2001.

[61] H.G.M. Edwards, M.G. Sibley, B. Derham and C. Heron, J. Raman Spectrosc., 35: 746–753, 2004.

[62] H.G.M. Edwards, L.F.C. de Oliveira and H.D.V. Prendergast, Analyst, 129: 134–138, 2004.

[63] W.L. Clowes, Royal Navy: A History from the Earliest Times to 1900, (1st Edition), Chatham Publishers, London, 1996.

[64] H.G.M. Edwards, P. Wyeth, D.W Farwell, N.F. Nikhassan, and P. Garside, J. Raman Spectrosc., 37: 1193–1200, 2006.

[65] H.G.M. Edwards, D.W Farwell and D. Webster, Spectrochimica Acta Part A, 53: 2383–2389, 1997.

[66] H.G.M. Edwards, E. Ellis, D.W. Farwell and R.C. Janaway, J. Raman spectrosc., 27: 663–668, 1996.

[67] R.H. Brody, H.G.M. Edwards and A.M. Pollard, Analytica Chim. Acta, 427: 223–232, 2001.

[68] M. Shimoyama, H. Maeda, H. Sato, T. Ninomiya and Y. Ozaki, Appl. Spectrosc., 51: 1154–1158, 1997.
[69] H.G.M. Edwrads, N.F. Nikhassan and N. Arya, J. Raman Spectrosc., 37: 353–360, 2006.
[70] H.G.M. Edwards, "Forensic Applications of Raman Spectroscopy to the Nondestructive Analysis of Biomaterials and their Degradation", in Forensic Geoscience: Principles, Techniques and Applications, K. Pye and D.J. Croft (Eds.), Geological Society Special Publication 232, Geological Society (London) Publishing House, Bath, UK, 159–170, 2004.
[71] P. Vandenabeele, H.G.M. Edwards and L. Moens, Chem. Revs., 107: 675–686, 2007.
[72] P. Vandenabeele, K. Castro, M.D. Hargreaves, L. Moens, J.M. Madriagara and H.G.M. Edwards, Analytica Chim. Acta, 588: 108–116, 2007.

CHAPTER 6

REAL-TIME VIBRATIONAL SPECTROSCOPY AND ULTRAFAST STRUCTURAL RELAXATION

Takayoshi Kobayashi

Contents

Abstract

In the first part of this chapter, the details of the key techniques are described for utilizing the full capability of noncollinear optical parametric amplifier (NOPA) such as group-velocity matching, pulse-front matching (PFM), and idler angular-dispersion compensation. By these careful consideration and designing, the first sub-10-fs pulse generation tunable in the wide ranges of both visible and near-infrared (NIR) was succeeded. Extension of this technique to a full-bandwidth operation has led to as short as 4.7-fs visible pulses, which is also the first sub-5-fs light source by other method as the conventional continuum-compression scheme. In the second part, application of the light source is described. Utilizing the developed sub-5-fs NOPA system and a multi-channel detector, polydiacetylene (PDA) dynamics of exciton in a conjugated polymer, PDA, was studied. Time-resolved absorbance changes (ΔA) probe at 128 wavelengths at the same time were analyzed by Fourier transformation (FT) of the real-time traces of ΔA with detailed probes wavelength dependence. From the amplitude of molecular vibration induced impulsively by the sub-5-fs pulses from the NOPA, the contributions of wave-packets in the ground state and excited state were separated from the phase of the ΔA trace induced by the wavepacket motion.

Keywords: NOPA; group-velocity mismatch; PFM; pulse-front tilting; ultrabroadband chirped mirror; FRAC; FROG; flexible mirror; BBO crystal; SHG FROG; wavepacket; C—C stretching mode; C≡C stretching mode; free exciton; geometrical relaxation; difference absorption spectrum; global fitting; C≡C stretching; Born–Oppenheimer approximation; internal conversion; transverse relaxations; longitudinal relaxation; Franck–Condon state; Franck–Condon factor; Fourier power; self-trapped exciton; geometrical relaxation

1. ULTRASHORT PULSE LASER

1.1. Introduction

One of the most progressive investigations recently performed in ultrafast science for the applications to spectroscopy especially ultrafast dynamics in molecular systems is the generation of extremely short pulses in a sub-5-fs regime in the ultraviolet (UV), visible, and near-infrared (NIR) spectral ranges. The continuum compression of mode-locked Ti: sapphire laser systems has approximately 4.5-fs duration [1,2], and recently approximately 5-fs pulse was generated directly from Ti: sapphire oscillators [3,4]. Even though the spectra of these pulses are extended in the ultrabroad range from approximately 600 to 1000 nm, the major fraction of the intensity is located in a relatively limited NIR region around 800 nm; therefore, the applicability to a time-resolved spectroscopy was very much restricted in terms of available spectral range. Because most of interesting materials in photophysics, photochemistry, and photobiology have various electronic transition energies over the wide UV, visible, and NIR ranges, an extremely short pulse source with broadband is highly required for the investigation of such varieties of material systems. Even shorter pulses have been reported in sub-femtosecond time regime. However, applications of such pulses are oriented to internal transitions because they have spectrum in far UV to soft X-ray regime. Such internal transitions are expected not to have inhomogeniety, and hence the dynamics of such transition between inner-shell states in atoms and molecules can be obtained from the Fourier transform (FT) of the stationary electronic transition associated with absorption and/or emission. Therefore, such sub-femtosecond or attosecond pulses are not necessarily useful for dynamics study. It is expected to be more frequently applied to high-power laser physics than relaxation dynamics in atomic and molecular systems.

Noncollinear phase-matched optical parametric conversion has been recently attracting a great deal of attention of scientists in wide range of research field as a novel method of ultrashort pulse generation [5–10]. The noncollinear geometry can realize the group-velocity matching between the signal and the idler in a very broad spectral range, which is equivalent to an achromatic phase matching with the spectral angular dispersion of the idler, and the broadest gain bandwidth is attained. In 1995, Gale et al. [5] applied this property to a synchronously-pumped optical parametric oscillator and gave the first success to generate as short as 13-fs pulses, which are tunable from 580 to 680 nm with an approximately 2-nJ pulse energy. After this study, noncollinear optical parametric amplifiers (NOPAs) based on this scheme are being exploited to generate even shorter pulses [6–10]. The outstanding advantages such as tunability outside the

Ti: sapphire 800-nm region, μJ-level pulse energy, and flexible pulse width and bandwidth endow a NOPA with ability of being one of the most useful light source for time-resolved spectroscopy with an extremely high time resolution.

In this chapter, the details of the key techniques are described for utilizing the full capability of NOPA such as group-velocity matching, pulse-front matching (PFM), and idler angular-dispersion compensation. By these careful consideration and designing, the first sub-10-fs pulse generation tunable in the wide ranges of both visible and NIR was succeeded [9]. Extension of this technique to a full-bandwidth operation has led to as short as 4.7-fs visible pulses, which is also the first sub-5-fs light source by other method than the conventional continuum-compression scheme [10].

In Section 1.2, the broadband property of group-velocity-matched NOPA is described. The problem of pulse-front tilting in the noncollinear interaction is discussed, and the PFM geometry essential for the transform-limited (TL) pulse generation, are introduced in Section 1.3. The visible sub-5-fs pulse generation is presented in Section 1.4. The sub-4-fs pulse generation is described in Section 1.5. In the following Section 2, we show an example of applications to the ultrafast spectroscopy of conjugated polymers. The NOPA is well exhibiting the outstanding capability for real-time observation of the ultrafast molecular dynamics.

1.2.   Group-velocity matching by noncollinear geometry

The gain bandwidth of parametric interaction is limited not by the group-velocity mismatched (GVM) between the pump and the signal but by that between the signal and the idler. It is inversely proportional to the GVM between the signal and the idler [11]. The noncollinear phase matching (NCPM) with an angular dispersion of the signal and idler has been studied for spectral broadening [12]. Takeuchi and Kobayashi [13] reported a broadband generation of idler pulses by focusing a white-1ight continuum with a large convergence angle into the crystal to be phase-matched noncollinearly among the wide range of the continuum. The bandwidth was increased to as broad as $1300\,cm^{-1}$ in the NIR. We used the noncollinearly generated signal and idler just around the pump axis. It is after these few years that the NCPM was investigated to eliminate the signal–idler GVM with arrangement of the signal and idler with larger noncollinear angles. These are equivalent to a well-known achromatic phase matching [14], and an ultrabroadband frequency division of the pump photon is able to be achieved [6,15]. In the case of down-conversion, the condition is *automatically* satisfied only by appropriately arranging the crystal angle θ [6].

In the case of OPA, the signal is amplified with a fixed crystal angle. This is able to be achieved with noncollinear angle α between signal and pump as shown in Figure 1. The bandwidth can then be described by the following equation [7,11].

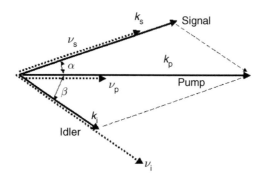

Figure 1 Geometrical configuration of the wave vectors in a noncollinear OPA. The wave vectors of the pump (k_p), signal (k_s), and idler (k_i) are shown in the gain crystal. The α and β are internal noncollinear angles. The group velocities of the pump (v_p), signal (v_s), and idler (v_i) are also shown by dashed lines.

$$\Delta\nu_{\text{parametric}} = \frac{0.53}{\left|\dfrac{1}{v_s} - \dfrac{1}{v_i\cos(\alpha+\beta)}\right|c}\sqrt{\frac{\Gamma}{l}}, \tag{1}$$

Here, c, v_s, v_i, l, Γ, and β are the light velocity, group velocity of signal, group velocity of idler, crystal length, coupling constant, and idler noncollinear angle, respectively. The suffixes $j = p$, s, and i denote the pump, signal, and idler, respectively. In the most general case of $v_s < v_i$, the noncollinear geometry can realize the GV matching by choosing α, and the bandwidth is only limited by the mismatch between signal and idler due to the GV–dispersion [11]. Figure 2 shows the phase-matching condition in a type I β-BaB$_2$O$_4$ (BBO) NOPA pumped at 395 nm. The gain bandwidth reaches the maximum value of 160 THz in the case for the angle between the optical axis and the laser beam in the crystal $\theta = 31.5°$ and noncollinear angle $\alpha = 3.7°$. The idler is generated with large angular dispersion to be phase-matched with the broad signal spectrum. This outstanding property was firstly discovered by Gale et al. [5] and is being widely investigated. The signal–idler GV-matched interaction allows the generation of pulses shorter than 5 fs in the spectral range below the pump frequency. The configuration of NOPA system is shown in Figure 3. This NOPA could generate sub-20-fs pulses in the visible angle. A NOPA with PFM configuration can generate sub-8-fs pulses tunable with central wavelength from 550 to 690 nm. The spectral shapes and fringe-resolved autocorrelation traces of the tunable output pulses from the NOPA system shown in Figure 3 are exhibited in Figure 4 [9].

1.3. Pulse-front matching

An ultrashort pulse is very thin in space. Spatial thickness of a pulse with a 10-fs duration is only 3 μm. Thus, the time-space coupling across the cross section by some distortion of propagation may give a tilted pulse front from the wave front.

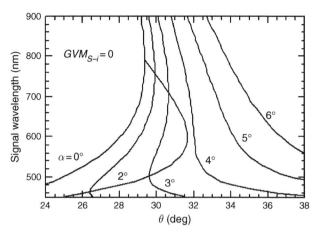

Figure 2 Theoretical phase-matching curves of θ in a type-I β-BaB$_2$O$_4$ (BBO) optical parametric amplifier (OPA) pumped at 395 nm with several different values of the signal noncollinear angle α (0°, 2°, 3°, 4°, 5°, and 6°). The signal branch and idler branch are shown. Curve exhibiting the condition for zero GVM between signal and idler is indicated by $GVM_{s-i} = 0$.

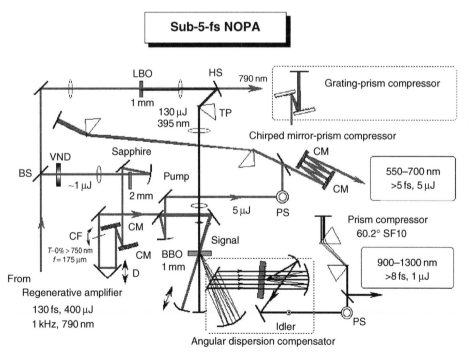

Figure 3 Experimental setup for the phase-matched noncollinearly optical parametric amplifier (NOPA). BS, beam sampler; SHG, second-harmonic generator; HS, fundamental and second-harmonic separator; D, variable optical delay line; NF, notch filter centered at 800 nm; PS, periscope for rotating the polarization of the signal. The cone-like parametric fluorescence with the minimized dispersion (see text) is illustrated with the external cone angle α_{ext}. Also illustrated is the crystal axes x, y, and z.

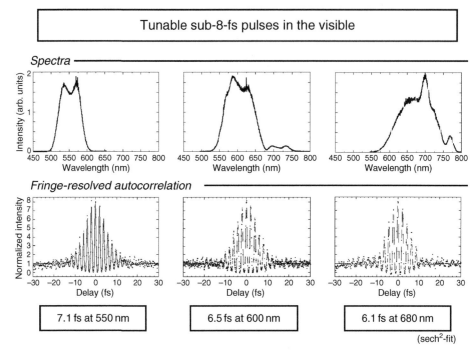

Figure 4 Output spectra and autocorrelation traces of the sub-8-fs tunable pulses between 550 and 690 nm from a noncollinearly optical parametric amplifier (NOPA). Tuning can be made by simply changing the delay in the variable optical delay line (D) in Figure 3.

This is called as pulse-front tilting, resulting in undesired pulse-width broadening and spectral lateral walk-off [16,17]. The dispersion of the exit angle ε is related to the tilt angle γ as follows,

$$\frac{d\varepsilon}{d\lambda} = -\frac{\tan\gamma}{\lambda}. \tag{2}$$

Here λ is the wavelength of the seed to be amplified. In the case of NOPA, the noncollinear interaction causes a slanted gain volume in the signal beam, resulting in the generation of a tilted signal by $\gamma = \alpha$ [18]. This was the origin of the difficulty in the generation of TL pulses in the initial stage of the study to construct NOPAs [6].

Tilted pumping is the geometry required essentially for pulse compression below 10 fs. The tilt-pumped OPA was firstly demonstrated by Danielius et al. [19] in 1996 to control the group velocity of the pump for efficient amplification in a collinear OPA. We utilized this scheme for PFM. The pump beam passes through a prism with incident and exit angles ϕ_1 and ϕ_2, respectively, and the pulse-front is tilted by γ_{prism} just after the prism which is given by [16]

$$\tan \gamma_{\text{prism}} = -\lambda_p \frac{d\phi_2}{d\lambda_p} = -\frac{\sin \alpha_{\text{apex}}}{\cos \phi_1' \cos \phi_2} \lambda_p \frac{dn}{d\lambda_p}, \tag{3}$$

where n is the refractive index of the prism material, α_{apex} is the prism apex angle, and ϕ_1' is the angle of refraction. The following telescope composed of lenses with focal lengths f_1 and f_2 recollimates the spectral lateral walk-off and images the tilted front at the crystal position with a longitudinal magnification factor f_1/f_2. The internal tilt angle γ_{int} is then reduced by refraction by v_p/c, so the PFM condition $\gamma_{\text{int}} = \alpha$ is given by

$$\tan \alpha = \frac{v_p}{c} \frac{f_1}{f_2} \tan \gamma_{\text{prism}}. \tag{4}$$

Because the spectral lateral walk-off in the horizontal direction is insignificant in the present experimental condition, the beam is recollimated both horizontally and vertically. Because the spot size is reduced by the lateral magnification factor f_2/f_1, the parameters of the telescope are practically determined for the efficient amplification. Thus, a prism apex angle and incident angle are the most essential parameters for optimization to satisfy the condition given by Equation (4). This PFM interaction gives maximum spatial overlap between the pump and the signal and dramatically suppresses the spatial chirp, resulting in the generation of nontilted spatially coherent signal pulses, which are compressible to the TL [9].

1.4. Generation of sub-5 fs

To obtain even shorter pulse, chirped mirrors were used [10]. The second harmonics (SHs) (150 fs, 100 μJ) of a Ti: sapphire regenerative amplifier (Clark-MXR CPA-1000, 400 μJ, 120 fs, 1 kHz at 790 nm) pumps a 1-mm-thick BBO crystal (type-I, 30° z-cut). A single–filament continuum is noncollinearly amplified with the noncollinear internal angle α of 3.7° and the crystal angle θ of 31.5° for the signal–idler group–velocity matching. The tilted-pump geometry with an $\alpha_{\text{apex}} = 45°$ fused-silica (FS) prism with $\phi_1 = 49°$ and a telescope ($f_1 = 200$ and $f_2 = 71$ mm) satisfies the PFM condition of Equation (4) in the BBO, and the nontilted signal amplification is attained with elimination of the spatial chirp [9]. Diameters of both beams at the BBO are about 1.5 mm (corresponding to the pump intensity of about 50 GW/cm^2). Both pump and signal beams are reflected back to the crystal by concave mirrors located in the confocal configuration and the inverse imaging gives the PFM interaction again also in the second-stage amplification with a negligible spectral change. The resulting output pulse energy is 6–7 μJ.

Because the seeded continuum has a chirp, the OPA is a system of chirped-pulse amplification [20]. To utilize the full bandwidth, the chirp rate must be small enough for all the spectral components in the bandwidth to interact with the pump pulses in a BBO crystal. In this chapter, we use relatively thick sapphire plate (2 mm) and a low-frequency cut-off filter before amplification for ultrastable operation for versatile spectroscopic applications. Pre-compression of the seed [21] is then required for full-bandwidth operation in our setup, and after amplification, a main compressor compensates the residual part of chirp. The total pulse compressor is composed of ultrabroadband chirped mirrors (UBCMs; Hamamatsu Photonics Hamamatsu, Japan), a 45°-FS prism pair, air, and 0.5-mm-thick Cr-coated broadband beam splitters in a fringe-resolved autocorrelator (FRAC). The UBCMs are specially designed to have appropriate wavelength dependence for the accurate phase compensation over 200 THz in the visible range. A UBCM pair with one round trip is used for pre-compression, and another UBCM pair with three round trips and prism pair are for main compression. Figure 5 shows the wavelength dependence of the group delay (GD) of each component and the total system. The parameters of the compressor are determined to cancel the measured GD across the whole spectral range. The best compression is attained in the case of a 1-m separation and a 6.0-mm intraprism path length (IPL) of the prism pair at 650 nm and the net four round trips of the UBCM pairs. The throughput of the compressor is about 80%, which is determined by Fresnel loss at the prism surfaces. The final pulse energy is hence about 5 μJ.

The spectral shape of the signal after the compressor depends on the positions of the delay lines, which are determined to maximize the bandwidth. The FWHM is as broad as 240 nm (150 THz) extending from 535 to 775 nm (Figure 6) with the corresponding TL pulse width of 4.4 fs. The expected phase at the entrance of the 10-μm-thick BBO crystal in the FRAC is obtained by frequency integration of the sum of the measured GD of the pulse just after amplification and the calculated GD of the compressor. The deviation of the phase is only within $\pm \pi/4$ radian over the whole spectral range.

The pulse shape is measured by the FRAC with low dispersion, where both arms are perfectly balanced by the same beam splitters and the fine delay is on-line calibrated. Figure 7a and b shows the FRAC traces from −15 to +15 fs and from −40 to +40 fs, respectively, of the signal after optimizing the compressor. The $sech^2$-fit pulse width is as short as 3.5 fs. Complex Fourier transformation of the spectrum (DIFT) shown in Figure 7c gives the intensity profile with a nearly TL 4.7-fs FWHM, and the calculated FRAC trace agrees very well with the experimental results even in the side wings on both sides. Fitting the trace with parameters of GD dispersion (GDD) and third-order dispersion with inclusion of the oscillating phase of the UBCMs also estimates 4.7 ± 0.1-fs duration [10]. Even though there remains some uncertainty especially in the leading and trailing regions of the intensity profile deduced by this method [1], the FWHM determined by the central part of the pulse is surely well below 5 fs. A frequency-resolved optical gating was measured and formed to be consistent with the above result [22]. As far as we know, this highly coherent 2.2-cycle pulse was the first sub-5-fs pulse in the visible region.

One of the outstanding advantages of NOPA is its flexibility of the pulse width in the range longer than 5 fs by selecting some spectral portion by increasing the seed chirp. For example, a tunable sub-10-fs operation is straightforward without the chirped mirrors or by inserting a glass plate before the crystal. The tunable TL pulses with a 6- to 7-fs duration were generated from 550 to 700 nm by simply changing the delay line of the seed without any further alignment [9] as explained earlier in this chapter (Figure 4).

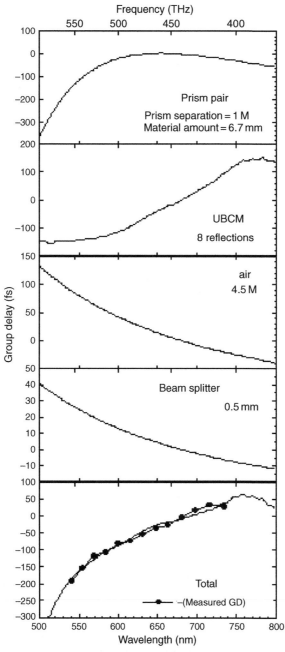

Figure 5 Group delay (GD) property of the compressor. The GD of each compressor system – the prism pair, the ultrabroadband chirped mirror (UBCM) pair with prism four round-trips eight reflections, the pair, air with a 4.5-m path length, and the beam splitter in the fringe-resolved autocorrelator (FRAC) – are shown. Also shown is the measured GD of the signal (full circles) with the sign reversed.

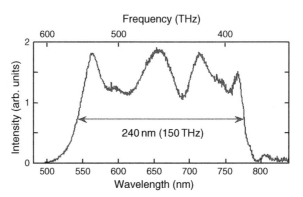

Figure 6   Spectrum of the amplified signal pulses from noncollinearly optical parametric amplifier (NOPA) under a full-bandwidth operation after the prism pair, ultrabroadband chirped mirror (UBCM), and the beam splitter. The bandwidth extends to 150 THz (240 nm).

1.5.   Generation of sub-4-fs pulse

In this subsection, we report modifications in the layout of a double-pass NOPA, which allow the generation of a visible near-IR spectrum that covers 300 nm at its FWHM. We further demonstrate compression of the resulting signal by means of a computer-controlled flexible mirror. The principal traits of double-pass sub-5-fs NOPA are described elsewhere [10]. The schematic of the experimental layout which can generate even shorter pulse as short as sub-4 fs is shown in Figure 8. The system is pumped by 120-fs pulses with 150-µJ energy from a regenerative amplifier (Clark-MXR, model CPA-1000) seeded by a fiber oscillator model (IMRA, a new Femtolite). We have implemented several important improvements that enabled further extension of the spectral bandwidth and furnished additional means of spectral shaping compared with the NOPA described in Shirakawa et al. [10].

Firstly, a pair of 45° FS prisms has been introduced to pre-compress the seed pulse, the white light continuum generated in a 2-mm sapphire plate. An adjustable razor blade behind the inner prism is used to remove the intense spectral components that are close to the fundamental 790-nm light. On the contrary, the insertion depth of the inner prism adjusts the cut-off wavelength on the blue side of the spectrum. The use of non-Brewster prisms with a more acute apex angle helps reducing higher order phase distortion [1]. However, material cubic dispersion of the prism dominates the phase properties of the prism pre-compressor even in such prisms. Consequently, the spectral components at 500 and 750 nm are roughly synchronized while 600-nm light is retarded with respect to them.

Secondly, because of the abundant intensity of the SH light, we have chosen to stretch the pump pulse in time by down-chirping it in a 100-mm block of fused silica. As a result, the peak intensity of the pump pulse is decreased below the damage threshold of the BBO crystal that is used in the NOPA. This allowed switching to a confocal configuration that greatly improved the output mode pattern. At the same

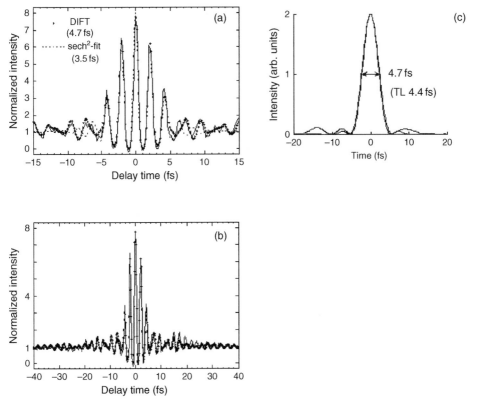

Figure 7 (a) Fringe-resolved autocorrelation (FRAC) trace of the compressor output for $|\tau| < 15$ fs and (b) for $|\tau| < 40$ fs (τ, delay time). Measured (solid curve), FT-fit (cross, see text), and sech^2-fit (dashed curve) traces are shown. The sech^2-fit pulse width is 3.5 fs, whereas the calculated and FT-fit pulse widths are 4.7 fs and 4.7 ± 0.1 fs, respectively. (c) Intensity profile of the compressed pulse. The transform-limited (TL) pulse width determined by the spectrum is 4.4 fs (thin solid curve), whereas the calculated (thick solid curve) and FT-fit to FRAC with phase parametric (dashed curve) pulse widths are 4.7 and 4.7 ± 0.1 fs, respectively.

time, the lengthened time envelope of the pump pulse dramatically lowered the requirement for an accurate pre-compression of the seed light.

Thirdly, the angular dispersion of the pump is utilized in such a way that various frequency components of the SH spectrum intersect with the seed beam in the NOPA BBO crystal at slightly different angles, which helps broadening the effective phase-matching bandwidth. Conversely, we find this effect to be another significant effect than the tilted pump geometry and PFM which was emphasized previously [10]. Although tilted-pulse pumping is essential for increasing the interaction length and pulse overlap in longer crystals [18], it plays a secondary role in the case of a 1-mm BBO crystal used in this work. Figure 9 illustrates the idea to obtain the enhancement of the phase-matching bandwidth. Because adjacent pump wavelengths correspond to slightly off-set phase-matching curves, the resulting

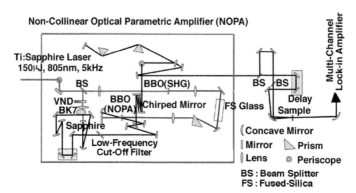

Figure 8 Schematic of the visible sub-4-fs pulse generator. BS, beam sampler; harmonic separator; TP, prism for pulse-front tilting; L1, L2, lenses for the telescope; SMs spherical mirrors ($\gamma = 100\,\mathrm{mm}$); VND, variable neutral-density filter; WSM, spherical mirrors ($\gamma = 120\,\mathrm{mm}$); CF, cut-off filter; D, optical delay line; PS, periscope; P1, P2, 45° fused-silica prisms; TBS, thin beam splitter; CCMs, corner-cube mirrors; BPF, band-pass filter; PMT, photomultiplier; PD, photodiode.

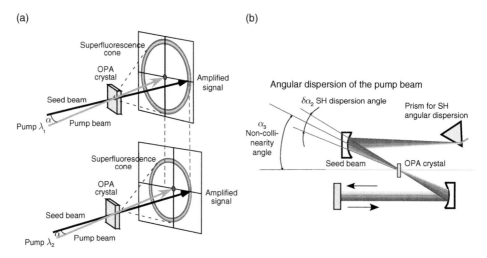

Figure 9 (a) Schematics of noncollinearly optical parametric amplifier (NOPA) with angularly dispersed pump beam with typical wavelength components of λ and λ_2 ($<\lambda_1$), with different incident angles to the OPA nonlinear crystal. They exhibit both superfluorescent cones and amplified seed beams resulting in the signal beams by the pump beams. (b) Schematic configuration of angular dispersive pump system composed of a prism and concave and flat mirrors for two-stage pumping.

bandwidth that can be simultaneously amplified in a given direction of the seed light is extended by use of a broadband SH pump. To maximize the usable SH bandwidth, we have employed a relatively thin, 0.4-mm, BBO SHG crystal that ensures a 30% frequency-doubling efficiency and provides a 6-nm-wide pump spectrum. The FWHM of the angular phase-matching was computed according to Dmitriev et al. [23] and represents the case of low efficiency of parametric frequency conversion [24]. An additional broadening can be obtained by manipulating the incidence angle for various pump frequencies. The actual conditions of our experiment is as follows. A Brewster-angled FS prism is used to introduce angular dispersion in the pump beam. The SH light is subsequently focused onto the BBO crystal by an $R = -200$ mm mirror placed at an 80 cm distance from the prism (Figure 8). Analogous ideas about phase-matching extension have been previously implemented in achromatic frequency doubling [25,26] and a multipass OPA [27].

The improvements listed above have led to the generation of a smooth parametrically amplified spectrum, which corresponds to a 3.5-fs FWHM pulse duration if ideally compressed.

Amplitude-phase characterization of both chirped and compressed pulse was carried out by SHG FROG in a very thin BBO wedge. The crystal thickness at the point used in the measurement was estimated to be about 5 μm. To reverse the effect of spectral filtering, a post-experiment correction [28] has been applied to all measured FROG traces. Recently, crystal angle dithering has been suggested as a way to solve the problem of the insufficient phase-matching bandwidth in an SHG FROG measurement [29]. Nevertheless, we avoided using this method in our work as it is very difficult to ensure that the axis, around which the crystal is being tilted, coincides precisely with the beam intersection. This concern becomes vital in the measurement of the compressed (<2-μm-long) NOPA pulses, in which crystal dithering can easily result in the equally unwanted time-delay filtering of the FROG/autocorrelation trace.

The pulse compressor consists of a pair of chirped mirrors CM1, 2 (Figure 8), and a grating dispersion line with a flexible mirror [30,31] (OKO Technologies Amsterdam, The Netherland) positioned in the focal plane. The combination of the grating dispersion line and the chirped mirrors has been designed through a dispersive ray-tracing analysis aiming to match the spectral phase of the chirped parametrically amplified pulse, which was obtained from the FROG measurement. The total throughput of the pulse shaper is less than 12% because of the low diffraction efficiency of the grating, which limits the energy of the compressed pulses to approximately 500 nJ.

The FWHM pulse duration corresponding to the optimal grating–spherical mirror separation and the "switched off" state of the flexible mirror (no bias voltage applied to the actuators) is 5.3 fs.

Provided the actual deflection of the membrane is well-calibrated as a function of applied voltage, the task of attaining the ultimate pulse compression becomes straightforward [30]. In practice, however, the perfect calibration of the membrane deflection is very cumbersome. Therefore, we have relied on a feedback with iterative optimization based on SHG FROG [30,32]. To maximize the SH signal at every wavelength, our algorithm employs Brent's search method [33], looped over 13 independent actuator settings. The automated iterative phase adjustment is typically complete within several minutes. An example of a FROG measurement before and after the described optimization is presented in Figure 10a and c and Figure 10b and d, respectively.

The retrieved spectral phase shows almost negligible deviations from a flat phase throughout most of the bandwidth. It must be noted, however, that the 5-μm thickness of the BBO crystal used for pulse diagnostic was not sufficient to cover the whole bandwidth at once. Abrupt spectral variations of SH conversion efficiency, regardless of any FROG data corrections, lead to a total loss of information from both the high- and the low-frequency wings of the spectrum, as is evident from the comparison of the measured and retrieved spectra in Figure 10e. The truncated SH bandwidth introduces system error on the FROG trace [34], causes the inversion algorithm to distort the real pulse shape, and generally results in a poor convergence. The obtained FROG error was 0.0068 on a 128 × 128 matrix for the trace in Figure 10b. Considering this experimental uncertainty, we believe that the resulting FWHM pulse duration is 4 fs within a 10% error margin. The accuracy is estimated from numerical simulations [34] and the

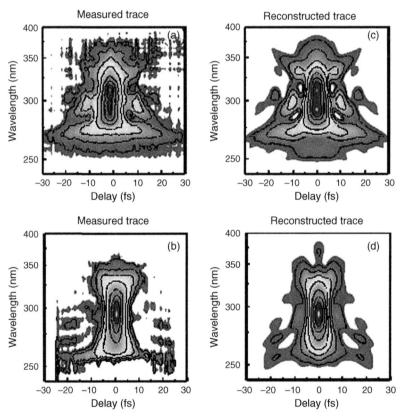

Figure 10 Overview of pulse-shaping results. (a and b) Measured second harmonic generation frequency-resolved optical gating (SHG FROG) traces before and after adaptive phase correction, respectively. Corresponding retrieved traces are displayed in (c) and (d). Contour lines in (a)–(d) are drawn at values 0.02, 0.05, 0.1, 0.2, 0.4, 0.6, and 0.8 of the FROG peak intensity. (e) Shaded area, fundamental spectrum measured at the crystal location in the FROG apparatus; open circles, spectrum recovered by the FROG retrieval algorithm; dashed-dotted curve, spectral phase before shaping; dashed curve, the optimized phase. (f) Initial (solid curve) and optimized (shaded area) temporal intensity profiles; dashed curve, temporal phase of the optimized pulse.

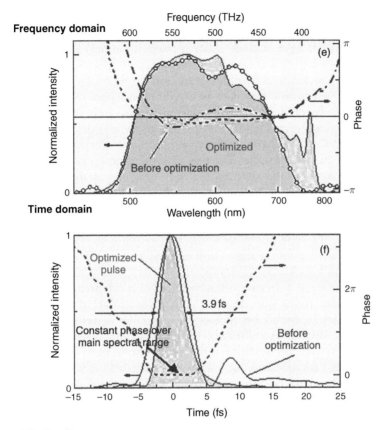

Figure 10 (*Continued*)

dispersion of FROG inversion results of the experimental data. It can be expected that a further reduction of the pulse duration after a modification of the dispersion line.

2. ULTRAFAST SPECTROSCOPY

2.1. Introduction

It has been a long dream of molecular physicists and chemists to visualize the structure of molecules during its reaction and vibration that can be a gateway of chemical reaction. Some specific molecular vibrational mode may be associated with the redistribution of electrons in the reacting molecules resulting in a configurational change or bond relocation. For such purposes, ultrashort laser pulses can be used as a "flash of camera" to pursue the nuclear wavepacket motions on the potential energy surfaces (PESs) to reveal photochemical reaction pathways [35–39]. Nuclear wavepacket can help the visualization of atomic motion during the chemical reaction or molecular vibration even though too much analogy may introduce misunderstanding the processes taking place. For example, there is a situation in which one nucleus exists in two (or more) distant places at the same time.

Nuclear motions induced by photoexcitation through vibronic coupling modulate in turn the electronic structure with vibrational periods [40,41]. This phenomenon can be observed as electronic spectral changes in both shape and intensity during the vibrational periods [42,43]. In case the vibration belongs to the gate mode of chemical reaction, the spectrum provides the electronic structure in the "transition state" in a broad sense. The reason of mentioning as "in a broad sense" is that "transition state" is the state where the curvature of the potential curve is negative, whereas in the statement above it means all of the states between two metastable intermediate states. Short pulses to be used must have broad enough spectral width to cover the spectral range of interest and short enough width to real-time-resolve the vibrational motions. This requires sub-5 fs FT-limited pulse [44] to time-resolve most strongly and frequently coupled stretching modes of single, double, and triple C—C bonds and C—H bond with frequencies in the range of 1000–3000 cm^{-1}. In case when materials in condensed phase are the targets of the study, there is an issue due to complication induced by inhomogeneous broadening, which is found for almost all molecular systems in the condensed phase, either in a solid, in a solid matrix, or in a crystalline phase. In case of solution and liquid phase system, the situation is slightly different. In an extremely short time scale, shorter than the correlation time of polarizability of the solvent molecules, the system is considered to be inhomogeneous, but longer than this time scale it becomes homogeneous.

In our previous study [45], we could partly solve the above three difficulties by utilizing the sub-5 fs pulse as pump and probe sources [42] with a spiky peak in the spectra and a multichannel detector. Especially, we tackled the third problem by taking advantage of the spectral feature of the spiky structure that enables femtosecond partial "hole burning" [46] to eliminate partial inhomogeneity and detect vibrational structure buried in part in the inhomogeneity. The "hole" spectrum reveals vibrational progression in the molecule with vibronic coupling [47] even though the spectrum is not a true "hole" spectrum because the burning pulse has much broader spectrum than the homogenous spectrum. However, this method is very useful and enabled us to track each vibrational peak in bleaching, gain, and induced absorption spectra.

As an example of the usefulness, the systematic study of time-resolved measurement of one of the most popular conjugated polymers, polydiacetylene (PDA), is described in detail in this chapter.

Molecular vibration of several modes in blue-phase polydiacetylene-3-butoxycarbonylmethylurethane (PDA-3BCMU) was real-time observed by 5-fs pump–probe measurement. The methodologies utilized include (i) "a peak-tracking" method applied to the time-resolved spectrum of the polymer. This provides valuable information of detailed dynamics in the polymer through 1-fs step sequence of electronic absorption and gain spectra and (ii) global fitting after the singular value decomposition of the time trace of the difference absorption spectra induced by pump pulses. (iii) The contribution of the vibration wavepackets in the ground state and in the excited state in the signal was also separated.

By applying these methods, the following results were clarified. The peak of $0 \rightarrow 0$ transition from the ground state in which vibration excitation does not participate oscillates with the frequencies of C—C stretching vibration modes. The oscillations of energy and intensity of a peak of induced absorption to higher excited state show different behaviors both in the vibration amplitude and phase. The results provided us comprehensive information to determine the relative locations of the multidimensional potential hypersurfaces of the ground, the lowest excited, and the higher excited states in the configuration space. The precise analysis of the time evolution of the difference absorption spectra has revealed the thermalization processes due to the energy redistribution among vibration levels of various quantum numbers and vibration modes in the polymer chains in the excited and in the ground states.

The C=C stretching mode in the ground state starts to oscillate π-out-of-phase with the C≡C stretching mode. The structure of PDA-3BCMU in the geometrically relaxed state is not pure butatriene-type but mixed with acetylene-type. The frequencies of C=C and C≡C stretching modes were determined by singular value decomposition method to be 1472 ± 6 and $2092 \pm 6 \, \text{cm}^{-1}$, respectively. The double and triple bond stretching frequencies in the ground state are 1463 ± 6 and $2083 \pm 6 \, \text{cm}^{-1}$, respectively.

Even though quantitative values of the parameters mentioned above are expected to vary from one PDA to the other, main characteristic features are expected to be common in all blue-phase PDAs because of their similarity in the electronic spectrum with a typical exciton peak around $2 \, \text{eV}$ and phonon side band due to C—C stretching modes.

Spectroscopic and electrical properties of conjugated polymers have been studied extensively because of their unique properties due to their low dimensionality, and they are considered to be model compounds of quasi-one-dimensional electronic systems. However, the primary photophysical phenomena in the relaxation kinetics of π-conjugated polymers are not yet fully understood despite their potential applications to electronics, optoelectronics, and photonics [48,49]. The examples are flexible conductors, light-emitting diodes, and all-optical switches [50]. Elucidating the early electron-lattice dynamics in the vibrational nonequilibrium after photoexcitation in these systems encourages engineering-appropriate materials for the applications.

Among many existing polymers, PDAs have especially attracted a lot of scientists' interest, because PDAs have many interesting features from various viewpoints. They have several phases named according to their colors, that is, blue, yellow, and red phases. They can also have various morphologies, that is, single crystals, cast films, Langmuir films, and solutions [51–54]. The ultrafast optical responses in PDAs have been intensively investigated using femtosecond spectroscopy and picosecond and femtosecond time-resolved Raman spectroscopies [9,37,49,55–65]. It is also because of interest from the viewpoint of fundamental physics as they are model materials of one-dimensional system with outstanding characteristic features due to their low dimensionality. It includes ultrafast relaxation resulting in the outstanding ultrafast large optical nonlinearity helped by the excitonic nature. The ultrafast large optical nonlinearity has been studied by nonlinear spectroscopy [56]. They exhibit several nonlinear optical effects such as phonon-mediated optical Stark effects [66], phonon-mediated hole burning [67], inverse Raman scattering [49], optical Stark effect [64], and Raman gain process [64]. These features are deeply related to the formation of localized nonlinear excitations such as polarons and a self-trapped exciton (STE) formed via a strong coupling between electronic excitations and lattice vibrations. STE is sometimes called exciton polaron or neutral bipolaron [55–57,68–78].

Electronic states in PDA are well characterized theoretically by the Pariser–Parr–Pople–Peierls (PPPP) model [79], which includes both the electron–electron (correlation) interaction (Pariser–Parr–Pople) and the electron–lattice (Peierls) coupling, capturing essential electronic properties of the π-electron system. One important consequence of the electronic correlation is the reversal of the energy-level positions of the first excited 1^1B_u and the second 2^1A_g states [80]. Because PDAs are spatially centrosymmetric and thus shows C_2 symmetry, the wave functions possess mirror plane and centroinversion symmetries. The group notation for mirror plane symmetries is A and B for the symmetric and the antisymmetric cases, respectively. Inversion symmetries are labeled g for the symmetric and u for the antisymmetric. The ground state is therefore labeled $1A_g$, and the first optically active dipole has to be the $1B_u$ state. The wave functions are either even (A_g) or odd (B_u) under inversion. Recent experimental and theoretical studies have revealed that the lowest excited singlet state in a blue-phase PDA to be an optically forbidden 2^1A_g state lying approximately 0.1 eV below an 1^1B_u free exciton (FE) state [65,81]. Blue-phase PDAs usually have much smaller fluorescence yield than red-phase PDAs. From various experimental results and discussion, this feature is well summarized in the following way [48]. First, photoexcited ^1B_u FE generated by photoexcitation relaxes to the nonfluorescent ^1A_g state, which lies below the ^1B_u exciton in such systems with long conjugation lengths. The nonthermal ^1A_g state relaxes to the bottom of the potential curve and then thermalizes. The time constants of the relaxation and thermalization are about 60 fs and a few picoseconds, respectively. Finally, ^1A_g relaxes to the ground state with a decay time of 1.5 ps at room temperature, even before full thermalization including bulk interchain thermalization, which takes place only after several tens of picoseconds.

2.2. Ultrafast dynamics in PDA

The properties of the broad bandwidth and the shortest duration of NOPA are quite useful for such real-time observation of ultrafast vibrational dynamics and structural change in photochemical reactions

From previous extensive studies [9,37,49,55–65], the initial changes in the electronic absorption spectra and their ultrafast dynamics in a femtosecond region after photoexcitation of PDA are explained in terms of the geometrical relaxation (GR) of a FE to a STE within 100 fs together with the internal conversion (IC) from ^1B_u state to ^1A_g state. The STE is well established to be a geometrically relaxed state with admixture of butatriene-type configuration $(-CR=C=C=CR'-)_n$ from an acetylene-type chain $(=CR-C\equiv C-CR'=)_n$ [58,59] Here, R and R' represent substituted side groups attached to the main chain. There are PDAs with various combinations of the substituting groups. All of the stretching vibrations of carbon atoms are considered to be coupled to the photogenerated FE and induce various nonlinear optical processes different from those in most of inorganic semiconductors [56,57,64,65,74,75,82]. Recent experimental and theoretical studies have shown that the lowest excited singlet state in a blue-phase PDA to be an optically forbidden 2^1A_g state. It is because the blue-phase PDAs have long enough conjugation lengths for the 2^1A_g state to be substantially stabilized. This is because of the 2^1A_g state with two-triplet character, and it has large repulsive interaction between the two triplet "components" because of the Pauli exclusion principle. These repulsive interactions relaxed in a long chain and as in ordinary organics "triplet" state is more stabilized than a corresponding singlet state. This 2^1A_g state is lying approximately 0.1 eV below a strongly allowed 1^1B_u–FE state, which provides characteristic intense blue color and metallic reflectance [68–70,83–86]. The IC mentioned above is then explained to take place along with self-trapping [74,75]. However, the detailed dynamical processes of IC and GR have not yet been fully characterized [74,75]. Recent progress in femtosecond-pulsed lasers has enabled to study molecular dynamics on a 10-fs timescale [37,60]. In the previous works by Bigot and others, a wavepacket motion of C=C stretching mode with a period of approximately 23 fs was found in the photon-echo and transient bleaching signals of PDA-DCAD (poly (1,6-di (n-carbazolyl)-2,4 hexadiyne)) films by using 9- to 10-fs pulses [58,59].

The real-time observation of the GR in PDA has been enabled by the recent development of sub-5-fs visible pulse generation based on NOPA system, which satisfies all of the PFM, phase matching, and group-velocity-matching conditions [87–89] as described in Section 1. Utilizing compressors such as a prism pair, a grating pair, chirped mirror pairs, and a deformable mirror, the shortest visible-NIR pulses as short as 4 fs were obtained with a broad featured structure as described in Section 1 [87–89]. These features are excellently suited for pump–probe spectroscopy The trace of the delay-time dependence of the normalized difference transmittance $\Delta T(t)/T$ induced by an ultrashort pump pulse or more appropriately the absorbance change ΔA calculated from $\Delta T/T$ is called a "(vibration) real-time spectrum," which means a spectrum not in a frequency domain but in a time domain probing the real-time behavior of molecular vibration amplitude. By using time-resolved analysis of the FT of the real-time spectrum, the dynamic features of self-trapping, IC, and coupling between stretching and bending modes in the relaxed state in a PDA have been elucidated using sub-5-fs pulses [65].

However, these experiments have a remaining problem of the possible ambiguity in the assignment of the pump–probe signals to either the ground state or the excited state, because the ultrashort laser pulse with a wide enough spectrum can drive the coherent vibrations in both ground and excited states. This prevents us from the well-defined discussion of the dynamics of the wavepacket after being photogenerated. In this study, we could attribute the origins of the oscillation signals in pump–probe traces either to the ground or to the excited state by utilizing broad spectral information of the time traces, which has been realized by the extremely stable sub-5-fs NOPA developed in the author's group.

2.3. Experimental

2.3.1. Sample

The sample used in this is a cast film of blue-phase PDA–3BCMU (poly [4,6–docadiyn–1, 10–diolbis (*n*–butoxycarbonylmethylurethane)]) on a glass substrate. PDA–3BCMU has side groups of

$$R{=}R'={-}(CH_2)_3OCONHCH_2COO(CH_2)_3CH_3,$$

in the backbone chain structure of $({=}RC{-}C{\equiv}C{-}CR'{=})_n$. PDA–3BCMU is one of the well-known soluble PDAs. The samples were prepared by the following method.

Oxidative coupling of 4–pentyn–1–ol (purchased from Tokyo Kasei Tokyo, Japan) by Hay's method [90] was used to obtain 4,6–decadiyn–1,10–diol with the yield of 79%. The diol compound reacted with butyl isocyanatoacetate at 23°C using triethylamine in THF resulting in the production of 3BCMU (monomers) with the yield greater than 98%. The monomer film of 3BCMU was irradiated by solid-state polymerization of an appropriate dose of ^{60}Co γ-ray irradiation. The total dose was approximately 150 kGy. The crystal of 3BCMU after the ^{60}Co γ-ray irradiation process was washed with methanol to remove the unreacted monomers. The obtained polymer (PDA–3BCMU) was dried under vacuum.

A sample PDA–3BCMU film was prepared by a doctor blade method from chloroform solution of about 0.1 wt%. By the doctor blade method, the solution of polymer was dropped onto a washed glass plate, stretched by a glass rod, and then dried at room temperature.

2.3.2. Laser system

Light source of sub–5–fs spectroscopy was a NOPA described in Section 1 of this chapter. The output pulses with a 4.7–fs width at a 5–kHz repetition rate from the NOPA seeded by a white-light continuum were used as both pump and probe pulses [9,44]. The continuum was generated by focusing the regenerative amplified Ti : sapphire laser pulses onto a 2–mm thick sapphire plate The duration of the NOPA output pulses was reduced with a 5–fs pulse compressor system [9,44,89–91] composed of a prism pair and chirp mirrors. The Ti : sapphire laser system is a commercially supplied regenerative amplifier (model Spitfire; Spectra-Physics California, USA), of which pulse duration, central wavelength, repetition rate, and average output power are 100 fs, 790 nm, 5 kHz, and 800 mW, respectively. The spectra of the pulses covered from 520 to 750 nm with a nearly constant phase, indicating that the pulses are FT limited. The pulse energies of the pump and probe were about 35 and 5 nJ, respectively. All the measurements were performed at room temperature (295 $\pm$ 1 K).

2.3.3. Data analysis

Difference absorption spectra (ΔA) were calculated by Equation (5) shown below from the normalized difference transmission spectra ($\Delta T/T$) at pump–probe delay times from −50 to 1949 fs with every 1-fs step.

$$\Delta A = -\log_{10}\left(1 + \frac{\Delta T}{T}\right) \tag{5}$$

The difference transmittances probed in a wavelength range from 541 to 740 nm (corresponding to photon energy 1.68–2.28 eV) were detected simultaneously using a multichannel lock-in amplifier with a 300 grooves/mm grating monochromator for polychromatic usage. The spectral resolution was about 3.6 nm (10 meV) limited by the monochromator.

2.4. Results and discussion

2.4.1. Peak tracking analysis

The spectra of excitation and probe pulse laser and absorption of the blue-phase PDA-3BCMU are both shown in Figure 11. The vibration energy levels of C—C double and triple bond stretching are indicated in the graph. The low lying electronic energy states are depicted in the figure together with transitions associated with excitation, IC between the excited states, induced absorption, and relaxation to the ground state. Typical example of experimental data of difference absorption spectra is displayed two dimensionally in Figure 12. Several examples of the observed traces of normalized difference transmittance at probe delay time of $t(\Delta T(t)/T)$ of the blue-phase PDA-3BCMU sample at several energies of probe photon were shown in Figure 13. All of the trace curves show signals of a finite size of $\Delta T(t)/T$ at negative delay times and sharp and intense peaks around zero probe delay time. The former is due to the perturbed free induction decay (FID) process associated with the third-order non-linearity induced by the sequential interaction of probe–pump–pump fields modified with molecular vibrations. The latter signals are due to pump–probe coupling induced

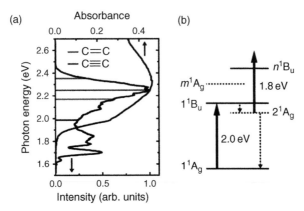

Figure 11 (a) Spectra of the excitation pulses and the absorption of PDA. The former and the latter have scales bottom and top, respectively. The lines denote the energy of 1^1B_u state and its vibration levels. (b) Schematic energy diagram of the essential electronic states: ground state (1^1A_g), excited state (1^1B_u), higher excited state (m^1A_g), and geometrically relaxed state (2^1A_g), from which optical transition is allowed to another excited state (n^1B_u) of PDA relevant to our experiment. The corresponding transition energies are also described. Dashed lines indicate radiationless transitions.

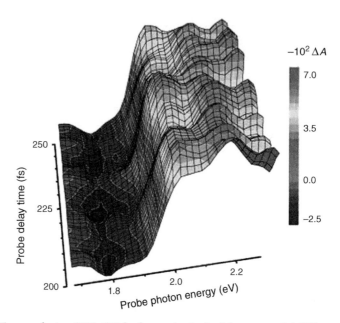

Figure 12 Time evolution (200–250 fs after excitation) of the sequential difference absorption spectra of polydiacetylene-3-butoxycarbonylmethylurethane (PDA-3BCMU) with resolution of ≤1 fs (vertical axis) and ≤10 meV (horizontal axis) obtained by the pump–probe measurement. Positive and negative difference absorbances are caused by the induced absorption from 2^1A_g state and the absorption saturation in 1^1A_g state, respectively. (See color plate 17).

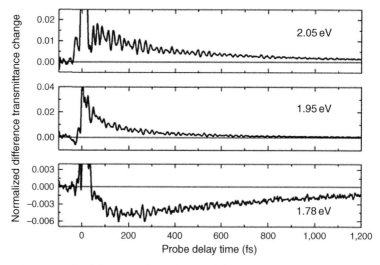

Figure 13 Pump–probe delay time dependence of the normalized difference transmittance ($\Delta T(t)/T$) on the probe delay time at three-probe-photon energies of 2.05, 1.95, and 1.78 eV.

by the nonlinear process of the pump–probe–pump time ordering. There is another contribution from the interference between the scattered pulses and the probe pulses of which duration are elongated in the polychromator. The details of the vibrational modulation observed at longer delay times than 100 fs are free from the distortions due to the above two mechanisms.

Here, the before-mentioned "peak-tracking" analysis developed recently by the author's group is to be explained. This is to study the peak energies of the transient electronic transition during molecular vibrations. This method was applied to demonstrate the oscillations of transition peak energies in the blue-phase PDA–3BCMU film [56], where 3BCMU denotes the side-chain of the conjugated polymer. Polymers composed of π-electrons along the one-dimensional carbon chain [92], including PDAs as a typical example, have large ultrafast electron–lattice interaction [67] and thereby exhibit ultrafast optical response [49]. Nonlinear optical properties in polymers including polyacetylene and PDAs can mostly be explained in terms of four essential electronic states model proposed in Mazumder et al. [93] and the ground state as shown in Figure 11b. Pump pulses with spectrum having the highest peak at 2.25 eV, which is also the energy of maximum absorbance in the stationary absorption spectrum, predominantly excite the allowed $1^1A_g \rightarrow 1^1B_u$ transition coupled by C≡C stretching (Figure 11). This creates reduced difference absorption spectrum (DAS) due to the population-hole in the ground state caused by depletion. The DAS has a relatively well-defined vibronic structure including the purely electronic zero-phonon $(0 \rightarrow 0)$ peak around 2.0 eV [34,42]. The bandwidth of each peak increases with delay time. This is caused by the thermalization in the ground state, and it indicates that each peak in the bleaching spectra corresponds to a well-defined vibronic exciton–exciton transition. Another clear peak of induced absorption $(2^1A_g \rightarrow n^1B_u)$ around 1.8 eV [49,56] also appears in the positive DAS (Figure 12). In the figure, the positive DAS is the downward signal.

These well-separated features due to the partial "hole burning" enable us to apply the "peak-tracking" method to the transition peaks. This method provides detailed information about the mechanism of the oscillating signals corresponding to all the vibronic transition separately in DAS. For each vibronic transition, the DAS signal can be analyzed in terms of two contributions: modulations of the transition energy and the transition probability. They appear in DAS as "vertical" and "horizontal" modulations with a molecular vibration frequency, Ω [94]:

$$\Delta A(\omega) \cong \Delta A_0(\omega) + \left(\frac{\delta(\mu^2)}{\mu^2} \Delta A(\omega) + \delta\omega \frac{d\Delta A(\omega)}{d\omega} \right) \cos(\Omega t + \theta). \quad (6)$$

Here, $\Delta A_0(\omega)$ is the nonoscillating component of the absorbance change $(\Delta A(\omega))$, and $\delta(\mu^2)/\mu^2$ and $\delta\omega$ correspond to the probability and energy modulations, respectively. The modulation frequency, Ω, and the initial phase, θ, are corresponding to those of the molecular vibration mode. The wavepacket motion along the potential curve causes the transition energy modulation [43]. Decomposed signal provides the correct time evolutions of transition energies and intensities with phase information and thus enable us to determine the relative locations of multiple PESs in the different electronic states.

The dynamics of the absorbance change was analyzed by the global fitting method, which is to be discussed later in detail to be classified into two components of short life (60 ± 3 fs) and long life (890 ± 40 fs) [59,69]. Initial relaxation process of $1^1B_u \rightarrow 2^1A_g$ (60 ± 3 fs) is observed as the decay of induced emission (IE) ($1^1B_u \rightarrow 1^1A_g$).

At longer delay time than 100 fs when the relaxation to 2^1A_g is nearly fully completed since the corresponding decay time is about 60 fs, the peak-tracking analysis clarified the dynamic behavior of the sequential spectra, which are composed of the saturation of the stationary absorption ($1^1A_g \rightarrow 1^1B_u$), and the induced absorption ($2^1A_g \rightarrow n^1B_u$), which includes oscillations in the slopes between the peaks. The modulation amplitudes of transition peak energies can be related to the strengths of vibronic couplings involved in the electronic transition. By using these relations, important spectroscopic parameters such as stabilization energies and Huang–Rhys factors [95] can be obtained in every direction of normal coordinates in the multidimensional configuration space. These parameters provide the displacement of the potential hypersurface minimum of the final state with respect to that of the initial state of the corresponding transition.

The excitonic zero-phonon transition peak near 2.0 eV ($1^1A_g \rightarrow 1^1B_u$) and an IA peak near 1.8 eV ($1^1A_g \rightarrow 1^1B_u$ transition) were tracked in Figure 14. Each peak and its slopes on both sides show energy oscillation following "sine" and "cosine" functions corresponding to the wavepacket motions in the ground and the excited states [30], respectively. The existence of two intense peaks in Figure 14c indicates that the two modes of C=C (1460 cm^{-1}) and C≡C (2080 cm^{-1}) stretching are strongly coupled with $1^1A_g \rightarrow 1^1B_u$ transition. However, the signal intensity of C≡C stretching is dramatically reduced in $2^1A_g \rightarrow n^1B_u$ transition (Figure 11b) with remaining strong peak of C=C stretching mode. This means that the

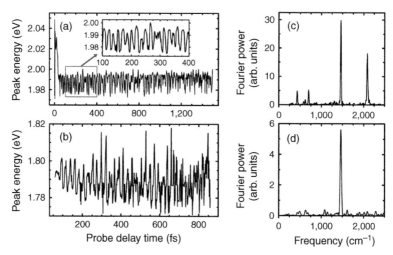

Figure 14 Dynamic behaviors of transition peaks and their oscillation frequencies. Energy oscillations of the zero-phonon transition peak (a) $1^1A_g \rightarrow 1^1B_u$ (~1.99 eV) and an induced absorption peak (b) $2^1A_g \rightarrow n^1B_u$ (~1.79 eV), and their Fourier power spectra calculated for the data of (c) 100–1500 fs and (d) 100–860 fs, respectively. The inset in (a) is the peak motion with an expanded delay-time scale from 100 to 400 fs.

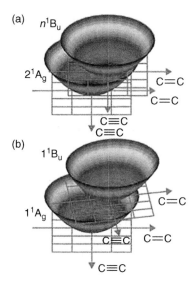

Figure 15 Schematic potential energy surfaces of the two electronic states involved in (a) $2^1A_g \rightarrow n^1B_u$ and (b) $1^1A_g \rightarrow 1^1B_u$ transitions.

potential-minimum shift between 2^1A_g and n^1B_u is largest in the C=C stretching direction, resulting in the impulsive starting of the vibrational mode. Schematic two-dimensional PESs obtained in the C=C and C≡C stretching coordinates are plotted in Figure 15. In this way, we can determine the structure of multidimensional potential hypersurfaces of molecular system. The dimensionality we are able to reach can be the number of vibration modes to be deleted and analyzed using the FTs of the real-time vibration spectra. As mentioned above, we can determine the displacement of more than two electronic states in case we can separate the electronic transition in the pump–probe experiment.

As described above, we picked up a specific transition among several peaks observed in the spectra for the characterization of the dynamic behavior of the corresponding vibronic transition. Applying the peak-tracking method, we have decomposed the two-dimensional modulation spectra into the transition-peak energy and the transition-intensity modulations. This method provides a new insight into the vibration amplitude distributions in the spectral features in DAS. It can be used as a general method to determine the multidimensional potential hypersurfaces in molecular systems. Because there is no limitations of the systems to be applied if sub-5-fs spectroscopy data can be obtained, it can be used by various systems including complicated systems such as organic polymers and biomolecules with multiple electronic states experiencing ultrafast dynamics and chemical reactions.

In case the photon energy of the laser pulse is in resonance with the electronic transition of the molecular systems being studied, femtosecond pulses with a broad spectrum excite multivibration levels of several modes simultaneously in phase in the relevant excited electronic state. The wavefunction of thus generated vibronic state is given by the product of the corresponding electronic wavefuntion and the product of the linear combinations of corresponding vibration levels of the corresponding modes being covered by the laser spectrum under the Born–Oppenheimer approximation. In such a way, ultrashort pulse creates wavepacket on the multimode multidimensional potential surfaces in the excited states. At the same time, such short pulses can generate coherent molecular vibrations in the ground state by the impulsive stimulated Raman scattering (ISRS). Identification and/or separation of the observed transmittance signals due to these two wavepackets are sometimes difficult. Vibration phase analysis of the wavepackets is useful to distinguish these two effects. However, the phase profile along the probed photon energy is complicated because of the inhomogeneity of electronic transition energy to the excited states, which may be sensitive to the local environment surrounding the excitations and the phase relaxation induced by the electronic interaction. One more factor that makes it even more difficult for the phase profile to be interpreted is that the vibration phase to be determined by the analysis is sensitive to the frequency of the vibration mode. Because of the homogeneous and inhomogeneous broadening of the vibration frequency, there is a limit of the precise determination of vibration frequency of each mode. It is also limited by the finite length of probe delay time spanning and finite lifetime of the corresponding excited state in case when the modulations are due to the excited state wavepacket. Because of these limitations, the frequency determined by the Fourier analysis may differ from a true value by 0.5%, namely $5\,\mathrm{cm}^{-1}$ for a $1000\,\mathrm{cm}^{-1}$ mode, which has a corresponding 33 fs period from the time-trace data extending to the probe delay of 2000 fs. In such a case, the phase obtained can be different by 15% of 2π radian estimated by extrapolation from the center of probe delay time that is 1000 fs.

One more difficulty in the precise determination of the phase especially in the case of high frequency modes stems on the difficulty in the precise determination of time zero position due again to the finite pulse duration. This is most serious in the cases of high frequency modes. For example, a mode with a frequency of $3333\,\mathrm{cm}^{-1}$ has a vibration period of 10 fs. In case the zero probe delay time position has 1 fs error, it causes $\pi/10$ error in the phase determination.

There is an even more difficult problem in the case of PDA studied in this chapter. There is a well-known GR process that takes place very rapidly just after photoexcitation to the excited electronic state. This is because of the barrier-free feature along the potential curve between the extended free-exciton state and STE expected in a one-dimensional system. To extract the well-defined signal of vibration coherences, we have to exclude the data during the initial GR process when the very strong coupling between the electronic excitation and molecular vibration is exerting on the electronic transition probability and its frequency dependence.

To discuss the effects of vibration modes to the electronic transitions including the ground state absorption, IE, and induced absorption, it is of vital importance to identify the modulations to the possible candidates of electronic transitions. In the case of PDA being studied, they are the 1^1A_g ground state $\rightarrow 1^1B_u$ exciton state, 1^1B_u exciton state $\rightarrow m^1A_g$ state, and 2^1A_g state $\rightarrow n^1B_u$ state. For the identification, it is needed to investigate the dependence of the dephasing time of the observed modes on the probe-photon energy to confirm the contributions of the vibrational coherences. The bleaching signal due to the ground-state depletion corresponds to the transition between 1^1A_g state and 1^1B_u exciton state. The induced absorption signals due to the absorption from the excited states to the higher excited states are expected to be composed of two transitions. They are the transition from the 1^1B_u exciton state to m^1A_g state and that from 2^1A_g state to n^1B_u state. Signals due to these two transitions are expected to have different decay dynamics [56].

There are several relaxation processes in such conjugated polymers as PDA after ultrashort pulse photoexcitation. One of them is the *thermalization* process that takes place among intrachain molecular vibration modes and interchain vibrational coupling including lattice vibration. They can appear in both signals because of the ground state and the excited states. The dynamics can be introduced by the deviation of population distribution of vibration levels in the Franck–Condon excited state generated just after the excitation from that of the thermal equilibrium state. Same thermalization process takes place due also to the deviation of the population distribution in the ground state depopulated by excitation, resulting in the hole burning and repopulated later on by the IC. There is another decay dynamics due to the radiationless relaxation of *IC* from 1^1B_u excited state to the 2^1A_g state associated with the *GR*. The above-mentioned decay dynamics are all classified as population dynamics. This process is also called "longitudinal" relaxation. This is because the relaxation is described in terms of the diagonal elements of density matrix describing the multilevel and/or multistate systems. The terminology comes from the analogy to the spin systems which are described with spin density matrix. The fields studying such systems are discussing about magnetic resonances of electrons (ESR) and nuclei (NMR).

There are also decay dynamics classified as phase relaxation processes in both vibration and electronic coherences. Oscillating component of the transient spectra reveals the "transverse" relaxation processes of the vibration coherences; this terminology is again from the magnetic resonance research fields such as ESR and NMR. The electronic coherence is in general difficult to be observed in ordinary pump–probe experiment. It is because in the electronically resonant case, the electronic polarization oscillates with optical frequency equal to the relevant frequency corresponding to the electronic transition. Therefore, it cannot be time-resolved with optical pulse of which duration is (much) longer than the period of the pulsed light determined mainly by the carrier frequency of the pulse. Even in nonresonant case, the situation is the same. By the interaction with ultrashort pulsed lights with the molecular system under nonresonant condition, the electronic system is forced to oscillate with the optical (carrier) frequency of the pulse. After the interaction is over, the electronic system starts to oscillate with its eigen frequency; nonoscillating component of the transient spectra reveals the longitudinal relaxation processes of the vibrational and electronic population distributions.

Clarification of the dynamics from both viewpoints of the above-mentioned transverse and longitudinal relaxations is required to understand the early photophysics in the electronically delocalized system. The electronic longitudinal relaxation is generally expected to appear as the global change over relatively broad spectral range in the difference absorption spectra in such a manner that the decay profile is common in the spectral range of the relevant electronic state. This common decay profile is found to be satisfied in case when the initial phase relaxation is over and thermalization is terminated, resulting in the constant population distribution of vibration levels that is maintained during the electronic longitudinal relaxation that is taking place.

In some cases, on the contrary, there can also be blue or red shift of the weight of mass of the spectrum in case when the initial state or the final state of the corresponding transition is lowered in energy because of the relaxation in the molecular geometrical configuration or dielectric relaxation in the media such as host matrix or solvent. Although the vibration energy redistribution among various modes in the polymer is expected to show several types of change in the features of the difference absorption spectra, they can be broadening or sharpening in case the initially distributed Franck–Condon state depending on the pump laser spectrum and the Franck–Condon factor has narrower or broader distribution than that in the equilibrium state. In previous experimental studies performed in the author's group, the transient absorption spectrum was measured with a 100-fs resolution, which revealed that the spectrum of IA ($2^1A_g \rightarrow n^1B_u$) is shifted to higher energy associated with the thermalization in 2^1A_g state in 0.5–1 ps [62–65]. However, these experiments used pump pulses with a narrow spectral width centered at 1.97 eV corresponding to the excitonic absorption peak around 2.0 eV of blue-phase PDA-3BCMU. The pump spectrum covers only the zero-phonon transition, and hence the coherent vibration cannot be generated. In this chapter, sub-5-fs visible pulses were used as a pump and a probe. Such extremely short pulses can drive wavepacket motions associated with molecular vibrational modes with even higher frequency than 3000 cm^{-1}. Their motions are along the potential hypersurfaces of the ground and in the excited states because of the possible vibronic coupling to many modes in both absorption process and ISRS process. In addition to the generation of vibrational coherence, nonequilibrium population distributions are also established both in the ground and the excited states by the pump pulses with a wide spectrum covering multiple vibration levels of CC stretching mode in the electronic excited state. The pump pulses induce population depletion in the ground state as well as the generation of population in the excited state. In the author's group, time-resolved measurement with a probe delay time step between 0.2 and 1.0 fs is performed. Even 1-fs-step difference absorption spectra can have enough time resolution to reveal both the vibration dynamics such as amplitude relaxation and instantaneous frequency change. The long step time as in the case of 1-fs step, the amplitude of the molecular vibration mode with high frequency will be reduced but frequency itself can be determined. The study of frequency modulation is very difficult to be studied by the conventional time-resolved vibration spectroscopy. Ultrafast vibrational thermalization processes in conjugated polymers can even more easily studied because the dynamics is much slower than the vibration period. This was enabled by the extremely stable sub-5-fs NOPA system developed in our group.

2.4.2. Bleaching and induced absorption spectra

We smoothed out the oscillating components from DAS by taking sequential convolutions of the data with rectangular functions of 3–101 fs time widths to investigate the nonoscillating spectral change due to molecular vibration. Then the smoothed signal can be attributed to the population dynamics of the vibration levels in the electronic states. Figures 16 and 17 show the nonoscillating DAS in 200–1900 fs. The initial population distribution in the excited state is determined by the absorbed laser intensity.

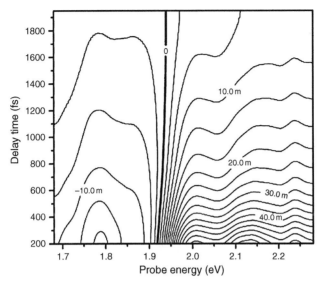

Figure 16 Two-dimensional difference absorption spectrum (DAS) of polydiacetylene-3-butoxycarbonylmethylurethane (PDA-3BCMU) after subtracting the oscillating components. Contour map of the equal absorbance change curves are shown with a 3.3-m OD unit.

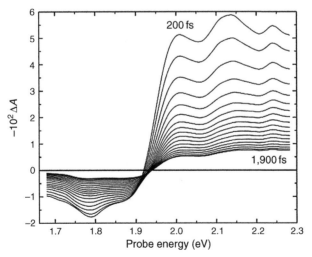

Figure 17 Difference absorption spectrum (DAS) of polydiacetylene-3-butoxycarbonylmethylurethane (PDA-3BCMU) after subtracting the oscillating components extracted from Figure 16 at delay times between 200 and 1900 fs with an interval of 100 fs.

Figure 18 shows the normalized spectra of the time derivative of DAS. The spectrum is obtained by taking the difference between the two successive difference spectra (difference–difference spectra (DDAS)) with delay time step of 1 fs. DDAS reflect the speed of the instantaneous absorbance change at each delay time and are

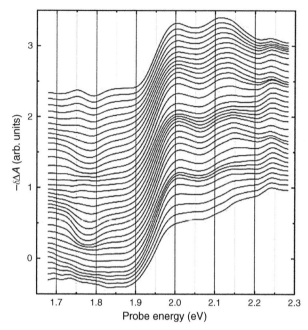

Figure 18   Time derivative of the difference absorbance in Figure 16. The spectrum is obtained by taking the difference between the two successive difference spectra (difference–difference spectra (DDAS)) with delay time step of 1 fs. DDAS is normalized at each delay time (200–1900 fs, 50 fs step).

attributed to the reduced population in the excited state, in other words, the recovered population in the ground state. The spectral changes in their shapes reveal that the two processes of the vibrational energy redistribution and the IC process from the 2^1A_g state to the ground state take place concurrently. The early bleaching spectrum with a vibration progression suggests that a "hole" is induced in the inhomogeneously broadened absorption spectrum. The transient bleaching spectrum undertakes spectral change due to the vibration thermalization through the intra- and interchain interactions via the low-frequency bending and torsion modes of the polymer chains. The initially prepared bleaching spectrum with a smaller inhomogeneity than that of the stationary absorbance relaxes to the thermalized spectrum with a feature similar to that of the stationary spectrum as shown in Figure 19. The recovery of the population in the ground state also refills the "hole" with a relatively large zero-phonon peak in the bleaching spectrum. These thermalization processes cause the spectral blue shift observed in the bleaching (negative ΔA) spectral region.

2.4.3.   Vibration thermalization in the ground and the excited states
The nonoscillating spectra in Figures 16 and 17 also reveal that a positive ΔA peak due to induced absorption grows up around 1.85–1.90 eV with increasing delay time. Figure 20 shows the time trace of the energies of $\Delta A = 0$ and $d(\Delta A)/dt = 0$

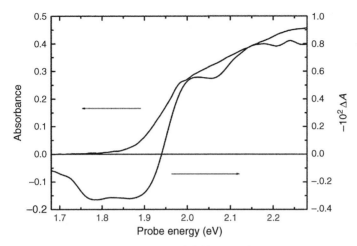

Figure 19   Stationary absorption spectrum and difference absorption spectrum with a quasi-thermalized feature at 1800 fs.

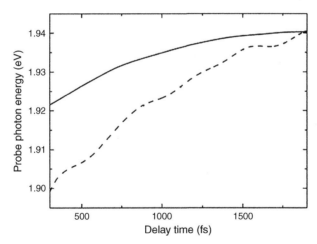

Figure 20   Energy shifts of $\Delta A = 0$ points (solid line) and intersection points between the two spectra separated by 50 fs in delay time (dashed line).

points in the ΔA spectra. The blue shift of the zero-ΔA energy, which have a single-exponential decay constant of about 0.7 ps, can be ascribed to the spectral changes of both the induced absorption ($2^1A_g \rightarrow n^1B_u$) and the bleaching ($1^1A_g \rightarrow 2^1B_u$). The probe energy dependence of the annihilation time of 2^1A_g exciton determined by biexponential fitting also exhibits a higher value of about 1.1 ps at 1.85 eV due to the induced absorption growth around 1.85–1.90 eV than 0.5 ps at 2.0 eV because of the bleaching decay at zero-phonon transition energy (2.0 eV).

Thermalization processes from the nonequilibrium states both in the ground 1^1A_g and in the excited 2^1A_g states were revealed in the long-term energy shifts of peaks and slopes in DAS. Seventeen transition energies with characteristic features in the spectra were investigated, which include 3 peaks, 4 valleys, 9 slopes, and 1 zero ΔA point as shown in Figure 21. Each peak showed a blue or red shift due to thermalization in 1^1A_g and 2^1A_g states depending on probe photon energy in

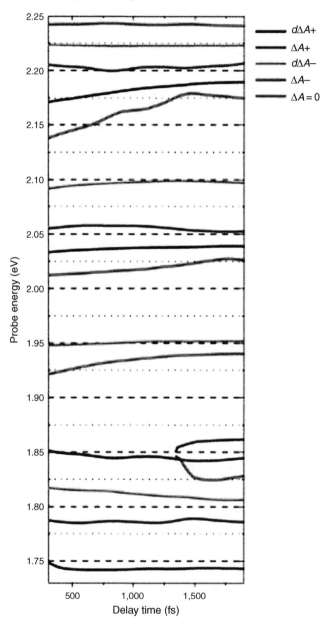

Figure 21 Energy shifts of the peaks and the slopes extracted from the difference absorption spectrum in Figure 16. Energies of slopes were determined by the maximum or minimum energies of the photon-energy derivative of the spectrum at each delay time. $+$ and $-$ denote the positive and negative signs of the peaks and the gradient peaks, respectively.

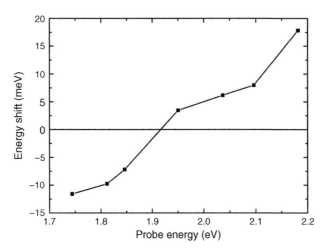

Figure 22　Energy-shift quantities of gradient peaks (slopes) during 1300 fs from 200 to 1500 fs.

Figure 21. The thermalization time in which population distribution among vibrational levels with distributed quantum numbers in various modes comes to thermal distribution within the main chains is estimated to be about 0.5 ps from the time constant of the spectral shift. The value of the time constant determined here is consistent with our previous papers [45,49,56,57,59].

Quantities of energy shifts of slopes in Figure 11 during 1.3 ps from 0.2 to 1.5 ps are shown in Figure 12, which shows that there are two regions with positive and negative energy shifts. The bandwidths of the vibrational peaks in the DAS are broadened with increasing delay time. The results may indicate the spectral diffusion in the 2^1A_g excited state. The zero crossing point in Figure 22 is around 1.9–eV, which is lower than $0 \rightarrow 0$ transition energy by about 100 meV. This energy difference is consistent with the amount of the Stokes shift in literature [82].

2.4.4. Singular value decomposition for the analysis of mode dependence of vibronic coupling in the excited state and the ground state

The two–dimensional normalized difference transmittance $\Delta T(t)/T$ of the blue–phase PDA-3BCMU sample is displayed in Figure 12 (probe-photon energy and probe delay time). Several examples of $\Delta T(t)/T$ traces at several probe-photon energies were shown in Figure 13. All of the traces have signals of finite size at negative delay times and sharp and several extremely intense peaks around zero probe delay time. The former is due to the perturbed FID process associated with the third-order nonlinearity of the sequential interaction of probe–pump–pump fields modified with coherent molecular vibrations. The latter signals are clearly appearing to have several peaks in the signal probed at all the three probe-photon energies of 2.05, 1.95, and 1.78 eV. Among the three, the negative-time signal sizes are small at 2.05 and 1.78 eV, but they have oscillation features while the one at 195 eV has much larger size but lacks clear vibration features. This is due to the fact that at this probe-photon energy, relative size of the vibration feature at positive delay time is weaker than the other two photon energies. All the signals at other wavelengths with detectable signal sizes have similar features. They are due to pump–probe coupling induced by the nonlinear process of the pump–probe–pump time ordering of the interacting fields. There is another source of artificial signal because of the interference between the scattered pulses and the probe pulses both of which durations are elongated by dispersion in the polychromator. The details of the vibrational modulation observed at longer delay times than 100 fs are free from the signal distortions due to the above two mechanisms, which are irrelevant to the mechanism of the vibronic coupling. Figure 23 shows the measured normalized difference transmittance spectra at a few probe delay times between 200 and 1100 fs. In the region above the probe-photon energy of about 1.95 eV, the normalized difference transmittance is positive assignable mainly to photobleaching of the 1^1B_u–FE absorption peaked around 2.0 eV. This feature in the lowest energy region can be due also to IE in the shorter time range than 100 fs as discussed earlier in this chapter. The normalized difference transmittance signals in the probe-photon energy region below 1.90 eV are negative due dominantly to the photo-induced absorption transition from the geometrically relaxed 2^1A_g state to the higher excited n^1B_u state. The latter state is one of the four essential states invoked to explain the characteristic features in the nonlinear spectra [56,78]. However, accompanying bleaching can modify the intensity and spectral shape. As described earlier in this chapter, the 1^1B_u–FE state decays within 100 fs into the geometrically relaxed 2^1A_g state, which is also consistent with several previous papers [49,82,96].

The power spectrum of FT of $\Delta T(t)/T$ at 128 probe-photon energies probed by using the lock-in amplifier is shown two dimensionally in Figure 24. The abscissa

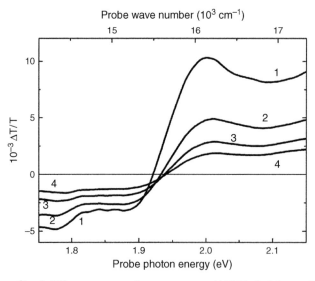

Figure 23 Normalized difference transmittance spectra ($\Delta T/T$); 1, delay at 200 fs; 2, 500 fs; 3, 800 fs; 4, 1100 fs.

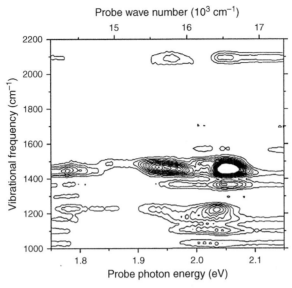

Probe wave number (10^3 cm^{-1})

Figure 24 Two-dimensional Fourier power spectra of the $\Delta T/T$ traces.

and ordinate of the two-dimensional description are vibrational frequency and probe-photon energy, respectively. For the calculation, the signal data in the probe delay time from 400 to 900 fs were used to avoid the interference effects taking place near the zero-delay time and weak signal at longer delay. FT was performed after high-pass step-function filtering with a cut-off frequency at 1000 cm^{-1} to avoid if any slow decay due to electronic relaxation. In the wide range of probe-photon energy, three peaks were commonly observed at 1220, 1460, and 2080 cm^{-1} corresponding to the C—C, C=C, and C≡C stretching modes, respectively [97,98]. The probe-photon energy dependencies of the Fourier power of these modes are shown in Figure 25. It is clearly seen that all of the three modes have peaks around 1.78, 1.92–1.96, and 2.03–2.05 eV in common even though the intensity distribution profile is substantially different among them.

The vibrational modes in the pump–probe signal were more deeply investigated by analysis with a linear prediction singular value decomposition (LP-SVD) method. By the LP-SVD, single or multiple mode(s) of damped oscillation were extracted from the data such as $y(t) = A\exp(-t/\tau)\cos(\Omega t + \theta)$, where A is the initial amplitude of the signal modulation of $\Delta T(t)/T$ due to the molecular vibration, τ is the decay time of the amplitude of the modulation signal, Ω is the relevant mode frequency, and θ is the initial phase of the molecular vibration just after excitation. These parameters for the vibrational modes of C—C, C=C, and C≡C were separately extracted by the LP-SVD method from the 200–900 fs data after rectangular frequency filtering of the Fourier power spectra in 1150–1290, 1395–1535, and 2010–2150 cm^{-1} ranges at observed probe-photon energies independently. The amplitudes at 300 fs of the modes are shown in Figure 26. The spectrum of amplitude of C—C stretching mode in Figure 26a does not resemble the

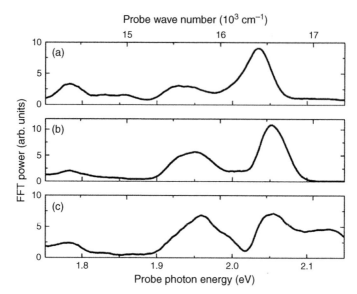

Figure 25 Fast Fourier power spectra of the modes of (a) C—C, (b) C=C, and (c) C≡C stretching extracted from the $\Delta T/T$ traces.

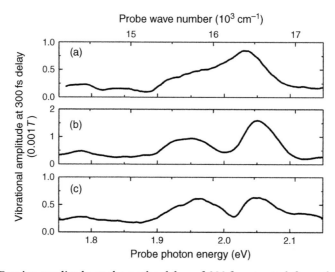

Figure 26 Fourier amplitude at the probe delay of 300 fs extracted from the normalized difference transmittance by the method of linear prediction singular value decomposition (LP-SVD) for the modes of (a) C—C, (b) C=C, and (c), C≡C stretching.

FFT-power spectrum of C—C mode in Figure 25a. The vibration signal of C—C stretching mode was not intense enough for the mode signal to be extracted out from the real-time data to analyze the decay time and initial phase precisely. The Fourier-power spectra show other modes than carbon–carbon stretching with weaker intensities, and they are presumably out-of-plane bending modes [65], but they are also difficult to be analyzed for precise detailed discussion because of insufficient intensity. Therefore, from here on, we concentrate on only C=C and C≡C stretching modes extracted by the LP-SVD method from the pump–probe signals.

The fitted initial phase of the C=C and C≡C modes are shown as phase spectra in Figure 27. This kind of phase data cannot be obtained from other time-resolved spectroscopy such as infrared absorption and Raman scattering, which have been developed in the last two decades and were recognized to be very useful for the identification of the molecular structure in the excited states after photoexcitation of molecules and intermediate species in the photochemical processes. However, the data of the initial vibration phases of multimode molecular vibration are expected to provide invaluable information about the mechanisms of the chemical reactions and excited state dynamics. This method of extraction of phase information from the real-time data will be an essential and powerful spectroscopic method to describe the full chemical reaction pathways in this 21st century [39].

To fully utilize the information of the phases, the following simple model shown in Figure 28 was proposed to understand systematically the features of the phase spectra in the figure. In Figure 28a and c, transition from the ground state to 1^1B_u–FE state and that from the geometrically relaxed 2^1A_g state to n^1B_u state, respectively, are indicated on the potential curves of PDA. The FWHM of pulse of the NOPA output used as a pump is much shorter than the oscillation periods of

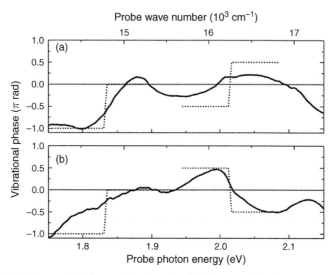

Figure 27 Initial phase (solid curve) of each vibration mode of normalized difference transmittance by the method of linear prediction singular value decomposition (LP-SVD) for the modes of (a) C=C, and (b) C≡C stretching. The expected phase (dotted line) of vibration from the model of the wavepacket motion in the ground state (1.95–2.07 eV) and in the excited state (1.76–1.88 eV).

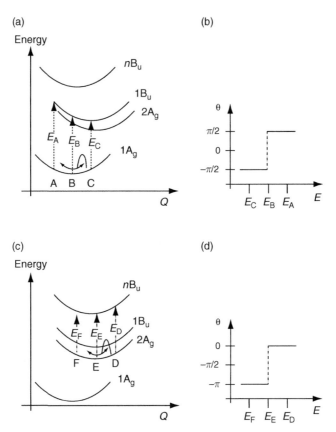

Figure 28   (a) Transition from the ground states on potential energy curves of polydiacetylene (PDA). Point B is located at the bottom of the curve of the ground states. (b) The model phase of vibration by the motion of the wavepacket in the ground state. E_X is the energy of transition from position X. (c) Transition from the excited states on potential energy diagram of PDA. Point E is located at the bottom of the curve of the ground states. (d) The model phase of vibration by the motion of the wavepacket in the excited state.

C=C (23 fs) and C≡C (16 fs) vibrations. Therefore, it can generate vibrational wavepackets impulsively in the ground 1^1A_g state and 1^1B_u–FE state. The short pulse width of the NOPA system developed in our laboratory is thus essentially important to study the dynamical behavior of systems with such high frequencies as C≡C stretching. There are a few groups in the world where utilizing NOPA output for real-time spectroscopy with substantially broader pulse durations and the data taken by using such systems are always suffering from the Fourier amplitude reduction or even disappearance of the high-frequency modes with respect to lower frequency modes. This makes discussion of the vibronic coupling difficult or even impossible.

A wavepacket just after being generated in the ground state at the bottom indicated by point B in Figure 28a starts to oscillate on the potential curve along

B → A → B → C → B → ... (or B → C → B → A → B → ...). In accordance with the oscillatory motion, absorption intensity increases when the wavepacket is located at the probe-photon frequency that corresponds to the vertical transition energy at the position of the wavepacket following the Franck–Condon principle based on the adiabatic decoupling between the electronic and the nuclear motions. Therefore, the oscillation of wavepacket results in the modulation of normalized difference transmittance with the vibration frequency.

If the wavepacket photoproduced at B starts to oscillate following the sequence B → A → B → ..., the transmittance at probe-photon energy E_A starts to decrease initially, and at the same time, the transmittance at E_C starts to increase. Here, E_X is the transition energy from position X (=A, B, C, D, E, or F) on the lower state to the corresponding Franck–Condon point of the potential curve of the upper state. In this case, the phase of the molecular vibration of $\Delta T(t)/T$ is $\pi/2$ at probe-photon energies higher than E_B, and it is $-\pi/2$ at that lower than E_B as shown in Figure 28b. While in case when the wavepacket starts to move to the reverse direction (B → C → B → ...) to the previous case, the phase is $-\pi/2$ for the probe-photon energies higher than E_B and $\pi/2$ at that lower than E_B.

Figure 27a shows that the phases of the C=C stretching mode determined in the probe spectral region between about 1.9 and 2.0 eV are negative while those between about 2.0 and 2.1 eV are positive. These features can be explained using Figure 28a and b, where E_B is 2.0 eV, which corresponds to the 1^1B_u-FE absorption peak. The phase of the C=C stretching mode at the probe-photon energy of 2.02 eV is clearly positive and is about 0.3π or at least far from the zero-crossing point of 1.975 eV. While, on the contrary, that of the C≡C stretching mode at the same probe-photon energy in Figure 27b is close to zero or at least close to the zero-crossing point of 2.015 eV. This energy (2.02 eV) corresponds to the photon energy shifted by the C=C stretching energy (0.18 eV) from 2.2 eV of the peak of the probe which is expected to give a $\pi/2$ phase shift. Therefore, it can be concluded that the phases around 2.02 eV in C=C are shifted to positive due to the contribution of wavepacket generated by the stimulated Raman gain process. The phases of C≡C stretching are positive from about 1.93 to 2.015 eV and are negative between 2.015 and 2.15 eV. It is thus concluded that the wavepacket of C≡C mode generated on the ground-state potential curve initializes its motion to the reverse direction to that of the C=C mode. This means that the changes of C=C and C≡C bond lengths upon photoexcitation to the Franck–Condon state is opposite to each other. The C≡C bond length is expected to be elongated by the change to the GR resulting in the butatriene-like configuration. The length of C=C bond is considered to be shortened from the above discussion. This is probably due to the redistribution of the π electrons of the triple bonds that increases not only C—C single bonds but also C=C double bonds neighboring to the single bonds.

The phases of C=C mode probed at about 1.95 and 2.05 eV are neither $\pi/2$ nor $-\pi/2$, which can be explained in the following way. At these two spectral positions, the signals of photoinduced absorption are expected to be relatively strongly modulated by the wavepacket of C=C mode in the excited state because of the strong excitonic peak at 2.00 eV. Then the phases of C=C mode are considered to be 0π as expected for a wavepacket in the excited state. Therefore, it is concluded that the vibration signal due to the ground state and that due to the excited state are coexisting. Therefore, observed values of the phases are expected to correspond to the weighted averages of the two contributions, as these two contributions are incoherent with each other.

From the result of the phases at 1.95 and 2.05 eV having opposite signs to each other, it can be concluded that the vibrational amplitude peaks at 1.95 and 2.05 eV are due to the modulation of the intensity of 1^1B_u-FE absorption (resonant at 2.00 eV) by the motion of the vibrational wavepacket in the ground state produced by the stimulated Raman scattering (SRS) process.

Figure 28c represents the diagram of photoinduced absorption from the "geometrically relaxed 2^1A_g state." Here, it is not known that the state 2^1A_g forms or does not form an exciton and hence the word "self-trapped exciton" was carefully avoided to be used. The pump pulse generates wavepacket not only in the ground state but also in the 1^1B_u-FE state. The photoexcited wavepacket in the 1^1B_u-FE state soon (<100 fs) after photogeneration relaxes into the geometrically relaxed 2^1A_g state. In the author's group, this relaxation time was determined to be 60 ± 20 fs [99]. This 2^1A_g state has a different geometrical configuration due to its property of the so-called two-triplet state. The oscillation of the wavepacket on the geometrically relaxed 2^1A_g state potential curve modulates the probe signal. A wavepacket is produced at point F in Figure 28c at the beginning, and then makes a start to move as $F \rightarrow E \rightarrow D \rightarrow E \rightarrow F \ldots$, where point E is located at the bottom of the potential curve along the corresponding stretching-mode coordinate associated with the geometrically relaxed 2^1A_g state.

The expected phases of the oscillation of the photoinduced absorption signal in the case of $E_D > E_F$ are shown in Figure 18b, that is, they are zero, $-\pi/2$, and $-\pi$ at the probe-photon energy of E_D, E_E, and E_F, respectively, in the same way as in the case of the ground state.

The absorption spectrum associated with the transition of from the geometrically relaxed 2^1A_g state to n^1B_u states has a peak around 1.8 eV [64]. At first, we consider the case when E_E is 1.83 eV. Then the phases of the C=C mode in Figure 27a correspond to those in the model shown in Figure 28d. The phases of C≡C-mode molecular vibration probed at 1.77 and 1.87 eV are $-\pi$ and 0π, respectively. These are the same as those of C=C-mode vibration. Therefore, it is concluded that not only the vibration wavepacket of C=C mode but also that of C≡C mode is generated in the geometrically relaxed 2^1A_g state. This clearly demonstrates that the full geometrically relaxed butatriene-type structure is not formed, but it can still be described in the form of acetylene-type structure with relatively small amount of configuration mixing of the butatriene-type structure in terms of quantum chemical resonance structure, as discussed previously by one of the present authors [56].

Figure 29 shows the C=C stretching-mode frequencies extracted by LP-SVD from the real-time traces at many probe-photon between 1.78 and 2.08 eV. From the previous discussion, the signal due to the wavepacket motion in the ground state appears strongly at the probe-photon energies of 2.04–2.07 eV and 1.92–1.95 eV, which correspond to E_A and E_C regions, respectively, in Figure 28. The modulated signal due to the motion of the wavepacket in the geometrically relaxed 2^1A_g state appears strongly at the probe-photon energies of 1.87–1.90 eV and 1.79–1.81 eV, indicated by E_D and E_F regions in the figure. The wavepacket motion in the

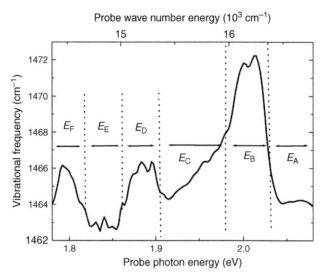

Figure 29 The probe-photon energy dependence of vibration frequency of C=C stretching mode. E_A and E_C are energy regions in which the pump–probe signal is modulated by the motion of the wavepacket mainly in the ground state. E_D and E_F are energy regions in which the pump–probe signal is modulated mainly by the motion of the wavepacket in the excited state. In E_B and E_E, the signal is modulated by the motion of the wavepacket only in the excited and ground state, respectively.

geometrically relaxed 2^1A_g state and that of the ground state appear nearly exclusively at 2.00 and 1.84 eV, respectively, corresponding to E_B and E_E, respectively. Figure 19 shows that the vibration frequencies determined for the probe-photon energy regions of E_B, E_D, and E_F are higher than those of E_A, E_C, and E_E regions. Hence, it can be concluded that the frequency of C=C stretching in the geometrically relaxed 2^1A_g state is close to that determined for E_B probe-photon energy. Therefore, the frequency of the C=C stretching in the excited state is concluded to be $1472 \pm 6 \, \text{cm}^{-1}$ and that in the ground state which is $1463 \pm 6 \, \text{cm}^{-1}$ concluded for the data for E_E probe-photon energy region. The frequency of C≡C stretching mode in the geometrically relaxed 2^1A_g state is determined to be $2092 \pm 6 \, \text{cm}^{-1}$, whereas that in the ground state is $2083 \pm 6 \, \text{cm}^{-1}$. The errors of the differences are smaller than those of their absolute values, because the errors in the frequency determination are mainly due to the imperfect step length of the delay stage. It can then be concluded that each of the frequencies of the C=C and C≡C stretching modes in the excited state is higher by about $10 \pm 2 \, \text{cm}^{-1}$ than each of those in the ground state.

REFERENCES

[1] A. Baltuska, Z. Wei, M.S. Pshenichnikov, D.A. Wiersma, R. Szipöcs, Appl. Phys. B, 65: 175–188, 1997.
[2] M. Nisoli, S. Stagira, S. De Silvestri, O. Svelto, S. Sartania, Z. Cheng, M. Lenzner, Ch. Spielmann, F. Krausz: Appl. Phys. B, 65:189–196, 1997.

[3] U. Morgner, F.X. Kartner, S.H. Cho, Y. Chen, H.A. Haus, J.G. Fujimoto, E.P. Ippen, V. Scheuer, G. Angelow, T. Tschudi: Opt. Lett., 24:411–413, 1999.

[4] D.H. Sutter, G. Steinmeyer, L. Gallmann, N. Matuschek, F. Morier-Genoud, U. Keller, V. Scheuer, G. Angelow, T. Tschudi: Opt. Lett., 24:631–633, 1999.

[5] G.M. Gale, M. Cavallari, T.J. Driscoll, F. Hache: Opt. Lett., 20:1562–1564, 1995.

[6] A. Shirakawa, and T. Kobayashi: Appl. Phys. Lett. 72:147–149, 1998; IEICE Trans. Electron., E81-C:246–253, 1998.

[7] T. Wilhelm, J. Piel, E. Riedle: Opt. Lett., 22:1494–1496, 1997.

[8] G. Cerullo, M. Nisoli, S. De Silvestri: Appl. Phys. Lett., 71:3616–3618, 1997; Opt. Lett., 23:1283–1285, 1998.

[9] A. Shirakawa, I. Sakane, T. Kobayashi: Opt. Lett., 23:1292–1294, 1998.

[10] A. Shirakawa, I. Sakane, T. Kobayashi: in *Ultrafast Phenomena XI*, T. Elsaesser et al. ed. (Springer-Verlag, Berlin, 1998), p. 54; A. Shirakawa, I. Sakane, M. Takasaka, T. Kobayashi: Appl. Phys. Lett., 74:2268–2270, .

[11] R. Danielius, A. Piskarskas, A. Stabinis, G.P. Banfi, P. Di Trapani, R. Righini: J. Opt. Soc. Am. B, 10:2222–2232, 1993.

[12] R.L. Byer, S.E. Harris, Phys. Rev., 168:1064–1068, 1968.

[13] S. Takeuchi, T. Kobayashi, J. Appl. Phys., 75:2757–2760, 1994.

[14] O.E. Martinez: IEEE J. Quantum Electron., QE-25:2464–2468, 1989.

[15] V. Krylov, A. Kalintsev, A. Rabane, D. Erni, U.P. Wild: Opt. Lett., 20:151–153, 1995.

[16] Zs. Bor, B. Rácz: Opt. Commun., 54:165–170, 1985.

[17] O.E. Martinez: Opt. Commun., 59:229–232, 1986.

[18] P. Di Trapani, A. Andreoni, C. Solcia, P. Foggi, R. Danielius, A. Dubietis, A. Piskarskas: J. Opt. Soc. Am. B, 12:2237–2244, 1995.

[19] R. Danielius, A. Piskarskas, P. Di Trapani, A. Andreoni, C. Solcia, P. Foggi, Opt. Lett., 21:973–975, 1996.

[20] A. Dubietis, G. Jonusauskas, and A. Piskarskas, Opt. Commun., 88:437–440, 1992.

[21] G.M. Gale, F. Hache, M. Cavallari, IEEE J. Sel. Top. Quantum Electron., 4:224–229, 1998.

[22] A. Baltuska, M.S. Pshenichnikov, D.A. Wiersma, Opt. Lett., 23:1474–1476, 1998.

[23] V.G. Dmitriev, G.G. Gurzadyan, and D.N. Nikogosyan, *Handbook of Nonlinear Optical Crystals*, 3rd edition (Springer-Verlag, Berlin, 1999), p. 96.

[24] The effective phase-matching width can be additionally broadened in the regime of strong pump depletion.

[25] B.A. Richman, S.E. Bisson, R. Trebino, M.G. Mitchell, E. Sidick, and A. Jacobson, Opt. Lett., 22:1223–1225, 1997.

[26] B.A. Richman, S.E. Bisson, R. Trebino, E. Sidick, and A. Jacobson, Opt. Lett., 23:497–499, 1998.

[27] T.S. Sosnowski, P.B. Stephens, and T.B. Norris, Opt. Lett., 21:140–142, 1996.

[28] G. Taft, A. Rundquist, M. Murnane, I. Christov, H. Kapteyn, K. DeLong, D. Fittinghoff, M. Krumbügel, J. Sweetser, and R. Trebino, IEEE J. Select. Topics Quantum Electron., 2:575–585, 1996.

[29] P. O'Shea, M. Kimmel, X. Gu, and R. Trebino. Opt. Express., 7:342–349, 2000.

[30] E. Zeek, K. Maginnis, S. Backus, U. Russek, M. Murnane, G.R. Mourou, H. Kapteyn, and G. Vdovin, Opt. Lett., 24:493–495, 1999.

[31] M. Armstrong, P. Plachta, E. Ponomarev, and R.J.D. Miller, Opt. Lett., 26:1152–1154, 2001.

[32] T. Brixner, M. Strehle, and G. Gerber, Appl. Phys. B, 68:281–284, 1999.

[33] W.H. Press, S.A. Teukolsky, W.T. Vetterling, and B.P. Flannery, *Numerical recipes in C*, 2nd edition (Cambridge University Press, New York, 1996), p. 402.

[34] A. Baltuška, M.S. Pshenichnikov, and D.A. Wiersma, IEEE J. Quantum Electron., 35:459–478, 1999.

[35] T.S. Rose, M.J. Rosker, A.H. Zewail, J. Chem. Phys., 91:7415–7436, 1989.

[36] M.H. Vos, F. Rappaport, J.-C. Lambry, J. Breton, J.-L. Martin, Nature, 363:320–325, 1993.

[37] Q. Wang, R.W. Schoenlein, L.A. Peteanu, R.A. Mathies, C.V. Shank, Science, 266:422–424, 1994.

[38] S. Pedersen, J.L. Herek, and A.H. Zewail, Science, 266:1359–1364, 1994.

[39] T. Kobayashi, T. Saito, H. Ohtani, Nature, 414:531–534, 2001.

[40] S. Pedersen, L. Banares, A.H. Zewail, J. Chem. Phys., 97:8801–8804, 1992.

[41] H. Torii, M. Tasumi, Chem. Phys. Lett., 260:195–222, 1996.

[42] Y.J. Yan, S. Mukamel, Phys. Rev. A, 41:6485–6504, 1990.

[43] A.T.N. Kumar, F. Rosca, A. Widom, P.M. Champion, J. Chem. Phys., 114:701–724, 2001.

[44] A. Shirakawa, I. Sakane, M. Takasaka, T. Kobayashi, Appl. Phys. Lett., 74:2268–2270, 1999.

[45] M. Ikuta, Y. Yuasa, T. Kimura, H. Matsuda, and T. Kobayashi, Phys. Rev. B, 70:214301–214306, 2004.

[46] S.A. Kovalenko, J. Ruthmann, N.P. Ernsting, J. Chem. Phys., 109:1894–1900, 1998.

[47] J. Friedrich, J.D. Swalen, D. Haarer, J. Chem. Phys., 73:705–711, 1980.

[48] T. Kobayashi , IEICE Trans. Fundam., E-75A:38–45, 1992.

[49] M. Yoshizawa, Y. Hattori, T. Kobayashi, Phys. Rev. B, 47:3882–3889, 1993.

[50] R.W. Carpick, D.Y. Sasaki, M.S. Marcus, M.A. Eriksson, and A.R. Burns, J. Phys.: Condens. Matter, 16:R679, 2004.

[51] G. Wegner , Macromol. Chem., 145, 85, 1971.

[52] G.N. Patel, R.R. Chance, and J.D. Witt, J. Chem. Phys., 70:4387–4392, 1979.

[53] D. Bloor, in *Polydiacetylenes* D. Bloor and R.R. Chance ed. (Martinus Nijhoff Publishers, Dordrecht, Netherlands, 1985).

[54] K. Tashiro, K. Ono, Y. Minagawa, M. Kobayashi, T. Kawai, and K. Yoshino, J. Polym. Sci., Part B: Polym. Phys., 29:1223–1233, 1991.

[55] M. Yoshizawa, T. Kobayashi, H. Fujimoto, and J. Tanaka, J. Phys. Soc. Jpn., 56:768–780, 1982.

[56] T. Kobayashi, M. Yoshizawa, U. Stamm, M. Taiji, and M. Hasegawa, J. Opt. Soc. Am. B, 7:1558–1578, 1990.

[57] M. Yoshizawa, K. Nishiyama, and T. Kobayashi, Chem. Phys. Lett., 207:461–467, 1993.

[58] J.Y. Bigot, T.A. Pham, and T. Barisien, Chem. Phys. Lett., 259:469–474, 1996.

[59] T.A. Pham, A. Daunois, J.C. Merle, J. Le Moigne, and J.Y. Bigot, Phys. Rev. Lett., 74:904–907, 1995.

[60] G. Cerullo, G. Lanzani, M. Muccini, C. Taliani, and S. De Silvestri, Phys. Rev. Lett., 83:231–234, 1999.

[61] A. Vierheilig, T. Chen, P. Walther, W. Kiefer, A. Materny, and A.H. Zeweil, Chem. Phys. Lett., 312:349–356, 1999.

[62] M. Yoshizawa, M. Taiji, and T. Kobayashi, IEEE J. Quantum Electron., 25:2532–2539, 1989.

[63] M. Yoshizawa, A. Yasuda, and T. Kobayashi, Appl. Phys. B, 53:296–307, 1991.

[64] M. Yoshizawa, Y. Hattori, and T. Kobayashi, Phys. Rev. B, 49:13259–13262, 1994.

[65] T. Kobayashi, A. Shirakawa, H. Matsuzawa, and H. Nakanishi, Chem. Phys. Lett., 321:385–393, 2000.

[66] G.J. Blanchard, J.P. Heritage, A.C.V. Lehmen, M.K. Kelly, G.L. Baker, and S. Etemad, Phys. Rev. Lett., 63:887–890, 1989.

[67] W.B. Bosma, S. Mukamel, B.I. Greene, and S. Schmitt- Rink, Phys. Rev. Lett., 68:2456–2459, 1992.

[68] A.J. Heeger, S. Kivelson, J.R. Schrieffer, and W.P. Su, Rev. Mod. Phys., 60:781–850, 1988.

[69] W.P. Su, J.R. Schrieffer, and A.J. Heeger, Phys. Rev. Lett., 42:1698–1701, 1979.

[70] W.P. Su, J.R. Schrieffer, and A.J. Heeger, Phys. Rev. B., 22:2099–2111, 1980.

[71] Z. Vardeny, Physica, 127B:338, 1984.

[72] L. Rothberg, T.M. Jedju, S. Etemad, and G.L. Baker, Phys. Rev. Lett., 57:3229–3232, 1985.

[73] B.I. Greene, J.F. Mueller, J. Orenstein, D.H. Rapkine, S. Schmitt-Rink, and M. Thakur, Phys. Rev. Lett., 61:325–328, 1988.

[74] T. Kobayashi, M. Yasuda, S. Okada, H. Matsuda, and H. Nakanishi, Chem. Phys. Lett., 267:472–480, 1997.

[75] J. Kinugusa, S. Shimada, H. Matsuda, H. Nakanishi, and T. Kobayashi, Chem. Phys. Lett., 287: 639644, 1998.

[76] F. Zerbetto , J. Phys. Chem., 98:13157–13161, 1994.

[77] J. Swiatkiewicz, X. Mi, P. Chopra, and P.N. Prasad, J. Chem. Phys., 87:1882–1886, 1987.

[78] L.X. Zheng, R.E. Benner, Z.V. Vardeny, and G.L. Baker, Synth. Metals, 49/50:313–320, 1992.

[79] A. Race, W. Barford, and R.J. Bursill, Phys. Rev., B 67:245202–245210, 2003.

[80] M.Y. Lavrentiev and W. Barford, Phys. Rev. B, 59:15048–15055, 1999.

[81] B. Lawrence, W.E. Torruellas, M. Cha, M.L. Sundheimer, G.I. Stegeman, J. Meth, S. Etemad, and G. Baker, Phys. Rev. Lett., 73:597–600, 1994.

[82] M. Yoshizawa, A. Kubo, and S. Saikan, Phys. Rev. B, 60:15632–15635, 1999.

[83] W.P. Su, J.R. Schrieffer, and A.J. Heeger, Phys. Rev. Lett., 68:1148–1151, 1982.

[84] S. Abe, J. Yu, and W.P. Su, Phys. Rev. B, 45:8264–8271, 1992.

[85] S. Abe, M. Schreiber, W.P. Su, and J. Yu, Phys. Rev. B, 45:9432–9435, 1992.

[86] K. Pakbaz, C.H. Lee, A.J. Heeger, T.W. Hagler, and D. McBranch, Synth. Met., 64:295–306, 1994.

[87] A. Shirakawa, and T. Kobayashi, Appl. Phys. Lett., 72:147–149, 1998.

[88] G. Cerullo, M. Nisoli, S. Stagira, S. De Silvestri, G. Tempea, F. Krausz, and K. Ferencz, Appl. Phys. B, 70:S253, 2000.

[89] A. Baltuška, and T. Kobayashi, Appl. Phys. B, 75:427–443, 2002.

[90] G. Wenz, and G. Wegner, Makromol. Chem. Rapid Commun., 3:231, 1982.

[91] T. Kobayashi, and A. Shirakawa, Appl. Phys. B, 70:S239–246, 2000.

[92] S. Mukamel, A. Takahashi, H.X. Wang, and G. Chen, Science, 266:250–254, 1994.

[93] S. Mazumder, D. Guo, and S.N. Dixit, Synth. Met., 57:3881–3888, 1993.

[94] H. Kano, T. Saito, and T. Kobayashi, J. Phys. Chem. B, 105:413–419, 2001.

[95] K. Huang, and A. Rhys, Proc. R. Soc. London A, 204:406, 1950.

[96] M. Turki, T. Barisien, J.Y. Bigot, and C. Daniel, J. Chem. Phys., 112:10526–10537, 2000.

[97] Z. Iqbal, R.R. Chance, and R.H. Baughman, J. Chem. Phys., 66:5520–5525, 1977.

[98] D. Grando, S. Sottini, and G. Gabrielli, Thin Solid Films, 327–329:336–340, 1998.

[99] Y. Yuasa, M. Ikuta, and T. Kobayashi, Phys. Rev. B, 72:134302–134309, 2005.

CHAPTER 7

APPLICATIONS OF COHERENT RAMAN SPECTROSCOPY

BACTERIAL SPORE DETECTION VIA OPTIMIZED PULSE CONFIGURATION

Dmitry Pestov, Alexei V. Sokolov, *and* Marlan O. Scully

Contents

Abstract

We explore the possibilities of using coherent Raman spectroscopy for real-time detection of biohazards. By exciting vibrational coherence on more than one Raman transition simultaneously, we are aiming for a robust and definitive means to obtain a molecule-specific signal and use it for species identification. In particular, we concentrate on detection of dipicolinic acid (DPA), known to be a marker molecule for bacterial spores. We consider time- and frequency-resolved techniques for coherent Raman spectroscopy and adapt those for this particular application.

Keywords: coherent anti-Stokes Raman scattering; FAST CARS; hybrid CARS; detection; spore; endospore; bacterial; four-wave mixing; dipicolinic acid; calcium dipicolinate; sodium dipicolinate; biohazard; CSRS

1. INTRODUCTION

The anxiety caused by the distribution of anthrax spores through the United States Postal Service in September and October 2001 and by subsequent false-alarm encounters with suspicious powder-like substances was further aggravated by the long time required for their analysis. The need for real-time detection methods, local and remote, became apparent. It triggered a large-scale search for suitable techniques in different fields of expertise, including biology, chemistry, nano-engineering, and laser spectroscopy.

The most prominent biological methods are polymerase chain reaction (PCR) and immunoassays. PCR employs primers to separate organic-specific nucleic acid sequences and polymerases to amplify the segment until it is detectable. Its speed depends on the marker used, but generally the complete analysis can be performed in a few hours. For example, the reported detection limit for PCR based on bacterial pagA gene is $\sim 10^3$ spores in 2–3 h [1]. Immunoassays, which rely on competitive binding of bioagents with their antibodies, can detect 10^5 spores in 15 min, but their false-positive rate is unacceptably high [2].

Some molecules, being just building blocks of more advanced structures, can still serve as unique markers for their hosts, reducing the detection problem to the recognition of those molecular signatures. To this end, calcium dipicolinate (CaDPA) is known to be a marker molecule of bacterial endospores, accounting for up to 15% of their dry weight [3,4]. Relatively fast methods have been developed to chemically extract CaDPA and then detect it by means of fluorescence or photoluminescence. In the latter case, the formation of the terbium (Tb^{3+}) dipicolinate complex leads to the enhanced luminescence compared with terbium alone and has been found to detect 10^3 *Bacillus subtilis* spores (a harmless analog for *Bacillus anthracis*) per milliliter in 5–7 min [5]. Photoluminescence and fluorescence methods, however, lack specificity [6].

Raman spectra, on the contrary, are excellent molecular fingerprints. The spontaneous inelastic scattering of photons on molecular vibrations with a gain or loss of vibrational quanta (the process referred to as Raman effect) provides a characteristic optical signature of molecular composition and structure. In particular, Raman spectra of bacterial spores are dominated by vibrations of aforementioned CaDPA [7–9], resulting in the reported detection limit of as few as 10^6 spores in 1 s (13 min for good quality spectra, see Ref. [10]). Unfortunately, the efficacy of conventional spontaneous Raman spectroscopy is hampered by low efficiency of the Raman scattering process.

The scattering efficiency can be enhanced by the use of molecular coherence. In coherent anti-Stokes Raman scattering (CARS) spectroscopy [11,12], the molecules are put into coherent oscillations by a pair of preparation pulses, pump (with the carrier frequency ω_1) and Stokes (ω_2). These macroscopic oscillations of the molecular polarization lead to the efficient scattering of the probe photons (ω_3). The generated beams of blue- (anti-Stokes) and red-shifted (Stokes) light, whose propagation directions are determined by the phase-matching conditions [13] for the corresponding wavevectors

$$\vec{k}_{\text{CARS}} = \vec{k}_3 + (\vec{k}_1 - \vec{k}_2), \quad \vec{k}_{\text{CSRS}} = \vec{k}_3 - (\vec{k}_1 - \vec{k}_2), \tag{1}$$

are referred to as CARS and coherent anti-Stokes Raman scattering (CSRS), respectively.

Phase-assisted buildup of the generated signal over the excitation volume is the key property that differentiates the coherent process from spontaneous Raman scattering. It results in the quadratic dependence of the scattering efficiency on the number of participating molecules. Between the two frequency-shifted components of the scattered radiation, the preference is usually given to CARS because it is offset from the fluorescence that might be produced by the probe pulse ($\omega_{CARS} > \omega_3 > \omega_{CSRS}$).

The immediate benefits of the indicated (coherent) approach are blemished by the well-known issue of dealing with other four-wave mixing (FWM) processes, triggered by the overlapped laser pulses. Scattering of the probe photons off the induced optical grating due to the instantaneous electronic response and off-resonant vibrational modes does not contribute toward species-specific signal but creates a confusing background. This work is a quest for a robust experimental technique that overcomes this limitation, i.e., provides means for suppression of the nonresonant (NR) FWM and adequate discrimination between the informative CARS signal and the left-over background, even for such analytes as powders and spores.

2. FAST CARS

The idea to utilize coherent Raman spectroscopy for spore detection was originally discussed in the paper entitled "FAST CARS: Engineering a laser spectroscopic technique for rapid identification of bacterial spores", included in the following pages. There, the emphasis is made on the use of femtosecond laser pulses, which offers several benefits.

First, the short pulse duration ($\sim$100 fs), which is smaller than a typical dephasing time of laser-induced molecular vibrations ($\sim$1 ps), allows for monitoring of the free-induction-decay dynamics of excited molecules while effectively suppressing the NR FWM contribution. This is because the later process is virtually instantaneous and happens only when the three laser pulses are overlapped.

The second advantage is the large spectral bandwidth of the preparation pulses. Multiple Raman modes can be accessed without tuning the laser frequencies. Such multiplicity is essential for specificity of the technique. The larger the Raman frequency span covered, the more information is gained for the analysis from a single measurement.

Finally, the spectral amplitude and phase of the laser pulses can be used for further refinement of the output. Pulse shaping has been already utilized in various coherent control schemes, where the process outcome is manipulated through constructive and destructive interferences of possible excitation paths [14–17]. In particular, the specificity can be gained through the mode-selective excitation of Raman transitions, i.e., at the preparation stage, or through the shaping of the probe pulse. Silberberg et al. have recently demonstrated the use of periodic modulation [18] and step-like jumps [19–21] of the pulse phase for this purpose.

REPRINT: M.O. Scully *et al.*, *Proc. Natl. Acad. Sci. USA* **99**, 10994 (2002)

FAST CARS: Engineering a laser spectroscopic technique for rapid identification of bacterial spores

M. O. Scully[†‡§¶‖], G. W. Kattawar[††], R. P. Lucht[†**], T. Opatrný[†,††], H. Pilloff[†], A. Rebane[‡‡], A. V. Sokolov[††], and M. S. Zubairy[†‡§§]

[†]Institute for Quantum Studies, Departments of [‡]Physics, [§]Electrical Engineering, and [**]Mechanical Engineering, Texas A&M University, College Station, TX 77843; [¶]Max-Planck-Institut fur Quantenoptik, D-85748 Garching, Germany; [††]Department of Theoretical Physics, Palacky University, CZ-77146 Olomouc, Czech Republic; [‡‡]Department of Physics, Montana State University, Bozeman, MT 59715; and [§§]Department of Electronics, Quaid-i-Azam University, Islamabad 45320, Pakistan

This contribution is part of the special series of Inaugural Articles by members of the National Academy of Sciences elected on May 1, 2001.

Contributed by M. O. Scully, May 15, 2002

Airborne contaminants, e.g., bacterial spores, are usually analyzed by time-consuming microscopic, chemical, and biological assays. Current research into real time laser spectroscopic detectors of such contaminants is based on e.g., resonance fluorescence. The present approach derives from recent experiments in which atoms and molecules are prepared by one (or more) coherent laser(s) and probed by another set of lasers. However, generating and using maximally coherent oscillation in macromolecules having an enormous number of degrees of freedom is challenging. In particular, the short dephasing times and rapid internal conversion rates are major obstacles. However, adiabatic fast passage techniques and the ability to generate combs of phase-coherent femtosecond pulses provide tools for the generation and utilization of maximal quantum coherence in large molecules and biopolymers. We call this technique FAST CARS (femtosecond adaptive spectroscopic techniques for coherent anti-Stokes Raman spectroscopy), and the present article proposes and analyses ways in which it could be used to rapidly identify preselected molecules in real time.

There is an urgent need for the rapid assay of chemical and biological unknowns, such as bioaerosols. Substantial progress toward this goal has been made over the past decade. Techniques such as fluorescence spectroscopy (1, 2) and UV resonant Raman spectroscopy (3–7) have been successfully applied to the identification of biopolymers, bacteria, and bioaerosols.

At present, field devices are being engineered (1) that will involve an optical preselection stage based on, e.g., fluorescence radiation as in Fig. 1. If the fluorescence measurement does not give the proper signature then that particle is ignored. Most of the time the particle will be an uninteresting dust particle; however, when a signature match is recorded, then the particle is selected for special biological assay (see Fig. 1*b*). The relatively simple fluorescence stage can very quickly sort out some of the uninteresting particles whereas the more time-consuming bio-tests will be used for only the "suspects."

The good news about the resonance fluorescence technique is that it is fast and simple. The bad news is that although it can tell the difference between dust and bacterial spores it cannot differentiate between spores and many other organic bioaerosols (see Fig. 1*c*).

However, despite the encouraging success of the above-mentioned studies, there is still interest in other approaches to, and tools for, the rapid identification of chemical and biological substances. To quote from a recent study (8):

> "Current [fluorescence-based] prototypes are a large improvement over earlier stand-off systems, but they cannot yet consistently identify specific organisms because of the similarity of their emission spectra. Advanced signal processing techniques may improve identification."

Resonant Raman spectra hold promise for being spore specific as indicated in Fig. 2*b*. This is the good news, the bad news is that the Raman signal is weak and it takes several minutes to collect the data of Fig. 2*b*. Because the through-put in a set-up such as that of Fig. 1*b* is large, the optical interrogation per particle must be essentially instantaneous.

The question then is: Can we increase the resonant Raman signal strength and thereby reduce the interrogation time per particle? If so, then the technique may also be useful in various detection scenarios.

The answer to the question of the proceeding paragraph is a qualified yes. We can enhance the Raman signal by increasing the coherent molecular oscillation amplitude R_0 indicated in Fig. 2*c*. In essence this means maximizing the quantum coherence between vibrational states $|b>$ and $|c>$ of Fig. 2*a*.

Our point of view derives from research in the fields of laser physics and quantum optics that have concentrated on the utilization and maximization of quantum coherence. The essence of these studies is the observation that an ensemble of atoms or molecules in a coherent superposition of states represents, in a real sense, a new state of matter aptly called phaseonium (9–11).

In particular, we note that matter in thermodynamic equilibrium has no phase coherence between the electrons in the molecules making up the ensemble. This is discussed in detail later. When a coherent superposition of quantum states is involved, things are very different, and based on these observations, many interesting and counterintuitive notions are now a laboratory reality. These include lasing without inversion (12–15), electromagnetically induced transparency (16, 17), light having ultra-slow group velocities on the order of 10 m/sec (18–23), and the generation of ultra-short pulses of light based on phased molecular states (24, 25).

Another emerging technology central to the present paper is the exciting progress in the area of femtosecond quantum control of molecular dynamics originally suggested by Judson and Rabitz (26). This progress is described and reviewed in the articles by Kosloff *et al.* (27), Warren *et al.* (28), Gordon and Rice (29), Zare (30), Rabitz *et al.* (31), and Brixner *et al.* (32). Other related work on quantum coherent control includes the quantum interference approach of Brumer and Shapiro (33), the time-domain (pump-dump) technique proposed by Tannor *et al.* (34), and the stimulated Raman adiabatic passage (STIRAP) approach of Bergmann *et al.* (35) to generate a train of coherent laser pulses. The preceding studies teach us how to produce pulses having arbitrary controllable amplitude and frequency time depen-

---

Abbreviations: CARS, coherent anti-Stokes Raman spectroscopy; FAST CARS, femtosecond adaptive spectroscopic techniques for CARS; STIRAP, stimulated Raman adiabatic passage; DPA, dipicolinic acid.

[‖]To whom reprint requests should be addressed. E-mail: scully@physics.tamu.edu.

REPRINT: M.O. Scully *et al.*, *Proc. Natl. Acad. Sci. USA* **99**, 10994 (2002)

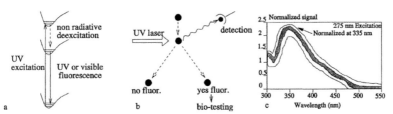

Fig. 1. (*a*) UV excitation radiation promotes molecules from ground state to an excited-state manifold. This excited-state manifold decays to the ground state via nonradiative processes to a lower manifold, which then decays via visible or UV fluorescence. (*b*) A scenario in which a UV laser interacts with dust particles and biospheres of interest. When, for example, a bacterial spore is irradiated, fluorescence will be emitted signaling that this particular system is to be further tested. In principle, uninteresting particles are deflected one way; but when fluorescence takes place, the particles are deflected in another direction and these particles are then subjected to further biological tests. (*c*) The shaded area displays the signal range for the fluorescence spectrum of a number of biological samples, *Bacillus subtilis, Bacillus thuringiensis, Escherichia coli,* and *Staphylococcus aureus.* It is not possible to distinguish between the different samples based on such a measurement (see ref. 2 for more details).

dence. Indeed the ability to sculpt pulses by the femtosecond pulse shaper provides an important new tool for all of optics [see the pioneering works by Heritage *et al.* (36), Weiner *et al.* (37), Wefers and Nelson (38), and Weiner (39)].

A promising approach is to use learning algorithms so that knowledge of the molecular potential energy surfaces and matrix elements between surfaces are not needed *a priori*; however, by using a pulse shaper coupled with a feedback system, complex spectra can be revealed.

Thus, we now have techniques at hand for controlling trains of phase-coherent femtosecond pulses so as to maximize molecular coherence. This process allows us to increase the Raman signal while decreasing the undesirable fluorescence background, which has much in common with the CARS spectroscopy (40) of Fig. 3, but with essential differences as we now discuss.

The presently envisioned improvement over ordinary CARS is based on enhancing the ground-state molecular coherence. However, we note that molecules involving a large number of degrees of freedom will quickly dissipate the molecular coherence among these degrees of freedom. This difficulty is well known and is addressed in the present work from several perspectives. First of all, when working with ultra-short pulses, we have the ability to generate the coherence on a time scale that is small compared with

the molecular relaxation time. Furthermore, we are able to tailor the pulse sequence in such a way as to mitigate and overcome key limitations in the application of conventional coherent anti-Stokes Raman spectroscopy (CARS) to trace contaminants. The key point is that we are trying to induce maximal ground-state coherence, as opposed to the usual situation within conventional CARS where the ground-state coherence is not a maximum as is shown later in this paper. With FAST CARS (femtosecond adaptive spectroscopic techniques applied to CARS) we can prepare the coherence between two vibrational states of a molecule with one set of laser pulses and use higher frequency visible or UV to probe this coherence in a coherent Raman configuration. This process will allow us to capitalize on the fact that maximally coherent Raman spectroscopy is orders of magnitude more sensitive than incoherent Raman spectroscopy.

Having stated our goals and our approach toward attaining these goals, we emphasize that the present article represents essentially an engineering endeavor. We propose to draw heavily on the ongoing work in quantum coherence and quantum control as mentioned earlier. For example, the careful experiments and analysis of the Würzburg group on the generation and probing of ground-state coherence in porphyrin molecules (41) by femtosecond-CARS (fs-CARS) are very germane to our considerations. However, ground-state coherence is not maximized in these experiments. In

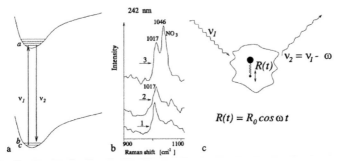

$$R(t) = R_0 \cos \omega t$$

Fig. 2. (*a*) Resonant Raman scattering in which radiation ν_1 excites the atom from |c> to |a> and the Stokes radiation is emitted taking the molecule from |a> to |b>. (*b*) Detail of UV resonance Raman spectra of spores of *B. megaterium* (trace 1), *Bacillus c.* (trace 2), and calcium dipicolinate (trace 3), all excited at 242 nm; adapted from ref. 4 (see also Fig. 6). (*c*) A more physical picture of Raman scattering in which a single diatomic molecule, consisting of a heavy nucleus and a light atom, scatters incident laser radiation at frequency ν_1. The vibrational degrees of freedom associated with the diatomic molecule are depicted here as occurring with amplitude R_0 oscillating at frequency ω. The scattered radiation from this vibrating molecule is at frequency $\nu_2 = \nu_1 - \omega$ for the Stokes radiation.

REPRINT: M.O. Scully *et al.*, *Proc. Natl. Acad. Sci. USA* **99**, 10994 (2002)

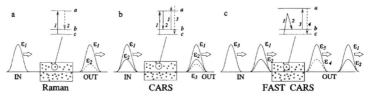

Fig. 3. (a) Ordinary resonant Raman spectroscopy in which a drive laser of amplitude ε_1 generates a weak signal field having an amplitude ε_2. The incident signal consists of one pulse at ν_1, and the pulse structure after interaction with the molecular medium consists of two pulses at ν_1 and ν_2. (b) The coherent Raman process associated with CARS is depicted in which two fields at frequency $\{\nu_1\}$ and $\{\nu_2\}$ are incident with amplitudes ε_1 and ε_2. The third radiated anti-Stokes signal field at frequency ν_3 is indicated. (c) FAST CARS configuration in which maximal coherent Raman spectroscopy is envisioned. The preparation pulses ε_1 and ε_2 prepare maximum coherence between states $|b>$ and $|c>$. Next the probe laser ε_3 interacts with this oscillating molecular configuration and the anti-Stokes radiation is generated.

another set of beautiful experiments (42) they investigate the selective excitation of polymers of diacetylene via fs-CARS. They control the timing, phase, and frequency (chirp) content of their preparation pulses. In these experiments it was necessary to focus attention on the evolution of the excited-state molecular dynamics. We hope to avoid this complication as is explained later.

Perhaps closest to our approach is the recent joint work of the Garching Max-Planck and Würzburg groups (43). Their paper is a prime example of a FAST CARS experiment. However, they concentrate on producing highly excited states of the "vibrational motion of a certain bond." The application of their technique to the production of maximum coherence between states $|b>$ and $|c>$ of Fig. 2a in a specific vibrational mode of their molecule would be of great interest to us and is underway.

Finally, we want to draw the reader's attention to the useful collection of articles in a recent special issue of the *Journal of Raman Spectroscopy* dedicated to fs-CARS (44). Likewise the recent work of Silberberg and coworkers (45, 46) in which they show that it is possible to excite one of two nearby Raman levels, even when they are well within the broad fs pulse spectrum, is another excellent example of the power of the FAST CARS technique.

The present work focuses on utilization of a maximally phase-coherent ensemble of molecules, i.e., molecular phaseonium, to enhance Raman signatures. This will be accomplished via the careful tailoring of a coherent pulse designed to prepare the molecule with maximal ground-state coherence. Such a pulse is a sort of "melody" designed to prepare a particular molecule. Once we know this molecular melody, we can use it to set that particular molecule in motion and this oscillatory motion is then detected by another pulse; this is the FAST CARS protocol depicted in Fig. 3c.

In the next section, the status of Raman spectroscopy applied to biological spores is reviewed. Then, we compare various types of Raman spectroscopy with an eye to the recent successful applications of quantum coherence in laser physics and quantum optics. Then, we present several experimental schemes for applying these considerations to the rapid identification of macromolecules, in general, and biological spores, in particular. Finally, we propose several scenarios in which FAST CARS could be useful in the rapid detection of bacterial spores. The various appendices referred to are to be found in our extended paper, which is published as supporting information on the PNAS web site, www.pnas.org. As stated earlier, the present article is an engineering science analysis of a promising approach to the problem of bacterial spore detection.

Pico-Review of Raman Spectroscopy Applied to Bacterial Spores

The bacterial spore is an amazing life form. Spores thousands of years old have been found to be viable. One textbook (48) reports that "endospores trapped in amber for 25 million years germinate when placed in nutrient media."

A key to this incredible longevity is the presence of dipicolinic acid (DPA) and its salt calcium dipicolinate in the living core that contains the DNA, RNA, and protein as shown in Fig. 4.

A major role of the calcium DPA complex seems to be the removal of water, as per the following quote (49): "The exact role of these [DNA] chemicals is not yet clear. We know, for instance, that heat destroys cells by inactivating proteins and DNA and that this process requires a certain amount of water. Since the deposition of calcium dipicolinate in the spore removes water ... it will be less vulnerable to heat."

Hence, one of the major components of bacterial spores is DPA and its ion as depicted in Fig. 5. Calcium dipicolinate can contribute up to 17% of the dry weight of the spores. A definitive demonstration (3) of this conjecture was made by comparing the 242-nm excitation spectra of calcium dipicolinate with spore suspensions of *Bacillus megaterium* and *Bacillus cereus*. From Fig. 6, it is seen that good matches were noted at 1,017 cm^{-1} with further matches being found at 1,396, 1,446, and 1,607 cm^{-1} peaks of the calcium dipicolinate.

As has been shown in previous work by, e.g., Carmona (3) and Nelson and coworkers (4–7), the presence of DPA and its calcium salt gives us a ready-made marker for endospores. As has been mentioned earlier and as will be further discussed later, this is the key to Raman fingerprinting of the spore.

We note, however, that fluorescence spectroscopy was one of the first methods used for detection of bacterial taxonomic markers and is still used for detection where high specificity is not required. This technique is an important addition to the tool kit of scientists and engineers working in this area.

A possible FAST CARS protocol is as follows: First, we obtain size and fluorescence information. If this information is consistent

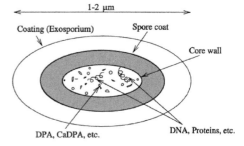

Fig. 4. Sketch of spore indicating that the DPA and its salts, e.g., Ca-DPA, are contained in the core and are in contact with the spore-specific DNA ribosomes and cell proteins.

REPRINT: M.O. Scully *et al.*, *Proc. Natl. Acad. Sci. USA* **99**, 10994 (2002)

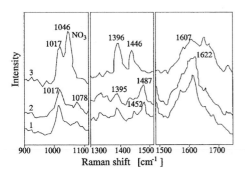

Fig. 5. (*a*) DPA [2,6-pyridinedicarboxylic acid, $C_5H_3N(COOH)_2$]. (*b*) The Ca^{2+} DPA complex.

with the presence of a particular bacterial spore we could then automatically perform a FAST CARS analysis sensitive to DPA so as to further narrow the number of suspects.

It is important to note that just as resonant Raman is some 10^6 times more sensitive than nonresonant, coherent Raman yields a much stronger signal than ordinary incoherent Raman spectroscopy. This makes it possible to collect the Raman spectra much more rapidly via FAST CARS, which is very important in the ultimate scheme of things.

We will generate quantum coherence in macromolecules by working with the now available femtosecond pulse trains in which phase coherence exists between the individual pulses. In this way, one can enhance coherent Raman signatures. The utilization of "molecular music" to generate maximal phase coherence holds promise for the identification and characterization of macromolecules and biomolecules.

Overview of Raman Spectroscopy

Raman scattering is an inelastic scattering of electromagnetic fields off vibrating molecules. The origin of Raman scattering dates back to a theoretical paper in *Naturwissenschaften* by A. Smekal in 1923 entitled (translated) "The quantum theory of dispersion" (50). It was followed by another paper in a 1923 *Physical Review* (by A. Compton) entitled "A quantum theory of the scattering of x-rays by light elements" (51). Some historians feel that these two papers gave C. V. Raman the idea for the experiments that were performed with K. S. Krishnan and led to the discovery of the effect in more than 60 liquids. Raman and Krishnan published their results entitled "A new type of secondary radiation" in *Nature* on March 28, 1928 (52). It was soon followed by the landmark paper of G. Landsberg and L. Mandelstam who found the same effect in quartz and published a paper entitled (translated) "A novel effect

of light scattering in crystals," which appeared on July 13, 1928 in *Naturwissenschaften* (53). By the end of 1928 dozens of papers had already been published on the Raman effect.

Raman scattering is an optical phenomenon in which there is a change of frequency of the incident light. Light with frequency v_1 scatters inelastically off the vibrating molecules such that the scattered field has frequency $v_2 = v_1 \pm \omega_{bc}$, where ω_{bc} is the frequency of the molecular vibrations. The field with down-shifted frequency $v_2 = v_1 - \omega_{bc}$ is called Stokes field, whereas the frequency up-shifted radiation is called the anti-Stokes.

There are two basic Raman processes: the so-called spontaneous and stimulated Raman scattering. Spontaneous scattering occurs if a single laser beam with intensity below a certain threshold illuminates the sample. In condensed matter, in propagating through 1 cm of the scattering medium, only approximately 10^{-6} of the incident radiation is typically scattered into the Stokes field (see, e.g., ref. 54). Stimulated scattering that occurs with a very intense illuminating beam is a much stronger process in which several percent of the incident laser beam can be converted into the other frequencies.

For spontaneous Stokes scattering, the intensity of the scattered field is roughly proportional to the length traveled by the incident field in the medium. On the other hand, the stimulated process becomes dominant and the scattered field intensity can increase exponentially with the medium length.

The resonant Raman process (appearing when the frequency of the incident radiation coincides with one of the electronic transitions) is much richer than the nonresonant, and we now turn to a discussion of the resonant problem. Resonant Raman radiation is governed by the oscillating dipole between states $|a>$ and $|b>$ (Stokes) and/or $|a>$ and $|c>$ (anti-Stokes) in the notation of Fig. 3 and Table 1. In the Stokes case, the steady-state coherent oscillating dipole $P(t)$, divided by the dipole matrix element $\rho_{ab} = e <a|r|b>$, is the important quantity. That is $\rho_{ab}(t) \equiv P(t)/\rho_{ab}$ is

$$\rho_{ab} = -i[\Omega_2(n_a - n_b) - \Omega_1\rho_{bc}]/[\gamma_{ab} - i(\omega_{ab} - v_2)], \quad [1]$$

where the Raman coherence is ρ_{bc}. In Eq. **1**, ω_{ab} is the transition frequency between the electronic states a and b, n_a and n_b are the populations of levels a and b, v_2 is the frequency of the generated field, and the other quantities are defined in the legend of Table 1.

The main advantage of resonant Raman scattering is that the signal is very strong—up to a million times stronger compared with the signal of nonresonant scattering (4, 5). It is also very useful that only those Raman lines corresponding to very few vibrational modes associated with strongly absorbing locations of a molecule show this huge intensity enhancement. On the other hand, the resonance Raman spectra may be contaminated with fluorescence. However, this problem can be avoided by using UV light so that most of the fluorescence appears at much longer wavelengths than the Raman scattered light and is easily filtered out.

Fast Cars

Generation of Atomic Coherence. The purpose of this section is to demonstrate the utility of pulse shaping as a mechanism for

Fig. 6. Shown are UV resonance Raman spectra of spores of *B. megaterium* (trace 1), spores of *B. cereus* (trace 2), and calcium dipicolinate (trace 3) in three spectral regions. All samples are excited at 242 nm. Figure adapted from ref. 4.

(Fig. 6 graph labels: Intensity vs Raman shift [cm^{-1}]; peaks labeled 1046, 1017, NO$_3$, 1017, 1078, 1396, 1446, 1395, 1487, 1452, 1607, 1622; x-axis values 900 1000 1100 1300 1400 1500 1500 1600 1700; trace labels 1, 2, 3)

REPRINT: M.O. Scully *et al.*, *Proc. Natl. Acad. Sci. USA* **99**, 10994 (2002)

Table 1. Comparison of different Raman spectroscopic techniques as derived in *Appendix A*, which is published as supporting information on the PNAS web site

Process	Raman coherence ρ_{cb}	Dipole coherence ρ_{ab}
Raman (Weak drive)	$i\dfrac{\Omega_2\Omega_1^*}{\gamma_{bc}\Delta}$ 10^{-5}	(incoh.) $-\dfrac{\Omega_2\,|\Omega_1|^2}{\Delta\,\Delta\gamma_{bc}}$ 10^{-9}
Resonant Raman (Weak drive)	$-\dfrac{\Omega_2\Omega_1^*}{\gamma_{ac}\gamma_{bc}}$ 10^{-2}	(incoh.) $-\dfrac{\Omega_2\,|\Omega_1|^2}{\gamma_{ab}\,\gamma_{ac}\gamma_{bc}}$ 10^{-3}
Raman (Strong drive)	(max. coh.) $\dfrac{i}{4}\sqrt{\dfrac{\gamma_1}{\gamma_{bc}}}$ 10^{-3}	(max. coh.) $i\dfrac{1}{4}\dfrac{\Omega_2}{\Delta}\sqrt{\dfrac{\gamma_1}{\gamma_{bc}}}$ 10^{-6}
Resonant Raman (Strong drive)	(max. coh.) $\dfrac{1}{2}$ 10^0	(max. coh.) $\dfrac{i\Omega_1}{2\gamma_{ab}}$ 10^{-1}

The density matrix element ρ_{bc} governs the amplitude of coherent vibration, whereas the element ρ_{ab} is proportional to the electronic polarization responsible for emission of radiation. $\Omega_{1,2}$ are the Rabi frequencies, Δ is the detuning of the electronic transition, γ_{ab}, γ_{ac} are the decay rates of the optical transitions, γ_{bc} is the decoherence rate of the vibrational states, and γ_1 is the decay rate from level b to c. The approximated values (shown in the lower right corner) were obtained for $\gamma_{ab} \approx \gamma_{ac} \approx \gamma_{bc} \approx 10^{12}\mathrm{s}^{-1}$, $\gamma_1 \approx 10^6\mathrm{s}^{-1}$, $\Delta \approx 10^{15}\mathrm{s}^{-1}$, and $\Omega_{1,2} \approx 10^{11}\mathrm{s}^{-1}$ for weak driving and $\Omega_{1,2} \approx 10^{12}\mathrm{s}^{-1}$ for strong driving. Note that $\Omega \approx 10^{11}\mathrm{s}^{-1}$ corresponds to a 10-ns pulse with 0.1 mJ energy focused on a square millimeter spot if the electronic transition dipole moment is $\wp \approx 10^{-19}\mathrm{C} \times 10^{-10}\mathrm{m}$.

generating maximal coherence. The Raman signal is optimized at the condition of maximal molecular coherence. When in this state, each of the molecules oscillates at a maximal amplitude, and all molecules in an ensemble oscillate in unison. Here we discuss several methods for the preparation of maximal coherence state.

Adiabatic rapid passage via chirped pulses. A particularly simple and robust approach to the generation of the maximal coherence is to use a detuning $\delta\omega$, which is largely independent of inhomogeneous broadening and variations in matrix elements (Fig. 7).

Such multilevel molecular system can be described in terms of an effective two-by-two Hamiltonian (55). Diagonalization of this Hamiltonian (see *Appendix C*, which is published as supporting information on the PNAS web site) allows us to analyze the evolution of the system by drawing analogies to two-state systems. If the excitation is applied resonantly ($\Delta\omega = 0$), such that the initial state of the system (the ground state $|c\rangle$ is projected onto the new basis formed by the eigenvectors $|+\rangle$ and $|-\rangle$, the system undergoes a sinusoidal Rabi flopping between states $|b\rangle$ and $|c\rangle$. In this situation one can choose to apply a $\pi/2$ pulse to create the maximal coherence $|\rho_{bc}| = 0.5$. Alternatively, one can apply an excitation at a finite detuning $\Delta\omega$, to allow all population, which is initially in the ground state, to follow the eigenstate $|+\rangle$ adiabatically.

Fractional STIRAP. In an all-resonant Λ scheme (Fig. 8, with $\delta\omega = \Delta\omega = 0$) maximal coherence can be prepared between the levels b and c in a fractional STIRAP set up by a counterintuitive pulse

sequence (35, 56, 57), such that the population of the upper state a is always zero and fluorescence from this state is eliminated. This can be accomplished via a counterintuitive sequence of two pulses at frequencies ω_{ab} and ω_{ac}. Under the condition of adiabatic passage, the molecule in the initial state $|b\rangle$ is transformed into a coherent state $(|b\rangle - |c\rangle)2^{-1/2}$.

The principle behind a STIRAP process is the adiabatic theorem as applied to the time-varying Hamiltonian $H(t)$. If the system at time t_0 is in an eigenstate of $H(t_0)$, and the evolution from t_0 to t_1 is sufficiently slow, then the system will evolve into the eigenstate of $H(t_1)$. The three-level atomic system driven by two fields has three eigenstates, one of which is a linear superposition of only the lower levels b and c. The time-dependent amplitudes of this eigenstate depend on the pulse shapes of the fields at frequencies ω_{ab} and ω_{ac}. Thus, by an appropriate pulse shaping, it should be possible to prepare a maximally coherent superposition of states b and c as shown in Fig. 8. The expressions for the Hamiltonian and the corresponding eigenstates are given in *Appendix D*, which is published as supporting information on the PNAS web site.

Comparing different schemes for the preparation of maximal coherence, we note that the required laser power is much lower for the all-resonant scheme, but in the case of biomolecules, UV lasers are required. The far-detuned scheme will work with more powerful infrared lasers, up to the point of laser damage. As for the comparison of adiabatic and nonadiabatic regimes, we should note

REPRINT: M.O. Scully *et al.*, *Proc. Natl. Acad. Sci. USA* **99**, 10994 (2002)

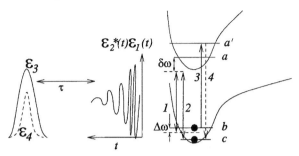

Fig. 7. Energy level schematics for a three-level system to generate maximum coherence between the levels $|b\rangle$ and $|c\rangle$ via fields ε_1 and ε_2. These fields are off-resonant with the electronic detuning $\delta\omega$ and possibly also with the Raman detuning $\Delta\omega$, which can vary in time, thus chirping the pulses. After preparing the coherence ρ_{bc} with fields $\varepsilon_{1,2}$ the probe field ε_3 gives rise to the anti-Stokes field ε_4.

that the adiabatic scheme may turn out to be more robust, because it does not rely on a particular pulse area and works for inhomogeneous molecular ensembles and nonuniform laser beams.

Femtosecond pulse sequences. In a series of beautiful experiments K. Nelson and coworkers (58) have generated coherent molecular vibration via a train of femtosecond pulses. They nicely describe their work as: "Timed sequences of femtosecond pulses have been used to repetitively 'push' molecules in an organic crystal . . . , in a manner closely analogous to the way a child on a swing may be pushed repetitively to reach oscillatory motion."

An interesting aspect of this approach is the fact that the individual pulses need not be strong. Only the collective effect of many weak pulses is required. This may be helpful if molecular "break-up," caused by strong ε_1 and ε_2, is a problem. This will be further discussed elsewhere.

Adaptive Evolutionary Algorithms. So far we described how one-photon and two-photon resonant pulse sequences can be used to produce a coherent molecular superposition state. The idea is that once this state is created a delayed pulse can be applied to produce Raman scattering, which will bear the signature of the molecular system. The Raman signal is expected to be optimized when the molecular coherence is maximal. In general, however, things are too complicated to enable us to work with simple predetermined laser pulses. For large biomolecules the level structure is not only very complex, but usually unknown. We now consider how search algorithms can be used to find the optimal pulse sequence for a complicated molecule with an unknown Hamiltonian. This ap-

proach will eventually lead to an efficient generation of "molecular fingerprints."

To achieve this goal we will need to (*i*) use a technique for preparation of complex-shaped pulse sequences, and (*ii*) find the particular pulse sequences, required for the excitation of the particular biomolecules and the production of spectral signatures, which will allow one to distinguish (with certainty) the target biological agent from any other species.

Pulse-shaping techniques already exist; they are based on spectral modification. First, a large coherent bandwidth is produced by an ultra-short pulse generation technique[¶¶] (59–61). Then, the spectrum is dispersed with a grating or a prism, and each frequency component is addressed individually by a spatial light modulator [a liquid crystal array (62, 63) or an acoustic modulator (64)]. This way, individual spectral amplitudes and phases can be adjusted independently. Finally, the spectrum is recombined into a single beam by a second dispersive element and focused onto the target. This technique allows synthesis of arbitrarily shaped pulses right at the target point and avoids problems associated with dispersion of intermediate optical elements and windows.

A particular shaped pulse sequence can be represented by a three-dimensional surface in a space with frequency-amplitude-phase axes. Each pulse shape, which corresponds to a particular

---

¶¶The shortest optical pulses generated to date (5–6 fs) are obtained by expanding the spectrum of a mode-locked laser by self-phase modulation in an optical waveguide, and then compensating for group velocity dispersion by diffraction grating and prism pairs.

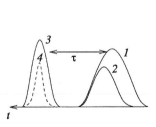

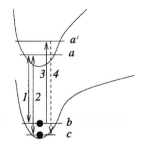

Fig. 8. Energy-level schematics for the generation of maximum coherence between the levels $|b\rangle$ and $|c\rangle$ via fractional STIRAP by counterintuitive pulses 1 and 2. After a time delay of τ the pulse resonant with $|a'\rangle \rightarrow |b\rangle$ transition produces a signal at $\omega_{a'c}$.

REPRINT: M.O. Scully *et al.*, *Proc. Natl. Acad. Sci. USA* **99**, 10994 (2002)

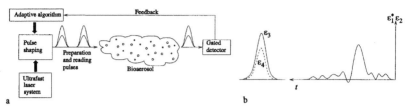

Fig. 9. (*a*) Experimental setup for the implementation of adaptive techniques. (*b*) Depiction of amplitude of possible optimized Raman preparation pulse sequence $\varepsilon_1^* \varepsilon_2$. Not indicated is the fact that the reading pulse ε_3 can also be profitably considered as a learning algorithm variable.

three-dimensional surface, produces a molecular response. The problem is to find the optimal shape. The search space is too large to be scanned completely. Besides, many local optima may exist in the problem. The solution is offered by global search algorithms (such as adaptive evolutionary algorithms) (32, 65). In this approach the experimental output is included in the optimization process. This way, the molecules subjected to control are called on to guide the search for an optimal pulse sequence within a learning loop (26). With the proper algorithm, automated cycling of this loop provides a means of finding optimal pulse shapes under constraints of the molecular Hamiltonian and the experimental conditions. No prior knowledge of the molecular Hamiltonian and the potential energy surfaces is needed in this case.

This adaptive technique was developed for coherent control of chemical reactions (32). The idea is that the pulses can be optimized to produce desired chemical products. In our problem we want to optimize Raman generation. In this case both preparation and reading pulses can be adaptively shaped to maximize the signal. Fig. 9 shows schematics for the experimental setup that implements these ideas.

Generated spectra will be different for different molecular species. And our task is not only to maximize Raman generation, but also to identify spectral patterns characteristic of particular species and maximize the difference in the spectrum produced by the target biomolecule from spectra produced by any other biomolecules. The key idea here is to apply the same adaptive algorithms to learn these optimal molecular fingerprints or perhaps better said molecular melody.

We note that the complexity of the molecular level structure is not so much a problem as a solution to a problem. We should take advantage of the richness of the molecular structure, and the infinite variety of possible pulse shapes, to distinguish different species.

Possible FAST CARS Measurement Strategies for Detection of Bacterial Spores

Having presented the FAST CARS concept in some detail we now return to the question of its application to fingerprinting of macromolecules and bacterial spores. Some aspects of the technique seem fairly simple to implement and would seem to hold relatively immediate promise. Others are more challenging but will probably be useful at least in some cases. Still other applications, e.g., the stand-off detection of bioaerosols in the atmosphere present many open questions and require careful study. In the following we discuss some simple FAST CARS experiments that are underway and/or being assembled in our laboratories.

Preselection and Hand-Off Scenarios.

At present, field devices are being engineered that will involve an optical preselection stage based on, e.g., fluorescence tagging. If the fluorescence measurement does not match the class of particles of interest then that

particle is ignored. When many such particles are tested and a possible positive is recorded, the particle is subjected to special biological assay. Such a two-stage approach can substantially speed up the detection procedure. The relatively simple fluorescence stage can very quickly sort out many uninteresting scattering centers whereas the more sophisticated Raman scattering protocol will be used only for the captured suspects.

The properly shaped preparation pulse sequence will be determined by, e.g., the adaptive learning algorithm approach. The amplitude and phase content of the pulse that produces maximum oscillation may be linked to a musical tune. Each spore will have a song that results in maximum Raman coherence. A correctly chosen melody induces a characteristic response of the molecular vibrations—a response that is as unique as possible for the bacterial spores to be detected. Playing a melody rather than a single tone is a generalization that enables us to see a multidimensional picture of the investigated object. We note that the optimization can (and frequently will) include not only the preparation pulses 1 and 2 (see Fig. 9b), but also the probe pulse 3, in particular, its central frequency and timing. Analysis of the response to such a complex input is a complicated signal processing problem. Various data mining strategies may be used in a way similar to speech analysis.

However, taking into account the fact that we work with femtosecond pulses chained in picosecond to nanosecond pulse trains, the whole analysis can be very short. In particular, if we recall the long sampling time of the complete fluorescence spectra of ref. 6 being ≈ 15 min, our estimation of a microsecond analysis is a very strong argument for the chosen approach.

Possible Further Raman Characterization.

After a suspect particle has been targeted, it may be subject to a whole variety of investigative strategies. Raman scattering off a flying particle can be very fast, but not necessarily the most accurate method. It will be very useful to pin the particle on a fixed surface and cool it down to maximize the decoherence time T_2 so that the characteristic lines are narrowed down. The particle can be deflected by optical means (laser tweezers, laser ionization, etc.) and attached to a cooled conducting surface. Cooling to liquid helium temperature should enable us to enhance the dephasing time from $T_2 \leq 10^{-12}$ sec at room temperature to perhaps a few picoseconds (or more) at a few degrees Kelvin.

Possible Spore-Specific FAST CARS Detection Schemes.

We conclude with some speculative observations for long-range (stand-off) measurements. The chemical state of DPA in the spore is of special interest to us because the stuff we hang on the DPA molecule will determine its characteristic Raman frequency. To this end, we quote from an article (66) by Murrell on the chemical composition of spores: "When DPA is isolated from spores it is nearly always in the Ca-CDPA chelate but sometimes

REPRINT: M.O. Scully *et al.*, *Proc. Natl. Acad. Sci. USA* **99**, 10994 (2002)

as the chelate of other divalent metals [e.g., Zn, Mn, Sr, etc.] and perhaps as a DPA-Ca amino complex."

Thus, because each different type of spore would have its own unique mixture of metals and amino acids, it may be the case that the finer details of the Raman spectra would contain spore specific fingerprints. This conjecture is supported by Fig. 2*b* where the difference between the DPA Raman spectra of the spores of *B. cereus* and *B. megaterium* is encouraging.

The extent to which the DPA Raman spectra are sensitive to its environment is an open question. That we might be able to achieve spore specific sensitivity is consistent with the well-known fact that substituents, e.g., NO_2 experience a substantial shift of their vibrational frequencies when bound in different molecular configurations. Furthermore, recent NMR experiments (47) show spore-specific fingerprints.

Clearly there are many opportunities and open questions implicit in the FAST CARS molecular melody approach to real-time

spectroscopy. However it plays out, this combination of quantum coherence and coherent control is a rich field.

M.O.S. notes: "My friend and mentor Vicky Weisskopf used to say 'The best way into a new problem is to bother people.' This is faster than searching the literature and more fun. I would like to thank my colleagues for allowing me to be a bother and especially my coauthors who have suffered the most!" This paper is dedicated to the memory of Prof. Viktor Weisskopf, premier physicist and scientist–soldier who stood by his adopted country in her hour of need. We thank R. Allen, Z. Arp, A. Campillo, K. Chapin, R. Cone, A. Cotton, E. Eisenstadt, J. Eversole, M. Feld, J. Golden, S. Golden, T. Hall, S. Harris, P. Hemmer, J. Laane, F. Narducci, B. Spangler, W. Warren, G. Welch, S. Wolf, and R. Zare for valuable and helpful discussions. We gratefully acknowledge the support from the Office of Naval Research (Contracts N000140210808 and N000140210741), the Air Force Research Laboratory (Rome, NY), Defense Advanced Research Projects Agency-QuIST, Texas A&M University Telecommunication and Informatics Task Force (TITF) Initiative, and the Welch Foundation.

1. Seaver, M., Eversole, J. D., Hardgrove, J. J., Cary, W. K. & Roselle, D. C. (1999) *Aerosol Sci. Tech.* **30,** 174–185.
2. Cheng, Y. S., Barr, E. B., Fan, B. J., Hargis, P. J., Jr., Rader, D. J., O'Hern, T. J., Torczynski, J. R., Tisone, G. C., Preppernau, B. L., Young, S. A. & Radloff, R. J. (1999) *Aerosol Sci. Tech.* **30,** 186–201.
3. Carmona, P. (1980) *Spectochim. Acta* **36A,** 705–712.
4. Manoharan, R., Ghiamati, E., Dalterio, R. A., Britton, K. A., Nelson, W. H. & Sperry, J. F. (1990) *J. Microbiol. Methods* **11,** 1–15.
5. Nelson, W. H. & Sperry, J. F. (1991) in *Modern Techniques in Rapid Microorganism Analysis,* ed. Nelson, W. H. (VCH, New York), pp. 97–143.
6. Ghiamati, E., Manoharan, R., Nelson, W. H. & Sperry, J. F. (1992) *Appl. Spectrosc.* **46,** 357–364.
7. Manoharan, R., Ghiamati, E., Chadha, S., Nelson, W. H. & Sperry, J. F. (1993) *Appl. Spectrosc.* **47,** 2145–2150.
8. Committee of the Institute of Medicine and the National Research Council (1999) *Chemical and Biological Terrorism: Research and Development to Improve Civilian Medical Response* (National Academy Press, Washington, DC), p. 90.
9. Scully, M. O. (1991) *Phys. Rev. Lett.* **67,** 1855–1858.
10. Scully, M. O. (1992) *Phys. Rep.* **219,** 191–201.
11. Scully, M. O. & Zubairy, M. S. (1997) *Quantum Optics* (Cambridge Univ. Press, Cambridge, U.K.).
12. Kocharovskaya, O. & Khanin, Y. I. (1988) *Pis'ma Zh. Eksp. Teor. Fiz.* **48,** 581–586.
13. Harris, S. E. (1989) *Phys. Rev. Lett.* **62,** 1033–1036.
14. Scully, M. O., Zhu, S. Y. & Gavrielides, A. (1989) *Phys. Rev. Lett.* **62,** 2813–2816.
15. Zibrov, A. S., Lukin, M. D., Nikonov, D. L., Hollberg, L., Scully, M. O., Velichansky, V. L. & Robinson, H. G. (1995) *Phys. Rev. Lett.* **75,** 1499–1502.
16. Boller, K.-J., Imamoğlu, A. & Harris, S. E. (1991) *Phys. Rev. Lett.* **66,** 2593–2596.
17. Field, J. E., Hahn, K. H. & Harris, S. E. (1991) *Phys. Rev. Lett.* **67,** 3062–3065.
18. Hau, L. V., Harris, S. E., Dutton, Z. & Behroozi, C. H. (1999) *Nature (London)* **397,** 594–598.
19. Kash, M. M., Sautenkov, V. A., Zibrov, A. S., Hollberg, L., Welch, G. R., Lukin, M. D., Rostovtsev, Y., Fry, E. S. & Scully, M. O. (1999) *Phys. Rev. Lett.* **82,** 5229–5232.
20. Kocharovskaya, O., Rostovtsev, Y. & Scully, M. O. (2001) *Phys. Rev. Lett.* **86,** 628–631.
21. Fleischhauer, M. & Lukin, M. D. (2000) *Phys. Rev. Lett.* **84,** 5094–5097.
22. Liu, C., Dutton, Z., Behroozi, C. H. & Hau, L. V. (2001) *Nature (London)* **409,** 490–493.
23. Phillips, D. F., Fleischhauer, A., Mair, A., Walsworth, R. L. & Lukin, M. D. (2001) *Phys. Rev. Lett.* **86,** 783–786.
24. Liang, J. Q., Katsuragawa, M., Kien, F. L. & Hakuta, K. (2000) *Phys. Rev. Lett.* **85,** 2474–2477.
25. Sokolov, A. V., Walker, D. R., Yavuz, D. D., Yin, G. Y. & Harris, S. E. (2001) *Phys. Rev. Lett.*, e-Print Archive, http://link.aps.org/abstract/PRL/v87/e033402.
26. Judson, R. S. & Rabitz, H. (1992) *Phys. Rev. Lett.* **68,** 1500–1503.
27. Kosloff, R., Rice, S. A., Gaspard, P., Tersigni, S. & Tannor, D. J. (1989) *Chem. Phys.* **139,** 201–220.
28. Warren, W. S., Rabitz, H. & Dahleh, M. (1993) *Science* **259,** 1581–1589.
29. Gordon, R. J. & Rice, S. A. (1997) *Annu. Rev. Phys. Chem.* **48,** 601–641.
30. Zare, R. N. (1998) *Science* **279,** 1875–1879.
31. Rabitz, H., de Vivie-Riedle, R., Motzkus, M. & Kompa, K. (2000) *Science* **288,** 824–828.
32. Brixner, T., Damrauer, N. H. & Gerber, G. (2001) *Adv. At. Mol. Opt. Phys.* **46,** 1–54.

33. Brumer, P. & Shapiro, M. (1986) *Chem. Phys. Lett.* **126,** 541–546.
34. Tannor, D. J., Kosloff, R. & Rice, S. A. (1986) *J. Chem. Phys.* **85,** 5805–5820.
35. Bergmann, K., Theuer, H. & Shore, B. W. (1998) *Rev. Mod. Phys.* **70,** 1003–1025.
36. Heritage, J. P., Weiner, A. M. & Thurston, R. N. (1985) *Opt. Lett.* **10,** 609–611.
37. Weiner, A. M., Heritage, J. P. & Kirschner, E. M. (1988) *J. Opt. Soc. Am. B* **5,** 1563–1572.
38. Wefers, M. M. & Nelson, K. A. (1995) *Opt. Lett.* **20,** 1047–1049.
39. Weiner, A. M. (2000) *Rev. Sci. Instrum.* **71,** 1929–1960.
40. Demtröder, W. (1981) *Laser Spectroscopy* (Springer, Berlin).
41. Heid, M., Schlucker, S., Schmitt, U., Chen, T., Schweitzer-Stenner, R., Engel, V. & Kiefer, W. (2001) *J. Raman Spectrosc.* **32,** 771–784.
42. Chen, T., Vierheilig, A., Waltner, P., Heid, M., Kiefer, W. & Materny, A. (2000) *Chem. Phys. Lett.* **326,** 375–382.
43. Zeidler, D., Frey, S., Wohlleben, W., Motzkus, M., Busch, F., Chen, T., Kiefer, W. & Materny, A. (2002) *J. Chem. Phys.* **116,** 5231–5235.
44. Kiefer, W. (2000) *J. Raman Spectrosc.* **31,** 1–144.
45. Oron, D., Dudovich, N., Yelin, D. & Silberberg, Y. (2002) *Phys. Rev.* e-Print Archive, http://link.aps.org/abstract/PRL/v88/e063004.
46. Dudovich, N., Oron, D. & Silberberg, Y. (2002) *Phys. Rev. Lett.* e-Print Archive, http://link.aps.org/abstract/PRL/v88/e123004.
47. Leuschner, R. K. G. & Lillford, P. J. (2001) *Int. J. Food Microbiol.* **63,** 35–50.
48. Black, J. (2002) *Microbiology: Principles and Explorations* (Wiley, New York).
49. Talaro, K. & Talaro, A. (1999) *Foundations in Microbiology* (William C. Brown Publishers, Dubuque, IA).
50. Smekal, A. (1923) *Naturwissenschaften* **11,** 873–875.
51. Compton, A. (1923) *Phys. Rev.* **21,** 483–502.
52. Raman, C. V. & Krishnan, K. S. (1928) *Nature (London)* **121,** 501–502.
53. Landsberg, G. & Mandelstam, L. (1928) *Naturwissenschaften* **16,** 557–558.
54. Boyd, R. W. (1992) *Nonlinear Optics* (Academic, Boston).
55. Harris, S. E. & Sokolov, A. V. (1997) *Phys. Rev. A* **55,** R4019–R4022.
56. Vitanov, N. V., Suominen, K.-A. & Shore, B. W. (1999) *J. Phys. B* **32,** 4535–4546.
57. Jain, M., Xia, H., Yin, G. Y., Merriam, A. J. & Harris, S. E. (1996) *Phys. Rev. Lett.* **77,** 4326–4329.
58. Weiner, A. M., Leaird, D. E., Wiederrecht, G. P. & Nelson, K. A. (1990) *Science* **247,** 1317–1319.
59. Fork, R. L., Brito-Cruz, C. H., Becker, P. C. & Shank, C. V. (1987) *Opt. Lett.* **12,** 483–485.
60. Baltuska, A., Wei, Z., Pshenichnikov, M. S. & Wiersman, D. A. (1997) *Opt. Lett.* **22,** 102–104.
61. Nisoli, M., DeSilvestri, S., Svelto, O., Szipocs, R., Ferencz, K., Spielmann, C., Sartania, S. & Krausz, F. (1997) *Opt. Lett.* **22,** 522–524.
62. Weiner, A. M. (1995) *Prog. Quant. Electr.* **19,** 161–237.
63. Baumert, T., Brixner, T., Seyfried, V., Strehle, M. & Gerber, G. (1997) *Appl. Phys. B* **65,** 779–782.
64. Hillegas, C. W., Tull, J. X., Goswami, D., Strickland, D. & Warren, W. S. (1994) *Opt. Lett.* **19,** 737–739.
65. Assion, A., Baumert, T., Bergt, M., Brixner, T., Kiefer, B., Seyfried, V., Strehle, M. & Gerber, G. (1998) *Science* **282,** 919–922.
66. Murrell, W. (1969) in *The Bacterial Spore,* eds. Gould, G. & Hurst, A. (Academic, New York), Vol. 1, pp. 215–273.

3. HYBRID TECHNIQUE FOR CARS

Working toward a realization of the FAST CARS spore detection scheme, we have carried out many experiments [22–28] and calculations [29–31]. With the timing between the laser pulses as the initial hint (time-resolved CARS), we eventually have developed a comprehensive spectroscopic toolset that utilizes both time and frequency domains (Figure 1; a generic

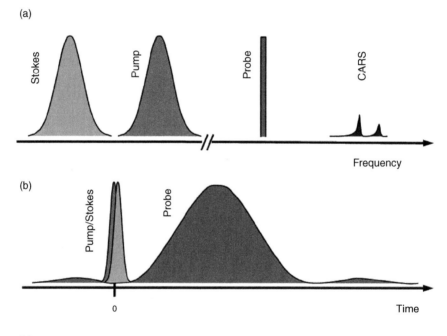

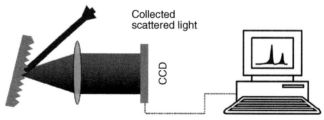

Figure 1 Schematic layout of hybrid coherent anti-Stokes Raman scattering (CARS) technique in the frequency (a) and time (b) domains. Instantaneous coherent broadband excitation of several characteristic molecular vibrations (here we imply the excitation of two Raman modes) and subsequent probing of these vibrations by an optimally-shaped time-delayed narrowband laser pulse help suppress nonresonant (NR) four-wave mixing (FWM) and spectrally discriminate the broadband NR background from the vibrationally resonant CARS signal; (c) multichannel, spectrally-resolved detection of the generated signal allows for robust, fluctuation-insensitive acquisition of the CARS spectrum.

setup layout is given in Figure 2). We refer to it as hybrid CARS [32], although the same approach was also introduced by Stauffer et al. as fs/ps CARS [33], and back in 1980s, by Zinth et al. as short excitation and prolonged interrogation (SEPI) technique [34–36].

In the optimized scheme, we utilize ultrashort transform-limited pulses for uniform broadband Raman excitation but use a tailored narrowband (~1 ps long) laser pulse for probing rather than a broadband one. We also use a multi-channel detector to simultaneously record the anti-Stokes signal at all optical frequencies within the band of interest. These modifications allow for spectral discrimination between the narrowband resonant contribution and broadband NR background and extraction of the CARS signal even at zero probe delay. Furthermore, combining the spectrally-resolved acquisition with the time-delayed probing provides the means to suppress the interfering FWM and associated noise, i.e., offers further optimization of the scheme along the lines of time-resolved CARS.

The basis for this work and the first experimental results were reported in the *Science* article, enclosed in the next pages. Note that later work [37], based on the introduced concept, has yielded single-shot, "on the fly" spore detection (see Figure 3).

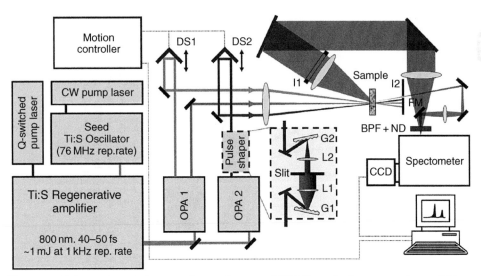

Figure 2 Schematics of the experimental setup. OPA1,2, optical parametric amplifiers; DS1,2, computer-controlled delay stages; I1,2, irises; FM, flip mirror; BPF+ND, bandpass and neutral-density filters; CCD, charge coupled device; inset: home-made pulse shaper. G1,2, ruled gratings (600 grooves/mm, Edmund Optics); L1,2, bi-convex lenses ($f = 20$ cm).

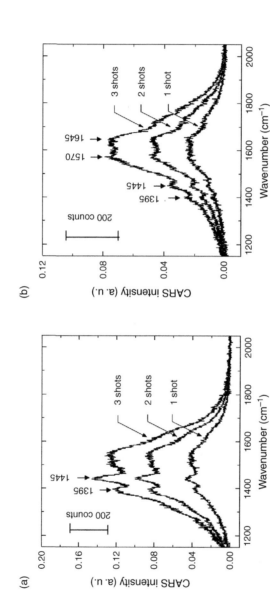

Figure 3 CARS spectra of *Bacillus subtilis* spores obtained through acquisition of the scattered radiation over one, two, and three laser shots. One shot corresponds to firing of two synchronized preparation pulses and a delayed probe pulse. Parameters used: (a) pump wavelength $\lambda_1 = 1.25\ \mu m$, 4 μJ/pulse; Stokes $\lambda_2 = 1.54\ \mu m$, 4 μJ/pulse; probe $\lambda_3 = 805.8$ nm, 3 μJ/pulse; (b) pump and probe pulses are the same as in (a), but Stokes $\lambda_2 = 1.56\ \mu m$, 4 μJ/pulse. The full width at half maximum (FWHM) of the probe pulse is 30 cm^{-1}. Its delay relative to the preparation pulses is 1.3 ps. The scattered CARS photons are collected with a 2-inch lens ($f = 10$ cm) put at the focal distance from the illuminated spores.

REPRINT: D. Pestov *et al.*, *Science* **316**, 265 (2007)

Optimizing the Laser-Pulse Configuration for Coherent Raman Spectroscopy

Dmitry Pestov,[1]* Robert K. Murawski,[1,2] Gombojav O. Ariunbold,[1] Xi Wang,[1] Miaochan Zhi,[1] Alexei V. Sokolov,[1] Vladimir A. Sautenkov,[1] Yuri V. Rostovtsev,[1,2] Arthur Dogariu,[2] Yu Huang,[2] Marlan O. Scully[1,2]

We introduce a hybrid technique that combines the robustness of frequency-resolved coherent anti-Stokes Raman scattering (CARS) with the advantages of time-resolved CARS spectroscopy. Instantaneous coherent broadband excitation of several characteristic molecular vibrations and the subsequent probing of these vibrations by an optimally shaped time-delayed narrowband laser pulse help to suppress the nonresonant background and to retrieve the species-specific signal. We used this technique for coherent Raman spectroscopy of sodium dipicolinate powder, which is similar to calcium dipicolinate (a marker molecule for bacterial endospores, such as *Bacillus subtilis* and *Bacillus anthracis*), and we demonstrated a rapid and highly specific detection scheme that works even in the presence of multiple scattering.

The Raman vibrational spectrum of molecules provides an excellent fingerprint for species identification. Because of the Raman effect, lower-frequency (Stokes) radiation is emitted (Fig. 1A) when light irradiates a molecule. The signal is weak, but with the advent of powerful lasers, spontaneous Raman spectroscopy became a useful technique.

When the molecules are put into coherent oscillation by a pair of preparation pulses (pulses 1 and 2 in Fig. 1B) and a third pulse is scattered off of this coherent molecular vibration, a strong anti-Stokes signal (pulse 4) is generated (Fig. 1B). This process is called coherent anti-Stokes Raman scattering (CARS) [(*1–3*) and references therein].

Unfortunately, CARS from the molecules of interest is frequently obscured by the nonresonant (NR) four-wave mixing (FWM) signal from other molecules (Fig. 1C), or even from the same molecules, because of contributions from multiple off-resonant vibrational modes and the instantaneous electronic response. This unwanted NR FWM is often much stronger than the resonant signal because there are usually many more background molecules than target molecules. Fluctuations of the NR background can, and frequently do, completely wash out the CARS signature (*4*). A variety of methods, including polarization-sensitive techniques (*5*) and heterodyne (*6, 7*) and interferometric (*8–11*) schemes, have been developed to increase the signal-to-background ratio. However, these methods do not work well in the presence of strong multiple scattering in rough samples because scattering randomizes

spectral phases and polarization. For this reason, applications of CARS for rapid detection and recognition of strongly scattering media, such as anthrax spores, have often been deemed impractical.

In our recent work, we have developed techniques for maximizing the coherent molecular oscillation and minimizing the NR background. A sequence of femtosecond pulses is used so that pulses 1 and 2 prepare a coherent molecular vibration (corresponding to a particular mode of oscillation) and a third time-delayed probe pulse is scattered off the oscillation, yielding the anti-Stokes signal. By delaying the third pulse relative to the first two, the NR FWM signal is eliminated. The combination of shaped preparation pulses and an ultrashort time-delayed probe pulse maximizes the signal and lessens the background contribution. We call this approach femtosecond adaptive spectroscopic technique via CARS (FAST-CARS) (*12*). The definitive fingerprint information is retrieved from the probe-delay dependence, but the delay tuning slows down the acquisition and makes the technique vulnerable to fluctuations.

In the present work, we used ultrashort (~50 fs) transform-limited pulses 1 and 2 for uniform broadband Raman excitation and a tailored narrowband (several hundred femtosecond long) probe. We used a multichannel detector to simultaneously record the anti-Stokes signal at all optical frequencies within the band of interest. These modifications allowed us to discriminate between the resonant contribution and the NR background, as explained below, and to extract the CARS signal even at zero probe delay. Furthermore, combining this method with the probe-pulse delay provided a means to suppress the interfering FWM and associated noise.

This combination of broadband preparation and frequency-resolved multichannel acquisition (*13–16*) with time-delayed narrowband probing yields a very sensitive and robust technique that allows us to identify bacterial endospores, such as anthrax, in real time. We refer to this technique as hybrid CARS for short, and we used it to study the Raman signature of endospores, which is dominated by the contribution from the dipicolinic acid (DPA) molecules.

In order to provide better insight into the advantages of the proposed scheme, we review here a few important aspects of the theory behind CARS, which has been extensively described elsewhere (*1–3, 8, 11, 14, 17*). In general, the third-order polarization induced by the pump, Stokes, and probe pulses can be split into resonant and NR contributions. If we assume that there are no one-photon resonances involved, the NR part can be attributed to the instantaneous electronic response and a sum over the contributions from far-detuned Raman transitions. The resonant part of the third-order polarization is attributed to the Raman transitions of interest. The NR component, $\chi_{NR}^{(3)}$, of the nonlinear susceptibility is usually frequency-insensitive within the considered spectral band and can be treated as a constant. The resonant component $\chi_R^{(3)}$, under the assumption of Lorentz-shape Raman lines, can be presented as $\chi_R^{(3)}(\omega_1 - \omega_2) = \sum_j A_j \Gamma_j / [\Omega_j - (\omega_1 - \omega_2) - i\Gamma_j]$, where ω_1 and ω_2 are the pump and Stokes laser

[1]Institute for Quantum Studies and Departments of Physics and Chemical Engineering, Texas A&M University, College Station, TX 77843, USA. [2]Applied Physics and Materials Science Group, Engineering Quadrangle, Princeton University, Princeton, NJ 08544, USA.

*To whom correspondence should be addressed. E-mail: dmip@neo.tamu.edu

Fig. 1. Level diagram and schematic of different scattering processes on simple molecules. In this example, CO is a target molecule and SO is a background molecule. **(A)** Incoherent Raman scattering (pulse 2) was derived from laser pulse 1 scattering off the CO molecule. **(B)** CARS signal 4 was derived from probe

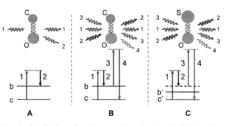

pulse 3 scattering off the CO molecular vibration, coherently prepared by pulses 1 and 2. **(C)** One of the possible channels for the NR background generation in SO. c, the ground state of the CO molecule; b, the target vibrational state of the CO molecule; c', the ground state of the background molecule; b', an off-resonant vibrational state of the background molecule.

REPRINT: D. Pestov *et al.*, *Science* **316**, 265 (2007)

frequencies, j is the summation index over all covered Raman transitions, A_j is a constant related to the spontaneous Raman cross section of the jth Raman transition and molecular density, Γ_j denotes the jth Raman line half-width, Ω_j denotes the jth vibrational frequency (17), and $i = \sqrt{-1}$. The summation is held over all of the Raman transitions involved (17).

In the frequency domain (8), the third-order polarization $P^{(3)}(\omega)$ can be written as

$$P^{(3)}(\omega) = P^{(3)}_{NR}(\omega) + P^{(3)}_R(\omega) = \int_0^{+\infty} d\Omega \times [\chi^{(3)}_{NR} + \chi^{(3)}_R(\Omega)]E_3(\omega - \Omega) \times S_{12}(\Omega),$$ where $P^{(3)}_{NR}$ is the NR component of the polarization, $P^{(3)}_R$ is the resonant component, $E_3(\omega)$ is the spectral amplitude of the probe pulse, and $S_{12}(\Omega) \equiv \int_0^{+\infty} d\omega' \times E_1(\omega')E_2^*(\omega' - \Omega)$ is the convolution of the pump and Stokes field amplitudes, $E_1(\omega)$ and $E_2(\omega)$. The signal arising from the nonlinear response of the medium is proportional to $|P^{(3)}(\omega)|^2$, so the spectra generally have complex shapes caused by the interference between both resonant contributions from different vibrational modes and the NR background. A straightforward analogy with the spontaneous Raman spectra can be made only for well-separated lines with no NR background. Otherwise, the direct fit of the recorded CARS spectrum is required for Raman spectrum retrieval (18).

The convolution of the pump and Stokes spectra, $S_{12}(\Omega)$, enters the two parts of the third-order polarization on equal grounds. It defines a Raman frequency band covered by the preparation pulses and is maximized for transform-limited ones; i.e., pulses with the constant spectral phase. The difference between the two contributions comes from the susceptibility and can be enhanced by the use of a properly shaped probe. One way to proceed is to modify the spectral phase of the probe pulse, as it was demonstrated by Oron *et al.* (9). Another way, which seems to be more robust and straightforward, is to shape its spectral amplitude, $|E_3(\omega)|$, as we did in our experiment. If a narrowband probe is applied together with the broadband transform-limited preparation pulses, the NR contribution inherits a smooth, featureless profile of $S_{12}(\Omega)$ with some characteristic width $\Delta\omega_{12}$, whereas the resonant part generates a set of narrow peaks (one for each excited vibrational mode) whose width is determined either by the Raman linewidth or the probe spectral width, $\Delta\omega_3$, whichever is greater.

The amplitude ratio between the resonant signal and the NR background at a Raman-shifted frequency is also affected by the spectral width of the probe pulse. At the zero probe delay, the ratio is inversely proportional to the square of the probe spectral width when the last one is in between the Raman linewidth and the width of the pump-Stokes convolution profile (i.e., $\Gamma \ll \Delta\omega_3 \ll \Delta\omega_{12}$) and the probe-pulse energy is fixed. This ratio saturates at the limits

and leads to a superior but finite signal-to-background ratio for the optimum probe width on the order of the Raman linewidth.

The present measurement strategy combines the benefits of frequency-resolved CARS signal discrimination against the NR FWM (pointed out above) with the NR background suppression {as in time-resolved CARS $[(19–22)$ and references therein]}. Indeed, when the probe delay is adjustable, further optimization is possible. In the plane of two parameters (the probe-pulse duration and its delay), the resonant response peaks for both parameters on the order of the inverse Raman linewidth. However, the NR FWM at the Raman-shifted frequency is maximized for zero probe delay, and the probe-pulse duration is matched to the time span of the pump-Stokes convolution profile.

As mentioned above, we can eliminate the NR background by just delaying the probe pulse, which gives a theoretically unlimited signal-to-background ratio. Unfortunately, this approach does not properly optimize the resonant contribution, and we can end up with no detectable signal at all. We suggest the simultaneous use of the two parameters (the probe-pulse duration and its delay) to achieve close-to-optimal resonant response with reasonable suppression of the NR background. The actual optimal values of the parameters depend on the Raman linewidth, the sensitivity of the setup used, and the relative strength of the resonant and NR susceptibilities.

Proper tailoring of the probe pulse can help to reduce the contribution of the NR background for probe delays comparable to the pulse length. For example, a rectangular-like spectrum gives a sinc-

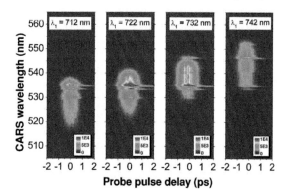

Fig. 2. CARS spectrograms recorded on NaDPA powder at different pump wavelengths. The CARS spectrum is shown as a function of the probe-pulse delay for the pump wavelength $\lambda_1 = 712, 722, 732,$ and 742 nm, respectively (left to right). The other parameters are: pump, full width at half maximum (FWHM) ~12 nm, 2 µJ per pulse; Stokes, $\lambda_2 = 803$ nm, FWHM ~32 nm, 3.9 µJ per pulse; probe, $\lambda_3 = 577.9$ nm, FWHM ~0.7 nm, 0.5 µJ per pulse. The integration time was 1 s per probe-delay step. 1E4 $\equiv 1 \times 10^4$.

Fig. 3. Cross sections of the CARS spectrograms from Fig. 2 for two probe delays, **(A)** 0 ps and **(B)** 1.5 ps. The wavelengths within the observed range were transferred into the Raman shift, relative to the probe central frequency. The integration time was 1 s: 0.5 s for the signal and 0.5 s for the background acquired for the delayed Stokes pulse. The absolute frequencies of the Raman transitions in NaDPA, observed in the CARS experiment and spontaneous Raman measurements, are summarized in Table 1. CCD, charge-coupled device.

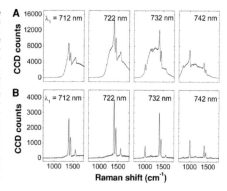

REPRINT: D. Pestov *et al.*, *Science* **316**, 265 (2007)

squared temporal profile of the probe-pulse intensity, which goes as $[\sin(\Delta\omega_3 t/2)/(\Delta\omega_3 t/2)]^2$ with the time t. Putting the preparation pulses in one of its nodes would result in the effective suppression of the NR background.

We compromise between the resolution, signal strength, and the extent of the NR background suppression. On a single-shot basis, the spectral resolution is usually determined by the probe bandwidth. However, this is not an intrinsic limit, and much better resolution can be achieved by recording the anti-Stokes spectrum while varying the probe-pulse delay, if the measurements are not overwhelmed by the fluctuations.

As mentioned above, we applied the developed technique in the context of the spore detection problem. A marker molecule for bacterial spores is calcium dipicolinate (CaDPA), which accounts for 10 to 17% of the bacterial spore dry weight (*12*). We focus here on NaDPA, which is easier to make and is a good surrogate for CaDPA. The spontaneous Raman spectrum of NaDPA exhibits a similar set of strong Raman lines as that of CaDPA. Both differ somewhat from the Raman spectrum of DPA itself (*23*). The important point is that although endospores are fairly complex in structure, their Raman spectra are dominated by several vibrational modes of CaDPA.

The details on our implementation of the hybrid CARS technique and the setup schematics can be found online (*24*). The CARS spectra of NaDPA powder as a function of the probe-pulse delay are shown in Fig. 2. The spectrograms were taken at different pump wavelengths to cover the whole spectral-fingerprint region of the molecule (800 to 1700 cm^{-1}). Streaklike horizontal lines are the signature of excited NaDPA Raman transitions. The broadband pedestal is the NR background. As expected, the tuning of the pump wavelength spectrally shifts the NR FWM but leaves the position of the resonant lines untouched. Also, the resonant and NR contributions exhibit different dependencies on the probe delay. The magnitude of the NR background is determined by the overlap of the three laser pulses and follows the probe-pulse profile. However, a relatively long decay time of the Raman transitions under consideration favors their long-lasting presence and makes them stand out when the probe is delayed.

The cross sections of the spectrograms at two different probe delays are given in Fig. 3. The integration time for each of those is only 1 s. When the three pulses are overlapped (zero delay), the resonant contribution is severely distorted by the interference with the NR FWM. Delaying the probe by 1.5 ps—that is, putting the preparation pulses into the first node of the sinc-shaped probe pulse—improves the signal-to-background ratio by at least one order of magnitude. In this case, the NR background suppression is limited by multiple scattering that scrambles the timing between the preparation and probe photons.

The absolute frequencies of the observed Raman transitions calculated from the retrieved peak positions and the probe wavelength are summarized in Table 1. Comparison with the data from spontaneous Raman measurements shows a remarkably good match.

Extracted CARS contributions from our first measurements on *Bacillus subtilis* spores (a surrogate for anthrax), in which we maximized the signal rather than the signal-to-background ratio, are summarized in Fig. 4. The Raman peaks are not normalized on the strength of the excitation and thus have an imprint of the pump-Stokes spectral convolution function, which sweeps through the Raman band from 800 to 1700 cm^{-1} while the pump wavelength is tuned. We assign the Raman transitions in the band (Table 1) and compare the retrieved line positions with the known positions from spontaneous Raman measurements (*25, 26*). Within the estimated experimental uncertainty of 15 cm^{-1}, the values are in good agreement. The data shown in Fig. 4 were acquired at zero probe delay over 2 min, although the Raman lines stand out from the background even after a few seconds of integration. Under similar experimental conditions, the signal arising from spontaneous Raman scattering is weaker by a few orders of magnitude and typically requires a longer integration time than the CARS signal.

To place the present work in context, our approach comes from the superposition of two well-known techniques developed over the past few decades and employed for combustion diagnostics [(*1, 21, 22*) and references therein] and chemically selective microscopic biological imaging (*27, 28*). Multifrequency acquisition has been implemented in so-called broadband or multiplex CARS (*14–17*), where together with the multichannel detection, a combination of narrowband pump-probe and broadband Stokes pulses is used to address a wide range of vibra-

Table 1. The observed Raman peaks and their calculated absolute frequencies for NaDPA and *B. subtilis* spores. The third column lists the frequency values from spontaneous Raman measurements.

Peak (nm)	CARS Raman shift (cm^{-1})	Spontaneous Raman shift (cm^{-1})
	NaDPA powder	
529.7	1575	1572*
533.5	1440	1442*
534	1395	1395*
540.8	1187	1189*
541.8	1153	1152*, 1157*
543.9	1082	1087*
546.2	1004	1007*
551.9	815	817*, 827*
	B. subtilis spores	
527.8	1643	1655†, 1624‡
529.9	1568	1572†, 1581‡
531.2	1524	1539‡
533.6	1437	1445†, 1447‡
535.2	1381	1395†, 1396‡
537.3	1308	1280‡
539.3	1239	1245†
540.7	1191	1192‡
543.9	1082	
546.4	998	1001†, 1013† / 981‡, 1018‡
551.8	819	822†

*See figure S2 in (*24*).   †From (*25*).   ‡From (*26*).

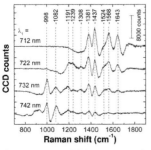

CCD counts

$\lambda_1 =$ 712 nm
722 nm
732 nm
742 nm

-998 -1082 -1191 -1239 -1308 -1381 -1437 -1524 -1568 -1643

8000 counts

800 1000 1200 1400 1600 1800
Raman shift (cm^{-1})

Fig. 4. CARS on *B. subtilis* spores at zero probe delay. The resonant contribution was retrieved by fitting the NR background with a smooth curve and subtracting it from the total acquired signal. The sample used was a pullet of spores fixed in a rotating sample holder. The pump wavelength, λ_1, was varied from 712 to 742 nm. Other parameters were the same as those for Fig. 2. The integration time was 2 min. The comparisons between the retrieved Raman frequencies and the available spontaneous Raman data are given in Table 1.

REPRINT: D. Pestov *et al.*, *Science* **316**, 265 (2007)

tional frequencies. In this degenerate scheme, the NR background has been addressed by means of the polarization-sensitive and interferometric techniques mentioned above.

A delayed probe has been used in time-resolved CARS (*19–22*). That technique uses ultrashort pulses for preparation and probing. Its ultimate source of species-specific information is multimode interference in the probe-delay signal profile, generally referred to as quantum beats (*19*). Time-resolved CARS eliminates the NR contribution by delaying the probe pulse, but the technique still remains vulnerable to fluctuations. It has been successfully applied to polycrystalline and opaque solids (*29*) to observe vibrational dephasing of single excited Raman transitions. However, the use of the multimode interference pattern for species recognition requires the ability to record high-quality quantum-beat profiles over a relatively large probe-delay span and therefore is challenging in the presence of scattering and fluctuations.

In our scheme, a generalized broadband or multiplex CARS technique is combined with background suppression by means of an optimal sequence of coherent excitation and time-delayed probe pulses. A schematic overview, pointing out the similarities and differences of the relevant established schemes and the one introduced here, is available online [figure S3 in (*24*)]. In short, we diverge from the conventional broadband CARS arrangement and deal with the probe and the two preparation pulses, pump and Stokes, separately. By adjusting the probe-pulse delay and its

spectral width, we suppress the NR background, as in time-resolved CARS, but keep the advantages of the frequency-resolved multiplex CARS spectroscopy. The experimental data demonstrate the efficacy of the ultrafast broadband excitation and time-variable narrowband probing, whereas the described implementation supports the versatility of the technique.

References and Notes
1. P. R. Régnier, J. P.-E. Taran, *Appl. Phys. Lett.* **23**, 240 (1973).
2. J. W. Nibler, G. V. Knighten, in *Raman Spectroscopy of Gases and Liquids*, A. Weber, Ed. (Springer-Verlag, New York, 1979), pp. 253–299.
3. T. G. Spiro, *Biological Applications of Raman Spectroscopy* (Wiley, New York, 1987).
4. W. M. Tolles, R. D. Turner, *Appl. Spectrosc.* **31**, 96 (1977).
5. J. L. Oudar, R. W. Smith, Y. R. Shen, *Appl. Phys. Lett.* **34**, 758 (1979).
6. Y. Yacoby, R. Fitzgibbon, B. Lax, *J. Appl. Phys.* **51**, 3072 (1980).
7. E. O. Potma, C. L. Evans, X. S. Xie, *Opt. Lett.* **31**, 241 (2006).
8. D. Oron, N. Dudovich, D. Yelin, Y. Silberberg, *Phys. Rev. A* **65**, 043408 (2002).
9. D. Oron, N. Dudovich, D. Yelin, Y. Silberberg, *Phys. Rev. Lett.* **88**, 063004 (2002).
10. S. H. Lim, A. G. Caster, S. R. Leone, *Phys. Rev. A* **72**, 041803 (2005).
11. T. W. Kee, H. Zhao, M. T. Cicerone, *Opt. Express* **14**, 3631 (2006).
12. M. O. Scully *et al.*, *Proc. Natl. Acad. Sci. U.S.A.* **99**, 10994 (2002).
13. D. Klick, K. A. Marko, L. Rimai, *Appl. Phys.* **20**, 1178 (1981).
14. A. Voroshilov, C. Otto, J. Greve, *J. Chem. Phys.* **106**, 2589 (1997).
15. H. Kano, H. Hamaguchi, *Appl. Phys. Lett.* **85**, 4298 (2004).
16. G. I. Petrov, V. V. Yakovlev, *Opt. Express* **13**, 1299 (2005).
17. Y. R. Shen, *The Principles of Nonlinear Optics* (Wiley, New York, 1984).
18. H. A. Rinia, M. Bonn, M. Muller, *J. Phys. Chem. B* **110**, 4472 (2006).
19. R. Leonhardt, W. Holzapfel, W. Zinth, W. Kaiser, *Chem. Phys. Lett.* **133**, 373 (1987).
20. A. Materny *et al.*, *Appl. Phys. B* **71**, 299 (2000).
21. P. Beaud, H.-M. Frey, T. Lang, M. Motzkus, *Chem. Phys. Lett.* **344**, 407 (2001).
22. R. P. Lucht, S. Roy, T. R. Meyer, J. R. Gord, *Appl. Phys. Lett.* **89**, 251112 (2006).
23. P. Carmona, *Spectrochim. Acta A* **36**, 705 (1980).
24. The details of the experimental setup, a schematic overview of the established schemes (time-resolved CARS and multiplex/broadband CARS), and the technique introduced in this work are available as supporting material on *Science* Online.
25. A. P. Esposito *et al.*, *Appl. Spectrosc.* **57**, 868 (2003).
26. W. H. Nelson, R. Dasari, M. Feld, J. F. Sperry, *Appl. Spectrosc.* **58**, 1408 (2004).
27. M. D. Duncan, J. Reintjes, T. J. Manuccia, *Opt. Lett.* **7**, 350 (1982).
28. A. Zumbusch, G. R. Holton, X. S. Xie, *Phys. Rev. Lett.* **82**, 4142 (1999).
29. X. Wen, S. Chen, D. D. Dlott, *J. Opt. Soc. Am. B* **8**, 813 (1991).
30. We thank J. Laane and K. McCann for their generous assistance with spontaneous Raman measurements and gratefully acknowledge support from the Office of Naval Research (award N00014-03-1-0385), the Defense Advanced Research Projects Agency, NSF (grant PHY-0354897), an award from the Research Corporation, and the Robert A. Welch Foundation (grants A-1261 and A-1547).

Supporting Online Material
www.sciencemag.org/cgi/content/full/316/5822/268/DC1
Materials and Methods
Figs. S1 to S3
References

19 December 2006; accepted 26 February 2007
10.1126/science.1139055

4. COHERENT VERSUS INCOHERENT RAMAN SCATTERING

The instantaneous Raman excitation of molecular vibrations is a step that brings ultrafast coherent Raman spectroscopy to another level of complexity, as compared with the ordinary Raman measurements. A fair question to ask is how much one benefits from it. We have addressed this question by comparative experimental analysis of spontaneous and coherent Raman scattering on liquids and opaque solids.

In particular, we have made such a comparison for two strong Raman lines, $992 \, \mathrm{cm}^{-1}$ and $1031 \, \mathrm{cm}^{-1}$, of pyridine. Spontaneous Raman and CSRS spectra of the excited vibrational modes, acquired at the laser setup in Figure 2 with a 200-μm cuvette of pyridine, are shown in Figure 4. It was found that for the given interaction length and reasonably high energies of the preparation pulses ($\sim 1 \, \mu J/$ pulse, $\sim 50 \, \mathrm{fs}$; the estimated peak intensity $\sim 4 \times 10^{11} \, \mathrm{W/cm}^2$), the number of generated CSRS photons exceeds the number of spontaneous Raman photons by a factor 10^5. For further details on the experiment, please refer to Ref. [38].

The observed 10^5-fold gain in the efficiency of coherent Raman scattering over incoherent one comes entirely from the phase-locked accumulation of the signal over the active volume. For similar excitation strength but the interaction length $\sim 1 \, \mu m$, typical for microscopy, one would have the enhancement factor in the order of 500 [39]. The last estimate agrees with the measurements on the powder of sodium dipicolinate (NaDPA), Figure 5, where the coherent buildup of the

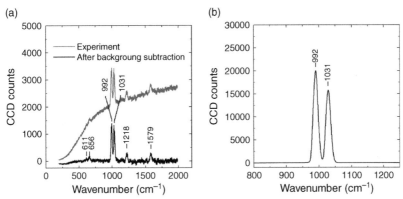

Figure 4 Spontaneous versus coherent Raman scattering: (a) spontaneous Raman spectrum of pyridine. Probe pulse parameters: $\lambda_3 = 577.9$ nm, full width at half maximum (FWHM) ≈ 19 cm^{-1}, average power is 0.13 mW (0.13 μJ/pulse, 1 kHz rep. rate). Integration time is 3 min. The estimated collection angle is $\sim 0.004 \times 4\pi$. The smooth background profile is determined by the transmission of the bandpass filters set in front of the spectrometer. Cuvette path length is 200 μm. (b) CSRS spectrum of pyridine. The two Raman lines, 992 cm^{-1} and 1031 cm^{-1}, are selectively excited via a pair of ultrashort pulses. The probe delay relative to the preparation pulses is fixed and equal to 1.8 ps. Pump: $\lambda_1 = 738$ nm, FWHM ~ 260 cm^{-1}, 0.72 μJ/pulse; Stokes: $\lambda_2 = 802$ nm, FWHM ~ 480 cm^{-1}, 1.36 μJ/pulse. Probe parameters and the sample are the same as in spontaneous Raman measurements. With 10^4 attenuation by a set of neutral-density filters, integration time is 1 s.

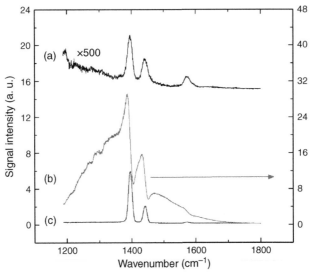

Figure 5 Spontaneous versus coherent Raman scattering on strongly scattering solids: (a) Spontaneous Raman spectrum recorded on sodium dipicolinate (NaDPA) powder with probe pulses at $\lambda_3 = 805.5$ nm (FWHM ≈ 14 cm^{-1}) and average power 1.7 mW (1.7 μJ/pulse, 1 kHz rep. rate). The signal is normalized on the integration time (4 s) and multiplied by a factor of 500; (b) normalized CSRS spectrum, obtained with the same probe pulse and ultrashort preparation pulses. Pump $\lambda_1 = 1.24$ μm, 10 μJ/pulse; Stokes $\lambda_2 = 1.54$ μm, 10 μJ/pulse. The probe pulse delay is zero, i.e., it is overlapped with the preparation pulses. (c) Same as in (b) but the probe pulse delay is 1.85 ps. The pump and Stokes pulses are put on the node of the shaped probe pulse to suppress the interfering nonresonant (NR) background.

frequency-shifted radiation is mitigated by a strong scattering within the sample. The experiment with NaDPA powder is similar to the one with pyridine but the laser pulse wavelengths are shifted into IR domain ($\lambda_1 = 1285$ nm, $\lambda_2 = 1565$ nm, $\lambda_3 = 805.5$ nm). Also, the frequency-shifted photons are collected in the back-scattering geometry for both CSRS and spontaneous Raman measurements.

From the curves (a) and (c) in Figure 5, one gets the ratios between the peaks at 1395, 1442, and 1572 cm^{-1} in CSRS and spontaneous Raman spectra to be 484, 274, and 38, respectively. Obviously, they depend on the frequency difference between the pump and Stokes laser light. The ratios are also affected by the energies of the excitation pulses, which were chosen close to the laser-induced damage threshold. The sample was continuously rotated to avoid its heat-assisted damage.

The results suggest three-orders-of-magnitude enhancement in the photon flux as a figure-of-merit for coherent Raman spectroscopy of NaDPA powder, as an example of highly scattering medium. Note though that the three-orders-of-magnitude increase of the acquisition time will not result in spontaneous Raman spectrum with similar signal-to-noise ratio. This is because the noise amplitude also increases with the acquisition time, T_{acq}. In particular, shot noise scales as $\sqrt{T_{acq}}$.

5. SUMMARY

The obtained results convey the utility of coherent Raman spectroscopy for detection of biohazards, such as *Bacillus anthracis*, as well as other applications requiring chemical specificity and acquisition speed. The proper choice of the laser pulse configuration yields efficient suppression of the optical background associated with parasitic nonlinear processes and allows taking advantage of coherent excitation of molecular vibrations. The performed experiments demonstrate single-shot-detection capability of the devised CARS technique. However, a lot of work still needs to be done for its implementation as a single, low-cost field device.

ACKNOWLEDGMENTS

We are grateful to our co-workers and collaborators for their invaluable contribution to the project, stimulating discussions, and continuous support. In particular, we would like to thank the team members at Texas A&M University (Miaochan Zhi, Xi Wang, Robert K. Murawski, Gombojav O. Ariunbold, Yuri V. Rostovtsev, Vladimir A. Sautenkov, George R. Welch) and Princeton University (Arthur Dogariu and Yu Huang). This work was sponsored by the Office of Naval Research under Award No. N00014-03-1-0385, Defense Advanced Research Projects Agency, the National Science Foundation (Grant No. PHY-0354897), an Award from Research Corporation, and the Robert A. Welch Foundation (Grants No. A-1261 and A-1547).

REFERENCES

[1] A. Fasanella, S. Losito, R. Adone, F. Ciuchini, T. Trotta, S. A. Altamura, D. Chiocco, and G. Ippolito, Journal of Clinical Microbiology **41**: 896–899, 2003.

[2] D. King, V. Luna, A. Cannons, J. Cattani, and P. Amuso, Journal of Clinical Microbiology **41**: 3454–3455, 2003.

[3] J. F. Powell, The Biochemical Journal **54**: 210–211, 1953.

[4] B. D. Church and H. Halvorson, Nature **183**: 124–125, 1959.

[5] P. M. Pellegrino, N. F. Fell, and J. B. Gillespie, Analytica Chimica Acta **455**: 167–177, 2002.

[6] Y. S. Cheng, E. B. Barr, B. J. Fan, P. J. Hargis, D. J. Rader, T. J. O'Hern, J. R. Torczynski, G. C. Tisone, B. L. Preppernau, S. A. Young, and R. J. Radloff, Aerosol Science and Technology **30**: 186–201, 1999.

[7] P. Carmona, Spectrochimica Acta Part a-Molecular and Biomolecular Spectroscopy **36**: 705–712, 1980.

[8] A. P. Esposito, C. E. Talley, T. Huser, C. W. Hollars, C. M. Schaldach, and S. M. Lane, Applied Spectroscopy **57**: 868–871, 2003.

[9] E. Ghiamati, R. Manoharan, W. H. Nelson, and J. F. Sperry, Applied Spectroscopy **46**: 357–364, 1992.

[10] S. Farquharson, L. Grigely, V. Khitrov, W. Smith, J. F. Sperry, and G. Fenerty, Journal of Raman Spectroscopy **35**: 82–86, 2004.

[11] W. M. Tolles and R. D. Turner, Applied Spectroscopy **31**: 96–103, 1977.

[12] W. M. Tolles, J. W. Nibler, J. R. Mcdonald, and A. B. Harvey, Applied Spectroscopy **31**: 253–271, 1977.

[13] Y. R. Shen, The principles of Nonlinear Optics, John Wiley & Sons, Inc., Hoboken, New Jersey 2003.

[14] A. Assion, T. Baumert, M. Bergt, T. Brixner, B. Kiefer, V. Seyfried, M. Strehle, and G. Gerber, Science **282**: 919–922, 1998.

[15] D. Meshulach and Y. Silberberg, Physical Review A **60**: 1287–1292, 1999.

[16] V. V. Lozovoy, B. I. Grimberg, E. J. Brown, I. Pastirk, and M. Dantus, Journal of Raman Spectroscopy **31**: 41–49, 2000.

[17] R. A. Bartels, T. C. Weinacht, S. R. Leone, H. C. Kapteyn, and M. M. Murnane, Physical Review Letters **88**: 033001, 2002.

[18] N. Dudovich, D. Oron, and Y. Silberberg, Nature **418**: 512–514, 2002.

[19] D. Oron, N. Dudovich, D. Yelin, and Y. Silberberg, Physical Review A **65**: 043408, 2002.

[20] D. Oron, N. Dudovich, and Y. Silberberg, Physical Review Letters **90**: 213902, 2003.

[21] D. Oron, N. Dudovich, D. Yelin, and Y. Silberberg, Physical Review Letters **88**: 063004, 2002.

[22] G. Beadie, J. Reintjes, M. Bashkansky, T. Opatrny, and M. O. Scully, Journal of Modern Optics **50**: 2361–2368, 2003.

[23] G. Beadie, M. Bashkansky, J. Reintjes, and M. O. Scully, Journal of Modern Optics **51**: 2627–2635, 2004.

[24] M. Mehendale, B. Bosacchi, E. Gatzogiannis, A. Dogariu, W. S. Warren, and M. O. Scully, Journal of Modern Optics **51**: 2645–2653, 2004.

[25] D. Pestov, M. C. Zhi, Z. E. Sariyanni, N. G. Kalugin, A. A. Kolomenskii, R. Murawski, G. G. Paulus, V. A. Sautenkov, H. Schuessler, A. V. Sokolov, G. R. Welch, Y. V. Rostovtsev, T. Siebert, D. A. Akimov, S. Graefe, W. Kiefer, and M. O. Scully, Proceedings of the National Academy of Sciences of the United States of America **102**: 14976–14981, 2005.

[26] A. Dogariu, Y. Huang, Y. Avitzour, R. K. Murawski, and M. O. Scully, Optics Letters **31**: 3176–3178, 2006.

[27] Y. Huang, A. Dogariu, Y. Avitzour, R. K. Murawski, D. Pestov, M. C. Zhi, A. V. Sokolov, and M. O. Scully, Journal of Applied Physics **100**: 124912, 2006.

[28] M. C. Zhi, D. Pestov, X. Wang, R. K. Murawski, Y. V. Rostovtsev, Z. E. Sariyanni, V. A. Sautenkov, N. G. Kalugin, and A. V. Sokolov, Journal of the Optical Society of America B-Optical Physics **24**: 1181–1186, 2007.

[29] Z. E. Sariyanni and Y. Rostovtsev, Journal of Modern Optics **51**: 2637–2644, 2004.

[30] G. Beadie, Z. E. Sariyanni, Y. V. Rostovtsev, T. Opatrny, J. Reintjes, and M. O. Scully, Optics Communications **244**: 423–430, 2005.
[31] Y. V. Rostovtsev, Z. E. Sariyanni, and M. O. Scully, Physical Review Letters **97**: 113001, 2006.
[32] D. Pestov, R. K. Murawski, G. O. Ariunbold, X. Wang, M. C. Zhi, A. V. Sokolov, V. A. Sautenkov, Y. V. Rostovtsev, A. Dogariu, Y. Huang, and M. O. Scully, Science **316**: 265–268, 2007.
[33] B. D. Prince, A. Chakraborty, B. M. Prince, and H. U. Stauffer, The Journal of Chemical Physics **125**: 044502, 2006.
[34] W. Zinth, Optics Communications **34**: 479–482, 1980.
[35] W. Zinth, M. C. Nuss, and W. Kaiser, Chemical Physics Letters **88**: 257–261, 1982.
[36] W. Zinth, M. C. Nuss, and W. Kaiser, Optics Communications **44**: 262–266, 1983.
[37] D. Pestov, X. Wang, G. O. Ariunbold, R. K. Murawski, V. A. Sautenkov, A. Dogariu, A. V. Sokolov, and M. O. Scully, Proceedings of the National Academy of Sciences of the United States of America **105**: 422–427, 2008.
[38] D. Pestov, G. O. Ariunbold, X. Wang, R. K. Murawski, V. A. Sautenkov, A. V. Sokolov, and M. O. Scully, Optics Letters **32**: 1725–1727, 2007.
[39] G. I. Petrov, R. Arora, V. V. Yakovlev, X. Wang, A. V. Sokolov, and M. O. Scully, Proceedings of the National Academy of Sciences of the United States of America **104**: 7776–7779, 2007.

CHAPTER 8

High-Resolution Laboratory Terahertz Spectroscopy and Applications to Astrophysics

Stephan Schlemmer, Thomas Giesen, Frank Lewen, *and* Gisbert Winnewisser

Contents

Abstract

High-resolution terahertz (THz) spectroscopy is a vivid field of current research due to its importance for astrophysics and molecular physics. New THz telescope facilities like ALMA, Herschel, and SOFIA make use of accurate ($\Delta \nu < 1\,MHz$) laboratory line positions in the frequency range up to 2 THz to identify interstellar molecules and to derive their astrophysical abundance. This chapter describes current laboratory activities to provide the necessary transition frequencies for molecular rotation and ro-vibration in the range from 300 GHz up to 3 THz. The continuous development and fabrication of monochromatic submillimeter (sub-mm) wave radiation sources is a key to recent successes of THz spectroscopy in laboratory and space. High-resolution spectrometers, based on backward wave oscillators, multiplier techniques, and laser side band generation, are specified, and examples of molecules studied at the Cologne laboratories are presented. General features of the instruments such as frequency accuracy and sensitivity are described in detail. Applications concern the study of light hydrides, complex molecules, radicals, and molecular ions. These transient species are very important molecules in the astrophysical environment because of their chemical activity. In the laboratory, they are experimentally as well as theoretically demanding. Combination of

supersonic jets and ion traps to THz spectroscopy is discussed as two recent developments in our laboratory to meet the challenge.

Keywords: terahertz; sub-mm wavelength; laboratory spectroscopy; astromolecules;
laser side band spectrometer; multiplier spectrometer; ion-trap; supersonic jet

1. INTRODUCTION

The development of powerful monochromatic radiation sources for centimeter (cm) and millimeter (mm) wavelengths in the 1940s led to a new era in
spectroscopy which allowed for precise studies of gas phase molecules with
unbiased accuracy. The rotational energy states and transitions of many molecules have been studied since then under laboratory conditions, and thus
supporting the development of radio-astronomy. More than 140 molecules
have been identified to date in the interstellar medium by the means of
radio-astronomy [1]. Our knowledge of the chemical composition and physical
conditions of interstellar clouds is mainly due to high-resolution spectroscopic
observations in the cm- and mm-wavelength region. For reviews, see Winnewisser
[2–5] and Herbst [6–8].

It is noteworthy that at least half of the luminous power of the Milky Way
galaxy is emitted at sub-mm wavelengths [9,10] which is largely because of the fact
that the spectral energy distribution (SED) according to Planck's law peaks at these
wavelengths for the cold environment between stars. Consequently, this spectral
range has been discovered in recent years by mostly ground-based telescopes and is
the target of a new generation of telescopes, such as the atacama path finder
experiment (APEX), which is operational since 2006 [11], the airborne telescope
Stratospheric Observatory For Infrared Astronomy (SOFIA) [12], and the HIFI and
PACS receivers onboard the Herschel satellite [13], both ready for operation in
2009. Finally, ALMA, a set of up to 66 12-m telescopes, currently built in the
Chilean Atacama desert at high altitudes, will give an unprecedented view on the
chemical inventory of the universe at sub-mm wavelength with uttermost spectral
and spatial resolution [14].

The sub-mm wavelengths from 1 mm to 100 microns correspond to frequencies
of 300 GHz–3 THz. In the terahertz (THz) region, numerous small to medium-size
molecules have characteristic spectral lines, either by means of (i) pure rotational or
(ii) low-lying ro-vibrational transitions. Another interesting aspect is related to the
fact that the photon energy $E = h\nu$, and thus the excitation energy of low-lying
vibrational modes is of the order of the thermal energy, $E = kT$. As a consequence
these modes are excited in the warm gas of star forming regions and even in cold
interstellar molecular clouds with a typical temperature of 10 K. Because of this
circumstance, (iii) rotational transitions of vibrationally excited species are of general interest although the spectra are getting much more complicated. Moreover,
lines from (iv) isotopologues can contribute significantly to interstellar spectra. This

becomes obvious for carbon, where the second most prominent isotope ^{13}C has a relative abundance to the main isotope ^{12}C in the few percent range. Moreover, in the cold environments in space isotopic enrichment due to the incorporation of deuterium instead of hydrogen can play an important and interesting role. Therefore, spectra of isotopologues are becoming a very sensitive tool for astrophysicists.

To exploit the wealth of spectral information coming especially from line surveys of present day [15,16] and future THz telescopes, laboratory spectra of the molecules of interest have to be supplied. The new instruments will provide high-resolution data up to some 2 THz. In addition, PACS onboard Herschel will provide lower resolution data up to some 5 THz. Therefore, line predictions based on laboratory spectra have to be available over the same spectral range for the interpretation of the observations [1,17–19]. The sensitivity of the telescope instruments is reaching the so-called confusion limit, where the baseline is no longer given by the noise of the receiver but by the emission signal from overlapping molecular lines. As a consequence, all spectral contributions (i) − (iv) have to be considered also in the laboratory approach.

The spectral resolution at which these data are needed for astronomy is largely given by the Doppler width, which is typically a few MHz at a transition frequency of 1 THz. In practice, the spectral resolution has to be in the order of $\Delta\nu/\nu = 10^{-7}$. Fourier transform IR spectrometers, which have the necessary frequency coverage, however, do not have sufficient resolution. Figure 1 shows the spectrum of H_2S_2 at 978 GHz. The lower trace spectrum has been obtained using the high-resolution Fourier transform spectrometer at Giessen University [20]. The upper trace spectrum has been recorded with a highly monochromatic radiation source, a backward wave oscillator (BWO) [21]. The Doppler-limited line width is 1.5 MHz, which is easily resolved by a BWO spectrometer of a few kHz spectral resolution. On the contrary, the FT spectrometer with 2 m optical path length reaches a spectral resolution of 2.5×10^{-3} cm^{-1} or 75 MHz which is not sufficient to resolve the Doppler-limited profile.

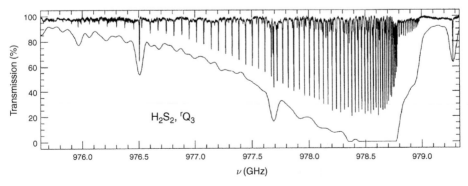

Figure 1 $^rQ_3$ branch of the rotational spectrum of H_2S_2 at 978 GHz. Upper trace: fully resolved spectrum using a free running BWO as radiation source. Lower trace: high-resolution FTIR spectrum taken at Giessen university, [20].

The focus of this chapter is on high-resolution gas phase laboratory spectroscopy in the THz region. In the following sections, the current development and the use of absorption spectrometers and their technical specifications will be discussed. Special emphasis will be given to recent applications, and examples of future experimental developments will be explained.

2.   HIGH-RESOLUTION TERAHERTZ SPECTROMETERS

High-resolution laboratory spectroscopy is carried out mostly in absorption. The spectrometers consist of a bright, monochromatic, and tunable radiation source, an absorption cell, and a sensitive detector. In our laboratory, usually liquid He cooled Bolometers with a noise equivalent power in the order of 10^{-12} W/Hz$^{1/2}$ are used for robust detection of even weak absorption THz signals.

Much effort is put in the development and operation of the radiation sources to largely cover the spectral region desired. The key characteristics of the radiation sources are as follows: (i) spectral purity defined by the single sideband phase noise, (ii) tuning range, (iii) continuous wave operation, (iv) output power of the source to match the sensitivity of the detector to record even small absorption signals, (v) the beam quality, (vi) the dynamic range of the combination source/detector which enables the detection of large as well as weak signals alike, and (vii) the accuracy of the determination of a line center. Of course, in practice, also the scanning speed, the handling, and the costs play an important role in designing the instrument. In this chapter, we only consider radiation sources that meet the constraints on the resolution as this is the prime target for the astrophysical application. The spectrometers are discussed in the following sections, guided by the use of the corresponding radiation source. An overview list of the sources used in our laboratory is shown in Table 1. Also the molecules studied in recent years are listed with the corresponding experimental frequency range of investigation.

A number of sophisticated techniques has been developed to produce tunable monochromatic radiation in the THz region. For review, see Siegel [22] and references therein. Only few monochromatic light sources give direct access to the THz region. BWO tubes have been widely used for frequencies from 50 GHz to 1.5 THz. A more recent, promising development is the quantum cascade laser (QCL). Originally operating in the mid-IR region, it has now reached operation in the far-infrared (FIR) region. Successful development of a QCL for 1.4 THz radiation has been reported by Walther et al. [23] and successfully locked to a precise frequency standard by a phase locked loop (PLL) loop at Cologne.

Other approaches to THz radiation rely on (i) frequency up–conversion of cm- and mm–wave generators by multiplier devices, (ii) on sideband generation combining high-frequency sources and tunable low-frequency sources, or (iii) on frequency down-conversion techniques, such as laser mixing.

Table 1 Terahertz radiation sources: astrophysically relevant molecules investigated recently at Cologne

Spectrometer	THz–Source	Multiplication Factor	Molecules	Range/THz	Refs
Fundamental Mode	BWO		aGg'-$C_2H_4(OH)_2$	0.05–0.37	[113]
			CD_3OD	0.06–0.23	[114]
			gGg'-$C_2H_4(OH)_2$	0.08–0.58	[115]
			CD_3OD	0.08–0.91	[116]
			CH_3CN	0.09–0.81	[117]
			$C^{17}O, ^{13}C^{17}O$	0.10–0.98	[34]
			$^{39,41}K^{35,37}Cl$	0.17–0.93	[118]
			C_4H_4	0.18–0.79	[119]
			$NaCl, Na^{37}Cl$	0.20–0.93	[120]
			Kr (Rydbg.)	0.24–0.38	[121]
			SiS	0.40–0.93[a]	[122]
			CH_2	0.43–0.60	[123]
			$HCS^+, DCS^+, HC^{34}S^+$	0.46–0.94	[124]
			PH_2	0.46–0.95	[125]
			H_2CS	0.56–0.93	[126]
			$HCOOH, H^{13}COOH$	0.84–0.99[a]	[127]
			CH_2	0.93–0.96	[128]
Multiplier	Planar Schottky[a]	×8	HC_5N	0.06–0.28	[129]
(BWO)[b]			HCN	0.09–1.59[c]	[53]
	Planar MoMed	×3	CCC	1.96	[83]
	P. Cont. Whisker membrane	×3			
(All solid state)[b]	Planar Schottky	×72	HSOH	1.2–1.4[d]	[130]
			H_2D^+, D_2H^+	1.3–1.5[d]	[110]
Superlattice	100 GHz	×(2n+1)	DC_3N	0.07–0.90[a,e]	[131]
(BWO)[b]	250 GHz	×(2n+1)	ND_2H, D_2O	1.0–2.7	[55]
			D_2O	0.08–2.7[a,e]	[58]
			ND_2H	0.08–2.58	[59]

Table 1 *(Continued)*

Spectrometer	THz-Source	Multiplication Factor	Molecules	Range/THz	Refs
Sideband	FIR laser + BWO[f]	-	DNC	0.68–1.98[c]	[69]
			H^{13}CN, H^{13}C^{15}N, HC^{15}N	0.08–0.95, 2.0[c]	[36]
			CH$_2$	1.91–1.96	[123]
			NH	1.9–2.0	[132]
			D^{13}CN,DC^{15}N,D^{13}C^{15}N	0.08–2.0[c]	[35]
			HSOH	0.09–1.9[c]	[133]
			CCC	1.9–2.0	[67]
			H$_2$CO	0.83–0.96, 1.76–2.01[c]	[68]
			HC^{14}N	0.05–0.98, 1.9–2.0[c]	[134]
			HCN	0.79–0.98, 1.78–1.97[a,c]	[135]
			D$_2$O	0.08–2.7[a,c]	[58]

[a] With additional measurements from other laboratories.
[b] Sources in parantheses denote the pump medium.
[c] Small tuning range around center frequency.
[d] Commercial multiplier chain, Virginia Diodes Inc.
[e] Frequency range includes measurements with fundamental BWO.
[f] BWO, backward wave oscillator; FIR, far-infrared.

In the following section, we present BWOs (Section 2.1) as monochromatic high-frequency radiation sources operating in the fundamental mode, Schottky multipliers (2.2) and superlattice (SL) multipliers (2.3) as examples for frequency multiplication, as well as laser sideband generation (2.4) using BWOs for broadband tunability. The selection is based on current use in our laboratory. More techniques are discussed in Ref. [24].

2.1. BWOs, powerful radiation sources for terahertz spectroscopy

One of the most powerful sources for monochromatic radiation up to above 1 THz is the BWO. High-frequency accuracy, broadband tunability, and high output power are the main features that make BWOs well suited radiation sources for many spectroscopic applications. Monochromatic electromagnetic radiation is generated from velocity-modulated electrons, interacting with a slow-wave structure. Electrons from a heated electrode are accelerated in a strong electrical field of several kV. An axial homogenous magnetic field of the order of 1 T confines the electrons traveling at some 10% of speed of light to a narrow beam with current densities as high as $300 \, A/cm^2$. A periodic structure of some tens of µm lattice constant set to an electrostatic potential of 20 V reduces the phase velocity of electromagnetic waves radiated by the electron beam below the vacuum speed of light. In resonance with the eigenmodes of the periodic structure, the electrons supply energy to stimulate the electromagnetic modes to oscillate. The oscillation frequency depends on (i) the electron velocity and (ii) the spacing of the slow-wave structure. A higher acceleration voltage causes a faster oscillation and thus a higher output frequency. BWOs have large frequency tuning ranges of 10–20% of their center frequency. Tubes with various slow-wave structures have been designed to cover the spectral range from 36 GHz to 1.4 THz [25] with output powers of some tens of mW at low and medium frequencies (300–600 GHz) to less than 1 mW at the high-frequency end. For a more detailed description of the principals, see Refs [26–28] and references therein. BWOs are commercially available from Istok, Moscow region, Russia. The high spectral resolution of BWOs is demonstrated in Figure 1 as discussed above. A set of BWO tubes is used in our laboratory to cover the spectral range up to 1 THz. Recent studies on a large number of astrophysically relevant molecules with BWO-based spectrometers are listed in Table 1.

2.1.1. Principle of phase locking

Frequency accuracy is a main issue for precise THz spectroscopy. Doppler-limited lines of small molecules at room temperature are in the order of a few hundred kHz in width. Molecules in interstellar clouds can be as cold as a few Kelvin with line profiles of a few 10 kHz, where additional line features, such as hyperfine structure components, are partly resolved. In the laboratory, sub-Doppler spectroscopy can be applied to resolve spectral features of a few 10 kHz which requires a frequency accuracy of $\Delta\nu/\nu = 10^{-9}$.

In general, microwave tube oscillators are inherently noisy and unstable in frequency. Without further frequency stabilization, the line width of BWO

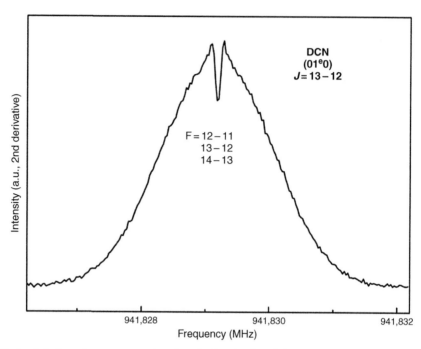

Figure 2 Sub-Doppler spectrum showing the Lamb-dip of the $J = 13 - 12$ transition of DCN at 942 GHz, [35].

radiation in the free running mode is typically 10–50 MHz, which is mainly due to phase noise, instabilities of the voltage supply, and inhomogeneities of the magnetic field. The spectroscopic accuracy of a BWO is drastically improved by several orders of magnitude and to below 1 kHz in a phase-locked stabilization mode. The technique of phase locking is well established in mm-wave spectroscopy and has been applied to low-frequency BWOs as early as 1970 [29]. In 1993, the Cologne spectroscopy group reported the first PLL BWO at 1 THz [30–32]. A review on phase-locked BWO has been published by Krupnov [33]. The high–resolution possibilities of BWO applications have been demonstrated by Lamb-dip spectra for $C^{17}O$ at 449 GHz [34], DCN up to 942 GHz [35] (Figure 2), $H^{13}CN$ at 604 GHz [36], and $D^{13}C^{15}N$ at 764 GHz [35]. In favorable cases, underlying hyperfine structure transitions could be resolved with $\Delta\nu$ as low as 30 kHz with a corresponding accuracy of 1–5 kHz.

2.2. Schottky multipliers and their applications

Since the early days of high-resolution microwave spectroscopy, frequency multiplication was and still is an important experimental method to extend microwave tunability and accuracy to higher frequencies. In general, the non-linear current–voltage relation (I–V curve) or the non-linear capacitance–voltage relation (C–V curve) of the corresponding semiconductor material is used to generate higher order harmonics of the input wave. Fast development of communication technology in the 1990s lead to a significant improvement of high-frequency devices. Current developments at THz frequencies mainly concern the preparation for the new telescopes mentioned above [22,37,38].

Today, BWOs with sufficient output power, Gunn diodes, and commercial synthesizers are used as fundamental frequency sources for input of multiplier devices. Whisker-contacted diodes have been the most common multipliers in this frequency range. They are more and more replaced by planar Schottky devices such as air-bridged diodes or devices on thin membranes [39–41]. Amplification of the corresponding mm-wave sources supplies enough power for multiplication purposes up to input frequencies of some 120 GHz. Advanced state-of-the-art Schottky diode devices for multiplication are based on GaAs semiconductors. Present fabrication processes of these microdevices allow for high-conversion efficiencies at rather high frequencies. These devices are designed by optimizing the semiconductor doping concentration, the break down voltage, heat conductivity, the anode diameter size, and the epitaxial layer thickness. Also, low series resistance, low parasitic capacities, and high cut-off frequencies in excess of 3–5 THz are optimization goals. New mechanical and electrical designs for the multiplier blocks have been developed. Mainly cross guide waveguides or inline waveguide blocks, either mechanically milled or electro-formed, are used to achieve broadband response, low standing wave ratios, and a good idler matching within the operation band [42,43].

In summary, the development of multiplier configurations is subject of current research [22,44–48], see below for applications in spectroscopy. However, multipliers and chains of multipliers up to 1.9 THz are commercially available from various commercial suppliers [41,49,50] and are ready for use as local oscillators in astronomy. The use of such chains up to frequencies above 2 THz in molecular spectroscopy has recently been described thoroughly by Drouin et al. [51]. Today, multiplier spectrometers can be considered the work horses in high-resolution THz spectroscopy because these instruments combine most criteria defined above: high spectral resolution, large scanning range, and sufficient power for spectroscopic applications to name the most important factors.

In Cologne, various THz diode multiplier sources, see list given in Table 1, have been developed, tested, and applied for spectroscopy. Figure 3 shows the principal setup of a multiplier spectrometer. In the case shown, the spectrometer is fed by the fundamental of a BWO source [52]. Part of the input radiation is split off for the phase lock circuit while most of the power is used for the generation of harmonics in the multiplier device. The characteristics of specific multiplier devices used in our laboratory are given in the references listed in Table 1.

The planar Schottky diode used in Figure 3 allowed for a tuning range from 0.2 up to 1.6 THz. In this range, a number of *high-J* rotational transitions of HCN have been recorded [53]. Figure 4 displays the lines between 1 and 1.6 THz referring to the 4th and 8th harmonic of the BWO fundamental operating from 180 to 265 GHz. These lines clearly show the broadband applicability of the multiplier spectrometer. Another new type of multiplier is the SL device which has led to new records in high-frequency applications as will be described in the following section.

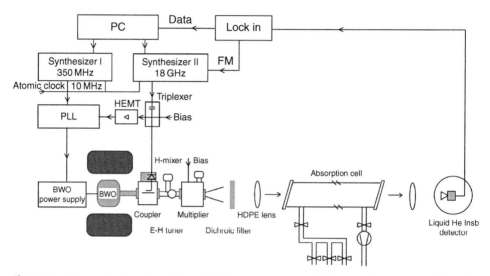

Figure 3   Schematic drawing of a multiplier spectrometer. The setup here uses a Schottky diode multiplier. Similar configurations use the new superlattice (SL) devices.

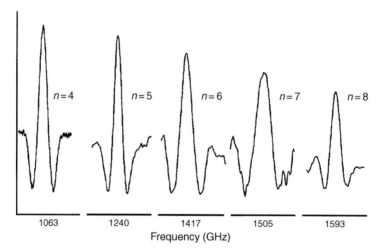

Figure 4   Operation of a Schottky diode multiplier spectrometer. Rotational transitions of HCN in the range of 1–1.6 THz.

2.3.   Superlattice multipliers and their applications

The SL device consists of a periodic arrangement of thin layers that are fabricated by molecular beam epitaxy [54]. The electron energy is restricted to minibands with a specific miniband gap. The transport of electrons leads to a negative differential resistance. Above a threshold value, the I–V characteristic of SL devices is highly non-linear and antisymmetric, making it suitable for high-frequency multiplication

[55]. The understanding and design of SL devices is an active field of research [56]. The devices used in our applications are made up of 18 monolayers GaAs and 4 monolayers AlAs with a total number of 18 periods (thickness 112 nm) [57].

These SL multipliers have been used in combination with two continuously tunable BWO tubes as input radiation sources, delivering output powers of 10–60 mW throughout the entire frequency range from 78 to 118 GHz for SL I and from 180 to 260 GHz for SL II. For test purposes, spectra of the rotational lines of CO have been recorded. Because of the SL's antisymmetric I–V curve, only the odd harmonics of the input frequency were produced. Absorption up to the $J = 11 \leftarrow 10$ transition have been detected. Careful calibration has allowed to derive the output powers of the SL device from the measured line intensities [55]. Figure 5 shows the relative intensities of the spectrometer as a function of frequency. Interestingly, the power drop corresponds to 0.053 ± 0.004 dB/GHz, which is even smaller than the value derived for Schottky diodes, which reach typical values of 0.08 dB/GHz.

Overall, the measurements with our SL spectrometer demonstrate that (i) the SL-operated sources are producing higher order odd harmonics in sufficient supply for laboratory spectroscopy. (ii) High-frequency transitions for D_2O [58] and ND_2H [59] in the range from 0.7 to 3.1 THz with multiplication factors up to 13 have been used. Figure 6 shows a rotational transition for D_2O at 2.7 THz. Another absorption line of D_2O has recently been detected at 3.1 THz. To our knowledge, this is the highest frequency transition recorded in high resolution using a multiplier. (iii) Very high line accuracies can be achieved, which has been shown for various lines of ND_2H. Based on SL spectra, its spectroscopic constants have been improved [59]. In addition, (iv) broadband scans are feasible and have

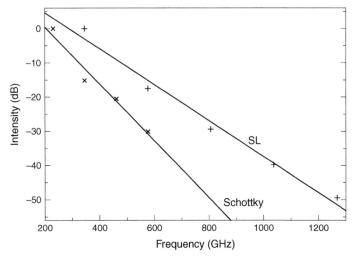

Figure 5   Relative intensities determined from rotational transitions in CO in dependence of the superlattice (SL) harmonic associated with this measurement. Measurements performed with a Schottky multiplier are shown for comparison.

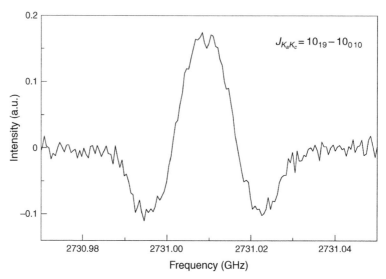

Figure 6 Rotational transition of D_2O at 2.7 THz taken with superlattice II (SL II) configuration.

been demonstrated for methanol using the SL I setup, [55]. Finally, (v) the SL-based spectrometers prove to be rather robust because of an intrinsic current limitation of the SL device [55]. Currently we set up an all solid state SL-based spectrometer which covers the frequency range up to 1 THz at full computer control. SL devices are also used for heterodyne detection of THz signals in the same spectrometer. As a result remote data aquisition for a large number of molecules at high frequencies is becoming reality.

2.4. Cologne sideband spectrometer for THz applications

An alternative approach to generate THz radiation is the FIR laser sideband technique. Radiation from an optically pumped FIR gas laser is superimposed by radiation of a tunable monochromatic microwave source on a fast, non-linear mixing device to generate sidebands at the sum and difference frequencies of the two radiation sources. FIR gas lasers, when operated with different gain media, such as CH_3F, CH_3OH, and $HCOOH$, have thousands of laser lines in the range of 0.2 to over 10 THz. Strong laser lines at a few 10 mW mixed with microwave radiation from a Yttrium Iron Garnet (YIG) oscillator (2–4 GHz) [60], a frequency multiplied synthesizer (up to 75 GHz) [61], or a klystron (22–114 GHz) [62] have been used to generate tunable laser sideband signals up to 3 THz on a GaAs–Schottky barrier diode. The microwave is either coupled to the mixer coaxially or via waveguides. The separation of sidebands from the fundamental laser radiation is achieved by a diplexer in combination with a Farbry–Perot interferometer or by a grating monochromator. Frequency tuning of the laser sidebands is achieved by tuning the frequency of the microwave radiation source.

In 1997, the Cologne group presented a FIR sideband laser system using a PLL-locked BWO (280–380 GHz) as tunable microwave source [63]. The high frequency of the BWO leads to sidebands well separated from the carrier frequency, which can easily be extracted by a grating monochromator. Both the optical coupling of BWO radiation onto the mixer and the clearly separated sidebands have substantially simplified the use of laser sideband techniques for THz applications. A schematic overview of the layout of the Cologne sideband spectrometer for THz applications (COSSTA) is presented in Figure 7. Aside from the traditional absorption cell and the hot carrier liquid-He-cooled, magnetically tuned InSb detector, COSSTA consists of three essential units: (i) the frequency-stabilized FIR laser system, (ii) the evacuated optics and mixer arrangement, which serves to superimpose the BWO and FIR beams, and (iii) the BWO phase lock assembly.

The tunability of the system is basically determined by the frequency coverage of the BWOs, which is about 100 GHz at 330 GHz center frequency. An instantaneous frequency scanning capability of 10 GHz has been demonstrated by recording the $^rQ_{13}$ branch of the near-prolate asymmetric-top molecule DSSD in the frequency range between 1872 and 1882 GHz. Figure 8 displays a fraction of the spectrum that shows contributions from DSSD and the minor isotopologue DS^{34}SD. The frequency stability of this sideband spectrometer is limited solely by the instability of the FIR laser frequency, which is typically in range of 15 MHz. To overcome this limitation, an analog automatic frequency control (AFC) loop was developed to synchronize the FIR laser frequency of 1.626 THz with the 15th harmonic of a Gunn oscillator (InP Gunn, 108 GHz), which is phase-locked to a stable reference frequency [64]. The sub-mm and FIR radiation are coupled

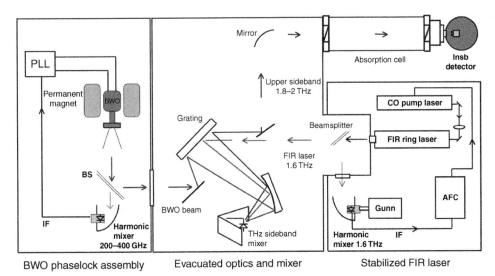

Figure 7 Schematic drawing of the Cologne sideband spectrometer for terahertz applications (COSSTA). The far-infrared (FIR) laser as well as the backward wave oscillator (BWO) output used for sideband generation are stabilized to reach very high-resolution radiation in the range 1.75 to 2.01 THz.

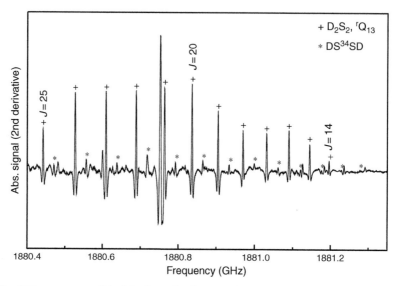

Figure 8   THz spectrum of the $^rQ_{13}$ branch of DSSD observed with COSSTA.

quasi-optically onto a whisker-contacted Schottky diode for generating the side-band power. The different frequencies are multiplexed with a low-loss, blazed grating (Figure 7). Finally, the sideband beam passes the absorption cell and is focused onto the liquid-He-cooled InSb bolometer. With a 1T6 Schottky diode (University of Virginia), sideband power levels of $1.6\,\mu W$ were reached. To avoid power losses introduced by water-vapor absorption, a vacuum box for the optical system was installed. The entire sideband beam and the main part of the laser beam is guided inside the vacuum box. Various applications of COSSTA to molecular spectroscopy can be found in Table 1 and in Refs [37,65–69].

 ## 3.   FUTURE TRENDS IN TERAHERTZ SPECTROSCOPY

3.1.   Supersonic jet-spectroscopy

Supersonic jets are nowadays widely in use for spectroscopic investigations of reactive, short living molecules and molecular ions [70], weakly bound van der Waals complexes [71], and large biomolecules [72]. In the adiabatic expansion of a buffer gas, the rotational temperature of seeded molecules drops to a few Kelvin, and even temperatures below 1 K are feasible. For details on jet techniques, see the excellent book by Scoles [73].

Because of the low temperatures in a supersonic expansion, the molecular partition functions are strongly reduced in favor over standard gas cell experiments at room temperature, leaving only the lowest rotational energy level to be thermally populated. Adiabatically cooled molecules have thus significantly simplified rotational spectra with strongly enhanced line intensities which compensate for the usually short optical path lengths of the probing beam through a jet.

A number of techniques have been developed over the last two decades to introduce reactive short living molecules or species with almost zero vapor pressures into an adiabatically expanding carrier gas. Two important techniques are (i) laser ablation of solid targets and (ii) plasma–induced techniques such as electrical discharge or electron bombardment [74] applied to precursor gases diluted in a non–reactive buffer gas [75]. Both techniques have been applied to produce species that are of particular interest for radio–astronomy.

Laser ablation of a graphite rod placed in front of a supersonically expanding jet leads to the formation of cold carbon clusters of various sizes. Pure carbon chains are expected to play an important role in the formation process of interstellar organic molecules. The Berkeley spectroscopy group has applied UV–laser ablation technique in a supersonic jet to study a number of carbon clusters by means of absorption spectroscopy in the mid- and far-infrared regions [76]. Laser ablation has also been used to produce small silicon–carbon clusters [77,78]. Since 1995, these studies have been continued at Cologne and led to the detection of C_8 and C_{10} in the IR wavelength range [67,79–82]. The lowest bending modes of carbon chain molecules lie in the THz region and are thus of great interest for observations with SOFIA and the Herschel satellite. Although progress has been made over the last years, to date only the lowest bending mode of C_3 has been studied. Figure 9 shows the Q(10) transition of C_3 at 1.9 THz measured with the Cologne Supersonic Jet

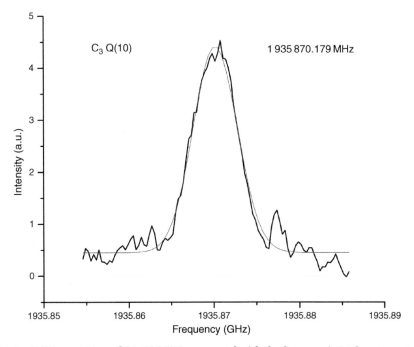

Figure 9 Q(10) transition of C_3 at 1.9 THz measured with the Supersonic Jet Spectrometer for Terahertz Applications (SuJeSTA). This instrument combines the formation of transient species using either laser ablation or pulsed discharges, jet cooling and multipass absorption spectroscopy in the tetrahertz range.

Spectrometer for Terahertz Applications (SuJeSTA). Here, the pulsed jet with carbon clusters from a laser ablation source has been combined with a THz spectrometer operating at 1.9 THz. The radiation source consists of a phase-locked BWO at 635 GHz, whose output is tripled in a Schottky multiplier [83].

While laser ablation has been primarily used for the production of pure clusters, discharge nozzles have been applied by several groups to produce highly unsaturated carbon chains [84–87]. A number of radicals and radical ions have been studied in this manner over a wide frequency range from a few GHz to the far- and mid-IR region and up to the optical spectral region although spectral data in the THz region are still sparse.

Spectroscopy of molecules in supersonic jets usually suffers from short optical path lengths. In a number of applications, this disadvantage over spectroscopy in static cells has been overcome by applying multipass techniques such as multi-reflection optics or optical resonators to increase the path of the probing beam through the jet. Most significant examples to be mentioned are Fourier transform microwave (FTMW) spectrometers. For example, more than 60 new carbon- and silicon-containing molecules have been detected in this manner by Pat Thaddeus and collaborators at Harvard University. Many of these molecules were later detected in space, making use of the precise and distinct laboratory data.

For technical reasons, the frequency range of FTMW spectrometers is constrained, and operation is usually limited to about 50 GHz. To our knowledge, the only type of cavity-enhanced mm-wave spectrometer above 50 GHz is the ORO-TRON, developed at Troitsk Institute for Spectroscopy, Russia. OROTRONs for continuous wave (cw) operation have been developed over the last decade and successfully used in our laboratories for spectroscopy in the range of 70–180 GHz [89]. This development has extended the FTMW operational range by more than two octaves at comparable sensitivity. The highest frequency generated by an OROTRON has been 360 GHz but in pulsed mode operation. Measurements of extremely weak "forbidden" rotational transitions of SiH_4 [89] and rare isotopic species of OCS in natural abundance, for example $^{18}O^{13}C^{34}S$ [90], demonstrate the high sensitivity of the instrument and allow to determine the minimal detectable absorption coefficient (MDAC) to be better than 3×10^{-10}/cm [91].

3.2. Spectroscopy in ion traps

Traditional absorption spectroscopy has to fight for the detection of a small signal change, ΔI, on top of a large signal, I_0. Despite this difficulty, it is a rather successful method as demonstrated in the preceding sections. In radio-astronomy, the emission signal from molecules is detected against the small 2.7 K radiation background. Provided rather sensitive detection schemes like in present day heterodyne techniques [92], emission spectroscopy is an interesting alternative in terms of sensitivity. Over the last couple of years, we developed a rather different spectroscopic method to reach even higher sensitivities for the spectroscopy of molecular ions. This approach belongs to the group of action spectroscopy and happens in an electrodynamic ion trap. Details are given in a number of publications focusing on the IR spectroscopy of molecular ions [93–96].

Multipole ion traps, like quadrupole and ocotpole, have been developed to study ion–molecule reactions [97]. In our 22-pole ion trap apparatus mass selected ions are stored over long periods of time (ms up to 100 s) [98,99]. Here they interact with a neutral collision partner. In our method, we promote an ion–molecule reaction by the excitation of the ion in the trap. For that purpose, the light from a laser or from some other radiation source is guided onto the axis of the trap, where it interacts with the stored ions and enhances the formation of product ions.

One example concerns the astrophysically relevant $C_2H_2^+$ cation. In collisions with H_2, it is largely unaffected as its reaction to form $C_2H_3^+ + H$ is slightly endothermic. The only reaction which happens in rather cold environments, like in the cold trap or in space, is radiative association, where the collision complex $C_2H_4^+$ is stabilized via the emission of a photon. Although this is an unlikely event, this process has been studied in detail in the 22-pole ion trap at low temperatures. In our experiments, we excited the ν_3 C–H stretching vibration and the ν_5 cis-bending vibration of $C_2H_2^+$ via various IR laser sources [94,96,100]. The additional energy in the ion leads to a substantial increase of the hydrogen abstraction channel forming $C_2H_3^+$. For the spectroscopic purposes, the number of $C_2H_3^+$ products are detected as a function of the excitation wavelength. Figure 10 shows the result of such a laser-induced reaction (LIR) experiment. It clearly shows the rovibrational lines of $C_2H_2^+$. It has to be pointed out that only a few hundred to some thousand ions are necessary to record such a LIR spectrum. This ultra-high sensitivity is reached due to the fact that ions can be detected with almost 100% efficiency. Because of this, action spectroscopy in traps has become a rather active field of research. Especially, photodissociation of complexes [101,102] and larger biomolecules [103–105] but also photodetachment of anions [106] have been employed recently. In our setup, the ions of interest are already formed in a trap.

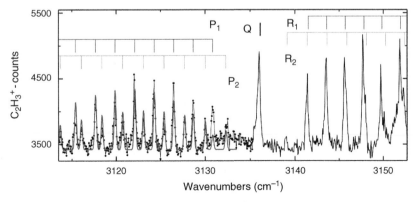

Figure 10 Rotationally resolved laser-induced reaction (LIR) spectrum of the antisymmetric CH stretching vibration of $C_2H_2^+$. $C_2H_3^+$ products monitor the excitation of the parent ions. For the P branch, a simulated spectrum is shown as a solid line, assuming a rotational temperature of $T = 90$ K.

Therefore, primary ions like $C_2H_2^+$ but also secondary ions like H_3^+, CH_5^+ and so on can be prepared and spectra taken in the 22-pole ion trap [93–97,100,107,108].

Another example of our method concerns the spectroscopy of the H_2D^+ molecule [109]. In this case, we promote the formation of H_3^+ in collisions with H_2. This is an endothermic reaction ($\Delta E = 232$ K) which is at the center of interest to understand isotopic enrichment in interstellar molecules. We used excitation of higher overtones and combination bands to enhance the reactivity. The observed high-resolution line positions for H_2D^+ and D_2H^+ were compared to state-of-the-art theoretical predictions including the treatment of the break-down of the Born–Oppenheimer approximation for these light species. Mode specific differences between theoretical predictions and experiment have been found and are discussed in Ref. [109]. Because the endothermicity is rather small, it is possible to promote the hydrogenation reaction by rotational excitation only. First results from this very exciting extension of our method to pure rotational spectroscopy in the THz regime have already been published [110]. Based on these laboratory experiments a search for the lowest transition in *para*-H2D+ will be started using the APEX telescope. Such observations will be most important for understanding the coldest region in the universe.

4.   CONCLUSIONS

Terahertz spectroscopy is a very active field of research. It is largely supported by the needs of astronomy, and many laboratories are collecting data which are supplied via the JPL and CDMS line catalogues for use in astrophysical applications. Moreover, as seen even in the very limited selection of examples in this chapter, THz spectroscopy is an exciting field of research on its own, covering a large number of interesting aspects of molecular physics ranging from molecular structure to molecular reactivity.

In recent years, many technical difficulties have been overcome, and this chapter demonstrates the accomplishment of high-resolution and substantial scanning range of existing instruments. However, the field is still facing several challenges. Small dipole moments and large partition functions have excluded the study of large sets of important molecules. Therefore, sensitivity is one of the key aspects which needs improvement. One example concerns the polycylcic aromatic hydrocarbons (PAHs). They are observed in space via the very prominent infrared bands with characteristic features for the various vibrational modes, [111] and references therein. Although responsible for a considerable fraction of carbon in space, not a single PAH has been identified to date. Dedicated laboratory emission THz spectra at high-resolution might set the starting shot for understanding interstellar carbon chemistry in much greater detail. THz spectra of complex molecules in general are considered a key instrument in astrophysics which could ultimately answer the question of forming prebiotic molecules in space.

Efficient formation, cooling and sensitive detection of transient molecules as demonstrated in our jet as well as ion trap experiments will lead to studies of otherwise inaccessible species and frequency ranges. The rotational spectrum of a molecule such as CH_5^+ can be considered as one of the holy grails of spectroscopy [112]. The ultimate sensitivity of ion action spectroscopy in traps will perhaps allow us to finally determine the high-resolution FIR spectra of only a few hundred molecular ions. It is the aim of our work to further develop the methods to get a handle on these and other exciting molecules.

ACKNOWLEDGMENTS

This work has been supported by the Deutsche Forschungsgemeinschaft (DFG) via grant SFB494. The authors thank the many students and postdocs who contributed to this work.

REFERENCES

[1] Cologne Database for Molecular Spectroscopy (CDMS): http://www.cdms.de.

[2] G. Winnewisser, C. Kramer, Spectroscopy between the stars, Space Sci. Rev. 90 (1999) 181–202.

[3] G. Winnewisser, E. Herbst, Interstellar-Molecules, Rep. Prog. Phys. 56 (1993) 1209–1273.

[4] G. Winnewisser, Spectroscopy in the terahertz region, Vib. Spectrosc. 8 (1995) 241–253.

[5] G. Winnewisser, Spectroscopy among the stars, Fresenius J. Anal. Chem. 355 (1996) 571–575.

[6] E. Herbst, Gas-grain models of low-mass star formation, in: R. I. Kaiser, P. Bernath, Y. Osamura, S. Petrie, A. M. Mebel (Eds.), Astrochemistry – From Laboratory Studies to Astronomical Observations, Vol. 855 of American Institute of Physics Conference Series, 2006, pp. 260–271.

[7] E. Herbst, H. M. Cuppen, Interstellar chemistry special feature: monte carlo studies of surface chemistry and nonthermal desorption involving interstellar grains, Proc. Nat. Acad. Sci. U.S.A. 103 (2006) 12257–12262.

[8] E. Herbst, Chemistry in the ISM: the ALMA (r)evolution, Astrophys. Space Sci. 313 (2008) 129–134.

[9] T. G. Phillips, J. Keene, Submillimeter astronomy, in: Proc. IEEE, Vol. 80, 1992, pp. 1662–1678.

[10] D. T. Leisawitz, W. C. Danchi, M. J. DiPirro, L. D. Feinberg, D. Y. Gezari, M. Hagopian, W. D. Langer, J. C. Mather, S. H. Moseley, M. Shao, Scientific motivation and technology requirements for the SPIRIT and specifications of far-infrared/submillimeter space interferometers, in: Proc. SPIE, Vol. 4013, Munich, Germany, 2000, pp. 36–46.

[11] C. M. Walmsley, C. Bertout, Special Letters edition: First science with APEX, Astron. Astrophys. 454 (2006) 1–683.

[12] The Stratospheric Observatory for Infrared Astronomy (SOFIA): http://www.sofia.usra.edu/ http://www.dsi.uni-stuttgart.de.

[13] Herschel space observatory: http://www.esa.int/esaSC/120390_index_0_m.html.

[14] Atacama Large Millimeter Array (ALMA): http://www.eso.org/projects/alma/.

[15] P. Schilke, D. J. Benford, T. R. Hunter, D. C. Lis, T. G. Phillips, A line survey of Orion-KL from 607 to 725 GHz, Astrophys. J. Suppl. Ser. 132 (2001) 281–364.

[16] C. Comito, P. Schilke, T. G. Phillips, D. C. Lis, F. Motte, D. Mehringer, A molecular line survey of Orion KL in the 350 micron band, Astrophys. J. Suppl. Ser. 156 (2005) 127–167.

[17] H. S. P. Müller, F. Schlöder, J. Stutzki, G. Winnewisser, The Cologne Database for Molecular Spectroscopy, CDMS: a useful tool for astronomers and spectroscopists, J. Mol. Struct. 742 (2005) 215–227.

[18] Jet Propulsion Laboratory (JPL): http://spec.jpl.nasa.gov/.

[19] H. M. Pickett, R. L. Poynter, E. A. Cohen, M. L. Delitsky, J. C. Pearson, H. S. P. Müller, Submillimeter, millimeter, and microwave spectral line catalog, J. Quant. Spectrosc. Radiat. Transfer 60 (1998) 883–890.

[20] G. M. Plummer, G. Winnewisser, M. Winnewisser, J. Hahn, K. Reinartz, HSSH revisited – the high-resolution Fourier-Transform spectrum of the ground-state between 30 and 90 cm^{-1}, J. Mol. Spectrosc. 126 (1987) 255–269.

[21] M. Liedtke, R. Schieder, K. M. T. Yamada, G. Winnewisser, S. P. Belov, A. F. Krupnov, The $^{r}Q_3$ branch of HSSH at 980 GHz – anomalous K-doubling of the $K_a = 3$ levels, J. Mol. Spectrosc. 161 (1993) 317–321.

[22] P. H. Siegel, Terahertz technology, IEEE Trans. Microwave Theory Tech. 50 (2002) 910–928.

[23] C. Walther, M. Fischer, G. Scalari, R. Terazzi, N. Hoyler, J. Faist, Quantum cascade lasers operating from 1.2 to 1.6 THz, Appl. Phys. Lett. 91 (2007) 131122.

[24] F. C. De Lucia, Spectroscopy in the Terahertz Spectral Region, in: D. Mittleman (Ed.), Sensing with Terahertz Radiation, Springer, Heidelberg, 2003, pp. 39–115.

[25] A. N. Korolev, S. A. Zaitsev, I. I. Golenitskij, Y. V. Zhary, A. D. Zakurdayev, M. I. Lopin, P. M. Meleshkevich, E. A. Gelvich, A. A. Negirev, A. S. Pobedonostsev, V. I. Poognin, V. B. Homich, A. N. Kargin, Traditional and novel vacuum electron devices, IEEE Microwave and Guided Wave Letters, Trans. Electron Devices 48 (2001) 2929–2937.

[26] B. Levush, T. Antonsen Jr., A. Bromborsky, W. R. Lou, D. Abe, S. Miller, Y. Carmel, J. Rodgers, V. Granatstein, W. Destler, Relativistic backward-wave oscillators: theory and experiment, IEDM 91 (1991) 775.

[27] B. Levush, T. M. Antonsen Jr., A. Bromborsky, W.-R. Lou, Y. Carmel, Theory of relativistic backward-wave oscillators with end reflections, IEEE Trans. Plasma Sci. 20 (1992) 263–280.

[28] J. W. Gewartowski, H. A. Watson, Principles of electron tubes, D. Van Nostrand Company, INC., New York, 1965.

[29] A. F. Krupnov, L. I. Gershtein, Sovjet Physics – Pribory 6 (1970) 143–144.

[30] G. Winnewisser, A. F. Krupnov, M. Y. Tretyakov, M. Liedtke, F. Lewen, A. H. Saleck, R. Schieder, A. P. Shkaev, S. V. Volokhov, Precision broad-band spectroscopy in the terahertz region, J. Mol. Spectrosc. 165 (1994) 294–300.

[31] S. P. Belov, M. Liedtke, T. Klaus, R. Schieder, A. H. Saleck, J. Behrend, K. M. T. Yamada, G. Winnewisser, A. F. Krupnov, Precision-measurement of the $^{r}Q_2$ branch at 700 GHz and the $^{r}Q_3$ branch at 980 GHz of HSSH, J. Mol. Spectrosc. 166 (1994) 489–494.

[32] F. Lewen, R. Gendriesch, I. Pak, D. G. Paveliev, M. Hepp, R. Schieder, G. Winnewisser, Phase locked backward wave oscillator pulsed beam spectrometer in the submillimeter wave range, Rev. Sci. Instr. 69 (1998) 32–39.

[33] A. F. Krupnov, Phase lock-in of mm/submm backward wave oscillators: development, evolution, and applications, Int. J. Infrared Millimeter Waves 22 (2001) 1–18.

[34] G. Klapper, L. A. Surin, F. Lewen, H. S. P. Müller, I. Pak, G. Winnewisser, Laboratory precision measurements of the rotational spectrum of $^{12}C^{17}O$ and $^{13}C^{17}O$, Astrophys. J. 582 (2003) 262–268.

[35] S. Brünken, U. Fuchs, F. Lewen, Š. Urban, T. Giesen, G. Winnewisser, Sub-Doppler and Doppler spectroscopy of DCN isotopomers in the terahertz region: ground and first excited bending states $(\nu_1\nu_2\nu_3) = (0\ 1^{e,f}\ 0)$, J. Mol. Spectrosc. 225 (2004) 152–161.

[36] U. Fuchs, S. Brünken, G. W. Fuchs, S. Thorwirth, V. Ahrens, F. Lewen, Š. Urban, T. F. Giesen, G. Winnewisser, High resolution spectroscopy of HCN isotopomers: $H^{13}CN$, $HC^{15}N$, and $H^{13}C^{15}N$ in the ground and first excited bending vibrational state, Z. Naturforsch. 59a (2004) 861–872.

[37] I. Mehdi, E. Schlecht, G. Chattopadhyay, P. H. Siegel, THz local oscillator sources: performance and capabilities, in: T. G. Phillips, J. Zmuidzinas (Eds.), Millimeter and Submillimeter Detectors for Astronomy., Vol. 4855 of Proc. SPIE, 2003, pp. 435–446.

[38] P. H. Siegel, THz instruments for space, IEEE Trans. Antennas Propag. 55 (2007) 2957–2965.

[39] A. Grüb, A. Simon, V. Krozer, H. L. Hartnagel, Future developments of terahertz Schottky barrier mixer diodes, Electrical Engineering (Archiv für Elektrotechnik) 77 (1993) 57–59.

[40] F. Lewen, D. G. Paveljev, B. Vowinkel, J. Freyer, H. Grothe, G. Winnewisser, Planar schottky diodes for THz applications, in: Proceedings of the 4th International Workshop on Terahertz Electronics, Erlangen, 1996.

[41] H. Y. Xu, G. S. Schoenthal, J. L. Hesler, T. W. Crowe, R. M. Weikle, Nonohmic contact planar varactor frequency upconverters for terahertz applications, IEEE Trans. Microwave Theory Tech. 55 (2007) 648–655.

[42] N. R. Erickson, High efficiency submillimeter frequency multipliers, in: IEEE/MTT-S Intl. Mic. Symp. Digest, 1990, pp. 1301–1304.

[43] P. Zimmermann, Solid-state oscillators for the THz-range, in: Proceeding of the 8th International Conference on Terahertz Electronics, Darmstadt, 2000.

[44] P. H. Siegel, R. P. Smith, M. C. Graidis, S. C. Martin, 2.5 THz GaAs monolithic membrane-diode mixer, IEEE Trans. Microwave Theory Tech. 47 (1999) 596–604.

[45] J. Bruston, E. Schlecht, A. Maestrini, F. Maiwald, S. C. Martin, R. P. Smith, I. Mehdi, P. H. Siegel, J. Pearson, Development of 200 GHz to 2.7 THz multiplier chains for submillimeter-wave heterodyne receivers, in: Proceedings of the SPIE-International Symposium on Astronomical Telescope and Instrumentation, 2000, pp. 27–31.

[46] G. Chattopadhyay, E. Schlecht, J. S. Ward, J. J. Gill, H. H. S. Javadi, F. Maiwald, I. Medhi, An all-solid-state broad-band frequency multiplier chain at 1500 GHz, IEEE Trans. Microwave Theory Tech. 52 (2004) 1538–1547.

[47] A. Maestrini, J. Ward, J. Gill, H. Javadi, E. Schlecht, G. Chattopadhyay, F. Maiwald, N. R. Erickson, I. Mehdi, A 1.7-1.9 THz local oscillator source, IEEE Microwave Compon. Lett. 14 (2004) 253–255.

[48] F. Maiwald, S. Martin, J. Bruston, A. Maestrini, T. Crawford, P. H. Siegel, 2.7 THz waveguide tripler using monolithic membrane diodes, Microwave Symposium Digest, 2001 IEEE MTT-S International 3 (2001) 1637–1640.

[49] Virginia Diodes Inc., 979 Second Street SE, Suite 309, Charlottesville, VA 22902.

[50] RPG Radiometer Physics GmbH, Birkenmaarstraße 10, 53340 Meckenheim, Germany: http://www.radiometer-physics.de/rpg/html/Home.html.

[51] B. J. Drouin, F. W. Maiwald, J. C. Pearson, Application of cascaded frequency multiplication to molecular spectroscopy, Rev. Sci. Instr. 76 (2005) 093113.

[52] F. Maiwald, F. Lewen, B. Vowinkel, W. Jabs, D. G. Paveljev, M. Winnewisser, G. Winnewisser, Planar Schottky diode frequency multiplier for molecular spectroscopy up to 1.3 THz, IEEE Microwave And Guided Wave Letters 9 (1999) 198–200.

[53] F. Maiwald, F. Lewen, V. Ahrens, M. Beaky, R. Gendriesch, A. N. Koroliev, A. A. Negirev, D. G. Paveliev, B. Vowinkel, G. Winnewisser, Pure rotational spectrum of HCN in the terahertz region: use of a new planar Schottky diode multiplier, J. Mol. Spectrosc. 202 (2000) 166–168.

[54] R. A. Davies, M. J. Kelly, T. M. Kerr, Tunneling between two strongly coupled superlattices, Phys. Rev. Lett. 55 (1985) 1114–1116.

[55] C. P. Endres, F. Lewen, T. F. Giesen, S. Schlemmer, D. G. Paveljev, Y. I. Koschurinov, V. M. Ustinov, A. E. Zhucov, Application of superlattice multipliers for high-resolution terahertz spectroscopy, Rev. Sci. Instr. 78 (2007) 043106.

[56] K. F. Renk, A. Rogl, B. I. Stahl, Semiconductor-superlattice parametric oscillator for generation of sub-terahertz and terahertz waves, J. Lumin. 125 (2007) 252–258.

[57] D. G. Paveliev, Y. Koshurinov, N. Demarina, V. Ustinov, A. Zhukov, N. Maleev, A. Vasilyev, A. Baryshev, P. Yagoubov, N. Whyborn, Temperature dependence of the radiation power emitted by a superlattice subject to a high-frequency electric field, in: Infrared and Millimeter Waves and 12th International Conference on Terahertz Electronics, 2004. Digest of the 2004 Joint 29th International Conference, 2004, pp. 279–280.

[58] S. Brünken, H. S. P. Müller, C. P. Endres, F. Lewen, T. Giesen, B. Drouin, J. C. Pearson, H. Mäder, High resolution rotational spectroscopy on D_2O up to 2.7 THz in its ground and first excited vibrational bending states, Phys. Chem. Chem. Phys. 9 (2007) 2103–2112.

[59] C. P. Endres, H. S. P. Müller, S. Brünken, D. G. Paveliev, T. F. Giesen, S. Schlemmer, F. Lewen, High resolution rotation-inversion spectroscopy on doubly deuterated ammonia, ND_2H, up to 2.6 THz, J. Mol. Struct. 795 (2006) 242–255.

[60] G. Piau, F. X. Brown, D. Dangoisse, P. Glorieux, Heterodyne-detection of tunable FIR sideband, IEEE J. Quantum Electron. QE-23 (1987) 1388–1391.

[61] G. A. Blake, K. B. Laughlin, R. C. Cohen, K. L. Busarow, D. H. Gwo, C. A. Schmuttenmaer, D. W. Steyert, R. J. Saykally, The Berkeley tunable far infrared-laser spectrometers, Rev. Sci. Instr. 62 (1991) 1701–1716.

[62] P. Verhoeve, E. Zwart, M. Versluis, M. Drabbels, J. J. ter Meulen, W. L. Meerts, A. Dymanus, D. B. McLay, A far infrared-laser sideband spectrometer in the frequency region 550-2700 GHz, Rev. Sci. Instr. 61 (1990) 1612–1625.

[63] F. Lewen, E. Michael, R. Gendriesch, J. Stutzki, G. Winnewisser, Terahertz laser sideband spectroscopy with backward wave oscillators, J. Mol. Spectrosc. 183 (1997) 207–209.

[64] E. Michael, F. Lewen, R. Gendriesch, J. Stutzki, G. Winnewisser, Frequency lock of an optically pumped FIR ring laser at 803 and 1626 GHz, Int. J. Infrared Millimeter Waves 20 (1999) 1073–1083.

[65] R. Gendriesch, F. Lewen, G. Winnewisser, J. Hahn, Precision broadband spectroscopy near 2 THz: frequency-stabilized laser sideband spectrometer with backward-wave oscillators, J. Mol. Spectrosc. 203 (2000) 205–207.

[66] R. Gendriesch, F. Lewen, G. Winnewisser, H. S. P. Müller, Far-infrared laser-sideband measurements of the amidogen radical, NH_2, near 2 THz with microwave accuracy, J. Mol. Struct. 599 (2001) 293–304.

[67] R. Gendriesch, K. Pehl, T. Giesen, G. Winnewisser, F. Lewen, Terahertz spectroscopy of linear triatomic CCC: High precision laboratory measurement and analysis of the ro-vibrational bending transitions, Z. Naturforsch. 58a (2003) 129–138.

[68] S. Brünken, H. S. P. Müller, F. Lewen, G. Winnewisser, High accuracy measurements on the ground state rotational spectrum of formaldehyde (H_2CO) up to 2 THz, Phys. Chem. Chem. Phys. 5 (2003) 1515–1518.

[69] S. Brünken, H. S. P. Müller, S. Thorwirth, F. Lewen, G. Winnewisser, The rotational spectra of the ground and first excited bending states of deuterium isocyanide, DNC, up to 2 THz, J. Mol. Struct. 780-781 (2006) 3–6.

[70] H. Linnartz, T. Motylewski, F. Maiwald, D. A. Roth, F. Lewen, I. Pak, G. Winnewisser, Millimeter wave spectroscopy in a pulsed supersonic slit nozzle discharge, Chem. Phys. Lett. 292 (1998) 188–192.

[71] M. Hepp, R. Gendriesch, I. Pak, Y. A. Kuritsyn, F. Lewen, G. Winnewisser, M. Brookes, A. R. W. McKellar, J. K. G. Watson, T. Amano, Millimetre-wave spectrum of the Ar-CO complex: the $K = 2 \leftarrow 1$ and $3 \leftarrow 2$ subbands, Mol. Phys. 92 (1997) 229–236.

[72] T. Imasaka, D. S. Moore, T. Vo-Dinh, Critical assessment: Use of supersonic jet spectrometry for complex mixture analysis – (IUPAC technical report), Pure Appl. Chem. 75 (2003) 975–998.

[73] G. Scoles, D. Bassi, U. Buck, D. C. Laine (Eds.), Atomic and molecular beam methods, Oxford University Press, Oxford, New York, 1988.

[74] H. Verbraak, D. Verdes, H. Linnartz, A systematic study of ion and cluster ion formation in continuous supersonic planar plasma, Int. J. Mass Spectrom. 267 (2007) 248–255.

[75] H. Linnartz, Planar plasma expansions as a tool for high resolution molecular spectroscopy, Physica Scripta 69 (2004) C37–C40.

[76] A. van Orden, R. J. Saykally, Small carbon clusters: spectroscopy, structure, and energetics, Chem. Rev. 98 (1998) 2313–2357.

[77] A. van Orden, R. A. Provencal, T. F. Giesen, R. J. Saykally, Characterization of silicon-carbon clusters by infrared-laser spectroscopy – the ν_1 band of SiC_4, Chem. Phys. Lett. 237 (1995) 77–80.

[78] A. van Orden, T. F. Giesen, R. A. Provencal, H. J. Hwang, R. J. Saykally, Characterization of silicon-carbon clusters by infrared-laser spectroscopy – The ν_3 (Σ_u) band of linear Si_2C_3, J. Chem. Phys. 101 (1994) 10237–10241.

[79] T. F. Giesen, A. O. van Orden, J. D. Cruzan, R. A. Provencal, R. J. Saykally, R. Gendriesch, F. Lewen, G. Winnewisser, Interstellar detection of CCC and high-precision laboratory measurements near 2 THz, Astrophys. J. 551 (2001) L181–L184.

[80] P. Neubauer-Guenther, T. F. Giesen, U. Berndt, G. Fuchs, G. Winnewisser, The Cologne Carbon Cluster Experiment: ro-vibrational spectroscopy on C_8 and other small carbon clusters, Spectrochim. Acta, Part A 59 (2003) 431–441.

[81] T. F. Giesen, U. Berndt, K. M. T. Yamada, G. Fuchs, R. Schieder, G. Winnewisser, R. A. Provencal, F. N. Keutsch, A. van Orden, R. J. Saykally, Detection of the linear carbon cluster C_{10}: Rotationally resolved diode-laser spectroscopy, Chem. Phys. Chem. 2 (2001) 242.

[82] P. Neubauer-Guenther, T. F. Giesen, S. Schlemmer, K. M. T. Yamada, High resolution infrared spectra of the linear carbon cluster C_7: The ν_4 stretching fundamental band and associated hot bands, J. Chem. Phys. 127 (2007) 014313.

[83] M. Philipp, U. U. Graf, A. Wagner-Gentner, D. Rabanus, F. Lewen, Compact 1.9 THz BWO local-oscillator for the GREAT heterodyne receiver, Infrared Phys. Technol. 51 (2007) 54–59.

[84] A. E. Boguslavskiy, J. P. Maier, Gas phase electronic spectra of the carbon chains C_5, C_6, C_8, and C_9, J. Chem. Phys. 125 (2006) 094308.

[85] P. Birza, T. Motylewski, D. Khoroshev, A. Chirokolava, H. Linnartz, J. P. Maier, CW cavity ring down spectroscopy in a pulsed planar plasma expansion, Chem. Phys. 283 (2002) 119–124.

[86] M. C. McCarthy, W. Chen, M. J. Travers, P. Thaddeus, Microwave spectra of 11 polyyne carbon chains, Astrophys. J. Suppl. Ser. 129 (2000) 611–623.

[87] C. D. Ball, M. C. McCarthy, P. Thaddeus, Cavity ringdown spectroscopy of the linear carbon chains HC_7H, HC_9H, $HC_{11}H$, and $HC_{13}H$, J. Chem. Phys. 112 (2000) 10149–10155.

[88] B. S. Dumesh, V. D. Gorbatenkov, V. G. Koloshnikov, V. A. Panfilov, L. A. Surin, Application of highly sensitive millimeter-wave cavity spectrometer based on Orotron for gas analysis, Spectrochim. Acta, Part A 53 (1997) 835–843.

[89] B. S. Dumesh, V. P. Kostromin, F. S. Rusin, L. A. Surin, Highly sensitive millimeter-wave spectrometer based on an Orotron, Meas. Sci. Technol. 3 (1992) 873–878.

[90] L. A. Surin, B. S. Dumesh, F. S. Rusin, G. Winnewisser, I. Pak, Doppler-free two-photon millimeter wave transitions in OCS and CHF_3, Phys. Rev. Lett. 86 (2001) 2002–2005.

[91] L. A. Surin, Intracavity millimeter wave spectroscopy of molecules in excited vibrational states, Vib. Spectrosc. 24 (2000) 147–155.

[92] K. Jacobs, Recent progress in submillimeter and terahertz heterodyne receivers with super-conducting mixers, in: R. Titz, H.-P. Röser (Eds.), SOFIA – Workshop: Astronomy and Technology in the 21st Century, Wissenschaft & Technik Verlag, Berlin, 1998.

[93] S. Schlemmer, T. Kuhn, E. Lescop, D. Gerlich, Laser excited N_2^+ in a 22-pole ion trap: experimental studies of rotational relaxation processes, Int. J. Mass Spectrom. 187 (1999) 589–602.

[94] S. Schlemmer, E. Lescop, J. von Richthofen, D. Gerlich, M. A. Smith, Laser induced reactions in a 22-pole ion trap: $C_2H_2^+ + h\nu_3 + H_2 \rightarrow C_2H_3^+ + H$, J. Chem. Phys. 117 (2002) 2068–2075.

[95] O. Asvany, P. Kumar, B. Redlich, I. Hegemann, S. Schlemmer, D. Marx, Understanding the infrared spectrum of bare CH_5^+, Science 309 (2005) 1219–1222.

[96] S. Schlemmer, O. Asvany, T. Giesen, Comparison of the cis-bending and C–H stretching vibration on the reaction of $C_2H_2^+$ with H_2 using laser induced reactions, Phys. Chem. Chem. Phys. 7 (2005) 1592–1600.

[97] D. Gerlich, Ion-neutral collisions in a 22-pole trap at very low energies, Phys. Scr. T59 (1995) 256–263.

[98] D. Gerlich, Inhomogeneous RF fields: a versatile tool for the study of processes with slow ions, in: C. Ng, M. Baer (Eds.), Adv. Chem. Phys.: State-Selected and State-to-State Ion-Molecule Reaction Dynamics, Vol. LXXXII, Wiley, New York, 1992, pp. 1–176.

[99] D. Gerlich, Experimental investigation of ion-molecule reactions relevant to interstellar chemistry, J. Chem. Soc. Faraday Trans. 89 (1993) 2199–2208.

[100] O. Asvany, T. Giesen, B. Redlich, S. Schlemmer, Experimental determination of the ν_5 cis-bending vibrational frequency and Renner-Teller structure in ground state $(X^2\Pi_u)$ $C_2H_2^+$ using laser induced reactions, Phys. Rev. Lett. 94 (2005) 073001.

[101] A. A. Adesokan, G. M. Chaban, O. Dopfer, R. B. Gerber, Vibrational spectroscopy of protonated imidazole and its complexes with water molecules: ab initio anharmonic calculations and experiments, J. Phys. Chem. A 111 (2007) 7374–7381.

[102] D. W. Boo, Y. T. Lee, Infrared spectroscopy of the molecular hydrogen solvated carbonium ions, $CH_5^+(H_2)_n$ (n = 1 − 6), J. Chem. Phys. 103 (1995) 520–530.

[103] O. V. Boyarkin, S. R. Mercier, A. Kamariotis, T. R. Rizzo, Electronic spectroscopy of cold, protonated tryptophan and tyrosine, J. Am. Chem. Soc. 128 (2006) 2816–2817.

[104] U. J. Lorenz, N. Solca, J. Lemaire, P. Maître, O. Dopfer, Infrared spectra of isolated protonated polycyclic aromatic hydrocarbons: protonated naphthalene, Angew. Chem. Int. Ed. 46 (2007) 6714–6716.

[105] A. Dzhonson, D. Gerlich, E. J. Bieske, J. P. Maier, Apparatus for the study of electronic spectra of collisionally cooled cations: para-dichlorobenzene, J. Mol. Struct. 795 (2006) 93–97.

[106] S. Trippel, J. Mikosch, R. Berhane, R. Otto, M. Weidemüller, R. Wester, Photodetachment of cold OH^- in a multipole ion trap, Phys. Rev. Lett. 97 (2006) 193003.

[107] S. Schlemmer, O. Asvany, Laser induced reactions in a 22-pole ion trap, in: J. Phys. Conf. Ser., Vol. 4, 2005, pp. 134–141.

[108] J. Mikosch, H. Kreckel, R. Wester, R. Plašil, J. Glosík, D. Gerlich, D. Schwalm, A. Wolf, Action spectroscopy and temperature diagnostics of H_3^+ by chemical probing, J. Chem. Phys. 121 (2004) 11030–11037.

[109] O. Asvany, E. Hugo, F. Müller, F. Kuhnemann, S. Schiller, J. Tennyson, S. Schlemmer, Overtone spectroscopy of H_2D^+ and D_2H^+ using laser induced reactions, J. Chem. Phys. 127 (2007) 154317.

[110] O. Asvany, O. Ricken, H. S. P. Müller, M. C. Wiedner, T. F. Giesen, S. Schlemmer, High resolution rotational spectroscopy in a cold ion trap: H_2D^+ and D_2H^+, Phys. Rev. Lett. 100 (2008) 223004.

[111] A. G. G. M. Tielens, The physics and chemistry of the interstellar medium, Cambridge University Press, Cambridge, New York, 2005.

[112] E. T. White, J. Tang, T. Oka, CH_5^+: the infrared spectrum observed, Science 284 (1999) 135–137.

[113] D. Christen, H. S. P. Müller, The millimeter wave spectrum of aGg ethylene glycol: the quest for higher precision, Phys. Chem. Chem. Phys. 5 (2003) 3600–3605.

[114] L.-H. Xu, H. S. P. Müller, F. F. S. van der Tak, S. Thorwirth, The millimeter wave spectrum of perdeuterated methanol, CD_3OD, J. Mol. Spectrosc. 228 (2004) 220–229.

[115] H. S. P. Müller, D. Christen, Millimeter and submillimeter wave spectroscopic investigations into the rotation-tunneling spectrum of gGg' ethylene glycol, $HOCH_2CH_2OH$, J. Mol. Spectrosc. 228 (2004) 298–307.

[116] H. S. P. Müller, L. H. Xu, F. van der Tak, Investigations into the millimeter and submillimeter-wave spectrum of perdeuterated methanol, CD_3OD, in its ground and first excited torsional states, J. Mol. Struct. 795 (2006) 114–133.

[117] M. Simecková, Š. Urban, U. Fuchs, F. Lewen, G. Winnewisser, I. Morino, K. M. T. Yamada, Ground state spectrum of methylcyanide, J. Mol. Spectrosc. 226 (2004) 123–136.

[118] M. Caris, F. Lewen, H. S. P. Müller, G. Winnewisser, Pure rotational spectroscopy of potassium chloride, KCl, up to 930 GHz and isotopically invariant analysis of KCl and NaCl, J. Mol. Struct. 695 (2004) 243–251.

[119] S. Thorwirth, H. S. P. Müller, H. Lichau, G. Winnewisser, G. C. Mellau, The submillimeter wave spectrum of the C_4H_4 isomer vinylacetylene, J. Mol. Struct. 695 (2004) 263–267.

[120] M. Caris, F. Lewen, G. Winnewisser, Pure rotational spectroscopy of sodium chloride, NaCl, up to 930 GHz, Z. Naturforsch. 57a (2002) 663–668.

[121] M. Schäfer, M. Andrist, H. Schmutz, F. Lewen, G. Winnewisser, F. Merkt, A 240-380 GHz millimetre wave source for very high resolution spectroscopy of high Rydberg states, J. Phys. B: At., Mol. Opt. Phys. 39 (2006) 831–845.

[122] H. S. P. Müller, M. C. McCarthy, L. Bizzocchi, H. Gupta, S. Esser, H. Lichau, M. Caris, F. Lewen, J. Hahn, C. D. Esposti, S. Schlemmer, P. Thaddeus, Rotational spectroscopy of the isotopic species of silicon monosulfide, SiS, Phys. Chem. Chem. Phys. 9 (2007) 1579–1586.

[123] S. Brünken, H. S. P. Müller, F. Lewen, T. F. Giesen, Analysis of the rotational spectrum of methylene (CH_2) in its vibronic ground state with an Euler expansion of the Hamiltonian, J. Chem. Phys. 123 (2005) 164315.

[124] L. Margulès, F. Lewen, G. Winnewisser, P. Botschwina, H. S. P. Müller, The rotational spectrum up to 1 THz and the molecular structure of thiomethylium, HCS^+, Phys. Chem. Chem. Phys. 5 (2003) 2770–2773.

[125] L. Margulès, E. Herbst, V. Ahrens, F. Lewen, G. Winnewisser, H. S. P. Müller, The phosphidogen radical, PH_2: terahertz spectrum and detectability in space, J. Mol. Spectrosc. 211 (2002) 211–220.

[126] A. Maeda, I. R. Medvedev, M. Winnewisser, F. C. De Lucia, E. Herbst, H. S. P. Müller, M. Koerber, C. P. Endres, S. Schlemmer, High-frequency rotational spectrum of thioformaldehyde, H_2CS, in the ground vibrational state, Astrophys. J. Suppl. Ser. 176 (2008) 543–550.

[127] M. Winnewisser, B. P. Winnewisser, M. Stein, M. Birk, G. Wagner, G. Winnewisser, K. M. T. Yamada, S. P. Belov, O. I. Baskakov, Rotational spectra of cis-HCOOH, trans-HCOOH, and trans-$H^{13}COOH$, J. Mol. Spectrosc. 216 (2002) 259–265.

[128] E. A. Michael, F. Lewen, G. Winnewisser, H. Ozeki, H. Habara, E. Herbst, Laboratory spectrum of the 1_{11}-2_{02} rotational transition of CH_2, Astrophys. J. 596 (2003) 1356–1362.

[129] C. D. Esposti, L. Bizzocchi, P. Botschwina, K. M. T. Yamada, G. Winnewisser, S. Thorwirth, P. Förster, Vibrationally excited states of HC_5N: millimeter-wave spectroscopy and coupled cluster calculations, J. Mol. Spectrosc. 230 (2005) 185–195.

[130] O. Baum, M. Koerber, T. Giesen, and S. Schlemmer, The $^rQ_4$ branch of HSOH at 1.4 THz, J. Mol. Spectrosc. (2008) in prep.

[131] H. Spahn, H. S. P. Müller, T. F. Giesen, J.-U. Grabow, M. E. Harding, J. Gauss, S. Schlemmer, Rotational spectra and hyperfine structure of isotopic species of deuterated cyanoacetylene, DC_3N, Chem. Phys. 346 (2008) 132–138.

[132] F. Lewen, S. Brünken, G. Winnewisser, M. Simecková, Š. Urban, Doppler-limited rotational spectrum of the NH radical in the 2 THz region, J. Mol. Spectrosc. 226 (2004) 113–122.

[133] G. Winnewisser, F. Lewen, S. Thorwirth, M. Behnke, J. Hahn, J. Gauss, E. Herbst, Gas-phase detection of HSOH: synthesis by flash vacuum pyrolysis of di-tert-butyl sulfoxide and rotational-torsional spectrum, Chem. Eur. J. 9 (2003) 5501–5510.

[134] S. Thorwirth, H. S. P. Müller, F. Lewen, S. Brünken, V. Ahrens, G. Winnewisser, A concise new look at the l-type spectrum of $H^{12}C^{14}N$, Astrophys. J. 585 (2003) L163–L165.

[135] Z. Zelinger, T. Amano, V. Ahrens, S. Brünken, F. Lewen, H. S. P. Müller, G. Winnewisser, Submillimeter-wave spectroscopy of HCN in excited vibrational states, J. Mol. Spectrosc. 220 (2003) 223–233.

Selective Detection of Proteins and Nucleic Acids with Biofunctionalized SERS Labels

S. Schlücker *and* W. Kiefer

Contents

Abstract

Surface-enhanced Raman scattering (SERS) plays an increasingly important role for the selective detection of proteins and nucleic acids. Significant advantages of SERS over existing labelling approaches include, for example, its multiplexing capacity for simultaneous target detection, target quantification and sensitivity. Using SERS as a readout strategy requires a Raman label in conjunction with a SERS substrate for signal enhancement and a biomolecule (e.g. an antibody) for specific binding to the corresponding target molecule. This modular principle offers many opportunities for different implementations. Here, various experimental configurations for the selective detection of proteins and nucleic acids by SERS are summarized. Because noble metal nanoparticles are central to bioanalytical SERS applications, methods for the encapsulation of nanoparticle Raman label conjugates are discussed. Such agents will probably also play a central role for target localization. SERS microscopy as a novel Raman imaging technique may therefore become widely used.

Keywords: surface-enhanced Raman scattering; surface-enhanced resonance Raman scattering; nanoparticles; gold; silver; plasmon resonance; Raman label; Raman dye; Raman reporter molecule; SERS label; biofunctionalization; biospecific binding; molecular recognition; selective detection; ultrasensitive detection; labelling; multiplexing; quantification; seeded growth; encapsulation; silica shell; glass shell; laser excitation; laser spectroscopy; bioanalytics; biomolecules; target molecules; immunoassay; ELISA; gene probe; immunohistochemistry; Raman microscopy; SERS microscopy; imaging; image contrast; localization; staining

1. INTRODUCTION

The selective detection of proteins and nucleic acids is usually achieved by labelled antibodies and oligonucleotides, respectively. A variety of different labelling and subsequent readout methodologies is available, including electronic absorption and fluorescence spectroscopy as widely used optical techniques. Surface–enhanced Raman scattering (SERS) plays an increasingly important role in this field of biological and biomedical research [1].

A SERS-based selective detection of proteins or DNA/RNA requires the following components (see Table 1): a Raman label [e.g. a Raman dye such as rhodamine 6G (R6G)] with a characteristic vibrational signature, a SERS substrate (e.g. Ag nanoparticle) for signal enhancement and a biomolecule (e.g. an antibody) for specific binding to the corresponding target molecule. In the following, the term SERS label refers to the combination of a SERS substrate and a Raman label (Table 1).

Describing the combination of the three components in Table 1 as a biofunctionalized SERS label is only one of several options; terminology certainly depends on the perspective and the individual background of researchers which may come

Table 1 Biofunctionalized SERS labels for selective protein and DNA/RNA detection comprise a Raman label (e.g. R6G) for the characteristic label signature, a SERS substrate (e.g. Ag nanoparticle) for Raman signal enhancement and a biomolecule (e.g. antibody) for target-specific binding

	Component	Examples	Function
SERS label	Raman label	Raman dyes (e.g. Rhodamine 6 G)	Characteristic Raman signature
	SERS substrate	Silver and gold nanoparticles	Raman signal enhancement
	Biomolecule	Antibodies, oligonucleotides	Selective target recognition

Table 2 Multiplexing capacity of different labelling approaches. In addition to the labels, the underlying physical processes are specified. SERS labels offer unique multiplexing capacities compared with existing techniques based on electronic absorption or emission spectroscopies

Number of Different Individual Labels	~1–3	~1–3	~3–10	~10–100
Label Type	Dyes	Molecular Fluorophores	Quantum Dots	SERS labels
Detection by	Electronic Absorption	Fluorescence emission		Raman scattering

from diverse scientific disciplines ranging from material science and spectroscopy to biology and medicine. Synonymous terms for an antibody-functionalized SERS label are antibody- and Raman label-functionalized nanoparticle or immuno-SERS label or SERS-labelled antibody.

SERS offers significant advantages over existing labelling approaches. A very important advantage is its unique multiplexing capacity, i.e. the potential for the simultaneous detection of numerous SERS labels (see Table 2); the line widths of vibrational Raman bands are typically two orders of magnitude or more narrower in comparison with the broad emission profiles of molecular fluorophores: this avoids or minimizes spectral overlap between different labels. A second advantage is that the simultaneous excitation of distinct SERS label signatures requires only a single laser line. Furthermore, quantification of target concentrations, an important aspect in bioanalytical applications, is achieved via SERS label signal intensities in the corresponding spectra. In contrast to fluorescence spectroscopy and microscopy, photo bleaching is generally not critical. Finally, image contrast in SERS microscopy can be maximized by using red to near-infrared laser excitation for which autofluorescence of biological specimens is minimized.

The first two sections of this chapter summarize different approaches to the selective detection of proteins (Section 2) and nucleic acids (Section 3) by SERS. Then various methods for the encapsulation of nanoparticle Raman label conjugates are discussed (Section 4). Finally, microscopic applications with SERS labels are presented (Section 5).

2. PROTEIN DETECTION WITH SERS LABELS

The selective recognition of a target protein by its corresponding antibody is shown schematically in Figure 1. The antigen–binding sites of the antibody (Figure 1 left) selectively recognize the target from a pool containing also other molecules (Figure 1 middle). The resulting antigen–antibody complex (Figure 1 right) can be detected by the signal of the label attached to the antibody. Discrimination of

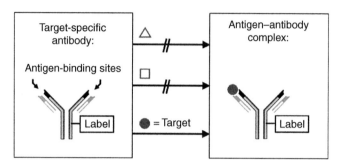

Figure 1   Selective target recognition by its antibody: the antigen-binding sites of the antibody bind (left) selectively to the target. The antibody–antigen complex (right) is detected by the signal of the label attached to the antibody.

antigen–antibody complexes from unbound labelled antibodies, both exhibiting the characteristic signal of the label, requires target immobilization. This is achieved, for example, in a so-called sandwich immunoassay depicted in Figure 2. First, unlabelled capture antibodies are immobilized to a substrate (Figure 2 top left). The capture antibody selectively binds target molecules from the solution (Figure 2 middle left). Addition of a labelled detection antibody yields an antigen "sandwich" between the matched pair of antibodies (Figure 2 bottom right). Finally, unbound detection antibodies are removed by washing. The label signal is then directly related to the amount of antigen, which is essential for quantification.

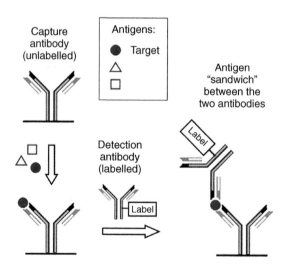

Figure 2   Steps in a sandwich immunoassay. The unlabelled capture antibody is conjugated to a substrate (top left) and binds selectively the target from a solution containing also other molecules. After immobilization of the antigen by the capture antibody (bottom left), a labelled detection antibody is added. It binds to a different recognition site of the antigen, forming a sandwich comprising the antigen and the antibody pair. The sandwich complex (bottom right) is detected by the signal of the label.

Various label types can be used, including dyes, molecular fluorophores and quantum dots (cf. Table 2). As shown in Table 1, the selective detection of proteins by SERS requires antibodies for specific protein binding, a Raman label or Raman-active reporter molecule and a SERS-active substrate. Many different possibilities of combining these components to a single functional unit exist. The Raman label can, for example, be directly attached to the antibody, forming a Raman-labelled antibody or immuno-Raman conjugate (Section 1.1); in a different implementation, the antibody is labelled with an enzyme, which converts a substrate molecule into the actual Raman label (Section 1.2). In combination with different SERS-active substrates such as metal films or colloids, various realizations for protein detection by SERS exist. A very effective implementation of biospecific binding in conjunction with SERS signal generation is the binding of antibody and Raman label molecules to metal nanoparticles (Sections 2.3 and 2.4).

2.1. Immunosandwich assay for protein detection employing silver films

The first SERS-based sandwich immunoassay was presented in 1989 by Rohr, Cotton and co-workers [2] (Figure 3). In this configuration, a roughened silver film acts as the SERS substrate. First, the silver surface is coated with the capture antibody (Figure 3a) and then overcoated with proteins such as bovine serum albumin (BSA) as a blocking agent. Then the corresponding antigen – in this application thyroid-stimulating hormone (TSH) – is added and immobilized by specific binding to the capture antibody (Figure 3b). The last step is the incubation with the Raman-active immuno-conjugate, i.e. a dye-labelled detection antibody which also specifically binds to the antigen (Figure 3c). Surface-enhanced resonance Raman scattering (SERRS) from the Raman reporter molecule, p-dimethylaminoazobenzene (DAB), was detected in the range of 4–60 μIU/mL. In this early study, the authors also explicitly state that generation of the SERRS signals is not restricted to silver films and electrodes: colloids of silver and other metals can serve as the SERRS-active substrates, too.

2.2. Enzyme immunosandwich assay for protein detection utilizing silver nanoparticles

Enzyme-based assays such as enzyme-linked immunosorbent assay (ELISA) are very popular in clinical diagnostics because many samples can be analyzed simultaneously under the same conditions in a relatively short time. ELISA technology utilizes enzyme-labelled detection antibodies and involves several washing steps. In conventional ELISA approaches, the enzyme converts a colourless substrate into a coloured product that can be quantitatively determined by electronic absorption spectroscopy. The first SERS-based ELISA was presented by Dou, Ozaki and co-workers [3] in 1997. In their system (Figure 4), an enzyme (peroxidase, POD) is covalently bound to the detection antibody (anti-mouse IgG).

(a)

(b)

(c)

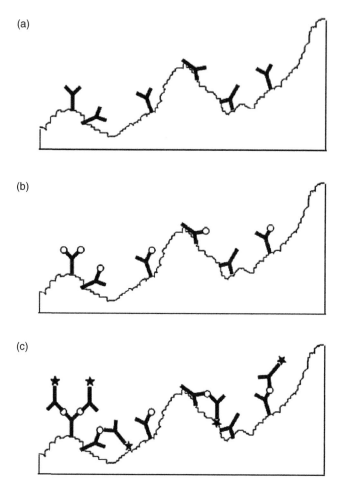

Figure 3 First SERS sandwich immunoassay employing silver films. (a) Unlabelled capture antibodies (Y) are adsorbed to a roughened silver film as the SERS substrate; (b) target proteins (O) are immobilized onto the surface by specific binding to the capture antibody; (c) detection antibodies conjugated to Raman labels (★) also bind specifically to the target proteins. SERS from the Raman labels is detected, confirming the presence of the target protein. Adapted from Ref. [2].

The oxidation–condensation reaction of o-phenylenediamine as a substrate with hydrogen peroxide is catalyzed by POD, yielding azoaniline as the reaction product. Azoaniline is then adsorbed to the surface of the silver colloid: this stable dye molecule gives rise to intense SERRS signals with very strong Raman bands assigned to N=N and C=C stretching modes. The concentration of the antigen is determined indirectly via the SERRS signal of the enzymatic reaction product azoaniline. The detection limit of this enzyme immunoassay method was found to be about 10^{-15} mol/mL; this sensitivity is one order of magnitude higher compared with that in the first SERRS study.

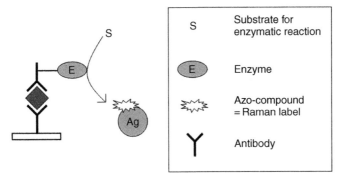

Figure 4   SERS sandwich immunoassay utilizing enzymes and silver nanoparticles. The detection antibody is labelled with an enzyme (E), which catalyzes the conversion of the substrate (S) into an azo-compound. After adsorption of the azo-compound on the surface of silver nanoparticles, the characteristic SERS signal of this Raman label is observed. Adapted from Ref. [3].

2.3.   Immunosandwich assay for protein detection with Raman-immunogold conjugates

The first concept in which both Raman label and antibody molecules are conjugated to a metal nanoparticle (Figure 5) was presented by Porter and co-workers [4] in 1999. The obvious advantage of this strategy is that in these biofunctionalized nanoparticles the capabilities for SERS signal generation and biospecific binding are both incorporated into a single functional unit. In contrast to the preceding configurations, the Raman reporter molecules are directly attached to the surface of the metal nanoparticles. The three-step assay involves the immobilization of capture antibodies on a gold surface, the selective immobilization of antigen molecules via specific antigen–antibody binding and finally the antigen detection by SERS with Raman reporter-labelled immunogold (Figure 5). Using near-infrared excitation (785 nm) allowed to generate efficiently the SERS signal while minimizing fluorescence contributions. Colloidal gold (30 nm) was used as the SERS substrate, which also allows to implement a thiol/disulfide platform for

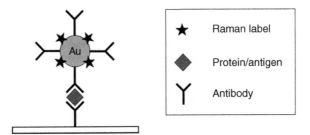

Figure 5   SERS sandwich immunoassay with Raman-labelled immuno-gold probes, i.e. SERS-labelled antibodies. Both Raman label molecules and antibodies are directly attached to the surface of the gold nanoparticle. Adapted from Ref. [4].

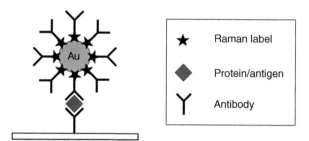

Figure 6 SERS sandwich immunoassay with Raman-labelled immuno-gold probes, i.e. SERS-labelled antibodies. In this modified approach (cf. Figure 5), the antibodies are attached to the Raman label molecules and not directly to the gold nanoparticle, providing a higher surface coverage with Raman reporter molecules. Adapted from Ref. [5].

Raman label conjugation; in this study, thiophenol, 2-naphthalenethiol and 4-mercaptobenzoic acid were employed: all molecules form Au–S bonds, leading to a stable monolayer on the gold surface via self-assembly. The sensitivity of this immunogold Raman methodology is around 10^{-9} mol/L. Some drawbacks arise from this configuration: in addition to the thiol compounds used as Raman reporter molecules, antibodies must adsorb to the surface. Because a complete self-assembled monolayer (SAM) with Raman labels cannot be formed, the SERS signal is lower than maximum. Additionally, other (bio)molecules can adsorb onto the gold surface, increasing the risk of unspecific binding and the appearance of unwanted additional SERS signals.

Porter and co-workers [5] presented a significant improvement in detection capabilities in 2003, employing modified SERS labels (Figure 6). In the new labelling strategy, the antibodies are covalently bound to the Raman reporter molecules, i.e. they are not directly bound to the gold surface. This allows a high surface coverage with Raman labels. Gold nanoparticles (30 nm) covered with a monolayer of 4-nitro-2-mercaptobenzoic acid as an intrinsically strong Raman scatterer gave very intense SERS signals upon 632.8 nm laser excitation. Prostate-specific antigen (PSA) could be determined in such a SERS sandwich immunoassay at the femtomolar level ($\sim$1 pg/mL).

The same concept was applied for the rapid and highly sensitive detection of viral pathogens such as feline calicivirus from cell culture media [6]. Various experimental parameters such as salt concentration and binding buffer were optimized in order to minimize non-specific binding and maximize antigen-binding efficiency. The dynamic range of this assay is 10^6–2.5×10^8 viruses/mL, with a limit of detection of 10^6 viruses/mL.

2.4. Protein detection with nanostructures prepared by seeded growth of silver on gold

Mirkin and co-workers [7] have developed Raman dye-labelled nanoparticles for protein detection by modifying probes originally designed for DNA and RNA detection

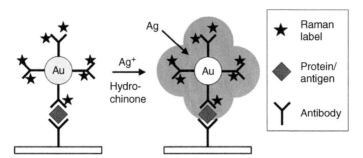

Figure 7 Protein detection with SERS-active nanostructures. Raman-labelled immuno-gold probes bind specifically to the protein target. Subsequent electroless silver deposition in the presence of Ag$^+$ and hydrochinone leads to the formation of SERS-active nanostructures [7,8]. Different configurations can be used, for example, a direct binding immunoassay [7] or a sandwich immunoassay [8] as shown here.

(see Section 3.3). In this concept, small gold nanoparticles (13 nm) labelled with Raman-active dyes and target-specific binding molecules are combined with a particle-initiated Ag developing technique (Figure 7). Seeded growth of silver on gold nanoparticles is performed in the presence of Ag(I) and hydrochinone: this electroless silver deposition is necessary for producing SERS-active nanostructures at locations where immunogold probes are immobilized (Figure 7).

Two different nanoparticle probes for selective protein detection, based on either protein–small molecule (type I) or protein–protein (type II) interactions, were developed; both types are water-soluble and very stable, have characteristic Raman spectroscopic signatures and specific target recognition properties [7].

In type I particle probes, designed for probing protein–small molecule interactions, the gold nanoparticle is functionalized with a molecule comprising the following units: a thiole group for binding to the gold surface, the Raman-dye, a hydrophilic deoxyribonucleotide oligomer of adenine and the small molecule (e.g. biotin) at the other end which binds specifically to the protein target [7]. The specificity of this approach was demonstrated for the following three small molecules: biotin, digoxigenin and dinitrophenyl. For SERS detection, the dyes Cy3, Cy3.5 and Cy5 – linking the thiol and A$_{20}$ (oligoA) units within the alkylthiol-capped oligoadenotides – were employed as Raman labels. The protein counterparts are the corresponding monoclonal antibodies against the set of three unrelated small molecules.

In type II particle probes, designed for probing protein–protein interactions, antibodies instead of small molecules are used (see Figure 7): the gold nanoparticles are covered with antibodies, alkylthiol-capped oligoadenotides containing the Raman label and BSA for passivating the particle surface [7]. This system was also tested for three proteins: mouse immunoglobulin G (IgG), ubiquitin and human protein C in combination with their respective antibodies. Various surfaces can be used for target immobilization in such a direct binding assay: in addition to glass chips, polymer substrates such as nitrocellulose work allow the application also in Western blotting experiments.

In addition to these direct binding assays [7], the silver developing technique has also been applied for protein detection in a sandwich immunoassay format (Figure 7) in which the antigen is immobilized by a capture antibody. Polyclonal antibodies against hepatitis B virus surface antigen were immobilized on silicon or quartz substrates [8]. After capture of the target antigen, the Raman-labelled detection antibody–nanoparticle conjugate was added. These probe-labelling immunogold particles contain 4–mercaptobenzoic acid as the Raman label and monoclonal antibodies against hepatitis B virus surface antigen; both Raman label and detection antibody molecules, together with BSA, are directly adsorbed on the surface of the silver nanoparticles. After specific binding of the Raman-labelled immunogold probes to the captured antigen molecules, silver staining is carried out for producing SERS-active nanostructures (Figure 7). For the concentration range 1–40 µg/mL, a linear calibration model was established, the detection limit being 0.5 µg/mL [8].

3. NUCLEIC ACID DETECTION WITH SERS LABELS

Labelled oligonucleotides play a central role for the selective detection of DNA and RNA in the same manner as labelled antibodies do for protein detection (Section 2). Again, various SERS substrates can be employed: solid SERS substrates (Section 3.1), nanoparticles (Section 3.2) and SERS-active nanostructures prepared by seeded growth (Section 3.3).

3.1. Gene probe for DNA detection employing solid SERS substrates

The first DNA gene probe based on SERS label detection was reported by Vo-Dinh and co-workers in 1994 [9]. A schematic diagram of the hybridization and detection steps is depicted in Figure 8. First, a nitrocellulose probe containing single stranded DNA fragments of interest is prepared (Figure 8a). Oligonucleotides labelled with cresyl fast violet (CFV), also called SER-gene probe, are added to these DNA fragments. Hybridization takes place for complementary DNA sequences of single strands (Figure 8b), in this study between 18 deoxyribonucleotide oligomers of thymine, $p(dT)_{18}$, and adenine, $p(dA)_{18}$, respectively. The SERS-labelled probes that hybridized to the DNA oligomers attached on nitrocellulose were recovered by washing (Figure 8c) and transferred to the SERS-active substrate (silver on alumina-coated glass slide) for spectral analysis employing 620 nm excitation (Figure 8d). The specificity of the SERS gene probe is demonstrated by showing that only complementary oligonucleotides hybridize, while control sequences such as $p(dC)_9$ do not. In this early study, the authors already explicitly mention the enormous multiplexing potential of Raman/SERS-based approaches compared with, for example, chemiluminescent or fluorescent labels: even in the case of a considerable spectral overlap between Raman label signatures, it should be possible to find up to 100 different labels for the simultaneous detection of distinct gene targets. Of course,

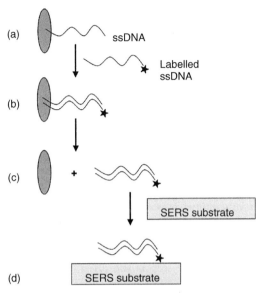

Figure 8 Gene probe for DNA detection employing solid SERS substrates. (a) Single-stranded DNA (ssDNA) is immobilized on a support matrix. (b) Hybridization takes place upon specific binding of complementary Raman-labelled ssDNA. (c) The hybridization product is removed from the support matrix and (d) transferred to the solid SERS substrate for detection. Adapted from Ref. [9].

the multiplexing capacity of Raman labels is not restricted to nucleic acid detection, but can also be applied to protein detection and any other area of targeted research in general.

Subsequent studies by Vo-Dinh and co-workers [10] extended the SERS gene probe concept also to the utility of CFV-labelled primers in polymerase chain reaction (PCR) amplification of target DNA sequences. Synthetic DNA templates from the *gag* gene region of the human immunodeficiency virus type 1 (HIV1) were used as a model system in order to demonstrate the applicability of the SERS gene probe for HIV detection. Cancer gene detection [11], including the use in a microarray platform [12], is another application of this technology. The combination of SERS-labelled probes with a spatially resolved data acquisition presents a new technology platform for genomics and biomedical analysis [13].

3.2. DNA detection by SERRS employing aggregated silver colloids

Graham, Smith and co-workers [14] presented an approach to the selective detection of DNA at ultralow concentrations by SERRS in 1997 (Figure 9). An organic polyamine (spermine) instead of inorganic salts was used for the aggregation of the silver colloids in order to achieve high SERS signal levels; spermine interacts with DNA to balance the negative charge of the phosphate groups and is effective for controlled aggregation of the colloidal particles (Figure 9) as documented

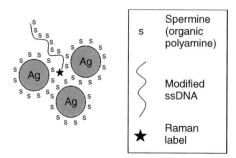

Figure 9 DNA detection with aggregated colloidal silver and modified DNA probes. An excess of spermine, an organic polyamine, increases surface-attraction of DNA and allows controlled aggregation of silver nanoparticles [14]. The DNA probe itself is also chemically modified, which leads to an additional improvement of surface attraction of the DNA.

by optical absorption/extinction spectroscopy and transmission electron microscopy (TEM) photographs. The second step is the chemical modification of DNA by incorporating propargylamino-modified nucleosides. Both procedures improve the surface attraction of the DNA, which is a prerequisite for obtaining SERRS signals from DNA labelled with Raman reporter molecules (see Figure 9). The dye hexa-chloro-6-carboxyfluorescein (HEX) in combination with aggregated silver colloids provides the required SERRS sensitivity as low as 8×10^{-13} mol/L. Also a second oligonucleotide probe was synthesized: instead of using propargyl amine groups, the positively charged dye R6G was used to adsorb directly to the colloid surface [15]. A direct comparison of the HEX- and R6G-modified oligonucleotide probes at the same concentration shows that the R6G probe yields a higher signal-to-noise ratio by a factor of about 10. Various parameters for maximum SERRS signals were investigated: for example, the optimum spermine (polyamine) concentration was found to be 0.1 mol/L. Additionally, the signal stability over time was investigated, demonstrating that a maximum occurs 60 s after the addition of spermine and that a constant signal level is obtained after 210 s [15]. Subsequently, both probes (R6G and HEX) were used to illustrate the ability of SERRS to discriminate different oligonucleotide probes in mixtures: this approach does not require separation procedures and can be applied to oligonucleotides in varying proportions [16].

Multiplexed genotyping by SERRS was first demonstrated for cystic fibrosis, a genetically transmitted disease [17]. This requires allele-specific SERRS-active primers for usage in the PCR. The separation of unincorporated primers is achieved by a biotin probe and streptavidin–coated magnetic beads. The last step is the denaturation of the immobilized hybridization product: it releases the SERRS-labelled single-stranded DNA (ssDNA), which can then be detected by SERRS.

Using labelled oligonucleotides and suspensions of silver colloid, quantitative detection over wide concentration ranges was achieved [18]. Commercially available dyes such as HEX, Cy3, Cy5, R6G and others can be used, enabling a comparison of the detection limit for both fluorescence and SERRS. Typically, limits of detection for SERRS were found to be lower by three orders of magnitude or more; this is partly due to the higher detection efficiency of Raman instrumentation. Quantitative and simultaneous detection of five different oligonucleotides was demonstrated by dual-wavelength SERRS, i.e. at two different excitation wavelengths (514.5 nm and 632.8 nm) matching the distinct electronic absorptions of the corresponding Raman dyes [18]. Detection limits were in the order of $10^{-11}-10^{-12}$ mol/L.

In addition to commercially available dyes, specifically designed and synthesized Raman labels such as benzotriazole derivates can be used. Benzotriazoles are aromatic heterocycles with a high affinity for different metals including silver [19], and their azo derivatives are excellent SERRS dyes. Functionalization of benzotriazole azo dyes allows conjugation to biomolecules. In case of DNA detection, phosphoramidites are convenient because they can be incorporated into any oligonucleotide probe using solid-phase synthesis. Maleimide and carboxylic acid derivatives have also been synthesized in order to make benzotriazoles available as SERRS labels for other biomolecules. In the so-called SERS beacons approach, benzotriazole dyes are used in combination with fluorophores [19].

3.3. DNA/RNA detection with nanostructures prepared by seeded growth of silver on gold

Mirkin and co-workers [20] have developed novel SERS substrates by seeded growth of silver on gold nanoparticle probes (cf. Section 2.4 for protein detection). Specifically, gold nanoparticles labelled with Raman-active dyes and target-specific binding oligonucleotide sequences are combined with a particle-initiated Ag developing technique (Figure 10). The multiplexed (6-plex) detection of DNA was demonstrated in 2002, employing a three-component sandwich assay in a micro-array format. The chip spotted with 15-nucleotide capture strands (Figure 10 left) was coated with a 30-nucleotide target sequence in a phosphate-buffered saline

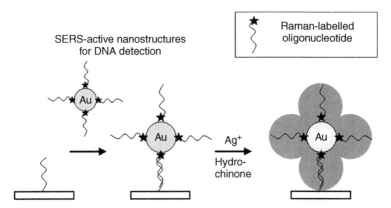

Figure 10   DNA detection with SERS-active nanostructures. Raman-labelled oligonucleotide-gold probes bind specifically to the DNA target. Subsequent electroless silver deposition in the presence of Ag^+ and hydrochinone leads to the formation of SERS-active nanostructures [20].

(PBS) solution. The presence of these specific target DNA strands was probed with 13 nm diameter Au nanoparticles modified with Raman label molecules and complementary alkylthiol-capped oligonucleotide strands (Figure 10 middle). Six commercially available dyes (Cy3, tetramethyl rhodamine, texas red, Cy3.5, R6G and Cy5), which can be incorporated into oligonucleotides through standard automated DNA synthesis, were used as Raman reporters. The six types of Raman-labelled and oligonucleotide-modified Au nanoparticles were used for probing a series of six statistically unique 30- to 36-nucleotide sequences, covering targets such as the HIV and the ebola virus. The selectivity of the oligonucleotide-nanoparticle probes was demonstrated in a microarray experiment containing mixtures of the different targets; no false-positives or false-negatives were observed, as documented by the corresponding flatbed scanner images of the Ag-developed microarrays (Figure 10 right) employing false colour images for the distinct Raman spectra of the six labels. The multiplexing capacities of composite Raman labels on gold nanoparticles for DNA microarrays were later demonstrated in the context of a random-array approach, employing glass beads exposing capture DNA strands on their surface [21].

Also the application of this oligonucleotide–nanoparticle based detection system for differentiating single-nucleotide polymorphisms (SNPs) was demonstrated [20]. Two RNA targets were chosen that can bind to the same capture-strand DNA but have a single-base mutation in the probe-binding regions. Additionally, it was also shown how mixtures can be analyzed in a semiquantitative fashion; for the two targets, a linear correlation between the target ratio and the Raman intensity ratio of the corresponding probes was observed.

These applications of oligonucleotide-modified nanoparticles for DNA and RNA detection stress the enormous potential of nanoparticle-based probes for biodiagnostic research.

4. ENCAPSULATION OF SERS LABELS

SERS labels comprise nanoparticles and Raman labels for signal generation in combination with specific binding molecules for target recognition. Because the chemical and physical properties of nanoparticles are strongly influenced by their environment, encapsulation is an important way for stabilizing them. Different materials such as silica (Section 4.1), proteins (Section 4.2) and synthetic polymers (Section 4.3) can be employed for encapsulation. The first publications in this field appeared in 2003 [22,23].

4.1. Silica encapsulation of SERS labels

Silica encapsulation of metal and semiconductor nanoparticles has been known for many years to the material science community, offering various advantages. In these core-shell particles, the silica shell protects the core both mechanically and chemically, while optical properties of the core are maintained and can be exploited. The glass shell also allows further functionalization, employing a variety of established conjugation protocols. Mulvaney, Natan and co-workers [22] were the first to apply the concept of glass encapsulation to Raman-labelled single silver and gold nanoparticles (Figure 11). In contrast to the encapsulation of, for example, metal nanoparticles, the Raman label molecules must be added before the initial formation of the glass shell. Co-adsorption of 3-amino-n-propyltrimethoxysilane (APTMS) and the Raman dye results in a vitreophilic SERS label. Then, a thin silica shell is grown by the addition of active silica. Silica growth was achieved with the Stöber method, using tetraethyl orthosilicate (TEOS). TEM images confirm the formation of a silica shell, whose thickness can be controlled by the amount of TEOS added. Two different Raman labels on noble metal nanoparticles were silica-encapsulated: *trans*-1,2-bis(4-pyridyl)ethylene (BPE) and 4-mercaptopyridine (MP). Various experiments documented the physical and chemical properties of the corresponding glass-coated nanoparticles. Below sample damage threshold, the SERS signal intensity of the particles is proportional to the laser excitation power. The position of the plasmon band is not influenced by the solvent, demonstrating the protective nature of the glass coating. A much more dramatic experiment demonstrating the same point is the dissolution of the glass-coated particles in *aqua regia*: the protective glass shell increases the etching time from 15 s or less for a bare colloidal Au sample to 3 h for a 20 nm and 1 h for a 8 nm glass shell, respectively.

In the same year, Doering and Nie [23] presented a very similar approach to glass-coated nanoparticles. They found that organic molecules with an isothiocyanate group or multiple sulfur atoms are compatible with silica encapsulation. The preparation procedure comprises the addition of the Raman label onto the gold nanoparticle surface, the addition of 3-mercapto-n-propyltrimethoxysilane (MPTMS), silica deposition by using sodium silicate and finally condensation of remaining silicate onto the existing shell.

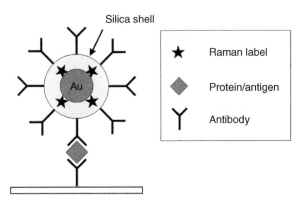

Figure 11 Silica encapsulated and biofunctionalized SERS labels comprising gold nanoparticle, Raman label, silica shell and antibodies [22].

Glass-coated, gold core-shell nanoparticles with crystal violet (CV), R6G, rhodamine B and 2-mercaptopyridine as Raman labels were applied for the detection of human IgG antigen detection, achieving a sensitivity of 4.9 ng/mL [24].

In addition to the encapsulation of single noble metal nanoparticles described above, nanoparticle assemblies can be glass coated. Aggregates of silver or gold nanoparticles have emerged as effective SERS substrates because for molecules at junctions the intensity of Raman scattering can be higher by several orders of magnitude. A very elegant procedure for preparing silica-encapsulated silver nano-particle assemblies (Figure 12) [25] uses spherical silica nanoparticles with diameters of 180–210 nm as a template (Figure 12a). The silica nanoparticles are then functionalized with MPTMS. From these MPTMS-treated silica nanoparticles, Ag-embedded silica particles are prepared by using the polyol method: silver nanoparticles are formed and deposited onto the thiol-coated silica core (Figure 12b). The next step is the addition of the Raman label and MPTMS, followed by the silica encapsulation of the Raman labelled, Ag-embedded silica particles by sodium silicate and TEOS (Figure 12c). Functionalization of the silica coated, Ag-embedded silica particles is achieved with APTMS, yielding primary amino groups at the surface to which spacer groups and, finally, the antibody can be conjugated (Figure 12d). A linear SERS signal dependence was found in the concentration range 0.5–10 mg/mL, with a limit of detection of ~0.5 mg/mL. The overall concept combines the advantages of nanoparticle assemblies for intense SERS signals with the benefits of a silica shell; in addition to the increased mechanical and physical stability, the silica shell also offers a variety of conjugation chemistries for further biofunctionalization of the outer surface.

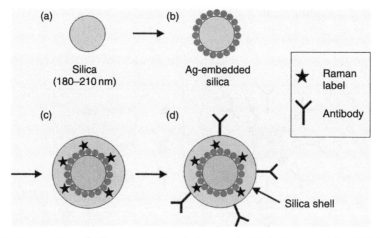

Figure 12 Preparation of silica encapsulated silver-embedded silica particles [25]. (a) Silica spheres (180–210 nm) serve as a template; (b) silver nanoparticles are deposited onto the silica particle; (c) addition of the Raman label and a SiO$_2$-precursor leads to a silica encapsulated SERS label; (d) the silica shell can be biofunctionalized, for example, with antibodies.

4.2. Protein encapsulation of SERS labels

Nanoparticles or nanoparticle aggregates can also be encapsulated by a protein shell. Composite organic–inorganic nanoparticles (COINs) [26,27] represent nanoparticle assemblies prepared by the controlled aggregation of silver nanoparticles (total diameter ca. 50 nm) induced by the addition of the Raman label molecules (Figure 13). An encapsulation layer formed by cross-linked BSA (~10 nm thick) is produced by the addition of BSA and glutaraldehyde to the colloidal COIN solution (Figure 13). Subsequent treatment with lysine or sodium borohydride removes excess aldehyde groups on the COIN surface. Finally, COIN biofunctionalization with antibodies (Figure 13) is achieved by activating the resulting carboxylic acid groups with carbodiimides such as EDC or DCC.

COINs with Raman label molecules such as 8-azaadenine, 9-aminoacridine and methylene blue were prepared. Incorporating these three distinct Raman labels, either as individual Raman labels or as mixtures into a single COIN, each with two possible intensity levels, all 37 possible signatures could be obtained. Based on a general formula from stochastic theory, it is estimated that millions of signatures could be generated with less than 50 Raman labels whose main peaks differ by 15 cm^{-1} within the wavenumber range 500–2000 cm^{-1} [26]. The limit of detection for a direct binding assay employing with PSA was 1 ng/mL [27].

4.3. Polymer encapsulation of SERS labels

In many areas, synthetic polymers are common materials for encapsulation. Polymer encapsulation has, for example, been applied to the aggregated silver nanoparticles described in Section 3.2, employing divinylbenzene as the monomer and 2,2′-azobis(isobutyronitrile), AIBN, as a radical starter. The polymer core of the beads itself can be further functionalized by the addition of a secondary polymer coating with free carboxylate groups or other groups, e.g. for attaching DNA to the bead surface. The overall size of the polymer-encapsulated SERRS beads is 1.0–1.5 μm [28].

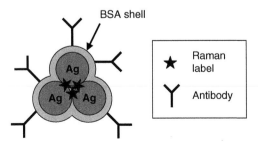

Figure 13 Composite organic–inorganic nanoparticles (COINs) are prepared via a controlled, Raman-label induced aggregation of silver nanoparticles, which are subsequently coated with a shell of bovine serum albumin (BSA) and finally biofunctionalized with antibodies [26,27].

5. MICROSCOPIC APPLICATIONS OF SERS LABELS

The applications described in Sections 2 and 3 deal with the detection of proteins and nucleic acids from solution (cf. Figures 1 and 2). This requires the immobilization of the target molecules onto a substrate, for example a modified glass slide or chip, prior to detection. For the localization of targets in cells and tissue, the immobilization step is usually not required. In tissue specimens from biopsies, for example, the antigens are intrinsically immobilized in the cellular matrix (see Figure 14); for formalin-fixed and paraffin-embedded samples, techniques of paraffin removal and antigen retrieval have to be applied.

5.1. Introduction to Raman microscopy with SERS labels

Microscopy with SERS labels aims at the selective localization of target molecules. The image contrast is based on the Raman signature of the corresponding label. This requires a spatially resolved detection of the sample. Mapping approaches with point or line focus illumination in combination with a xy-translation stage are commonly used. False colour images can be generated from the spatially resolved Raman spectra; image contrast is usually based on peak heights (intensities) or peak areas (integrated Raman intensities). Very often the term Raman microspectroscopy is used in order to stress this inherently spectrally resolved detection [29].

For antigen localization, either labelled primary or secondary antibodies can be used. Employing labelled secondary antibodies has the amplification advantage of achieving high signal intensities because multiple secondary antibody molecules can bind to the primary unlabelled antibody. This approach has, for example, been used in conjunction with malachite green labelled gold particles with subsequent silver staining [30]. Antigen quantification, however, is difficult because of the

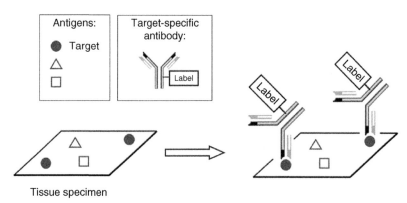

Figure 14 Antigen localization in tissue specimens with labelled antibodies (immunohistochemistry). **In contrast to antigen detection from solution (cf. Figure 2), the antigens are already immobilized within the cellular matrix.**

amplification step. If both quantitative and multiplexed detection in cells and tissue specimens have to be addressed, the use of SERS-labelled primary antibodies is preferred.

5.2. SERS microscopy for protein localization

The Raman spectrum of a SERS-labelled primary antibody is depicted in Figure 15 [5,31]. The SERS distance dependence leads to a high selectivity: because the electromagnetic field enhancement is rapidly decaying with increasing distance from the surface of the gold nanoparticle, only bands from the aromatic Raman label unit are observed (Figure 15 top). By contrast, no contributions from the antibody such as the amide vibrations are detected. This type of SERS-labelled antibodies has first been used in a sandwich immunoassay for protein detection at the femto-molar level by the Porter group [5] and later been applied to SERS microscopy (μSERS) [31]. The barcode representation of the Raman spectrum (Figure 15 bottom) illustrates the high multiplexing capacity of SERS-labelled antibodies. Many other Raman labels with distinct spectral signatures exist: either they are already commercially available or they have to be specifically designed and synthesized.

The proof of principle for SERS microscopy was first demonstrated in 2006 by Schlücker and co-workers [31]. The localization of PSA in tissue specimens of biopsies from patients undergoing prostatectomy was achieved with SERS-labelled primary anti-PSA antibodies in combination with 632.8 nm laser excitation (Figure 16). Generally, red to near-infrared excitation reduces autofluorescence and therefore improves image contrast. In the incubated prostate tissue section, SERS from

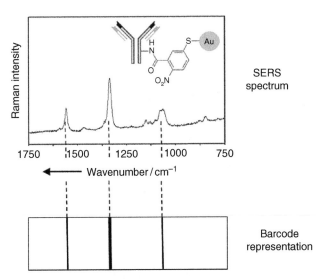

Figure 15 Raman spectrum of a SERS-labelled antibody (top) together with a barcode representation (bottom). The Raman spectrum exhibits only few selected bands from the Raman marker unit, an aromatic compound with a nitro group, because of the SERS distance dependence. Adapted from Refs. [5] and [31].

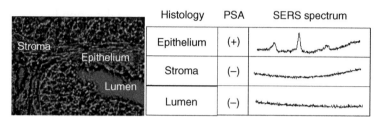

	Histology	PSA	SERS spectrum
	Epithelium	(+)	
	Stroma	(−)	
	Lumen	(−)	

Figure 16 Proof of principle for SERS microscopy [31], demonstrated by the localization of prostate-specific antigen (PSA) in biopsies from patients with prostate cancer. Prostate tissue (left) contains different histological classes in which PSA is either abundant (+) or not (−) (right). SERS-labelled anti-PSA antibodies are detected in the PSA-(+) epithelium (right), showing the characteristic Raman signals of the SERS label. Locations in the PSA-(−) stroma and lumen serve as negative controls at which no spectral contributions of the SERS-labelled antibody are detected (right). In the stroma, only minimal tissue autofluorescence is observed upon 632.8 nm laser excitation, while in the lumen only glass substrate contributions are detected (right). (See color plate 18).

aromatic Raman labels attached to hollow gold/silver spheres was detected only in the PSA-(+) epithelium (Figure 16); the PSA-(−) stroma and lumen served as controls, showing either broad autofluorescence contributions in the stroma or glass substrate contributions in the lumen (Figure 16) [31].

Further applications appeared soon after. Glass-coated, Ag-embedded silica spheres (see Section 4.1) were used for the localization of two targets (HER2 and CD10) on cellular membranes [25]. Also COINs (see Section 4.2) were applied to PSA localization in tissue specimens [27]. The issue of steric hindrance was analyzed in a simultaneous two-COIN staining for PSA antigen: each of the two distinct Raman labels was conjugated separately to anti-PSA antibody, yielding two distinct biofunctionalized SERS labels with anti-PSA antibody molecules on their surface. The characteristic Raman signatures from both COINs could be detected at almost every location in the epithelium, suggesting that steric hindrance from COINs

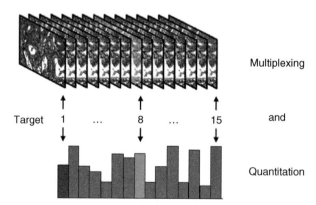

Figure 17 SERS microscopy has the potential for the detection of multiple targets (multiplexing) and target quantitation, two important aspects for tissue diagnostics. (See color plate 19).

(diameter 60–100 nm versus laser beam diameter 800 nm or 0.8 μm) does not represent a major problem.

SERS microscopy is a novel and very promising approach to immunohisto-chemistry, in particular because of its immense multiplexing capacity for simultaneous protein localization in combination with target quantification (see Figure 17). We expect that numerous applications for the analysis of cells and tissue specimens will follow within the next years. Cancer diagnostics will certainly be an important field where this innovative methodology offers significant advantages over existing approaches. In 2008, the first in-vivo application of SERS for tumor diagnostics has been demonstrated by Nie and co-workers [32]. As with every new methodology, however, the future will show how this Raman microspectroscopic technique will establish itself especially in the life science community.

REFERENCES

[1] D.A. Stuart, A.J. Haes, C.R. Yonzon, E.M. Hicks and R.P. Van Duyne, IEE Proc. Nanobio-technol., 152: 13–32, 2005.

[2] T.E. Rohr, T. Cotton, N. Fan and P.J. Tarcha, Anal. Biochem., 182: 388–398, 1989.

[3] X. Dou, T. Takama, Y. Yamaguchi, H. Yamamoto, and Y. Ozaki, Anal. Chem., 69: 1492–1495, 1997.

[4] J. Ni, R.J. Lipert, G.B. Dawson and M.D. Porter, Anal. Chem., 71: 4903–4908, 1999.

[5] D.S. Grubisha, R.J. Lipert, H.-Y. Park, J. Driskell and M.D. Porter, Anal. Chem., 75: 5936–5943, 2003.

[6] J.D. Driskell, K.M. Kwarta, R.J. Lipert, M.D. Porter, J.D. Neill and J.F. Ridpath, Anal. Chem., 77: 6147–6154, 2005.

[7] Y.C. Cao, R. Jin, J.-M. Nam, C.S. Thaxton and C.A. Mirkin, J. Am. Chem. Soc., 125: 14676–14677, 2003.

[8] S. Xu, X. Ji, W. Xu, X. Li, L. Wang, Y. Bai, B. Zaho and Y. Ozaki, Analyst, 129: 63–68, 2004.

[9] T. Vo-Dinh, K. Houck and D.L. Stokes, Anal. Chem., 66: 3379–3383, 1994.

[10] N.R. Isola, D.L. Stokes and T. Vo-Dinh, Anal. Chem., 70: 1352–1356, 1998.

[11] T. Vo-Dinh, L.R. Allain and D.L. Stokes, J. Raman Spectrosc., 33: 511–516, 2002.

[12] L.R. Allain and T. Vo-Dinh, Anal. Chim. Acta, 469: 149–154, 2002.

[13] T. Vo-Dinh, D.L. Stokes, G.D. Griffin, M. Volkan, U.J. Kim and M.I. Simon, J. Raman Spectrosc., 30: 785–793, 1999.

[14] D. Graham, W.E. Smith, A.M.T. Linacre, C.H. Munro, N.D. Watson and P.C. White, Anal. Chem., 69: 4703–4707, 1997.

[15] D. Graham, B.J. Mallinder and W.E. Smith, Biopolymers, 57: 85–91, 2000.

[16] D. Graham, B.J. Mallinder and W.E. Smith, Angew. Chem. Int. Ed., 39: 1061–1063, 2000.

[17] D. Graham, B.J. Mallinder, D. Whitcombe, N.D. Watson and W.E. Smith, Anal. Chem., 74: 1069–1074, 2002.

[18] K. Faulds, F. McKenzie, W.E Smith and D. Graham, Angew. Chem. Int. Ed., 46: 1829–1831, 2007.

[19] D. Graham, K. Faulds and W.E. Smith, Chem. Commun., 42: 4363–4371, 2006.

[20] Y.C. Cao, R. Jin and C.A. Mirkin, Science, 297: 1536–1540, 2002.

[21] R. Jin, Y.C. Cao, C.S. Thaxton and C.A. Mirkin, Small, 2: 375–380, 2006.

[22] S.P. Mulvaney, M.D. Musick, C.D. Keating and M.J. Natan, Langmuir, 19: 4784–4790, 2003.

[23] W.E. Doering and S. Nie, Anal. Chem., 75: 6171–6176, 2003.

[24] J.-L. Gong, J.-H. Jiang, H.-F. Yang, G.-L. Shen, R.-Q. Yu and Y. Ozaki, Anal. Chim. Acta, 564: 151–157, 2006.

[25] J.-H. Kim, J.-S. Kim, H. Choi, S.-M. Lee, B.-H. Jun, K.-N. Yu, E. Kuk, Y.-K. Kim, D.H. Jeong, M.-H. Cho and Y.-S. Lee, Anal. Chem., 78: 6967–6973, 2006.

[26] X. Su, J. Zhang, L. Sun, T.-W. Koo, S. Chan, N. Sundararajan, M. Yamakawa and A.A. Berlin, Nano Lett., 5: 49–54, 2005.

[27] L. Sun, K.-B. Sung, C. Dentinger, B. Lutz, L. Nguyen, J. Zhang, H. Qin, M. Yamakawa, M. Cao, Y. Lu, A.J. Chmura, J. Zhu, X. Su, A.A. Berlin, S. Chan and B. Knudsen, Nano Lett., 7: 351–356, 2007.

[28] A.F. McCabe, C. Eliasson, R.A. Prasath, A. Hernandez-Santana, L. Stevenson, I. Apple, P.A.G. Cormack, D. Graham, W.E. Smith, P. Corish, S.J. Lipscomb, E.R. Holland and P.D. Prince, Faraday Discuss., 132: 303–308, 2006.

[29] S. Schlücker, M.D. Schaeberle, S.W. Huffman and I.W. Levin, Anal. Chem, 75: 4312–4318, 2003.

[30] D.A. Stuart, A. Haes, A.D. McFarland, S. Nie and R.P. Van Duyne, Proc. SPIE - Int. Soc. Opt. Eng., 5327: 60–73, 2004.

[31] S. Schlücker, B. Küstner, A. Punge, R. Bonfig, A. Marx and P. Ströbel, J. Raman Spectrosc., 37: 719–721, 2006.

[32] X. Qian, X.-H. Peng, D.O. Ansari, Q. Yin-Goen, G.Z. Chen, D.M. Shin, L. Yang, A.N. Young, M.D. Wang and S. Nie, Nat. Biotechn., 26: 83–90, 2008.

SURFACE-ENHANCED RAMAN SCATTERING SPECTROSCOPY

ELECTROMAGNETIC MECHANISM AND BIOMEDICAL APPLICATIONS

Tamitake Itoh, Athiyanathil Sujith, *and* Yukihiro Ozaki

Contents

Abstract

Surface-enhanced Raman scattering (SERS) spectroscopy is a promising tool in analytical science because of the enormous enhancement factors that have increased the detection limits of a wide variety of molecules to single-molecule level. From many earlier experiments, it has been concluded that SERS enhancement is determined by electromagnetic (EM) field enhancement than chemical enhancement. The EM and chemical enhancement factors, specificity of vibrational spectra, and the insensitivity to the aqueous environment increase the significance of SERS to study complex biological systems. Also, the development of diverse SERS probes made this technique a practical analytical tool for biomedical applications. This chapter consists of two sections. In the first section, we discuss details of a microspectroscopic system that allowed mechanistic investigations of SERS by simultaneous measurements of plasmon resonance Rayleigh scattering spectra and SERS spectra from single silver (Ag) nanoaggregates and a direct demonstration of relationships between plasmon resonance and SERS. The second section explains biomedical applications of SERS with reference to cellular probing, biological imaging, and pathogen detection.

Keywords: surface-enhanced Raman scattering spectroscopy; SERS; plasmon resonance Rayleigh scattering; electromagnetic field enhancement; SERS and bioanalysis; SERS and cell imaging

1.   INTRODUCTION

Raman scattering is extensively used as a powerful tool for the determination of molecular structures and nature of bonding in molecules because it allows selective structural, surface processes, interface reaction, and kinetic investigations of various kinds of molecules from gaseous molecules, liquid molecules, polymers, and biomolecules [1(a),(b)]. Despite such advantages, Raman scattering suffers the disadvantage of extremely poor efficiency because of its very low cross section (e.g., 10^{-30} cm^2 per molecule). Resonance Raman effect is a powerful effect to enhance the cross section. Surface-enhanced Raman scattering (SERS) spectroscopy [2(a),(b)], first reported by Jeanmaire and Van Duyne [2(c)] and Albrecht and Creighton [2(d)], yields stronger enhancement of cross section by several orders of magnitude (up to 10^6). SERS offers the possibility of overcoming many of the problems in conventional Raman spectroscopy. It is a Raman spectroscopic technique that provides greatly enhanced Raman signal from Raman-active analyte molecules that have been adsorbed onto certain specially prepared metal surfaces [typically, silver (Ag), or gold (Au)]. The magnitude of enhancement in the cross section depends on (i) the chemical nature of the adsorbed molecules, (ii) the roughness of the surface, and (iii) the optical properties of the adsorbent [3]. The electromagnetic (EM) and chemical enhancements are the two primary mechanisms operating in SERS. EM is dependent on the presence of metal surface's roughness features, whereas chemical enhancement involves changes in the adsorbate electronic states due to chemisorption of the analyte [4,5]. The total SERS enhancement is determined by EM rather than the chemical effects [6].

This chapter consists of two parts. Part I deals with recent development of studies of EM mechanism of SERS, and Part II is concerned with biomedical applications of SERS, which are one of the most important and interesting applications.

2.   PART I: EXAMINATION OF ELECTROMAGNETIC MECHANISM OF SURFACE-ENHANCED RAMAN SCATTERING

SERS has attracted great interest in analytical science because of the enormous enhancement factors that have increased the detection limits of a wide variety of molecules to the single-molecule level [7–11]. The SERS EM model describes single-molecule sensitivity at interparticle junctions and sharp edges in Ag and Au nanoaggregates based on the 4th power of a local EM field enhancement factor M [9–17]. The realization of SERS enhancement factors $|M|^4$ of up to 10^{14} has made single-molecule sensitivity realistic. In other words, two-fold EM enhancement processes are important for verifying SERS enhancement factors that enable single molecules to be detected; in these processes, the first enhancement is due to coupling between incident photons and plasmons and the second enhancement is due to coupling between SERS photons and plasmons [18–24].

Recent experimental investigations regarding the origin of SERS have been mainly discussed in terms of the two-fold EM enhancement processes [7–10,12–24]. Contribution from chemical enhancement to a SERS signal is also possible [25–27]. From investigations using scanning near-field optical microscopy (SNOM) and scanning electronic microscopy (SEM), it has been revealed that the SERS enhancement is related to nanoparticle junctions in metal nanoaggregates [22,28]. Also, recent calculations of EM field at internanoparticle junctions in Au/Ag nanoaggregates supported enormous enhancement of SERS by coupling to plasmon resonance [9,12–16]. Therefore, exploration of relationships between SERS and plasmon resonance of metal nanoaggregates would be crucial for identifying ideal conditions for SERS measurements. Nonetheless, experimental investigations in collective ensemble measurements of relationships between SERS and plasmon resonance are difficult because of size and shape dependencies of the plasmon resonance energy of metal nanoaggregates [23,24]. In the collective ensemble measurements, plasmon resonance bands become averaged, and the relationships are difficult to be discussed because of inhomogeneous broadening of plasmon resonance bands. The difficulties associated with ensemble averaging were circumvented by developing a dark-field microspectroscopic system for correlated measurements of plasmon resonance and SERS [24].

In this section, we discuss details of the microspectroscopic system that allows simultaneous measurements of plasmon resonance Rayleigh scattering spectra and SERS spectra from a single common Ag nanoaggregate and direct demonstration of relationships between plasmon resonance and SERS [24,29]. Specifically, we have identified three kinds of relationships among excitation polarization, excitation energy, and plasmon resonance energy [24]. The microspectroscopic system and the correlated spectral measurements would be important to find optimum conditions of SERS detection and to explore mechanistic details of SERS.

2.1. Experiments

2.1.1. Preparation of SERS-active Ag nanoaggregates

A colloidal solution of Ag nanoparticles was prepared following a literature method [23,24]. Steps involved in the preparation of samples for spectromicroscopic analyses are summarized here. An aqueous solution of a mixture of NaCl (10 mM) and rhodamine 6G (R6G, 6.4×10^{-9} M) was added to a colloidal solution (9.6×10^{-11} M) of Ag nanoparticles. This mixture was incubated for 5 min at room temperature and spin-coated on a slide glass. From correlated plasmon imaging and SEM imaging, we found that SERS-active Ag nanoaggregates are composed of several Ag nanoparticles of $\sim$40 nm diameter. The Ag nanoaggregates were covered using a thin layer of a NaCl (3.5 M) aqueous solution for obtaining stabilized SERS signals [20].

2.1.2. Microspectroscopic setup of plasmon resonance Rayleigh scattering and SERS

The experimental setup for measuring Rayleigh scattering spectra of single-Ag nanoaggregates is shown in Figure 1 [24]. In an inverted optical microscope (IX70; Olympus, Tokyo, Japan), a collimated unpolarized white light beam from a 100-W halogen lamp was introduced into a sample surface through a dark-field condenser lens. Rayleigh scattering light from a single bright spot, likely a single-Ag nanoaggregate, was collected using an objective lens O1 (LCPlanFl, 60X, N.A. 0.7; Olympus) and detected either using a digital camera (CCD1) (COOLPIX5000; Nikon, Tokyo, Japan) for plasmon and SERS imaging or a polychromator (Pro-275; Acton, Tokyo, Japan) by connecting to a charge-coupled device (CCD2) (DV434-FI; Andor, Tokyo, Japan) for spectral measurements. We detected Rayleigh scattering light from a single-Ag nanoaggregate and minimized background Rayleigh scattering light by selective measurement on a sample area of $1.5\,\mu m$ diameter (shown in an open circle in Figure 2a) using a pinhole ($300\,\mu m$ radius) set in the image plane of the inverted microscope [30]. The white light spectral shape of the 100 W halogen lamp was compensated by normalizing the Rayleigh scattered light intensity using Rayleigh scattering from a frost plate (DFQ1-30C02-240; Sigma, Tokyo, Japan); the frost plate was found to have uniform Rayleigh scattering efficiency within the detection range of the spectrometer.

The experimental setup for SERS spectral measurement of single-Ag nanoaggregates is also shown in Figure 1. The inverted optical microscope is common to the detections of Rayleigh scattering and SERS signals. Excitation lasers used are an Ar ion laser (L1) (2016–05; Spectra Physics, Tokyo, Japan), second harmonics (532 nm) of a LD YAG laser (L2), a Kr ion laser (L3) (643R–AP–A01; Melles Griot, Tokyo, Japan), and a He–Ne laser (L4) (05-LHP-151; Melles Griot). A set of mirrors (M1–M3) on position-controlled ports are used for selecting 457, 488, 514, 532, 567, and 633 nm laser wavelengths for SERS excitations. A laser beam is passed through a polarizer (P1) and a quarter-wave plate (W) and reflected by a half mirror (HM) into the dark-field condenser lens before focusing on the sample surface. We adjust the focal point of the laser beam to that of white light by using two convex lenses (L1 and L2). This arrangement is very helpful for simple detection of both the Rayleigh scattering and the SERS signals without additional optics. SERS signal from a single bright spot (Figure 2b) is collected using the common objective lens (O1), passed through a holographic notch filter (N) [HNF(457.8, 488. 0,514. 5, 532, 568.2, and 632.8)-1.0; Kaiser Optic Systems, Inc., Ann Arbor, MI, USA], and detected using the CCD camera in the same way as Rayleigh scattering detection. The excitation laser power is $100\,mW/cm^2$ at the sample surface. We selected SERS signal from a single-Ag nanoaggregate and minimized the contribution from background fluorescence to SERS by using a pinhole.

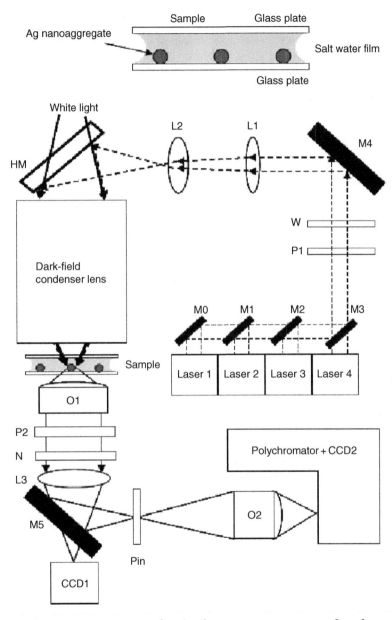

Figure 1 Microspectroscopy setup for simultaneous measurement of surface-enhanced Raman scattering (SERS) spectroscopy and plasmon resonance Rayleigh scattering from single-Ag nanoaggregates. An aqueous solution of NaCl and Ag nanoaggregates adsorbed with rhodamine 6G (R6G) were sandwiched between the two glass plates. M0–M5 are mirrors; P1 and P2 are polarizers; W is a 1/4 wave plate; L1–L3 are convex lenses; HM is a half mirror; O1 and O2 are objective lenses; N is a notch filter; Pin is a pinhole; and CCD1 and CCD2 are charge-coupled devices.

(a)   (b)

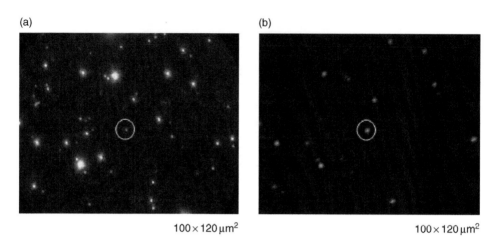

100×120 µm²   100×120 µm²

Figure 2 Plasmon resonance Rayleigh scattering (a) and surface-enhanced Raman scattering (SERS) spectroscopy (b) images of Ag nanoaggregates dispersed on a glass plate. White circles in (a) and (b) correspond to areas selected by a pinhole shown in Figure 1. (See color plate 20).

2.2. Results and Discussion

2.2.1. Polarization dependence of SERS induced by anisotropy of plasmon resonance

To understand optical correlation between excitation polarization and SERS, we investigated polarization dependencies of plasmon resonance Rayleigh scattering and SERS for a single-Ag nanoaggregate. For this, we calculated both the optical far-field corresponding to a Rayleigh scattering spectrum and the near-field corresponding to a SERS spectrum of an Ag nanoaggregate in which two nanoparticles (diameter ∼20 nm) are in contact. This two-nanoparticle system is the simplest SERS-active system [9,12,13,16,21,28]. For the calculation, we used a finite difference time domain (FDTD) method (PLANC-FDTD Ver.6.2; Information and Mathematical Science Laboratory Inc., Tokyo, Japan). We compared polarization dependencies among the calculated Rayleigh scattering spectra, experimental Rayleigh scattering spectra, and SERS spectra of an Ag nanoaggregate.

The calculated far-field spectra are shown in Figure 3A. When the incident light polarization is parallel to the long axis of a nanoaggregate, the Rayleigh scattering spectra show two distinctive plasmon resonances centered at 355 and 630 nm. On the contrary, only one plasmon resonance centered at 350 nm is significant when the polarization of incident light is perpendicular to the long axis of the nanoaggregate. Calculated near-field images corresponding to the 350 and 630 nm laser excitations with excitation polarizations perpendicular and parallel to the long axis of the Ag nanoaggregate are shown in Figure 3b. The color gradient in the calculated images are proportional to the SERS EM enhancement factor that is defined as the 4th power of the ratio of incident EM field (E_{in}) and enhanced EM field (E_{loc}), $|E_{loc}/E_{in}|^4$ [9,10]. When the excitation polarization is parallel to the long axis, an enhanced near-field is observed at a particle junction. On the contrary,

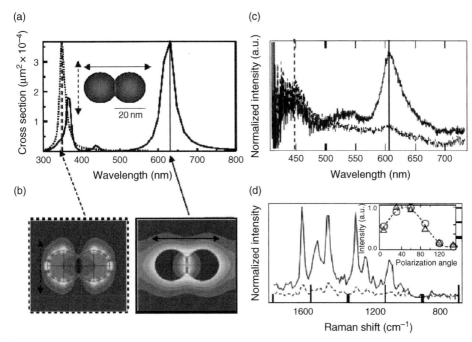

Figure 3 (a) Calculated Rayleigh scattering spectra of an Ag nanoaggregate shown in the inset: solid and dotted curves are spectra with excitation polarization parallel to and perpendicular to long axis of the Ag nanoaggregate, respectively. (b) Calculated near-field images of the Ag nanoaggregate with excitation polarization perpendicular (left panel) to and parallel (right panel) to the long axis of the Ag nanoaggregate. Values of color scale are 4th power of ratios of incident EM field (E_{in}) and enhanced EM field (E_{loc}), $|E_{loc}/E_{in}|^4$. (c) Experimental Rayleigh scattering spectra of a Ag nanoaggregate with orthogonal excitation polarizations. (d) Experimental surface-enhanced Raman scattering (SERS) spectra corresponding to E with orthogonal excitation polarizations. Polarization angles are the same for solid curves in (c) and (d) and dotted curves in (c) and (d). Inset of (d): intensity of the plasmon resonance maxima at 610 nm in (c) and SERS maxima in (d) represented as open circles and open triangles respectively with respect to polarization angle.

enhancement is negligible when the excitation polarization is perpendicular to the long axis. The polarization-angle dependence of the plasmon resonance band at 630 nm (Figure 3a) is well fitted with a $\cos^2\theta$ curve (data not shown). Therefore, this resonance band is likely a dipole. Also, we confirmed that the polarization-angle dependence of near-field intensity at the nanoparticle junction can be fitted with the same $\cos^2\theta$ curve (data not shown) as in the case of the plasmon resonance at 630 nm. The common $\cos^2\theta$ curve revealed that the enhanced near-field at the nanoparticle junction is contributed by the dipole of plasmon resonance with the maximum at 630 nm.

Typical Rayleigh scattering spectra from a SERS-active Ag nanoaggregate are shown in Figure 3c. The Rayleigh scattering spectra indicated by solid and dotted curves correspond to orthogonal excitation polarization angles. Interestingly, for a given excitation polarization, the experimental (Figure 3c) and calculated (Figure 3a) curves show only one major plasmon resonance band; band positions are not exactly the same though. Generally, the plasmon resonance maximums at 610, 530, and 445 nm in the experimental spectrum (Figure 3c, solid curve) are comparable to the plasmon resonance maximums at 630, 440, and 360 nm in the calculated spectrum (Figure 3a, solid curve). Furthermore, the polarization-angle dependencies of the plasmon resonance maximums at 610 and 530 nm (inset of Figure 3d) and 630 and 440 nm (data not shown) were also fitted with a common $\cos^2\theta$ curve; the $\cos^2\theta$ curve reveals that the plasmons are dipoles. From the identical spectral shapes and polarization dependencies of the calculated and experimental plasmon resonance bands, we considered that the experimental maxima at 610 and 530 nm (Figure 3c) correspond to the calculated maxima at 630 and 440 nm (Figure 3a), which are longitudinal plasmons parallel to the long axis of a Ag nanoaggregate.

We examined the polarization-angle dependence of SERS signals from the identical Ag nanoaggregate and found that the polarization-angle dependence of SERS is similar to that of the plasmon resonance maximum at 610 nm (Figure 3c) [24]. The SERS signal detected from a single-Ag nanoaggregate yields the maximum (solid curve in Figure 3d) and minimum (dotted curve in Figure 3d) intensities when the excitation polarization is parallel and perpendicular to the plasmon resonance band, respectively. The polarization-angle dependence of the SERS intensity is illustrated in the inset of Figure 3d. The polarization-angle dependence of both the plasmon and the SERS signals can be fitted with a common $\cos^2\theta$ curve. This fitting strongly supports that SERS is contributed by an enhanced EM field that is coupled to the plasmon resonance at 610 nm. This relation between SERS and plasmon resonance leads us to conclude that the SERS intensity becomes the optimum when the excitation polarization angle is parallel to the longitudinal plasmon mode in Rayleigh scattering spectrum [29].

2.2.2. Optimal SERS excitation wavelength dependence on plasmon resonance maximum

Recently, excitation-wavelength dependence of SERS was examined by using arrays of Au nanoparticles with a homogeneous plasmon resonance band, and it was found that the wavelength of the plasmon resonance maxima affects the excitation-wavelength dependence of SERS [24(a)]. Here, we identified optimum excitation wavelength for SERS on the basis of relations between the excitation-wavelength dependence of SERS and the wavelength of plasmon resonance maxima of single-Ag nanoaggregates. For this, we selected four single SERS-active Ag nanoaggregates with different plasmon resonance maxima and examined relations between the excitation-wavelength dependence of SERS and the plasmon resonance wavelength of individual nanoaggregates. From polarization dependencies (results not shown) of these plasmon resonance maxima and SERS maxima, we confirmed that the plasmon dipoles are coupled to SERS and directly investigated the relationships between SERS and plasmon resonance without considering inhomogenity caused by overlapping of many dipoles and multipoles in Ag nanoaggregates. Plasmon resonance bands selected from the four Ag nanoaggregates with intensity maxima at 587, 622,

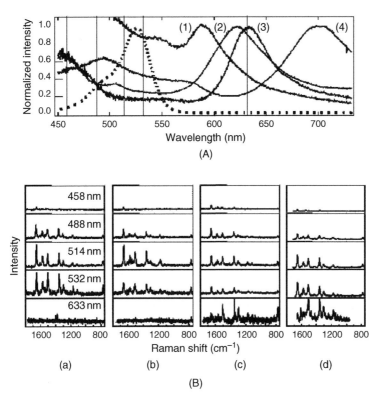

Figure 4 (A) Solid curves (1), (2), (3), and (4): plasmon resonance Rayleigh scattering spectra of four single-Ag nanoaggregates; dashed curve: absorption spectrum of rhodamine 6G (R6G). The vertical solid lines correspond to excitation laser wavelengths (458, 488, 514, 532, and 633 nm). (B) Panels (a), (b), (c), and (d): surface-enhanced Raman scattering (SERS) spectra of four SERS-active single-Ag nanoaggregates measured by exciting at different wavelengths. The SERS bands in (B) [panels (a), (b), (c), and (d)] correspond to plasmon resonance bands in (A) [(1), (2), (3), and (4)].

634, and 700 nm are shown in Figure 4A (solid curves). The excitation laser wavelengths used were 458, 488, 514, 532, and 633 nm and are indicated by vertical lines in Figure 4A. The dotted curve in Figure 4A depicts an absorption spectrum of R6G. The excitation wavelength dependencies of SERS signals from the four Ag nanoaggregates are shown in Figure 4B. Plasmon resonance of the nanoaggregates corresponding to the SERS panels a, b, c, and d are curves 1, 2, 3, and 4 in Figure 4A.

It has been known that for a colloidal solution of SERS-active Ag nanoaggregates, the spectral shape of excitation-wavelength dependence of SERS intensity is similar to the absorption spectrum of R6G [31]. In particular, an aqueous solution of R6G has both the optical absorption maximum and the spectral shape of excitation-wavelength dependence of the SERS intensity around 525 nm [31]. Indeed, from single nanoaggregate SERS investigations, we revealed that a definitive similarity between the absorption spectrum and the excitation wavelength dependency of SERS is not valid from nanoaggregate to nanoaggregate. A typical example is shown in Figure 4B [panels (a) and (b)], where under the 633 nm excitation characteristic SERS bands are absent for two nanoaggregates with higher energy plasmon resonance bands. On the contrary, under the same excitation conditions, characteristic SERS bands of R6G are clearly identified for other two nanoaggragtes with relatively low plasmon resonance energy [panels (c) and (d)]. This comparison of the excitation-wavelength dependence and plasmon resonance bands suggested that the excitation wavelength for efficient SERS shifts to the longer wavelength region as the plasmon resonance maximum shifts to the longer wavelength region. The red-shifted excitation wavelength likely shifts the EM field toward a longer wavelength that produces the plasmon resonance. It is noted that the excitation wavelength for efficient SERS is different from the plasmon resonance maximum as shown in Figure 4B. For example, panel (b) in Figure 4B shows that the SERS intensity is maximized when excited at 514 nm although the corresponding plasmon resonance maximum is located at 622 nm. These observations suggested that the ideal excitation wavelength for SERS is not always determined by molecular absorption maximum or plasmon resonance maximum but by spectral overlap between electronic absorption spectrum of a molecule and spectrum of EM filed intensity, which is enhanced by coupling between particle plasmon and incident laser light. We consider that the optimum excitation-wavelength for SERS is, in general, between a plasmon resonance maximum and a molecular absorption maximum.

2.2.3. Variations in SERS spectra induced by plasmon resonance maxima

SERS spectra are different for each nanoaggregate [7–11,23–26]. These differences provide difficulties in characterizing molecules by SERS. The origin of the differences has been attributed to variations in chemical interactions between adsorbed molecules and surface metal atoms [11,25] or to temperature variations of SERS-active molecules [26]. We have investigated the contribution of plasmon resonance maxima to the above differences. In these investigations, we have considered that SERS spectra are varied by plasmon resonance band shape because SERS EM model predicts that SERS photon radiation is mediated by plasmon dipole radiation [9,12,13,16,23(a)].

To explore the origin of the plasmon-resonance dependence of spectral variations in SERS, we measured SERS and plasmon resonance bands for three Ag nanoaggregates on which rhodamine 123 (R123) molecules were absorbed. Typical plasmon resonance bands (dotted curve) and SERS bands (solid curves) from the three Ag nanoaggregates obtained with the 568-nm laser excitation are shown in Figure 5. Interestingly, the SERS bands near the plasmon resonance maximum were selectively enhanced. The selective enhancement of SERS bands are presented in Figure 5a–c. It can be seen from Figure 5a that a plasmon resonance maximum at 570 nm and SERS bands around 570 nm, which corresponds to a Raman shift of 400 cm^{-1}, are larger than other SERS bands. Figure 5b shows that a plasmon resonance maximum at 590 nm and SERS bands around 570 and 620 nm are equally enhanced. In Figure 5c, a plasmon resonance maximum at 620 nm and SERS bands around 620 nm, which corresponds to a Raman shift of 1500 cm^{-1},

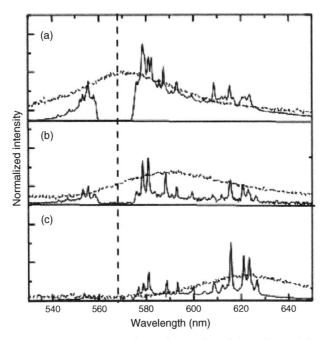

Figure 5 Plasmon resonance maxima dependence of surface-enhanced Raman scattering (SERS) spectra from three Ag nanoaggregates. Plasmon resonance Rayleigh scattering spectra are indicated by dotted curves and SERS spectra by solid curves.

are stronger than other SERS bands. The observed selective enhancement of SERS bands is consistent with SERS EM model [7–9], which points out SERS as a two-fold enhanced Raman scattering; first, enhancement from incident light and second, enhancement from scattered light. The second enhancement is rephrased as radiation of Raman scattering photon is mediated by plasmon dipole radiation. Thus, we consider that SERS spectral envelope is modulated by spectral envelope of plasmon resonance Rayleigh scattering. Therefore, the observed spectral modulation of SERS dictates a consideration that a SERS spectrum is a convolution between normal Raman spectrum and plasmon resonance Rayleigh scattering spectrum.

In this way, our investigations directly demonstrated three kinds of relationships between plasmon resonance and SERS [24,29]. We identified the relationships among excitation polarization, excitation energy, and plasmon resonance energy, and we reached the following three specific conclusions: (i) SERS bands have the same polarization dependence as that of a longitudinal mode plasmon resonance band, suggesting that excitation polarization set parallel to the longitudinal plasmon mode can provide maximum SERS intensity; (ii) SERS intensity is dependent on plasmon resonance maximum and thus larger overlap between molecular absorption band and plasmon resonance band can provide higher SERS intensity; and (iii) a spectral shape of SERS is reliant on plasmon resonance energy, and it is very likely that a SERS spectrum can be modulated by plasmon resonance band shape.

3. PART II: BIOMEDICAL APPLICATIONS

SERS has been used for bioapplications from the first stage of its invention. Thanks to the demonstration of single-molecule SERS; in recent years, biomedical applications of SERS have become a subject with a variety of challenges and opportunities for practitioners of chemistry, biology, and medicine [32–37]. Cotton, Schultz, and Van Duyne [38] demonstrated very early in the history of SERS that there was great potential for probing small quantities of biological materials using SERS. Their works demonstrated that cytochrome C and myoglobin could be adsorbed to roughened silver electrodes and SERS spectra could be acquired. SERS measurements for biological and medical specimens often use near-infrared (NIR) lasers, which can reduce the risk of damaging the specimens by applying high power [1]. The high specificity of vibrational spectra, the reduction of fluorescence, and the insensitivity to the aqueous environment increase the significance of SERS to study complex biological systems. SERS studies have also been performed on living cells [39–41]. In these experiments, colloidal silver or gold particles were incorporated inside the cells, and SERS was applied to monitor the intracellular distribution of drugs in the whole cell and to study the antinum or drugs/nucleic acid complexes. The universal biocompatibility of these types of noble metal nanoparticles has contributed to the increasing interest in utilizing them as vehicles for drug and gene delivery [42,43].

This section is devoted to the biomedical applications of SERS spectroscopy. Special attention has been given to the applications of SERS in cellular probing, biological imaging, and pathogen detection. The discussion here is based on selected works from scientific literatures, which are noteworthy in the current world, and also these reports provide a promising way to further development in this field.

3.1. Cellular Probing

SERS is now extensively used for cell detection [44] and probing of cellular compartments [45] due to the possibility of detailed investigations of specific biochemical components in a cell [46]. It overcomes the detection time from $\sim$300 to $\sim$1 s and spatial detection limit from $\sim$400 to $\sim$100 nm compared with normal Raman spectroscopy [3,47,48]. The background fluorescence, photoinduced degradation, and overheating of living cells are also considerably reduced by the surface enhancement [3,47]. This enhancement is a useful tool for detecting or tracking different known biomolecules with SERS-active nanoparticles instead of using different fluorescent tags, which may have a problem of overlapping of spectra and nonuniform fluorophores photobleaching rates that leads to several potential complications [49].

Gold and silver nanoparticle probes have been considered as the ideal tools for the targeted studies in living biological systems, in particular, the investigations of small morphological structures in cells [50]. Gold nanostructures can serve as nanosensors by delivering enhanced Raman spectroscopic signatures of biological molecules and structures in their environment [51]. Delivery of nanoparticles into cellular interior, as well as routing of the particles or targeting of cellular compartments, can be achieved in various ways, depending not only on the nature of the experiments but also on the type of cell line and physicochemical particle parameters, such as size, shape, and surface functionalization [52–54]. These methods include fluid phase uptake from the cultural medium and mechanical methods such as micro injection. The in vivo molecular probing of cellular compartments by measuring SERS spectra from endosomes in living individual epithelial cell line IRPT and macrophage cells J774 were reported [45]. Figure 6a shows transmission electron micrograph of the fluid phase endocytotic uptake of an individual gold nanoparticle by an IRPT cell. The cell membrane (arrowheads, labeled m) encloses the particle (arrowheads, labeled p), thereby forming a vesicle in the cytoplasm (scale bar: 500 nm). A schematic of SERS measurements inside the endosomal compartment is shown in Figure 6b. This study suggested that intracellular investigations based on SERS enable sensitive, controlled molecular probing with extremely high lateral resolution from nanometer volumes.

Typical SERS spectra measured with 1-μm spot size within the $30 \times 30 \, \mu m^2$ area at different places on a living intestinal epithelial cell HT29 monolayer incubated with colloidal gold are shown in Figure 7 [6]. All the spectra measured at different places are distinct, reflecting the inhomogeneity of the cell. The $1120 \, cm^{-1}$ band can be assigned to an O—P—O DNA backbone vibration. Raman bands around $1300 \, cm^{-1}$ are associated with adenine and guanine vibrations. The spectral region around $1000 \, cm^{-1}$ is dominated by phenyl alanine ($1004 \, cm^{-1}$) and a C—O DNA backbone vibration at $980 \, cm^{-1}$. The absence of a strong SERS band at $735 \, cm^{-1}$, which is due to the adenine ring breathing mode, indicates that native, and not denatured, DNA contributes to the spectra. SERS spectra also show the conformationally sensitive guanine bands around $650/670 \, cm^{-1}$. The strong SERS band at $650 \, cm^{-1}$ in the upper spectrum is very likely due to the superposition of a strong tyrosine vibration and a guanine one.

(a)   (b)

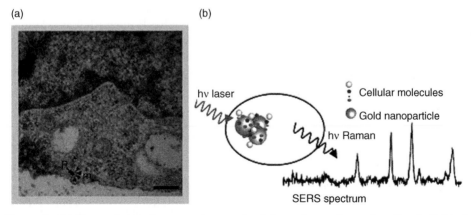

Figure 6 (a) Transmission electron micrograph of the endocytotic uptake of an individual gold nanoparticle by an IRPT cell. (b) A schematic of surface-enhanced Raman scattering (SERS) spectroscopy measurements inside the endosomal compartment [45].

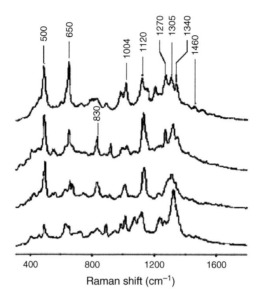

Figure 7 Surface-enhanced Raman scattering (SERS) spectra measured at different spots on a living intestinal epithelial cell HT29 monolayer incubated with colloidal gold [6].

The detection of cellular components in intact single cells can be improved by Raman labels such as dyes. Raman labels are small chromophoric molecules that are placed at biological sites as Raman reporter groups which may resemble closely a natural biological component or it may have no biological counterpart and be placed at the site simply to probe the properties of that site [55]. This can offer several advantages: use of longer excitation wavelengths that are suitable for living cells, lack of energy transfer, and presence of lower autofluorescence background [56]. In addition, the Raman labels have narrow peaks and can be excited by a single wavelength, which allows the simultaneous detection of multiple components [56]. The localization of anti-12-LO antibody in bovine coronary artery endothelial cells (BCAECs) has been studied using SERS technique with a Raman label cresyl violet (CV) [56]. To determine the localization of 12-LO in single BCAEC cells, the SERS spectra were detected at different points horizontally along the x or y plane across the center of the cell as shown in Figure 8a. Figure 8b shows the contour plot of the intensity of CV as a function of the detection points across the cell. The intensity of CV at $592 \, \text{cm}^{-1}$ was detected mainly in the cytosolic region of the cells. It was not detected on the cellular membrane or the nucleus.

The technique of using nanoparticles only for SERS measurements is often difficult to obtain reproducible results. Moreover, the signal will be dominated by the species that adsorb most readily to the nanoparticle surface or those that are in high concentration leading to problems of an overwhelming background [57]. These problems can be solved by developing functionalized nanoparticle sensors that employ nanoparticles that have been coated with a molecule that will bind to the analyte of interest. The functionalized nanoparticle probes have several advantages over non-functionalized probes. (i) The functional group adds a degree of specificity to the

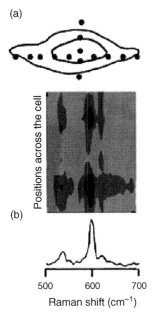

Figure 8 (a) Surface-enhanced Raman scattering (SERS) spectra detection at different points horizontally along the *x* or *y* plane across the center of the cell. (b) The contour plot of the intensity of cresyl violet (CV) as a function of the detection points across the cell [56].

sensor by providing a specific interaction with the target analyte. (ii) The analyte molecule does not need to be Raman active or have a particularly large Raman cross section and (iii) the surface is coated by the functional molecule, interfering molecules cannot adsorb to the particle surface, and therefore the background is reduced [57].

Nanoparticle-based pH sensors using 4-mercaptobenzoic acid (4-MBA) was developed by Talley et al. [58]. A Raman spectrum of 4-MBA changes according to the state of the acid group is shown in Figure 9. As the pH is reduced and the acid group becomes protonated, the COO^- stretching mode at $1430\,cm^{-1}$ decreases in intensity. The strong ring breathing modes at 1075 and $1590\,cm^{-1}$ are not affected by the change in pH. The 4-MBA–coated nanoparticle sensors show a pH response in the range of 6–8 which is ideal for biological measurements. To demonstrate the feasibility of utilizing these nanoparticle sensors in living cells, the 4-MBA–functionalized nanoparticles were incorporated into Chinese hamster ovary (CHO) cells by passive uptake. A representative SERS spectrum of the incorporated nanoparticles, shown in Figure 10, illustrates that the functionalized nanoparticles retain their functionality and are not overwhelmed by a large background when they are placed into a biological matrix. The spectrum indicates that the pH surrounding the nanoparticle is below 6, which is consistent with the particles being located inside a lysosome (pH ~ 5) [59]. Although CHO cells are not normally considered phagocytotic, they have been shown to uptake latex beads as large as 1 μm in diameter [60]. These studies also showed that the phagocytosed latex particles were localized in lysosomes once they were internalized.

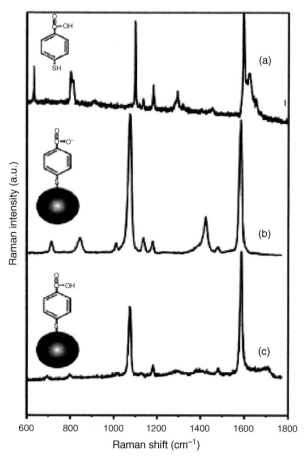

Figure 9 (a) The Raman spectrum of solid 4-mercaptobenzoic acid (4-MBA) and (b) the surface-enhanced Raman scattering (SERS) spectra of 4-MBA attached to silver nanoparticles at pH 12.3 and (c) pH 5.0. The insets to the left of each spectra illustrate the dominant state of the molecule under the conditions described above [58].

SERS is a promising tool to monitor neurotransmitter release at the single-cell level; it is a sensitive technique that provides structural information about the released compounds and spatial information about their release sites [61].

A study by Dijkstra et al. [61] demonstrated that depolarization-evoked neuro-transmitter (noradrenaline, adrenaline and dopamine) secretion by rat pheochro-mocytoma (PC12) cells can be spatially resolved by SERS using silver colloids. Figure 11a shows gold nanoparticle adsorbed by PC12 cell. Nomarski-DIC micro-scopy with reflection confocal laser scanning microscopy (CLSM) revealed that the colloidal particles were primarily situated outside the cells. Arrowheads in the figure indicate intracellular locations in two different cells. Figure 11c illustrates the spatial distribution of the correlation coefficients over one particular cell (see spots in the dashed rectangle in Figure 11b), which varies between −0.46 and 0.82. The spectra are

(a)   (b)

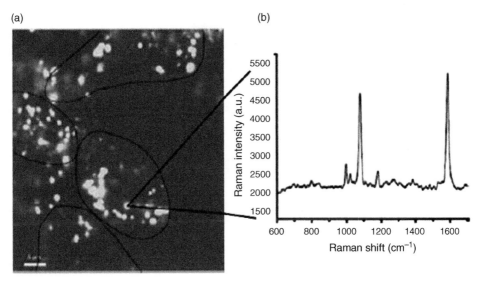

Figure 10 (a) A confocal image of Chinese hamster ovary (CHO) cells with 4-mercaptobenzoic acid (4-MBA) nanoparticle sensors incorporated into the cells. The cells are outlined in black to facilitate viewing the low contrast of the cells against the bright nanoparticles. (b) The surface-enhanced Raman scattering (SERS) spectrum of one of the nanoparticle sensors (indicating that the pH around the nanoparticle sensor is <6) [58].

distributed over five spatially isolated positions (D–H), which can be considered to be individual release sites, because the dilution upon the release is considerable [62]. All the spectra of a single event were averaged, and the correlation coefficients with the three catecholamine reference spectra were determined (E–H). The spectrum of position E had a good correlation with the spectra of both noradrenaline and adrenaline; the signal-to-noise ratio does not allow discriminating between their very similar reference spectra. The spectra of positions F–H correlate strongly with the spectrum of dopamine due to a significant spectral structure around 1320 cm^{-1}. Most events are located on the cell plasma membrane, but the position of spectrum H seems to be adjacent to the cell, which is most likely due to a small-cell movement during the measurements; the cell image was recorded directly before the Raman measurements.

3.2. Biological Imaging

The observation of molecular events inside living cells may dramatically improve our understanding of basic cellular processes and our knowledge of the intracellular transport and the fate of therapeutic agents. Raman imaging is better nondestructive analytical tool than fluorescence microscopy because it yields highly compound-specific information for chemical analysis and has great potential for high-throughput analysis and direct imaging [63]. Because of the low efficiency of Raman scattering, the role of noble metal nanoparticles

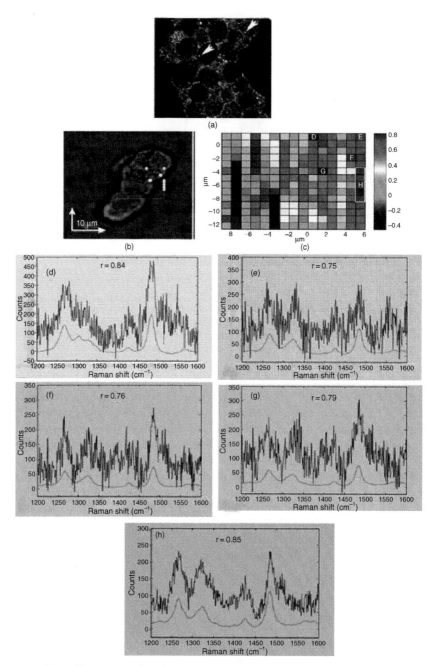

Figure 11 (a) Gold nanoparticle adsorbed by PC12 cell. (b) Cells showing the measured area and (c) spots showing the spatial distribution of the correlation coefficients over one particular cell (dashed rectangle). Corresponding surface-enhanced Raman scattering (SERS) spectra and their similarities with neurotransmitters is also given. [61]. (See color plate 21).

enhancement in the scattering offers a great promise toward the biological Raman imaging of living cells. It has been demonstrated that SERS imaging, by contrast, has the potential to provide a much stronger signal and hence shorter integration times necessary for living cell imaging applications [46].

3.2.1. Imaging of individual cells

For monitoring and imaging of individual living cells, Vo-Dinh et al. [64] developed an optical silica nanofiber with Ag-coated tip by thermal vacuum deposition. 2,4-dinitrobenzoic acid (DNBA) was chosen as a Raman reporter for cellular imaging because of the strong cross section of its symmetric NO_2 stretching mode. The Ag nanoparticles are functionalized with DNBA anti-EGF antibodies. This functionalized nanoparticle has been used for the incubation of the cells to be imaged (CHO cells). The intracellular responses are measured by inserting the nanofiber tip inside the cell. Global SERS images and spectra of CHO cells incubated with SERS nanoprobes functionalized with DNBA and anti-EGF are shown in Figure 12a and b, respectively. The SERS image was acquired at the major SERS peak at $1333 \, cm^{-1}$ using a 633-nm He–Ne laser for excitation. The images indicate that the SERS signal is inhomogeneously distributed over the cell surface. There are two possible explanations for the observed variations in the SERS signal intensities within the cell: (i) differing enhancement because of the SERS-active site resulting from nanostructure (silver clusters) and (ii) the heterogeneous distribution of the laser within the global imaging field of view of the silver nanoparticles that are functionalized with DNBA and anti-EGF. The image in Figure 12a was obtained with the Raman shift set at $1333 \, cm^{-1}$, which showed correlation with DNBA's characteristic peak shown in the SERS spectrum (Figure 12b).

3.2.2. Differentiating cancer cells

SERS has recently been successfully applied to differentiating cancerous cells with normal cells. For example, to obtain a highly sensitive cellular image of living normal HEK293 cells and HEK293 cells expressing PLCγ1 using the SERS technique, functional nanoprobes based on Au/Ag core-shell nanoparticles, conjugated with monoclonal antibodies, were used by Lee et al. [65]. PLCγ1 is a protein whose abnormal expression may be associated with tumor development. Schematic illustration of silver-coated gold nanoprobes with R6G for the SERS imaging of cancer cells is given in Figure 13.

Functionalized nanoprobes, conjugated with secondary antibodies, were only attached to the markers on cancer cells. On specific binding, nanoprobes were washed out using a buffer solution. After washing, the remaining nanoprobes were selectively attached to the cancer markers by antibody–antibody interaction. Each Raman spectrum was measured by moving down the laser spot from the top of the cell with an interval of 3 μm, but any characteristic SERS signal of R6G was not observed form the normal cells [Figure 14a and b]. Only a few noise signals caused by fluorescence were observed in the spectra.

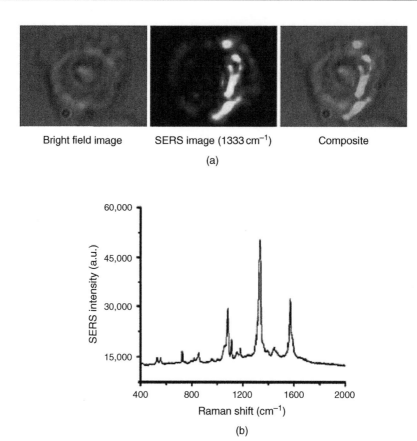

Figure 12   (a) Global surface-enhanced Raman scattering (SERS) spectroscopy images and spectra of Chinese hamster ovary (CHO) cells incubated with nanoparticles functionalized with 2,4-dinitrobenzoic acid (DNBA) and antibody to epidermal growth factor. The brightfield image (left), the total SERS image (middle), and the composite image (right) of a CHO cells are shown. (b) The SERS image was acquired at a SERS intensity signal of 1333 cm^{-1} using a 633 nm He−Ne laser for excitation [64].

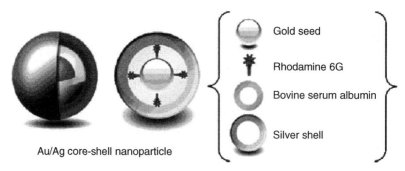

Figure 13   Schematic illustration of silver-coated gold nanoprobes for the surface-enhanced Raman scattering (SERS) spectroscopy imaging of cancer cells [65].

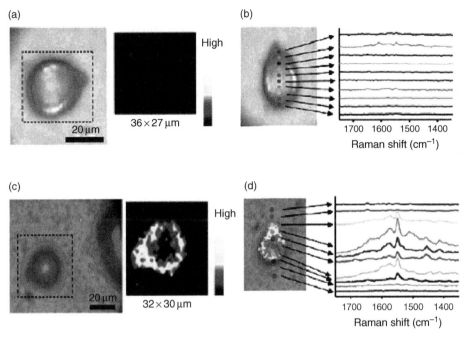

Figure 14 (a) Normal cell-dark field image and surface-enhanced Raman scattering (SERS) spectroscopy image (b) Raman measurements. (c) single cancer cell: brightfield image (left), SERS image (right). (d) Overlay image of brightfield and SERS and Raman mapping for single cancer cell. The spots in (b) and (d) indicate the laser spots across the middle of the cell along the y axis [65].

The SERS image for the same PLCγ1-expressing HEK293 cells is displayed in Figure 14c. In this figure, the image colors for the SERS spectra were displayed by a color-decoding method using the strongest Raman peak of R6G at 1650 cm^{-1}. The darker the color of the bar, the lower the concentration of SERS nanoprobes attached to the cell. Figure 14d shows a combined image of the bright field and SERS images. It shows the distribution of the PLCγ1 markers on the cell membrane. For a clear understanding, the Raman depth-profiling spectra were also measured and are displayed in Figure 14d. Here, the R6G reporter-labeled SERS spectrum on the cell membrane looks somewhat different from the SERS spectrum of pure R6G. In particular, one strong extra Raman peak was observed at 1542 cm^{-1}. However, the other Raman peaks of Raman reporter R6G in HEK cells are well matched with those of pure R6G. In this work, SERS signals of cell components were measured along with SERS signals of Raman reporter indocyanine green (ICG). The relative contributions of ICG and cell components depend on the co-adsorption of both kinds of molecules. According to the experimental data, however, the intensity of the strongest Raman peak at 1650 cm^{-1}, which was used for Raman mapping, was very much consistent with the intensity of antibody interactions between nanoparticles and biomarkers. Thus, it is believed that SERS imaging with the antibody-conjugated metal nanoprobes can clearly distinguish between cancerous and noncancerous cells.

3.2.3. Imaging of proteins

An important frontier in molecular imaging is the utilization of the enormous sensitivity of SERS to achieve specific detection of cellular proteins in vivo [66]. Hu et al. [67] demonstrated imaging of membrane proteins on cells using a cyano-labeled SERS probe. Figure 15c shows an optical image of labeled HeLa cells expressing the transmembrane domain of the platelet-derived growth factor receptor (TM) containing an acceptor peptide (AP) tag that was subsequently labeled with a ketone analog of biotin7 and Ag-NP-1 (silver nanoparticles with a ligand containing cyano group as a Raman reporter 1 shown in Figure 15a). Its

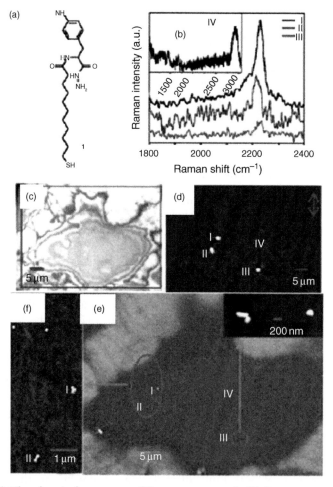

Figure 15   (a) The chemical structure of Raman reporter 1; (b) Raman spectra of the CN vibration mode extracted from positions I, II, and III of the cell shown in the optical image (c). Inset of (b) is a cellular Raman spectrum taken from spot IV of the same cell. (d) Raman intensity map of the CN band of the same cell and (e) the corresponding scanning electronic microscopy (SEM) image. Inset in (e) shows the nanoparticles in the lower right circle. (f) The group of nanoparticles as shown in the large oval of (e) [67].

corresponding Raman intensity map from the CN vibration band is depicted in Figure 15d. In the Raman map, three intense spots (I, II, and III) are identified, and their corresponding Raman spectra are shown in Figure 15b. Both spectra I and II exhibit an obvious CN band at $2230\,cm^{-1}$. A much weaker CN stretching band was observed at spot III. Intensity contrasts between the three SERS hot sites and regions without Ag-NPs are approximately 3 orders of magnitude. Inset of Figure 15b is a typical cellular Raman spectrum obtained from spot IV of the same cell but in a region without Ag-NP-1 (Figure 15d). This cell spectrum consists of the following bands: 1660 (amide I), 1447 (CH_2, CH_3), 1285 and 1350 (amide III, NH, CH), and 2990 (CH) cm^{-1}. [68]. SEM studies (Figure 15e and f) proved that strong SERS signals observed at spots I, II, and III were generated by tetramer, trimer, and dimer nanoparticles, respectively. It is fortunate that many cell surface receptors function as clusters which may help to facilitate nanoparticle aggregation and orientation in cellular and tissue imaging experiments.

3.3. Pathogen detection

A pathogen or infectious agent is a biological agent that causes disease or illness to its host. New investigations in the detection of pathogens are getting more attention because of its important role in the human health care, veterinary medicine, and bioterrorism prevention [69]. Bacteria and virus are important pathogens causing threat to human life. This section discusses the application of SERS toward the detection of bacteria and virus.

3.3.1. Bacteria detection

Bacteria are important biological systems, and their association with decease, infection, and recently, bio-terror threats, makes their detection of prime interest. Figure 16 shows the processed SERS spectra of different bacteria obtained by R.M. Jarvis and R. Goodacre [70]. The gram-negative facultative anaerobic rods Eco17, *Escherichia coli*; kp59, *Klebsiella pneumoniae*; and kox104, *Klebsiella oxytoca* show similar spectra with some shifts in the region of 480–800 cm^{-1}. cf109, *Citrobacter freundii*, exhibits sharp peak at $677\,cm^{-1}$ attributed to an extracellular L-cysteine-rich polypeptide toxin produced by the organism. The spectra of EntC90, *Enterococcus spp.*, and pm65, *Proteus mirabilis* (EntC90, gram positive and pm65, gram negative), should be different when considering their cell envelope biochemistry. However, they seem to be similar to each other due to the presence of N-acetyl-D-glucaosamine in their cell envelope which can individually show an intense SERS peak at $730\,cm^{-1}$. The potential application of SERS bacterial studies lies in the field of bioterrorism. SERS can detect dipicolinic acid (DPA) in the low concentration range quantitatively, which is a marker for bacterial spores including those of *Bacillus anthracis* (anthrax), and therefore, SERS could be used as a transduction method for sensitive anthrax pore detection [71]. Figure 17 shows the SERS spectra of varying concentrations of DPA in sulfate-aggregated colloid with CNS^- internal standard.

Recently, there are new substrates and procedures for the detection of bacterial cells for the fast and selective SERS response. Zhang et al. [72] developed a rapid

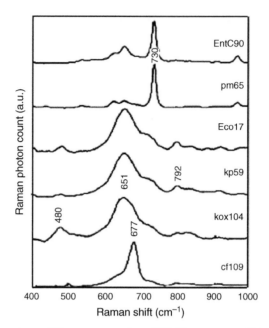

Figure 16 Surface-enhanced Raman scattering (SERS) spectrum of Ent90, *Enterococcus spp.*; pm65, *Proteus mirabilis*; Eco17, *Escherichia coli*; kp59, *Klebsiella pneumoniae*; kox104, *Klebsiella oxytoca*; and cf109, *Citrobacter freundii* [70].

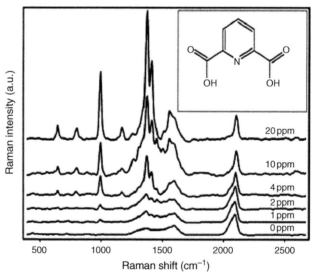

Figure 17 The surface-enhanced Raman scattering (SERS) spectra of varying concentrations of dipicolinic acid (DPA) in sulfate-aggregated colloid with CNS⁻ internal standard. Inset: structure of DPA [71].

detection protocol suitable for use by first-responders to detect anthrax spores. The speed and sensitivity of this method is provided by silver film over nanosphere substrates (AgFON). A limit of detection (LOD) of approximately 2.6×10^3 spores, below the anthrax infectious dose of 10^4 spores, has been achieved within 11 min. A schematic illustration of the fabrication of alumina-modified AgFON substrates is given in Figure 18.

The selectivity and discrimination of the SERS technique in bacteria can be assured by using a specific antibody to the model bacterium, *Escherichia coli* [73]. The new detection procedure offers a signal enhancement by 20-fold compared to conventional Raman spectroscopy. The SERS detection procedure, schematically depicted in Figure 19, consisted of three steps: (i) sorption of protein A (a highly stable cell surface receptor produced by *Staphylococcus aureus* strains) onto silver nanoparticles; (ii) immuno-reaction between the antibody and the protein A-conjugated silver nanoparticles; and (iii) bacterial sorption on silver immuno-nanoparticles.

3.3.2. Virus detection

In the area of viral pathogen detection, the most used techniques include electron microscopy, fluorescent antibody labeling of frozen tissue sections, enzyme-linked immunosorbent assay (ELISA), polymerase chain reaction (PCR), DNA hybridization, virus isolation, and serologic testing. These methodologies, however, often lack the sensitivity, specificity, speed, cost, versatility, portability, and throughput

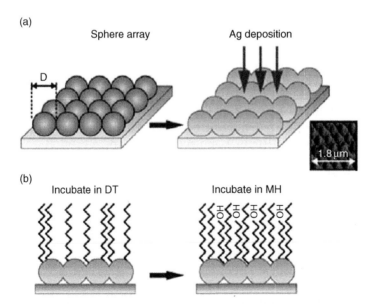

Figure 18 (a) Schematical illustration of the fabrication of AgFON substrates. The inset shows an atomic force microscope (AFM) image of an AgFON substrate. (b) The AgFON surface was incubated in a solution of 1 mM decanethiol (DH) in ethanol for 45 min and then transferred to 1 mM mercapto-1-hexanol (MH) in ethanol for at least 12 h [72].

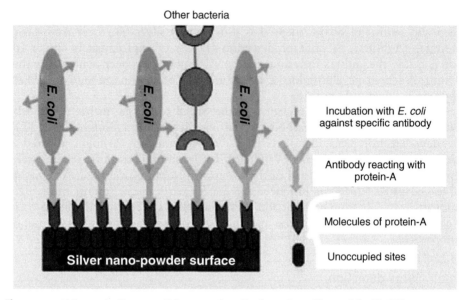

Figure 19   Schematic diagram of the procedure for detection of bacterial cells [73].

sought for such applications [74]. Here, the role of SERS is very important, which can provide all these requirements.

Figure 20 shows the SERS spectra of (i) adenovirus (Ad), (ii) rhinovirus (rhino), and (iii) human immuno-viruses (HIV) obtained by Shanmukh et al. [75]. The most prominent spectral features observed in the spectra are at 654, 730, 1247, and $1326\,cm^{-1}$ corresponding to guanine, the adenine ring vibration, thymine, and adenine, respectively. The Raman bands between 1580 and $1700\,cm^{-1}$ can be attributed to carbonyl groups on the amino acid side chains and the amide I vibration while the spectral region near $1000\,cm^{-1}$ has bands due top phenylalanine (1001 and $1030\,cm^{-1}$). Notable differences in the SERS HIV spectra compared with Ad or rhino are the shifted spectral positions of the guanine band ($643\,cm^{-1}$) and the adenine band ($719\,cm^{-1}$) and the presence of a band at $1523\,cm^{-1}$ in the HIV spectra that is absent in the spectra of the other two viruses. Based on the differences in the SERS spectra, it was possible to distinguish between the three viruses investigated in this study. This result highlights the potential of SERS as a tool to rapidly detect (30–50 s) and identify different viral pathogens in trace amounts (the concentrations of the purified virus samples were $\sim 5 \times 10^8$ PFU/mL) in diminutive specimen volumes in real time. This study used a silver nanorod SERS array fabricated by oblique angle deposition (OAD) method that acts as an extremely sensitive SERS substrate with enhancement factors of greater than 10^8.

These days, many researchers are interested in developing a direct nucleic acid-based test which can detect the presence of DNA sequences related to HIV [76–78]. Infection with the human immuno-deficiency virus type 1 (HIV1) results in a uniformly fatal disease. Unfortunately, standard HIV serologic tests, including

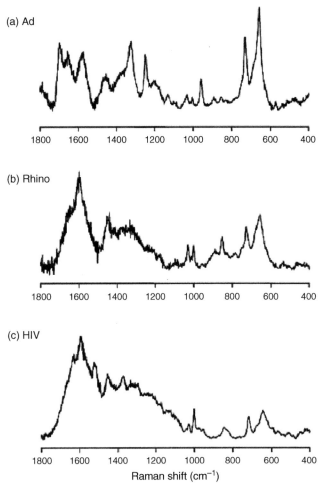

Figure 20 Surface-enhanced Raman scattering (SERS) spectra of (a) adenovirus (Ad), (b) rhinovirus (rhino), and (c) and human immuno-viruses (HIV) [75].

the ELISA and the western blot assay, are not useful in the diagnosis of HIV infection during early infancy because of the confounding presence of transplacentally derived maternal antibody in the infant's blood [63]. There is direct nucleic acid-based test that detects the presence of HIV viral sequences using SERS. The gene detection procedure uses different Raman labels such as cresyl fast violet (CFV), rodamine B, R123, and R6G [63]. For a detailed study based on the multiplexed SERS detection of the different oligonucleotide targets, using different Raman labels, refer the work of Cao et al. [49]

A SERS gene probe that has been developed by Vo-Dinh's [79] group can selectively detect HIV DNA. Figure 21A shows the SERS detection of the CFV-labeled target DNA strand, which has been hybridized with the surface-bound capture DNA probe. The signal persists even after the rigorous rinsing of the

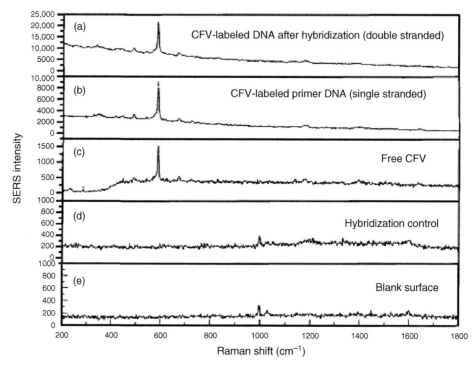

Figure 21 Demonstration of SERGene technique for the detection of the HIV1 gag gene using hybridization experiments [79].

sample. The detection of a nonhybridized, CFV-labeled single strand of DNA is also shown in curve Figure 21b. In this case, the DNA strand is the CFV-labeled forward primer sequence, which has no complementarity with the surface-bound capture probe sequence. This sample was simply applied to the polystyrene plate and coated with silver; no rinsing step was performed. For standard free CFV comparison, note the strong CFV SERS signal at $590\,\mathrm{cm}^{-1}$ (Figure 21c). There does not appear to be any major alteration in the CFV spectrum as a result of being bound to single- or double-stranded DNA. To the contrary, no CFV signals were observed following the hybridization and rinsing steps when no complementarity existed between the capture probe sequence and the single-stranded, CFV-labeled DNA. Because of the lack of complementarity, the labeled primer was effectively removed from the plate during the ensuing rinsing step. The blank illustrated in Figure 21c demonstrates that the silver-coated polystyrene plate exhibits no major spectral background.

3.4. Conclusion

In this section, we outlined recent advances in the applications of SERS spectroscopy and imaging to both biological and medical fields. The development of diverse SERS probes has made this technique as a practical analytical tool for biomedical applications. The potentiality of SERS in the detailed investigation of specific biochemical components of cells and direct nucleic acid-based tests at the single-molecule level yields considerable promise in medical diagnostics. SERS direct imaging techniques can improve our knowledge of molecular events inside living cells. New developments in this field also have made possible rapid detection of pathogens, which plays important roles in human health care and bioterrorism prevention.

REFERENCES

[1] (a) J. R. Ferraro, K. Nakamoto and C. W. Brown, "Introductory Raman Spectroscopy", Second Edition, Elsevier, San Diego, USA, 2003.
   (b) B. Schrader (ed.), "Infrared and Raman Spectroscopy-Methods and Applications" Wiley-VCH, Weinheim 1995.
[2] (a) R. Aroca, "Surface-Enhanced Vibrational Spectroscopy", John Wiley & Sons, Chechester, 2006.
   (b) K. Kneipp, M. Moskovits, and H. Kneipp (Eds.), "Surface-enhanced Raman Scattering: Physics And Applications", Topics Appl. Phys. Springer, Berlin, 2006.
   (c) D. L. Jeanmaire and R. P. Van Duyne, J. Electroanal. Chem., 84: 1–20, 1977
   (d) M. G. Albrecht and J. A. Creighton, J. Am. Chem. Soc., 99: 5215–5218, 1977.
[3] K. Kneipp, H. Kneipp, I. Itzkan, R. R. Dasari and M. S. Feld, J. Phys.: Condens. Matter, 14: R597–R624, 2002.
[4] P. Kambhampati, C. M. Child, M. C. Foster and A. Champion, J. Chem. Phys., 108: 5013–5026, 1998.
[5] M. J. Weaver, S. Zou and H. Y. H. Chan, Anal. Chem., 72: 38A–47A, 2000.
[6] K. Kneipp, A. S. Haka, H. Keipp, K. Badizadegan, N. Yoshizawa, C. Boone, K. E. Shafer-Peltier, J. T. Motz, R. R. Dasari and M. S. Feld, Appl. Spectrosc., 56: 150–154, 2002.
[7] K. Kneipp, Y. Wang, H. Kneipp, L. T. Perelman, I. Itzkan, R. Dasari and M. S. Feld, Phys. Rev. Lett., 78: 1667–1670, 1997.
[8] S. M. Nie and S. R. Emory, Science, 275: 1102–1106, 1997.
[9] H. Xu, E. J. Bjerneld, M. Käll, and L. Borjesson, Phys. Rev. Lett., 83: 4357–4360, 1999.
[10] A. M. Michaels, M. Nirmal and L. E. Brus, J. Am. Chem. Soc., 121: 9932–9939, 1999.
[11] S. Habuchi, M. Cotlet, R. Gronheid, G. Dirix, J. Michiels, J. Vanderleyden, F. D. Schryver and J. Hofkens, J. Am. Chem. Soc., 125: 8446–8447, 2003.
[12] M. Inoue and K. Ohtaka, J. Phys. Soc. Jpn., 52: 3853–3864, 1983.
[13] M. Moskovits, Rev. Mod. Phys., 57: 783–826, 1985.
[14] F. J. García-Vidal and J. B. Pendry, Phys. Rev. Lett., 77: 1163–1166, 1996.
[15] E. Hao and G. C. Schatz, J. Chem. Phys., 120: 357–366, 2004.
[16] H. Xu, X. H. Wang, M. P. Persson, H. Q. Xu, M. Käll and P. Johansson, Phys. Rev. Lett., 93: 243002-1–243002-4, 2004.
[17] B. Pettinger, J. Chem. Phys., 85: 7442–7451, 1986.
[18] M. Kerker, D. S. Wang and H. Chew, Appl. Opt., 19: 4159–4174, 1980.
[19] J. Gersten and A. Nitzan, J. Chem. Phys., 73: 3023–3037, 1980.
[20] D. A. Weitz, S. Garoff, J. I. Gersten and A. Nitzan, J. Chem. Phys., 78: 5324–5338, 1983.
[21] M. Inoue and K. Ohtaka, Phys. Rev. B, 26: 3487–3490, 1982.
[22] A. D. McFarland, M. A. Young, J. A. Dieringer, and R. P. VanDuyne, J. Phys. Chem., B, 109: 11279–11285, 2005.
[23] (a) T. Itoh, K. Yoshida, V. Biju, Y. Kikkawa, M. Ishikawa and Y. Ozaki, Phys. Rev. B, 76: 085405-1–085405-5, 2007
   (b) T. Itoh, V. Biju, M. Ishikawa, Y. Kikkawa, K. Hashimoto, A. Ikehata and Y. Ozaki, J. Chem. Phys., 124: 134708-1–134708-6, 2006.
[24] (a) T. Itoh, K. Hashimoto, A. Ikehata and Y. Ozaki, Chem. Phys. Lett., 389: 225–229, 2004.
   (b) T. Itoh, K. Hashimoto and Y. Ozaki, Appl. Phys. Lett., 83: 2274–2276, 2003.
   (c) T. Itoh, K. Hashimoto, A. Ikehata and Y. Ozaki, Appl. Phys. Lett., 83: 5557–5559, 2003.

[25] A. Weiss and G. Haran, J. Phys. Chem. B, 105: 12348–12354, 2001.

[26] Y. Maruyama, M. Ishikawa and M. Futamata, J. Phys. Chem. B, 108: 673–678, 2004.

[27] A. Otto, I. Mrozek, H. Grabhorn and W. Akemann, J. Phys.: Condens. Matter., 4: 1143–1212, 1992.

[28] K. Imura, H. Okamoto, M. K. Hossain and M. Kitajima, Nano Lett., 6: 2173–2176, 2006.

[29] T. Itoh, Y. Kikkawa, K. Yoshida, K. Hashimoto, V. Biju, M. Ishikawa and Y. Ozaki, J. Photochem. Photobiol. A, 183: 322–328, 2006.

[30] (a) T. Itoh, T. Asahi and H. Masuhara, Appl. Phys. Lett., 79: 1667–1669, 2001.
     (b) T. Itoh, T. Asahi and H. Masuhara, Jpn. J. Appl. Phys., 41: L76–L78, 2002.

[31] P. Hildebrant and M. Stockburger, J. Phys. Chem., 88: 3391–3395, 1984.

[32] J. M. Reyes-Goddard, H. Barr and N. Stone, Photodiagn. Photodyn. Ther., 2: 223–233, 2005.

[33] S. Farquharson, A. D. Gift, C. Shende, P. Maksymiuk, F. E. Inscore and J. Murran, Vib. Spectrosc., 38: 79–84, 2005.

[34] G. Breuzarda, O. Piota, J.-F. Angibousta, M. Manfaita, L. Candeilb, M. Del Riob and J.-M. Millota, Biochem. Biophys. Res. Commun., 329: 64–70, 2005.

[35] V. P. Drachev, M. D. Thoreson, V. Nashine, E. N. Khaliullin, D. Ben-Amotz, V. J. Davisson and V. M. Shalaev, J. Raman Spectrosc., 36: 648–656, 2005.

[36] R. M. Jarvis, A. Brookerb and R. Goodacre, Faraday Discuss., 132: 281–292, 2006.

[37] F. Yan and T. Vo-Dinh Sens. Actuators B, 121: 61–66, 2007.

[38] T. M. Cotton, S. G. Schultz and R. P. Van Duyne, J. Amer. Chem. Soc., 102: 7960–7962, 1980.

[39] H. Morjani, J. F. Riou, I. Nabiev, F. Lavelle and M. M. Manfait, Cancer Res., 53: 4784–4790, 1993.

[40] M. Manfait, H. Morjani and I. Nabiev, J. Cell. Pharmacol., 3: 120–125, 1992.

[41] I. Nabiev, H. Morjani and M. Manfait, Eur. Biophys. J., 19: 311–316, 1991.

[42] K. K. Sandhu, C. M. McIntosh, J. M. Simard, S. W. Smith and V. M. Rotello, Bioconjug. Chem., 13: 3–6, 2002.

[43] G. Han, C. C. You, B. J. Kim, R. S. Turingan, N. S. Forbes, C. T. Martin and V. M. Rotello, Angew. Chem. Int. Ed. Engl., 45: 3165–3169, 2006.

[44] R. M. Jarvis and R. Goodacre, Anal. Chem., 76: 40–47, 2004.

[45] J. Kneipp, H. Kneipp, M. McLaughlin, D. Brown and K. Kneipp, Nano Lett., 6: 2225–2231, 2006.

[46] L. Zeiri, B. V. Bronk, Y. Shabtai, J. Eichler and S. Efrima, Appl. Spectrosc., 58: 33–40, 2004.

[47] C. Eliasson, A. Loren, J. Engelbrektsson, M. Josefson, J. Abrahamsson and K. Abrahamsson, Spectrochim. Acta A: Mol. Biomol. Spectrosc., 61: 755–760, 2005.

[48] K. Kneipp, H. Kneipp, I. Itzkan, R. R. Dasari and M. S. Feld, Chem. Rev., 99: 2957–2976, 1999.

[49] Y. C. Cao, R. Jin and C. A. Mirkin, Science, 297: 1536–1540, 2002.

[50] K. Kneipp, H. Kneipp and J. Kneipp, Acc. Chem. Res., 39: 443–450, 2006.

[51] W. R. Premasiri, D. T. Moir, M. S. Klempner, N. Krieger, G. Jones and L. D. Ziegler, J. Phys. Chem. B, 109: 312–320, 2005.

[52] J. Rejman, V. Oberle, I. S. Zuhorn and D. Hoekstra, Biochem. J., 377: 159–169, 2004.

[53] W. J. Arlein, J. D. Shearer and M. D. Caldwell, Am. J. Physiol., 44: R1041–R1048, 1998.

[54] A. G. Tkachenko, H. Xie, Y. L. Liu, D. Coleman, J. Ryan, W. R. Glomm, M. K. Shipton, S. Franzen and D. L. Feldheim, Bioconjug. Chem., 15: 482–490, 2004.

[55] P. R. Carey, J. Raman Spectrosc., 29: 861–868, 1998.

[56] K. Nithipatikom, M. J. McCoy, S. R. Hawi, K. Nakamoto, F. Adar and W. B. Campbella, Anal. Biochem., 322: 198–207, 2003.

[57] C. E. Talley, T. R. Huser, C. W. Hollars, L. Jusinski, T. Laurence and S. M. Lane, 'Nanoparticle Based Surface-Enhanced Raman Spectroscopy, UCRL-PROC-208863', NATO Advanced Study Institute: Biophotonics Ottawa, Canada, 2005.

[58] C. E. Talley, L. Jusinski, C. W. Hollars, S. M. Lane and T. Huser, Anal. Chem., 76: 7064–7068, 2004.

[59] B. Alberts, D. Bray, D. J. Lewis, M. Raff, K. Roberts and J. D. Watson, "Molecular Biology of the Cell", Garland Publishing: New York, 1994.

[60] M. Fukasawa, F. Sekine, M. Miura, M. Nishijima and K. Hanada, Exp. Cell Res., 230: 154–162, 1997.
[61] R. J. Dijkstra, W. J. J. M. Scheenen, N. Dama, E. W. Roubos and J. J. ter Meulen, J. Neurosci. Meth., 159: 43–50, 2007.
[62] R. H. Chow, L. von Ruden and E. Neher, Nature, 356: 60–63, 1992.
[63] T. Vo-Dinh, F. Yan and M. B. Wabuyele, J. Raman Spectrosc., 36: 640–647, 2005.
[64] T. Vo-Dinh, P. Kasili and M. Wabuyele, Nanomed. Nanotechnol. Biol. Med., 2: 22–30, 2006.
[65] S. Lee, S. Kim, J. Choo, S. Y. Shin, Y. H. Lee, H. Y. Choi, S. Ha, K. Kang, and C. H. Oh, Anal. Chem., 79: 916–922, 2007.
[66] G. R. Souza, D. R. Christianson, F. I. Staquicini, M. G. Ozawa, M. G, E. Y. Snyder, R. L. Sidman, J. H. Miller, W. Arap and R. Pasqualini, Proc. Natl. Acad. Sci. U.S.A, 103: 1215–1220, 2006.
[67] Q. Hu, L.-L. Tay, M. Noestheden and J. P. Pezaki, J. Am. Chem. Soc., 129: 14–15, 2007.
[68] (a) G. U. Puppel, F. F. M. De Mul, C. Otto, J. Greve, M. Robert-Nicoud, D. J. Arndt-Jovin and T. M. Jovin, Nature, 347: 301–303, 1990.
     (b) W. L. Peticolas, T. W. Patapoff, G. A. Thomas, J. Postlewait and J. W. Powell, J. Raman Spectrosc., 27: 571–578, 1996.
[69] B. B. Chomel, J. Vet. Med. Educ., 30: 145–147, 2003.
[70] R. M. Jarvis and R. Goodacre, Anal. Chem., 76: 40–47, 2004.
[71] S. E. J. Bell, J. N. Mackle and N. M. S. Sirimuthu, Analyst, 130: 545–549, 2005.
[72] X. Zhang, N. C. Shah and R. P. Van Duyne, Vib. Spectrosc., 42: 2–8, 2006.
[73] G. Naja, P. Bouvrette, S. Hrapovic and J. H. T. Luong, Analyst, 132: 679–686, 2007.
[74] J. D. Driskell, K. M. Kwarta, R. J. Lipert, and M. D. Porter, J. D. Neill and J. F. Ridpath, Anal. Chem., 77: 6147–6154, 2005.
[75] S. Shanmukh, L. Jones, J. Driskell, Y. Zhao, R. Dluhy and R. A. Tripp, Nano Lett., 6: 2630–2636, 2006.
[76] Y. C. Cao, R. Jin, J. –M. Nam, C. S. Thaxton and C. A. Mirkin, J. Am. Chem. Soc., 125: 14676–14677, 2003.
[77] J. Johanson, K. Abravaya, W. Caminiti, D. Erickson, R. Flanders, G. Leckie, E. Marshall, C. Mullen, Y. Ohhashi, R. Perry, J. Ricci, J. Salituro, A. Smith, N. Tang, M. Vi and J. Robinson, J. Virol. Methods, 95: 81–92, 2001.
[78] S. M. H. Abanto, M. H. Hirata, R. D. C. Hirata, E. M. Mamizuka, M. Schmal and S. Hoshino-Shimizu, J. Clin. Lab. Anal., 14: 238–245, 2000.
[79] N. R. Isola, D. L. Stokes and T. Vo-Dinh, Anal. Chem., 70: 1352–1356, 1998.

CHAPTER 11

SPECTROSCOPY AND BROKEN SYMMETRY

P.R. Bunker *and* Per Jensen

Contents

Abstract

We outline how symmetry principles are used to understand high-resolution molecular spectra. The transformation operations involved are believed to commute with the molecular Hamiltonian; the most important operations of this kind are permutations of electrons, permutations of identical nuclei, coordinate inversion (or space inversion), and time reversal.

The constraints on the molecular wavefunction imposed by symmetry have important spectroscopic implications, and we present a broad general picture of the current experimental work done to test the truth of these implications. These experiments use highly precise and/or sensitive molecular-spectroscopy techniques to look for the breakdown of the Pauli exclusion principle, the appearance of "missing" levels, the effects of parity violation (the breakdown of space inversion symmetry), and the effects arising from the breakdown of time reversal symmetry. The reason for studying these symmetry breakdowns is to test our understanding of the form of the complete molecular Hamiltonian and its inherent symmetry.

Keywords: symmetry; molecule; molecular spectra; electron permutations; identical-nuclei permutations; coordinate inversion; space inversion; time reversal; Pauli exclusion principle; chirality; symmetry breakdown; parity violation; Molecular Symmetry group

1. INTRODUCTION

In a geometrical sense, the symmetry of a molecule is described by a point group, and this is obtained by determining the number and type of rotational symmetry axes and reflection planes of the molecule in its equilibrium configuration. However, the use of symmetry in the analysis of high-resolution molecular spectra involves using transformation operations that are believed to commute with the molecular Hamiltonian, such as permutations of electrons, permutations of identical nuclei, coordinate inversion (or space inversion), and time reversal [1,2]. The use of these operations has important spectroscopic implications. In recent years, highly precise and highly sensitive spectral measurements are being made that test the truth of these implications by looking for the breakdown of the Pauli exclusion principle, the appearance of "missing" levels, the effects of parity violation (the breakdown of space inversion symmetry), and the measurement of effects arising from the breakdown of time reversal symmetry. The reason for studying these symmetry breakdowns is to test our understanding of the form of the *complete* molecular Hamiltonian and its symmetry.

Currently, the complete molecular Hamiltonian derives from the Standard Model of theoretical physics [3,4]. The Standard Model incorporates the unification of the electromagnetic and weak forces; this results in the existence of the weak neutral current interaction between electrons and nuclei on top of the usual electromagnetic interaction. When we include both interactions, we have the *electroweak* Hamiltonian, which is the complete Hamiltonian that we are interested in testing.

Although rather new in molecular spectroscopy, parity violation and the breakdown of time reversal symmetry have long been the object of study in the particle physics community. A precise experimental way of determining the extent of parity violation by studying β-decay was suggested by Lee and Yang [5] and observed by Wu et al. [6] in the 1950s. In the context of this chapter, it is of interest to quote Shanmugadhasan [7]: "In the 1940s, whenever the occasion arose, Dirac used to argue forcefully against the requirement of space-reflection invariance and of time-reflection invariance for physical theories. Dirac [8] recorded his thoughts in 1949. In 1959, in introducing Dirac to an audience at the National Research Council in Ottawa, G. Herzberg quoted this prophetic paragraph of Dirac's." Thus, although Herzberg was well aware of the breakdown of these symmetries, he, quite reasonably, did not think it necessary to mention the fact in his 1960s textbook [9] on molecular electronic spectroscopy because the effects seemed totally negligible. A modern text on molecular spectroscopy should probably not make this omission, because the current level of precision and sensitivity in molecular spectroscopy is very close to that required to see such broken symmetry effects.

Sections 2, 3, and 4 introduce the various symmetries that are involved in the study of high-resolution molecular spectra. We begin with electron permutation symmetry and the symmetrization postulate. We then summarize the definition of the Molecular Symmetry (MS) group; this group involves permutations of identical nuclei and the space-inversion operation E^* (called "P" in the theoretical physics community). After that we discuss the consequences of time reversal symmetry.

The last three sections are concerned with how the possible breakdown of these symmetries is being probed in modern spectroscopy experiments.

Within electromagnetic theory, an important type of broken symmetry is that which involves the interaction of molecular states that differ in their nuclear spin symmetry. These are generally termed *ortho–para* interactions. We discuss this subject in Section 3.1.

Our aim is to introduce the reader to the current work being done in this field using highly precise and/or sensitive molecular spectroscopy experiments, and we present a broad general picture. We apologize to the many authors whose works we have not referenced, but it would have overpowered the text if we had included all possible references. We encourage the reader to delve into the still rather large number of references that we do quote in order to discover other relevant papers on the subject.

2. THE SYMMETRIZATION POSTULATE

In the Standard Model of theoretical physics, electrons are fundamental particles and they are identical to each other. This means that for an atom or a molecule any permutation of the space and spin coordinates of the electrons commutes with the Hamiltonian and is therefore a *symmetry operation*. For a system containing n electrons, the symmetry group of all possible $n!$ electron permutations is called the *symmetric group* and is denoted $\mathbf{S}_n$ [10].

The class structure of a symmetric group is easy to determine because all permutations of the same shape (i.e., consisting of the same number of independent transpositions, independent cycles of three, independent cycles of four, etc.) are in the same class. For example, in the group $\mathbf{S}_5$ there are seven classes, and the elements in these classes have the following shapes:

$$E, (xx), (xx)(xx), (xxx), (xxx)(xx), (xxxx), \text{ and } (xxxxx). \tag{1}$$

The number of classes in the symmetric group $\mathbf{S}_n$ is given by the *partition number* of n, i.e., the number of ways of writing n as the sum of integers. For example, for $n = 5$ we can write

$$5 = 1 + 1 + 1 + 1 + 1, 2 + 1 + 1 + 1, 2 + 2 + 1, 3 + 1 + 1, \\ 3 + 2, 4 + 1, \text{ or } 5, \tag{2}$$

so that the partition number of 5 is 7. The parallel between this partitioning of the number 5 and the shapes of the permutations in each class in Equation (1) is obvious. Because the number of irreducible representations in a group is equal to the number of classes, we can immediately deduce the number of irreducible representations in $\mathbf{S}_n$ once we have the partition number of n. For example, the group $\mathbf{S}_5$ has seven irreducible representations; the character table of the symmetric group $\mathbf{S}_5$ is given in Table 1.

Table 1   The character table[a] of the symmetric group $\mathbf{S_5}$

	E	(12)	(12)(34)	(123)	(12)(345)	(1234)	(12345)
	1	10	15	20	20	30	24
$\mathbf{D}^{(0)}$:	1	1	1	1	1	1	1
$\overline{\mathbf{D}}^{(0)}$:	1	−1	1	1	−1	−1	1
$\mathbf{D}^{(1)}$:	4	2	0	1	−1	0	−1
$\overline{\mathbf{D}}^{(1)}$:	4	−2	0	1	1	0	−1
$\mathbf{D}^{(2)}$:	5	1	1	−1	1	−1	0
$\overline{\mathbf{D}}^{(2)}$:	5	−1	1	−1	−1	1	0
$\mathbf{D}^{(3)} = \overline{\mathbf{D}}^{(3)}$:	6	0	−2	0	0	0	1

[a] One representative element in each class is given, and the number written below each element is the number of elements in the class.

A common feature of all symmetric groups is that they have only two one-dimensional irreducible representations: the *symmetric* representation $\mathbf{D}^{(0)}$ having character $+1$ for all permutations, and the *antisymmetric* representation $\overline{\mathbf{D}}^{(0)}$ having character $+1$ for even permutations such as (12)(34), (123), and (12345), and character -1 for odd permutations such as (12), (12)(345), and (1234). All other representations of $\mathbf{S}_n$, for n greater than 2, have more than one dimension. The representations of the symmetric group occur in pairs $[\mathbf{D}^{(k)}, \overline{\mathbf{D}}^{(k)}]$ such that

$$\overline{\mathbf{D}}^{(k)} = \mathbf{D}^{(k)} \otimes \overline{\mathbf{D}}^{(0)}, \tag{3}$$

and the representations $\mathbf{D}^{(k)}$ and $\overline{\mathbf{D}}^{(k)}$, which are related in this way, are said to be *dual* or *associate*. Equation (3) implies that

$$\overline{\mathbf{D}}^{(k)} \otimes \mathbf{D}^{(k)} \supset \overline{\mathbf{D}}^{(0)}. \tag{4}$$

It can happen that a representation is self-dual, i.e.,

$$\overline{\mathbf{D}}^{(k)} = \mathbf{D}^{(k)}. \tag{5}$$

Such a representation has character zero for all odd permutations.

Presuming that the Hamiltonian for an n-electron system commutes with all the operations of $\mathbf{S}_n$, it follows that the wavefunctions of an n-electron atom or molecule can be symmetry classified according to the irreducible representations of $\mathbf{S}_n$. From the time-dependent Schrödinger equation, it can be shown that this symmetry classification does not change with time, i.e., it is *conserved* (see Section 14.2 of Ref. 11). Electrons are such that only states of symmetry $\overline{\mathbf{D}}^{(0)}$ occur, and because of symmetry conservation this has always been the case; it is fundamental to the nature of electrons. In fact, it has been found that all particles having half-integral spin (i.e., fermions) have the same symmetry property. This is called the *Pauli exclusion principle*. Although the Pauli exclusion principle is consistent with the laws of quantum mechanics, it does not follow from them, and has to be considered as an extra postulate [12]. In quantum chemical calculations of electronic wavefunctions, one only considers *antisymmetric* electronic basis functions; that is, one only combines orbital and spin basis functions that transform in the symmetric group according to the dual of each other so that their product transforms as $\overline{\mathbf{D}}^{(0)}$, or one uses the more standard technique of constructing *Slater determinants* (which, by necessity, transform as $\overline{\mathbf{D}}^{(0)}$) from one-electron space and spin wavefunctions.

As one moves from the hydrogen atom to heavier and heavier elements, the Pauli exclusion principle leads to a periodically recurring outer orbital structure, and because it is the outer orbital structure that largely determines the chemical properties, these will also periodically repeat. Thus, the Pauli exclusion principle leads to the periodic system of the elements.

In contrast to the situation for fermions, it has been found that for particles with integral spin (i.e., bosons), such as ^{12}C and ^{16}O nuclei, only symmetric states (i.e., states of symmetry $\mathbf{D}^{(0)}$) occur. The consequences of the permutation symmetry properties of nuclei will be discussed later in this chapter after we show how permutations of identical nuclei get involved in considerations of molecular symmetry by introducing the MS group.

In summary, the quantum mechanical treatment of atoms and molecules containing identical particles makes the *symmetrization postulate:*

> The states that occur in nature are symmetric with respect to identical boson exchange and antisymmetric with respect to identical fermion exchange.

States not represented by wavefunctions of this allowed symmetry are *completely forbidden.* Because this is based on experimental observation, it immediately encourages two lines of research: (i) the search for particles that do not obey the symmetrization postulate, and (ii) the search for forbidden states involving well-known particles such as electrons, protons, or ^{12}O nuclei. Molecular spectroscopy experiments aimed at (ii) are discussed in Section 5.

3. THE MOLECULAR SYMMETRY GROUP

The MS group was introduced by Longuet-Higgins [1] and is the subject of two text books [2,11], so we will only give a brief overview in order to set the context for the way that the permutations of identical nuclei and the inversion operation are used in the study of molecular symmetry.

For any molecule, once we know its chemical formula, we can appreciate the structure of its Complete Nuclear Permutation Inversion (CNPI) group. This group consists of the following operations: all possible permutations, including the identity E, of the space and spin coordinates of identical nuclei in the molecule; the space-fixed inversion operation E^*, which inverts the space coordinates of all particles in the molecule (nuclei and electrons) but which does not affect the spins; and the product of each of the possible identical nuclei permutations with E^*. If we presume that these elements commute with the molecular Hamiltonian, then we can use the irreducible representations of the CNPI group to symmetry label the energy levels. These symmetry labels could be used, for example, to determine which levels interact with each other (they would have the same symmetry labels) and to

Table 2 The character table[a] of the CNPI group $C_{2v}(M)$ for H_2O

	E	(12)	E^*	$(12)^*$
A_1:	1	1	1	1
A_2:	1	1	-1	-1
B_1:	1	-1	-1	1
B_2:	1	-1	1	-1

[a] The operation (12) interchanges the two protons in H_2O, E^* is the inversion operation, and $(12)^*$ represents the product (12) $E^* = E^*$ (12). Each element is in a class of its own.

determine optical selection rules (the product of their symmetries would contain the symmetry of the dipole moment operator). As an example of a simple CNPI group, we give in Table 2 the character table for $C_{2v}(M)$, the CNPI group of H_2O. This group is isomorphic to C_{2v}, the point group of an H_2O molecule at equilibrium.

For molecules with two or more versions (non-superposable equilibrium structures that differ only in the numbering of identical nuclei [13]) separated from each other by insuperable potential energy barriers, the symmetry labeling obtained using the CNPI group would be rather unwieldy. In this circumstance, a proper understanding of the symmetry and relevant degeneracies of the molecular energy levels is obtained using a subgroup of the CNPI group called the Molecular Symmetry group. This subgroup consists only of feasible elements of the CNPI group [1,2,11]. For nonlinear molecules in isolated electronic states having no observable tunneling between versions (a so-called "rigid" nonlinear molecule), the MS group is isomorphic to the point group of the equilibrium structure of the molecule, and the symmetry labeling of vibrational and electronic (vibronic) states is the same using either group. However, whereas the elements of the point group only transform vibronic variables, the elements of the MS group also transform the rotational and nuclear spin variables. This means that the MS group can be used to label rotational and nuclear spin states, as well as vibrational and electronic states.

The MS group can be used equally well to classify the states of nonrigid molecules, i.e., molecules with observable tunnelings between versions, and the states of molecules that have strong spin–orbit coupling. Extended MS groups have been developed so that the rotational, vibrational, and electronic states of linear molecules, and the rotational, contorsional, vibrational, and electronic states of coaxially nonrigid molecules, can be classified in the same group as used for the rovibronic and nuclear spin states.

3.1. Nuclear spin and *ortho–para* interactions

When identical nuclei in a molecule are interchanged, the symmetrization postulate (Section 2) applies to the complete molecular wavefunction Ψ_{total}. This wavefunction can be expressed as the product

$$\Psi_{total} = \Psi_{trans}\Psi_{int},$$

(6)

where the wavefunction Ψ_{trans} describes the translational motion and the "internal" wavefunction Ψ_{int} describes all degrees of freedom other than the translation (i.e., electron orbital motion, electron spin, vibration, rotation, and nuclear spin). Ψ_{trans} is unchanged by permutations of identical nuclei, and so the symmetrization postulate applies also to Ψ_{int}.

The contribution to the molecular energy from Hamiltonian terms involving the nuclear spins is extremely small in comparison with the contributions from electron orbital motion, electron spin, vibration, and rotation. Consequently, the *hyperfine* effects originating in the nuclear spins can often be neglected. This approximation is consistent with an internal wavefunction in the form

$$\Psi_{\text{int}} = \Psi_{\text{rve}}\Psi_{\text{ns}}, \tag{7}$$

where the rovibronic function Ψ_{rve} describes electron orbital motion, electron spin, vibration, and rotation, while Ψ_{ns} is a nuclear spin function. With this approximation, the molecular energy does not depend on Ψ_{ns}.

In Equation (7), the functions Ψ_{rve} and Ψ_{ns} must be combined to produce Ψ_{int} functions that comply with the symmetrization postulate. This is done by using the symmetry properties of the functions in the appropriate MS group. As an example, we consider the water molecule. Because H_2O has only one version in the sense defined above, its MS group is identical to its CNPI group $C_{2v}(M)$, whose character table is given in Table 2. The two protons in $H_2^{16}O$ have spin $I = 1/2$ and are fermions, while the ^{16}O nucleus has $I = 0$. To comply with the symmetrization postulate, allowed Ψ_{int} wavefunctions for $H_2^{16}O$ have B_1 or B_2 symmetry in $C_{2v}(M)$ (Table 2); these two irreducible representations describe wavefunctions with a sign change under the proton-exchange operation (12). A proton has two possible spin functions, $\alpha = |I = 1/2, m_I = 1/2\rangle$ and $\beta = |I = 1/2, m_I = -1/2\rangle$, where I is the spin quantum number and m_I is the spin projection, in units of $\hbar$, on the Z axis of an XYZ axis system fixed in space. The ^{16}O nucleus has one possible spin function, $\gamma = |I = 0, m_I = 0\rangle$. For an $H_2^{16}O$ molecule, we can construct three spin functions, $\Psi_{\text{ns}}^{(A)} = \alpha\alpha\gamma$, $\beta\beta\gamma$, and $(\alpha\beta\gamma + \beta\alpha\gamma)/\sqrt{2}$ (where, in the products of three one-nucleus spin functions, the first factor is the spin function for the proton labeled 1 and the second factor is the spin function for the proton labeled 2), with A_1 symmetry in $C_{2v}(M)$ (the nuclear spins are unchanged under E^*), and a single function $\Psi_{\text{ns}}^{(B)} = (\alpha\beta\gamma - \beta\alpha\gamma)/\sqrt{2}$ with B_2 symmetry.

For $H_2^{16}O$, we can combine a Ψ_{rve} wavefunction of B_1 (B_2) symmetry in $C_{2v}(M)$ with any one of the three $\Psi_{\text{ns}}^{(A)}$ functions to produce an allowed Ψ_{int} wavefunction of B_1 (B_2) symmetry. A Ψ_{rve} wavefunction of A_1 (A_2) symmetry can only be combined with the one $\Psi_{\text{ns}}^{(B)}$ spin function to produce an allowed Ψ_{int} wavefunction of B_2 (B_1) symmetry. In the approximation where the molecular energy does not depend on Ψ_{ns}, the Ψ_{rve} states of B_1 or B_2 symmetry have a *nuclear-spin degeneracy* of 3, or a *nuclear-spin statistical weight factor* $g_{\text{ns}} = 3$, whereas Ψ_{rve} states of A_1 or A_2 symmetry have $g_{\text{ns}} = 1$. States with $g_{\text{ns}} = 3$ are called *ortho* states, whilst states with $g_{\text{ns}} = 1$ are called *para* states. For $H_2^{16}O$, the *ortho* states have a total nuclear spin $I_{\text{total}} = 1$ while the *para* states have $I_{\text{total}} = 0$. The spin statistical weight factors are manifest in the intensities of

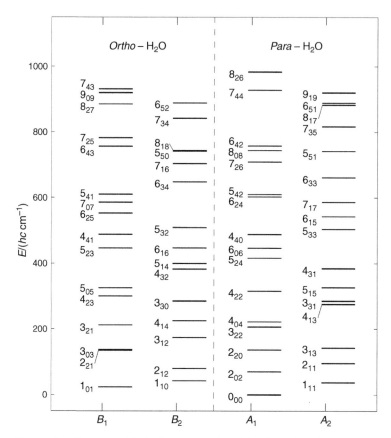

Figure 1 Term value diagram with the lowest rotational levels in the vibrational ground state of $H_2^{16}O$. The levels are labeled by $J_{K_aK_c}$ [2,11] and by their rovibronic $C_{2v}(M)$ symmetry label.

molecular transitions observed under conditions where the hyperfine effects are not resolved.[1] Figure 1 is a term value diagram showing the lowest rotational levels in the vibrational ground state of $H_2^{16}O$; the levels are labeled by $J_{K_aK_c}$ [2,11] and by their $C_{2v}(M)$ symmetry label.

Many molecules are like H_2O in that they have two possible nuclear-spin modifications (i.e., two possible values of g_{ns}). In general, for such molecules we refer to the Ψ_{rve} states with the larger value of g_{ns} as *ortho* states, and to those with the smaller value of g_{ns} as *para* states. For an isolated molecule, whose states can be taken to be exact *ortho* and *para* states [in that the corresponding wavefunctions are given exactly by Equation (7)], *ortho–para conversion* is a forbidden process because transitions induced by interactions with electromagnetic fields (i.e., light) have *ortho–ortho* and *para–para* selection rules. A molecule with exact *ortho* and *para* states can only convert between these states in collisions with other molecules or with surfaces. In reality, however, the states of an isolated molecule will not be exact

---

[1] The calculation of spin statistical weight factors is discussed, for example, in Chapter 8 of Ref. 2, in Chapter 9 of Ref. 11, and in Ref. 14.

ortho and *para* states. The hyperfine-structure terms in the molecular Hamiltonian, which describe the contribution of the nuclear spins to the molecular energy, can couple *ortho* and *para* states for which the wavefunction products $\Psi_{rve}\ \Psi_{ns}$ in Equation (7) have the same symmetry in the MS group of the molecule in question. This leads to internal wavefunctions in the form

$$\Psi_{int}^{(o-p)} = c^{(o)}\,\Psi_{rve}^{(o)}\Psi_{ns}^{(o)} + c^{(p)}\,\Psi_{rve}^{(p)}\Psi_{ns}^{(p)},\tag{8}$$

where the superscript "(o)" and "(p)" label wavefunctions and expansion coefficients c associated with exact *ortho* and *para* states, respectively. The hyperfine-structure terms give rise to extremely small contributions to the molecular energy, and so the mixing of *ortho* and *para* states expressed in Equation (8) is normally very weak. Thus, a typical internal wavefunction $\Psi_{int}^{(o-p)}$ will have $|c^{(o)}| \gg |c^{(p)}|$, so that it is essentially an *ortho* state with a small admixture of *para* character, or it has $|c^{(p)}| \gg |c^{(o)}|$, so that it is essentially an *para* state with a small admixture of *ortho* character. Between such states, *ortho–para* transitions become possible. For example, a predominantly *ortho* state $\Psi_{int}^{(o-p)}$ with $|c^{(o)}| \gg |c^{(p)}|$ is connected by strong electric-dipole transitions (with line strengths proportional to $|c^{(o)}|^2$) to pure (or predominantly) *ortho* states and by weak *ortho–para* transitions (with line strengths proportional to $|c^{(p)}|^2$) to pure (or predominantly) *para* states.

Ortho–para transitions have been observed, for example, for CH_4 molecules in magnetic and electric fields [15–17] and in the ν_3 fundamental band of SF_6 [18] (see Section 4 of Ref. 19 for a discussion in terms of MS group theory). Raich and Good [20] (see also p. 473 of Ref. 2) have made a rough calculation of the probability per unit time for the electric dipole *ortho–para* transition $(J,F) = (1,1)$ $\leftarrow$ $(0,0)$ in the vibrational ground state of the H_2 molecule. This transition steals intensity from allowed rovibronic transitions. Their estimate is that the radiative lifetime of the $F = 1$ hfs component of the $J = 1$ level in the vibrational ground state of the H_2 molecule is 5×10^{12} years, which is about two orders of magnitude greater than the age of the universe. A more recent calculation [K. Pachuki and J. Komasa, Phys Rev A 77: 030501 (R) 2008] obtains a lifetime of 1.6×10^{13} years.

As discussed in Section 17.7 of Ref. 2, the so-called g/u mixing in homonuclear diatomic molecules is a further example of *ortho–para* interaction. It is likely to be observed if a g and a u vibronic state[2] are close in energy. If a g and a u electronic state share the same dissociation limit, the conditions for g/u mixing will be fulfilled in the energy region immediately below this limit. This type of mixing was observed for the first time in the I_2 molecule [21]. It has also been observed in Cs_2 [22] and in H_2^+ (see Refs. 23–25 and references therein).

Since in general *ortho–para* mixing is weak, we can, to a good approximation, consider the *ortho* and *para* modifications of a given molecule as two different molecules. These two molecules will have different properties. For example, the *ortho* and *para* forms of H_2 are well studied and they are characterized by different values of specific heat, boiling-point

---

[2] Wavefunctions of a homonuclear diatomic molecule have g symmetry if they generate the character $+1$ under $(12)^*$, where (12) is the interchange of the two nuclei in the molecule, and u symmetry if they generate the character -1 under $(12)^*$ (see, for example, Ref. 2).

temperature, heat of vapor formation, etc. One can ask whether, for other molecules, it would be possible to obtain a pure sample of one spin modification, *ortho* say, and observe the conversion to the *para* form as the molecules approach *ortho/para* equilibrium. Studies of this kind have been carried out by Chapovsky, Hermans, and co-workers (see Ref. 26 and references therein), chiefly for the CH_3F molecule. They use the *laser-induced drift* (LID) technique [26] to obtain an enhanced concentration of *ortho* and *para* molecules, and absorption spectroscopy to monitor the ensuing conversion. It is found that the mixing of *ortho* and *para* states by hyperfine effects, as expressed in Equation (8), is the dominating mechanism that governs *ortho–para* conversion. So-called "direct *ortho–para* transitions," involving collisions with other molecules and with surfaces, can be neglected [26]. In order for a significant mixing of *ortho* and *para* states to occur, it is necessary that pure *ortho* and *para* basis states, obtained in the approximation of the hyperfine-structure terms of the molecular Hamiltonian being neglected and described by the wavefunction products $\Psi_{rve}^{(o)}\Psi_{ns}^{(o)}$ and $\Psi_{rve}^{(p)}\Psi_{ns}^{(p)}$ in Equation (8), be close in energy. One reason that Chapovsky and Hermans [26] chose $^{12}CH_3F$ and $^{13}CH_3F$ for their studies, apart from these molecules being well suited for the LID technique, is the fact that they have such favorable *gateway states*. The importance of gateway states in spin conversion processes was suggested in an early, pioneering paper by Curl, Kasper, and Pitzer [27]. Based on the ideas of this paper, Chapovsky [28] developed the *quantum relaxation* model, which describes the process of *ortho–para* conversion as resulting from molecular collisions in which the nuclear spins are not directly flipped. What happens in the collision is that a molecule in a pure *ortho* state (say) is converted to a gateway state with a wavefunction in the form given in Equation (8). The gateway state has appreciable *ortho–para* mixing so that in Equation (8), both $|c^{(o)}|$ and $|c^{(p)}|$ are significant. Time-dependent quantum mechanics describes a molecule in such a state as oscillating between the *ortho* and *para* components of its wavefunction. When the molecule has a high probability of being in the *para* component, another collision can convert it to a pure *para* state.

The quantum relaxation theory [28,29] explains, for example, measurements of *ortho–para* conversion rates in formaldehyde H_2CO [30,31], methyl fluoride CH_3F [26,32], and ethylene C_2H_4 [33]. In the case of CH_3F, the parameters in the hyperfine Hamiltonian responsible for the mixing of the gateway states in these cases have been determined, and they are in satisfactory agreement with *ab initio* values [34].

In 1988, Konyukhov et al. [35] reported the enrichment of H_2O spin modifications by adsorption on the surface of corundum ceramics at room temperature. In 2002, Tikhonov and Volkov [36] and Vigasin et al. [37] reported similar enrichments obtained in an experiment involving water vapor flow through a chromatography medium. In these experiments, the *ortho/para* ratio was monitored by rotational spectroscopy; the transitions monitored lie between 36 and 38 cm^{-1}. Tikhonov and Volkov [36] froze the spin-modification enriched water obtained in their chromatography experiment and thus apparently produced *ortho* and *para* enriched ice, analogous to separately solidified *ortho* and *para* H_2 (see Ref. 38 and references therein). Veber et al. [39], however, have attempted to reproduce the H_2O results of Refs. 35–37 and failed to do so. It is suggested that the signals detected in Refs. 35–37 may originate in microwave radiation-induced desorption of water molecules, and there is no *ortho* and *para* enrichment. A fairly recent spectroscopic study concerned with *ortho–para* conversion in water is by Nela et al.

[40], who used laser-induced fluorescence to investigate overtone states of water in the 3000–4000 cm^{-1} region. They observed rich collision-induced spectra but found that no *ortho–para* conversion takes place: The spin states are conserved in the collisional processes.

Spin conversion processes of molecules in matrices have been investigated by Momose and co-workers [41,42] for the case of CH_4 molecules in *para*-hydrogen matrices and by Michaut et al. [43] for the case of H_2O molecules in solid argon matrices.

3.2. Missing levels

The $^{12}C^{16}O_2$ molecule has the CNPI (or MS) group $\mathbf{C}_{2v}(M)$, just as H_2O.[3] Because, however, ^{12}C and ^{16}O nuclei both have $I = m_I = 0$, this molecule has one possible spin function Ψ_{ns} only, and this one function is invariant under the symmetry operations (12) (which now describes the interchange of the two ^{16}O nuclei) and $\hat{E}^*$, so that it has A_1 symmetry in $\mathbf{C}_{2v}(M)$. The ^{16}O nuclei are bosons so that to comply with the symmetrization postulate, an allowed Ψ_{int} wavefunction for $^{12}C^{16}O_2$ has A_1 or A_2 symmetry in $\mathbf{C}_{2v}(M)$ (Table 2). Such a Ψ_{int} wavefunction can be obtained by combining a Ψ_{rve} wavefunction of A_1 or A_2 symmetry with the one available Ψ_{ns} function; the corresponding levels have $g_{ns} = 1$. From Ψ_{rve} wavefunctions of B_1 or B_2 symmetry $\mathbf{C}_{2v}(M)$, we cannot produce allowed Ψ_{int} wavefunctions; the corresponding levels have $g_{ns} = 0$ and are *missing rotational levels* that do not occur in nature. As a result, for CO_2 molecules in their ground vibronic state, rotational levels having odd J value are missing.

The molecules CO_2, SO_3, BH_3, and NH_3 all have in common that some of their rotational levels are missing. In a short-hand notation, we can write the spin-statistical weight factors of $^{12}C^{16}O_2$ as

$$\Gamma_{rve}^{sw} = 1A_1 \oplus 1A_2 \oplus 0B_1 \oplus 0B_2, \tag{9}$$

where the coefficient of each irreducible representation is the g_{ns}-value for Ψ_{rve} states of that symmetry species.

The molecules $^{32}S^{16}O_3$ and $^{11}BH_3$ are planar at equilibrium. They both have the MS group $\mathbf{D}_{3h}(M)$, whose character table is given in Table 3 with the three ^{16}O nuclei in $^{32}S^{16}O_3$ and the three protons in $^{11}BH_3$ labeled as 1, 2, 3. In the short-hand notation of Equation (9), we obtain for $^{32}S^{16}O_3$ (^{32}S has $I = 0$)

$$\Gamma_{rve}^{sw} = 1A_1' \oplus 1A_1'' \oplus 0A_2' \oplus 0A_2'' \oplus 0E' \oplus 0E'' \tag{10}$$

and for $^{11}BH_3$ (^{11}B has $I = 3/2$)

$$\Gamma_{rve}^{sw} = 0A_1' \oplus 0A_1'' \oplus 16A_2' \oplus 16A_2'' \oplus 8E' \oplus 8E'' \tag{11}$$

---

[3] Because the equilibrium structure of CO_2 is linear (while that of H_2O is bent), the group $\mathbf{C}_{2v}(M)$ used for H_2O is renamed $\mathbf{D}_{\infty h}(M)$ when used for CO_2, and its irreducible representations are relabeled as Σ_g^+ (or $+s$) $= A_1$, Σ_u^+ (or $+a$) $= B_2$, Σ_g^- (or $-a$) $= B_1$, and Σ_u^- (or $-s$) $= A_2$. See Table A-18 in Ref. 2. For simplicity, we use the H_2O notation for CO_2 here.

Table 3 The character table^a of the CNPI (and MS) group $\mathbf{D}_{3h}(M)$ for SO_3, BH_3, and NH_3

	E	(123)	(12)	E^*	$(123)^*$	$(12)^*$
	1	2	3	1	2	3
A_1' :	1	1	1	1	1	1
A_1'' :	1	1	1	-1	-1	-1
A_2' :	1	1	-1	1	1	-1
A_2'' :	1	1	-1	-1	-1	1
E' :	2	-1	0	2	-1	0
E'' :	2	-1	0	-2	1	0

^{a} One representative element in each class is given, and the number written below each element is the number of elements in the class.

Ammonia $^{14}NH_3$ is nonplanar at equilibrium, but the effects of tunneling between two equivalent equilibrium geometries can be observed, and the appropriate MS group is therefore $\mathbf{D}_{3h}(M)$, just as for $^{11}BH_3$. The g_{ns}-values are (^{14}N has $I=1$)

$$\Gamma_{rve}^{sw} = 0A_1' \oplus 0A_1'' \oplus 12A_2' \oplus 12A_2'' \oplus 6E' \oplus 6E'' \qquad (12)$$

As a result, in its ground vibronic state, levels of ammonia having $K=0$ and even J value are missing, as are half of the levels having K a multiple of 3 (see Table 15-1 and Figure 15-5 of Ref. 2).

3.3.   Parity and chiral molecules

If we neglect, in the molecular Hamiltonian, the very small terms originating in the weak interaction force, the resulting approximate Hamiltonian (the electromagnetic Hamiltonian) commutes with the inversion operation E^*, and so E^* is a symmetry operation in the approximation of the weak interaction force being ignored.[4] The CNPI group for a general molecule contains E^* as does, for example, $\mathbf{C}_{2v}(M)$, the CNPI group for H_2O whose character table is given in Table 2. Generally there are two nondegenerate irreducible representations Γ_{CNPI}^+ and Γ_{CNPI}^- of the CNPI group allowed by Fermi–Dirac and/or Bose–Einstein statistical formulas. These representations are $\Gamma_{CNPI}^+ = B_2$ and $\Gamma_{CNPI}^- = B_1$ for H_2O and $\mathbf{C}_{2v}(M)$ as explained in Section 3.1. In general, the allowed representations $\Gamma_{CNPI}^\pm$ have characters $\chi^\pm[E^*] = \pm 1$ (where the signs are correlated) under E^* so that an internal wavefunction Ψ_{int}, associated with Γ_{CNPI}^+ or Γ_{CNPI}^-, satisfies the equation

$$E^*\Psi_{int} = \chi^\pm[E^*]\Psi_{int} = \pm\Psi_{int}. \qquad (13)$$

The *parity* of the state in question is unambiguously determined by the symmetry of the wavefunction Ψ_{int} in the CNPI group. Wavefunctions belonging to Γ_{CNPI}^+ (with character $+1$ under E^*) have *positive parity*, and wavefunctions belonging to Γ_{CNPI}^- (with character -1 under E^*) have *negative parity*. Because, in Γ_{CNPI}^+ or

[4] When we consider the effects of the weak interaction force, *parity violation* occurs, and this is discussed in Section 6 below.

Γ_{CNPI}^-, the transformation properties under the effect of any permutation of identical nuclei is fixed by the statistics, the only symmetry distinction that the CNPI group can make for the complete internal wavefunctions Ψ_{int} is that of parity. Thus, if we use the CNPI group to label these wavefunctions, we could in fact just as well label them by a parity label only.

Normally we use the MS group, rather than the CNPI group, for labeling the complete internal states Ψ_{int}. If the MS group contains E^* (an obvious example of this is H_2O for which the CNPI and MS groups are identical), there are two irreducible representations of the MS group, Γ_{MS}^+ and Γ_{MS}^-, allowed by Fermi–Dirac and/or Bose–Einstein statistics. When we now let $\chi^{\pm}[E^*] = \pm 1$ denote the characters under E^* of $\Gamma_{MS}^{\pm}$ (where the signs are correlated), Ψ_{int} satisfies Equation (13). Consequently, we obtain nondegenerate complete internal states, each one with a definite parity determined by the symmetry (Γ_{MS}^+ or Γ_{MS}^-) of Ψ_{int} in the MS group.

A molecule is *chiral* or *optically active* when its MS group contains neither E^* nor any permutation-inversion operation P^*. A chiral molecule has two distinguishable forms of its equilibrium structure that are the mirror images of each other, and the two forms are separated by an insuperable energy barrier on the potential energy surface.[5] The two forms, called *enantiomers*, are physically distinguishable, and there is no observed tunneling between the two potential energy minima supporting the two enantiomers. The two enantiomers rotate the plane of polarization of linearly polarized light in opposite directions. An often quoted example of a chiral molecule is a substituted methane molecule with four different substituents, CHFClBr say. For this molecule, no permutations of identical nuclei are possible, and E^* is unfeasible, so the MS group is the trivial one-element group $\{E\}$, which indeed contains neither E^* nor any permutation-inversion operation. As another example of a chiral molecule, we consider the skew-symmetric chain molecule ClOOCl, whose two enantiomers [44], labeled **(a)** and **(b)**, are drawn in Figure 2. If we rotate form **(a)** so that its nuclei 2, 3, and 4 cover the corresponding nuclei in form **(b)**, then the two Cl nuclei labeled 1 will be on opposite sides of the plane defined by the nuclei 2, 3, and 4. The two forms **(a)** and **(b)** are therefore physically distinguishable. These forms are connected by internal rotation of the two ClO moieties about the O–O axis. The barriers to this rotation [44] are insuperable so that E^* (which interconnects the two forms) is unfeasible. The MS group is $\mathbf{G}_2 = \{E, (12)(34)\}$, where $(12)(34)$ is the simultaneous interchange of the two Cl and the two O nuclei, and the molecule is chiral.[6] The MS group $\mathbf{G}_2$ has the irreducible representations

$$
\begin{array}{c|cc}
\mathbf{G}_2 & E & (12)(34) \\
\hline
A & 1 & 1 \\
B & 1 & -1
\end{array}
\tag{14}
$$

---

[5] An alternative definition of chirality merely requires the presence of two distinguishable equilibrium structures that are mirror images of each other. An insuperable energy barrier between the forms is not required with this "static" definition of chirality, and so the forms may not be separable. Our definition requires separability and, hence, an insuperable energy barrier.

[6] The molecule HOOH, for example, has structures similar to those of ClOOCl. For this molecule, the barriers to internal rotation are superable on the time scale of a typical spectroscopic experiment. Consequently, the MS group is $\{E, (12)(34), E^*, (12)(34)^*\}$ and, according to our definition of chirality, HOOH is not chiral.

(a)   (b)

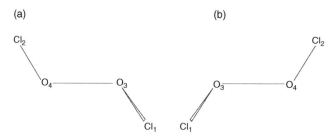

Figure 2 The two enantiomers of the molecule ClOOCl. In both displays, the Cl nucleus labeled 1 is above the plane of the page.

In $^{35}Cl_2\,^{16}O_2$, the two ^{35}Cl nuclei are fermions (with $I = 3/2$) and the two ^{16}O nuclei are bosons. Thus, all allowed internal states $\Psi_{int}^{(a)}$ for form **(a)** and all allowed internal states $\Psi_{int}^{(b)}$ for form **(b)** have B symmetry in G_2; for a chiral molecule, there is only one allowed irreducible representation in the MS group. In general, the internal states of a chiral molecule are doubly degenerate because the wavefunctions occur in pairs $(\Psi_{int}^{(a)}, \Psi_{int}^{(b)})$ of the same energy. The phase factors of the wavefunctions can be chosen such that $\Psi_{int}^{(b)} = E^*\Psi_{int}^{(a)}$ and, because $(E^*)^2 = E$, $\Psi_{int}^{(a)} = E^*\Psi_{int}^{(b)}$. That is, the states $\Psi_{int}^{(a)}$ and $\Psi_{int}^{(b)}$ cannot be assigned a definite parity. It is suggested above that by using the CNPI group for the symmetry classification of the internal states, we can obtain states of definite parity. In the CNPI group, the wavefunctions $\Psi_{int}^{(a)}$ and $\Psi_{int}^{(b)}$ generate the reducible representation $\Gamma_{CNPI}^+ \oplus \Gamma_{CNPI}^-$, where Γ_{CNPI}^+ and Γ_{CNPI}^- are defined above, and the corresponding symmetrized wavefunctions are

$$\Psi_{int}^{(\pm)} = \frac{1}{\sqrt{2}}\left(\Psi_{int}^{(a)} \pm \Psi_{int}^{(b)}\right), \tag{15}$$

where (with correlated signs) $E^*\Psi_{int}^{(\pm)} = \pm\Psi_{int}^{(\pm)}$ so that the states $\Psi_{int}^{(\pm)}$ have definite parities. However, in each of these states there is equal probability of finding the chiral molecule in the forms **(a)** and **(b)**, and so the CNPI group classification with definite parity is not appropriate for a single enantiomer, for example for a molecule known to be in the **(a)** form and described by the wavefunction $\Psi_{int}^{(a)}$.

The MS group for the molecule PH_3 is [2,11]

$$C_{3v}(M) = \{E, (123), (132), (12)^*, (13)^*, (23)^*\}, \tag{16}$$

where the nuclei are labeled as indicated in Figure 3. This group, whose character table is given in Table 4, does not contain E^*, but it does contain the three permutation-inversion operations $\{(12)^*, (13)^*, (23)^*\}$. The PH_3 molecule has two versions of the equilibrium structure as shown in Figure 3. These versions

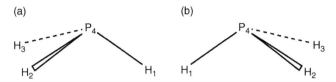

Figure 3 The two versions of the molecule PH_3. In both displays, proton 1 and the phosphorus nucleus are in the plane of the page, proton 2 is above this plane, and proton 3 is below it.

are connected by E^* and separated by an insuperable energy barrier on the potential energy surface. Unlike the enantiomers of a chiral molecule, the versions are not physically distinguishable; they only differ by the labeling of the nuclei. So, even though the E^* operation interconverts versions separated by an insuperable energy barrier, PH_3 and molecules of this type are not chiral. For such molecules there are always two MS group symmetry labels, Γ_{MS}^+ and Γ_{MS}^-, allowed for the Ψ_{int}; for PH_3 they are A_1 and A_2 (Table 4). The internal states $\Psi_{int}^{(a)}$ localized at form **(a)** and the internal states $\Psi_{int}^{(b)}$ localized at form **(b)** (Figure 3) must all belong to Γ_{MS}^+ or Γ_{MS}^- in the MS group and, just as for an optically active molecule, they do not have a definite parity. Normally, we consider only the internal states associated with one of the forms given in Figure 3, form **(a)** say, and ignore those associated with form **(b)**. We can do this because we know that when the potential energy barrier between **(a)** and **(b)** is insuperable, each internal state $\Psi_{int}^{(b)}$ for form **(b)** has the same energy as an internal state $\Psi_{int}^{(a)}$ for form **(a)** and can be obtained (for a suitable choice of phase factors) as $\Psi_{int}^{(b)} = E^* \Psi_{int}^{(a)}$, just as for an optically active molecule. We can now, just as for an optically active molecule, introduce the "parity states" $\Psi_{int}^{(\pm)}$ defined by Equation (15). These states are symmetrized in the CNPI group as discussed above. Unlike the situation for optically active molecules, however, there is no problem with using the parity states for describing the molecule. This is because for PH_3 and the type of molecule it represents, the forms **(a)** and **(b)** cannot be distinguished experimentally and we never encounter a molecule known to be, for example, in form **(a)**. When the potential energy barrier between **(a)** and **(b)** is insuperable, the internal states are doubly degenerate and we may just as well describe them by the wavefunctions $\left(\Psi_{int}^{(+)}, \Psi_{int}^{(-)} \right)$ as by the equivalent

Table 4 The character table[a] of the MS group $C_{3v}(M)$ for PH_3

	E 1	(123) 2	(12)* 3
A_1:	1	1	1
A_2:	1	1	-1
E :	2	-1	0

[a] One representative element in each class is given, and the number written below each element is the number of elements in the class.

wavefunctions $\left(\Psi_{\text{int}}^{(\text{a})}, \Psi_{\text{int}}^{(\text{b})} \right)$. If the energy barrier between the (a) and (b) forms becomes superable (because we design an improved experiment capable of resolving the energy splittings resulting from the *inversion tunneling* through it), the appropriate wavefunctions for the new situation are $\left(\Psi_{\text{int}}^{(+)}, \Psi_{\text{int}}^{(-)} \right)$. These wavefunctions now describe states of slightly different energy; the previously unresolved splitting is given approximately as

$$\left| E_{\text{int}}^{(-)} - E_{\text{int}}^{(+)} \right| = 2 \left| \langle \Psi_{\text{int}}^{(\text{a})} | \hat{H}_{\text{int}} | \Psi_{\text{int}}^{(\text{b})} \rangle \right|, \tag{17}$$

where $\hat{H}_{\text{int}}$ is the molecular electromagnetic Hamiltonian.

The complete internal states of PH_3 can be of symmetry A_1 or A_2 (but not of symmetry E) in $\mathbf{C}_{3v}(M)$. We can use the technique of *reverse correlation* [45] to determine that in the CNPI group $\mathbf{D}_{3h}(M)$ (Table 3), internal states of symmetry A_1 in $\mathbf{C}_{3v}(M)$ apparently split into states of symmetry $A_1' \oplus A_2''$, and that internal states of symmetry A_2 split into states of symmetry $A_2' \oplus A_1''$. However, the spin-statistical weight factors for PH_3 in $\mathbf{D}_{3h}(M)$, written in the short-hand notation of Section 3.1, are (^{31}P has $I = 1/2$)

$$\Gamma_{\text{rve}}^{\text{sw}} = 0A_1' \oplus 0A_1'' \oplus 8A_2' \oplus 8A_2'' \oplus 4E' \oplus 4E''. \tag{18}$$

Thus, internal states of symmetry A_1' or A_1'' in $\mathbf{D}_{3h}(M)$ are forbidden by the rules of Fermi–Dirac statistics and, as a consequence, internal states of symmetry A_1 (A_2) in $\mathbf{C}_{3v}(M)$ become internal states of symmetry A_2'' (A_2') in $\mathbf{D}_{3h}(M)$ if there is inversion tunneling. These $\mathbf{C}_{3v}(M)$ states are not subject to any further splitting from the tunneling. The eigenstates of the tunneling PH_3 molecule have definite parities: A_2'' states have $-$ parity, and A_2' states have $+$ parity. A Ψ_{rve} state of A_1 (A_2) symmetry in $\mathbf{C}_{3v}(M)$ gives rise to internal states Ψ_{int} of A_1 (A_2) symmetry in $\mathbf{C}_{3v}(M)$ and therefore of A_2'' (A_2') symmetry in $\mathbf{D}_{3h}(M)$. Thus, we say [11] that Ψ_{rve} states of nontunneling PH_3 with A_1 (A_2) symmetry in $\mathbf{C}_{3v}(M)$ have *incipient* $-$ $(+)$ parity, because this is the unique parity of the state that arises if there is inversion tunneling. Nontunneling PH_3 states of E symmetry in $\mathbf{C}_{3v}(M)$ give rise to internal states of A_1 and A_2 symmetry in $\mathbf{C}_{3v}(M)$ and these states in turn give rise to internal states of A_2'' and A_2' symmetry, respectively, in $\mathbf{D}_{3h}(M)$. The initially degenerate E states have incipient double parity (i.e., $+$ and $-$); they will be split by inversion tunneling and also by nuclear spin hyperfine effects.

4. TIME REVERSAL SYMMETRY

The operation $\hat{\theta}$ of reversing all linear and angular momenta, including spin angular momenta, is called time reversal. In this section, we summarize some important points about the use of this operation in atomic and molecular spectroscopy. Much more detailed discussions exist in the literature [10,46–48].

For a spin-free system, the operation of time reversal is simply that of complex conjugation $\hat{K}$, as we can see by considering the time-dependent Schrödinger equation

$$i\hbar \frac{\partial \Psi_n(\mathbf{q}, t)}{\partial t} = \hat{H}\Psi_n(\mathbf{q}, t). \tag{19}$$

If we apply the operation $\hat{K}$ to both sides, remembering that $\hat{H}$ is real, we get

$$i\hbar \frac{\partial \Psi_n(\mathbf{q}, t)^*}{\partial(-t)} = \hat{H}\Psi_n(\mathbf{q}, t)^*. \tag{20}$$

From this equation, we can see that complex conjugation accomplishes time reversal because the state $\Psi_n(\mathbf{q}, t)^*$ evolves in positive time in the same way that $\Psi_n(\mathbf{q}, t)$ would evolve in negative time. Probability density is unchanged because it is proportional to $|\Psi_n(\mathbf{q}, t)|^2$, and for a stationary state, the energy is unchanged because $\hat{H}\Psi_n(\mathbf{q}, t) = E\Psi_n(\mathbf{q}, t)$ implies $\hat{H}\Psi_n(\mathbf{q}, t)^* = E\Psi_n(\mathbf{q}, t)^*$, but we must bear in mind that any externally applied magnetic field in $\hat{H}$ must be explicitly reversed to incorporate the effect of $\hat{\theta}$. A linear and unitary operator $\hat{U}$ satisfies $\hat{U}(c\Psi) = c(\hat{U}\Psi)$ and $|\langle \hat{U}\Psi | \hat{U}\Psi \rangle| = |\langle \Psi | \Psi \rangle|$. The complex conjugation operation $\hat{K}$ is *antilinear* and *antiunitary* because it satisfies $\hat{K}(c\Psi) = c^*(\hat{K}\Psi)$ and $|\langle \hat{K}\Psi | \hat{K}\Psi \rangle| = |\langle \Psi | \Psi \rangle|$.

Including spin in the Hamiltonian means that the time reversal operation is no longer simply the operation of complex conjugation, but it can be shown that it is the product of a unitary operation (which we call $\hat{U}$) and the complex conjugation operation $\hat{K}$, i.e.,

$$\hat{\theta} = \hat{U}\hat{K}. \tag{21}$$

The product of a unitary and an antiunitary operation is antiunitary, so the time reversal operation $\hat{\theta}$ is antiunitary. Presuming that $\hat{\theta}$ is a symmetry operation (i.e., that it commutes with the Hamiltonian), we find that $\hat{\theta}\Psi$ and Ψ must have the same energy. Labeling a complete atomic or molecular eigenfunction Ψ by the total angular momentum quantum number F and the projection quantum number m_F, it can be shown that, with the appropriate choice of phase (see Equation (3.73) in Ref. 48):

$$\hat{\theta}\Psi(F, m_F) = i^{2m_F}\Psi(F, -m_F). \tag{22}$$

A pair of states having a nonzero value of F, $\Psi(F,m_F)$ and $\Psi(F,-m_F)$, which differ only in that the direction of the total angular momentum is reversed, are interconverted by the time reversal operator and they have the same energy. This is so even if the atom or molecule is placed in an external electric field because the direction of an external electric field is unaffected by time reversal.

It is awkward to include $\hat{\theta}$ in a symmetry group of the Hamiltonian because it is antiunitary. However, it can be done, but one cannot form representations as one can with a group of unitary operations. Instead one must form *corepresentations* [10]. In the study of gas-phase molecules, it turns out that including $\hat{\theta}$, and then using corepresentations, does not lead to any new symmetry labels on the levels. Time reversal symmetry can be responsible for extra degeneracies, and it can help in determining whether certain matrix elements vanish [49]. The extra degeneracies come about in the circumstance that in the original (unitary) symmetry group there are two irreducible representations, R and R^* say, that are the complex conjugate of each other. As a result of time reversal symmetry, an energy level of symmetry R will always coincide with a level of symmetry R^* [50].

5. THE BREAKDOWN OF THE SYMMETRIZATION POSTULATE

The symmetrization postulate was introduced in Section 5. There is a vast literature on the subject both of its truth and of its possible violation [12,51–60]. It is very difficult to construct a quantum field theory that is relativistically invariant if the symmetrization postulate is abandoned, but it can be done (see Refs. 59 and 60 and references therein). This make it very worthwhile to try to find states that are forbidden by this postulate in the spectra of atoms and molecules. However, we have no idea of the extent of the violation that might occur, if at all, and so the search is rather a shot in the dark. The main searches that have been made concern looking for exchange-symmetric states for electrons, and for exchange-antisymmetric states for ^{16}O nuclei. The first of these would involve the breakdown of the Pauli exclusion principle.

5.1. The breakdown of the Pauli exclusion principle

A simple example is provided by the helium atom, which is a two electron system; electronic states that are symmetric with respect to electron exchange are forbidden. Such states are termed *paronic* states [54]. We might consider that the two electrons in some or all helium atoms have a small admixture of symmetric states. For the helium atoms like this, the two electrons would have to be distinguishable; the Pauli exclusion principle being only approximately satisfied if the particles are only approximately identical. For this to be possible, we would have to postulate that electrons have a property, as yet unknown, that makes them distinguishable, and maybe just a few exotic electrons are distinguishable from the rest as a result. This was conjectured by Fermi [61], but he predicted that such a state of affairs would probably have had drastic consequences on the properties of the elements over the time of their existence and so it was very unlikely. An alternative possibility for the existence of paronic states involves the idea of para-statistics [53–55]. At a more fundamental level, string theory does not have particles as fundamental entities and requires something more general than the symmetrization postulate. The generalization of the symmetrization postulate to strings might lead to greater insights as to whether it can be violated by particles.

Drake [62] points out that regardless of how paronic states of the helium atom might arise, in the nonrelativistic LS coupling approximation, the electronic wavefunctions for paronic helium would be identical to those for normal helium except that the singlet and triplet spin functions would be interchanged so as to form symmetric states. Thus, the nonrelativistic energies of the normal and paronic states of the helium atom are identical. By considering the spin-dependent terms in the Breit interaction and the anomalous magnetic moment in the quantum electrodynamic corrections, the paronic energy shift of each paronic state away from its reference normal helium state has been calculated [62]. Using these results, a search was made for the paronic partner of the 389 nm 2^3S_1–3^3P_1 transition of helium, in an atomic beam laser induced fluorescence experiment [63]; the paronic transition is predicted in Ref. [62] to occur 535 MHz higher in frequency. This paronic line was not seen, and from the sensitivity of the experiment, an upper limit to the violation of 5×10^{-6} was obtained. The He experiment is the most direct and simple-to-interpret experiment that has been performed so far to test for the breakdown of the Pauli exclusion principle. Several other experimental tests, that do not involve atomic or molecular spectroscopy measurements, have all yielded a null result [64–68].

5.2. The appearance of "missing" levels

In Section 3.1, nuclear spin was discussed and the concept of "missing" rotational levels, forbidden by the symmetrization principle, was introduced [see Equation (9)–(12)]. In the same vein as looking for Pauli excluded electronic states, one can look for missing rotational energy levels. Usually it is possible to calculate the energies that such states would have by interpolation using the energies of allowed states and the appropriate effective Hamiltonian. But we have no idea how sensitive our spectroscopy experiment has to be in order to detect them, and so far none have been seen [69–73]. The proceedings of a conference on this matter has been published [74], and the subject has been reviewed by Tino [75].

There can be missing rotational levels either because of the exchange symmetry of bosons such as D or ^{16}O nuclei, or because of the exchange antisymmetry of fermions such as H or ^{13}C nuclei. The most sensitive search so far [73] is that for the missing R(25) line in the ν_3 fundamental infrared band (00^01)–(00^00) of the $C^{16}O_2$ molecule [73] whose wavenumber can be predicted as 2367.265 cm^{-1}. An infrared frequency-modulated difference-frequency spectrometer was used to take the absorption spectrum. Using the calculated line strength of this line, and that of the nearby allowed R(80) "marker" line of the (02^21)–(02^20) hot band at 2367.230 cm^{-1}, together with the observed line amplitudes and a calculation of the RMS noise, leads to an upper limit for the violation parameter $\beta^2/2$ [54] of 1.7×10^{-11}.

In Ref. 76 experiments are proposed that will test the symmetrization principle for three identical nuclei by looking for evidence of missing levels in the spectra of $S^{16}O_3$, NH_3, and BH_3. The latter two molecules would enable one to test the breakdown of nuclear fermion antisymmetry for protons. There is also the interesting fact that the symmetric group S_3, appropriate for the permutation symmetry of three particles, has a degenerate irreducible representation as well as the usual one-dimensional symmetric and antisymmetric representations. Can states with this degenerate symmetry exist? Finally we should mention that many papers have been written concerning the possibility of testing the symmetrization principle for photons, which are bosons (see Refs. 75,77,78 and references therein).

To make more progress, it is necessary to develop more sensitive techniques [75]. As pointed out in Ref. [73], greatly increased sensitivity in this type of spectroscopic test could be obtained if the NICE-OHMS technique [79] were extended to the infrared region. Also four of the authors of Ref. 73, along with two others, have developed a comb-referenced difference-frequency spectrometer for cavity ring down spectroscopy in the 4.5 μm region [80], and this would be ideal for making a more sensitive infrared test without the need of a nearby marker line. Among other sensitive spectroscopic techniques that could be applied is that of ionization in which the ion(s) produced by irradiation at sufficient frequency to ionize the upper missing level of a missing transition could be monitored while tuning another radiation through the predicted frequency of that missing transition [11].

6. THE BREAKDOWN OF INVERSION SYMMETRY

In Section 3.3 parity was introduced as the eigenvalue of the inversion operation E^*. The operation E^* commutes with the electromagnetic Hamiltonian and eigenstates of that Hamiltonian have a definite parity, except for the states of an enantiomer of a chiral molecule for which the parity label does not apply. As mentioned in the Introduction, the complete molecular Hamiltonian within the Standard Model is the electroweak Hamiltonian. The electroweak Hamiltonian does not commute with E^* and it mixes states of opposite parity; this is called parity violation. The (very small) effect of parity violation has been observed in atomic spectroscopy (see Refs. 81–83 and references therein), but not yet in molecular spectroscopy. In atomic spectroscopy, parity violation is studied by observing very weak parity violating transitions between states that nominally have the same parity. In molecular spectroscopy, parity violation is studied because of its particular effect for chiral molecules.

For the enantiomers of a chiral molecule, the electronic energies (within the Born–Oppenheimer approximation) at equilibrium are not exactly the same if we use the electroweak Hamiltonian. Also the shapes of the potential energy surfaces around the two minima differ slightly. As a result the rovibronic energies of the two enantiomers are not identical when we allow for parity violation. These facts were first pointed out by Rein [84] and Letokhov [85]. Theoretical calculations of the parity-violating energy difference ΔE_{PV} between the equilibrium energies of enantiomers were first made by Rein and coworkers [86,87], and by Mason and Tranter [88–90]. Recent *ab initio* work [91–95] shows that these earlier estimates of ΔE_{PV} were too low by an order of magnitude. For chiral molecules containing first row atoms, ΔE_{PV} is calculated to be of the order of $10^{-13}\,\mathrm{cm}^{-1}$, but it depends strongly on the charges of the nuclei involved [96]. No experiment has yet been able to measure the parity-violating energy difference between enantiomers. In Ref. 97 the transition frequencies of the two enantiomers of CHFClBr at around 9.3 μm are found to be equal to within 13 Hz. CHFClI is a more favorable example [98]. There is a truly astounding number of review papers on the subject of the effect of parity violation for chiral molecules, and this is probably because there is yet to be any experimental observation of it. We direct the interested reader to the references cited in this section, and to the papers that they cite. One reason for the considerable effort being expended on both the theoretical calculation and the experimental determination of the parity-violating energy difference for chiral molecules is its possible implication in biological homochirality (see Refs. 99–101 and references therein).

7. THE BREAKDOWN OF TIME REVERSAL SYMMETRY

Although parity violation was observed in the 1950s [6], it was not until 1998 that particle physicists made the first direct observation of the breakdown of time reversal symmetry T [102]. However, because there is strong evidence that CPT invariance[7] is a perfect symmetry operation (see Ref. 103 and references therein), a previous observation of CP breakdown [104] is equivalent to the observation of T breakdown. A small amount of CP (or T) breakdown, such as that seen in Ref. [104], can be accommodated within the Standard Model of theoretical physics but this theory needs to be extended (somehow) in order to accommodate the amount of CP violation that is required to explain the matter/antimatter asymmetry of the universe.

The explanation of why the universe is made up chiefly of matter, rather than equally of matter and antimatter, is one of the great mysteries of physics [105]. The most plausible explanation for this *baryogenesis* is that the imbalance was dynamically generated, rather than an initial condition, and this requires, among other things, that there was CP violation during the first few seconds after the Big Bang [106]. The mechanism for CP violation is not well understood and we cannot with confidence calculate it [107]. As a result, the measurement of the extent of CP (or T) violation in a variety of systems is important in order to test any suggested extensions of the Standard Model.

One family of such measurements involves trying to determine the electric dipole moment (EDM) of particles such as the neutron or electron or of a diamagnetic atom [108]. A finite EDM for any of these systems would mean that there was CP violation beyond the Standard Model because EDMs caused by CP violation within the Standard Model are negligible [107,109]. This work was pioneered by Purcell and Ramsey [110,111], who were particularly concerned with the measurement of the EDM of the neutron. The current upper limit for the magnitude of the EDM of the neutron[8] is $2.9 \times 10^{-26}\,e\,\mathrm{cm}$ [112], and as an atomic example, the current upper limit for the magnitude of the EDM of the $^{199}\mathrm{Hg}$ atom is $2.1 \times 10^{-28}\,e\,\mathrm{cm}$ [113]. As regards the measurement of the EDM of the electron, the current upper limit on its magnitude is $1.6 \times 10^{-27}\,e\,\mathrm{cm}$ from a magnetic resonance experiment on $^{205}\mathrm{Tl}$ [114]. However, it appears that high-resolution molecular spectroscopy is particularly appropriate for the task of improving on the determination of the EDM of the electron [115,116].

---

[7] CPT is the sequential operation in any order of charge conjugation C (the change of each particle into its antiparticle), space inversion P (called E^* in Section 3), and time reversal T.

[8] $1\,e\,\mathrm{cm} \approx 1.602\ 18 \times 10^{-21}\,\mathrm{C\,m} \approx 4.803\ 22 \times 10^{8}\,\mathrm{D}$.

The determination of the electron EDM using high-resolution molecular spectroscopy involves, in essence, the measurement, in the presence of an applied external electric field, of the difference in energy between two states that only differ in the sign of m_F, the projection of the total angular momentum F along the field direction. As pointed out in Section 4, in the absence of T violation such states would be degenerate. The measurement of a nonzero energy splitting for such levels would measure the extent of T violation and can be analyzed to give the EDM of the electron. The best system for making such a measurement is a polar diatomic molecule containing one heavy atom such as YbF [115] or PbO [116], because the effective electric field within such a molecule is greatly enhanced over the value of the applied external field because of the strong electron orbital polarization. The calculation of the effective electric field requires an electronic structure calculation [117,118], and values in the 1–100 GV/cm range, about 10^6 times the applied external field, are obtained; for some examples, see Table 1 in Ref. 115

None of the experiments aimed at measuring an EDM have yet managed to obtain a nonzero result. However, the precision of the measurements has lead to the upper limits quoted above. As regards the measurement of the electron EDM using molecular spectroscopy, significant improvement in precision for the PbO and YbF experiments are imminent [119,120], and a realistic goal is to measure the electron EDM with a precision of the order of $10^{-29}\,e\,\mathrm{cm}$. The reduction of the upper limits on the magnitudes of the EDM for electrons, neutrons, and diamagnetic atoms places constraints on any extension of the Standard Model [109]. The actual measurement of a nonzero EDM would surely lead to significant insights into the symmetry of the universe.

ACKNOWLEDGMENTS

We are very grateful to Jean Cosléou and Patrice Cacciani for reading the manuscript and giving us helpful comments. The work of PJ was supported in part by the Deutsche Forschungsgemeinschaft and the Fonds der Chemischen Industrie.

REFERENCES

[1] H. C. Longuet-Higgins, Mol. Phys., 6: 445, 1963.
[2] P. R. Bunker and P. Jensen,, 'Molecular Symmetry and Spectroscopy' (2nd Edition, 2nd printing), NRC Research Press, Ottawa, 2006.
[3] M. K. Gaillard, P. D. Grannis and F. J. Sciulli, Rev. Mod. Phys., 71: S96, 1999.
[4] G. Kane, Sci. Am., 15: 4, 2005.
[5] T. D. Lee and C. N. Yang, Phys. Rev., 104: 254, 1956.
[6] C. S. Wu, E. Ambler, R. W. Hayward, D. D. Hoppes. and R. P. Hudson, Phys. Rev., 105: 1413, 1957.
[7] S. Shanmugadhasan, 'Dirac as Research Supervisor and other Remembrances'. In Tributes to Paul Dirac, J.G. Taylor, ed., Adam Hilger: Bristol, 1987.

[8] P. A. M. Dirac, Rev. Mod. Phys., 21: 392, 1949 (see especially the paragraph below Equation (1) on page 393).

[9] G. Herzberg,, 'Molecular Spectra and Molecular Structure, III. Electronic Spectra and Electronic Structure of Polyatomic Molecules,' Krieger, Malabar, Florida, 1991.

[10] E. P. Wigner, 'Group Theory,' (English translation by J. J. Griffin), Academic Press, New York, 1959.

[11] P. R. Bunker and P. Jensen, 'Fundamentals of Molecular Symmetry,' Taylor and Francis, Oxford and Philadelphia, 2004.

[12] A. M. Messiah and O. W. Greenberg, Phys. Rev., 136: B248, 1964.

[13] R. G. A. Bone, T. W. Rowlands, N. C. Handy and A. J. Stone, Mol. Phys., 72: 33, 1991.

[14] P. Jensen and P. R. Bunker, Mol. Phys., 97: 821, 1999.

[15] I. Ozier, P. Yi, A. Khosla and N. F. Ramsey, Phys. Rev. Lett., 24: 642, 1970.

[16] I. Ozier, Phys. Rev. Lett., 27: 1329, 1971.

[17] W. M. Metano and I. Ozier, J. Chem. Phys., 72: 3700, 1980.

[18] J. Bordé, Ch. J. Bordé, C. Salomon, A. Van Leberghe, M. Ouhayoun and C. D. Cantrell, Phys. Rev. Lett., 45: 14, 1980.

[19] P. R. Bunker and P. Jensen, Mol. Phys., 97: 255, 1999.

[20] J. C. Raich and R. H. Good, Jr., Astrophys. J. 139: 1004, 1964.

[21] J. P. Pique, F. Hartmann, R. Bacis, S. Churassy and J. B. Koffend, Phys. Rev. Lett., 52: 267, 1984.

[22] H. Weickenmeier, U. Diemer, W. Demtröder and M. Broyer, Chem. Phys. Lett., 124: 470, 1986.

[23] R. E. Moss, Chem. Phys. Lett., 206: 83, 1993.

[24] P. R. Bunker and R. E. Moss, Chem. Phys. Lett., 316: 266, 2000.

[25] A. D. J. Critchley, A. N. Hughes and I. R. McNab, Phys. Rev. Lett., 86: 1725, 2001.

[26] P. L. Chapovsky and L. J. F. Herman, Annu. Rev. Phys. Chem., 50: 315, 1999.

[27] R. F. Curl, J. V. V. Kasper and K. S. Pitzer, J. Chem. Phys., 46: 3220, 1967.

[28] P. L. Chapovsky, Phys. Rev. A, 43: 3624, 1991.

[29] P. Cacciani, J. Cosléou, F. Herlemont, M. Khelkhal, C. Boulet and J.-M. Hartmann, J. Mol. Struct., 780–781: 277, 2006.

[30] G. Peters and B. Schramm, Chem. Phys. Lett., 302: 181, 1999.

[31] C. Bechtel, E. Elias and B. F. Schramm, J. Mol. Struct., 741: 97, 2005.

[32] P. Cacciani, J. Cosléou, F. Herlemont, M. Khelkhal and J. Lecointre, Phys. Rev. A, 69: 032704, 2004.

[33] Z.-D. Sun, K. Takagi and F. Matsushima, Science, 310: 1938, 2005.

[34] C. Puzzarini, J. Cosléou, P. Cacciani, F. Herlemont and M. Khelkhal, Chem. Phys. Lett., 401: 357, 2005.

[35] V. K. Konyukhov, V. I. Tikhonov and T. L. Tikhonova, Kratk. Soobshch. Fiz., 9: 12, 1988.

[36] V. I. Tikhonov and A. A. Volkov, Science, 296: 2363, 2002.

[37] A. A. Vigasin, A. A. Volkov, V. I. Tikhonov and R. V. Shchelushkin, Dokl. Phys., 47: 842, 2002.

[38] A. Chijioke and I. F. Silvera, Phys. Rev. Lett., 97: 255701, 2006.

[39] S. L. Veber, E. G. Bagryanskaya and P. L. Chapovsky, J. Exp. Theor. Phys., 102: 76, 2006.

[40] M. Nela, D. Permogorov, A. Miani and L. Halonen, J. Chem. Phys., 113: 1795, 2000.

[41] M. Miki and T. Momose, J. Low. Temp. Phys., 26: 661, 2000.

[42] T. Momose, H. Hoshina, M. Fushitani and H. Katsuki, Vib. Spectrosc., 34: 95, 2004.

[43] X. Michaut, A.-M. Vasserot and L. Abouaf-Marguin, Vib. Spectrosc., 34: 83, 2004.

[44] M. Quack and M. Willeke, J. Phys. Chem. A, 110: 3338, 2006.

[45] J. K. G. Watson, Can. J. Phys., 43: 1996, 1965.

[46] M. Tinkham, 'Group Theory and Quantum Mechanics,' McGraw-Hill, New York, 1964.

[47] A. Abragam and B. Bleaney, 'Electron Paramagnetic Resonance of Transition Ions,' Clarendon Press, Oxford, 1970.

[48] R. G. Sachs, 'The Physics of Time Reversal,' University of Chicago Press, Chicago, 1987.

[49] J. K. G. Watson, J. Mol. Spectrosc., 50: 281, 1974.

[50] J. T. Hougen, J. Chem. Phys., 41: 899, 1964.

[51] W. Pauli, Phys. Rev., 58: 716, 1940.

[52] R. P. Feynman, Phys. Rev., 76: 749, 1949.

[53] H. S. Green, Phys. Rev., 90: 270, 1953.

[54] O. W. Greenberg and R. N. Mohapatra, Phys. Rev. Lett., 59: 2507, 1987; Phys. Rev. Lett., 61: 1432(E), 1987; Phys. Rev. Lett., 62: 712, 1987.

[55] O. W. Greenberg, Phys. Rev. Lett., 64: 705, 1991 Phys. Rev., D, 43: 4111, 1991.

[56] M. V. Berry and J. M. Robbins, Proc. R. Soc. London, Ser. A, 453: 1771, 1997.

[57] J. Twamley, Nature, 389: 127, 1997.

[58] A. S. Wightman, Am. J. Phys., 67: 742, 1999.

[59] O. W. Greenberg and J. D. Delgado, Phys. Lett., A, 288: 139, 2001.

[60] O. W. Greenberg and A. K. Mishra, Phys. Rev., D, 70: 125013, 2004.

[61] E. Fermi, 'Le Ultime Particelle Costitutive Della Materia,' Scientia, 55: 21, 1934.

[62] G. W. F. Drake, Phys. Rev., A, 39: 897, 1989.

[63] K. Deilamian, J. D. Gillasp and D. E. Kelleher, Phys. Rev. Lett., 74: 4787, 1995.

[64] E. Ramberg and G. A. Snow, Phys. Lett., B, 238: 438, 1990.

[65] C. Kurdak, et al., Surface Science, 361/362: 705, 1996.

[66] Y. M. Tsipenyuk, A. S. Barabash, V. N. Kornoukhov and B. A. Chapyzhnikov, Radiat. Phys. Chem., 51: 507, 1998.

[67] E. Baron, R. N. Mohapatra and V. L. Teplitz, Phys. Rev., D, 59: 036003, 1999.

[68] R. Arnold, et al., Eur. Phys. J., A, 6: 361, 1999.

[69] R. C. Hilborn, Bull. Am. Phys. Soc., 35: 982, 1990.

[70] G. M. Tino, Nuovo Cimento Soc. Ital. Fis., 16D: 523, 1994.

[71] M. de Angelis, G. Gagliardi, L. Gianfrani and G. M. Tino, Phys. Rev. Lett., 76: 2840, 1996.

[72] R. C. Hilborn and C. L. Yuca, Phys. Rev. Lett., 76: 2844, 1996.

[73] D. Mazzotti, P. Cancio, G. Giusfredi, M. Inguscio and P. De Natale, Phys. Rev. Lett., 86: 1919, 2001.

[74] R .C. Hilborn and G. M. Tino, ed. 'Spin-Statistics Connection and Commutation Relations,' AIP Conference Proceedings No. 545, AIP, New York, 2000.

[75] G. M. Tino, Phys. Scr., T95: 62, 2001.

[76] G. Modugno and M. Modugno, Phys. Rev., A, 62: 022115, 2000.

[77] D. DeMille, D. Budker, N. Derr and E. Deveney, Phys. Rev. Lett., 83: 3978, 1999.

[78] R. C. Hilborn, Phys. Rev., A, 65: 032104, 2002.

[79] J. Ye, L.-S. Ma and J. L. Hall, J. Opt. Soc. Am., B, 15: 6, 1998.

[80] D. Mazzotti, P. Cancio, A. Castrillo, I. Galli, G. Giusfredi and P. De Natale, J. Opt. A: Pure Appl. Opt., 8: S490, 2006.

[81] C. S. Wood, S. C. Bennett, D. Cho, B. P. Masterson, J. L. Roberts, C. E. Tanner and C. E. Wieman, Science, 275: 1759, 1997.

[82] C. S. Wood, S. C. Bennett, J. L. Roberts, D. Cho and C. E. Wieman. Can. J. Phys., 77: 7, 1999.

[83] V. A. Dzuba, V. V. Flambaum and M. S. Safronova, Phys. Rev., A, 73: 022112, 2006.

[84] D. W. Rein, J. Mol. Evol., 4: 15, 1974.

[85] V. S. Letokhov, Phys. Lett., A, 53: 275, 1975.

[86] D. W. Rein, R. A. Hegstrom and P. G. H. Sandars, Phys. Lett., A, 71: 499, 1979.

[87] R. A. Hegstrom, D. W. Rein and P. G. H. Sandars, J. Chem. Phys., 73: 2329, 1980.

[88] S. F. Mason and G. E. Tranter, Mol. Phys., 53: 1091, 1984.

[89] S. F. Mason and G. E. Tranter, Proc. R. Soc. London, Ser. A, 397: 45, 1985.

[90] G. E. Tranter, Mol. Phys., 56: 825, 1985.

[91] A. Bakasov, T.-K. Ha and M. Quack, in 'Chemical Evolution: Physics of the Origin and Evolution of Life'. Proceedings of the Fourth Trieste Conference on Chemical Evolution, J. Chela-Flores and F. Raulin ed., Kluwer, Dordrecht, 1996.

[92] P. Lazzeretti and R. Zanasi, Chem. Phys. Lett., 279: 349, 1997.

[93] R. Zanasi and P. Lazzeretti, Chem. Phys. Lett., 286: 240, 1998.

[94] A. Bakasov, T.-K. Ha and M. Quack, J. Chem. Phys., 109: 7263, 1998; 110: 6081, 1999.

[95] A. Bakasov and M. Quack, Chem. Phys. Lett., 303: 547, 1999.

[96] J. K. Laerdahl and P. Schwerdtfeger, Phys. Rev., A, 60: 4439, 1999.

[97] Ch. Daussy, T. Marrel, A. Amy-Klein, C. T. Nguyen, Ch.J. Bordé and Ch. Chardonnet, Phys. Rev. Lett., 83: 1554, 1999.

[98] P. Soulard, et al., Phys. Chem. Chem. Phys., 8: 79, 2006.

[99] P. Frank, W. A. Bonner and R. N. Zare, 'On the One Hand but not the Other: The Challenge of the Origin Survival of Homochirality in Prebiotic Chemistry.' Chapter 11 in 'Chemistry for the 21st Century,' E. Keinan and I. Schechter ed., Wiley-VCH, Weinheim, 2000.

[100] J. K. Laerdahl, R. Wesendrup and P. Schwerdtfeger, Phys. Chem. Chem. Phys., 1: 60, 2000.

[101] R. Berger and M. Quack, Phys. Chem. Chem. Phys., 1: 57, 2000.

[102] A. Angelopoulos, et al., Phys. Lett. B, 444: 43, 1998.

[103] A. Angelopoulos, et al., Phys. Lett. B, 444: 52, 1998.

[104] J. H. Christenson, J. W. Cronin, V. L. Fitch and R. Turlay, Phys. Rev. Lett., 13: 138, 1964.

[105] M. Dine and A. Kusenko, Rev. Mod. Phys., 76: 1, 2004.

[106] A. D. Sakharov, Sov. Phys. JETP Letters, 5: 24, 1967.

[107] E. D. Commins, J. D. Jackson and D. P. DeMille, Am. J. Phys., 75: 532, 2007.

[108] N. Fortson, P. Sandars and S. Barr, Phy. Today, 56: No. 6, 33, 2003.

[109] S. J. Huber, M. Pospelov and A. Ritz, Phys. Rev., D, 75: 036006, 2007.

[110] E. M. Purcell and N. F. Ramsey, Phys. Rev., 78: 807, 1950.

[111] J. H. Smith, E. M. Purcell and N. F. Ramsey, Phys. Rev., 108: 120, 1957.

[112] C. A. Baker et al., Phys. Rev. Lett, 97: 131801, 2006.

[113] M. V. Romalis, W. C. Griffith, J. P. Jacobs and E. N. Fortson, Phys. Rev. Lett, 86: 2505, 2001.

[114] B. C. Regan, E. D. Commins, C. J. Schmidt and D. DeMille, Phys. Rev. Lett, 88: 071805, 2002.

[115] J. J. Hudson, B. E. Sauer, M. R. Tarbutt and E. A. Hinds, Phys. Rev. Lett, 89: 023003, 2002.

[116] M. G. Kozlov and D. DeMille, Phys. Rev. Lett, 89: 133001, 2002.

[117] N. S. Mosyagin, M. G. Kozlov and A. V. Titov, J. Phys. B. At. Mol. Opt. Phys., 31: L763, 1998.

[118] A. N. Petrov, A. V. Titov, T. A. Isaev, N. S. Mosyagin and D. DeMille, Phys. Rev. A, 72: 022505, 2005.

[119] D. Kawall, F. Bay, S. Bickman, Y. Jiang and D. DeMille, Phys. Rev. Lett, 92: 133007, 2004.

[120] M. R. Tarbutt et al., Phys. Rev. Lett, 92: 173002, 2004.

BROADBAND MODULATION OF LIGHT BY COHERENT MOLECULAR OSCILLATIONS

A.M. Burzo *and* A.V. Sokolov

Contents

Abstract

In recent years, generation of ultrashort light pulses has opened a new window of opportunities for accessing processes occurring on an electronic timescale. Coherent molecular modulation is one method that in the past has proven to produce collinear mutually coherent spectral sidebands that extend in frequency from the infrared to the far ultraviolet, giving promise for compression of optical subcycle pulses and for non-sinusoidal field synthesis. Here, we review the basic fundamental theoretical concepts of the molecular modulation and describe experimental demonstrations. In addition, we discuss recent developments that emerged from the need to improve the initial concept of molecular modulation: in particular, we emphasize the benefits of applying more input fields that drive different Raman transitions in the same molecular medium and the use of hollow fibers filled with molecular gases for improving the efficiency of generation without the need of high-power input lasers.

Keywords: molecular modulation; Raman generation; subfemtosecond pulses

1. INTRODUCTION

The generation, characterization, and applications of subfemtosecond pulses are the goals of an emerging field, attoscience, which is attracting a lot of efforts in many laser laboratories. Today, the search for obtaining shorter (subfemtosecond or attosecond) laser pulses is a thriving field with the ultimate goal of controlling ultrafast electronic processes in atoms, molecules, and solids, in the same way as in recent past femtosecond pulses allowed access to molecular processes in femtochemistry [1]. Subfemtosecond pulses could be used to observe rapid motion of inner-shell electrons, which are tightly bound to the nuclei, could also provide information about fast ionization processes, and would allow precise control over high harmonic generation (HHG) [2–6]. Generation of subfemtosecond pulses with a spectrum centered around the visible region would open an additional unprecedented opportunity: the pulse duration would be shorter than the optical period and would allow subcycle field shaping. Present techniques produce pulses with shaped frequency and amplitude envelopes [7–9]; such shaped pulses already allow coherent control of molecular dynamics and chemical reactions [10–12]. Subcycle shaping will allow synthesis of waveforms where the electric field is an arbitrary predetermined function of time, not limited to quasi-sinusoidal oscillations.

According to the uncertainty principle in wave mechanics, $\Delta t \Delta \epsilon \geq \hbar/2\pi$, where Δt determines the extent of temporal confinement for a particle, $\Delta \epsilon$ gives the corresponding particle energy spread, and h is the Planck's constant. In optics, where photon energy is proportional to the frequency of light, the uncertainty principle is a simple recognition of Fourier transformation property that requires a wide frequency spectrum for the production of a short optical pulse. Furthermore, the so-called transform–limited pulse duration – the shortest duration allowed by the available spectrum – is reached when all spectral components have equal phase. When the duration of transform–limited pulses becomes comparable to the period of optical cycle, the phase of the individual frequency components (the so-called absolute phase) becomes an important parameter that needs to be controlled to achieve a prescribed temporal shape of the electric field.

Gradual development of the solid-state mode-locked (Ti : sapphire) laser technology resulted in pulse length reduction from 6 fs in 1987 to just under 4 fs in recent years [13–17]. As the pulse length approached the duration of a single optical cycle (which is on the order of 2 fs for visible light), researchers have learned to control the absolute phase of the pulse (also known as the carrier-envelope phase) [18–20]. However, by the end of the last century, it was obvious that ultrashort pulse production by the "traditional" techniques reached the "few-femtosecond barrier," and that new techniques were needed to break it. A step forward in ultrashort pulse generation was made in 2001. Workers in the field of HHG have measured subfemtosecond pulses in the soft X-ray spectral region [21–23]. The origins of this work date to the late 1980s, when HHG was discovered [24, 25], and early 1990s, when it was recognized that if these equidistant frequencies are locked in phase, a train of attosecond pulses (separated by half the laser period) would result [26–31]. Since 2001, a remarkable progress has been made toward production and characterization of ever-shorter pulses in this short-wavelength spectral region [32–34], including generation of phase-stabilized few cycle pulses in the extreme-ultraviolet regime [35]. The optical comb technology [36], which played a crucial role in allowing production of single well-controlled attosecond X-ray pulses [37], has also been recently extended to vacuum–ultraviolet region [38, 39]. The atto-second precision of the applied laser field shape has already allowed controlling electron localization in molecular dissociation [40] and multiphoton photoemission from metal surfaces [41]. The attosecond X-ray pulses have been used to study, for example, inner-shell spectroscopy [42]. HHG is a unique source of X-ray pulses, but by their very nature, these pulses are difficult to control because of intrinsic problems of X-ray optics; besides the conversion efficiency into these pulses is very low (typically 10^{-5}) [26–31].

An alternative approach (Raman-based) relies on broadening of the incident laser spectrum by the time-varying polarizability of vibrating molecules [43–46]. Impulsive Raman scattering has produced pulses as short as 3.8 fs in the near–ultraviolet region [47]. Recent experiments have shown that a weak probe pulse can be compressed by molecular oscillations, which are excited impulsively by a strong pump pulse [47–52]. Related work has shown that molecular wavepacket revivals can produce frequency chirp, which in turn would allow femtosecond pulse compression by normal group-velocity dispersion (GVD) in a thin output window [53, 54]. A disadvantage of the impulsive excitation method is that the generated Raman coherence is several orders of magnitude smaller than the maximal possible value. If the pump intensity is increased, the nonlinear response of the medium becomes substantial and ultimately limits the level of molecular excitation; as a result, the spectral broadening is far from maximal. Meanwhile, it has been shown that near-maximal molecular coherence can be achieved with adiabatic Raman excitation, resulting in collinear broadband Raman generation, which was termed molecular modulation [55–57]. As suggested by its name, the technique is based on creating a substantial modulation of the refractive index in an ensemble of coherently oscillating molecules. The coherence is produced by two (transform-limited) laser fields whose frequency difference is tuned close to a Raman transition. The resultant frequency-modulation sidebands (spaced by the Raman frequency) are mutually coherent, as demonstrated in gasses [58–60] and recently in Raman-active crystals [61]. The total bandwidth generated through this process extends from the infrared, through the visible, and into the ultraviolet region, that is, spans over more than $50,000 \, \text{cm}^{-1}$. Molecular modulation allows highly efficient production of single-cycle pulse trains with 0.5–2 fs pulse duration [62], because of a nearly 100% conversion efficiency of the driving fields into the generated sidebands [63].

The purpose of this chapter is to review the basics of the molecular modulation technique, with an emphasis on the characteristics that are relevant for optical pulse synthesis. An introductory theoretical overview is given in Section 2. The next section describes several experiments that have demonstrated the generation of a wide comb of Raman sidebands. Experimental evidence of their mutual coherence is provided in Section 4. To put the recent advances into context, other techniques that evolved from the necessity of improving the generation of the Raman frequency comb are presented in Sections 5 and 6. In particular, we emphasize the benefits of applying more input fields that drive different Raman transitions in the same molecular medium (Section 5) and the use of hollow fibers filled with molecular gases for improving the efficiency of generation without the need of high-power input lasers (Section 6). Section 7 concludes our review with a look at the prospective directions for this technique.

2. MOLECULAR MODULATION: BASIC PRINCIPLES

Basic principles of molecular modulation can be understood using a very simple analogy with the generation of new sidebands in an electro–optical crystal, such as lithium niobate. By applying a sinusoidal voltage to the crystal, the refractive index of the material is modulated with a specific modulation depth. When light at a frequency ω_m is sent through the medium, it experiences a time–varying phase shift, which gives rise to sidebands at frequencies $\omega_n = \omega + n\omega_m$, where n is an integer.

Instead of an electro–optical crystal, we use a molecular medium and apply two (or more) input fields that drive a Raman transition (often a vibrational one). The external sinusoidal voltage used in the case of the electro–optical crystal is replaced in our system by a pair of tunable nanosecond lasers with their frequency difference tuned close to the Raman transition frequency. Essentially, the driving fields excite molecules into a coherent superposition of vibrational states, causing the macroscopic refractive index of the medium to change at the applied difference frequency.

Physically, a large ensemble of molecules oscillates either in phase with the applied fields (phased state) or 180° out of phase (anti-phased) as determined by the sign of the detuning from the molecular resonance. A small Raman detuning is crucial for adiabatic preparation of a strong molecular excitation driven by the product intensities of the applied fields. If the coherence approaches its maximum value of 0.5, efficient sideband generation at frequencies $\omega_n = \omega + n\omega_m$ can occur (here n is an integer number, ω_m is the frequency difference for the applied lasers, and ω is the frequency of one of the driving fields).

The excited molecular coherence determines the extent to which the molecular response affects the laser pulse during propagation. It yields a time–varying perturbation in the index of refraction which in turn causes variations of phase and group velocities. Time–varying phase velocity leads to frequency modulation and results in an up or down conversion of the probe–laser pulse, or in a frequency chirp, depending on the timing of a short probe pulse with respect to the molecular motion [64]. Chirped pulses are then either stretched or compressed, depending on the sign of the chirp, by normal–GVD, possibly in the same medium [60].

3. BROADBAND GENERATION: EXPERIMENTAL RESULTS

Efficient collinear generation of a broadband comb of equidistant frequencies has been demonstrated in molecular gases [60]. Deuterium or hydrogen have typically been used in these experiments, partly because of the large Raman transition frequencies exhibited by these molecules and also because of the availability of the laser wavelengths that allows convenient access to these transitions.

The coherence is prepared by a pair of transform–limited nanosecond lasers, with frequency difference tuned around the vibrational Raman resonance. Laser intensities (typically of order of few GW/cm^2) are chosen such that the product of the corresponding one-photon Rabi frequencies exceeds the product of the detuning from the molecular electronic states and the Raman detuning. After the pulse overlap is achieved, the laser beams are loosely focused into a Raman cell (which is typically about 1 m in length), containing a low-pressure molecular gas. The cell is cooled to a temperature of 77 K to increase the population in the ground rotational state and to decrease the linewidth of the Raman transition [60]. When the frequency of one of the lasers is tuned through the Raman resonance, a bright well-collimated beam of white light is observed at the output of the cell. A prism is introduced into the path of the output light, and different colors corresponding to different frequencies are projected onto a white screen: up to 13 anti-Stokes sidebands and 2 Stokes sidebands are observed [60]. These sidebands range from $2.94\,\mu m$ to 195 nm in wavelength. The next (fourteenth) anti-Stokes sideband has a wavelength of 184 nm and is absorbed by air.

The dispersed output of the Raman cell is shown in Figure 1 for different Raman detunings and different D_2 pressures. The collinear nature of the generation process is clear at low pressures, where phase matching among different sidebands automatically occurs in the collinear geometry. For larger pressures, phase matching causes off-angle generation of the higher order sidebands. The spatial profiles of these sidebands show divergence from that of the driving fields, which is a signature of the cascaded nature of sideband generation process.

As we have pointed out in the introduction, the conversion efficiency for the generated sidebands can be very high. For example, Yavuz et al. have demonstrated 100% Raman conversion efficiency by driving the rotational transition at $587\,cm^{-1}$ in H_2 with a pair of nanosecond lasers (which happened to have incommensurate field frequencies). By tuning the frequency of one of the lasers to a positive detuning from the Raman resonance, that is making the molecules oscillate in phase with the driving force, they obtained a broadband comb of frequencies ranging from $1.32\,\mu m$ to 352 nm [63].

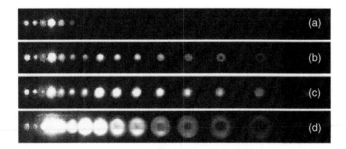

Figure 1 Broadband generation by vibrational molecular modulation in D_2 gas. The light beam emerging from the output of the molecular gas cell is dispersed by a prism and projected onto a screen. (a) $P = 71$ Torr and $\Delta\omega = -400\,MHz$, (b) $P = 71\,torr$ and $\Delta\omega = 100\,MHz$, (c) $P = 71\,torr$ and $\Delta\omega = 700\,MHz$, and (d) $P = 350\,torr$ and $\Delta\omega = 700\,MHz$. At low pressures, the Raman sidebands are emitted collinearly, with all sidebands having similar divergences. An increase in pressure results in a divergent structure for the higher order beams (reprinted from Ref. [57]). (See color plate 22).

Figure 2   Broadband generation by rotational molecular modulation in H_2 gas, driven by a pair of nanosecond lasers with frequencies equal to integer multiples of their frequency difference. The gas pressure is 270 torr and the temperature is 77 K. (See color plate 23).

In very recent experiments, we have used the same transition but made the frequencies of the driving fields to be integer multiples of their frequency difference. As a result, the absolute phases of the individual pulses in the resultant pulse train are now equal within one laser shot. This is a step toward realizing stable, reproducible waveforms with prescribed subcycle shape. The dispersed spectrum obtained in our experiment is shown in Figure 2. Detailed results will be presented elsewhere.

4.   AMPLITUDE AND FREQUENCY MODULATION AT 90 THZ

In the previous sections, we have reviewed recent approaches that use non-linear frequency conversion to create coherent output at different wavelengths. Their combined large bandwidth supports the generation of near single-cycle trains of pulses separated by a modulation period of the order of few femtoseconds.

The crucial element that allows the synthesis of various waveforms is the coherence among the generated sidebands. If the relative phases of these sidebands remain constant across their spatial and temporal profiles, manipulating their values can result in an amplitude-modulated (AM) or frequency-modulated (FM) light at a very high frequency (90 THz corresponding to fundamental vibrational transition in deuterium). A traditional way of viewing the FM or AM light is presented in Figure 3.

This mutual coherence among the phases of the sidebands was demonstrated in previous experiments in molecular gases driven by nanosecond pulses [60] and

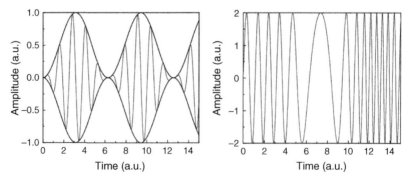

Figure 3   Amplitude-modulated (left) and frequency-modulated (right) wave.

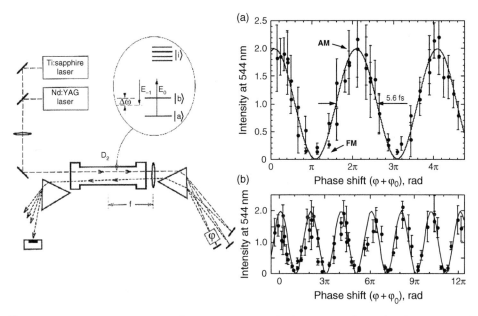

Figure 4 Experimental setup for generation and detection of amplitude-(AM) and frequency-modulated (FM) light in D_2 gas. Right, intensity of the generated sideband at 544 nm as a function of the relative phases among the three sidebands at 1.06 μm, 807 nm, and 650 nm. When the phases are such to produce AM light, the generation of 544 nm beam maximizes, whereas the lack of generation of 544 nm indicates FM light (reproduced from Ref. [58]).

recently in Raman-active crystals with femtosecond pumping [61]. The experimental setup is shown in Figure 4. Here, three sidebands are separated from the generated comb and then sent back into the Raman cell after the adjustment of their phases. When the phases of the three sidebands are adjusted such as to produce an FM waveform, no anti-Stokes generation at 544 nm is obtained. With a proper phase adjustment that maximizes the time-dependent intensity profile (AM regime), an efficient generation of the fourth sideband occurs. The results of this experiment are shown in Figure 4. The mutual coherence of the sidebands, which allows synthesis of AM and FM waveforms when only three sidebands are used, has also allowed the synthesis of single-cycle pulse trains (with the repetition rate equal to the molecular frequency) when a larger number of sidebands were used [59, 62], and in the future, this may lead to production of waveforms shaped with subfemtosecond (and subcycle) precision.

5. RAMAN GENERATION IN GASES WITH SEVERAL INPUT FIELDS

The scheme is not limited to applying two frequencies to a given medium. By mixing multiple media driven by several input fields, a multiplicative increase in the number of new frequencies (sidebands) can be achieved [65, 66].

When more fields are used to drive different transitions in the same medium, quantum interference effects among the probability amplitudes of the molecular states result not only in an increase in the spectral density of the generated comb of frequencies but also in a dramatic change of the efficiency of the generation [67]. The interference within the molecule itself resembles the electromagnetically induced transparency (EIT) process that occurs when, in a medium, the absorption of a probe-laser light driving a particular transition in atoms or molecules is cancelled out by turning on a second coupling field that is tuned to a neighboring transition. This effect has been a subject of numerous investigations because of its potentially large number of applications: from ultraslow light [68], laser cooling [69], and nonlinear optics [70] to light storage [71]. EIT has been demonstrated so far mostly in atoms, and only very recent experiments have shown that EIT is possible in a molecular system [72].

Here, we show that EIT-based ideas such as interference effects can be used to enhance the generation at a particular frequency by changing the (two-photon) detuning from the Raman resonance, or even to selectively access different degrees of freedom within the molecule itself.

The system which we consider is described in Figure 5. Two pairs of laser fields (pump and coupling) E_{p1}, E_{c2}, and E_{p2}, E_{c2} drive two Raman transitions in a molecular medium. To be specific, we assume that transition $a \rightarrow c$ corresponds to ro-vibrational Raman transition $v'' = 0$, $J'' = 0 \rightarrow v' = 1$, $J' = 2$ in molecular deuterium at 77 K, while $b \rightarrow c$ represents pure vibrational transition $v'' = 0$, $J'' = 2$

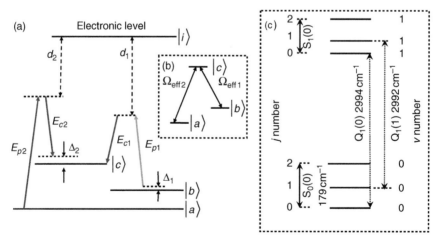

Figure 5   (a) The energy level schematic for establishing coherences ρ_{ac} and ρ_{bc}. Two Raman transitions are driven by two pairs of coupling and pump fields: fields E_{p1}, E_{c1} drive $b \rightarrow c$ Raman transition while transition $a \rightarrow c$ is driven by E_{p2}, E_{c2}. Transition $a \rightarrow b$ is Raman-allowed. The proposed quantum interference among transition probabilities amplitudes for the three-level system takes place in a standard electromagnetically induced transparency Λ scheme driven by two effective fields (b). Here, the (large) one photon detunings from electronic state $|i\rangle$ are indicated by d_1 and d_2. (c) The few of the most representatives transitions used in experiments with molecular deuterium.

$\rightarrow \upsilon' = 1$, $J' = 2$. The Raman transition $\upsilon'' = 0$, $J'' = 0 \rightarrow \upsilon' = 0$, $J' = 2$ is also allowed, as it follows from the Raman selection rules. Thus, a coherence between states $|a\rangle$ and $|b\rangle$ is set by the initial pairs of driving fields. For simplicity, we assume that only one electronic state $|i\rangle$ is present in our system. Inclusion of all other electronic levels can be done without affecting our results, but the physics of the problem can be rendered and understood easier in the simpler case presented here. The Hamiltonian of such a system is given in the dipole approximation by

$$H = H_0 + H_{\text{int}}, \tag{1}$$

where H_0 is defined by

$$H_0 = \hbar\omega_a|a\rangle\langle a| + \hbar\omega_b|b\rangle\langle b| + \hbar\omega_c|c\rangle\langle c| + \hbar\omega_i|i\rangle\langle i|. \tag{2}$$

The interaction Hamiltonian assumes the form

$$\begin{aligned}
H_{\text{int}} = 2\hbar\big[&\Omega_{p2}(t)\text{Cos}\left(\omega_{p2}t\right)|a\rangle\langle i| + \Omega_{p1}(t)\text{Cos}\left(\omega_{p1}t\right)|b\rangle\langle i| \\
+ &\Omega_{c1}(t)\text{Cos}(\omega_{c1}t)|c\rangle\langle i| + \Omega_{c2}(t)\text{Cos}\left(\omega_{c2}t\right)|c\rangle\langle i|\big] + c.c,
\end{aligned} \tag{3}$$

where $\Omega = E\mu/2\hbar$ is the Rabi frequency, E represents the electric field, and μ is the dipole moment. Here the driving fields are defined in terms of detunings (Figure 5) $\omega_{p1} = \omega_i - \omega_b - d_1$; $\quad \omega_{p2} = \omega_i - \omega_a - d_2$; $\quad \omega_{c1} = \omega_{p1} - (\omega_c - \omega_b + \Delta_1)$, and $\omega_{c2} = \omega_{p2} - (\omega_c - \omega_a + \Delta_2))$.

Following Refs. [55] and [73], we assume that one photon detunings from electronic level $|i\rangle$ are large compared to Rabi frequencies (d_1, d_2 are approximately four orders of magnitude larger than the effective Rabi frequencies). This implies that the probability amplitudes of electronic states are small compared to these one-photon detunings. After a unitary transformation of the initial Hamiltonian in the rotating reference frame, we find the following 3×3 effective Hamiltonian:

$$H_{\text{eff}} = -\hbar \begin{pmatrix}
\dfrac{\Omega_{p2}^2(t)}{(d_1 + d_2)} & 0 & \dfrac{\Omega_{c2}(t)\Omega_{p2}(t)}{(d_1 + d_2)} \\[2ex]
0 & \Delta_1 - \Delta_2 - \dfrac{\Omega_{p1}^2(t)}{(d_1 + d_2)} & \dfrac{\Omega_{c1}(t)\Omega_{p1}(t)}{(d_1 + d_2)} \\[2ex]
\dfrac{\Omega_{c2}(t)\Omega_{p2}(t)}{(d_1 + d_2)} & \dfrac{\Omega_{c1}(t)\Omega_{p1}(t)}{(d_1 + d_2)} & \Delta_2 + \dfrac{\Omega_{c1}^2(t) + \Omega_{c2}^2(t)}{(d_1 + d_2)}
\end{pmatrix} \tag{4}$$

The effective 3×3 Hamiltonian describes essentially a standard EIT Λ scheme driven by two effective fields, as illustrated in Figure 5. However, depending on the initial choice of driving fields, one can realize a V or $Ladder$ type of scheme. Evidence of such interference effects in an effective V scheme in molecular deuterium will be described in the next sections.

Next, we solve the time-dependent Schrödinger equation with the condition that all population is found initially in the ground state. We are assuming gaussian pulses with full width at half maximum (FWHM) pulse duration of 4 ns for E_{p1}, E_{c1}, E_{p2} and 1 ns for the second coupling field E_{c2} and peak intensities of $10^{10}\,\text{GW/cm}^2$.

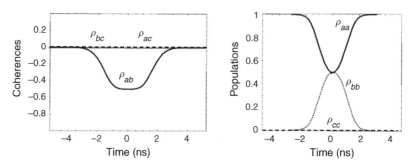

Figure 6   Time evolution of populations and coherences in the Λ scheme described in Figure 5a. At the peak pulse intensity corresponding to the time $t=0$, population of state $|a\rangle$ (solid line indicated in (a)) reaches 0.5 value. Population in state $|b\rangle$ (small dash line in (b)) increases from zero initial value to 0.5 at time $t=0$. Population of state $|c\rangle$ is virtually zero before and during the time that pulses are applied. Coherence ρ_{ab} is maximum while no coherence is established between initially driven transitions ρ_{ac} and ρ_{bc}. All field intensities are the same.

The resulting dynamics of the populations of each state are presented in Figure 6. Both transitions are driven on exact resonance, with two–photon detunings Δ_1 and Δ_2 equal to zero. Time $t=0$ corresponds to the peak intensity of applied pulses. One can see that populations of states $|a\rangle$ (solid line) and $|b\rangle$ (small dash line) reach 0.5, corresponding to a maximum value of coherence ρ_{ab}. Population of state $|c\rangle$ is virtually zero before and during the time that pulses are applied and therefore is essentially decoupled from the rest of the system. Population of state $|a\rangle$ returns to its maximum value of one after pulses are gone, indicating that adiabaticity condition is clearly satisfied.

Decoupling of state $|c\rangle$ effectively leads to cancellation of the driven ro-vibrational transition and to the enhancement of the pure rotational transition, with a maximum established coherence of 0.5 (see Figure 6).

We have demonstrated these quantum interference ideas in recent experiments with molecular deuterium. A few of its energy levels and their approximate energies are given in Figure 5c. We drove the fundamental vibrational transition $(Q_1(0))$ by two–transform–limited nanosecond laser pulses with wavelengths of 1064 and 807 nm. Their tunable frequency difference was approximately equal to the vibrational frequency of 2994.6 cm^{-1}. We observed (self–starting) stimulated rotational Raman generation corresponding to the $S_0(0)$ transition which is present even when one laser pulse is applied. We found that the stimulated rotational Raman generation is strongly affected by the vibrational Raman generation when both lasers are applied.

The two lasers were focused in the middle of a 1.2-m long liquid nitrogen-cooled cell. By cooling the cell to $T=77$ K, we increased the population in the $J=0$ state to 66% and reduced the linewidth of the transition. When dispersed by a pair of prisms onto a white screen, the generated light output showed a combined ro-vibrational comb. The comb transformed into a pure vibrational comb when the two photon detuning from vibrational Raman resonance was changed. Depending on the vibrational Raman detuning, we observed a complete suppression of the

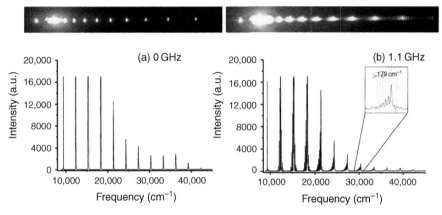

Figure 7 The spectrum generated in D_2 at two different vibrational Raman detunings. (a) A picture of a pure vibrational dispersed spectrum taken with a digital camera and the same spectrum recorded by the spectrometer at a zero detuning from vibrational Raman resonance (adapted from Refs [81] and [67]). (b) A ro-vibrational dispersed spectrum taken with a digital camera and the same spectrum recorded by the spectrometer at a 1.1 GHz detuning from vibrational Raman resonance. The 807-nm laser energy is 8 mJ/pulse, whereas the energy of the 1064-nm laser is 180 mJ/pulse, and the D_2 pressure is 300 Torr (adapted from Ref. [81]). (See color plate 24).

rotational generation presented in Figure 7a or a huge generation enhancement of several orders of magnitude (Figure 7b). The two spectra presented in Figure 7 were obtained for identical energies of the driving fields and for the same gas pressure, but at a different detuning from the vibrational Raman resonance. We monitored experimentally the dependence of all the rotational and vibrational sidebands intensities on the detuning from the vibrational Raman resonance and calculated them afterwards within the framework of our V-EIT-like interference. Figure 8 shows a comparison of our experimental and theoretical results. For details see Ref. [67].

The behavior of vibrational lines as a function of detuning is similar to the one observed, for example, in Ref. [60]. For negative detunings, vibrational generation is less efficient than for positive detunings. On-resonance generation is overall less efficient than off-resonance one [60]. Rotational generation around 1064 nm is completely suppressed around vibrational Raman resonance and features a several orders of magnitude asymmetry around the Raman resonance. Although the efficient generation occurs (surprisingly at a first glance) in the nonadiabatic regime, there is good agreement between our results and recent experimental observations of Gundry et al. [74]. These authors point out that increasing the two-photon Rabi frequency of the driving fields results in a more efficient generation although the nonadiabatic behavior of the system becomes more pronounced. In addition, larger two-photon Rabi frequency makes the generation efficiency dependence on detuning less pronounced [74] and explains the effective generation that occurs in our experiment at a large vibrational Raman detunings of few GHz (Figure 8). No enhancement of the rotational generation is observed, however, when

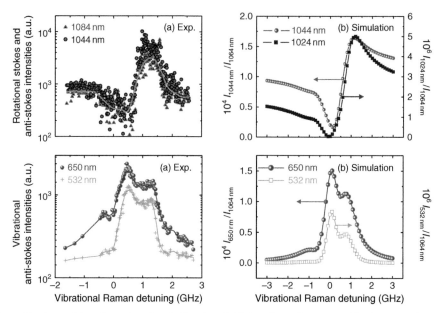

Figure 8   Intensities of rotational and vibrational sidebands as a function of the detuning from the vibrational Raman resonance. Experimental results are shown in (a), while (b) shows the results of our model simulations. In this simulation, all paramenters are taken to be the same as in experiment.

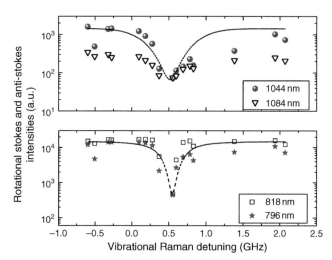

Figure 9   The intensities of first rotational Stokes and anti-Stokes of the two driving fields as a function of the vibrational Raman detuning, when the frequencies of the lasers were tuned around the frequency corresponding to $v'' = 0$, $J'' = 1 \rightarrow v' = 1$, $J' = 1$ Raman transition. The dashed line is meant to guide the eye. The pressure of the D_2 gas is 270 torr and the temperature is 77 K. No enhancement of the rotational generation originated from the Stimulated Raman Scattering (SRS) by 1064-nm laser is observed as the frequency of the second laser is tuned through the Raman resonance. This is essentially due to the fact that the transition $v'' = 0$, $J'' = 2 \rightarrow v' = 1$, $J' = 1$ is not Raman-allowed, and therefore the probability amplitudes of the three levels involved cannot interfere.

the Ti : sapphire laser frequency was tuned to 807.4 nm to drive $v'' = 0, J'' = 1 \rightarrow v' = 1, J' = 1$, as shown in Figure 9. The lack of enhancement of the rotational generation is due to the fact that the levels $v'' = 0, J'' = 2 \rightarrow v' = 1, J' = 1$ cannot be coupled via a Raman transition. This leads to impossibility of realizing a quantum interference among the transition probabilities of the levels involved.

So far, we have discussed molecular modulation in free space and with a molecular gas as a Raman medium. The use of optical fibers has certain advantages, which we discuss in the next section.

6. MOLECULAR MODULATION IN WAVEGUIDES

There is extensive literature demonstrating the use of hollow optical fibers for the enhancement of nonlinear processes. These fibers now play an important role in high-energy laser pulse compression [75], high-order harmonic generation [76,77], and generation of ultrashort pulses [78].

The use of hollow core fibers is a perspective tool to improve the efficiency of the sideband generation in the molecular modulation technique. There are several factors that affect sideband generation. In particular, to obtain the necessary coherence on the molecular transition, a high-enough intensity of driving fields in the order of several GW/cm^2 must be achieved. For example, for a free-space interaction length of 25 cm and a pulse length of 15 ns, the necessary driving laser pulse energy is about 100 mJ [79]. If we add to these requirements a good overlapping of the beam profiles and a long interaction length, it is not surprising that recently different experimental groups attempted the use of hollow core waveguides for Raman generation [79–81].

It is well known that focusing of laser beams in free space is restricted by Rayleigh range [82]. For a laser beam with a spot size w_0 at the beam waist, Rayleigh range is defined as the distance z_R at which the diameter of the spot size increases by a factor $\sqrt{2}$ and is given by $z_R = \pi w_0^2 / \lambda$, where λ is the wavelength. The intensity of laser beam is approximately constant over a range equal to twice the Rayleigh range. Focusing on a smaller size (and thus increasing the field intensity) will decrease the interaction length and, as a result, will reduce the efficiency of the generation process.

Hollow waveguides are free of the Rayleigh range constraint, which allows us to achieve high field intensity while maintaining a long interaction length and also a good quality of beam profile. However, losses limit the use of hollow waveguides [83]. They occur because of waveguide bending (proportional to $1/R$, where R represents the bending radius) and finite bore radius a (such losses are proportional to $1/a^3$).

For fused silica hollow waveguide, the lowest loss mode corresponds to H$_{11}$ mode. For this mode, the attenuation coefficient α is given by [84]

$$\alpha = \left(\frac{2.405}{2\pi} \right)^2 \frac{\lambda^2}{a^3} \operatorname{Re} \left(\frac{1}{2} \frac{v^2 + 1}{\sqrt{v^2 - 1}} \right), \tag{5}$$

where λ is the wavelength in the gas medium filling the waveguide and ν is the ratio of the refractive indices of the cladding and the core material. Rapid growth of losses with decrease of the bore radius limits the bore size of the waveguide that can be used efficiently. Nevertheless, waveguides remain a better alternative for Raman generation than focusing on free space.

GVD of the medium is usually a negative factor for efficient sideband generation because it requires phase matching. This translates into a requirement for group-velocity matching among the sidebands [85]. Dispersion consists of two contributions: dispersion of the fiber and dispersion of the filling gas. It is necessary to minimize the combined dispersion of the gas and the waveguide in the region of interest. This, for example, can be achieved by optimizing the gas pressure. Dispersion in a Raman gas media is given by [60]

$$n_{gas} = 1 + \frac{1}{2\pi\varepsilon_0 c}\hbar N\omega_q a_q \lambda_0,$$ (6)

where ε_0 is the permittivity of vacuum, N is the number of molecules in a unit volume, ω_q is the frequency of the qth sideband, λ_0 is the wavelength in vacuum, a_q is the dispersion coefficient for the generated qth sideband, and c is the speed of light in vacuum. For a waveguide, dispersion depends on both the propagating wavelength λ and the bore radius a:

$$n_{WG} = 1 - \frac{1}{2}\left(\frac{1.2\lambda^2}{c\pi^2 a}\right)^2.$$ (7)

Compensating the group-velocity contributions due to the media and the waveguide is achieved by adjusting the medium concentration for a given bore radius.

Next, we describe our recent experiment on sideband generation in a hollow fiber filled with deuterium [81]. We drive $v'' = 0, J'' = 2 \to v' = 0, J' = 4$ rotational transition in molecular deuterium at room temperature with two transform-limited nanosecond laser pulses at wavelengths 782.33 and 807.56 nm. Their tunable frequency difference is approximately equal to the transition frequency from the second excited rotational state of the ground vibrational state to the fourth excited rotational state, as shown in Figure 10b.

The two lasers were coupled into a 1-m long fused silica hollow fiber with 160 μm radius filled with deuterium. Despite the lower energies of the transmitted fields in the fiber (2.8 mJ/pulse for 782 nm and 1.6 mJ/pulse for 807 nm), approximately 20 sidebands were generated.

For comparison, the same experiment was repeated by focusing the two beams in free space using a Raman cell of same length at the same pressure. For optimizing the efficiency of the generation in free space, the beams were focused on a larger spot. Light produced in the fiber was dispersed by a pair of prisms and projected onto a white screen; the compared output light is shown in Figure 10 for both free space and waveguide cases. Figure 10c shows rotational spectrum in deuterium taken by a spectrometer. Because of the limited range of the spectrometer (650–900 nm), some of the generated sidebands are not seen.

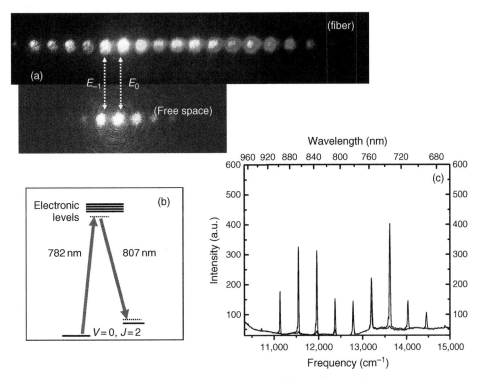

Figure 10 A pair of lasers (807 and 782 nm) drive the $v'' = 0, J'' = 2 \rightarrow v' = 0, J' = 4$ transition. The frequency difference of the driving lasers is slightly detuned from the Raman resonance to allow adiabatic preparation of molecules in a superposition state. (a) A comparison of the dispersed spectrum generated in a hollow fiber filled with D_2 gas and (b) in free space at a pressure of 704 torr. (c) Some of the generated sideband frequencies as measured by a spectrometer. The driving fields are indicated by E_{-1} and E_0. (See color plate 25).

The results obtained clearly demonstrate that the use of a hollow fiber improves efficiency of the Raman process. In the present experiment, because of the fiber, the interaction length was extended to more than twice the Rayleigh range.

A similar experiment on molecular modulation using a hollow fiber filled with molecular deuterium at 77 K was also done by the Stanford group [79]. They drove the fundamental vibrational transition of D_2 at 2994 cm^{-1} using laser pulse energies of several mJ coupled into a 22.5-cm long, 200 μm diameter hollow fused-silica fiber. The driving field wavelengths were 1064 and 807 nm. Twelve sidebands with wavelengths from 1.56 μm to 254 nm were generated. Compared with generation in free space with the same beam waist, the fiber extended the interaction length by a factor of six.

Figure 11 shows the generated energies in each sideband with the fiber present and with it removed. The Raman detuning is optimized to produce the widest bandwidth in both cases. The total energy in all sidebands drops by almost half when the fiber is present. This occurs mostly because of coupling and propagation losses associated with the hollow fiber. Despite fiber-related losses and at 20 times

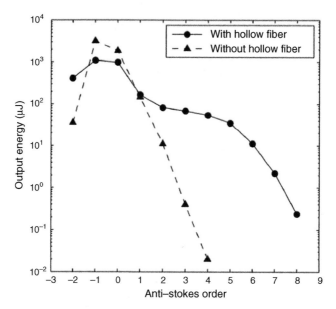

Figure 11   Energies of generated sidebands at the output of the gas cell with and without a hollow fiber. The input driving fields energies are 2.8 mJ/pulse for each laser. The two driving fields are labeled as −1 and 0 (reproduced from Ref. [79]).

lower incident energy, the generation efficiency in the hollow fiber is roughly equal to that in free space. This demonstrates fiber advantages in enhancement of molecular modulation.

Finally, we want to mention the recent experiment made in Japan on sideband generation in a hollow fiber by excitation of D_2 rotational transition $v'' = 0$, $J'' = 0 \rightarrow v' = 0$, $J' = 2$ [86]. The rotational transition was driven by ultraviolet 248- and 249.1-nm laser pulses from a KrF laser with a few mJ pulse energy. The inner diameter of the fiber was 124 μm and the length was 30 cm.

At the pressure of 30 kPa, 34 purely rotational Raman sidebands were obtained in the hollow fiber, where each of the spectral lines exhibits an intensity magnitude within 10% of the strongest line (Figure 12). In free space, only 15 sidebands were produced despite the large pump energy (70 mJ) because of the dispersion and broadening of the focused beam size due to diffraction.

A significant change in the spectrum was observed when the gas pressure in the fiber was increased to 60 kPa. The merged sidebands covered wavelengths from 220 to 660 nm and originated from a combination of rotational and vibrational transitions of the D_2 gas. This is explained by the fact that at high pressures, the vibrational Raman gain becomes larger than the rotational Raman gain, which yields generation of vibrational sidebands. The total number of Raman lines was approximately 320. With a proper phase control of the generated sidebands, this spectrum would correspond to 1-fs pulse generation.

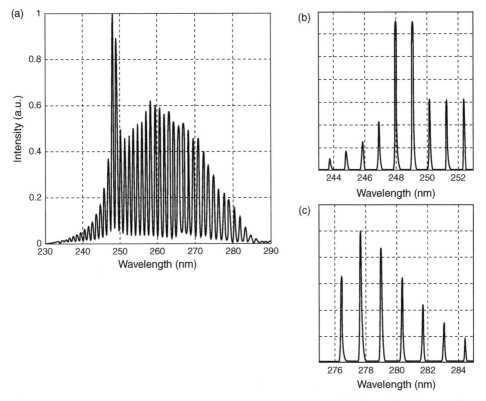

Figure 12 Rotational Raman generation in D_2 at pressure of 30 kPa: (a) the complete spectrum; (b) and (c) show regions of the same spectrum with higher resolution (reproduced from Ref. [86]).

7. SUMMARY

Advancing attosecond technology is the key for controlling the dynamics of different processes in many fields such as chemistry, physics, and biology. Here, we have reviewed a technique for subfemtosecond pulse synthesis and subcycle pulse shaping. The essence of the technique is the preparation of a macroscopic medium in a coherent superposition state. This preparation is accomplished by two laser pulses tuned close to a molecular Raman resonance. The resultant molecular motion modulates input light to produce a wide spectrum of discrete sidebands. The superposition of these spectral components with bandwidth more than three octaves allows subcycle shaping of the light waveform. We have shown different ways of improving the efficiency of the generation process, from using multiple input fields to the molecular ensemble to the use of hollow core fibers filled with gases. As more technologies are emerging today, a future way of lowering even more the input power necessary, while making the frequency comb more uniform, is the use of photonic crystal fibers that guide a broad bandwidth [87]. Implementations of molecular modulation in both gasses and solids possess distinct advantages that motivate us to pursue both. The ultimate goal is to use these Raman light sources to study ultrafast processes such as photoionization and HHG in coherent molecular ensembles.

The potential impact of this research derives not only from the promised discoveries in fundamental science but may also come from a variety of "more practical" applications of the new broadband Raman source. These applications are optical coherence tomography, ultrafast spectroscopy, precision metrology, to name a few. In addition, the technique of molecular modulation can be used to generate coherent radiation in the UV and IR spectral regions, where laser sources are not readily available.

ACKNOWLEDGMENTS

We gratefully acknowledge the contributions from our co-workers and collaborators from Stanford (D.D. Yavuz, D. R. Walker, M. Y. Shverdin, G. Y. Yin, S. Sensarn, and S. E. Harris), where this project began. Also, we thank P. Anisimov and A. Svidzinsky from Texas A&M University for fruitful discussions. In addition, we thank some of the authors named in this review for granting permission to reprint figures from their published work. This work has been supported by the National Science Foundation Grant No. PHY-0354897, an award from Research Corporation, the Texas Advanced Research Program Grant No. 10366-01-2007.

REFERENCES

[1] A. H. Zewail, J. Phys. Chem. A **104**, 5660 (2000).
[2] P. Corkum, Opt. & Photon. News **6**, 18 (1995).
[3] F. Krausz, T. Brabec, M. Schnürer, and C. Spielmann, Opt. & Photon. News **9**, 46 (1998).
[4] T. Brabec and F. Krausz, Rev. Mod. Phys. **72**, 545 (2000).
[5] F. Krausz, Phys. World **14**, 41 (2001).
[6] F. Krausz, Opt. & Photon. News **13**, 62 (2002).
[7] C. W. Hillegas, J. X. Tull, D. Goswami, D. Strickland, and S. D. Warren, Opt. Lett. **19**, 2018 (1994).
[8] A. M. Weiner, Prog. Quant. Electr. **19**, 161 (1995).
[9] T. Baumert, T. Brixner, V. Seyfried, M. Strehle, and G. Gerber, Appl. Phys. B **65**, 779 (1997).
[10] W. S. Warren, H. Rabitz, and M. Dahleh, Science **259**, 1581 (1993).
[11] S. A. Rice and M. Zhao, Optical Control of Molecular Dynamics. (Wiley, New York, 2000).
[12] M. Shapiro and P. Brumer, Adv. At. Mol. Opt. Phys. **42**, 287 (2000).
[13] R. L. Fork, C. H. Brito-Cruz, P. C. Becker, and C. V. Shank, Opt. Lett. **12**, 483 (1997).
[14] A. Baltuška, Z. Wei, M. S. Pshenichnikov, and D. A. Wiersman, Opt. Lett **22**, 102 (1997).
[15] M. Nisoli, S. DeSilvestri, O. Svelto, R. Szipöcs, K. Ferencz, C. Spielmann, S. Sartania, and F. Krausz, Opt. Lett **22**, 522 (1997).
[16] U. Keller, Nature **424**, 831 (2003).
[17] A. J. Verhoef, J. Seres, K. Schmid, Y. Nomura, G. T. L. Veisz, and F. Krausz, Appl. Phys. B **82**, 513 (2006).

[18] H. Telle, G. Steinmeyer, A. Dunlop, J. Stenger, D. Sutter, and U. Keller, Appl. Phys. B **69**, 327 (1999).

[19] D. J. Jones, S. A. Diddams, J. K. Ranka, A. Stentz, R. S. Windeler, J. L. Hall, and S. T. Cundiff, Science **288**, 635 (2000).

[20] A. Apolonski, A. Poppe, G. Tempea, C. Spielmann, R. H. Th. Udem, T. W. Hänsch, and F. Krausz, Phys. Rev. Lett. **85**, 740 (2000).

[21] M. Drescher, M. Hentschel, R. Kienberger, G. Tempea, C. Spielmann, G. A. Reider, P. B. Corkum, and F. Krausz, Science **291**, 1923 (2001).

[22] P. M. Paul, E. S. Toma, P. Breger, G. Mullot, F. Augé, P. Balcou, H. G. Muller, and P. Agostini, Science **292**, 1689 (2001).

[23] M. Hentschel, R. Kienberger, C. Spielmann, G. A. Reider, N. Milosevic, T. Brabec, P. Corkum, U. Heinzmann, M. Drescher, and F. Krausz, Nature **414**, 509 (2001).

[24] M. Ferray, A. L'Huillier, X. F. Li, L. A. Lomprk, G. Mainfray, and C. Manus, J. Phys. B: At. Mol. Opt. Phys. **21**, L31 (1988).

[25] X. F. Li, A. L'Huillier, M. Ferray, L. A. Lompré, and G. Mainfray, Phys. Rev. A **39**, 5751 (1989).

[26] G. Farcas and C. Toth, Phys. Lett. **168**, 447 (1992).

[27] S. E. Harris, J. Macklin, and T. Hänsch, Opt. Commun. **100**, 487 (1993).

[28] P. B. Corkum, N. H. Burnett, and M. Y. Ivanov, Opt. Lett. **19**, 1870 (1994).

[29] P. Antoine, A. L. 'Huiller, and M. Lewenstein, Phys. Rev. Lett. **77**, 1234 (1996).

[30] J. Schafer and K. C. Kulander, Phys. Rev. Lett. **78**, 638 (1997).

[31] I. P. Christov, M. M. Murnane, and H. C. Kapteyn, Phys. Rev. Lett. **78**, 1251 (1997).

[32] R. Kienberger and F. Krausz, Top. Appl. Phys **95**, 343 (2004).

[33] Y. Mairesse, A. de Bohan, L. J. Frasinski, H. Merdji, L. C. Dinu, P. Monchicourt, P. Breger, M. Kovacev, R. Taïeb, B. Carré, et al., Science **302**, 1540 (2004).

[34] G. Sansone, E. Benedetti, F. Calegari, C. Vozzi, L. Avaldi, L. P. R. Flammini, P. Villoresi, C. Altucci, R. Velotta, S. Stagira, et al., Science **314**, 443 (2006).

[35] J. Mauritsson, P. Johnsson, E. Gustafsson, A. L'Huillier, K. J. Schafer, and M. Gaarde, Phys. Rev. Lett. **97**, 013001 (2006).

[36] T. Udem, R. Holwartz, and T. W. Hänsch, Nature **416**, 233 (2002).

[37] A. Baltuška, T. Udem, M. Uiberacker, M. Hentschel, E. Goulielmakis, C. Gohle, R. Holwartz, V. S. Yakovlev, A. Scrinzi, T. W. Hänsch, et al., Nature **421**, 611 (2003).

[38] C. Gohle, T. Udem, M. Herrmann, J. Rauschenberger, R. Holzwarth, H. A. Schuessler, F. Krausz, and T. W. Hänsch, Nature **436**, 234 (2005).

[39] R. J. Jones, K. D. Moll, M. J. Thorpe, and J. Ye, Phys. Rev. Lett. **94**, 193201 (2005).

[40] M. F. Kling, C. Siedschlag, A. J. Verhoef, J. I. Khan, M. Schultze, T. Uphues, Y. Ni, M. Uiberacker, M. Drescher, F. Krausz, et al., Science **312**, 246 (2006).

[41] P. Dombi, F. Krausz, and G. Farkas, J. Mod. Opt. **53**, 163 (2006).

[42] M. Drescher, M. Hentschel, R. Kienberger, M. Uiberacker, A. S. V. Yakovlev, T. Westerwalbesloh, U. Kleineberg, U. Heinzmann, and F. Krausz, Nature **419**, 803 (2002).

[43] E. M. Belenov, A. V. Nazarkin, and I. P. Prokopovich, Pis'ma Zh. Eksp. Teor. Fiz. **55**, 223 (1992).

[44] S. Yoshikawa and T. Imasaka, Opt. Commun. **96**, 94 (1993).

[45] A. E. Kaplan, Phys. Rev. Lett. **73**, 1243 (1994).

[46] A. E. Kaplan and P. L. Shkolnikov, J. Opt. Soc. Am. B **13**, 347 (1996).

[47] N. Zhavoronkov and G. Korn, Phys. Rev. Lett. **88**, 203901 (2002).

[48] E. M. Belenov, P. G. Kryukov, A. V. Nazarkin, and I. P. Prokopovich, Zh. Eksp. Teor. Fiz **105**, 28 (1994).

[49] A. Nazarkin, G. Korn, M. Wittman, and T. Elsaesser, Phys. Rev. Lett. **83**, 2560 (1999).

[50] M. Wittmann, A. Nazarkin, and G. Korn, Phys. Rev. Lett. **84**, 5508 (2000).

[51] M. Wittmann, A. Nazarkin, and G. Korn, Opt. Lett. **26**, 298 (2001).

[52] V. P. Kalosha and J. Herrmann, Phys. Rev. Lett. **85**, 1226 (2000).

[53] V. Kalosha, M. Spanner, J. Herrmann, and M. Ivanov, Phys. Rev. Lett. **88**, 103901 (2002).

[54] R. A. Bartels, T. C. Weinacht, N. Wagner, M. Baertschy, C. H. Greene, M. M. Murnane, and H. C. Kapteyn, Phys. Rev. Lett. **88**, 013903 (2002).

[55] S. E. Harris and A. V. Sokolov, Phys. Rev. A **55**, R4019 (1997).

[56] S. E. Harris and A. V. Sokolov, Phys. Rev. Lett. **81**, 2894 (1998).

[57] A. V. Sokolov, D. R. Walker, D. D. Yavuz, G. Y. Yin, and S. E. Harris, Phys. Rev. Lett. **85**, 562 (2000).

[58] A. V. Sokolov, D. D. Yavuz, D. R. Walker, G. Y. Yin, and S. E. Harris, Phys. Rev. A **63**, 051801 (2001).

[59] A. V. Sokolov, D. R. Walker, D. D. Yavuz, G. Y. Yin, and S. E. Harris, Phys. Rev. Lett. **87**, 033402 (2001).

[60] A. V. Sokolov, M. Y. Shverdin, D. R. Walker, D. D. Yavuz, A. M. Burzo, G. Y. Yin, and S. E. Harris, J. Mod. Opt. **52**, 285 (2005).

[61] M. Zhi and A. V. Sokolov, New J. Phys. **10**, 025032 (2008).

[62] M. Y. Shverdin, D. R. Walker, D. D. Yavuz, G. Y. Yin, and S. E. Harris, Phys. Rev. Lett. **94**, 033904 (2005).

[63] D. D. Yavuz, D. R. Walker, G. Y. Yin, and S. E. Harris, Opt. Lett. **27**, 769 (2001).

[64] F. L. Kien, K. Hakuta, and A. V. Sokolov, Phys. Rev. A **66**, 023813 (2002).

[65] S. E. Harris, D. R. Walker, and D. D. Yavuz, Phys. Rev. A **65**, 021801 (2002).

[66] D. D. Yavuz, D. R. Walker, M. Y. Shverdin, G. Yin, and S. E. Harris, Phys. Rev. Lett. **91**, 233602 (2003).

[67] A. M. Burzo, A. V. Chugreev, and A. V. Sokolov, Phys. Rev. A **75**, 022515 (2007).

[68] L. V. Hau, S. E. Harris, Z. Dutton, and C. H. Behroozi, Nature **397**, 594 (1999).

[69] A. Aspect, E. Arimondo, R. Kaiser, N. Vansteenkiste, and C. C. Tannoudji, Phys. Rev. Lett. **61**, 826 (1988).

[70] M. Jain, H. Xia, G. Y. Yin, A. J. Merriam, and S. E. Harris, Phys. Rev. Lett. **77**, 4326 (1996).

[71] C. Liu, Z. Dutton, C. H. Behroozi, and L. V. Hau, Nature **409**, 490 (2001).

[72] F. Benabid, P. Light, F. County, and P. S. J. Russell, Opt. Express **13**, 5694 (2005).

[73] F. L. Kien, J. Q. Liang, M. Katsuragawa, K. Ohtsuki, K. Hakuta, and A. V. Sokolov, Phys. Rev. A **60**, 1562 (1999).

[74] S. Gundry, M. P. Ascombe, A. M. Abdulla, S. D. Hogan, E. Sali, J. W. G. Tish, and J. P. Marangos, Phys. Rev. A **72**, 033824 (2005).

[75] M. Nisoli, S. D. Silvestri, and O. Svelto, Appl. Phys. Lett. **68**, 2793 (1996).

[76] Y. Tamaki, Y. Nagata, M. Obara, and K. Midorikawa, Phys. Rev. A **59**, 4041 (1999).

[77] T. Pfeifer, R. Kemmer, R. Spitzenpfeil, D. Waltera, C. Winterfeldt, G. Gerber, and C. Spielmann, Opt. Lett. **30**, 1497 (2005).

[78] A. M. Zheltikov, Phys. Usp. **45**, 687 (2002).

[79] S. Sensarn, S. N. Goda, G. Y. Yin, and S. E. Harris, Opt. Lett. **31**, 2836 (2006).

[80] E. Sali, P. Kinsler, G. H. C. New, K. J. Medham, T. Halfmann, J. W. G. Tish, and J. P. Marangos, Phys. Rev. A **72**, 013813 (2005).

[81] A. M. Burzo, A. V. Chugreev, and A. V. Sokolov, Opt. Commun. **264**, 454 (2006).

[82] A. Siegman. Lasers. (University Science Books, Mill Valley, CA, 1986).

[83] M. Miyagi and S. Kawakami, Appl. Opt. **29**, 367 (1990).

[84] E. A. Marcantili and R. A. Schmeltzer, Bell Sys. Tech. J. **43**, 1783 (1964).

[85] A. Nazarkin, G. Korn, M. Wittman, and T. Elsaesser, Phys. Rev. A **65**, 041802 (2002).

[86] E. Takahashi, S. Kato, Y. Matsumota, and L. L. Losev, Opt. Express **15**, 2535 (2007).

[87] F. Couny, F. Benabid, and P. S. Light, Opt. Lett. **31**, 3574 (2006).

CHAPTER 13

Generalized Two-Dimensional Correlation Spectroscopy

Isao Noda

Contents

Abstract

The basic background of generalized two-dimensional (2D) correlation spectroscopy is described here to show how this technique can be applied to a very broad range of spectral analysis problems. In generalized 2D correlation spectroscopy, the underlying similarity or dissimilarity among systematic variations in spectroscopic signal intensities is examined. The signal variations are induced by an external perturbation or physical stimulus applied to the sample, and the pattern of the intensity variation is systematically analyzed by a simple cross-correlation technique. The correlation intensities thus obtained are displayed in the form of a 2D map defined by two independent spectral axes for further analysis. Such a 2D map is referred to as a 2D correlation spectrum, even though many of today's 2D correlation studies include applications in the field outside of spectroscopy, such as chromatography and microscopy. An illustrative example is given for the analysis of attenuated total reflection (ATR) IR spectra collected during the crystallization of biodegradable polymer poly(3-hydroxybutyrate-*co*-3-hydroxyhexanoate) to demonstrate the merits of this technique, such as the enhancement of spectral resolution, detection of coordinated changes among spectral signals, and determination of relative directions and sequential order of intensity variations.

Keywords: two-dimensional correlation spectroscopy; spectral analysis; biodegradable polymer

1. INTRODUCTION

The basic background of *generalized two-dimensional (2D) correlation spectroscopy* applicable to a broad range of spectral analysis problems [1–6] is described. 2D spectroscopy is a field of analytical technique, in which a spectral intensity is obtained as a function of two independent spectral variables, such as frequencies, wavenumbers, or wavelengths of an electromagnetic probe. The concept of 2D spectroscopy was first introduced several decades ago in the field of nuclear magnetic resonance (NMR) spectroscopy [7,8]. 2D spectra in NMR spectroscopy are typically obtained by applying a series of radio frequency pulses to a sample. Similar pulse-based 2D spectroscopy techniques are now practiced in the field of nonlinear optical spectroscopy, where a number of ultrafast optical pulses are used to produce 2D optical spectra.

Generalized 2D correlation spectroscopy, which have been developed parallel to multiple pulse-based 2D spectroscopy techniques, employs a somewhat different approach to constructing 2D spectra. It utilizes a form of cross-correlation for the analysis of spectral data obtained from a system under an external perturbation. The basic technique used in the generalized 2D correlation spectroscopy actually is applicable to the examination of a very broad range of experimental data obtained from different types of physical and analytical measurements, including not only applications in spectroscopy but also applications in chromatography and microscopy [4–6].

In a generalized 2D correlation spectroscopy study, the underlying similarity or dissimilarity among systematic variations in spectroscopic signals is examined. Signal variations are induced by an external perturbation or physical stimulus applied to the sample. The pattern of intensity variations is systematically analyzed by a simple

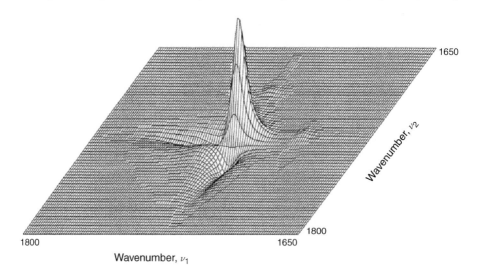

Figure 1 A pseudo 3D fishnet plot of a synchronous 2D IR correlation spectrum of a biodegradable polymer during the crystallization process.

cross-correlation technique. The correlation intensities thus obtained are displayed in the form of a 2D map defined by two independent spectral axes for further analysis. Such a 2D map is referred to as a *2D correlation spectrum*, even though many of today's 2D correlation studies include applications in the field outside of spectroscopy, such as chromatography and microscopy.

Figure 1 shows a typical 2D correlation spectrum, displayed in a pseudo 3D representation mode or a so-called fishnet plot. This particular 2D spectrum, defined by two independent IR wavenumber axes, shows the correlation among IR absorption bands of a biodegradable polymer undergoing a crystallization process from the melt. Peaks appearing over the 2D spectral plane provide detailed information about the complex dynamics of the phase transition process. It is noted that 2D correlation peaks can take either positive or negative intensities, and signs of correlation peaks play an important role in the interpretation of 2D correlation spectra. More detailed discussion on this particular 2D correlation spectrum will be given later.

2. GENEREALIZED 2D CORRELATION ANALYSIS

2.1. Perturbation-induced dynamic spectra

Figure 2 shows the basic scheme used for a generalized 2D correlation spectroscopy experiment. In a typical (1D) optical spectroscopy measurement, a selected electromagnetic probe, such as an IR or UV beam, is used to study the system of interest. Characteristic interactions between the probe and the system constituents are displayed in the form of a spectrum, which in turn is used for further examination of the state of the system. In a generalized 2D correlation experiment, an additional factor comes into play, which is the external perturbation applied to the system to somehow stimulate its constituents. The response of the system to the applied perturbation often manifests itself as measurable variations in spectral signals of the system. The portions of the spectral signals influenced by the given perturbation to undergo some intensity variations are by convention referred to as *dynamic spectra*.

In a typical generalized 2D correlation spectroscopy experiment, a series of perturbation-induced dynamic spectra are collected in a sequential order in an

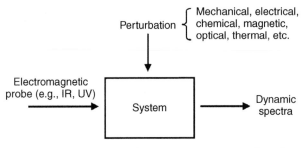

Figure 2   Application of an external perturbation to a system to induce dynamic spectra used in generalized 2D correlation spectroscopy.

alignment with the applied external perturbation. Dynamic spectra for a 2D correlation experiment may be collected, for example, as a function of time. Time-dependent evolution of transient spectral signals, arising from various constituents of the system affected by the perturbation, is thus monitored. Indeed, earlier development of 2D correlation spectroscopy was based primarily on the analysis of time-resolved spectra. Such a set of response signals may be readily analyzed by using a traditional statistical technique of time-series analysis [1]. So-called dynamic spectra actually need not be time-dependent responses, as long as they represent some systematic changes of spectral signals of a system in response to the given perturbation. Nowadays, it is common to collect dynamic spectra as a direct function of the essentially static measure of the imposed physical effect itself, such as temperature, pressure, concentration, stress, electrical field, and so on [4–6].

The conceptual scheme described in Figure 2 to induce dynamic spectra obviously is a very general one, as it does not even specify the physical nature or mechanisms with which the applied perturbation affects the system constituents. There are, of course, a number of different types of external perturbations, which could be used to stimulate a system of interest. For example, various molecular level changes may be induced by an electrical, thermal, magnetic, chemical, mechanical, or even acoustic excitation. Each perturbation affects the system in a unique and selective way, governed by the specific interaction between the macroscopic stimulus and microscopic or molecular level responses of individual system constituents. The type of physical information contained in dynamic spectra, therefore, is determined by the selection of specific perturbation method and analytical probe. Furthermore, the waveform of the applied perturbation also can be varied. For example, a simple sinusoidally varying stimulus [9–11], or a sequence of pulses, random noise, or even a set of static physical variables, like temperature or concentration, can be applied [1–6]. In principle, any analytical experiment that leads to the generation of systematic change in spectral signals becomes a good candidate for generalized 2D correlation analysis.

2.2.   2D Correlation analysis of dynamic spectra

The generalized 2D correlation formalism was evolved from the classical statistical theory of time-series analysis of multivariate signals [1]. Although the original concept of generalized 2D correlation is based on a sound and rigorous theoretical framework, it is somewhat complicated and cumbersome to implement. A much simplified practical treatment of the formal mathematical procedure is provided here to produce useful 2D correlation spectra based on the application of the discrete Hilbert transformation scheme [2,4,12]. For those who are interested in the detailed background, a complete description of the generalized 2D correlation theory is provided elsewhere [1,12].

Let us first consider a set of spectra $A(\nu, t)$ obtained for a sample of interest under the influence of an external perturbation. For convenience, the variable ν is taken as the IR wavenumber, although any other variable used in spectroscopy (e.g., Raman shift, UV-visible wavelength, or even X-ray diffraction angle) is equally applicable here. The variable t represents the effect of an external perturbation. For example, it can be the chronological time during the application of the perturbation. It can also be any reasonable measure of physical quantity associated with the type of external perturbation, such as temperature, pressure, concentration, voltage, and the like, depending on the specific experiment. Thus, spectra $A(\nu, t)$ can be, for example, a set of temperature-dependent IR spectra, time-resolved Raman spectra during a chemical reaction, UV spectra of solutions with varying concentrations, and so on.

For m discrete measurements of t-dependent spectra $A(\nu, t)$, we have

$$A_j(\nu) = A(\nu, t_j) \quad j = 1, 2, \ldots, m \tag{1}$$

Given the above, the *dynamic spectra* $\widetilde{A}_j(\nu)$ is formally defined as

$$\widetilde{A}_j(\nu) = A_j(\nu) - \bar{A}(\nu) \tag{2}$$

where $\bar{A}(\nu)$ is the *reference spectrum* of the system. It is common to set $\bar{A}(\nu)$ to be the *averaged spectrum* during the observation, as given by

$$\bar{A}(\nu) = \frac{1}{m} \sum_{j=1}^{m} A_j(\nu) \tag{3}$$

The selection of average spectrum as the reference may not always be the best option, so other choice of reference spectrum may also be used.

Given an arbitrary pair of two dynamic spectral signals, measured separately at wavenumbers ν_1 and ν_2, the *synchronous* and *asynchronous* 2D correlation intensities can be obtained as

$$\Phi(\nu_1, \nu_2) = \frac{1}{m-1} \sum_{j=1}^{m} \widetilde{A}_j(\nu_1) \cdot \widetilde{A}_j(\nu_2) \tag{4}$$

$$\Psi(\nu_1, \nu_2) = \frac{1}{m-1} \sum_{j=1}^{m} \widetilde{A}_j(\nu_1) \cdot \sum_{k=1}^{m} N_{jk} \cdot \widetilde{A}_j(\nu_2) \tag{5}$$

The term N_{jk} correspond to the element of the j-th raw and k-th column of the so–called discrete Hilbert–Noda transformation matrix [2,12], which is defined as

$$N_{jk} = \begin{cases} 0 & \text{if } j = k \\ \dfrac{1}{\pi(k-j)} & \text{otherwise} \end{cases} \tag{6}$$

We have so far assumed that the set of spectra is sampled at an equally spaced constant increment over the perturbation variable t. Slightly modified form of an expression is needed if spectra are sampled with varying increments instead [2,13].

The synchronous and asynchronous correlation intensity formally correspond to the real and imaginary part, respectively, of a complex cross-correlation function calculated along the external variable t between two spectral signal variations, measured at two different wavenumbers ν_1 and ν_2. The synchronous correlation intensity $\Phi(\nu_1, \nu_2)$ represents the overall similarity or coincidental nature between two signal variations. The asynchronous correlation intensity $\Psi(\nu_1, \nu_2)$, on the other hand, represents the out-of-phase nature of the signals. Because the complex cross-correlation function is calculated for an arbitrary pair of two spectral signals measured at ν_1 and ν_2, it can be viewed as a continuous function of two independent wavenumbers, i.e., 2D spectrum.

We should also point out a very intriguing possibility of generalized 2D correlation analysis called 2D *hetero-spectral correlation* [1,4,10,14]. In this scheme, two different types of spectral signals obtained by using different analytical probes are directly compared. Thus, a dynamic spectrum $\widetilde{A}(\nu, t)$ measured by one analytical technique (e.g., IR absorption) may be compared with another dynamic spectrum $\widetilde{B}(\mu, t)$ using a different probe (e.g., X-ray diffraction) [14]. The 2D hetero-correlation spectra are given by

$$\Phi(\nu, \mu)^{\text{hetero}} = \frac{1}{m-1} \sum_{j=1}^{m} \widetilde{A}_j(\nu) \cdot \widetilde{B}_j(\mu) \tag{7}$$

$$\Psi(\nu, \mu)^{\text{hetero}} = \frac{1}{m-1} \sum_{j=1}^{m} \widetilde{A}_j(\nu) \cdot \sum_{k=1}^{m} N_{jk} \cdot \widetilde{B}_j(\mu) \tag{8}$$

If there exists an underlying commonalty between the response pattern of individual system constituents, which are monitored by two different analytical probes under the same perturbation, one should be able to detect the correlation even between the different analytical signals.

3. PROPERTIES OF 2D CORRELATION SPECTRA

3.1. Synchronous spectra

The intensity of a synchronous 2D correlation spectrum $\Phi(\nu_1, \nu_2)$ given by Equation (4) represents the simultaneous or coincidental changes of signal intensity variations measured at ν_1 and ν_2 during the observation period along the externally defined variable t. Figure 3 shows a schematic example of a typical synchronous 2D correlation spectrum plotted as a contour map. The contour map representation of a 2D spectrum is much easier to navigate through for detailed analysis of fine features, compared with the pseudo 3D fishnet plot like the one used in Figure 1. On the other hand, a fishnet plot provides much better visual sense for the relative intensities of correlation peaks.

It can be easily seen that a synchronous 2D spectrum is a symmetric spectrum with respect to the main diagonal line corresponding to coordinates $\nu_1 = \nu_2$. Correlation peaks appear at both diagonal and off-diagonal positions. The intensity of peaks located at diagonal positions corresponds mathematically to the autocorrelation function (equivalent to the continuous form of statistical *variance*) of spectral signal intensity variations. Diagonal peaks are therefore often referred to as *autopeaks*. In the example spectrum shown in Figure 3, there are four distinct autopeaks

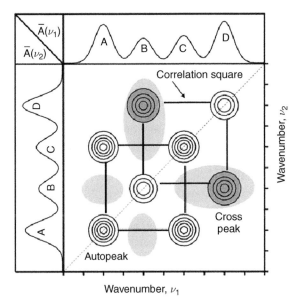

Figure 3 Schematic contour map of a synchronous 2D correlation spectrum. Shaded areas indicate negative correlation intensity regions.

located at the coordinates of the spectrum: A, B, C, and D. The magnitude of an autopeak intensity, which is always positive, represents the overall extent of spectral intensity variation observed at the specific wavenumber coordinate ν. Thus, a region of the spectrum which changes intensity to a greater extent under a given perturbation will show a stronger autopeak, while those remaining near constant develop little or no autopeak. In other words, an autopeak represents the overall susceptibility of the spectral signal intensity to change, when an external perturbation is applied to the system.

A *cross peak* located at an off-diagonal position of a synchronous 2D spectrum represents simultaneous or coincidental intensity changes of two different signals observed at coordinates ν_1 and ν_2. Such a synchronized change, in turn, suggests the possible existence of a coupled or closely related origin of the signal variations. It is often useful to construct a *correlation square*, joining the pair of cross peaks located at opposite sides of the main diagonal line drawn through the corresponding autopeaks, to show the existence of coherent variations of signal intensities at these coordinates. In the example spectrum, bands A and C are synchronously correlated, as well as bands B and D. Two separate synchronous correlation squares could, therefore, be constructed.

While the sign of an autopeak is always positive, the sign of a cross peak can be either positive or negative. For convenience, negative peaks (troughs) are indicated by shading. The sign of a synchronous cross peak becomes positive if the two spectral signals, measured at the wavenumbers ν_1 and ν_2 corresponding to the coordinates of the cross peak, are either increasing or decreasing together as functions of the external variable t. On the other hand, the appearance of a negative cross peak indicates that one of the signals is increasing while the other is decreasing. In the example spectrum of Figure 3, the signs of cross peaks at the coordinate (A, C) and (C, A) are positive, indicating that both intensities at A and C are either increasing or decreasing together. By contrast, the cross peak signs at the coordinate (B, D) and (D, B) are negative, indicating that intensity at one band is increasing, while the other is decreasing. The correlation intensity between band A and B is very small, suggesting that the patterns of intensity changes of these two bands are substantially different. The appearance of slightly negative correlation intensity indicates that the intensities at bands A and B are probably changing in the opposite direction.

3.2. Asynchronous spectra

Figure 4 shows an example of an asynchronous 2D correlation spectrum. The intensity of an asynchronous spectrum obtained by Equation (5) represents the not simultaneous changes of signal intensities at ν_1 and ν_2, occurring instead in a sequential or successive manner. The asynchronous spectrum has no autopeaks, consisting exclusively of cross peaks located at off-diagonal positions. In Figure 4, asynchronous correlations are observed for band pairs A and B, A and D, B and C, as well as C and D. Unlike a synchronous 2D spectrum, an asynchronous spectrum is anti-symmetric with respect to the main diagonal line. Thus, the sign of the asynchronous peak at (A, B) is opposite to the peak sign at (B, A). By extending lines from the spectral coordinates of the pair of cross peaks to corresponding diagonal positions, one can construct the asynchronous correlation square.

An asynchronous cross peak develops only if the intensities of two signals measured at ν_1 and ν_2 change out of phase (i.e., delayed or accelerated) with each

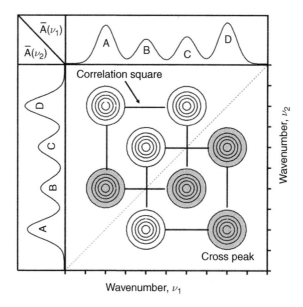

Figure 4 Schematic contour map of an asynchronous 2D correlation spectrum. Shaded areas indicate negative correlation intensity regions.

other. This feature is especially useful in differentiating overlapped bands arising from signals of different physical origins. For example, different signal contributions from individual components of a complex mixture, chemical functional groups experiencing different effects from some external field, or inhomogeneous materials comprised of multiple phases may all be effectively discriminated. Even if features are located close to each other, as long as the signatures or the pattern of signal intensity variations along the external variable t are substantially different, asynchronous cross peaks will develop between their corresponding coordinates.

Sign of an asynchronous cross peak can also be either positive or negative. The sign of asynchronous peak is used to determine the *sequential order* of signal changes. An asynchronous cross peak becomes positive if the intensity change along the variable t at the coordinate ν_1 occurs predominantly before the intensity change at ν_2. On the other hand, it becomes negative if the change occurs after ν_2. However, this sequential relationship is reversed if the synchronous correlation intensity at the same coordinate is negative, i.e., $\Phi(\nu_1, \nu_2) < 0$. An easier way to remember the sequential order rules (sometimes referred to as the *Noda's rules*) is to simply compare the signs of $\Phi(\nu_1, \nu_2)$ and $\Psi(\nu_1, \nu_2)$. If they are the same, the signal intensity variation measured at ν_1 occurs predominantly before that at ν_2. By contrast, if the signs are different, signal intensity variations at ν_1 occurs after ν_2. The example spectrum in Figure 4 indicates the signal intensity changes (either increase or decrease) at bands A and C occur before the changes at B and D.

4. APPLICATION EXAMPLE

An illustrative example of generalized 2D correlation spectroscopy applied to the study of a biodegradable polymer is discussed here. Figure 5a shows the molecular structure of poly(3-hydroxybutyrate-*co*-3-hydroxyhexanoate) or PHBHx, which is one of poly(hydroxyalkanoates) or PHA family [15–17]. It has been known that certain species of bacteria convert renewable carbon sources, like sugars and vegetable oils, to high molecular weight PHA copolymers and accumulate them within their cellular body as an energy storage medium. Figure 5b shows one of such bacterial species *Aeromodas caviae* magnified by 125,000 × , accumulating a very high content of PHBHx granules within their body. Because PHAs can be produced by a simple fermentation of readily available renewable resources, this class of copolymers is attracting keen public and industrial interest as a potential alternative source of polymeric materials. In particular, PHBHx exhibits uniquely rapid biodegradability under both aerobic and anaerobic conditions combined with a set of useful physical properties comparable with petroleum-based plastics. Their thermal behavior, especially the melt crystallization characteristics, is of great practical interest, because such material is now being seriously considered for their potential of replacing traditional plastics to be melt processed with conventional processing equipment.

Figure 6a shows the carbonyl stretching region of time- and temperature-dependent attenuated total reflection (ATR) IR spectra of a PHBHx copolymer,

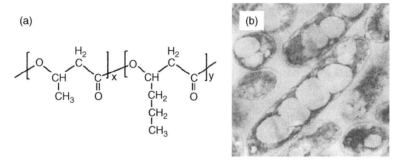

Figure 5 (a) Molecular structure of biodegradable copolymer poly(3-hydroxybutyrate-*co*-3-hydroxyhexanoate), PHBHx. (b) Granules of PHBHx accumulated within the cellular bodies of bacteria *Aeromonas caviae* (magnified by 125,000 ×).

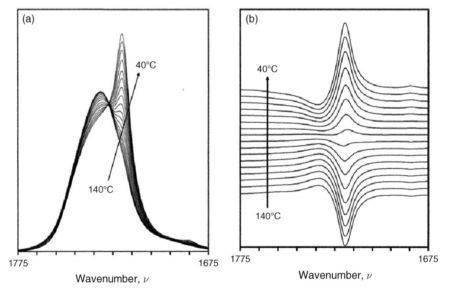

Figure 6 (a) Temperature-dependent attenuated total reflection (ATR) IR spectra of poly(3-hydoxybutyrate-*co*-3-hydroxyhexanoate) during the crystallization. (b) Corresponding dynamic spectra obtained by subtracting the temperature-averaged spectrum as the reference.

which is undergoing a dynamic crystallization process from the melt. This particular PHBHx sample had the comonomer composition of about 12 mol% of 3-hydro-xyhexanoate unit with the rest being 3-hydroxybutyrate. The weight average molecular weight was about 894,000. The sample was heated to 170°C (which was well above the melt temperature of this copolymer around 120°C) and kept for a minute to erase thermal memory, then quickly brought down to 140°C, and finally cooled gradually to 40°C at the rate of 3.5°C/min. IR spectra were continuously collected in situ, while the sample was cooled to undergo the crystallization. The corresponding dynamic spectra shown in Figure 6b were obtained by subtracting the temperature-averaged spectrum as the reference spectrum.

Some major features of the time-dependent IR ATR spectra in Figure 6a are easy to detect, as the PHBHx sample is being steadily cooled from the melt. The increase in the peak intensity at $1723\,cm^{-1}$, which is associated with the crystal growth of PHBHx copolymer, is accompanied by the decrease of a broad peak centered around $1740\,cm^{-1}$, which is attributed to the reduction of the amorphous component of PHBHx, by crystallization. A similar trend associated with the crystallization of PHBHx is observed more clearly in the dynamic spectra (Figure 6b), where the sign changes of these characteristic bands may be detected for difference spectra between the spectral traces above and below the melt temperature. It is, however, difficult to directly extract more detailed and subtle underlying evolution of the time- and temperature-dependent spectral features from the raw or simple difference spectra. 2D correlation spectra will be a powerful tool to effectively achieve such a task.

To construct 2D IR correlation spectra, Equations (4) and (5) were applied to the set of dynamic spectra (Figure 6b) representing the systematic IR intensity changes of the PHBHx sample during the crystallization. Figure 7 shows the synchronous 2D IR correlation spectrum of PHBHx displayed as a contour map. The pseudo 3D fishnet plot of the synchronous 2D correlation spectrum of this system has already been shown in Figure 1. Contour map representation is useful in locating the accurate coordinate of the peak maximum or minimum, whereas a fishnet plot often gives a better sense of relative intensities among various correlation peaks. The perturbation variable is the time-dependent temperature, which was decreasing at a constant rate of 3.5°C/min. The reference spectrum was

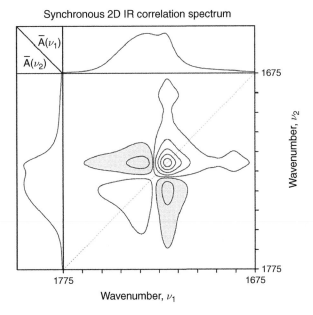

Synchronous 2D IR correlation spectrum

Figure 7 Synchronous 2D correlation spectrum of poly(3-hydoxybutyrate-*co*-3-hydroxyhexanoate) during the crystallization. Shaded areas represent negative intensity.

obtained by calculating the average over the temperature range of observation between 140°C and 40°C. The plot of the reference spectrum is placed at the top and side of the contour map. Negative correlation intensity areas of the 2D map are represented by the shading.

The synchronous 2D IR correlation spectrum of PHBHx shows several prominent features. The two autopeaks at the diagonal position around 1723 and 1740 cm^{-1} clearly reflect the dynamics of intensity changes of bands associated with the crystalline and amorphous component of PHBHx during the cooling process. A pair of broad negative cross peaks developed around (1723 and 1740 cm^{-1}) and (1740 and 1723 cm^{-1}) between the crystalline and amorphous band positions, indicating that the intensity of one band (crystalline) is increasing, while the other (amorphous) is decreasing during the crystallization of PHBHx from the melt. The result so far is consistent with the observation made from the original IR spectra. A small feature, visible in Figure 6, at 1685 cm^{-1} is also visible but much more clearly in the synchronous 2D spectrum, as a pair of cross peaks developed in the region at (1685 and 1723 cm^{-1}) and (1723 and 1685 cm^{-1}). From the sign of the cross peak, it can be deduced that the pattern of the change at 1685 cm^{-1} closely follows that of the crystalline band features around 1723 cm^{-1}, suggesting that the origin of this band is also associated with the crystalline component of PHBHx.

Even more intriguing results are obtained once we start analyzing the asynchronous 2D correlation spectrum (Figure 8). Asynchronous cross peaks appear

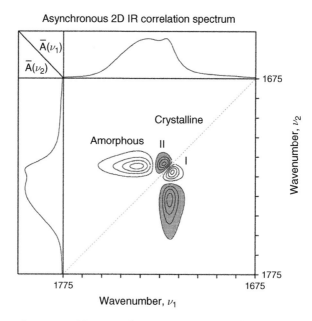

Figure 8 Asynchronous 2D correlation spectrum of poly(3-hydoxybutyrate-*co*-3-hydroxyhexanoate) during the crystallization. Shaded areas represent negative intensity.

between the band regions for amorphous ($1740\,\mathrm{cm}^{-1}$) and crystalline ($1722\,\mathrm{cm}^{-1}$) components, which have already been established by the analysis of the synchronous spectrum. The presence of asynchronous peaks unambiguously indicates that the disappearance of the amorphous component does not occur simultaneously with the appearance of the crystalline component. The signs of the synchronous and asynchronous cross peaks at the spectral coordinate (1722 and $1740\,\mathrm{cm}^{-1}$) are both negative. The negative peak signs at the same coordinate indicate the intensity change at ν_1 (i.e., the increase of the intensity at $1722\,\mathrm{cm}^{-1}$) seems to occur sooner at a higher temperature compared to the intensity change at ν_2 (i.e., the overall decrease of the intensity at $1740\,\mathrm{cm}^{-1}$) during the cooling.

The apparent discrepancy between the behavior of amorphous and crystalline intensities can be explained, once other asynchronous cross peaks shown in Figure 8 are taken into account. An additional asynchronous cross peak pair appears near the diagonal line at the spectral coordinate of (1724 and $1721\,\mathrm{cm}^{-1}$) and (1721 and $1724\,\mathrm{cm}^{-1}$), both within the crystalline region of the spectrum. It is reasonable to draw a conclusion that there are at least two distinct types of populations in the crystalline components of PHBHx copolymer, each represented by a slightly different characteristic IR frequency. The lower wavenumber band located at $1721\,\mathrm{cm}^{-1}$ most likely represents the well-formed PHBHx crystal with a high level of structural order, while the higher wavenumber band at $1724\,\mathrm{cm}^{-1}$ is probably ascribed to the crystalline components with more disorder or defects. During the crystallization of a semicrystalline polymer from the melt, one often observes the unhindered formation of the highly ordered primary crystals, followed by the formation of less ordered secondary crystals within constrained geometry already occupied by the primary crystals. The signs of the asynchronous cross peaks indicate that the growth of the intensity at $1721\,\mathrm{cm}^{-1}$ occurs later than that at $1724\,\mathrm{cm}^{-1}$. Thus, we assign the intensity change at $1721\,\mathrm{cm}^{-1}$ to the unhindered primary crystal (I) formation, and that at $1724\,\mathrm{cm}^{-1}$ to the formation of less ordered secondary crystal (II) under more constrained condition. The consumption of the amorphous component continues to a later stage to produce the secondary crystals, well after the completion of the primary crystals formation.

The seemingly simple illustrative example given above has actually demonstrated several key advantageous features commonly found in generalized 2D correlation spectra. By spreading highly overlapped peaks along the second dimension, the spectral resolution is clearly enhanced. Indeed, the identification of the existence of two separate crystalline bands at 1721 and $1724\,\mathrm{cm}^{-1}$ will be rather difficult from a cursory observation in this region of the original spectra (Figure 6). The presence of autopeaks confirms the dynamic variations of spectral band intensities by the external perturbation, i.e., gradual cooling of a molten polymer sample in our case. The signs of cross peaks also provide a very useful set of information. In the synchronous 2D correlation spectrum, peak signs could be used to discriminate and classify pertinent IR bands to those from growing crystalline components and disappearing amorphous components. Once the identification of certain dominant bands is established, the subsequent classification of the rest of minor bands becomes much easier. Thus, by using the appearance of a positive cross peak at (1685 and $1723\,\mathrm{cm}^{-1}$), one can easily identify $1685\,\mathrm{cm}^{-1}$ as one of the bands associated with the crystalline component. The sequential order information, such as the order of formation and consumption of polymer constituents during the crystallization process, can be readily obtained by combining the cross peak signs of synchronous and asynchronous 2D correlation spectra. The sequential order, in turn, provides a useful insight into the mechanistic understanding of the complex dynamics of crystallization process, which manifested only subtle changes in the spectral intensities of a set of original spectra.

5. CONCLUSIONS

Generalized 2D correlation spectroscopy is a powerful and versatile technique effective to a broad range of applications. The construction of 2D correlation spectra is relatively straightforward. One only needs a series of systematically varying spectra generated by applying an external perturbation to a system of interest during the measurement. The perturbation can take a form of physical or chemical reaction, change in temperature, pressure, or concentration, magnetic, electrical, or mechanical stimuli, and the like. The set of spectra are then converted to the synchronous and asynchronous 2D correlation spectrum representing, respectively, the similarity and dissimilarity of intensity variations. The spectral resolution is enhanced by spreading the overlapped bands along the second dimension. The 2D correlation spectra provide the rich information about the presence of coordinated changes among spectral signals, as well as the relative directions and sequential order of intensity variations. The technique should be a useful addition to the toolbox of analytical and experimental scientists.

REFERENCES

[1] I. Noda, Appl. Spectrosc., 47: 1329, 1993.
[2] I. Noda, A. E. Dowrey, C. Marcott, G. M. Story, and Y. Ozaki, Appl. Spectrosc., 54: 236A, 2000.
[3] I. Noda, in J. M. Chalmers and P. R. Griffiths (Eds.), Handbook of Vibrational Spectroscopy, vol. 3, p. 2113, John Wiley and Sons, Chichester, 2003.
[4] I. Noda and Y. Ozaki, Two-Dimensional Correlation Spectroscopy. Applications in Vibrational and Optical Spectroscopy, John Wiley and Sons, Chichester, 2004.
[5] I. Noda, Vib. Spectrosc., 36: 143, 2004.
[6] I. Noda, J. Molec. Struct., 799: 34, 2006.
[7] A. Bax, Two Dimensional Nuclear Magnetic Resonance in Liquids, Reidel, Boston, 1982.
[8] R. R. Ernst, G. Bodenhausen, and A. Wakaun, Principles of Nuclear Magnetic Resonance in One and Two Dimensions, Oxford University Press, Oxford, 1987.
[9] I. Noda, J. Am. Chem. Soc., 111: 8116, 1989.
[10] I. Noda, Appl. Spectrosc., 44: 550, 1990.
[11] C. Marcott, A. E. Dowrey, and I. Noda, Anal. Chem., 66: 1065A, 1994.
[12] I. Noda, Appl. Spectrosc., 54: 994, 2000.
[13] I. Noda, Appl. Spectrosc., 57: 1049, 2003.
[14] I. Noda, Chemtracts-Macromol. Chem., 1: 89, 1990.

[15] M. M. Satkowski, D. H. MElik, J.-P. Autran, P. R. Green, I. Noda, and L. A. Schechtman, in Y. Doi and A. Steinbüchel (Eds.), Biopolymers, vol. 3b, p. 231, Wiley-VCH, Weinheim, 2001.
[16] I. Noda, M. M. Satkowski, A. E. Dowrey, and C. Marcott, Macromol. Biosci., 4: 269, 2004.
[17] I. Noda, P. R. Green, M. M. Satkowski, and L. A. Schechtman, Biomacromolecules, 6: 580, 2005.

MICROWAVE SPECTROSCOPY

EXPERIMENTAL TECHNIQUES

Jens-Uwe Grabow *and* Walther Caminati

Contents

Abstract

For more than 15–20 years now, the field of microwave (MW) spectroscopy is progressing impressively: As will be seen in the subsequent chapter, this is partially by virtue of the experimental developments that combine jet-expansion sources with specific means of sample preparation for new chemical systems. But even more important, progress is due to advancements in experimental equipment, namely the rise of very sensitive time-domain techniques with high-resolution. Although structural studies continue to be a strength of rotational spectroscopy, many of the problems that became amenable for investigation in the recent years involve processes of intramolecular dynamics, such as conformational, tautomeric equilibria, and other large-amplitude motions as well as intermolecular vibrational energy redistribution and isomerization. Numerous interesting systems such as molecules with multiple internal motions, larger complexes, aggregates, biomolecules, and transient species can almost routinely be treated now. The studies also cover different kinds of intermolecular interactions, extending from hydrogen bonding and van der Waals interactions to the effects observed in quantum solvation. The development of pulsed excitation multiresonance techniques and the advent of real-time broadband microwave excitation and detection are impressively widening the capabilities of rotational spectroscopy to characterize the structure and dynamics of larger molecular species.

Keywords: microwave spectroscopy; rotational spectra; structure; dynamics; supersonic-jet expansion; Fabry–Pérot-type resonator; Fourier transform; coherence; time domain; impulse; fast passage; double resonance; triple resonance; transient absorption; transient emission; transient nutation; free-induction decay

1. INTRODUCTION

A large variety of techniques have been developed to record spectra that are associated with transitions between rotational states. For most molecules and molecular systems, these transitions occur in the microwave (MW) region, Table 1. The rotational transitions of molecular complexes, inorganic clusters, and heavier molecules typically lie in the centimeter (cm) wavelength range, those of lighter molecules in the millimeter (mm), and sub-millimeter (sub-mm) wavelength ranges of the microwave region. In most spectrometers, monochromatic oscillators provide the required radiation. Apart from multipliers for the generation of the frequencies at the higher end of the microwave region, Fourier-transform interferometers as well as far infrared (FIR) semiconductor laser, free-electron laser (FEL), and synchrotron light source absorption spectrometers have also been used to study spectra in the latter range.

Although molecular rotation spectra traditionally are associated with molecular structure determination, today's microwave spectroscopy is addressing a wide variety of basic and applied key problems in physical chemistry, molecular physics, and related fields: Questions on molecular structure, conformational and tautomeric conversion, chemical bonding, charge transfer, and internal dynamics are elucidated not only for isolated molecules: Over the years, the scope of rotational spectroscopy has widened from fundamental intramolecular observations to the interrogation of intermolecular interactions. Related studies extend from van der Waals complexes, electron donor–acceptor complexes, and hydrogen-bonded complexes to chiral recognition, molecular aggregation, and quantum solvation. Targeted systems are atmospheric species, interstellar compounds, biomolecules, transient/instable systems as well as metal, semiconductor, and ionic clusters. Consequently, the work includes laboratory studies of the radiative and collisional properties of molecules, clusters, radicals, and ions as well as field applications directed toward atmospheric sensing or radio astronomy.

Table 1 Approximate extension of the microwave region

Range	Wavelength (cm)	Wavenumber (cm^{-1})	Frequency (GHz)
cm wave	1.0–100.0	1.0–0.01	30.0–0.3
mm wave	0.1–1.0	10.0–1.0	300.0–30.0
sub-mm wave	0.01–0.1	100.0–10.0[a]	3000.0–300.0[b]

[a] Often referred to as part of the far-infrared (FIR) range.
[b] Terahertz (THz) range or T-rays are used in relation to generation with pulsed LASERs.

Microwave spectroscopy is facing a number of general challenges when studying rotational transitions. The quanta of radiation in this region of the electromagnetic spectrum are fairly low in energy and thus more difficult to detect than in infrared (IR) or ultraviolet/visible (UV/VIS) spectroscopy. At the same time, as the energy levels that take part in the transition are relatively close, the Boltzmann population difference is small, thus setting a limit to the population transfer that can occur during a molecular transition, which entails an even weaker signal. However, somewhat contrary to a widely held opinion, the background noise in the MW region is quite low, allowing for very sensitive detection.

In the late 1950s and early 1960s, Hewlett-Packard Inc. marketed a microwave spectrometer operating in the 10–40 GHz range. The complex and costly instrument experienced only modest commercial success at that time. Since then, to a good part due to huge advances in communication electronics, vast progress in performance, availability, and affordability of microwave instrumentation was made. Whereas the mm wave and sub-mm wave ranges of high-resolution microwave spectroscopy are still dominated by frequency-domain techniques, the cm wave range has been taken over by time-domain techniques. We shall discuss most of the basic modern methods that use microwave oscillators as coherent light sources as well as some more specialized ones that use arbitrary-waveform generation to overcome specific problems and open a wide field of new opportunities.

2. FREQUENCY-DOMAIN (FREQUENCY SCAN/CONTINUOUS WAVE) TECHNIQUES

In traditional microwave spectrometers, monochromatic radiation is passed through a sample cell, and the transmitted intensity is monitored [1]. Because of the presence of a sample, the radiation is modified. The transmission recorded as a function of the incident radiation frequency yields the spectrum.

The majority of recent frequency-domain spectroscopy (FDS) instruments are using backward wave oscillators (BWOs) as microwave radiation sources. Compared to the related mechanically tuned reflex klystrons, which have widely been used in the past, as well as mechanically tuned Gunn diode oscillators, BWOs have the advantage that they can be tuned electronically over an entire band. To ensure frequency stability, the oscillator usually is phase-locked to a stable signal generator typically operating at much lower frequency. An intermediate beat frequency between a harmonic of the low frequency and the microwave frequency is used to drive a feedback circuit adjusting the input voltage which is controlling the frequency of the microwave source. Broadband spectra can be taken by sweeping the low-frequency generator as the microwave source, being governed by the phase-locked loop (PLL) feedback circuit will maintain the intermediate frequency (IF). Different modulation techniques such as Stark effect modulation [2] in conjunction with phase-locked detection are often employed to eliminate cell background characteristics, thereby lowering the detection limit.

Microwave spectrometers following the traditional designs have been reviewed in the literature [1,3,4]. Here, we limit our detailed discussion of instruments to those steady-state techniques that are employed in the more recently developed frequency-domain microwave spectrometers, i.e., continuous wave (CW) techniques with swept frequency sources that can cover a broad frequency region.

2.1. Continuous radiation microwave excitation on stationary gases

Different types of static gas cells are employed, depending on the particular implementation of the microwave spectrometer; the main types for broadband spectrometers are waveguide cells and free-space cells. The implementation of a microwave experiment governs the choice of the cell type.

Plain waveguide cells, which might be coiled into a compact arrangement that can conveniently be temperature controlled, bear the disadvantage that the detected signal is altered by the cell's transmission characteristics. This can be very unpleasant, and to retrieve the signal, such cells usually require source modulation techniques. A better way to eliminate the background is the use of a strip-electrode cell, also known as Stark-effect modulation cell [2], where the resonance frequencies of the molecules can be modulated by the application of a high-voltage zero-based square-wave potential between the electrode and the waveguide. In the cm wave range, the Stark-effect modulation cell is often to be preferred. In the mm wave range, however, smaller waveguide dimensions are required for efficient single-mode propagation. This makes strip-electrode cells impractical even though most of the insulating dielectric material that is used to hold the strip-electrode half-way between the broad faces of the waveguide can be eliminated by cutting a groove in the wall of the guide itself. At smaller dimensions, however electrical breakdown of the sample gas starts to occur. Two accurately machined parallel metal plates can also be used to propagate microwave radiation [5]; horn antennae are used as microwave radiation feeds, here. Placed inside a suitable vacuum system, high-pumping rates can be maintained across the volume between the plates, which makes this type of cell useful for working with instable species.

Alternatively, the radiation emitted from a horn antenna at one end of a vacuum cell can be collimated by a conducting reflector or a dielectric lens and collected at the other end by a matching horn antenna which feeds the transmitted radiation into a suitable detector [6]. Often used in the sub-mm wave range, free-space systems, commonly placed inside a glass tube as vacuum cell, are also useful when instable species require a high pumping rate or when rapid metal-catalyzed decomposition occurs.

For the CW experiment, the molecular signal is obtained as power absorption of linearly polarized radiation with the power P at the detector [3]. For optically thin media, the amplitude of the molecular signal $|S_{ab}|$, being proportional to the power decrease dP at the end of the sample cell of length L, is given by

$$|S_{ab}| \propto \left(\frac{\Delta^1 N_{ab}\, \mu_{ab}^2\, \omega_{ab}\, L}{\varepsilon_0 c\, \hbar \Delta \nu_{ab}} \right) \times L(\omega_{ab} - \omega), \qquad (1)$$

where $L(\omega_{ab} - \omega)$ is an appropriate lineshape function, ε_0 is the electric field constant, $\hbar$ is Planck's constant, c is the vacuum speed of light, $\Delta^1 N_{ab}$ is the population difference per unit volume of the two levels involved in the transition $b \leftarrow a$ with the electric dipole matrix element μ_{ab} of the space-fixed z-component of the dipole operator, $\omega_{ab} = 2\pi\nu_{ab}$ is the angular resonance frequency, and $\Delta\nu_{ab}$ is the half width at half maximum (HWHM).

2.1.1. Phase-locked sweep

Microwave spectrometers following the PLL scheme are now reaching frequencies beyond 1 THz [7,8]. In these designs, a high-frequency BWO is phase-stabilized to the harmonic output of a continuously tunable mm wave synthesizer that delivers enough power for frequency multiplication. In Figure 1, a block diagram for a typical PLL spectrometer is shown; the crucial components are as follows: a tunable mm wave synthesizer (1), a planar Schottky diode multiplier mixer (3), a PLL feedback circuit (4), a high-frequency broadband tunable BWO (5), a liquid-He-cooled InSb detector (9), and a computer-controlled data-acquisition system (12). Locked to a harmonic of the synthesizer, continuous scanning over the entire range of the mm wave synthesizer can achieve hundreds of GHz at the BWO output. Available BWOs offer continuous tuning capability over one octave depending on the specification. In Figure 2, an example for a spectrum close to the upper frequency limit presently achievable with a broadband tunable BWO is given.

Very recently, it was shown that newly developed superlattice (SL) multipliers are competitive sources for the terahertz range: Being more efficient in the harmonic generation than Schottky diodes, they require only moderate input frequencies for the production of THz radiation. An SL multiplier [9] consists of a periodic arrangement of thin epitaxial layers. Because of the periodicity of the structure, the electron energy is restricted to bands with a specific band gap. The transport of the band electrons leads

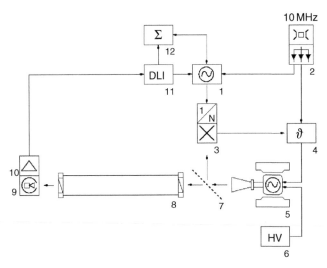

Figure 1 A backward wave oscillator (BWO) phase-locked loop (PLL) sub-mm wave spectrometer for use up to 1.2 THz: (1) mm wave synthesizer, (2) frequency standard and distribution amplifier, (3) planar Schottky-diode harmonic mixer, (4) PLL feedback circuit, (5) sub-mm wave BWO, (6) HV power supply, (7) beam splitter, (8) absorption cell, (9) liquid-He-cooled InSb detector, (10) low-noise amplifier, (11) digital lock-in amplifier, (12) experiment control and data acquisition. *Source*: Figure 1 of supplement to Ref. [103].

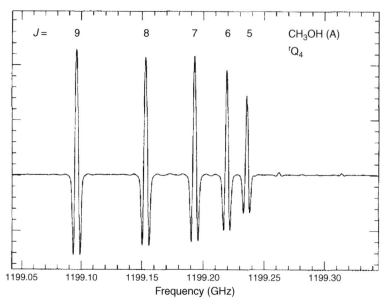

Figure 2   Spectrum of A-species low-J transitions constituting the head of the $^rQ_4$ ($\Delta J = 0$; $5 \leftarrow 4 = |K|$) branch of methanol in the $v_t = 0$ torsional state. (From Ref. [8]. Copyright 1995, Elsevier Inc.)

to a negative differential resistance. Above a threshold value, the current–voltage characteristic of an SL device is highly non-linear and anti-symmetric, making it suitable for high-frequency multiplication [10]. An SL multiplier BWO spectrometer setup [11] is shown in Figure 3. The BWO (6) is PLL stabilized with respect to a rubidium reference (2) with a frequency accuracy $\Delta f/f = 10^{-11}$. For the detection (11) of the SL (9) output signal below 2.33 THz, a magnetically tuned InSb hot electron bolometer is used, while a more sensitive Ga–doped Ge photoconductor is deployed for higher frequencies. Up to 2.7 THz can be generated at an input frequency of 250 GHz while very broad spectral ranges can be tuned by the SL; Figure 4 gives a spectral example. In the near future, a full solid-state THz spectrometer will be assembled where SL multiplication of 50 GHz radiation from a commercial synthesizer will be used to replace the BWO source used so far. The development of THz radiation sources is a very dynamic field of research. New instruments – based on novel devices such as SLs – are expected to outperform the widely used Schottky diode spectrometers in the (near) future. Considering the wide tuning range combined with the rather high frequencies reached, it is anticipated that SL multipliers will become the most interesting devices for THz spectroscopy.

2.1.2. Fast scan

There has been considerable recent interest in the development of spectroscopic gas sensors. From the beginning of microwave spectroscopy, its potential for the chemical analysis of gases that can be collected under low-pressure conditions has

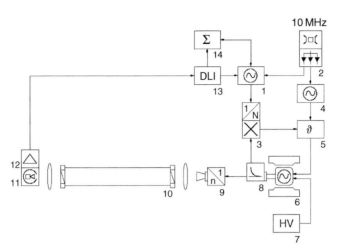

Figure 3 A superlattice (SL) multiplier backward wave oscillator (BWO) phase-locked loop (PLL) sub-mm wave spectrometer for use up to 2.7 THz: (1) cm wave synthesizer, (2) frequency standard and distribution, (3) DC-biased harmonic mixer, (4) mm wave synthesizer, (5) PLL feedback circuit, (6) sub-mm wave BWO, (7) HV power supply, (8) coupler, (9) SL multiplier, (10) absorption cell, (11) InSb hot electron bolometer for $f < 2.33$ THz or Ga-doped Ge-photoconductor for $f > 2.33$ THz, (12) low-noise amplifier, (13) digital lock-in amplifier, (14) experiment control and data acquisition. *Source*: Figure 2 of supplement to Ref. [103].

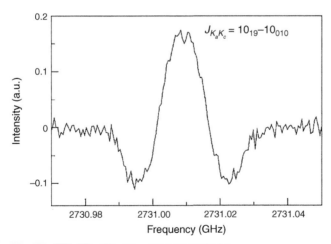

Figure 4 Spectrum of deuterium oxide recorded with a source-modulated backward wave oscillator (BWO) and a superlattice (SL) multiplier using the 11th harmonics. (Reused with permission from Ref. [11]. Copyright 2007, American Institute of Physics.)

been recognized [1]. Studying molecular systems in stationary situations offers the advantage that partition functions are more straightforward to calculate than in non-equilibrium situations, e.g., supersonic-jet expansions, where the thermal population is not easily obtained. If the partition function can be calculated or experimentally derived, the absolute absorption coefficients for rotational transitions can be precisely calculated from dipole moments, which can be accurately obtained from Stark-effect measurements. Subsequently, relating the recorded signal to concentrations is straightforward, representing an absolute quantitative calibration traceable to fundamental theory. As the Doppler broadening is small in the microwave region of the electromagnetic spectrum, a resolution unrivalled by other analytical techniques is obtained at low pressure. This resolution provides virtually absolute specificity for the detection of a molecule even in complex mixtures [12,13].

The recent developments in frequency-domain microwave spectroscopy of stationary gases are dominated by the introduction of fast scanning instruments that utilize free-running radiation sources in the mm and sub-mm range: The so-called fast-scan sub-mm spectroscopy technique (FASSST) [14] provides the sensitivity, speed, and extraordinary resolution capability, required for modern CW microwave spectroscopy on stationary samples to overcome the long-standing limitations that precluded its development for quantitative analytical purposes [15].

Based on electronically tunable sources of high spectral brightness and high spectral purity, typically BWOs, the FASSST methodology makes use of modern data-acquisition capabilities. This facilitates the replacement of complex analog phase-lock techniques with digital processing approaches for the frequency calibration of fast scanning radiation sources. Broad spectral coverage and simplicity of the system make the technique applicable to a wide variety of spectroscopic problems.

A block diagram of a typical FASSST spectrometer [16] is shown in Figure 5; its basic components are as follows: an electronically filtered swept DC voltage source (1) controlling the BWO frequency, a fundamental electronic oscillator (3) in the mm and/or sub-mm wave range with high-intrinsic spectral purity, an absorption cell (5), a reference gas cell (6), a resonator (10) generating the interference fringes used for frequency calibration, two liquid-He-cooled InSb detectors for the signal (7a) and reference (7b) channel, A low-noise amplifier (8) with adjustable frequency roll-off, and a computer-controlled data-acquisition system (12).

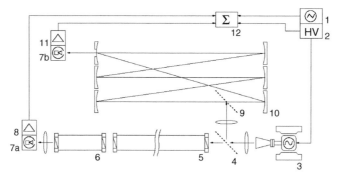

Figure 5 A backward wave oscillator (BWO) FASSST mm wave spectrometer: (1) Swept DC voltage source, (2) voltage-controlled HV power source, (3) mm wave BWO, (4) beam splitter, (5) sample gas cell, (6) reference gas cell, (7) liquid–He-cooled InSb detector, (8) adjustable bandpass low-noise amplifier, (9) beam splitter, (10) Fabry–Pérot resonator, (11) low-noise amplifier, (12) experiment control and data–acquisition. *Source:* Figure 3 of supplement to Ref. [103].

We will discuss the key characteristics of the spectrometer components: BWOs (3), which have traditionally been used in high-resolution phase-locked implementations [7], are available for all of the mm wave range and the sub-mm wave range up to ~1000 GHz. Operating in specific waveguide bands, several BWOs are needed to cover the entire frequency region. By applying a triangular waveform from an electronically filtered voltage source (1), the output frequency can be swept rapidly. Operated in free-running mode, they provide a short-time spectral purity of ~10 kHz. Thus, because the Doppler-limited line width in this frequency region is typically >100 kHz, the spectral absorption features are not significantly broadened. A sweep rate of 10–20 GHz/s freezes the slow fluctuations in the BWO's frequency output on the time scale of a sweep between two interference fringes. Furthermore, at high rates, the sweep across a single-absorption line results in a signal recovery frequency that largely eliminates the $1/f$-noise and thus the need for a superimposed high modulation frequency. A horn antenna is attached to the output waveguide of the BWO. A set of dielectric lenses is used to improve the collimation of the microwave beam.

To achieve an accurate frequency calibration, the well-defined line positions of a sample, e.g., sulfur dioxide SO_2 contained in a reference gas cell (6), are utilized. The known frequencies of the reference gas lines are used to accurately determine the mode spacing which is on the order of 5–10 MHz and the absolute frequency of the fringes generated with a folded Fabry–Pérot resonator (10). Interpolation between the positions of adjacent fringes, relative to the position of an unknown absorption line, yields the calculated line position. To enable heating or cooling of the absorption cell (5), the use of an aluminium pipe rather than a glass pipe is advantageous. In this case, care has to be taken against reflections from the aluminum cell wall to minimize the influence on the baseline signal.

Because, opposed to a widely held opinion, the background noise is very low in the mm wave and sub-mm wave range, very sensitive CW detectors can be utilized [17]. Sensitive liquid-He-cooled hot-electron InSb detectors (7) offer a frequency response $f = \sim 1$ MHz, depending on the specification. This corresponds to a maximum of $f/2 = \sim 5 \times 10^5 s^{-1}$ resolved spectral features while scanning. A typical Doppler-limited line width of 0.1–1 MHz then translates into a maximum scan rate of 50–500 GHz/s. The InSb bolometers are used in conjunction with low-noise preamplifiers (8, 11). In the case of the signal channel of the spectrometer, the amplifier (8) is supplemented with an adjustable bandpass amplifier. The low-frequency roll-off is set such that it transforms the Gaussian line shape absorption function approximately into its first derivative, thereby decreasing the baseline variations and eliminating most of the $1/f$ noise. The high frequency roll-off is set to decrease the high frequency noise. For the reference channel of the spectrometer, the amplifier (11) has bandpass characteristics extending from DC to 1 MHz.

The computer-controlled data acquisition (12) is triggered by the swept DC high-voltage source (1). The data-acquisition system digitizes the input from four sources: (i) the signal channel, (ii) the reference channel, (iii) the high-voltage ramp signal, and (iv) the trigger signal. The trigger channel information is used to identify the margins of the sweep. The initial estimate of the frequency range is obtained from the data of the high-voltage ramp channel by comparison to the stored BWO frequency–voltage characteristics. Because the frequency–voltage function depends critically on the temperature of the BWO, each sweep has to be calibrated separately using the following steps before individual sweeps are co-added: The spectral channel is convoluted with a first derivative Gaussian line shape to transform the signal of the spectral line shape function into an approximate second derivative line profile, thereby reducing the baseline signal further. The calibration lines can be identified in the spectrum by using cross-correlation techniques and are used to calculate the frequencies of the cavity fringes and the frequency position for each data point in the spectrum data array.

Finally, a linearized spectrum is calculated via interpolation. As an example, Figure 6 shows a portion of the spectrum of diethyl ether around 208 GHz [16] recorded at room temperature. Because the entire spectrum in the frequency range 108–366 GHz contains as much as $\sim 2.58 \times 10^6$ resolution elements (10 data points/ MHz), it is not possible to display a complete spectrum graphically. The transition frequency accuracy of the spectra obtained with the FASSST method and from the traditional PLL technique is quite comparable. However, comparison of the signal peak frequencies from FASSST and PLL measurements reveals that time delays and phase shifts caused by the data-acquisition hardware will not completely cancel out, if the sweep rate varies and if the widths of the sample and of the reference lines are not identical. Thus, the spectral positions of the second-derivative lineshapes can be systematically affected, and averaged up- and down sweeps are necessary to eliminate the deviations.

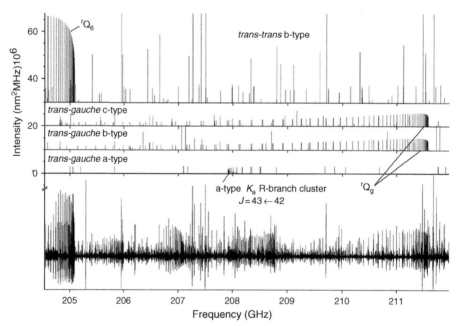

Figure 6   Part of the mm wave spectrum of diethyl ether obtained with a FASSST spectrometer for the frequency range of 108–366 GHz: The top four traces represent the theoretical predictions for both trans–trans and trans–gauche conformer transitions. The bottom trace shows the experimental spectrum. (From Ref. [16]. Copyright 2004, Elsevier Inc.)

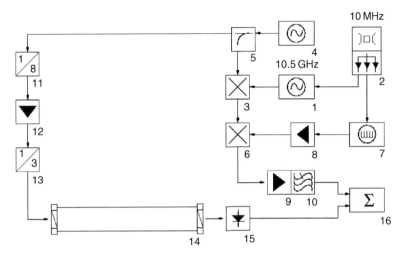

Figure 7 A solid–state FASSST mm wave spectrometer: (1) 10.5 GHz frequency reference, (2) frequency standard and distribution, (3) mixer, (4) cm wave VTO, (5) coupler, (6) mixer, (7) RF comb generator, (8) RF amplifier, (9) RF amplifier, (10) low-pass filter, (11) active transistor multiplier, (12) transistor power amplifier, (13) diode multiplier, (14) absorption cell, (15) Schottky–diode detector, (16) experiment control and data acquisition. *Source*: Figure 4 of supplement to Ref. [103].

Although very simple in comparison to PLL microwave spectrometers and systems based on FIR laser sources, a key element of the BWO implementation of the FASSST instrument is a large (∼40 m) Fabry–Pérot resonator [14]. The development of a compact FASSST spectrometer, shown in Figure 7, is accomplished with an all solid-state-devices-based approach using a scanning electronic system [18]. Here, a voltage-tunable oscillator (VTO) in the cm wave range (4) is multiplied by an active transistor multiplier (e.g., ×8) to produce mm wave radiation (11), which is subsequently amplified with a transistor power amplifier (12) and multiplied with a diode multiplier (e.g., ×3) to produce a signal of high spectral purity ($<1/10^7$) and high brightness in the mm wave and sub-mm wave region (13). After passing through the sample cell (14), the signal is detected with a Schottky diode (15). The high sweep rate (∼15 GHz/s) of the fully electronic frequency scan freezes any drift associated with the VTO. Fixed frequency points are provided by down-converting (3) the VTO frequency to <500 MHz and subsequent comparison (6) to a 10 MHz comb (7). With a ×24 overall multiplication of the VTO output, the reference points are spaced by 240 MHz at the measurement frequency, which is sufficient for an interpolation of the line positions to obtain the typical Doppler-limited measurement accuracy in this spectral range, i.e., <1/10 of the linewidth or ∼50 kHz, typically. To take advantage of the sensitivity due to the favorable molecular absorption strength in the mm wave range and fully utilize the speed and frequency coverage of the system, it is important to have bright sources that provide enough power within the Doppler linewidth to closely approach molecular saturation within the time of radiative interaction. Because of the advances in semiconductor technology, suitable solid-state amplifiers are

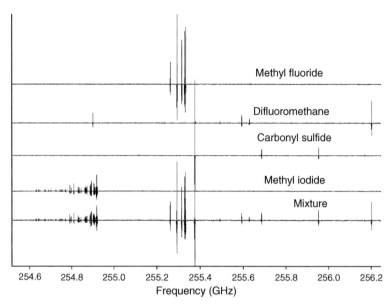

Figure 8 Spectra of four individual substances ($p_i = \sim 0.3$ Pa) and their mixture taken in ~ 0.1 s with a solid-state FASSST spectrometer for the frequency range of 246.8–261.2 GHz. (Reused with permission from Ref. [18]. Copyright 2005, American Institute of Physics.)

available in the mm wave and sub-mm wave spectral ranges. The information content and high specificity for analytical purposes, and the speed of the solid–sate FASSST system are illustrated in Figure 8, showing spectra taken within ~ 0.1 s.

2.2. Continuous radiation microwave excitation on supersonic expansions

The assignment of rotational spectra at room temperature can be very difficult or even virtually impossible if the spectral line density is very high. This situation is frequently encountered for heavy molecules, where the rotational constants are small and the spacing of the rotational transitions in the microwave spectrum becomes narrow. Consequently, the rotational population is distributed over many states, decreasing the intensity of a given transition. Furthermore, if conformational equilibria, vibrational satellites, or tunneling splittings due to large-amplitude motions are present, an easy assignment of the spectrum is usually prevented. The presence of nuclear hyperfine structure further complicates the spectrum, especially if the nuclear quadrupole coupling is either very large or results in a splitting that is on the same order as that of another pattern in the spectrum. In addition to the problem of congestion, the intensity is further distributed among the increasing number of levels from which transitions occur.

Finally, large molecules are typically not very volatile and heating of the sample is required to obtain a suitable gas–phase concentration. An even more severe spread of the rotational and vibrational level populations results as the partition function rapidly increases with temperature and the individual line intensity in the spectrum diminishes seriously.

2.2.1. Phase-locked sweep

Most of these problems can be overcome by the use of a gas jet into which the sample is seeded. Adiabatic cooling of the rotational and vibrational degrees of freedom in a supersonic-jet expansion greatly simplifies the rotational spectrum. While increasing the intensity of the low-J transitions, high-J transitions are eliminated along with most vibrational satellites and higher energy conformers. Thus, assignment usually becomes straightforward. The advantages of the jet-expansion technique in microwave absorption experiments have been demonstrated for an uncollimated free jet that propagates between two parallel plates to which an electric field for Stark-effect modulation is applied [19,20].

The jet expansion and mm wave absorption chamber of a spectrometer capable of Stark-effect modulation, as well as MW–MW or MW–RF double resonance (DR) experiments, is depicted in Figure 9 [21]. The vacuum chamber, 25 cm in diameter and 25 cm in length, is equipped with a $10''$ oil-diffusion pump (see Figure 9b) of 4000 l/s capacity. A liquid-N$_2$-cooled chevron baffle prevents back streaming into the vacuum chamber to avoid arcing of the high-voltage electrodes for Stark-effect modulation. Nozzles of typically 0.1–0.3 mm diameter are used with Ar as carrier gas at stagnation pressures of 10–100 kPa. These conditions result in a background pressure of $\sim 5 \times 10^{-4}$ kPa.

The vacuum chamber has four input ports for MW radiation to interact with the jet expanding along the x-axis (Figure 9a). A horn/dielectric lens system along the y-axis, which forms the absorption cell, focuses the radiation onto the x-axis (Figure 9a and b) with a beam waist of ~ 2 cm. The position of the jet source on this axis can be adjusted to optimize the mm wave absorption signal. Two metal plates, at a distance of 5 cm normal to the z-axis (Figure 9b), constitute the high-voltage electrodes for Stark-effect modulation of up to 3 kV at a frequency of 33 kHz. The Stark-effect electrodes can be removed to use a semiconfocal

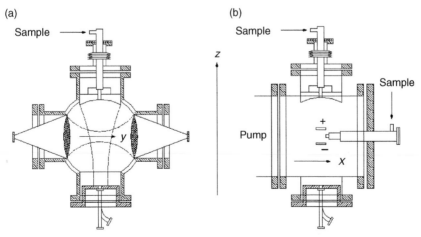

Figure 9 Supersonic-jet expansion mm wave absorption cell chamber with vertical semi-confocal Fabry–Pérot type resonator and horizontal conical horn/dielectric lens arrangement: (a) y,z cross section; (b) x,z cross section. *Source*: Figure 5 of supplement to Ref. [103].

Fabry–Pérot-type resonator along the z-axis, confining a resonant radiation field that crosses the x-axis. The resonator setup allows for a coaxial free jet expansion originating at the center of the spherical reflector, which can be electromechanically adjusted to be at a resonance position (Figure 9a). A mm wave synthesizer in the frequency range of 53–78 GHz is used as the radiation source. The transmitted signal is detected by Schottky diodes feeding the modulated DC conversion signal to a lock-in amplifier. The spectral example given in Figure 10 shows the syn-conformer of allyl alcohol [21] exhibiting μ_c-type lines, widely split because of the torsion of the hydroxyl group, preventing its earlier assignment in measurements at room temperature.

For MW–MW DR experiments, the horn/dielectric lens combination and the interferometer are used simultaneously, the latter usually providing the DR pump radiation. MW–RF DR experiments are performed by applying an amplitude-modulated radiofrequency (RF) signal of up to 150 MHz at the high-voltage electrodes.

If the stagnation pressure is sufficiently high, typically >100 kPa depending on the nature of the carrier gas and the molecule under investigation, the formation of molecular complexes occurs in the jet. To ensure optimal expansion conditions also at higher stagnation pressures, i.e., maintain the low background pressure for a given pump capacity, the nozzle diameter can be reduced. However, the required nozzle diameters can become impractically small. Therefore, in many cases, a pulsed

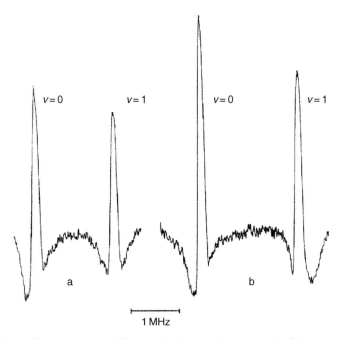

Figure 10 Ground state, $v = 0$, and first torsional excited state, $v = 1$, of the syn-conformer of allyl alcohol: (a) $J(K_a,K_c) = 8(0,8) \leftarrow 7(0,7)$ at 77287.98 and 77289.90 MHz, respectively; (b) $J(K_a,K_c) = 8(1,8) \leftarrow 7(1,7)$ at 76531.24 and 76533.64 MHz, respectively. (From Ref. [21]. Copyright 1995, Elsevier Inc.)

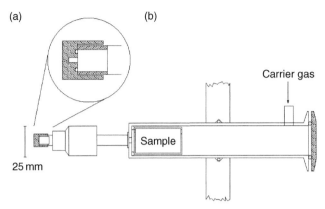

Figure 11 Heated source for continuous and pulsed supersonic-jet expansions: (a) pin-hole nozzle head; (b) electromagnetic valve assembly with sample reservoir. *Source*: Figure 6 of supplement to Ref. [103].

jet expansion is favorable [22] for the observation of molecular complexes. A source suitable for pulsed supersonic-jet expansions that can be electrically heated and temperature controlled is shown in Figure 11. Its head (Figure 11a) consists of a cap with a nozzle diameter of ~0.3 mm, which is screwed onto an unmodified (except for the thread) automobile fuel injector. The source is mounted on a stainless steel tube that can also be heated. It can hold a container for condensed phase samples (Figure 11b). Typical pulse frequencies are 5–10 Hz at 0.5–0.1 duty cycles. Typically, rotational temperatures of 5–10 K are obtained for expansions at sample mole fractions of ~1% in Ar. Higher expansion temperatures are accessible by using larger sample concentrations.

In CW experiments, the pulsed free jet expansion results in an amplitude modulation of the absorption signal with the pulse frequency. Stark-effect modulation can be used simultaneously in a doubly phase-locked detection scheme, virtually removing all baseline artifacts [22]: (i) lock-in detection at 33 kHz using a time constant sufficiently short to transmit the pulse modulation and (ii) lock-in detection of the transmitted signal at the pulse frequency using a time constant suitable for the speed of the frequency scan.

Supersonic-jet sources have also been combined with sub-mm wave spectrometers [23]. In the spectrometer shown in Figure 12, the jet is probed by the radiation of tunable BWOs (5) operating between 300 and 1200 GHz. The probing MW beam intersects the molecular jet six times in a multipass optical array (8), resulting in a ~30 cm overall absorption pathlength. The spectrometer can be operated in two modes: a PLL (4) mode for highly precise measurements over a small range of up to 500 MHz (typically 10–20 MHz) and a free-running fast-scanning mode, which covers a large frequency range of several GHz but with less accuracy. The fast-scanning mode is used for searching the absorption lines of spectroscopically little known or even completely unknown species, whereas the frequency-stabilized mode is used for highly precise measurements (10 kHz accuracy) on selected, identified lines and for long integration times to detect weak signals. Very recently,

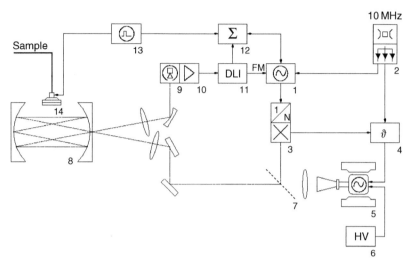

Figure 12 Supersonic-jet backward wave oscillator (BWO) phase-locked loop (PLL) sub-mm wave spectrometer with multipass optical array for use up to 1.2 THz: (1) mm wave synthesizer, (2) frequency standard and distribution, (3) harmonic mixer, (4) PLL feedback circuit, (5) sub-mm wave BWO, (6) HV power supply, (7) beam splitter, (8) multipass optical array, (9) liquid-He-cooled InSb detector, (10) low-noise amplifier, (11) digital lock-in amplifier, (12) experiment control and data acquisition, (13) pulse generator, (14) slit nozzle. *Source*: Figure 7 of supplement to Ref. [103].

supersonic-jet sources have also been used in combination with sub-mm wave spectrometers based on the SL multiplier BWO PLL implementation up to 2.7 THz [11].

2.2.2. Digital scan

Recently, there have also been considerable efforts to combine fast scanning mm wave and sub-mm wave absorption spectrometers with pulsed jets expanding collinearly with the propagation axis of the electromagnetic field [24]. While primarily directed toward the spectroscopic characterization of weakly bound complexes, many of these efforts are also aiming at general improvements in resolution, accuracy, and sensitivity. The pulsed supersonic expansion implementation of the FASSST shown in Figure 13 differs considerably from the stationary gas counterpart [25]. The source of the sub-mm wave radiation is a BWO (7) tunable by a high-voltage power supply (8) in the kV range with mV stability. The IF for the PLL is generated in a quasi-optical broadband sub-mm wave multiplier mixer (4) driven by the BWO (7) and fed by the signal of a highly stable phase-locked oscillator mm wave synthesizer (1, 2). The synthesizer is referenced to a 10 MHz Rb standard (3). The IF signal centered at 35 MHz is compared with the signal of an RF synthesizer (5) referenced to the same time standard (3). The frequency of the BWO (7) can then be rapidly digitally scanned (10 μs/step) in a range of ±5 MHz by incrementing the output of the synthesizer (5) within 30–40 MHz with

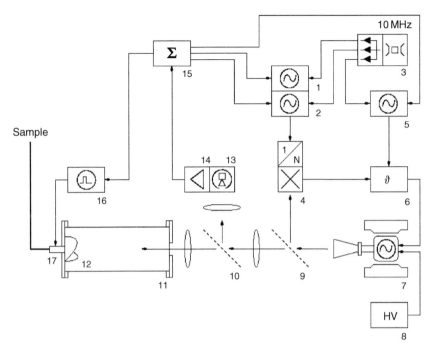

Figure 13 A collinear supersonic-jet expansion backward wave oscillator (BWO) phase-locked loop (PLL)/FASSST mm wave spectrometer: (1) RF synthesizer, (2) phase-locked mm wave BWO, (3) frequency standard and distribution, (4) harmonic mixer, (5) RF synthesizer, (6) PLL feedback circuit, (7) mm wave BWO, (8) HV power supply, (9) beam splitter, (10) wire-grid polarizer, (11) supersonic-jet expansion chamber, (12) roof-top reflector, (13) liquid-He-cooled InSb bolometer, (14) low-noise amplifier, (15) experiment control and data-acquisition, (16) pulse generator, (17) pin-hole nozzle. *Source*: Figure 8 of supplement to Ref. [103].

step sizes selectable from 10 Hz–100 kHz. The start of each fast frequency scan is synchronized with each supersonic-jet pulse (16). Continuous broadband scans of the BWO (7) frequency are accomplished by appropriate frequency steps of the mm wave synthesizer (1, 2) to sequentially acquire 10 MHz frequency segments. To achieve increased sensitivity, if required, fast frequency scans of several jet pulses can be averaged for each frequency segment in the computer-based data-acquisition and instrument control system (15).

A pulsed nozzle (17) with an orifice of typically 0.5 mm diameter is used, while a background pressure of $<2 \times 10^{-5}$ kPa is maintained. Directed through a wire-grid polarizer (10), the BWO radiation used for molecular excitation is focused on the orifice of the jet source, located in the center of a roof-top reflector (12), such that the field propagates collinearly with the supersonic-jet expansion. With the apex of the roof-top reflector (12) being rotated by 45° with respect to the polarization axis, the polarization direction of the reflected radiation is rotated by 90° relative to the polarization axis of the incident radiation. The backward propagating mm wave and sub-mm wave radiation is then efficiently reflected by

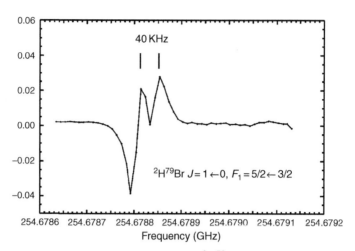

Figure 14   Spectrum of the $J = 1 \leftarrow 0$ transition of the $^2H^{79}Br$ isotopologue of hydrobromic acid, obtained with a collinear supersonic-jet expansion backward wave oscillator (BWO)/phase-locked loop (PLL) FASSST spectrometer for the frequency range of 236–540 GHz, with the deuterium hyperfine components $F_1, F = 5/2,3/2 \leftarrow 3/2,1/2;\ 5/2,7/2 \leftarrow 3/2,5/2$ partially resolved from the $F_1, F = 5/2,5/2 \leftarrow 3/2,1/2;\ 5/2,5/2 \leftarrow 3/2,5/2$ components. (Reused with permission from Ref. [25]. Copyright 2003, American Institute of Physics.)

the wire-grid polarizer (10) and subsequently focused onto a liquid-He-cooled InSb bolometer (13). The detector registers the absorption signal consisting of two Doppler-shifted components, one for the incident and one for the reflected radiation of the scanned BWO (7), which is resonant to a transition of the molecules traveling with the jet. A round-trip sweep then gives four frequency components whose arithmetic mean is the molecular transition frequency. Figure 14 demonstrates the resolution that can be achieved with a collinear supersonic-jet expansion BWO PLL/FASST spectrometer. From previous data [26], the splitting between the components in the observed doublet should be 37.0 kHz.

3.   TIME-DOMAIN (STATIONARY FREQUENCY/IMPULSE) TECHNIQUES

Instead of continuously passing monochromatic radiation through a sample cell and detecting the transmitted signal as a function of frequency, contemporary microwave spectrometers apply the radiation for a short period of time [27]. In the presence of a sample, a radiative response is induced. To obtain the spectrum, the response signal is recorded as a function of time and subjected to Fourier transformation (FT).

The trend in microwave spectroscopy toward more sophisticated experiments such as time-domain spectroscopy (TDS) instruments was initiated by the endeavor to detect the rotational spectra of ions or short-lived reaction products, which made the development of more sensitive detection systems mandatory. The goal of higher sensitivity in steady-state microwave spectroscopy has been achieved with superheterodyne detection systems together with very low-noise amplifiers. To fully exploit the capabilities of modern high-frequency electronics, additional measures such as the use of a balanced bridge system [28] and additional Stark-effect modulation for baseline stabilization had to be taken. In the detection scheme of a bridge system, the optimum noise figure may well be limited by source noise from the microwave oscillator due to imperfect balancing of the bridge. For high-resolution work, care has to be taken to minimize power and modulation broadening.

The time-dependent behavior of absorption and emission of two-level quantum mechanical systems makes it possible to measure rotational transitions in the time domain, analogous to the pioneering development of pulsed nuclear magnetic resonance (NMR) experiments. Transient experiments have been familiar in NMR for two decades before their routine introduction in MW spectroscopy. Normally, the relaxation times in magnetic resonance experiments are $10^6 - 10^7$ times longer than in gas-phase rotational systems. Thus, the radiation switching and speed of detection requirements in rotational spectroscopy are much more severe, and this has led, in part, to the long delay in developing transient experiments in rotational spectroscopy.

Because the detection of the signals takes place in the absence of any power from the microwave oscillator, the need for a balancing bridge is eliminated and an empty waveguide or any other suitable container can be used as sample cell. As a result, the signal-to-noise (S/N) ratio is unaffected by any source noise. Another advantage over steady-state measurements is the considerably increased resolution because of the total absence of any power or modulation broadening.

Microwave molecule resonant transient phenomena are centered on two main processes: *transient absorption*, which is normally called transient nutation (TN) in NMR experiments, and *transient emission*, which is normally called the free-induction decay (FID) in NMR experiments.

Transient absorption occurs when a two-level system is driven—in a time which is short compared to the relaxation processes—from stationary conditions characterized by the equilibrium population difference ΔN^0_{ab} and negligible polarization to a new state that features a macroscopic polarization $\boldsymbol{P}$ at a non-equilibrium population difference ΔN_{ab}. During this process, energy is taken from the radiation field to increase the energy of the molecular system. Thus, we refer to this process, which has involved a net absorption of energy, as transient absorption.

Transient emission occurs when the system is taken from a condition of interaction with the radiation, where the system is polarized and not in thermal equilibrium, to a condition, where the external radiation is either removed or at least taken far off-resonance—and thus out of interaction—with the molecular two-level system. Some of the energy stored in the molecules is subsequently released by spontaneous coherent emission.

The rates and behavior of the system as it changes toward a state of non-thermal equilibrium ΔN_{ab} and macroscopic polarization $\boldsymbol{P}$, and when it relaxes from this state after the radiation–molecule interaction is terminated, are determined by solving a system of coupled differential equations, the optical Bloch equations, thereby extracting the phenomenological relaxation constants T_1 and T_2, respectively. In NMR, T_1 is called the longitudinal or spin–lattice relaxation time. It represents the inverse of the rate at which energy is exchanged between the spin system and the environment. Similarly, because energy is stored in the form of a population difference in microwave rotational experiments, $1/T_1$ in this case is also the rate at which this stored energy is released to the environment. T_2 in NMR is called the transverse or spin–spin relaxation time. It represents the rate at which the individual spins lose coherence. Very often in NMR, the apparent T_2 contains a significant contribution due to inhomogeneity of the magnetic field. If each spin is subjected to a slightly different field at its local position, it will evolve in time slightly differently, leading to a loss of coherence. Obviously, this situation has its MW analog in the application of an inhomogeneous electric field to the sample. In addition, inhomogeneous broadening in microwave spectroscopy is caused by the Doppler effect, resulting from the fact that the interaction of the field with each dipole is slightly different depending on its individual velocity. In the experiments discussed here, the velocity dependence of the radiation molecule interaction is usually negligible, so that T_2 can be interpreted as a loss of coherent polarization because of molecular interactions.

The signal at the output of the super-heterodyne detector arises from a beat between the radiation field coherently emitted by the molecular system and the signal of the microwave oscillator. The signal obtained is subsequently Fourier transformed to give the spectrum of the transitions that were polarized. In comparison to absorption methods, Fourier-transform microwave (FT-MW) techniques offer the same advantages as experienced in FT-NMR.

3.1. Pulsed radiation microwave excitation of stationary gases

Early transient microwave experiments have been performed by bringing an ensemble of two-level quantum mechanical systems into or out of resonance with a radiation field by switching the electric field in a conventional Stark-effect modulation cell [29]. Here, we describe the much more suitable approach of sending a high-power radiation pulse through a waveguide cell. Although the principles of coherent transients have already been discussed in the early years of microwave spectroscopy [30], the technique of TDS as a tool for the investigation of molecular rotational spectra was established only about two decades later [31]. Because of the short relaxation times at usual gas sample pressures of 10^{-4}–10^{-2} kPa, which are typically on the µs or even sub-µs time scale, this achievement was only possible with the fast electronic devices becoming available at that time. Improvements, further developments, and numerous applications of FT-MW spectroscopy to rotational spectra of static gases were reported later on [32,33].

The majority of spectroscopic studies employing FT-MW techniques are based on a pulsed single-resonance interaction of molecular two-level systems with the microwave radiation field. The theoretical description of the phenomenon is based on the density matrix formalism that models the sample of rotating molecules as an ensemble of two-level systems. The time evolution of the gaseous molecular ensemble, interacting resonantly or near resonantly with the pulsed external MW radiation, was first evaluated in a treatment considering a homogeneous z-polarized field [34]. Extensions to wave propagation in rectangular and circular waveguides were reported later to account for lineshape effects due to inhomogeneous propagation mode profiles and wall collisions [35,36]. The analysis is based on the optical Bloch equations, which describe the time evolution of the diagonal and off-diagonal density matrix elements corresponding to the populations N_a, N_b of the two levels, and their two-level coherence, respectively. For an isolated two-level single-resonance experiment, the coherence term is proportional to the pulse-induced sample polarization P which, as follows from Maxwell's equations, is the source of the molecular field E of the emitted MW radiation.

The magnitude of the macroscopic oscillating dipole moment, i.e., the sample polarization and thus the amplitude of the corresponding microwave field emitted at the end of a MW pulse, depends on the coupling strength of the molecular dipoles with the external radiation [34]. For a transition b←a with the electric dipole matrix element μ_{ab} of the space-fixed z-component of the dipole operator, this coupling strength is characterized by the Rabi angular frequency $x = \mu_{ab}\varepsilon/\hbar$ where ε is the z-polarized electric field amplitude of the external radiation. The polarization achieved depends on x, the MW pulse length τ_p, and the off-set $\Delta\omega = \omega_{ab} - \omega_e$ from resonance. Effects from collisional damping and deviations from resonance during the pulse can be neglected for strong pulses ($x \gg \Delta\omega$) sufficiently short with respect to the relaxation times ($\tau_p \ll T_1, T_2$). For the period following the transient absorption, the amplitude $|S_{ab}| \propto (P_{LO}P_{ab})^{1/2}$ of the molecular signal after super-heterodyne detection with the local oscillator power P_{LO}, being proportional to the square root of the power flow $P_{ab}^{1/2}$ at the end of the sample cell is given by

$$P_{ab}{}^{1/2} \propto |S_{ab}| \propto \sqrt{\frac{ab}{\varepsilon_0 c}} \Delta^1 N_{ab}\mu_{ab}\omega_{ab}L \sin(x\tau_p) \exp\left(\frac{-t}{T_2}\right)$$
$$\times |\exp(i(\omega_{ab}t + \theta_{ab}))| \tag{2}$$

in the case of a rectangular waveguide with the cross-section ab and a length L. ε_0 is the electric field constant, c is the vacuum speed of light, $\Delta^1 N_{ab}$ is the population difference per unit volume of the two levels involved in the transition, and $\omega_{ab} = 2\pi\nu_{ab}$ is their angular resonance frequency. For a circular waveguide of radius R, $a \times b$ is replaced by $2\pi R^2$; propagation losses are neglected. According to Equation (2), the largest signal amplitude S_{ab} is obtained for a "$\pi/2$-pulse", defined by a pulse angle $x\tau_p = \pi/2$. For waveguides, the trigonometric function $\sin(x\tau_p)$ is replaced by a Bessel function of the first kind $J_1(\Theta)$ with a more complicated argument, having its first maximum at $\Theta = 1.871$, and only "effective $\pi/2$-pulses" can be achieved because of the electric field inhomogeneity due to the spatial distribution of the propagation mode. If more than one transition is in near resonance, the right-hand side of Equation (2) is replaced by the sum over the signal contributions from each transition. The phases of the individual signals are generally unknown, and thus amplitude spectra are normally displayed after FT. The exponential decay of the transient emission time-domain signal due to collision-induced relaxation corresponds to a pressure broadening of the frequency-domain line with a HWHM of $\Delta\nu_{ab} = 1/2\pi T_2$. At low pressures, other broadening mechanisms become important and the decay term in Equation (2) has to be modified accordingly. The significant decay terms need to be considered if the resolution capability of FT–MW on static gases is compared to steady-state techniques, such as Stark-effect modulation MW spectroscopy. Because, as shown below, the sensitivity of FT–MW spectroscopy is inherently larger, quite low sample pressures can often be used and pressure broadening becomes rather unimportant. Because of the lack of modulation and saturation broadening occurring in absorption techniques, the resolution is then only limited by Doppler and wall collision broadening, which depend on the frequency range and cell geometry, respectively [35].

The enhanced resolution of FT–MW spectroscopy translates into a higher precision of observed line center frequencies. For narrow line multiplets, however, the analysis of the amplitude frequency-domain spectra suffers from distortion effects, causing line position shifts as large as the HWHM, because of overlapping lines from signals with *a priory* unknown phases [37]. To recover the frequency information contained in the transient emission, parameters of a model function can be fit to the time-domain signal [38].

The TDS technique is inherently more sensitive than the FDS methods, i.e., the initial transient emission signal amplitude after a "$\pi/2$-pulse" is larger than any steady-state absorption signal [39]. The sensitivity enhancement is particularly pronounced if condition $S_p = xT_2 \gg 1$ is met, ensuring that an "effective $\pi/2$-pulse" $(x\tau_p = \pi/2, \tau_p \ll T_1, T_2)$ converts the initial population difference ΔN_{ab} efficiently into coherence ρ_{ab}. Comparing the resulting signal amplitude $S_{ab}(\mathrm{TDS})$ of the FT technique with the signal amplitude $S_{ab}(\mathrm{FDS})$ corresponding to the steady-state polarization [34], applicable to the CW absorption method, yields

$$\frac{S_{ab}(\mathrm{TDS})}{S_{ab}(\mathrm{FDS})} = \frac{1+S_p^2}{S_p} \tag{3}$$

if identical detection schemes, e.g., super-heterodyne detection, are employed in both experiments. As a consequence, at $S_p \gg 1$, low gas pressures can be employed and, if the incident MW power is sufficiently high, transitions with rather weak dipole matrix elements can still be investigated. The linear dependence of the signal on the dipole matrix element μ_{ab} in FT–MW spectroscopy evident from Equation (2) differs from the quadratic μ_{ab}^2 dependence obtained for CW absorption spectroscopy at low MW power with Equation (1). For the latter method, this means a severe sensitivity disadvantage for small transition dipole moments [32]. For very weak dipole moments, Stark-effect modulation spectroscopy, which is the most sensitive FDS technique to obtain rotational spectra of polar molecules, might not be applicable at all because of insufficient modulation. This breakthrough of TDS does not only extend to the observation of weakly polar species, it also makes the technique suitable for the observation of signals associated with transition dipole matrix elements μ_{ab}, which are small for quantum-mechanical reasons, e.g., the so-called "forbidden" transitions. Because no high-voltage electrode for modulation purposes, which would increase the cell attenuation, is needed for the TDS method, further sensitivity enhancement can be obtained by the use of longer sample cells.

Transient emission microwave signals have been obtained from pulse excitation of the sample by the methods of switching the microwave power [31] or frequency [40]. The power switching technique is achieved with fast pin-diode switches, capable of generating pulses of less than 10 ns duration. Such pulse lengths are sufficiently short with respect to rotational relaxation times, which are in the order of 10 μs at gas sample pressures of 10^{-4} kPa. The emission signals are rather weak, ranging from 10^{-12} W for the strongest transitions to less than 10^{-20} W at the detection limit [41]. For weakly polar species, high pulse power even beyond the W range is needed to create an "effective π/2-pulse." Super-heterodyne detection is necessary to record the weak signals and phase synchronous addition of the digitized transient emission signals from consecutive pulse excitations is performed to obtain a sufficient S/N ratio. For a given excitation bandwidth, typically on the order of several 10 MHz, according to the Nyquist theorem, the analog-to-digital (A/D) conversion rate must at least be equal to the bandwidth for a dual-channel vector or two-fold higher for single-channel scalar detection.

Covering the frequency range 1–40 GHz, various pulsed radiation designs of static gas FT-MW spectrometers in the cm wave range are employing coaxial (1–4 GHz) and rectangular or circular (4–40 GHz) waveguides [31–33]. A mm wave range implementation utilizes the frequency switching technique [40]. At such shorter wavelengths, free-space sample cells can also be used.

A state-of-the-art static gas waveguide FT-MW spectrometer depicted in Figure 15 comprises key components operating in the following way: A primary

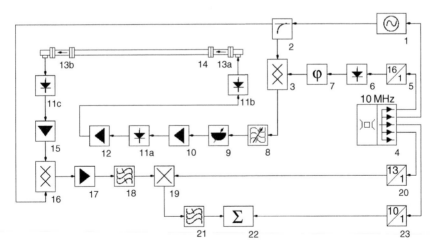

Figure 15 Static gas waveguide FT-MW spectrometer: (1) MW synthesizer, (2) MW directional coupler, (3) MW SSB mixer, (4) frequency standard & distribution amplifier, (5) RF multiplier, (6) RF GaAs switch, (7) RF bi-phase modulator, (8) MW YIG tunable bandpass filter, (9) MW variable attenuator, (10) MW solid-state amplifier, (11) MW pin-diode switch, (12) MW TWT amplifier, (13) MW isolator, (14) waveguide sample cell, (15) MW low-noise amplifier, (16) MW IR mixer, (17) RF amplifier, (18) RF bandpass filter, (19) RF double-balanced mixer, (20) RF multiplier, (21) RF low-pass filter, (22) experiment control and data acquisition, (23) RF multiplier. *Source*: Figure 9 of supplement to Ref. [103].

frequency source (1), typically a MW synthesizer or a MW oscillator phase-locked to a harmonic of an RF synthesizer, provides coherent radiation with an output power of typical 10 mW. The synthesizer, like all frequency signals and clock rates of the spectrometer, is referenced to a single 10 MHz time standard (4), either a high-quality atomic clock or a global-positioning system (GPS) receiver. The primary frequency source is used as the local oscillator at frequency $\nu_e - \delta$ for the super-heterodyne detection. An MW single-sideband (SSB) modulator (3), typically accepting a modulation signal $\delta = 160$ MHz, produces the frequency ν_e of the radiation field used for polarization of the molecular ensemble. An MW yttrium–iron–garnet (YIG) tunable bandpass filter (8) suppresses any leakage signals of the carrier frequency and residuals of unwanted sidebands. An MW solid-state high-power amplifier (10) is used to increase the power of the excitation signal. An MW pin-diode switch (11a) suppresses a residual external signal during detection of the transient emission. An optional MW very-high-power amplifier (12), e.g., a traveling wave tube amplifier (TWTA), further amplifies the external signal if necessary. An RF GaAs switch (6) introduces the power modulation of the signal radiation. An RF bi-phase modulator (7) is employed for 0°/180° phase-inversion-keying (PIK), which effectively reduces coherent leakage signals in an addition/subtraction scheme. An MW pin-diode switch (11b) blocks the considerable TWTA idle noise during detection of the transient emission. Two MW isolators (13) are necessary for the elimination of pulse reflections, which otherwise would bury the weak transient emission. Alternatively, in coaxial waveguides at low frequency, the ferrite isolators (13) and single-pole single-throw (SPST) pin-diode switches before (11b) and after (11c) the cell are replaced with single-pole double-throw (SPDT) pin-diode switches that guide the external radiation into a termination. The sample cell (14) is typically a rectangular waveguide of about 10 m length with a cross-section permitting only the propagation of the lowest transverse-electric TE_{10} mode for the frequency band specified. A bridge-type construction also allows the use of oversized rectangular waveguides [42]. Using circular waveguides, even longer cells (>20 m, typically) can be employed by exploiting the very-low-loss circular TE_{01} mode [43]. An MW pin-diode switch (11c) protects the detection system from the high-power polarization MW pulses. An MW low-noise front-end amplifier (15) boosts the weak transient emission, such that the signal level stays significantly above the thermal noise level after conversion losses and attenuation before a subsequent amplification stage is invoked. Thus, the initial S/N level at the input of the front-end amplifier is maintained. An MW image-rejection (IR) mixer (16) converts the signal to an IF signal band centered at the sideband frequency $\delta = 160$ MHz. An RF bandpass amplifier (17) prevents the signal band at δ from dropping to the thermal noise level. An RF double-balanced mixer (19), accepting a modulation signal of $\delta - 30$ MHz, is used to convert the IF band to the final signal band centered at 30 MHz. An RF low-pass filter (21), limiting the signal band to 50 MHz, serves as anti-alias filter for signals and noise outside the Nyquist range. Accordingly, an A/D conversion rate of 100 MHz, corresponding to a sampling interval of 10 ns, is required. An instantaneously re-triggerable transient recorder and real-time signal averager (22) is used for phase synchronous addition of the weak digitized signals following consecutive pulse excitations to obtain a sufficient S/N ratio. For maximum dynamic range, the input sensitivity of the A/D converter should be such that noise fluctuations are just flipping the logic state of the least significant bit. High-bit converters with a wide input range offer a better dynamic range but do not enhance the sensitivity. A workstation is used to control the experiment as well as to store, display, and analyze the data after FT. The entire sequence of MW and A/D-trigger pulses is initiated from the record and average system (22) utilizing a computer-controlled pulse generation circuit that is referenced to the 10 MHz time standard (4).

3.2.   MW-MW and MW-RF double resonance

Standard spectroscopic studies employing FT-MW techniques are based on a short single-resonance interaction of molecular two-level systems with the microwave radiation field. Spectra in a single-frequency domain of typically several 10 MHz width are obtained after FT of the transient molecular emission MW signal recorded in the time domain subsequent to the pulse-induced polarization of the gas sample. To obtain spectra with sufficient S/N, for the majority of molecules 10^6 transient emission signals are typically required to be added prior to FT. To cover broad frequency ranges, the MW synthesizer is appropriately stepped for sequential frequency segments. If the molecular sample is subjected to a second radiation field, the FT-MW technique can serve as a detector for molecular resonances introduced by its presence.

3.2.1.   One-dimensional DR spectroscopy

DR techniques were already introduced in the early period of impulse FT-MW spectroscopy on static gases as an extension to single-resonance methods. Such experiments were first only employed to facilitate an easy assignment of rotational transitions by simplifying the one-dimensional (1D) spectra. DR methods applying CW radiation for a second resonance were soon extended to experiments based on excitation pulse trains for both "signal" and "pump" resonance frequencies [44]. Various modulation schemes for molecular three-level systems were developed for this purpose, utilizing a second resonance not only in the MW [44] but also in the RF [45] range. In most cases, 0°/180° PIK or on/off-modulation of the excitation signal of the second resonance, while taking the difference from two consecutive sequences, is employed [46] to select transient emission signals belonging to three-level systems that share a common DR transition.

   The high sensitivity of pulsed DR methods employed in 1D FT-MW techni-ques is illustrated in Figure 16, showing the DR signal obtained for the $\Delta k = \pm 3$ "pump" and "signal" transitions of phosphoryl trifluoride [47], i.e., a three-level system where both transitions involved are only very weakly allowed.

   The transition frequency of the second resonance is not precisely obtained from 1D DR experiments. The quest of getting access to both the "signal" and the "pump" resonance frequencies at high precision led to the development of more advanced pulsed DR techniques. Various modulation techniques have been explored for short DR interaction, yielding dual frequency-domain spectra.

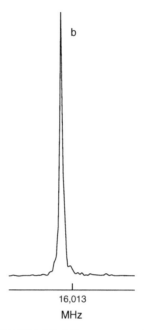

16,013
MHz

Figure 16 One-dimensional FT-MW MW-DR spectrum of phosphoryl trifluoride showing the J, K, $\Gamma = 58, 14, E \leftarrow 58, 11, E$ transition using the J, K, $\Gamma = 58, 11, E \leftarrow 58, 8, E$ transition for MW-DR modulation. The spectrum covers 2.5 MHz. (From Ref. [47]. Copyright 1992, Elsevier Inc.)

3.2.2. Two-dimensional DR spectroscopy

Further developments of pulsed DR techniques allow the simultaneous determination of both resonance frequencies by employing two-dimensional (2D) spectra. A variety of pulse sequences have been created for quite different applications. The various 2D FT-MW DR methods use DRs either in three-level systems [48] or four-level systems [49,50]. The latter DR application exploits collision-induced coupling between individual two-level systems. In general, utilization of 2D FT-MW DR experiments on three-level systems for spectroscopic studies has been employed most frequently.

To date, five different pulsed DR techniques for 2D FT-MW spectroscopy have been demonstrated. The choice of a particular pulse sequence is largely governed by the frequencies of the "pump" and "signal" transition in the three-level system studied [48]. If the excitation bandwidth (<50 MHz, typically) for the "signal" radiation covers the spectral position of the pump resonance, no second "pump" radiation source is required. Accordingly, a pulse sequence employing a single-excitation frequency has been developed to record 2D-autocorrelation spectra [51]. 2D-correlation techniques for well separated DR transitions utilize an excitation sequence with two "pump" pulses and one "signal" pulse [48].

Among the variety of 2D FT-MW DR techniques, the double-quantum (DQ) correlation sequence is the most important so far. As it retains the sensitivity at that of the "signal" transition, it allows for highly sensitive experiments employing "pump" frequencies that are significantly lower than the "signal" frequency. Accordingly, after its original development for MW–DR spectroscopy, this method was extended to RF–DR spectroscopy which is based on a "pump" transition in the RF range [52]. With the 2D FT–MW RF–DR technique, direct l-type doubling transitions of linear molecules [52], low-frequency rotational transitions of asymmetric-top molecules [53], direct nuclear quadrupole hyperfine transitions [54], and direct A_1–A_2 splitting transitions of symmetric-top molecules [55] are accessible.

The theoretical treatment underlying the pulse sequence experiment to obtain 2D FT–MW RF–DR spectra is based on the density matrix formalism applied to an ensemble of three-level systems [52]. As illustrated in Figure 17, the non-degenerate levels are denoted as (a), (b), and (c), where (b) is the level common to the "signal" and "pump" transitions and the "pump" level (c) either connected progressively (see Figure 17a, left), or regressively (see Figure 17b, left), to the "signal" level (a). The resonance off-set for the "signal" and "pump" transitions is given as $\Delta\omega_{ab} = \omega_{ab} - \omega_s$ and $\Delta\omega_{bc} = \omega_{bc} - \omega_p$, respectively, with ω_s and ω_p being the angular frequency of the external "signal" and "pump" z-polarized radiation field, assuming an electric field amplitude of ε_s and ε_p, respectively. The DQ transition c←a, which is not necessarily dipole-allowed, is then characterized by a resonance off-set $\Delta\omega_{ac} = \Delta\omega_{ab} + \Delta\omega_{bc}$ or $\Delta\omega_{ac} = \Delta\omega_{ab} - \Delta\omega_{bc}$ for the progressive or regressive scheme, respectively. The strengths of the electric dipole interaction are characterized by the Rabi angular frequencies $x_s = \mu_{ab}\varepsilon_s/\hbar$ and $x_p = \mu_{bc}\varepsilon_p/\hbar$, with the electric dipole matrix elements μ_{ab} and μ_{bc} of the space-fixed z-component of the dipole, respectively.

By repeatedly applying the pulse sequence shown in Figure 18, a 2D time-domain data array is obtained, which yields the spectrum in two frequency dimensions after 2D FT. Each sequence is assumed to start at thermal equilibrium

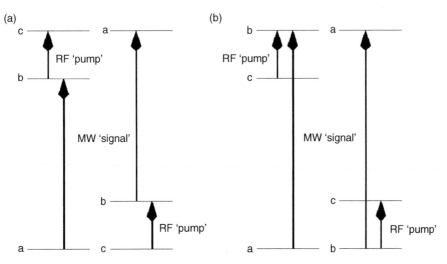

Figure 17 Progressive (a) and regressive (b) three-level system DR energy diagrams. *Source:* Figure H of supplement to Ref. [103].

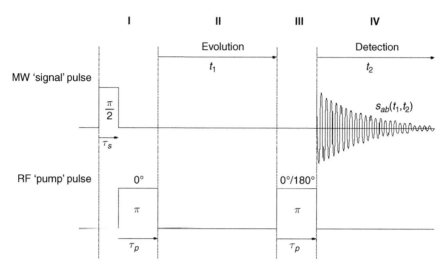

Figure 18 2D FT-MW DR pulse sequence: (I) preparation, (II) evolution, (III) mixing, and (IV) detection. See text for details. *Source*: Figure J of supplement to Ref. [103].

conditions. In the "preparation" period (*I*), the initial "signal" pulse of length τ_s converts the b←a population difference ΔN_{ab} into b←a coherence ρ_{ab} which, according to Equation (2), corresponds to maximum signal S_{ab} after an "effective $\pi/2$-pulse" defined as $x_s \tau_s = \pi/2$. The first "pump" pulse of length τ_p completely converts the b←a coherence ρ_{ab} into DQ c←a coherence ρ_{ac}, if an ideal "π-pulse" defined as $x_p \tau_p = \pi$ is applied. In the "evolution" period (II), the c←a DQ coherence ρ_{ac}, exhibiting an angular frequency of $\omega_{ac} = \omega_{ab} + \omega_{bc}$ or $\omega_{ac} = \omega_{ab} - \omega_{bc}$ for the progressive or regressive scheme, respectively, evolves freely. The delay time t_1 between two "pump" pulses is incremented for consecutive pulse sequences. In the "mixing" period (III), a second "pump" pulse fulfilling ideal "π-pulse" conditions converts DQ c←a coherence ρ_{ac} back into observable b←a coherence ρ_{ab}. As in 1D FT-MW experiments, see Equation (2), the b←a coherence ρ_{ab} is associated with a signal amplitude $|S_{ab}|$. In the "detection" period (IV), the resultant MW radiation at the end of the sample cell is recorded as a function of the recording time t_2. Compared to the 1D techniques, the phase of the transient emission exhibits an additional modulation with respect to the delay time t_1 between the two "pump" pulses:

$$|S_{ab}| \propto \sqrt{\frac{ab}{\varepsilon_0 c}} \Delta^1 N_{ab} \mu_{ab} \omega_{ab} L \sin\left(x_s \tau_s\right) \tag{4}$$
$$\times \exp\left(\frac{-(t_1 + t_2)}{T_2}\right) \exp\left(i(\omega_{ab} t_2 + \Delta\omega_{ac} t_1 + \theta_{ab})\right)|$$

with $\Delta\omega_{ac} = (\omega_{ab} - \omega_s) \pm (\omega_{bc} - \omega_p)$ for progressively (+) or regressively (−) connected three-level systems. As discussed in connection of Equation (2), at low pressures other broadening mechanisms than collision–induced relaxation, e.g., broadening resulting from the signal decay due to "Doppler-dephasing" [52], become important and the decay term in Equation (4) has to be modified accordingly. For multiple three-level systems excited by the radiation pulse, Equation (4) is replaced by the sum over the individual DR signals. Comparing Equations (2) and (4), it can be seen that, if the two "pump" pulses fulfill ideal "π-pulse" conditions, the signal amplitudes—apart from the additional $\exp(-t_1/T_2)$ damping term—are equal, and thus the 2D DQ method should in principle retain the sensitivity of 1D FT-MW spectroscopy. In reality, however, i.e., under actual experimental conditions, the Rabi angular frequencies x_s and x_p exhibit a spatial variation because of propagation mode inhomogeneities as well as additional dephasing due to the M-dependence of the transition dipole matrix elements of the individual M-sublevels. In consequence, the real-world sensitivity of 2D FT-MW spectroscopy for the "signal" transition may be reduced by one order of magnitude with respect to 1D FT-MW spectroscopy. Nevertheless, the sensitivity of 2D FT-MW spectroscopy for the "pump" transition is many orders of magnitude higher than that of direct 1D FT-RF spectroscopy, since—due to the linear frequency dependence of the Boltzmann population difference $\Delta^1 N_{ab} \propto \omega_{ab}$ at small energy separations and the size of standard waveguides $a \times b \propto \lambda_g^2 \propto (c/\nu_g)^2$ at their nominal frequency ν_g—its signal amplitude, see Equation (2), depends linearly on the transition frequency:

$$P_{ab}^{1/2} \propto |S_{ab}| \propto \sqrt{\frac{ab}{\varepsilon_0}}\ \Delta^1 N_{ab}\mu_{ab}\omega_{ab}L \propto \omega_{ab} \tag{5}$$

Because of incomplete conversion into and from DQ c←a coherence ρ_{ac}, additional molecular signals—not containing information about the pump transition frequency via $\Delta\omega_{ac}$—can occur in the 2D spectrum. To suppress such coherent perturbations of molecular origin specific to 2D FT-MW experiments, as well as phase-coherent instrumental artifacts also occurring in 1D FT-MW spectroscopy, various phase sequences that select dedicated coherence pathways have been developed [48,50]. In the case of the 2D FTMW RF-DR experiment discussed here, the DQ c←a coherence ρ_{ac} can be selected by a two-step phase sequence that inverts the "pump" excitation phase in every second repetition and takes the difference from two consecutive sequences [53].

For the most part, the general setup of a 2D FT-MW DR spectrometer is the same as that for 1D FT-MW spectroscopy (Figure 15). To perform the RF-DR experiment discussed here, the FT-MW spectrometer is modified to allow for additional RF pulses down to a frequency of a few MHz [54]. The key components of the bridge spectrometer shown in Figure 19 are as follows: An MW synthesizer (1), an MW SSB modulator (3), a YIG-tuned MW bandpass filter (6), an MW amplifier (8), MW pin-diode SPST switches (9), an MW TWTA (10), a reference cell (11), a sample cell (12), MW phase shifters (13), MW attenuators (14), an MW low-noise amplifier (15), an MW image-rejection mixer (16), an RF double-balanced mixer (19), a transient recorder and signal averager (22), an RF synthesizer (24), an RF bi-phase modulator (25), and an RF power amplifier (25). The RF radiation is generated using an RF synthesizer (24), referenced to the 10 MHz time standard (4). The source is followed by an RF bi-phase modulator (25) to allow for

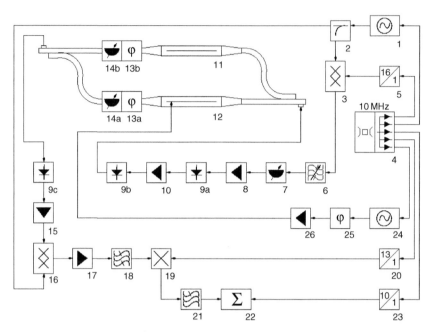

Figure 19 Static gas waveguide 2D FT-MW RF-DR spectrometer: (1) MW synthesizer, (2) MW directional coupler, (3) MW SSB mixer, (4) frequency standard and distribution, (5) RF multiplier, (6) MW YIG tunable bandpass filter, (7) MW variable attenuator, (8) MW solid-state amplifier, (9) MW pin-diode switch, (10) MW TWT amplifier, (11) reference gas cell, (12) sample gas cell, (13) MW phase shifter, (14) MW variable attenuator, (15) MW low-noise amplifier, (16) MW IR mixer, (17) RF amplifier, (18) RF bandpass filter, (19) RF double-balanced mixer, (20) RF multiplier, (21) RF low-pass filter, (22) experiment control and data acquisition, (23) RF multiplier, (24) RF synthesizer, (25) RF bi-phase modulator, (26) RF power amplifier. *Source*: Figure 10 of supplement to Ref. [103].

$0°/180°$ PIK, which effectively reduces the coherent perturbations of molecular origin in the addition/subtraction scheme discussed above. For pulse generation, the fast pulse modulation capabilities of the RF synthesizer output circuit or the RF GaAs switches are used. The pulse sequence is driven by the computer-controlled pulse generation circuit under control of the record and average system (22), which is also referenced to the 10 MHz time standard (4). An RF amplifier (26) increases the "pump" signal to a power level of up to 20 W before it is fed into the sample cell. Typically, oversized rectangular waveguides (11, 12) with high-voltage electrodes used for Stark-effect modulation are utilized. To minimize artifacts from the high-power MW radiation, the depicted bridge-type spectrometer is tuned by adjusting the MW phase shifters (13) and MW attenuators (14) to achieve destructive interference of the MW pulse after passing the waveguide cells (11, 12).

The frequency boundary ν_N of the ν_2 domain is set by the sampling interval of the A/D conversion according to the Nyquist theorem, $\nu_N = 1/2\Delta t$: Typically, the Nyquist boundary is $\nu_2(\text{max}) = 50$ MHz, corresponding to sampling intervals of $\Delta t_2 = 10$ ns. Here, the Nyquist range of the FT applied to the real part of the transient emission record only spans the positive frequency branch $0 \cdots +\nu_2(\text{max})$, unless dual-channel vector detection is performed. The pulse generator varies the delay time t_1 between the two "pump" pulses in increments as short as $\Delta t_1 = 20$ ns. The increment Δt_1 determines the Nyquist frequency boundary of the ν_1 domain after 2D FT, e.g., $\nu_1(\text{max}) = 1/2\Delta t_1 = 25$ MHz at $\Delta t_1 = 20$ ns. The Nyquist range of the 1D FT with respect to t_1 applied to the complex output of the 1D FT with respect to t_2 spans $-\nu_1(\text{max}) \ldots +\nu_1(\text{max})$. To limit the impact of field inhomogeneities on the sensitivity, the maximum evolution time is practically restricted to $t_1(\text{max}) < 10\ \mu s$.

The $2D(\nu_1,\nu_2)$ frequency domain is obtained by discrete 1D FT on the complex array in the $2D(t_1,\nu_2)$ mixed domain. Similar to single-resonance FT-MW spectroscopy, the phases of the individual signals are generally unknown and therefore the spectra are normally displayed as 2D contour or pseudo-3D surface amplitude plots. As encountered in 1D FT-MW spectroscopy, the precise analysis of the frequency domain spectra with narrow line multiplets suffers from distortion effects causing line position shifts. To recover the frequency information contained in the transient emission records, a model function $s(t_1,t_2)$ can be fit directly to the $2D(t_1,t_2)$ time domain, yielding the information on the MW transition frequency ω_{ab}, the $c \leftarrow a$ coherence off-set $\Delta\omega_{ac}$, and thus also the RF transition frequency ω_{bc}. Using this approach, it was found that the accuracies of the MW transition frequency and the RF transition frequency scale with the evolution-time increment range to detection-record length ratio, e.g., $1/10 = 5\ \mu s/50\ \mu s$. Alternatively,

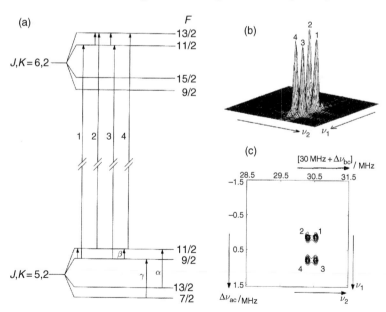

Figure 20   2D FT-MW RF-DR spectroscopy of the CF$_3{}^{79}$Br isotopologue of trifluoro bromo methane: (a) Energy level diagram showing the MW transitions J, K, $F = 6$, 2, 11/2$\leftarrow$5, 2, 9/2; 6, 2, 13/2$\leftarrow$5, 2, 11/2 at 25186.137 MHz; 25186.302 MHz, respectively, and RF transitions J, K, $F = 5$, 2, 11/2$\leftarrow$5, 2, 9/2; 6, 2, 13/2$\leftarrow$6, 2, 11/2 at 26.240 MHz; 26.004 MHz, respectively, involving two regressive (1,2) and two progressive (3,4) three-level systems. (b) Pseudo-3D(ν_1,ν_2) surface plot. (c) 2D(ν_1,ν_2) contour plot. (From Ref. [54]. Copyright 1993, Elsevier Inc.)

instead of analyzing the $2D(t_1,t_2)$ time domain directly, which often turns out to be rather time consuming, the $2D(t_1,\nu_2)$ mixed domain may be analyzed [53]. Fitting a theoretical expression $s(t_1,\nu_2)$ to this complex array yields the c←a coherence off-set parameter $\Delta\omega_{ac}$; the RF transition frequency ω_{bc} can be derived using the line position of the MW transition ω_{ab} in the ν_2 domain.

Figure 20 illustrates the application of 2D FT-MW spectroscopy [54], showing the RF-DR experiment on two regressive (1, 2) and two progressive (3, 4) three-level systems of the $CF_3{}^{79}Br$ isotopologue of trifluoro bromo methane (Figure 20a). The four corresponding transitions are visualized as a pseudo-$3D(\nu_1,\nu_2)$ surface (Figure 20b) and as a $2D(\nu_1,\nu_2)$ contour plot (Figure 20c). Though resonance frequencies are routinely obtained from a fit of a model function to either the $2D(t_1,t_2)$ time domain or $2D(t_1,\nu_2)$ mixed domain, connectivities of the three-level systems are much easier revealed visually from the plotted $2D(\nu_1,\nu_2)$ frequency domain, i.e., the 2D spectrum.

3.3.   Pulsed radiation microwave excitation of supersonic jets

If molecular species are prepared in the velocity-equilibrating conditions of a supersonic-jet expansion, Doppler- and collision-induced broadening become rather unimportant. Conventional waveguides, however, are not very well suited for a free expansion. Instead, resonators of appreciable size are used. With the molecular systems all traveling at the same speed on collisionless trajectories, the line width is ideally only governed by the duration of their stay in the active region of the resonator. Using an electromagnetic valve, this technique allows spectroscopy to be conducted on a pulsed jet of an appropriate gas mixture, which provides a sample of high number density. As well known, supersonic expansions of molecular samples in rare gas mixtures are rich in molecules of very low rotational and moderately low vibrational effective temperatures, generally providing a significant sensitivity advantage for transitions originating from low-energy rotational levels in the vibrational ground state.

Consequently, FT-MW spectroscopy on pulsed supersonic-jet expansions has become popular because of its unique ability to provide high resolution and high sensitivity simultaneously. Furthermore, in the course of the supersonic-jet expansion, which converts the enthalpy of the sample—determined by the thermal equilibrium conditions before the expansion—into the energy of a directed translation, weakly bound species are formed. Once existing, these species are isolated in the collisionless environment of the expanding gas and hence persist, however weak the binding. In the advent of the technique, which was originally developed as a new method of investigating the rotational spectra of weakly bound molecular complexes [56], studies on such systems have dominated the application of the method for a long time.

3.3.1.   Short stimulus FT-MW apparatus

After the early reports on FT-MW spectroscopy on pulsed supersonic-jet expansions [57], a number of significant advances in the technique were achieved especially in recent years [58–63]. Presently, this MW spectroscopic method serves as one of the most productive techniques and shall be presented in detail here. The state-of-the-art supersonic-jet expansion resonator FT-MW spectrometer discussed here provides further improvements in some significant points [64]. However, the principle design considerations concerning its high-frequency electronics are generally instructive for all FT-MW methods. This implementation is especially suited for the detection and characterization of larger, instable compounds—also in very small concentrations. A block diagram of the spectrometer, using a short MW radiation pulse for excitation, is presented in Figure 21; its principle of operation is as follows: A primary signal of frequency ν is created by a wobble generator (1). Employing an SPDT switch (2a), the output signal is either used for excitation or detection. In the first case, an MW pulse for sample polarization is created; the

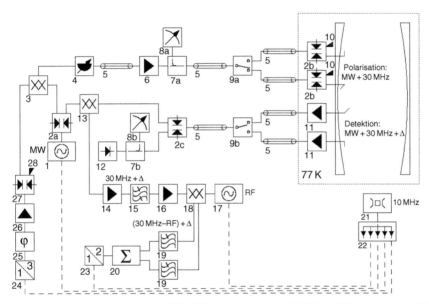

Figure 21 Supersonic-jet expansion FT-MW spectrometer: (1) MW wobble generator, 10 dBm output power; (2) MW pin-diode SPDT switch, 50 dB isolation, 3 dB insertion loss; (3) MW SSB modulator, 15 dB carrier suppression, 12 dB insertion loss; (4) MW variable attenuator, 0...71 dB attenuation; (5) MW coaxial cable, flexible; (6) MW power amplifier, 25 dB gain, 13 dBm output power; (7) MW directional coupler, 20 dB coupling, 1.2 dB insertion loss; (8) MW power sensor, 20...−70 dBm input power; (9) MW electromechanical switch, 60 dB isolation, 1.25 dB insertion loss; (10) MW termination, 50 Ω impedance; (11) MW low-noise amplifier, 1.1/2.0 dB noise figure at 77/300 K, 30 dB gain, 0 dBm output power; (12) MW diode detector, 0.5 mV/μW sensitivity; (13) MW IR demodulator, 18 dB image rejection, 14.5 dB insertion loss; (14) RF low-noise amplifier, 1.2 dB noise figure, 35 dB gain, 6 dBm output power; (15) RF bandpass filter, 2.0 MHz 3 dB-bandwidth, 2 dB insertion loss; (16) RF VGC amplifier, −13...17 dB gain, 3 dB output power; (17) RF signal generator, 10 dBm output power; (18) RF I/Q demodulator, 7 dB insertion loss; (19) RF low-pass filter, 6 MHz 3 dB-bandwidth; (20) PXI-based experiment control, data-acquisition, and data-analysis system; (21) frequency standard, 10 MHz output frequency, 10^{-11} frequency accuracy; (22) RF distribution amplifier, 13 dBm output power; (23) RF frequency doubler, 0 dB gain; (24) RF frequency tripler, 0 dB gain; (25) RF QPSK modulator, 4 × 90°; (26) RF VGC amplifier, −6...24 dB gain, 13 dB output power; (27) RF GaAs SPDT switch, 40 dB isolation, 0.9 dB insertion loss; (28) RF termination, 50 Ω impedance. Specifications are typical minimum requirements allowing for optimized performance. *Source*: Figure 3.3 of Ref. [64].

supply of MW power for the entire excitation path lasts only for the duration of the polarization pulse and therefore a risk of coherent perturbations because of signal leakage during detection is decreased. In the second case, the MW response of the sample is down-converted into a lower frequency band by taking the difference-frequency signal of the molecular and the primary radiation. Because the low-noise processing of signals in the RF range is superior to the processing of direct current (DC) signals, the MW source (1) is operated at a primary frequency lower than that of the radiative transition. The excitation signal at a frequency of typically $\nu + 30\,\text{MHz}$ is generated in an SSB modulator (3). Because the optimum polarization conditions depend on the transition dipole matrix element of the molecular system, a variable attenuator (4) adjusts the power in 1 dB steps to the level required for polarization. A power amplifier (6) compensates for the conversion loss to the sideband and, if necessary, allows for high excitation power required by small transition dipole matrix elements. Through a directional coupler (7a), a power sensor (8a) is monitoring the output power. Using an electromechanical switch (9a), the excitation signal is inserted into one of two signal paths being optimized for the lower or the upper frequency range of the spectrometer. SPDT switches (2b) perform pulse shaping and isolate the resonator from noise of the excitation electronics during the detection. The idle noise between excitation pulses is annihilated in a termination (10) at the second output of the pin-diode SPDT switch (2b); thus, an electrical short at the output of the amplifier arising from reflective pin-diodes does not occur.

The MW power of the signal paths for both frequency bands is supplied to the resonator by L-shaped antennae. The bent wire has approximately a length of $\lambda_{\min}/2$ with the value of $\lambda_{\min}$ corresponding to the highest polarization frequency $\nu_{\max}$ of each band. The direction of the antenna wire also determines the direction of the linearly polarized MW field. To achieve the most effective power transmission, the distance between the bent wire and the reflector surface can be electromechanically varied for impedance matching, i.e., reflection-free so-called critical coupling. The coupling coefficient β_1 of the exciting antenna becomes $\beta_1 = 1 + \beta_2$ after impedance matching, where the coefficient β_2 is that of the receiving antenna [63].

The sample response, i.e., the transient emission, is received by a second, diametrically opposite antenna. The direction of the polarized MW field, determined by the exciting antenna, is transferred onto the direction of the polarized molecular radiation field. Therefore, the bent wire of the receiving antenna, also of length $\lambda_{\min}/2$, has to be oriented in the same direction to absorb the radiation power. To obtain the maximum molecular signal, impedance matching is again facilitated by electromechanically varying the distance between the bent wire and the reflector surface. The coupling coefficient β_2 obtained for the receiving antenna at critical coupling is $\beta_2 = 1 + \beta_1$, whereas optimum conditions result for a vanishing coupling coefficient β_1, i.e., if the exciting antenna is removed. Because reflecting pin-diodes represent an electrical short, this is accomplished by directly connecting the pin-diode SPDT switch (2b) on the reflector's rear side to the antenna. With the short spatially close to the antenna, the coupling coefficient is forced to $\beta_1 = 0$ after the excitation pulse was applied; then optimum coupling of the receiving antenna is true for $\beta_2 = 1$. Critical coupling can still be obtained if the antennae are not located in the center of the reflector, i.e., near the maximum field amplitude. In fact, to facilitate a wideband coupling characteristic, the receiving antenna as well as the exciting antenna might not be placed on the resonator axis but on the perimeter of the beam waist w_0 of the corresponding boundary frequency ν_{max} [63].

The transient emission at the output of the receiving antenna is fed into a low-noise amplifier (11). To avoid degradation of the signal quality due to insertion losses, the amplifier is—without any protection—directly attached to the antenna output on the reflector's rear side. The recovery time of the unprotected amplifier after the exciting MW pulse is short enough to observe the transient emission without additional delay. The gain is chosen such that the power loss in the subsequent signal processing stages is somewhat less than the gain; thus, the S/N ratio remains unaffected [63]. A gain significantly in excess of the loss should be avoided, because even small signals might then cause power saturation in the downstream components, i.e., the dynamic range of the spectrometer will be reduced.

Using an electromechanical switch (9b), the molecular pulse response is taken from one of two signal paths being optimized for the lower or the upper frequency range of the spectrometer. A pin-diode SPDT switch (2c) passes the excitation signal to a diode detector (12), thus protecting the downstream receiver components from saturation. The diode detector is also used as signal monitor when the center frequency of the resonator is tuned to the excitation signal at frequency $\nu + 30\,\mathrm{MHz}$. During this tuning process, while the wobble generator (1) sweeps over a frequency segment centered at the excitation frequency, the pin-diode SPDT switches (2a,b) are set to supply CW power to the diode detector to capture the transmission curve. By changing the reflector distance electromechanically, the transmission curve of a selected "transverse electric magnetic" TEM_{plq} mode is centered at the excitation frequency. For optimum input at amplifier (11), the CW power is adjusted by the attenuator (4). Through a directional coupler (7b), the transmitted power is precisely monitored using a power sensor (8b).

After a short delay following the excitation pulse, the signal path is connected to the IR demodulator (13), which converts the molecular MW signal to an IF band at 30 MHz. The conversion loss is compensated by the low-noise amplifier (14). After conversion, the signal passes a bandpass filter (15). Its bandwidth is chosen such that the spectral width of the spectrometer [63], determined by the quality factor Q of the resonator, is not reduced, but noise outside the signal band is suppressed. An effective degradation of the S/N ratio after discrete FT, caused by projecting high-frequency noise onto an equivocal frequency scale according to the Nyquist theorem, is prevented. A voltage gain control (VGC) amplifier (16) is used to avoid saturation of the downstream components by strong molecular signals.

The IF signal is down-converted in an in-phase/quadrature-phase (I/Q) demodulator (18) utilizing the output of a signal generator (17), in such a way that the entire spectral range passing the bandpass filter is lying inside the Nyquist frequency range of the discrete A/D conversion. Higher mixing products are held off from the digitization by a low-pass filter (19). The absolute spectral position, depending on the output frequency of the signal generator (17) after final down-conversion, is arbitrary. Therefore, a super-heterodyne detection at a center frequency larger than half the width of the signal band as well as detection at DC might be performed. Discrimination between the negative and the positive frequency branches—required in the latter case—is facilitated by two–channel I/Q detection with subsequent complex FT performed by the PCI eXtensions for Instrumentation (PXI)–based experiment control, data–acquisition, and data–analysis system (20). Through a distribution amplifier (22), a 10-MHz reference frequency signal of a GPS receiver (21) or atomic clock is used for the generation of all analog and digital RF signals and system clock rates. By referencing all signal sources of the system to the same standard, the transient emission signal can repeatedly be generated with a recurring phase relation. This is necessarily the case if all clocks used in the experiment, as well as every IF of the excitation frequency after conversion in each of the demodulation stages, represent a harmonic of the clock that is used to derive the trigger signal. Because the weak transient emission is frequently far below the thermal noise level, phase-stable repetitions are essential for the correct addition of individual experiments to increase the S/N ratio.

The pulse sequence depicted in Figure 22 is generated using 32-bit counters sourced by a 20 MHz or 100 kHz base clock, depending on the required pulse duration. The clock of the A/D conversion is phase stabilized relative to a 10 MHz base clock. The signal of the 10 MHz reference standard (21) is used to generate these base frequencies either directly or after division of the 20 MHz output signal of a frequency doubler (23). For the generation of the 30 MHz sideband, a frequency tripler (24) is used. The trigger signal is derived from the 100 kHz basis such that all harmonics of this frequency can be used in the system. Besides the transistor–transistor logic (TTL) pulse sequence for the jet valve (SOURCE), laser photolysis, laser ablation, or flash lamp trigger (CHARGE), DC discharge or

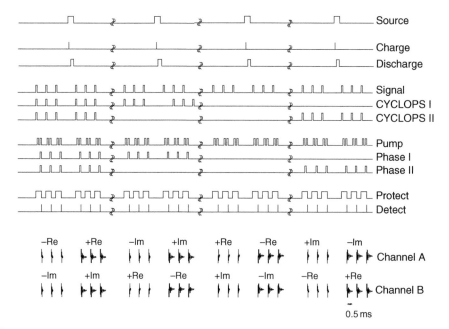

Figure 22 Supersonic-jet expansion FT-MW experiment sequence. See text for details.
Source: Figure 3.5 of Ref. [64].

Q-switch trigger (DISCHARGE), MW pulse (SIGNAL), DR pulse (PUMP), MW/DR protection (PROTECT), and transient recorder (DETECT), TTL control signals for the generation of the phase sequences are also required (CYCLOPS I, II). A quadrature phase shift keying (QPSK) modulator (25) is used to equalize the asymmetry of the input channels (CHANNEL A,B) in applying a suitable, ordered phase sequence, i.e., the cyclically ordered phase sequence (CYCLOPS). A VGC amplifier (26) compensates for the insertion loss of the QPSK modulator and adjusts the signal level to meet the SSB modulator (3) input power requirement. A GaAs SPDT switch (27) ensures that the RF power required for sideband generation is only supplied during the pulse generation, so that coherent leakage signals are prevented. Outside the polarization pulses, the power is annihilated in a termination (28) at the second output of the GaAs SPDT switch. For optional DR experiments, the relative phase between the DR pulses can be adjusted according to the requirements of a specific DR phase sequence (PHASE I,II). All TTL status signals for phase control are only active for the duration of the corresponding radiation pulse, so that remaining coherent leakage signals during the detection period are not subjected to the phase modulation and therefore are eliminated because of the channel rotation and sign alternation in the data-acquisition cycles.

By avoiding components that operate only in octave bands or are otherwise limited in their bandwidth, e.g., waveguides, isolators, circulators, power dividers, etc., a very broadband instrument can be constructed, such that—without any modifications—multioctave operation from 2–26.5 GHz is possible [62]. All essential external spectrometer components are integrated into the experiment control, data-acquisition, and data-analysis system either using the general purpose interface bus (GPIB) or local area network (LAN), all internal components are using the PXI. Using the platform-independent virtual instruments software architecture (VISA) and interchangeable virtual instruments (IVI) drivers, the entire spectrometer can be controlled via a graphical user interface (GUI) operated either interactively or completely automated [61].

3.3.2. Fabry–Pérot-type resonator

The electric dipole interaction of the molecular sample, expressed by the optical Bloch equations that describe the time evolution of a two-level system in resonance or near resonance with the external field, is caused by exposure to a standing wave field of the microwave radiation propagating in a TEM_{plq}-mode of a Fabry–Pérot-type resonator. Its geometric and electric parameters have a profound impact on the sensitivity of the spectrometer, especially at the low-frequency end. While cylindrical cavity designs utilizing the low-loss TE_{01}-mode have also been tried alternatively [65], considerable improvements were achieved by optimizing the parameters of the Fabry–Pérot-type resonator. Quite compact resonators [66] can successfully be employed in the cm wave range; rather extreme geometries [67] have been utilized to access the dm wave range with reasonably sized resonators. Also, operation throughout the mm wave range was sought after [68]. Further developments include cryogenic operation of the resonator spectrometer [63]. The resonator arrangement discussed here provides advantages in some significant points [64]. It is well suited for sensitive detections in the lower frequency range, important for the study of bigger or heavier species, i.e., molecular systems exhibiting larger moments of inertia.

The main component of the experimental setup shown in Figure 23 is a Fabry–Pérot-type resonator formed by two spherical mirrors (1) of equal curvature b' at distance

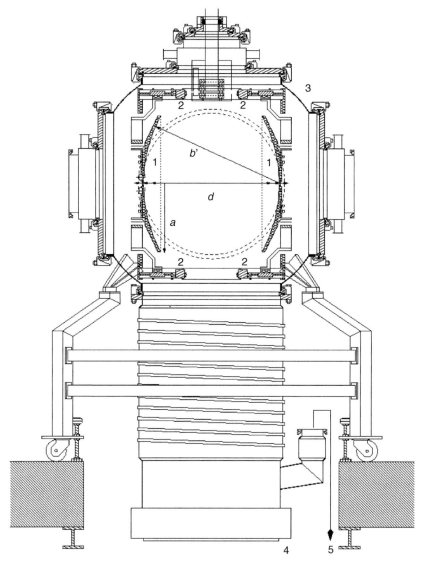

Figure 23 Supersonic-jet expansion Fabry–Pérot-type resonator chamber: (1) aluminum reflectors, conductivity $\sigma = 1.6 \times 10^7 \, 1/\Omega\text{m}$, $a' = 315\,\text{mm}$, $b' = 630\,\text{mm}$, $d = 530\ldots730\,\text{mm}$ distance; (2) linear translation stage, 100 mm travel, 0.5 μm resolution, 25 mm/s speed; (3) stainless steel six-way cross DN630 ISO-K, 1033 mm diameter spherical body; (4) oil diffusion pump DN630 ISO-F, 20,000 L/s pump capacity; (5) roots/two-stage rotary vane pump system, 210/65 m^3/h pump capacity. *Source*: Figure 3.6 of Ref. [64].

d, typically made of aluminum. A near-confocal arrangement with $g = 1 - d/b'$ and $(1 - g^2)^{1/2} > 0.987$ exhibits near optimum beam radii w'_s and w_0 at the reflectors and at its center, respectively [63]

$$w_s'^2 = \frac{d\lambda}{\pi} \times \sqrt{(1-g^2)}, w_0^2 = \frac{b'\lambda}{2\pi} \times \sqrt{(1-g^2)}, \tag{6}$$

resulting in lowest diffraction losses α_d for each path of the radiation propagating back and forth between the reflectors and a large active volume, respectively, at a given wavelength λ. Low diffraction loss is found for reflector radii a', large enough to give a Fresnel number

$$N = \frac{a'^2}{d\lambda} \times \sqrt{(1-g^2)} \tag{7}$$

of, at least, unity for near-confocal arrangements. Because aluminum is a good electrical conductor, the ohmic loss α_r due to the finite conductivity σ is very small. Therefore, resonator geometries with Fresnel numbers N somewhat larger than unity are required for negligible diffraction losses $\alpha_d << \alpha_r$. The unloaded quality factor Q_0 is then approximated as [63]

$$Q_0 = \frac{2\pi}{\lambda} \times \frac{d}{(\alpha_d + \alpha_r)} = d/2\delta, \delta = \sqrt{\frac{\lambda}{\pi c \mu \sigma}}, \tag{8}$$

where the skin depth δ is the distance within the conductor at which the amplitude of an electromagnetic wave has fallen to $1/e$ of its surface value. For non-ferromagnetic materials, the permeability μ is close to the permeability of vacuum μ_0. Values for Q_0 of up to 2×10^5 are obtained for Fabry–Pérot-type resonators. When critically coupling the resonator to the spectrometer with $\beta_1 = 0$ and $\beta_2 = 1$ as described above,

$$Q_L = \frac{Q_0}{1 + \beta_1 + \beta_2}, \tag{9}$$

a loaded quality factor of $Q_L = 10^5$ can be achieved. For a resonator critically coupled as given above, the amplitude of the transient emission is $P_{ab^{1/2}} \propto |S_{ab}| \propto (Q_0/4)^{1/2}$, while optimizing toward minimum transmission loss, i.e., $\beta_1 = \beta_2 >> 1$, decreases the signal to $|S_{ab}| \propto (Q_0/4\beta_2)^{1/2}$.

An electromechanical linear translation stage (2) is employed to establish the resonance condition for a TEM$_{plq}$ mode [63]

$$\frac{2d}{\lambda} = (q+1) + \frac{1}{\pi}(2p+l+1) \arccos(g), 0 \le g^2 \le 1, \tag{10}$$

where g is limited due to the stability condition of large resonators with circular reflectors of equal radius of curvature. Normally, the fundamental TEM$_{00q}$ modes, providing the largest active volume for the radiative interaction with the molecular systems, are utilized.

The aluminum reflectors can be cooled to a temperature of $T_r = 77$ K by flowing liquid N_2 through copper coils brazed to a copper plate bolted to its back side. The effective noise temperature T_n of the resonator is frequency dependent and its value is maintained at the reflector temperature T_r, only if diffraction losses are negligible, i.e., a Fresnel number $N > 1$ is required. To preserve this noise temperature at the input of the low-noise amplifier, the terminating and leading components of the exciting and receiving signal path are thermally connected to the back of the cold reflector.

The resonator chamber (3) is evacuated by an oil diffusion pump, backed by a three-stage mechanical pump system (4). To achieve the maximum nominal pump capacity, the oil diffusion pump is operated baffle-free without drawback on the MW experiment. Thus, jet-expansion rates of 20 Hz are feasible.

3.3.3. Pulsed supersonic-jet expansion

The use of an electromagnetic valve does not only limit the gas load for the pump system, thus reducing the required capacity, it also ideally combines with the pulsed excitation scheme of the FT-MW TDS technique. Skimming or collimation of the expansion to obtain a narrower line width leads to serious loss of intensity in the detected signals [69]. However, in comparison to the case where a gas mixture is propelled into the resonator perpendicularly to its symmetry axis, the resolution as well as sensitivity of supersonic-jet expansion FT-MW spectroscopy using Fabry–Pérot-type resonators could be significantly increased with the introduction of the coaxially oriented beam resonator arrangement (COBRA) [59,62]. This arrangement, illustrated in Figure 24, places the jet source in the center of one of the reflectors that form the resonator.

A diluted gas mixture, typically ≤1 percent sample seeded in a rare gas at a total pressure in the range 50–500 kPa, is expanded into the evacuated resonator chamber. The nozzle, in its simplest form, consists of a circular orifice of 0.5–2.0 mm diameter with an exit channel of 2 mm length conically widening to 4 mm. The properties of the emerging gas pulse are of paramount importance for the rotational spectra of the molecular systems detected by this method: It affects the effective translational, rotational, and vibrational temperatures of the species, as well as their spatial distribution and the nature of systems present in the jet.

The flow of the gas expanding through the nozzle into the evacuated resonator chamber is supersonic rather than effusive: As a result of the rapid adiabatic expansion, the molecular systems have very low effective rotational and vibrational temperatures and are traveling in a collisionless expansion with a speed v along radial paths. Before the expansion, the gas has a Maxwell–Boltzmann distribution of molecular velocities v_0 determined by the equilibrium temperature T_0. After the adiabatic expansion, there is a highly directed mass flow: Binary collisions during the initial stages of the expansion lead to the conversion of random motion of all species into directional translation. The internal energy of the molecular species is also transformed into directed kinetic energy. Accordingly, the rotational and vibrational energy of the seeded molecules fall rapidly, while the Maxwell–Boltzmann velocity distribution narrows and shifts very near toward the terminal value

$$v_\infty = \sqrt{\left(2k_B T_0 / m \times \gamma/\gamma - 1\right)}, \quad \gamma = c_p/c_v, \tag{11}$$

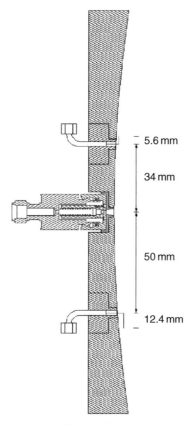

5.6 mm

34 mm

50 mm

12.4 mm

Figure 24 COBRA implementation: Supersonic-jet expansion source in the center of a reflector forming the Fabry–Pérot-type resonator; two perpendicular pairs of MW antennae pointing radially away from the axis with their location and shape optimized for operation in the upper and lower frequency band can be used. *Source*: Figure 3.4 of Ref. [64].

where c_p and c_v are the mean heat capacities at constant pressure and volume, respectively. k_B is the Boltzmann constant and m the mean mass of all species. The terminal velocity v_∞ corresponds to an effectively complete conversion of the enthalpy into directed kinetic energy for $T_0 \gg T_t \approx 0\text{K}$. To a very good approximation, the properties of the expansion are determined by the carrier gas alone, as it is present in vast excess. For a dilute mixture, m and γ can thus be set to the carrier gas values. The translational temperatures of supersonically expanded rare gases can be calculated [70]. For the conditions described here, translational temperatures $T_t < 1\text{ K}$ are predicted. The resulting narrowed velocity distribution corresponds to such small relative velocities that, once the molecules emerge from the high density region of the nozzle, they travel virtually collisionless.

Because there is a still large number of collisions while the gas emerges from the nozzle throat, rotational temperatures similar to T_t are expected, assuming a high efficiency of rotational–translational energy transfer. The vibrational temperatures of the species in the supersonic jet are also quite low. Without special measures, only in a few cases rotational transitions have been observed in low-lying vibrationally excited states.

The spatial distribution of the molecules emerging from the jet source is well described as a mass flow, in which the molecules all travel with the same speed v_∞ along radial paths at angles θ with respect to the symmetry axis [64]. At times t, restricted to the actual spatial presence of a transiting supersonic-jet with a pulse duration τ_M, the molecular number density $^1N(r,t)$ at distance r from the jet source is

$$^1N(r,t) = \frac{(b+1)N}{2\pi v_\infty \tau_M} \times \frac{\cos^b(\theta)}{r^2} \tag{12}$$

where the coefficient b varies from 1 to 3 for ideal effusive to ideal supersonic expansions, respectively; the total number N of species in the jet is determined experimentally from the consumption of a known gas mixture or, e.g., if the sample is produced in situ, by suitable diagnostic methods. As evident from the "peaking factor," $\kappa_p = (b+1)/2$ in Equation (12), the number density close to the expansion axis—in the COBRA arrangement coinciding with the active region of the resonator—is two-fold larger for supersonic jets than for effusive beams.

3.3.4. COBRA-transient emission signal

The interaction of the microwave radiation and the molecular systems within the supersonic-jet results in rotational coherence that establishes a polarized sample moving coaxially along the resonator axis. While the exciting MW field is relatively short-lived, the polarization of the molecular ensemble, due to the collisionless environment, is persistent. The time evolution is discussed in terms of the optical Bloch equations describing a two-level system in the absence of an external field. With very long relaxation times T_2, collisional broadening does not need to be considered. Instead, the expansion of the jet on the radial trajectories, reducing the number density of the molecular ensemble contributing to the molecular field in the resonator, is becoming the dominant mechanism for the decay of the signal. The lineshape, which is determined by the time-dependent number densities according to Equation (12), is still close to a Lorentzian in the frequency domain after FT.

According to Maxwell's equations, the polarized jet produces an electric field. The electrical field distribution $E(r,t)$ is obtained by projecting Maxwell's equations on the TEM_{plq} mode, i.e., by integrating the product of the respective expressions over the volume of the resonator. The molecular signal power as a function of time is finally obtained as the fraction of the total energy, stored by the field within the resonator volume, which is lost due to coupling to the receiver of the spectrometer [63]. The amplitude $|S_{ab}| \propto (P_{LO}P_{ab})^{1/2}$ of the molecular signal after super-heterodyne detection, being proportional to the square root of the power flow $P_{ab}^{1/2}$ at the receiving antenna, is approximated by

$$|S_{ab}| \propto \sqrt{\left(\frac{\pi w_0^2 \omega_{ab}}{4\omega_0 \delta}\right)} \frac{\kappa_p \Delta N_{ab}\mu_{ab}}{v_\infty^2 \tau_M \tau_E} \sin(x\tau_p) \tag{13}$$

$$\times \cos(k v_\infty t) |\exp(i(\omega_{ab}t + \theta_{ab}))|$$

where ΔN_{ab} represents the two-level population difference of the total number of species in the jet and τ_E denotes the expansion period from closing the jet source to the onset of the observation of the transient emission, and $k = \omega/c$ is the wavenumber of the radiation. During excitation, the spatial field distribution of the TEM_{plq} mode is projected as a spatial polarization pattern onto the expanding ensemble of molecules. If the electric field inhomogeneity of the TEM_{00q} mode propagating in the Fabry–Pérot resonator is taken into account, the trigonometric function $\sin(x\tau_p)$ is replaced by a Bessel function of the first kind $J_1(\Theta)$, exhibiting a more complicated argument with a first maximum at $\Theta = 1.871$, and only "effective $\pi/2$-pulses" can thus be achieved. During the transient emission, the spatial polarization distribution of the jet, which is expanding along the resonator axis, is projected on the TEM_{plq} mode that fulfills the resonance condition for electromagnetic field propagation in the Fabry–Pérot resonator. As a result of the jet expansion being directed along the resonator axis, the molecular field is modulated in time as given in Equation (13), resulting in

$$S_{ab}(t) \propto s' \exp\left(i((\omega_{ab} - kv_\infty)t + \theta'_{ab})\right) \tag{14}$$
$$+ s'' \exp\left(i((\omega_{ab} + kv_\infty)t + \theta''_{ab})\right),$$

and thus a Doppler doublet consisting of frequency components at $\nu_{ab}(1 - \nu_\infty/c)$ and $\nu_{ab}(1 + \nu_\infty/c)$ is observed in the frequency domain. The molecular resonance frequency is then recovered as the arithmetic mean of the components separated by $\Delta\nu_{ab} = 2\nu_{ab}\nu_\infty/c$. The linewidth (HWHM) of the individual components is on the order of 1.5 kHz, or $0.0000005 \text{ cm}^{-1}$; at an appreciable S/N ratio, a frequency accuracy of 150 Hz, or $0.00000005 \text{ cm}^{-1}$, is achieved for unblended lines. The sensitivity allows for routine observation of mono-deuterated asymmetric-top molecules in natural abundance [71]. Figure 25 demonstrates resolution and sensitivity of the cryogenic setup for an example of a transient species generated in situ. This extraordinary resolution provides supersonic-jet expansion FT-MW spectroscopy with a chemical specificity unrivalled by any analytical method [66,72].

3.4. External fields

When subjected to static homogeneous fields, either electric or magnetic, the $2J+1$ degeneracy of the spatial components with the quantum numbers $M_J = J, J-1, J-2, \ldots, -J$ of a rotational state with the total angular momentum quantum number J is lifted. The resulting splitting (with selection rules $\Delta M_J = 0, \pm 1$ depending on the orientation of the external field being either parallel or perpendicular to the MW field, respectively) in the rotational spectrum is called Stark effect or Zeeman effect, respectively.

The molecular systems can also be subjected to oscillating fields in addition to the high-frequency radiation that is resonant to the molecular transition being monitored. For electric fields varying in time, the so-called AC Stark effect is observed. If the molecular systems are subjected to a homogeneous or inhomogeneous pulsed or stationary field in the RF or MW range, which is resonant to a transition that shares an energy level with the "signal" transition, various DR effects modifying the transient emission signal occur.

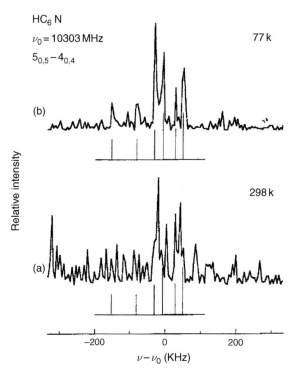

Figure 25 $J, K_a, K_c = 5, 0, 5 \leftarrow 4, 0, 4$ transition of the non–linear HC$_6$N isomer with ^{14}N nuclear quadrupole hyperfine structure, $\nu_0 = 10303.0\,\text{MHz}$: Measurements with resonator at 298 K (a) and 77 K (b). (Reused with permission from Ref. [63]. Copyright 2005, American Institute of Physics.)

3.4.1. Stark effect

The interaction of a polar molecular species possessing the dipole moment $\boldsymbol{\mu}$ with a homogeneous electric field $\boldsymbol{E}$ is described by the Hamiltonian

$$H = -\boldsymbol{\mu} \cdot \boldsymbol{E} \qquad (15)$$

Generally, the FT-MW technique does not rely on Stark-effect modulation. Thus, the investigation and exploitation of the Stark effect requires FT-MW spectrometers that are modified for such studies. Because the line width obtained with the TDS MW techniques is generally smaller than the one achieved by FDS MW methods, higher field homogeneity is required to exploit the resolution enhancement for, e.g., very precise determinations of the dipole moment μ.

In FT-MW spectroscopy of static gases with waveguides as sample cells, higher field homogeneity compared to that achieved with Stark-effect modulation cells used in absorption spectroscopy can be obtained with a configuration using an unusually thick high-voltage electrode. A thick "Stark septum," however, increases the attenuation of the MW radiation propagating in the waveguide cell resulting in a reduced sensitivity. Despite larger attenuation compared to a conventional Stark-effect modulation cell, it was still possible to determine the dipole moment of benzene-d_1 [73]. Also FT–MW spectrometers with a bridge-type waveguide cell have been modified for Stark-effect measurements of very weakly polar species such as allene-d_2 [74].

Despite all the differences in the various FT–MW techniques, the accuracy of Stark-effect measurements is always limited by the inhomogeneity of the employed electric field. Because the supersonic-jet expansion, as well as the MW radiation propagating in a low-loss TEM_{plq} mode of the Fabry–Pérot resonator, requires a substantially larger free volume than a static gas in a waveguide experiment, the impact of field inhomogeneities can become quite severe. Conventionally, Fabry–Pérot resonator spectrometers for the measurement of the Stark effect employ a pair of high-voltage electrodes that consist of rectangular or quadratic plates or grids oriented such that they face each other perpendicularly to the resonator axis [75]. Several approaches to decrease the field inhomogeneity were made: Obviously, a more homogeneous field can be achieved by increasing the ratio of the electrodes dimensions with respect to their distance [76]. However, the choice of electrode distance and extension is limited by the boundary conditions for low-loss propagation of the MW radiation, if the sensitivity of the spectrometer has to be maintained. This is especially important in the lower frequency range, because according to Equation (6), the spatial MW field distribution is significantly widened at larger wavelengths. Another approach employs an array of wires running parallel to the edges of the electrodes. The wires—connected in series by resistors acting as potential dividers—homogenize the electric field at the outer regions of the plate electrodes [77]. Such wires can only be placed parallel to the edges above and below the resonator axis. Additional sheets of metal placed at the edges of the plates can also enlarge the homogeneous region of the electric field between the Stark electrodes [78].

A very homogeneous electric field can be obtained if the reflectors of the Fabry–Pérot resonator themselves serve as Stark electrodes, when being mounted electrically insulated. Thus, the rear reflector can be set to a high electric potential of up to $U = 20\,kV$, whereas the front reflector, equipped with the MW antennae and the jet source, has to be kept at ground potential. As illustrated in Figure 26, the reflectors alone are not producing a sufficiently homogeneous electric field. However, in employing additional coaxial ring electrodes, i.e., utilizing the "coaxially aligned electrodes for Stark effect applied in resonators" (CAESAR) arrangement [79], the field is homogenized very efficiently. In this arrangement, the ring electrodes and the reflectors are connected via identical resistors. Guided by numerical simulations, the distances of the identical ring electrodes, with a diameter matching the reflector's diameter, are adjusted such that a constant potential gradient along the resonator axis is produced. The resulting electric field is homogeneous over a large volume. Inhomogeneities in radial directions occur sufficiently far away from the resonator axis, and thus, as the Gaussian TEM_{plq} modes confine the active volume of the resonator according to Equation (6), the inhomogeneous regions practically do not contribute to the observed molecular signal. The CAESAR setup thus provides the principle advantages of a COBRA spectrometer for Stark-effect experiments. Because the sensitivity is increased along with the resolution, narrow spectral splittings or shifts can now be observed also for species at low concentrations or with small dipole moments.

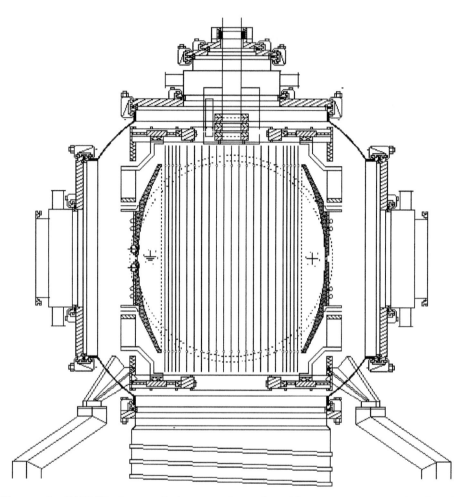

Figure 26 CAESAR Supersonic-jet expansion Fabry–Pérot-type resonator chamber for rotational Stark-effect measurements. *Source*: Figure 3.13 of Ref. [64].

3.4.2.   Zeeman effect

The measurement of the rotational Zeeman effect requires quite substancial instrumental efforts to obtain a sufficiently homogeneous magnetic field over the volume of the sample. Precision becomes very important to maintain resolution and sensitivity. The vast majority of FT-MW Zeeman studies were performed on stationary gases employing waveguide cells [80].

The interaction of a molecular species with a magnetic moment in a homogeneous magnetic field with flux density $\boldsymbol{B}$ is described by the Hamiltonian

$$H = \frac{-\mu_{\mathrm{N}}}{\hbar}\,\boldsymbol{B}\cdot\boldsymbol{g}\cdot\boldsymbol{J} + \frac{-1}{2}\,\boldsymbol{B}\cdot\boldsymbol{\chi}\cdot\boldsymbol{B} \qquad (16)$$

with the nuclear magneton μ_N, molecular g tensor, and the molecular susceptibility tensor χ. In the absence of contributions of large-amplitude motions, the g tensor can be written as the sum of positive nuclear and negative electronic contributions $g = g^n + g^e$. The susceptibility tensor χ can be written as the sum of a negative diamagnetic term that depends only on the electronic ground state and a positive paramagnetic term that contains a sum over excited electronic states $\chi = \chi^d + \chi^p$.

For molecular systems in a $^1\Sigma$ state, typical magnetic fields on the order of 1 T are necessary to observe the Zeeman effect. Species with electronic contributions to the angular momentum exhibit significantly larger magnetic moments, which are related to the orbital angular momentum L and the spin momentum S of the unpaired electrons:

$$\mu_L = \mu_B L, \ \mu_S = - g_e \mu_B S, \tag{17}$$

where μ_B is the Bohr magneton and $g_e = 2\mu_e / \mu_B = 2.0023$ is the g factor of the free electron. Typically, the Zeeman energy according to $H = -\mu \cdot B$ of species with unpaired electrons is four orders of magnitude more sensitive toward magnetic fields than the one for closed-shell molecules. As a result, observable spectral splittings are already induced by the Earth's magnetic field.

Although the Zeeman effect provides valuable information on the electronic structure of molecular systems, it is often desirable to avoid the Zeeman effect when studying a species with unpaired electrons. By eliminating the Zeeman splitting, the analysis of the spectra, which additionally exhibit a hyperfine splitting due to spin coupling, is simplified. Even more important, if unknown species in low concentrations are searched for, is the lower detection limit because the intensity of the unsplit transitions is increased. To eliminate the Earth's magnetic field, the so-called Helmholtz coils can be employed. An arrangement of three pairs of Helmholtz coils as shown in Figure 27 allows for the compensation of all spatial components. The

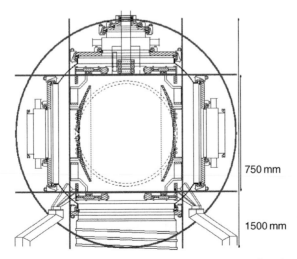

750 mm

1500 mm

Figure 27 Supersonic-jet expansion Fabry–Pérot-type resonator chamber with Helmholtz coil arrangement to suppress the rotational Zeeman effect of open-shell molecular species. *Source*: Figure 3.15 of Ref. [64].

magnetic flux density $B = \mu\mu_0 H$ produced by the pair of large diameter ring coils when subjected to an electric current I is quite homogeneous if their distance d equals the radius R. Along the coil axis y, one obtains

$$H = \left(\frac{IR^2}{2}\right)\left(1/\left(R^2+(d/2-y)^2\right)^{3/2}+1/\left(R^2+(d/2+y)^2\right)^{3/2}\right), \qquad (18)$$

and a similar dependence is obtained in the radial direction. If such a coil arrangement is used in combination with a Fabry–Pérot resonator, a sufficiently homogeneous field is achieved with such a pair of coils at a distance equal or larger than that of the resonator reflectors [81].

The Zeeman effect of species in a $^1\Sigma$ state can also be exploited using supersonic-jet expansion experiments when the Fabry–Pérot resonator is contained within a superconducting solenoid, which provides a highly homogeneous magnetic field along its axis at positions sufficiently distant from the ends [82].

3.4.3. Population effects

Early developed DR modulation schemes for molecular three-level systems utilize a second resonance in the MW or RF range. These DR techniques were first used on static gases [83] to facilitate an easy assignment of rotational transitions by simplifying the spectra. In this scheme, a continuous rather than a pulsed excitation field is used for DR modulation. When employing 0°/180° PIK or on/off-modulation on the excitation field of the second resonance for two consecutive experiments, transient emission signals belonging to three-level systems that share the DR transition can be selected from the spectrum by taking the difference of the time-domain signals from the two consecutive experiments prior to Fourier transform.

Similar techniques have later been developed for rotational spectroscopy on supersonic jets. Owing to the diminished occupation of higher energy rotational states after population transfer during adiabatic expansion, the rotational spectra are often already relatively sparse compared to those obtained from static gas samples. Still, DR modulation can be very valuable for the assignment of lines in congested regions and of otherwise complicated or difficult-to-predict spectra, as sufficiently homogeneous fields for Stark-effect measurements, which are traditionally utilized for this purpose, are more complicated to achieve in a COBRA spectrometer than in a WG experiment.

The first DR modulation supersonic-jet expansion FT-MW experiments were implemented using a perpendicular crossed-resonator arrangement [84], operated with radiation in the cm wave range of the MW region for both resonances. A drawback of this setup is the necessity to have two resonators tuned for the signal and pump radiation. Rapid frequency changes in either of the domains, enabling time-efficient broadband coverage to accelerate spectral assignment, are thus prevented. Alternatively to a resonator a high-power amplifier can be used to provide a suitable field strength for the pump radiation. This is overcome in DR modulation experiments that employ MW radiation in the mm wave range. Owing to the generally more favorable transition dipole moments at higher frequencies, much lower field amplitudes are needed for an efficient DR modulation, and consequently, resonator enhancement for the DR field is not required. Using a horn antenna, the mm wave radiation is simply fed into the resonator volume propagating perpendicularly to its axis [85], as illustrated in Figure 28. The setup allows for mm wave DR as well as cm wave DR, the latter with phase sequence capability. For mm wave DR, frequency modulation with an off-set of 2.34 MHz rather than 0°/180° PIK or on/off-modulation of the excitation field of the second resonance was utilized to observe the DR effect as a difference signal. When performing a broadband frequency survey in the mm wave frequency domain as shown in Figure 29, a doublet appears for each transition belonging to three-level systems that share the DR transition, as the mm wave radiation, at a separation of 2.34 MHz, comes into resonance twice. As displayed in Figure 30, reducing the mm wave frequency increments allows for high-resolution measurements of transitions beyond the spectral limit of the FT-MW spectrometer. Here, the FT-MW serves as a detector with its accessible frequency range indirectly extended to that of the DR source.

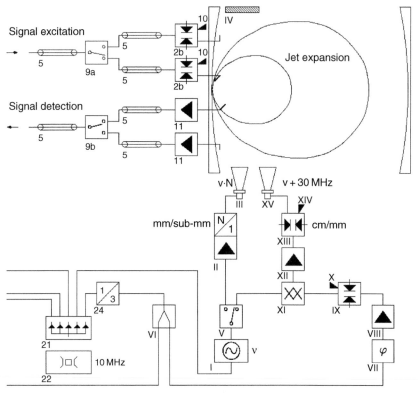

Figure 28 Supersonic-jet expansion FT-MW DR experiment with: mm/sub-mm wave DR excitation signal path: (I) cm wave synthesizer; (II) mm/sub-mm wave active multiplier; (III) mm/sub-mm wave horn antenna; (IV) metal reflector; cm/mm wave DR excitation signal path: (I) MW synthesizer; (V) MW electromechanical switch, (VI) RF power divider; (VII) RF QPSK modulator; (VIII) RF VGC amplifier; (IX) RF GaAs SPDT switch; (X) RF termination; (XI) MW SSB modulator; (XII) MW high-power amplifier; (XIII) MW pin-diode SPDT switch; (XIV) MW termination; (XV) MW horn antenna. Arabic numbers refer to the caption of Figure 21. *Source*: Figure 2 of Ref. [103].

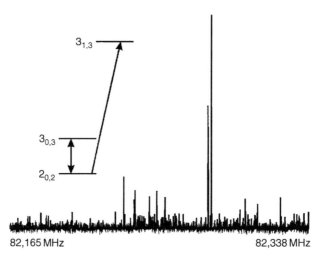

Figure 29   Broadband mm wave DR spectrum showing the J, K_a, $K_c = 3, 1, 3\leftarrow2, 0, 2$ "pump" transition of Ar−CO. With the "signal" pulse frequency fixed at that of the J, K_a, $K_c = 3, 0, 3\leftarrow2$, 0, 2 transition, the largest amplitudes of 1442 MW spectra from 10 added experiments are plotted as function of the mm wave "pump" frequency being varied in 120 kHz increments. Because of the modulation scheme, the transition appears as a doublet; see text for details. (Reused with permission from Ref. [85]. Copyright 1995, American Institute of Physics.)

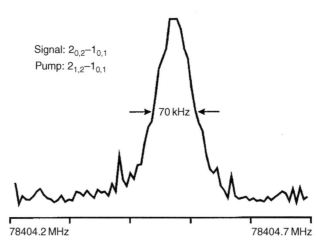

Figure 30   High-resolution mm wave DR spectrum showing the J, K_a, $K_c = 2, 1, 2\leftarrow1, 0, 1$ "pump" transition of Ar−CO. With the "signal" pulse frequency fixed at that of the J, K_a, $K_c = 2, 0, 2\leftarrow1, 0, 1$ transition, the largest amplitudes of 82 MW spectra from 10 added experiments are plotted as function of the mm wave "pump" frequency being varied in 6 kHz increments. (Reused with permission from Ref. [85]. Copyright 1995, American Institute of Physics.)

Extension of DR modulation FT-MW spectroscopy toward the sub-mm wave range, suitable for the investigation of low-lying intermolecular vibrational modes or of tunneling transitions, has also been pursued. This was accomplished in combining the TDS MW technique with broadband tunable and phase-stabilized BWOs [86]. Using an arrangement based on the same principles as the one shown in Figure 28, the FT-MW instrument is used to detect population changes N_b in the upper level of the "pump" transition b←c as a function of the sub-mm wave radiation frequency. Coherence changes ρ_{ab} can be neglected because of the short interaction region very close to the nozzle orifice, which ensures that the "pump" interaction period is essentially completed before the "signal" excitation pulse takes place. Strong DR effects are obtained for the MW "signal" transitions b←a or a←b, as the initial populations N_a, N_b of the two states are essentially zero because of the very low rotational temperatures after supersonic expansion. In fact, if a metal reflector is used so that the radiation traverses the resonator beam waist twice, the sub-mm wave excitation is sufficiently effective to achieve saturation that becomes visible as a dip on top of the line. As seen in Figure 31, this Lamb-dip-type effect results in a significant increase in the precision of frequency measurements, which was confirmed by combination differences [86]. The observation of effects on the MW "signal" transition from population changes in N_b, i.e., in the lower level of a mm wave or sub-mm wave "pump" transition c←b, however, is typically rather demanding with respect to the conditions required for appreciable DR excitation. In the near-IR, VIS, and near-UV regions [87], or if sufficient power and interaction time is provided in the mm wave range [88,89] as well as the IR region [90], depletion of the lower level is quite efficient and strong DR signals are obtained.

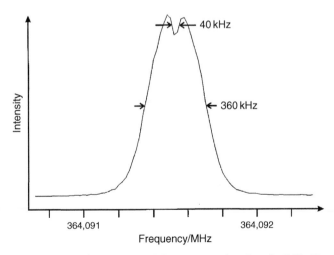

Figure 31 High-resolution sub-mm wave-DR spectrum showing the J, K_a, $K_c = 2, 0, 1 \leftarrow 0, 0, 0$; $\nu_{\text{bend}} = 1 \leftarrow 0$ "pump" transition of Ar−CO. With the "signal" pulse frequency fixed at that of the J, K_a, $K_c = 2, 0, 2 \leftarrow 1, 0, 1$; $\nu_{\text{bend}} = 1 \leftarrow 1$ transition, the largest amplitudes of 100 MW spectra from 10 added experiments are plotted as function of the sub-mm wave "pump" frequency being varied in 15 kHz increments. Five plots were averaged. (Reused with permission from Ref. [86]. Copyright 1998, American Institute of Physics.)

3.4.4.   Coherence effects

If, after a resonant "signal" transition b←a or a←b pulse excitation, a "pump" transition b←c or c←b of a molecular system interacts with coherent electromagnetic radiation, the coherence ρ_{ab} of the "signal" transition is modulated with the angular frequency $\Omega_s(r)$, dependent only on the Rabi angular frequency $x_p(r)$ of the "pump" transition and independent of the "signal" transition. At resonance, i.e., for a difference $\Delta\omega_{bc}=0$ between the applied, ω_p, and the resonant, ω_{bc}, angular frequencies, one obtains [91]

$$\Omega_s^2(r)=x_p^2(r)/4=\left(\mu_{bc}\,\varepsilon_p(r)/\hbar\right)^2/4. \tag{19}$$

Here, μ_{bc} is the electric dipole matrix element of the space-fixed z-component of the dipole operator and ε_p is the z-polarized electric field amplitude of the external radiation. When off-resonant, i.e., $\Delta\omega_{bc}\neq 0$, coherence modulation becomes much more complicated [92]. If the electric field $\varepsilon_p(r)$ is inhomogeneous, and also due to the spatial velocity distribution $v(r)$ of the supersonic-jet expansion, resulting in $\omega_p(r)$ from the Doppler effect, dephasing occurs and the molecular systems do not return to the initial coherence ρ_{ab}, i.e., the macroscopic polarization depletes. Because molecules dephased by DR interaction no longer contribute to the macroscopic polarization, complete depletion of the transient emission signal can in principle be achieved. This is in contrast to non-coherent DR phenomena, where a maximum depletion of half of the signal intensity cannot be exceeded.

For any inhomogeneous stationary field in the RF, MW, IR, VIS, or UV regions, applied after excitation of the "signal" transition, which is resonant to a transition that shares an energy level with the latter transition, the coherence associated with the transition can be destroyed and its transient emission signal then rapidly decays. This DR technique in combination with supersonic-jet FT-MW spectroscopy was first realized for the near-IR, VIS, and near-UV regions [87]. In the experiment, a light pulse from a tunable laser irradiates the sample molecules that are already polarized by a MW pulse, as illustrated in Figure 32. The frequency of the

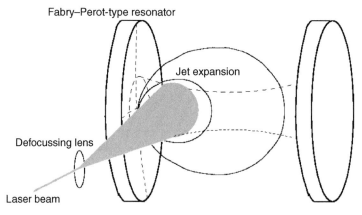

Figure 32   Supersonic-jet expansion = FT-MW (VIS, UV)-DR experiment. The laser beam is expanded with a concave lens to match the beam-waist of the MW field in the resonator. *Source*: Figure 3 of Ref. [103].

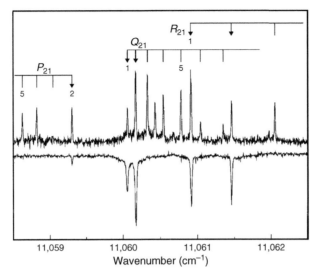

Figure 33 904 nm band of CCS: (upper trace) LIF spectrum; (lower trace) near-IR-DR spectrum. With the "signal" pulse frequency fixed at that of the J, $N = 3, 1 \leftarrow 1, 0$ transition, the largest amplitudes of 1100 MW spectra from 30 added experiments are plotted as function of the near-IR "pump" frequency being varied in 0.0003 nm increments. (Reused with permission from Ref. [87]. Copyright 2002, American Institute of Physics.)

laser is scanned, and when the laser is resonant to an electronic transition involving the upper or lower level of the MW transition, the "signal" coherence is reduced and a faster decline of the FID occurs during the interaction period. The electronic spectrum of transitions sharing a level of the "signal" transition is thus obtained by monitoring the MW transient emission signal as function of the laser frequency. As seen in Figure 33, the resulting FT-MW near-IR DR spectra resemble absorption spectra having a better S/N ratio and being less congested than the LIF spectra. Similar setups have later been utilized in the mm wave range of the MW region [88,89].

The versatility of the method is especially important for the IR region where direct detection of emission is difficult. Employing the FT-MW instrument as an IR detector, makes it possible to measure vibrational spectra of low-abundance species with excellent sensitivity [90]. The unrivalled spectral resolution of the FT-MW spectrometer provides fully resolved pure rotational spectra, even if the sample contains several different molecular species. By monitoring a selected pure rotational transition, it is then possible to obtain the vibrational spectrum of a single-molecular species without spectral overlap from other molecules, conformers, or isomers. This capability gives the technique a distinct advantage over fluorescence, ionization dip, and rotational coherence methods. In contrast to ionization detection methods—because FT-MW spectroscopy is non-destructive—the problems associated with fragmentation of larger species are avoided. The ability to monitor a single pure rotational transition is another major strength of FT-MW IR-DR spectroscopy as it greatly simplifies the rotational contour of the vibrational spectrum. Besides facilitating an easier assignment, weak spectral features, which are obscured by the rotational band contour even at the low-rotational temperatures found in supersonic-jet expansions, are revealed. The spectral simplification also allows for a determination of homogeneous spectral widths associated with internal vibrational energy redistribution (IVR) and dissociation, even when the width is in the order of the rotational spacing. This ability makes the method more suitable for dynamics studies than, e.g., techniques like cavity ring-down spectroscopy and slit-jet FTIR spectroscopy.

Population and coherence effects can be employed at the same time in triple-resonance (TR) experiments [93,94], essentially preserving the S/N ratio of the FT-MW experiment if efficient excitation is achieved. In the supersonic-jet FT-MW MW/IR-TR experiment, an IR laser pulse is directed into a plane-parallel multipass assembly with a multipass length matched to the MW resonator beam waist such that a "flock" of molecules, which can fill the active region of the resonator, is vibrationally excited. Using a slit-jet aligned along the laser beam path, a slab of vibrationally excited molecules drifts out of the multipass region and into the active region of the cavity. A second CW tunable MW source radiates into the active region through a standard gain horn. The FT-MW MW/IR-TR spectra are obtained by gating on the intensity value of the center of a monitored IR-excited state FT-MW signal in the frequency domain. The center height is recorded as the CW MW frequency is scanned. A resonance of the CW MW radiation with a transition that shares a level with the FT-MW transition is observed as a dip in the monitored signal. The symmetric splitting, also known as Javan doublet [91], arises from modulation of the "signal" transition ρ_{ab} coherence according to Equation (19). If the "pump" modulation is off-resonant, i.e., $\Delta\omega_{bc} \neq 0$, the splitting gets asymmetric, exhibiting different amplitudes and the line might even become a quartet where the pattern is inverted about ω_{bc} if the sign of $\Delta\omega_{bc} = \omega_{bc} - \omega_p$ is inverted [94]. The resulting FT-MW MW/IR-TR signal, obtained by taking the amplitude at the dip as a function of the CW MW frequency, resembles an absorption spectrum. Depending on the IR excitation lifetime, an observable lifetime broadening of the vibrationally excited-state FTMW signals may or may not occur.

4. TIME-DOMAIN (FREQUENCY RAMP/CHIRP) TECHNIQUES

The TDS technique has a further advantage over FDS steady-state techniques: The time required to record a spectrum in the transient excitation experiment is a priori independent of the width of the spectrum. In principle, this "multiplex advantage" translates into an improvement of the S/N ratio for a spectral band that is proportional to the square root of the spectral band to linewidth ratio [31]. However, unlike the situation in pulsed NMR spectroscopy [95], this S/N improvement was, until recently, not exploited in FT-MW spectroscopy. In particular, high-Q resonator techniques, or waveguide methods for the case of low-transition dipole moments, are still narrow-banded because of the rather limited excitation bandwidth typically achieved under the conditions of a "short stimulus" (SS) of the molecular systems to a stationary frequency radiation field.

4.1. Chirped radiation microwave excitation of supersonic jets

If the frequency of an electromagnetic field is swept through a molecular resonance in a short time compared to the relaxation time, the so-called fast passage (FP) excitation occurs. Even though the molecules are in resonance only for a very short time, a surprisingly large change in the population difference of the states in resonance and in the coherence of the two-level ensemble can be achieved, resulting in an appreciable oscillating macroscopic polarization. FP can be rationalized as transient absorption during the sweep, followed by transient emission after the sweep [96]. Comparing the resulting signal amplitude $S_{ab}(FP)$ of the frequency ramp technique with the signal amplitude $S_{ab}(SS)$ obtainable by the resonant stationary frequency method, i.e. for $x \gg \Delta\omega$, at a given geometry of a sample cell with flat frequency response, yields

$$ S_{ab}(FP)/S_{ab}(SS) = x\sqrt{\left(\frac{2\pi}{\alpha}\right)}/\sin\left(x\tau_p\right) \tag{20} $$

with the Rabi angular frequency $x = \mu_{ab}\,\varepsilon/\hbar$ of the transition b←a, characterized by the electric dipole matrix element μ_{ab}, the duration τ_p of the resonant radiation pulse, and the slope $\alpha = d(\Delta\omega)/dt$ of the frequency ramp of a z-polarized electric field ω_e with the amplitude ε at off-set $\Delta\omega = \omega_{ab} - \omega_e$ from the resonance frequency ω_{ab}. With $x^2/\alpha \ll 2\pi$, a *linear FP* according to $x/2\pi \ll \alpha/x = d(\theta(\Delta\omega))/dt$ has been assumed, i.e., in the Bloch vector model, the precession frequency is low compared to the changes $d(\theta(\Delta\omega))/dt$ in the direction of its precession axis [39]. This is consistent with a negligible population transfer $d(\Delta N_{ab})/dt \approx 0$. Still, it becomes clear from the Bloch vector model, that for $x/2\pi \approx \alpha/x$, i.e., if the precession frequency is similar to the frequency of reorientation $d(\theta(\Delta\omega))/dt$ of the precession axis upon FP, a population equilibration $\Delta N_{ab} \approx 0$ can also be obtained, resulting in maximum coherence. Any molecular transition b←a in the band covered by the frequency ramp $\omega_e(t)$, i.e., at frequencies within the interval $\omega_i \leq \omega_e(t) \leq \omega_f$, will be polarized. Because it is possible that the passage includes transitions that share a common level, DR effects have to be considered. However, for a *linear FP*, $d(\Delta N_{ab})/dt \approx 0$, and thus population effects that modify ΔN_{ab} are negligible. Similarly, coherence effects according to Equation (19) which might alter ρ_{ab}, i.e., the coherence from the first resonance, are minor in the *linear FP* regime unless $\mu_{bc} \gg \mu_{ab}$. Thus, FP across molecular resonances in a sufficiently short time with respect to the relaxation times, i.e., $1/\alpha \ll T_2^2$, can excite all transitions in a broad band using the TDS technique. In contrast, for the regime $x^2/\alpha \gg 2\pi$, i.e., a precession frequency which is high compared to changes $d(\theta(\Delta\omega))/dt$ in the direction of the precession axis, an *adiabatic FP* also known as "adiabatic rapid passage" (ARP) occurs and a population inversion, i.e., $\Delta N_{ab} = -\Delta N_{ab}^0$, at negligible coherence excitation, i.e., $\rho_{ab} \approx 0$, can be achieved [97]. The term *adiabatic* refers to the Bloch vector, which follows its precession axis smoothly; the effect is not adiabatic in the sense that no transitions occur. The *adiabatic FP* in its pure form is thus the "π-pulse" SS analog, whereas the *linear FP* produces results analogous to the "π/2-pulse" SS. While the $S_{ab}(FP)/S_{ab}(SS)$ ratio of a single experiment is typically an order of magnitude smaller than unity, the bandwidth $\omega_f - \omega_i$ of the frequency ramp "chirp" can be several orders of magnitude larger than the spectral coverage of a stationary frequency"impulse."

4.2.   Fast passage FT-MW apparatus

Even though more experiment repetitions might be necessary for the FP experiment to recover the S/N ratio of the SS method, in particular for low transition dipole moments, as no high-Q resonator can be employed, the "multiplex advantage" still yields a huge improvement in the S/N ratio if a broad spectral band is to be recorded in a limited time.

Exploitation of the "multiplex advantage" in early implementations of the FP experiments [96,98,99] was limited by the slope and coverage of the frequency ramp that could be provided with sufficient phase stability for repetition of the coherent experiments as well as the requirement for high-speed signal detection and the huge amount of signal processing. Owing to the enormous advances in digital electronics, the situation has changed, and one can now generate phase-coherent broadband MW frequency ramps as well as digitize sufficiently fast while simultaneously performing massive signal processing. The sensitivity and extreme broadband capabilities of the method, as shown in Figure 34, are impressive [100]. Even though the specifications

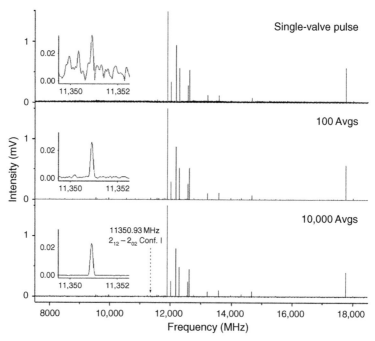

Figure 34   FP rotational spectrum of epifluorhydrin: 7.5 . . . 18.6 GHz broadband (11 GHz) spectra after 1, 100, and 10,000 added experiments taken at a repetition rate of 3.5 Hz. Conformer I is observed after one experiment with good S/N ratio, conformers II and III as well as the ^{13}C-species of conformer I are seen well after 100 experiments. The inset visualizes the S/N improvement of the weak J, K_a, $K_c = 2, 1, 2 \leftarrow 2, 0, 2$ transition of conformer I, marked on the bottom trace, on an expanded scale. (From Ref. [100]. Copyright 2006, Elsevier Inc.)

of key components required for FP experiments are quite demanding, the technical effort for building such an instrument can be significantly reduced if "very" rather than "ultra-" broadband operation is aimed for [101].

4.2.1. IMPACT spectrometer

The state-of-the-art supersonic-jet expansion FP FT-MW spectrometer discussed here provides advantages in some significant points [102]. The general design considerations on its high-frequency electronics are instructive for all FP methods [103]. Utilizing the surprisingly simple in-phase/quadrature-phase (I/Q) modulation scheme for direct SSB frequency translation, the design implements the "in-phase/quadrature-phase modulation passage-acquired coherence technique" (IMPACT). This implementation, with readily available components allowing for frequency coverage from 2.0–26.5 GHz in a single setup, is especially suited for the detection and characterization of larger, instable compounds—also in very small concentrations. A block diagram of the IMPACT spectrometer, using a short MW radiation frequency ramp for excitation, is presented in Figure 35; the principle of operation of the—compared to other FT-MW instruments—relatively straight forward setup is as follows: A primary signal with frequency ν is generated by the signal generator (1). With the pin-diode SPDT switch (2a), the output signal is either routed for excitation or for observation. In the first case, an MW frequency ramp for the FP excitation of the molecules is created; the entire excitation branch will only be supplied with microwave power for the duration of the polarization pulse, such that the risk of coherent perturbations by signal leakage during the detection is reduced. In the second case, the microwave response of the probe—taken as a difference-frequency signal between the molecular and the primary radiation—will be converted into a lower frequency band, the so-called base band. Similar to the COBRA FT-MW apparatus, the primary signal source is not used directly for the molecular excitation. The output signal of an SSB up-converter (4) is utilized instead. However, in contrast to the narrow-banded (<1 MHz) excitation by the COBRA apparatus, the SSB modulator is not used to generate a specific fixed frequency: Instead, the concept of a broadband excitation is realized via I/Q modulation at frequencies $\Delta\nu = -500 \ldots +500\,\text{MHz}$ utilizing an arbitrary-signal generator (3), such that a broadband (>1 GHz) SSB frequency ramp $\nu + \Delta\nu(t)$ is created for an excitation that is phase-invariantly repeatable. The power level necessary for the polarization of the molecular systems is adjusted via the amplitude of the modulation, so that a variable attenuator (4 of Figure 21) becomes obsolete. A high-power amplifier (6) not only compensates for the conversion loss but also provides high power levels, which becomes necessary because of the spectral width of 1 GHz, which is swept in $\tau_p = 10\mu s$ typically. The FP excitation is to be preferred over a fixed frequency pulse, which—at the same mean width and energy—would require the 10,000-fold power lasting for 1 ns, if it could be realized at all. Furthermore, the FP technique allows for a uniform excitation over the entire frequency range. The pin-diode SPDT switch (2b) isolates the molecular interaction zone during the detection from noise of the excitation electronics. The idle noise between the excitation pulses is eliminated in the terminating load (7) at the second output of the SPDT switch.

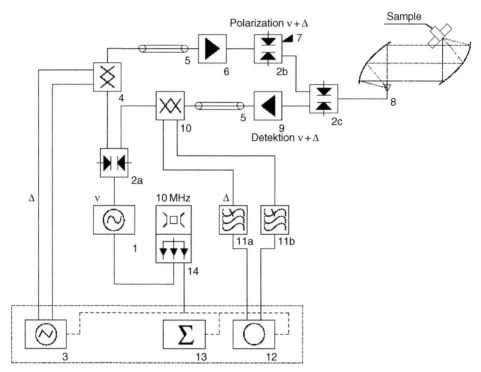

Figure 35 Supersonic-jet expansion IMPACT FT-MW spectrometer: (1) MW signal generator, 10 dBm output power; (2) MW SPDT switch, 50 dB isolation, 3 dB insertion loss; (3) RF arbitrary-signal generator, 2×1.25 GS/s generation, 15 bit resolution; (4) MW I/Q-up-converter, 15 dB carrier suppression, 12 dB insertion loss; (5) MW coaxial cable, flexible; (6) MW (very-) high-power amplifier, >27 dB gain, >30 dBm output power; (7) MW termination, 50 Ω impedance; (8) MW horn antenna with offset-parabola reflector, 25 dBi gain; (9) MW low-noise amplifier, 2.8 dB noise figure, 30 dB gain, 0 dBm output power; (10) MW I/Q down-converter, 18 dB image rejection, 14.5 dB insertion loss; (11) MW low-pass filter, DC-990 HMz, 2 dB insertion loss; (12) MW digitizer, $2 \times$ GS/s real-time acquisition; (13) PXIe-based experiment control, data-acquisition, and data-analysis system; (14) Rubidium frequency standard, 10 MHz output frequency, 5×10^{-10} frequency stability. *Source*: Figure 4 of Ref. [103].

After passing a pin–diode SPDT switch (2c), set for the excitation branch, the energy of the excitation signal is focussed via an antenna (8), in combination with an off-axis parabolic reflector, onto a metal plate that holds the molecular beam valve in its center. The maximum field amplitude for the polarization of the molecular systems is thus available in the region of the highest particle density. Because the parabolic reflector arrangement focuses the radiation beam onto the expanding jet and thus enhances the field amplitude ε, most efficient use is made of the available source power for molecular polarization. Containing signals within the range ν-500...ν + 500 MHz, the molecular IMPACT response – being projected onto the propagation mode of the incident and reflected radiation field – is then received by the same antenna/reflector combination (8) and passes the pin–diode SPDT switch (2c), now set for the detection branch. Besides setting the signal pathway, the switch protects the delicate MW receiver electronics from the harmful power levels of the frequency ramp excitation.

The molecular signal is passed to an extremely low-noise amplifier (9). The gain is chosen such that the insertion losses in the subsequent stages of the signal processing have just a negligible impact on the S/N ratio. An excess gain must be avoided, because power saturation of the subsequent components is provoked and thus the dynamic range of the spectrometer is reduced. In the SSB modulator (10), the broadband molecular IMPACT response is converted back into the base band around DC with $\Delta\nu = -500 \ldots +500$ MHz as an I/Q signal. By setting the primary frequency ν, any desired region of the spectrum within the range 2–26.5 GHz can be projected into this base band. The frequency conversion from and back to the base band in a single step minimizes destructive phase fluctuations (jitter). The low-pass filters (11) reject signals outside the Nyquist frequency limit according to the folding theorem of the A/D conversion performed by the transient recorder (12), which provides two-channel quadrature detection. With a complex FT implemented in the data-acquisition, data-control, and data-analysis system (13), the spectral composition of the transient emission is obtained.

Because the extremely weak molecular signal is usually far below the thermal noise level, phase-synchronized repetitions are essential to achieve an improvement in the S/N ratio by the phase-correct addition of single experiments. A signal provided from a rubidium frequency standard (14) is used for the generation of all analog frequency signals and to derive the digital system clock rates. Only if all signal sources used in the system are referenced to the same frequency standard, it becomes possible to generate the molecular signal with recurring phase. The relative arrangement of the molecular jet and the high-frequency field is very similar to a semiconfocal COBRA apparatus, but the parabolic reflectors are not forming a resonator in this setup. From the induced polarization of the velocity-equilibrated molecular jet – expanding coaxially with respect to the high-frequency incident and reflected radiation beam – a Doppler doublet is obtained again. Its components exhibit a very small spectral width similar to the FWHM $\cong 3$ kHz, found for the COBRA experiment, because the observation time is similar in this arrangement. Then, at good S/N ratios, the signal frequency can be determined as accurate as 300 Hz or 0.00000001 cm^{-1}.

4.2.2. WIDE-IMPACT spectrometer

The state-of-the-art supersonic-jet expansion wide-band FP FT–MW spectrometer discussed here maintains the IMPACT advantages while extended frequency coverage for single experiments is achieved [102]. The general design of its high-frequency electronics provides further opportunities for additional, more complex experiments [103]. This implementation is especially suited for the detection and characterization of compounds exhibiting dynamical processes, even in low concentrations. A block diagram of the WIDE-IMPACT spectrometer, using a short MW radiation frequency ramp for excitation, is presented in Figure 36. By wide-band I/Q modulation, it is possible to generate another impulse or CW signal without delay after FP excitation. Thus, 2D or 1D FP FT–MW MW–DR experiments are inherently feasible by the technique, whereas conventional spectrometers

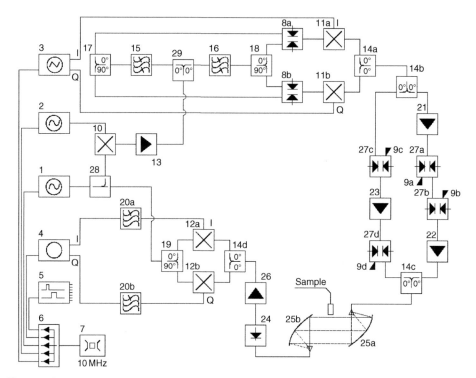

Figure 36 Supersonic-jet expansion WIDE-IMPACT FT-MW spectrometer: (1) MW phase-locked oscillator, f_0, +12 dBm; (2) MW phase-locked oscillator, f_1, +12 dBm; (3) arbitrary-waveform generator, $2 \times DC \ldots f_1$; (4) digital oscilloscope, $2 \times DC \ldots 2f_1$, real-time acquisition; (5) PXIe-based experiment control, data-acquisition, and data-analysis system; (6) distribution amplifier, 5×10 MHz; (7) frequency standard, 10 MHz, 5×10^{-10}; (8) MW SPDT-switch, $f_0 - f_1 \ldots f_0 + f_1$, speed: 20 ns, insertion loss <3.0 dB; (9) MW termination, $f_0 - 2f_1 \ldots f_0 + 2f_1$; (10) MW modulator, IF-in: f_1, LO-in: f_0, RF-out: $f_0 - f_1 \ldots f_0 + f_1$; (11) MW modulator, IF-in: $DC \ldots f_1$, LO-in: $f_0 - f_1 \ldots f_0 + f_1$, RF-out: $f_0 - 2f_1 \ldots f_0 + 2f_1$; (12) MW modulator, RF-in: $f_0 - 2f_1 \ldots f_0 + 2f_1$, LO-in: f_0, IF-out: $DC \ldots 2f_1$; (13) MW driver amplifier, frequency: $f_0 - f_1 \ldots f_0 + f_1$, +20 dBm, 27 dB; (14) MW in-phase power divider/combiner, $0°/0°$, $f_0 - 2f_1 \ldots f_0 + 2f_1$; (15) MW bandpass filter, $f_0 - f_1$, insertion loss: <1.0 dB, rejection: 30 dB; (16) MW bandpass filter, $f_0 + f_1$, insertion loss: <1.0 dB, rejection: 30 dB; (17) MW quadrature-phase power divider, $0°/90°$, $f_0 - f_1$; (18) MW quadrature-phase power divider, $0°/90°$, $f_0 + f_1$; (19) MW quadrature-phase power divider, $0°/90°$, f_0; (20) MW low-pass filter, $DC \ldots 2f_1$, insertion loss: <0.5 dB, rejection: 30 dB; (21) MW pre-amplifier, $f_0 - 2f_1 \ldots 18$ GHz, 22 dBm, 29 dB; (22) MW-pulsed high-power amplifier, $f_0 - 2f_1 \ldots 18$ GHz, 63 dBm, 46 dB; (23) MW power amplifier, $f_0 - 2f_1 \ldots f_0 + 2f_1$, 27 dBm, 27 dB; (24) MW high-power limiter, $f_0 - 2f_1 \ldots 18$ GHz, 1.5 kW, insertion loss: <3.0 dB; (25) MW horn antenna w/offset-parabola reflector, $f_0 - 2f_1 \ldots f_0 + 2f_1$, 25 dBi; (26) MW low-noise amplifier, $f_0 - 2f_1 \ldots f_0 + 2f_1$, noise figure: <3.0 dB, 27 dB; (27) MW SPDT-switch, $f_0 - 2f_1 \ldots f_0 + 2f_1$, speed: 20 ns, insertion loss <3.0 dB; (28) MW directional coupler, <−10 dB, f_0; (29) MW in-phase power divider/combiner, $0°/0°$, $f_0 - f_1 \ldots f_0 + f_1$. *Source:* Figure 5 of Ref. [103].

would need an additional coherent source. Because I/Q modulators for extreme frequency coverage are not yet commercially available, wide-band I/Q modulation has to be realized using a number of discrete components. Due to the larger number of components, the block diagram seems more complicated, although the complexity of the principal setup is unchanged. Unlike other FT-MW spectrometers, the WIDE-IMPACT apparatus does not even require a tuneable MW source; the particular implementation discussed here utilizes a fixed-frequency dual-sideband source assembly, thereby doubling the frequency coverage. The principle of operation of the MW-DR-capable FP setup is as follows: A primary signal of frequency $f_0 = 16.5$ GHz is generated by a phase-locked oscillator (1). The output signal, distributed via a directional coupler (28), is either used for excitation or for frequency translation of molecular signals. In the first case, an MW frequency ramp for an FP excitation of the molecules is created. In the second case, the microwave response of the probe – obtained from a difference-frequency signal between the molecular and the primary radiation f_0 – will be down-converted to the base band. The primary signal source (1) is not used directly for the molecular excitation but is modulated by the signal of a secondary phase-locked oscillator (2). Subsequently, one of the two sideband signals with frequencies $[f_0 - f_1]$ and $\{f_0 + f_1\}$ at the output of the modulator (10) is selected by the pin-diode SPDT switches (8) after passing the power divider (29) and filters (15, 16). The concept of a broadband excitation is realized via I/Q modulation at frequencies $\Delta f = -f_1 \ldots +f_1$ utilizing an arbitrary-signal generator (3), such that a broadband SSB frequency ramp $[f_0 - f_1 + \Delta f(t)]$ or $\{f_0 + f_1 + \Delta f(t)\}$ is created for a phase-invariantly repeatable excitation, each with a bandwidth of 10 GHz. The generation of such a frequency ramp by means of a voltage ramp via a voltage-controlled oscillator (VCO) does not provide the phase stability required for a coherence experiment. This becomes feasible only by using an arbitrary-waveform generator providing a two-channel analog bandwidth of $f_1 = 5.0$ GHz, which, via I/Q modulation of the carrier signal at $[f_0 - f_1]$ or $\{f_0 + f_1\}$, generates bands for the ranges $[f_0 - 2f_1 \ldots f_0]$ or $\{f_0 \ldots f_0 + 2f_1\}$, respectively, with a bandwidth of $2f_1 = 10.0$ GHz in each case. The particular type of frequency ramp generation via digital-to-analog (D/A) conversion, with a clock phase-locked to a common system reference for all spectrometer components, facilitates the overall phase-invariant repeatable coherence experiment. A wide vertical dynamic range of the generator allows for the compensation of the frequency response of the instrument as well as a power-transduced signal shaping for optimum polarization of the molecular systems studied in the coherence experiment. SSB generation via the I/Q method is realized by means of the modulators (11), the quadrature-phase power divider (17, 18), and the in-phase power divider (14a). Prior to this sideband generation, the power at the output of the amplifier (13) is increased to a level such that, after passing the in-phase power divider (29), the filters (15, 16), the quadrature-phase power divider (17), (18), and the pin-diode SPDT switches (8), sufficient power for either sideband is available at the modulators (11).

The SSB signal generated via I/Q modulation can be repeated at any time to phase-invariantly pass a frequency band, because phase stabilization is inherently warranted by the D/A clock since it is directly derived from the common system reference signal. Because the extremely weak molecular signal is frequently buried under the thermal noise, phase-synchronized repetitions are essential to achieve an improvement of the S/N ratio by phase-invariant addition of single experiments. A 10-MHz reference signal, being shared via a distribution amplifier (6) at the output of a frequency standard (7), is used to generate all analog and digital frequency signals and to derive the system clocks. By referencing all of the system's signal sources (1–3) and recorders (4) to the same frequency standard, it is possible to repeatedly generate and receive molecular signals at recurring identical phases excluding any drift. The long lasting transient emission, supersonic-jet experiments in the MW region allow for a very high frequency precision, and thus a very accurate reference such as a GPS receiver-disciplined rubidium standard is required.

The power level necessary for inducing a polarization of the molecular systems is adjusted via the amplitude of the modulation. A high-power amplifier (22) provides a very high output power of 2 kW, which is necessary because of the wide spectral width of 10 GHz, which is swept in $\tau_p = 10\,\mu s$ typically. This FP excitation is preferential to a fixed frequency pulse which – with the same mean width and energy – would translate to an unrealistic 100,000-fold power at 100 ps length. Furthermore, the FP technique can provide constant excitation conditions over the entire frequency range. Wide bands together with high output powers are provided by traveling wave tube (TWT) amplifiers. Two pin–diode switches (27a, b) are used in such a way that the input of the amplifier receives the signal from the pre-amplifier (21) only for the duration of the frequency ramp, thus reducing a risk of coherent perturbations by signal leakages during the detection. Outside the excitation period, the terminations (9a,b) avoid an impedance mismatch at the output of the pre-amplifier (21) and the input of the TWT amplifier (22). During the detection period, the gate control of a TWT pulse amplifier constitutes a very effective isolation of the molecular interaction zone from the noise of the excitation electronics. The idle noise between the excitation periods is eliminated in the termination resistor (7a) at the second output of the SPDT switch. Commercially available wideband TWT amplifiers with a range of 6.5–18.0 GHz cover the entire band of the lower sideband's frequency ramp $[f_0\text{-}2f_1\ldots f_0] = [6.5\ldots 16.5\,\text{GHz}]$. After switching the sideband, which is achieved by the pin-diode SPDT switches (8) within 20 ns, an additional 1.5 GHz frequency interval of the upper sideband's frequency ramp $\{f_0\ldots f_0 + 2f_1\} = \{16.5\ldots 26.5\,\text{GHz}\}$ below 18 GHz is accessible.

After passage of the wide-band frequency ramp, an impulse or CW signal at a given frequency can be generated by I/Q-modulation without any time lag: Thus, via two inphase-power dividers (14b, c), specific DR excitations with narrow frequency-ramp or fixed-frequency signals in the entire instrumental range $[f_0\text{-}2f_1\ldots f_0]\{\ldots f_0 + 2f_1\} = [6.5\ldots 16.5]\{\ldots 26.5\,\text{GHz}\}$ can be performed. Because the arbitrary-waveform generator (3) allows for complete amplitude and phase control, elaborate multi-impulse phase sequences are freely programmable. For the duration of the DR excitation, a pin-diode SPDT switch (27c) supplies the amplifier (23) with the input signal. The termination (9c) avoids an impedance mismatch at the input of the amplifier outside the DR excitation period. Because selective DR excitations do not require wide-frequency ramps, solid-state amplifiers with an output power below 1 W are sufficient. A second assembly of a pin-diode SPDT switch (27d) and a termination (9d) at the amplifier's output isolates the idle noise from the molecular interaction zone during the detection period. By exploiting the frequency agility of the I/Q modulation setup, 2D or 1D FP FT-MW MW-DR experiments are easily performed with a single MW source. Because the WIDE-IMPACT apparatus provides wide-band coverage in a single experiment, 2D FT-MW MW-DR experiment sequences with wide-band coverage in both dimensions now become time efficient and thus practicable, whereas conventional supersonic-jet expansion COBRA FT-MW spectroscopy can still be performed in the entire instrumental range by employing a Fabry–Pérot-type resonator instead of a sample cell or a free-space interaction assembly with flat frequency response.

Through an antenna/reflector assembly (25a), the energy of the excitation signal is collimated towards the expansion volume of the supersonic jet. The maximum field amplitude for the polarization of the molecular systems is thus available in the region of high particle density. Suitable wide-band double-ridge horn/off-axis parabolic reflector assemblies can cover the entire instrumental range $[f_0 - 2f_1 \ldots f_0]\{\ldots f_0 + 2 f_1\} = [6.5 \ldots 16.5]\{\ldots 26.5\,\text{GHz}\}$ at high directivity. Antennae with high directivity, i.e., antenna gain, are the prerequisite for an efficient excitation of the molecular sample as well as for the sensitive detection of the wide-band molecular IMPACT response in the instrumental frequency range. The transient emission, which is projected onto the propagation mode of the incident and reflected radiation field, is then received by a second antenna/reflector assembly (25b). The molecular signal at the output of the receiving antenna is passed to a very low-noise amplifier (26) with a noise figure of less than 3.0 dB, which is protected from the powerful excitation signals by a pin-diode power limiter (24). Being the first active element in the signal chain after antenna and limiter, the noise figure of the amplifier (26) is the most important factor for the achievable S/N ratio. To ensure that insertion losses in the subsequent signal-processing stages reduce the S/N ratio just negligibly, its gain is chosen to be about 30.0 dB. Because power saturation of the subsequent components is otherwise provoked and thus the dynamic range of the spectrometer is reduced, a higher gain should be avoided. The frequency translation via the assembly of the modulators (12), the quadrature-phase power divider (19), and the in-phase power divider (14d) converts the wide-band molecular IMPACT response back to an I/Q signal covering the DC–10.0 GHz band. When repeating the experiment, the I/Q demodulation of the transient emission with the output of the primary signal source (1) warrants the phase-invariance of the projection into the base band. An extension of the 12-GHz wide signal band of 6.5–18.0 GHz, determined by the frequency coverage of the TWT amplifier (22), to the spectral range of 18.0–26.5 GHz can be achieved by retro-fitting high-power components (21, 24) suitable for the latter frequency range. Then, the entire 20-GHz wide spectral range of $[f_0 - 2f_1 \ldots f_0]\{\ldots f_0 + 2f_1\} = [6.5 \ldots 16.5]$ $\{\ldots 26.5\,\text{GHz}\}$ can be projected into the base band span of $-2f_1 \ldots +2f_1 = -10\,\text{GHz} \ldots +10\,\text{GHz}$. The frequency conversion from and back to the base band in single modulation and demodulation steps, respectively, minimizes destructive phase fluctuations. The low-pass filters (20) reject signals outside the Nyquist frequency limit according to the folding theorem of the A/D conversion, which is performed by the digital oscilloscope (4) providing a two-channel quadrature detection. After addition of phase-invariantly repeated single experiments in the digital oscilloscope to improve the S/N ratio, a subsequent complex FT, and after possibly applying appropriate additional signal processing in the data-acquisition, data-control, and data-analysis system (5), the spectral composition of the transient emission is obtained. The system controls all instrument components, generates trigger as well as excitation sequences, and synchronizes them for phase-invariant experiment repetition.

4.2.3. Linear FP signal

Although the foremost benefit of FP FT-MW spectroscopy is the exploitation of the "multiplex advantage," it provides some additional benefits with respect to the spectral representation of the observed signals: Because, compared to the narrow-band resonator FT-MW experiment, the transient emission amplitudes of this wide-band free-space FT-MW experiment are significantly weaker, there is virtually no risk to saturate any suitable receiver components of the spectrometer. With the overall gain of the spectrometer adjusted such that the noise fluctuations just trigger the least-significant bits of the digital oscilloscope, a very large dynamic range previously only available by waveguide FT-MW is obtained. Its limit, i.e., the detection threshold, can be selected as desired and, as indicated in Figure 34, is determined by the number of experiments that are performed. What is even more important – because the excitation conditions over a wide range are consistent with those for a *linear FP* ensuring the validity of Equation (20) – is that the amplitudes of the transient emission signals are proportional to $\Delta N_{ab}\mu_{ab}^2$, i.e., the amplitudes show the same behavior as obtained with Equation (1) for absorption spectroscopy. This differs from the $\Delta N_{ab}\mu_{ab} \sin(\varepsilon\tau_{\mathrm{p}}\,\mu_{ab}/\hbar)$ dependency of the SS molecular response stated in Equation (2). Thus, even if a sample cell with flat frequency response – in particular no resonator setup – is employed and exact molecular resonance is carefully ensured, easily comparable SS excitation conditions are not straight-forward to maintain for signals exhibiting different transition dipole moments. Especially if assignments and/or the transition dipole moments are a priori doubtful or unknown, the *linear FP* response line intensities are more reliable. For narrow line multiplets though, the analysis of the amplitude frequency-domain spectra suffers from distortion effects causing line position shifts as large as the HWHM because of overlapping lines from signals with a priory unknown phases [37]. Being an implication inherent to all TDS methods, the reconstruction of the true frequency information encoded in the transient emission of narrow patterns can be facilitated by fitting the parameters of a model function to the time-domain signal [38].

Because of the favorable resonant interaction time τ_{p}, "$\pi/2$-pulse" SS conditions can usually be obtained even for less polar species. Then, Equation (20) becomes

$$S_{ab}(FP)/S_{ab}(SS) = \sqrt{\frac{x(2\pi)}{\alpha}} = \frac{\varepsilon\mu_{ab}}{\hbar}(2\pi/\alpha)^{1/2} \qquad (21)$$

The dependence on the transition dipole matrix element does not necessarily mean that FP FT-MW is less sensitive for the detection of less polar species, as the Rabi angular frequency $x = \mu\varepsilon/\hbar$ can be adjusted, while $\alpha/x^2 \gg 2\pi$ for a *linear FP* operation is maintained. To keep x constant, however, the power requirement with regard to the MW source is more demanding. For only weakly polar species, i.e., when "$\pi/2$-pulse" conditions cannot be established, Equation (20) reduces to

$$S_{ab}(FP)/S_{ab}(SS) = (2\pi/\alpha)^{1/2}/\tau_{\mathrm{p}} \qquad (22)$$

independent of the transition dipole matrix element μ_{ab}. As seen here, the sensitivity of both methods becomes equivalent if a constant slope $\alpha = \Delta(\Delta\omega)/\Delta t$ of the frequency ramp is chosen such that its time span Δt equals the impulse duration τ_{p} and its frequency span $\Delta(\Delta\nu) = \Delta(\Delta\omega/2\pi)$ covers the same spectral width $1/\tau_{\mathrm{p}}$ as the impulse. This equivalence persists if a series of n consecutive SS spectra is compared to a single FP spectrum: If the same MW source is used, the slope α increases with n, $S_{ab}(FP)$ drops with $n^{-1/2}$, and thus the number of experiments has to be increased by a factor of n to recover the S/N ratio. Because of this and from Equation (21), an m-fold "multiplex advantage" of the FP FT-MW technique only applies if MW sources are used that provide m times more power than needed for a "$\pi/2$-pulse" in the SS technique. The situation is different if a Fabry–Pérot-type resonator of spectral width $\delta\nu = \delta\omega/2\pi$ with a quality factor $Q_{\mathrm{L}} = \omega/\delta\omega$ resulting from critical coupling according to Equation (9) is used. Now the passive resonator gain enhances the transient emission signal with the amplitude $|S_{ab}(SS)| \propto Q_L^{1/2}$, resulting in the behavior described by Equation (13). The excitation field ε is enhanced with the same $Q_{\mathrm{L}}^{1/2}$ dependency, such that for identical SS polarization conditions, the output power requirement for the MW source decreases with $1/Q$. If maximum polarization is obtained in the FP experiment, the signal ratio of the two experiments is approximately $S_{ab}(FP)/S_{ab}(SS) \approx Q_L^{-1/2}$, and Q times more measurements have to be added to recover the S/N. Because Q SS measurements can be performed at the same total time in this regime, an m-fold "multiplex advantage" of the FP FT-MW technique is only achieved if the MW source provides enough power for an FP frequency ramp that covers $m \times Q$ times the spectral width $\delta\nu$ of the resonator mode at frequency ν. This is a bandwidth of $m \times Q \times \delta\nu = m \times \nu$, i.e., a rather wide range $\nu - m \times \nu/2 \ldots \nu + m \times \nu/2$, and therefore $m > 1$ is not realistic. However, the situation changes if the molecular spectrum exhibits an S/N ratio of n in single SS experiments covering the desired frequency range. The FP experiment now only needs Q/n^2 measurements to acquire a sufficient S/N ratio and an m-fold "multiplex advantage" of $m = n^2$ can be realized. In praxi, the advantage can even be much larger, because additional resonator setup times between SS measurements can become substantial, in particular if only a small number of experiments per frequency interval are required. Furthermore, it is not trivial to achieve broadband critical coupling for a Fabry–Pérot-type resonator, which amounts to an additional sensitivity advantage for the flat frequency response FP technique.

The development of A/D and D/A converters is one of the most dynamic fields of technical development, resulting in new devices such as the fast A/D recorder and D/A sources that are expected to outperform the still widely used analog sources and signal-processing devices. Considering the versatility combined with even wider coverage and higher frequencies soon to be reached, it is anticipated that FP FT-MW implementations, like WIDE-IMPACT discussed above, are becoming the most interesting instruments for FT-MW spectroscopy and will nothing less than revolutionize the field.

4.3. MW-(MW, mm, sub-mm) and MW-(MW, mm, sub-mm)/ (IR, VIS, UV) multiple resonance

The FP technique can easily be combined with additional static or oscillating external fields as demonstrated above for the established FT-MW methods. In this area, the broad frequency coverage also provides intriguing advantages. When applying static fields, i.e. in Stark effect measurements, precise frequency shifts are obtained for a large number of transitions at once, providing a "multiplex advantage" of a different kind: Instead of precisely determining the Stark effect for a molecular system on a limited number of lines for various electric fields, the frequency shifts of many lines measured at once can be evaluated for a single electric field.

Even more promising are the DR capabilities of the FP technique arising from the "multiplex advantage." Because of the drastically reduced time effort for wide frequency coverage in the MW domain, 2D DR spectroscopy, which required unrealistically long measurement times if broad coverage in both domains was desired, is now becoming feasible for many frequency regions of the second domain. This allows for a paradigm shift in MW spectroscopy: Being well established for determining the geometrical and electronic structure of molecular systems, including various effects of non-rigidity, rotational spectroscopy so far focused on problems that did not require a time-dependent treatment as mostly systems in the ground state or only moderately vibrationally excited states are accessible in conventional MW instruments. Even though the preparation of excited molecules by laser excitation opens up new vistas on dynamical effects to be elucidated by MW spectroscopy, its unrivaled resolving power and high sensitivity have largely not been exploited for studies on dynamical aspects, e.g., the kinetics of conformational isomerization [104].

In principle, the FP technique can be combined with DR excitation from almost any frequency region of the electromagnetic spectrum, exploiting the population and coherence effects discussed in the preceding section with the extended capabilities that arise from the fast production of broad 1D or 2D spectra. Arbitrary-waveform generation of the MW frequency ramp and impulses through direct I/Q modulation, which avoids subsequent mixing and multiplication stages, allows for exact signal shaping in amplitude and phase. Thus, cm wave DR excitation sequences of almost any complexity can easily be generated and provide the possibility for DR experiments that use even more elaborate sequences than those presented here for FT-MW with stationary gases. As outlined above for conventional narrow-band SS FT-MW spectroscopy, additional mm wave and sub-mm wave DR sources can now probe connectivities of resonances in MW spectra in a broadband fashion.

The most exciting perspective for FP FT-MW though, might be its suitability for the investigation of dynamic processes by rotational spectroscopy. Because the dynamic behavior manifests itself in the overall spectral envelope, the foremost prerequisite for such measurements is a fast wide-band coverage with reliable reproduction of line intensities ideally without sacrificing the sensitivity of narrow-band FT-MW methods. A drastically reduced acquisition time in the MW domain is particularly important if spectra are also taken in a second domain, i.e., if, rather than obtained at a given DR frequency, the laser excitation wavelength is also scanned to unravel intramolecular processes. Conformational isomerization, e.g., reveals itself in the composition of highly excited quantum states, which can be probed in detail by rotational spectroscopy. This becomes possible because the eigenfunctions of the Hamiltonian remain bound, i.e., the isomerizing molecular system still exhibits stationary wavefunctions with corresponding discrete energy levels. However, upon isomerization, the structurally delocalized nuclear wavefunction features characteristics of the involved conformations. To relate the dynamic process with the overall envelope of the rotational spectrum of the excited state, a time-dependent treatment is required: Interconversion between distinct species of different shape – characterized by rotational frequencies related to their different moments of inertia – constitutes an oscillation between the characteristic rotations that appears as coalescence of the spectral features, a well-known phenomenon in NMR spectroscopy [105,106]. The isomerization rate, corresponding to the conformational lifetime, which can be derived from the overall contour, serves as an upper limit, because IVR processes within a conformational species also contribute to spectral broadening.

Typically, the bandwidth of the IR-laser pulse is broad enough to excite multiple quantum states. Thus, FP FT-MW IR-DR spectra are not quantum-state resolved. To obtain state-resolved spectra, analogous to the narrow-band SS FT-MW TR method discussed in the preceding section, an FP FT-MW MW/IR-TR experiment can be employed [104]. State selection may be based on a level connectivity in any part of the entire MW region, by using either the cm wave impulse capabilities of the FP FT-MW setup or the additional mm and sub-mm wave sources. 2D spectra can then be obtained in two ways, i.e., the second dimension of the spectra can span different domains: by actively scanning the IR laser wavelength, 2D MW/IR spectra are generated or by incrementing the (MW, mm, sub-mm)-source frequency, 2D MW/(MW, mm, sub-mm) spectra are created. A supersonic-jet setup for FP FT-MW-detected (MW, mm, sub-mm)/tunable-IR TR interaction is illustrated in Figure 37: The fundamental output of a pulsed Nd : YAG laser (1) is used to pump an optical parametric oscillator/optical parametric amplifier (OPO/OPA) laser system (2). The "signal" output of the OPO, as well as part of the Nd : YAG laser light, is fed into a pulsed wavemeter (3) for wavelength control. The "idler" output of the OPO is used either directly or after difference-frequency mixing (DFM) with the Nd : YAG output. Using a periscope, the tunable mid-IR OPA output of the pulsed laser system, mounted on an optical table (4), is guided to a multipass arrangement, e.g., two concentric spherical mirrors of high reflectivity, intersecting with the supersonic-jet expansion. For an efficient overlap, a slit-nozzle may be employed. The mm or sub-mm sources (5) are based on a primary generator, phase-locked with the same frequency standard that is used by the FP FT-MW instrument, e.g., a setup as shown in Figure 36. The generator's cm wave output is multiplied, using active mm or sub-mm wave modules. If the primary generator, either internally or externally, allows for I/Q or phase modulation and subsequent signal modules are compatible with this, complex coherence experiments are also possible. Horn antennae (6) irradiate the expanding jet with the electromagnetic field. A computer-based system (7) controls all instrument components, generates trigger as well as excitation sequences, and synchronizes them for phase-invariant experiment repetition.

One can envisage the expansion of laser-excitation schemes for dynamic spectroscopy probed by FP FT-MW multiresonance techniques to higher energy

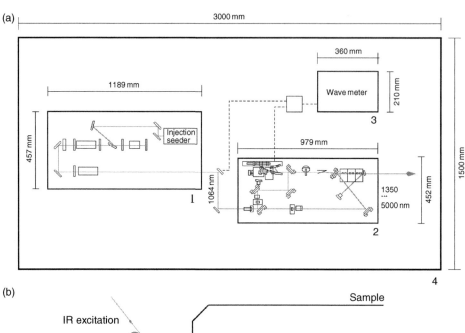

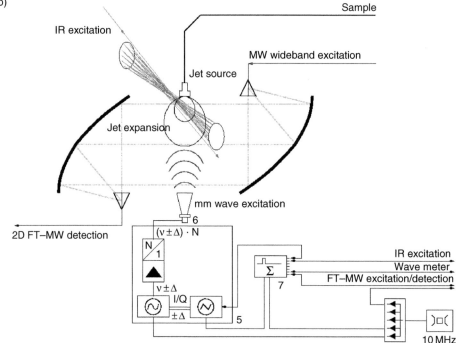

Figure 37 Supersonic-jet expansion FP 2D FTMW IR/mm wave-TR experiments: (a) tunable mid-IR-pulsed laser source: (1) Q-switched Nd : YAG laser, injection-seeded, 1064 nm, 1000 mJ, 20 Hz, (2) single-mode KTP/KTA OPO/OPA, 1350–5000 nm, ~10 mJ, 20 Hz, $\Delta\lambda$ <0.02 cm^{-1}, (3) pulsed wavemeter, $\Delta\lambda$ <0.02 cm^{-1}, (4) optical table; (b) supersonic-jet TR excitation assembly: (5) mm wave source, 60–325 GHz, (6) mm wave horn antennae, 60–325 GHz, >24 dBi, (7) PXIe-based experiment control system. For components not numbered, refer to Figure 36 and respective caption. *Source*: Figure 7 of Ref. [103]. (See color plate 26).

regions from the IR across VIS to the UV, in the near future accessing conformational dynamics at complete ergodicity and finally even above the barrier to bond-breaking structural isomerization.

ACKNOWLEDGEMENTS

We thank all microwavers who contributed over many years in countless productive co-operations. J.-U.G is indebted to his coworkers D. Banser, M. Rosemeyer, M. Schnell, J. Fritzsche, M. Hess, H. Saal, E. Locatelli, D. Dewald, M. Jahn, M. Vogt, D. Bremm, the electronic and mechanical workshop in Hannover as well as H. Dreizler, A. Guarnieri, H. Mäder, W. Stahl, D. Sutter, and P. Thaddeus. W.C. is thankful to P. G. Favero, R. D. Brown, and A. Bauder. J.-U.G is grateful for funds from the Bundesministerium für Bildung und Forschung (BMBF), the Deutsche Akademische Austauschdienst (DAAD), the Fonds der Chemie, the Land Niedersachsen, and explicitly for the crucial support provided by the Deutsche Forschungsgemeinschaft (DFG).

REFERENCES

[1] C.H. Townes, A.L. Schawlow, Microwave Spectroscopy, McGraw-Hill Dover Publications, Inc., New York, 1955.

[2] R.H. Hughes, E.B. Wilson, Phys. Rev., 71: 562, 1947.

[3] W. Gordy, R.L. Cook, Microwave Molecular Spectra, 3rd edition, John Wiley & Sons, New York, 1984.

[4] H.W. Kroto, Molecular Rotation Spectra, John Wiley & Sons, New York, 1975.

[5] J.S. Muenter, J. Chem. Phys., 48: 4544, 1968.

[6] R. Kewley, K.V.L.N. Sastry, M. Winnewisser, W. Gordy, J. Chem. Phys., 39: 2856, 1963.

[7] G. Winnewisser, A.F. Krupnov, M.Y. Tretyakov, M. Liedtke, F. Lewen, A.H. Saleck, A. Schieder, A.F. Shkaev, S.V. Volokhov, J. Mol. Spectrosc., 165: 294, 1994.

[8] S.P. Belov, G. Winnewisser, E. Herbst, J. Mol. Spectrosc., 174: 253, 1995.

[9] R.A. Davies, M.J. Kelly, T.M. Kerr, Phys. Rev. Lett., 55: 1114, 1985.

[10] L. Esaki, R. Tsu, IBM J. Res. Dev., 16: 61, 1970.

[11] C.P. Endres, F. Lewen, T.F. Giesen, S. Schlemmer, D.G. Paveliev, Y.I. Koschurinov, V.M. Ustinov, A.E. Zhucov, Rev. Sci. Instrum., 78: 043106, 2007.

[12] L.W. Hrubesh, Appl. Spectrosc., 32: 425, 1978.

[13] F. C. De Lucia, in D. Mittleman (ed.), Sensing with Terahertz Radiation, Springer, Berlin, 2003, p. 39.

[14] D.T. Petkie, T.M. Goyette, R.P.A. Bettens, S.P. Belov, S. Albert, P. Helminger, F.C. De Lucia, Rev. Sci. Instrum., 68: 1675, 1997.

[15] S. Albert, D.T. Petkie, R.P.A. Bettens, S.P. Belov, F.C. De Lucia, Anal. Chem., 70: 719A, 1998.

[16] I. Medvedev, M. Winnewisser, F.C. De Lucia, E. Herbst, E. Bialkowska-Jaworska, L. Pszczól-kowski, Z. Kisiel, J. Mol. Spectrosc., 228: 314, 2004.

[17] F.C. De Lucia, J. Opt. Soc. Am. B, 21: 1273, 2004.

[18] I.R. Medvedev, M. Behnke, F.C. De Lucia, Appl. Phys. Lett., 86: 154105, 2005.

[19] H.S. Zivi, A. Bauder, H.H. Günthard, Chem. Phys. Lett., 83: 469, 1981.

[20] R.D. Brown, J.G. Crofts, P.D. Godfrey, D. McNaughton, A.P. Pierlot, J. Mol. Struct., 190: 185, 1988.

[21] S. Melandri, W. Caminati, L.B. Favero, A. Millemaggi, P.G. Favero, J. Mol. Struct., 352/353: 253, 1995.

[22] S. Melandri, G. Maccaferri, A. Maris, A. Millemaggi, W. Caminati, P.G. Favero, Chem. Phys. Lett., 261: 267, 1996.

[23] T.F. Giesen, S. Brünken, M. Caris, P. Neubauer-Guenther, U. Fuchs, G.W. Fuchs, F. Lewen, in D. C. Lis, G. A. Blake, E. Herbst (eds.), Astrochemistry: Recent Successes and Current Challenges, Proceedings IAU Symposium No. 231, 2005, International Astronomical Union, 2006, p. 87.

[24] S.P. Belov, B.A. McElmurry, R.R. Lucchese, J.W. Bevan, I. Leonov, Chem. Phys. Lett., 370: 528, 2003.

[25] B.A. McElmurry, R.R. Lucchese, J.W. Bevan, I.I. Leonov, S.P. Belov, A.C. Legon, J. Chem. Phys., 119: 10687, 2003.

[26] F.A. Van Dijk, A. Dymanus, Chem. Phys., 6: 474, 1974.

[27] W.H. Flygare, T.G. Schmalz, Acc. Chem. Res., 9: 385, 1976.

[28] M.L. Unland, W.H. Flygare, J. Chem. Phys., 45: 2421, 1966.

[29] J.C. McGurk, H. Mäder, R.T. Hofmann, T.G. Schmalz, W.H. Flygare, J. Chem. Phys., 61: 3759, 1974.

[30] R.H. Dicke, R.H. Romer, Rev. Sci. Instrum., 26: 1, 1955.

[31] J. Ekkers, W.H. Flygare, Rev. Sci. Instrum., 47: 448, 1976.

[32] A. Bauder, in J. Durig (ed.),Vibrational Spectra and Structure, Vol. 20, Elsevier, Amsterdam, 1993, p. 157.

[33] H. Dreizler, Ber. Bunsen-Ges. Phys. Chem., 99: 1451, 1995.

[34] J.C. McGurk, T.G. Schmalz,, W.H. Flygare, in I. Prigogine, S. A. Rice (eds.), Advances in Chemical Physics, Vol. XXV, Wiley, New York, 1974, p. 1.

[35] H. Mäder, J. Quant, Spectrosc. Radiat. Transfer, 32: 129, 1984.

[36] T. Köhler, H. Mäder, Mol. Phys., 86: 287, 1995.

[37] I. Merke, H. Dreizler, Z. Naturforsch., 43a: 196, 1988.

[38] J. Haeckel, H. Mäder, Z. Naturforsch., 43a: 203, 1988.

[39] R.L. Shoemaker,in J. I. Steinfeld (ed.), Laser and Coherence Spectroscopy, Plenum, New York, 1979, p. 197.

[40] R. Schwarz, A. Guarnieri, J.-U. Grabow, J. Doose, Rev. Sci. Instrum., 63: 4108, 1992.

[41] J. Gripp, H. Mäder, H. Dreizler, J.L. Teffo, J. Mol. Spectrosc., 172: 430, 1995.

[42] P. Wolf, H. Mäder, Mol. Phys., 64: 43, 1988.

[43] V. Meyer, W. Jäger, R. Schwarz, H. Dreizler, Z. Naturforsch., 46a: 445, 1991.

[44] G. Bestmann, H. Dreizler, Z. Naturforsch., 38a: 452, 1983.

[45] W. Caminati, B. Vogelsanger, A. Bauder, J. Mol. Spectrosc., 128: 384, 1988.

[46] H. Ehrlichmann, J.-U. Grabow, H. Dreizler, N. Heineking, R. Schwarz, U. Andresen, Z. Naturforsch., 44a: 751, 1989.

[47] C. Styger, I. Ozier, A. Bauder, J. Mol. Spectrosc., 153: 101, 1992.

[48] B. Vogelsanger, A. Bauder, J. Chem. Phys., 92: 4101, 1990.

[49] D.A. Andrews, J.G. Baker, N.J. Bowring, Chem. Phys. Lett., 145: 505, 1988.

[50] B. Vogelsanger, A. Bauder, H. Mäder, J. Chem. Phys., 91: 2059, 1989.

[51] B. Vogelsanger, M. Andrist, A. Bauder, Chem. Phys. Lett., 144: 180, 1988.

[52] W. Jäger, J. Haekel, U. Andresen, H. Mäder, Mol. Phys., 68: 1287, 1989.

[53] W. Jäger, H. Krause, H. Mäder, M.C.L. Gerry, J. Mol. Spectrosc., 143: 50, 1990.

[54] H. Harder, H.-W. Nicolaisen, H. Dreizler, H. Mäder, J. Mol. Spectrosc., 160: 244, 1993.

[55] H. Harder, H.-W. Nicolaisen, H. Dreizler, H. Mäder, Chem. Phys. Lett., 214: 265, 1993.

[56] T.J. Balle, E.J. Campbell, M.R. Keenan, W.H. Flygare, J. Chem. Phys., 71: 2723, 1979.

[57] T.J. Balle, W.H. Flygare, Rev. Sci. Instrum., 52: 33, 1981.

[58] R.D. Suenram, F.J. Lovas, G.T. Fraser, J.Z. Gillies, C.W. Gillies, M. Onda, J. Mol. Spectrosc., 137: 127, 1989.

[59] J.-U. Grabow, W. Stahl, Z. Naturforsch., 45a: 1043, 1990.

[60] C. Chuang, C.J. Hawley, T. Emilsson, H.S. Gutowsky, Rev. Sci. Instrum., 61: 1629, 1990.

[61] U. Andresen, H. Dreizler, J.-U. Grabow, W. Stahl, Rev. Sci. Instrum., 61: 3694, 1990.

[62] J.-U. Grabow, W. Stahl, H. Dreizler, Rev. Sci. Instrum., 67: 4072, 1996.

[63] J.-U. Grabow, E.S. Palmer, M.C. McCarthy, P. Thaddeus, Rev. Sci. Instrum., 76: 093106, 2005.
[64] J.-U. Grabow, Habilitationsschrift, Universität Hannover, Hannover, 2004.
[65] V. Strom, H. Dreizler, D. Consalvo, J.-U. Grabow, I. Merke, Rev. Sci. Instrum., 67: 2714, 1996.
[66] R.D. Suenram, J.-U. Grabow, A. Zuban, I. Leonov, Rev. Sci. Instrum., 70: 2127, 1999.
[67] K.C. Etchison, C.T. Dewberry, K.E. Kerr, D.W. Shoup, S.A. Cooke, J. Mol. Spectrosc., 242: 39, 2007.
[68] W.F. Kolbe, B. Leskovar, Rev. Sci. Instrum., 56: 97, 1985.
[69] E.J. Campbell, L.W. Buxton, T.J. Balle, M.R. Keenan, W.H. Flygare, J. Chem. Phys., 74: 829, 1981.
[70] R.E. Smalley, L. Wharton, D.H. Levy, Acc. Chem. Res., 10: 139, 1977.
[71] H. Dreizler, U. Andresen, J.-U. Grabow, D.H. Sutter, Z. Naturforsch., 53a: 887, 1998.
[72] R.R. Bousquet, P.M. Chu, R.S. DaBell, J.-U. Grabow, R.D. Suenram, IEEE Sens. J., 5: 656, 2005.
[73] E. Fliege, H. Dreizler, Z. Naturforsch., 42a: 72, 1987.
[74] V. Meyer, D.H. Sutter, Z. Naturforsch., 49a: 725, 1993.
[75] E.J. Campbell, W.G. Read, J.A. Shea, Chem. Phys. Lett., 94: 69, 1983.
[76] D. Consalvo, Rev. Sci. Instrum., 69: 3136, 1998.
[77] T. Emilsson, H.S. Gutowsky, G. De Oliveira, C.E. Dykstra, J. Chem. Phys., 112: 1287, 2000.
[78] Z. Kisiel, J. Kosarzewski, B.A. Pietrewicz, L. Pszczolkowski, Chem. Phys. Lett., 325: 523, 2000.
[79] M. Schnell, D. Banser, J.-U. Grabow, Rev. Sci. Instrum., 75: 2111, 2004.
[80] A. Klesing, D.H. Sutter, Z. Naturforsch., 45a: 817, 1990.
[81] R.D. Suenram, G.T. Fraser, F.J. Lovas, K. Matsumura, J. Chem. Phys., 92: 4724, 1990.
[82] E.J. Campbell, W.G. Read, J. Chem. Phys., 78: 6490, 1983.
[83] H. Dreizler, E. Fliege, H. Mäder, W. Stahl, Z. Naturforsch., 37a: 1266, 1982.
[84] L. Martinache, S. Jans-Bürli, B. Vogelsanger, W. Kresa, A. Bauder, Chem. Phys. Lett., 149: 424, 1988.
[85] W. Jäger, M.C.L. Gerry, J. Chem. Phys., 102: 3587, 1995.
[86] V.N. Markov, Y. Xu, W. Jäger, Rev. Sci. Instrum., 69: 4061, 1998.
[87] M. Nakajima, Y. Sumiyoshi, Y. Endo, Rev. Sci. Instrum., 73: 165, 2002.
[88] K. Suma, Y. Sumiyoshi, Y. Endo, J. Chem. Phys., 121: 8351, 2004.
[89] Y. Sumiyoshi, H. Katsunuma, K. Suma, Y. Endo, J. Chem. Phys., 123: 054324, 2005.
[90] K.O. Douglass, J.E. Johns, P.M. Nair, G.G. Brown, F.S. Rees, B.H. Pate, J. Mol. Spectrosc., 239: 29, 2006.
[91] A. Javan, Phys. Rev., 107: 1579, 1957.
[92] U. Wötzel, W. Stahl, H. Mäder, Can. J. Phys., 75: 821, 1997.
[93] K.O. Douglass, J.C. Keske, F.S. Rees, K. Welch, H.S. Yoo, B.H. Pate, I. Leonov, R.D. Suenram, Chem. Phys. Lett., 376: 548, 2003.
[94] K.O. Douglass, F.S. Rees, R.D. Suenram, B.H. Pate, I. Leonov, J. Mol. Spectrosc., 230: 62, 2005.
[95] R.R. Ernst, W.A. Anderson, Rev. Sci. Instrum., 37: 93, 1966.
[96] J.C. McGurk, T.G. Schmalz, W.H. Flygare, J. Chem. Phys., 60: 4181, 1974.
[97] T.G. Schmalz, W.H. Flygare, in J. I Steinfeld (ed.), Laser and Coherence Spectroscopy, Plenum, New York, 1978, p. 125.
[98] F. Wolf, J. Phys. D: Appl. Phys., 27: 1774, 1994.
[99] V.V. Khodos, D.A. Ryndyk, V.L. Vaks, Eur. Phys. J. Appl. Phys., 25: 203, 2004.
[100] G.G. Brown, B.C. Dian, K.O. Douglass, S.M. Geyer, B.H. Pate, J. Mol. Spectrosc., 238: 200, 2006.
[101] G.S. Grubbs II, C.T. Dewberry, K.C. Etchison, K.E. Kerr, S.A. Cooke, Rev. Sci. Instrum., 78: 096106, 2007.
[102] J.-U. Grabow, Chem. Phys. Chem: to be published, 2008.

[103] J.-U. Grabow, Applikationsschrift, Hannover2008; Available on-line at: http://www.ft-mw.org/

[104] B.C. Dian, G.G. Brown, K.O. Douglass, B.H. Pate, Science, 320: 924, 2008.

[105] J.A. Pople, W.G. Schneider, H.J. Bernstein, High Resolution Nuclear Magnetic Resonance, McGraw-Hill, New York, 1959.

[106] H. S. Gutowsky, Chap. XLI Vol. 1, part IV in A. Weissberger (ed.), Technique of Organic Chemistry, 3rd ed. Interscience, New York, 1960.

CHAPTER 15

Microwave Spectroscopy
Molecular Systems

Walther Caminati *and* Jens-Uwe Grabow

Contents

Abstract

The development of Fourier transform microwave (FTMW) spectroscopy and — most notably — its combination with supersonic-jet expansion techniques allows for challenging investigations of chemically and physically interesting molecular systems. These are, among others, in situ prepared molecular species such as radicals, ions, and other transient species, generated by combining an electrical discharge, laser ablation, or laser photolysis, and highly dynamical systems including weakly bound molecular complexes. From these latter systems, information on nonbonding intermolecular interactions and on the internal dynamics is easily obtained. Many chemical problems, difficult to unravel with other techniques, were solved only by rotational spectroscopy experiments on supersonic jets. The widely unexplored interactions in the intermediate regime between bonding and nonbonding, molecular recognition, molecular aggregation, and many more are investigated with these techniques without having to rely on ab initio calculations – their support is of great help though, both in guiding spectroscopic searches and in interpreting the spectra.

The laboratory preparation and investigation of a large number of new transient species generated data on existing chemical systems, many of them being either of important astrochemical interest or crucial in the understanding of (photo-)chemical processes in the atmosphere. Because of the emerging new techniques, the conformational equilibria within amino acids and other biomolecules were precisely described.

The potential of MW spectroscopy to obtain this information is demonstrated by numerous investigations. Many interesting examples, most of them achieved in the last 15–20 years, are compiled here.

Keywords: Molecules; molecular systems; molecular complexes; van der Waals complexes; biomolecules; astrophysical molecules; nonpolar molecules; tautomeric equilibria; conformational equilibria; large amplitude motions; potential energy surfaces; molecular aggregation; molecular solvation; quantum solvation; oligomers; hydrogen bonding; noncovalent interaction; charge transfer; molecular recognition; transition metal complexes; quadrupole coupling; microwave spectroscopy; rotational spectra; structure; internal dynamics; supersonic jet expansion; Fourier transform spectroscopy

1. INTRODUCTION

In the past microwave (MW) spectroscopy has mostly been – and sometimes even by spectroscopists still is – considered as a tool to obtain molecular structures. Indeed, the rotational constants, which are the primary spectroscopic data related to MW spectra, directly depend on the mass distribution within a given molecular systems. Thus, information obtained from the rotational spectra was routinely related to bond lengths and angles. However, from the very beginning of rotational spectroscopy, the information encoded in MW spectra was used to obtain detailed chemical information, e.g., the electronic environment of a given atom. Very early also the effects of internal dynamics, producing tunneling splittings on the rotational energy levels, were investigated for prototype motions such as methyl group internal rotation and motions involving a W-shaped two minima potential.

The combination of Fourier transform MW (FTMW) spectroscopy with supersonic-jet expansions allowed studying the rotational spectra of weakly bound molecular complexes. Some of these molecular systems are dominated by large amplitude motions – sometimes complicating the spectra that much of their assignment was inhibited. New theoretical tools have been developed to fit the spectra and to relate the obtained data to the potential energy functions. It also became possible to size the intermolecular nonbonding interactions that lead to the preferred configuration of such a system. Although important spectroscopic and chemical information is still presently obtained by MW spectroscopy employing static gas cells, many chemical problems could be unraveled only with the help of rotational spectroscopy on supersonic jets. We will show several examples in the following sections. At the same time, it was possible to prepare and investigate in the laboratory a large number of transient new species, many of them also being of astrochemical interest. Because of the detailed and precise chemical information obtained from MW spectroscopic studies, many of these results are found in periodicals dedicated to a wide chemical audience.

With its capability to detect the spectra of heavier and larger molecules as well as molecular aggregates, to combine with in situ methods to prepare molecular species not otherwise obtainable, and to identify signals reaching the earth from interstellar space, rotational spectroscopy is not just reinforcing but strongly increasing its key role in delivering precise information on numerous chemical and physical objectives.

2. GENERAL ASPECTS

We will not treat the fundamental theoretical background of rotational spectroscopy, which is available from several text books devoted to MW spectroscopy, e.g., those written by Townes and Schawlow [1], Gordy and Cook [2], Sugden [3], or Kroto [4]. An exhaustive book on the rotational spectra of diatomic molecules has been published very recently [5]. The effects of large amplitude motions on rotational spectra are considered in detail in books written by Wollrabb [6] and Lister et al. [7], while their treatment within the framework of permutation inversion (PI) group theory is given by Bunker and Jensen [8].

Historically, the analysis of rotational spectra was performed employing numerous computer programs, often specifically written for molecular system by MW spectroscopists in their own laboratories. While – in a number of cases, notably those where molecular symmetry (MS) is of importance – there is still a considerable need for specialized programs, the vast majority of rotational spectra can now be analyzed by a very general computer program written by Pickett, the CALPGM suite of programs, which can treat 99 states and 9 interacting nuclei simultaneously [9], available at the Jet Propulsion Laboratory (JPL) spectroscopy group website [10]. Several more general programs for rotational spectroscopy are available at the "Programs for ROtational SPEctroscopy" (PROSPE) [11] and at the "The Cologne Database for Molecular Spectroscopy" (CDMS) [12] websites.

Among them, several programs deal with internal rotation – typically of methyl groups – up to three internal rotors. However, for low and intermediate barrier values, more specific programs, able to take into account higher order perturbation effects, have been developed by Ohashi [13], Kleiner [14], and Groner [15]. Severe perturbing terms associated with low/intermediate internal rotation barriers can also be handled by the spectral fitting program JB95 provided by Plusquellic [16], available from the National Institute of Standards and Technology (NIST) website [17] – as well as by Pickett's

CALPGM suite in employing the program IAMCALC for the calculation of Fourier coefficients needed by SPFIT and SPCAT in internal rotation calculations [18].

Specific models for complicated interactions have been developed, e.g., by Hougen [19] and Coudert, who developed a theoretical approach to model the spectrum of ethylene glycol [20], a molecule displaying a complicated isomerization large amplitude motion. Their first successful assignment of the spectrum led to its first detection in the interstellar medium.

Ab initio and density functional theory, implemented in programs such as Gaussian [21] or GAMESS [22], are widely used to predict the molecular equilibrium structure, the conformational and tautomeric preferences from the potential energy surface (PES), the dissociation energies of molecular complexes, molecular force fields, electric field gradients, etc., and to calculate trial values of the related spectroscopic parameters. Additional programs building upon quantum chemical calculations have been written: an example is GDMA, a Fortran 90 program for performing a Distributed Multipole Analysis (DMA) of wavefunctions calculated by Gaussian [23]. Using the formatted checkpoint file produced by the Gaussian, the program system calculates electric multipole moments at atom positions, or at other specified sites, to give a representation of the electrostatic field of a molecule. Build upon GDMA is ORIENT, a program for various calculations on the assembly of interacting molecules, such as in molecular complexes [24]. It uses a site–site potential specified by the user, including electrostatic, induction, repulsion, dispersion, and charge-transfer interactions if required. Both programs are available at Stone's website [25].

Very simple and thus useful in easily and quickly predicting minima in the PESs of molecular complexes are the programs RGDMIN and MIN16 [11]. RGDMIN predicts the geometry of RG...Molecule dimers from a simple distributed model for dispersive interaction [26,27]. The program minimizes the interaction energy between a rare gas (RG) atom and a molecule, assuming dispersive attraction is counterbalanced by hard sphere repulsion. MIN16 predicts the geometries of hydrogen-bonded dimers from the electrostatic model of Buckingham and Fowler [28,29], based on the assumption that the dominant contribution to the interaction energy comes from the electrostatic term. This can rather satisfactorily be modeled by using the DMA of the electron charge distribution, and the resulting attractive electrostatic term is counterbalanced by a van der Waals hard sphere repulsion term. The program searches for minima in intermolecular interaction energy between two molecules by reorienting the second around the first molecule.

One- and two-dimensional so-called flexible models for calculating the rotational energy levels $J = 0$, 1, 2 are available. These models are employed in the analysis of rovibrational problems exhibiting large amplitude motions: in supplying eigenvalues and wavefuntions, the flexible models are suited to reproduce the experimental observed energy-level difference as a function of a user-supplied potential energy function [30,31]. Specifically, these models relate to centrifugal distortion and rotational constants, measured in different vibrational states and observed vibrational frequencies, directly related to a parametrized function that describes the large amplitude motions within an arbitrary (nonlinear) molecule.

Several reviews have been written on MW spectroscopy topics, concerning mainly species of astronomical interest [32,33] and molecular complexes [34–37]. A bibliography on the rotational spectra of weakly bound complexes is compiled by Novick and available at the Wesleyan University website [38].

Before presenting specific molecular examples, we will briefly describe some of the situations and phenomena often encountered in rotational spectra.

2.1. Conformational equilibria

Rotational spectroscopy is one of the prime experimental techniques in discriminating among several conformers that can constitute a molecular system. One of the first conformational studies performed by MW spectroscopy concerned the investigation of *trans* and *gauche* normal propyl fluoride: in 1962 Hirota found the *gauche* form to be more stable than the *trans* by 2.0(13) kJ/mol [39] and presented a picture of the PES governing the molecular conformations. In 1982 Caminati reported the spectra of three conformers of 3-fluoropropan-1-ol [40], with one of them displaying an internal hydrogen bond resulting in a distorted chair conformation of the six-membered ring. The two non-hydrogen-bonded forms are 1.7 and 5.9 kJ/mol higher in energy. In 1994, the assignment of the rotational spectra of four conformers of *n*-chlorobutane was setting the largest number of conformers identified by rotational spectroscopy for one molecule at that time [41].

Today, roughly a decade after that Suenram and coworkers started to investigate conformational equilibria by supersonic-jet FTMW spectroscopy, the number of conformers which could be identified in a rotational spectrum of a large flexible molecule increased dramatically: indeed, 7, 13, and 15 conformers have been assigned for 1-hexene [42], 1-heptanal [43], and 1-octene [44], respectively. On-the-fly graphical assignment and real-time spectral analysis, as featured by the JB95 frontend [16], greatly facilitated the identification of the multi-conformer spectra. The conformational complexity of 1-heptanal is shown in Figure 1.

If spectra are recorded in absorption in a standard cell, the relative energies of two conformers can be evaluated in assuming a Boltzmann distribution [2]:

$$\Delta E_{0.0} = kT \ln \frac{I_1 \omega_2 \Delta \nu_1 \mu_{g,2}{}^2 \gamma_2 \nu_2{}^2}{I_2 \omega_1 \Delta \nu_2 \mu_{g,1}{}^2 \gamma_1 \nu_1{}^2} \tag{1}$$

where $\Delta E_{0.0}$ [$= E_{0.0}(2) - E_{0.0}(1)$] is the energy difference between the conformers in their rotational and vibrational ground states. I, ω, $\Delta \nu$, γ, μ_g, and ν are peak intensity, conformational degeneracy, line width at half height, line strength, dipole moment component ($g = a$ or b or c), and transition frequency, respectively, of the considered transition. Because intensity measurements are typically considerably less accurate than the corresponding transition frequencies and precise dipole moment components are not always readily available, energy differences are not evaluated at very high precision in many cases.

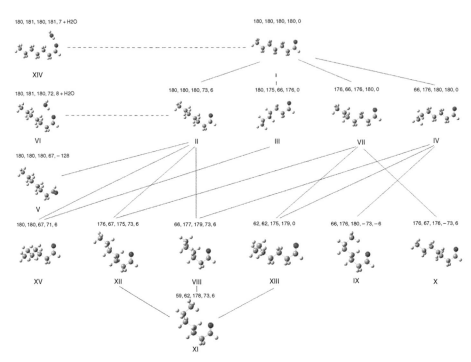

Figure 1 Global figure containing ab initio determined structures for all observed 1-heptanal conformers. Dihedral angles labeled are in degrees and in order of C4C5—C6C7, C3C4—C5C6, C2C3—C4C5, C1C2—C3C4, and OC1—C2C3. Lines connecting conformers indicate possible tunneling paths involving successive rotation of dihedral angles, starting from conformer *I* of *Cs* symmetry. (Reprinted with permission from Ref. [43].)

Intensity measurements in a supersonic jet are even more difficult, because rather undefined factors contribute to the intensity of the signals. However, approximations can still be used [45]:

$$\Delta E_{0.0} = kT_{\mathrm{conf}} \ln \frac{I_1 \omega_2 \mu_{g,2} \gamma_2 \nu_2^2}{I_2 \omega_1 \mu_{g,1} \gamma_1 \nu_1^2} + \left[(E_{\mathrm{rot}})_1 - (E_{\mathrm{rot}})_2 \right] \frac{T_{\mathrm{conf}}}{T_{\mathrm{rot}}} \tag{2}$$

where T_{conf} is the temperature in the bulk, prior to expansion, and T_{rot} is the "rotational" temperature in the jet.

Errors of 50–100 cm^{-1} are easily encountered with this technique. However, very recently, direct transitions between the three conformers of *n*-propanol [46] could be observed. Relying on those, the conformational energy difference does have the precision typical for MW spectroscopic measurements. Several systems exhibiting conformational equilibria will be considered in the next sections.

2.2. Tautomeric equilibria

Valuable information on tautomeric equilibria, present in many systems of chemical and biological interest, can be obtained via rotational spectroscopy. Keto–enolic equilibrium is the most prominent example of tautomerism. At a pioneering stage,

Baughcum and collaborators [47] analyzed the rotational spectrum of malonaldehyde and showed unambiguously that the molecule adopts C_s symmetry in an enolic form with a low barrier to proton tunneling. It was much later that the rotational spectrum of the totally symmetric tunneling A-state of the enolic form of acetylacetone could be assigned [48], indicating that the tautomeric equilibrium in the gas phase is again shifted toward the enolic species, but showing evidence for C_{2v} symmetry.

Another example is 2-hydroxypyridine, which can be considered the prototype system for the study of the keto–enolic equilibrium present in more complex systems such as nucleobases. In 2-hydroxypyridine, both the enolic and the ketonic forms were observed in the gas phase, and the stabilization energy of 2-hydroxypyridine with respect to 2-pyrimidone was determined from rotational spectroscopy to be 0.32(3) kJ/mol [49]. In the related system 4-hydroxypyrimidine, the ketonic form is more stable by 2.0(9) kJ/mol [50]. This may indicate that the stability of the ketonic with respect to the enolic form depends on the number of ring nitrogen atoms as well as the number of hydroxyl groups that can convert into ketonic groups. We will see several examples of tautomeric equilibria in the following sections, especially in the one devoted to "biomolecules".

2.3. Large amplitude motions

Some of the earliest MW spectroscopy measurements dealt with the rotational spectrum of ammonia and with its inversion motion tunneling [51]. Since that time, the large amplitude internal motions of many molecular systems were characterized from the tunneling splittings observed in their rotational spectra. Often, two or more motions were simultaneously generating tunneling splittings, sometimes with severe Coriolis couplings among them. Typical motions are (i) internal rotation of symmetric (generally methyl) groups; (ii) inversion of amino or imino hydrogens; (iii) internal rotation of light asymmetric groups (OH, SH, NH_2); (iv) ring puckering of (saturated) four- or (near saturated) five-membered rings; (v) pseudorotation. Even heavy atoms (or structural groups) can produce large splitting if their motions are characterized by low-barrier PESs as in many molecular complexes. In the following subsections, we will limit our examples to isolated molecules. The presence of internal motions in molecular complexes will be considered in Section 4.

2.3.1. Methyl group internal rotation

Numerosus molecules with methyl groups have been investigated, and the corresponding V_3 (or V_6) barriers to internal rotation have been determined. These barriers range from $V_6 = 8$ J/mol [52] to $V_3 = 17$–21 kJ/mol [53]. Higher expansion terms in the PES have often been taken into account when high torsionally excited states have been investigated, such as in the prototype case for methyl group internal rotation, acetaldehyde [54].

2.3.1.1. *Methyl group pure internal rotation*

Intermediate barriers and global fits of rotational transitions in various vibrationally excited states represent a difficult task. Acetaldheyde is the prototype molecule for the study of methyl group internal rotation, and one of the most investigated. For this molecule, Kleiner et al. [55] could fit several thousands of far-infrared and MW transitions of acetaldehyde, up to the fourth torsional state. The torsional rotational spectrum of methanol is very complicated, but a global fit of transitions in the ground and first excited torsional states was succesfully achieved by Xu and Hougen [56].

2.3.1.2. Two and three methyl group internal rotation

Groner developed a model for two rotating equivalent methyl groups within a planar molecular frame, which was successfully used to interpret the rotational spectrum of acetone in its first torsionally excited state (80 cm^{-1} above the ground state) [57]. The splittings between the four torsional substates are significantly larger than in the vibrational ground state, making assignment difficult. Eventually, 40 parameters of an effective rotational Hamiltonian for a molecule with two periodic internal motions were fit to frequencies of 571 transitions and 50 blends to near experimental precision.

Ohashi and collaborators could fit the supersonic-jet FTMW spectra of molecules with two- and three methyl rotors and ^{14}N quadrupole hyperfine structure. For N-methylacetamide, which possesses two different methyl tops with relatively low barriers to internal rotation, a global fit of 48 molecular parameters to 839 hyperfine components of 216 torsion–rotation transitions involving 152 torsion–rotation levels was performed and achieved a root mean square of 4 kHz, i.e., near experimental accuracy [58]. Set up in the principal axis system, the principal axis method (PAM), of the molecule and utilizing a single-step diagonalization procedure, a program using a free-rotor basis set for each top and a symmetric-top basis set for the rotational functions was written for this analysis. For N,N-dimethylacetamide [59], a molecule with three methyl groups, also Coriolis-like coupling parameters characterizing the interaction between internal rotation of methyl groups and the overall rotation were determined from internal-rotation tunneling splittings of the rotational transitions. The Coriolis-like coupling parameters permitted the determination of the barrier heights to internal rotation of the three methyl groups, which were found to be 677, 237, and 183 cm^{-1} for the C-methyl top, the trans-N-methyl top, and the cis-N-methyl top, respectively.

A considerable effort has been done by Grabow and collaborators in revealing concurrent tunneling pathways in molecules with three equivalent internal rotors leading to a multidimensional large-amplitude motion. They investigated with FTMW spectroscopy the series $(CH_3)_3XY$ (X = Si, Ge, Sn, Y = Cl, Br), whose central atoms systematically increase in size [60–63]. They found that the electrostatic repulsion, and thus top–top communication through space, is gradually reduced in the order Si > Ge > Sn, and that the relative contribution of the character of the chemical bond X—CH_3 (X = Si, Ge, Sn) becomes more important and leads to dramatic differences in the internal molecular dynamics. The appropriate PI group for the molecules of the series is G162, which consists of 162 elements and is not isomorphic with any point group. The multidimensional tunneling formalism provides a description of the internal rotation in terms of tunneling pathways. The experimental V_3 barriers within the series $(CH_3)_3XCl$ are 6.91(1), 4.449(5), and 1.774(6) kJ/mol for X = Si, Ge and Sn, respectively.

2.3.1.3. Interaction of methyl group internal rotation with other motions

An interesting example of interaction of a methyl group internal rotation with another motion is the coupling between the methyl group internal rotation with the proton tunneling in 2-methylmalonaldehyde. As shown in Figure 2, the hydrogen transfer in 2-methylmalonaldehyde is accompanied by the same tautomeric rearrangement of single and double bonds as in malonaldehyde [47], but it is accompanied in addition by an internal rotation of the methyl group by 60°, which is required to preserve the methyl group's lowest-energy orientation with respect to the C=C—C (or C—C=C) skeleton being bound to the center carbon. Chou and Hougen [64] used a tunneling-rotation Hamiltonian, developed on the basis of the G_m^{12} MS group, to perform global fits to the MW spectra of several isotopologues of 2-methylmalonaldehyde reported previously [65]. In this new analysis, the root-mean-square deviations could be reduced to the order of the measurements precision (0.12–0.10 MHz).

Methyl group internal rotation systematically interacts, through kinetic or potential energy couplings, with skeletal torsions. Then, the spectral splitting of the two symmetry species of the methyl torsion, the A and E states, in excited states of the skeletal torsion differs from the ground state splitting, and occasionally is even inverted. These effects can easily be reproduced in using Meyer's two-dimensional flexible model [30]. The inversion of the spectral sequence of the A- and E-substates in the first excited state of the skeletal torsion in methyl thiolfluoroformate is explained in terms of kinetics energy interaction [66], while, for example, the methyl torsion–skeletal torsion interaction in the case of methyl vinyl ketone was interpreted in terms of a potential energy gear-type coupling [67].

2.3.2. Hydroxyl group internal rotation

The OH group is the lightest top, and therefore its motions can generate tunneling splittings even at relatively high barriers, as observed for hydroxyl groups bound to aromatic rings. Large splittings have been observed, indeed, in the case of phenol [68] and 4-hydroxypyridine [69], despite V_2 barriers as high as 14.4 and 18.1 kJ/mol, respectively. Also some other interesting cases associated with the internal rotation of a hydroxyl group are worth being considered here.

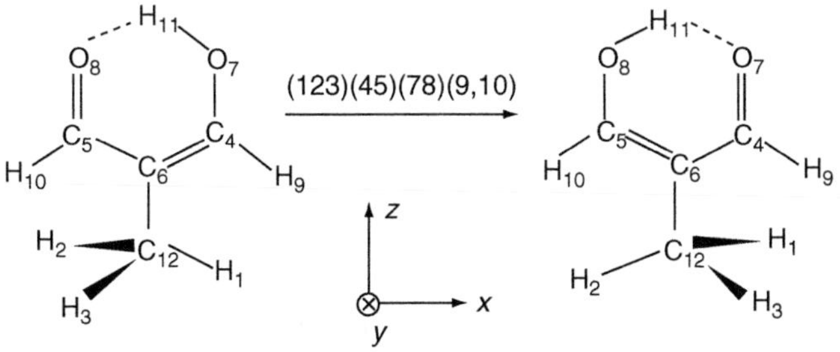

Figure 2 View of 2-methylmalonaldehyde showing the atom numbering and molecule-fixed axis system used in this paper. The permutation operation corresponding to the composite large-amplitude motion of hydrogen transfer and corrective internal rotation of the methyl group is also indicated. (Reused with permission from Ref. [64]. Copyright 2006, American Institute of Physics.)

2.3.2.1. *Hydroxyl group internal rotation in near spherical tops*

There are a couple of molecules, such as perchloric acid and t–butyl alcohol, which are near spherical tops possessing an internally rotating hydroxyl group. The rotational spectrum of $HClO_4$ was recently reported. Although no E torsional states, due to the spin-statistical weight arising from the oxygen atoms with spin $I = 0$ in the ClO_3 top, are present in the spectrum, the R-branches show three distinct groups of lines [70]: a relatively tight cluster of symmetric rotor-like transitions with $K = 3n$, a rather regular progression of transitions with $K = 3n + 2$ at higher frequency and a less regular group of transitions with $K = 3n + 1$ at lower frequency. Because the molecule is nearly spherical, the energy as a function of K is dominated by the K-dependent solutions of the Mathieu equation. This unusual energy-level distribution gives rise to numerous anomalous splittings and shifts due to avoided crossings within the K stacks as well as widely scattered μ_b transitions. Set up in the internal axis system, the internal axis method (IAM), the program IAMCALC as part of Pickett's CALPGM suite, was used to determine 68 parameters, reported in Table 1, from a huge number of measured transitions.

2.3.2.2. *Hydroxyl group internal rotation connecting transient chiral forms*

There is a considerable interest on the feasibility of observation of splittings in the spectra of enantiomers of chiral molecules [71]. Such a kind of splitting is easily observed in transient chiral forms connected through a rotation with respect to a single chemical bond. This is the case for the *gauche* forms of ethyl alcohol [72], allyl alcohol [73], and isopropanol [74], with splittings due to the tunneling across the barrier connecting the two equivalent "chiral" forms of 96748.816(7), 14168.15(14), and 46798.90(6) MHz, respectively.

Hirota suggests that, in the case of optically active rotational isomers, the Coriolis interaction may mix a symmetric and an antisymmetric level to a considerable extent, when these two levels happen to be nearly degenerate, such as in the case of isopropanol-OD. Then a rovibrational state of the *gauche* form may be expressed as

$$|v, r\rangle = |R\rangle [a|r_s\rangle + b|r_a\rangle] + |L\rangle [a|r_s\rangle - b|r_a\rangle] \qquad (3)$$

where $|R\rangle + |L\rangle$ and $|R\rangle - |L\rangle$ correspond to the symmetric and antisymmetric torsional states of *gauche*, respectively, and $|r_s\rangle$ and $|r_a\rangle$ represent the associated rotational wavefunctions. Hirota then foresees the observation of interesting relaxation processes, potentially supplying invaluable information on the interactions between two chiral molecules.

2.3.3. Other internal rotors

The FTMW spectrum of dinitrogen pentoxide (N_2O_5), a molecule consisting of two NO_2 rotors which have C_{2v} symmetry, has been reported [75]. This study has very interesting features: (i) PI group theoretical treatment including spin statistics which eliminates some levels; (ii) application of high-barrier tunneling matrix formalism; (iii) two rotors that are C_{2v}, which make the rotational spectrum difficult to treat because – unlike for C_{3v} – the moments of inertia change upon internal rotation; (iv) a very low

Table 1 Molecular parameters (MHz) for the ground torsional state of $HClO_4$ [70]

Operator	SPFIT i.d.[b]	$H^{35}ClO_4$ [a]
P^2	1	5276.778947(62)[c]
$\times (c1-1)$[d]	10000001	$-0.4731(53)$
$\times (c2-1)$	20000001	0.1446(53)
P^2	100	291.328036(100)
$\times (c1-1)$	10000100	0.2654(54)
$\times (c2-1)$	20000100	$-0.1690(53)$
$P_+^2 + P_-^2$	400	17.4464307(106)
$\times (c1-1)$	10000400	2.34743(265)
$\times (c2-1)$	20000400	$-0.0624(60)$
$\times (c3-1)$	30000400	2.03(59)E-03
$\times (c4-1)$	40000400	$-6.88(101)$E-03
$\times (c5-1)$	50000400	0.675(54)E-03
P_a^4	20	3.3668(42)E-03
$\times (c1-1)$	10000020	0.4998(43)E-03
$P^2 P_a^2$	11	$-2.7304(49)$E-03
$\times (c1-1)$	10000011	$-0.4775(50)$E-03
P^4	2	$-1.51567(160)$E-03
$\times (c1-1)$	10000002	0.01555(160)E-03
$P_a^3 \times s1$	110002010	$-2.396(231)$E-03
$P^2 P_a \times s1$	110002001	6.835(232)E-03
$P_a^3 \times s2$	120002010	3.949(113)E-03
$P^2 P_a \times s2$	120002001	$-3.790(113)$E-03
$P_a^s \times s1$	110002020	$-0.8806(272)$E-06
$P^2 P_a^3 \times s1$	110002011	0.7212(248)E-06
$P^4 P_a \times s1$	110002002	$-0.0534(65)$E-06
$P^2 \left(P_+^2 + P_-^2\right)$	401	$-0.028110(149)$E-03
$\times (c1-1)$	10000401	2.987(52)E-06
$\left[P_{a,}^2 \left(P_+^2 + P_-^2\right)\right]_+/2$	410	0.3751(41)E-03
$\times (c1-1)$	10000410	$-0.1147(46)$E-03
$P_+^4 + P_-^4$	500	$-0.01278(87)$E-03
$\times (c1-1)$	10000500	$-0.03892(87)$E-03
P_a^6	30	0.952(73)E-09
$P_a^4 P^2$	21	$-0.011244(199)$E-06
$\times (c1-1)$	10000021	$-5.330(135)$E-09
$P_a^2 P^4$	12	9.720(157)E-09
$\times (c1-1)$	10000012	4.197(137)E-09
P^6	3	0.2537(65)E-09
$P^4 \left(P_+^2 + P_-^2\right)$	402	0.14658(158)E-09
$\times (c1-1)$	10000402	0.26090(267)E-09
$P^2 \left[P_{a,}^2 \left(P_+^2 + P_-^2\right)\right]_+/2$	411	0.854(103)E-09
$\times (c1-1)$	10000411	4.417(76)E-09

(Continued)

Table 1 *(Continued)*

Operator	SPFIT i.d.[b]	H^{35}ClO$_4$[a]
$\left[P_{a,}^4\left(P_+^2 + P_-^2\right)\right]_+/2$	420	$-1.13(58)$E-09
$\times(c1-1)$	10000420	$1.59(45)$E-09
$P_+^6 + P_-^6$	600	$0.1305(36)$E-09
$P^2\left[P_{a,}^4\left(P_+^2 + P_-^2\right)\right]_+/2$	412	$0.05218(220)$E-12
$P_aP_b + P_bP_a$	6100	$-21.5991(79)$
$\times c1$	10006100	$7.8964(45)$
$\times c2$	20006100	$-0.5910(101)$
$\times c3$	30006100	$-9.06(94)$E-03
$\times c4$	40006100	$-0.01077(96)$
$\times c5$	50006100	$-2.890(130)$E-03
$P^2\left(P_aP_b + P_bP_a\right)$	6101	$-27397(46)$E-03
$\times c1$	10006101	$-0.06456(210)$E-03
$\left[P_{a,}^2\left(P_aP_b + P_bP_a\right)\right]+/2$	6110	$0.32037(98)$E-03
$\times c1$	10006110	$0.03552(247)$E-03
$P^2\left[P_{a,}^2\left(P_aP_b + P_bP_a\right)\right]_+/2$	6111	$2.610(43)$E-09
$\times c1$	10006111	$4.557(109)$E-09
$P^4\left(P_aP_b + P_bP_a\right)$	6102	$1.4447(113)$E-09
$\times c1$	10006102	$-0.289(69)$E-09
$\left[P_{a,}^4\left(P_aP_b + P_bP_a\right)\right]_+/2$	6120	$-7.102(157)$E-09
$\left[P_{a,}\left(P_+^3 + P_-^3\right)\right]_+/2^{1/2}$	6300	$-0.51701(209)$E-03
$\times c1$	10006300	$0.08066(152)$E-03
$\times c2$	20006300	$1.165(223)$E-06
$P^2\left[P_{a,}\left(P_+^3 + P_-^3\right)\right]_+/2^{1/2}$	6301	$2.3926(296)$E-09
$2\left[P_{a,}\left(P_+^5 + P_-^5\right)\right]_+/3^{1/2}$	6500	$0.3460(146)$E-09
$\times c1$	10006500	$0.785(34)$E-09
$2P^2\left[P_{a,}\left(P_+^5 + P_-^5\right)\right]_+/3^{1/2}$	6501	$-0.03773(139)$E-12
Δ_{E-A}[e]		$248706.1838(84)$

[a] Fit with $s=5.15$, $\rho=0.991029871$.
[b] The vibrational state identifiers are not shown.
[c] Numbers in parentheses are approximately 1σ uncertainties in units of the least significant figure.
[d] The preceding operator is multiplied by the function shown with $cn=\cos 2\pi n\rho K/3$ and $sn=\sin 2\pi n\rho K/3$.
[e] A–E splitting at $K=0$ obtained from the Fourier expansion of the energy giving $F=612937.4$ MHz and $V_3=236.911$ cm^{-1}.

barrier which also makes the two nitrogens dynamically equivalent; (v) ^{14}N-quadrupole coupling constants were utilized to determine the average orientation of the NO$_2$ rotors within the molecule, i.e., the N-O-N angle and the rotation angle of the C_{2v} rotors; (vi) it is an important atmospheric molecule, playing a role in the NO$_n$ photochemistry in the upper atmosphere.

For the two internal rotor states, rotational levels with $Ka + Kc$ even have $I_N = 0, 2$, while levels with $Ka + Kc$ odd have $I_N = 1$, where I_N is the resultant nitrogen nuclear spin. This observation establishes that the equilibrium configuration of the molecule has a twofold axis of symmetry.

2.3.4.   Ring motions

Saturated four-membered rings and saturated (or near saturated) five-membered rings are very floppy molecules characterized by several equivalent or similar energy minima. This is due to a subtle balance of electronic and steric interaction energy. In consequence, these molecular systems possess – slightly spaced in energy – two or more interacting vibrational levels. The rotational spectra show distinct features, typically doublets or multiplets of irregularly split lines. Substituted rings display typical *axial-equatorial* conformational equilibria.

2.3.4.1.   *Ring puckering*

Prototype molecules exhibiting ring-puckering motions are cyclobutane (a saturated four-membered ring) and cyclopentene (a five-membered ring with one double bond). Many four- and five-membered ring compounds showing the effects of this motion have been analyzed, obtaining very precise information on large amplitude motions that are represented by the two prototype molecules discussed here. Cyclobutane virtually has no dipole moment, but its rotational spectrum is observable from the dipole moment induced by $H \rightarrow D$ asymmetric isotopic substitution. The study of the mono- and of the 1,1 bi-deuterated species allowed for the determination of ring-puckering potential energy function (barrier $V = 505\,\mathrm{cm}^{-1}$), and for the associated structural relaxation [76,77]. An MW investigation of cyclopentene [78] allowed for the precise determination of the splittings of rotational states leading to the ring-puckering barrier, evaluated to be $V = 242\,\mathrm{cm}^{-1}$.

2.3.4.2.   *Interaction of ring puckering with butterfly motions*

A valuable application of MW spectroscopy is its power to unravel far-infrared (FIR) spectra and determine the PESs of two-ring molecules constituted by an aromatic ring fused to a five-membered ring, the latter being saturated apart from the bond shared with the aromatic ring. The corresponding FIR or single vibronic fluorescence (SVLF) spectra were correctly interpreted only after the MW spectra have been assigned. The prototype molecule of this kind, indan, combines a benzene and cyclopentene ring. Initial FIR and SVLF investigations lead to a very high barrier to ring puckering, $V = 1980\,\mathrm{cm}^{-1}$ [79], much higher than the barrier for cyclopentene. The assignment of the rotational spectrum of indan showed a tunneling splitting of the ground state $\Delta E_{0+0-} = 22.3\,\mathrm{MHz}$, facilitated a rapid re-assignment of the FIR and SVLF spectra and evaluation of the ring puckering barrier to be $V = 434\,\mathrm{cm}^{-1}$ [80], much closer to the value in isolated cyclopentene [78]. Basically, the histories of all other molecules of this type being investigated with these three techniques are similar: Indoline [81], phthalan [82], 1,3-benzodioxole [83], and coumaran [84]. A two-dimensional model was found, however, to be more appropriate to interpret the vibrational spacings [81,85].

2.3.4.3.   *Pseudo-rotation*

The prototype of the saturated five-membered rings, cyclopentane, is characterized by nearly-free pseudo-rotation, and several equivalent minima-separated by very low barriers-appear in the 2π range of the potential energy function. Because of the lack of a permanent dipole moment, precise information on this motion has not been obtained from the rotational spectrum though. While it was possible to investigate

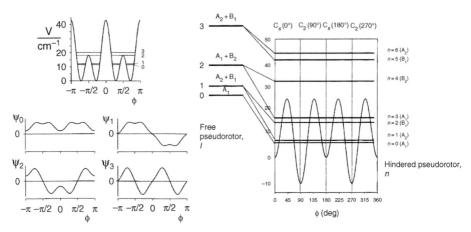

Figure 3 The pseudorotational energy level diagram and the empirical potential energy function for pseudorotation of tetrahydrofuran according to Refs. [87] (left) and [88] (right). None of them interpret the spectrum of perdeuterated tetrahydrofuran [90]. (Reused with permission from Refs. [87 and 88]. Copyright 1999 & 2003, American Institute of Physics.)

some nonpolar molecules such as cyclobutane [76] and cyclohexane [86] owing to the dipole moment induced by asymmetric deuteration, this has not been achieved for cyclopentane so far, probably because of the high density of pseudo-rotational states. However, rotational spectra have early been reported for polar molecules such as tetrahydrofuran [87,88] and 1,3-dioxolane [89] and have subsequently been interpreted in terms of a model assuming low barriers to pseudo-rotation. In the case of tetrahydrofuran, two different PESs, shown in Figure 3, could interpret several spectral spacings, but the one to the right, from Ref. [88], could take better into account the $V=2\leftarrow1$ spacing. However, none of the two models could satisfactorily explain the spectrum of perdeuterated tetrahydrofuran [90].

2.4. Chemical bond and nuclear quadrupole coupling: examples of severe interaction

The overall molecular rotation can couple to a nuclear spin. A nuclear spin $I>1/2$ gives rise to a nuclear quadrupole moment [2]; it is then possible to experimentally determine the quadrupole coupling tensor's diagonal elements χ_{gg}, with $g=a, b, c$ in the principal moment of inertia's axes system. They can be converted, in case of high MS, or when also off-diagonal coupling constants or enough structural data are available, to the quadrupole coupling contants in the quadrupole coupling tensor's principal axes system of the atom of interest, χ_x, χ_y, and χ_z. With $\chi_z=eq_zQ=e(d^2V/dz^2)Q$, etc., where e is the elementary charge and Q is the electric quadrupole moment of the nucleus. The coupling constant is directly related to the electric field gradient d^2V/dz^2 at the coupling nucleus. Because Q is independently available from atomic spectra, q can be calculated to obtain the electronic environment of the nucleus. Relations between quadrupole coupling constants and the electronic structure are given for several molecules in Ref. [2]. While the assignment in the cases of a single coupling nucleus or weak coupling is usually straight forward, several strongly coupling nuclei give rise to a complexity in the spectrum which challenges the analysis of the hyperfine structure.

2.4.1. A molecule with two iodine atoms

A molecule bearing an iodine nucleus exhibits a complicated rotational spectrum for three reasons: a large atomic mass, a very large nuclear quadrupole moment ($Q = -78.9\,fm^2$), and a large nuclear spin $I = 5/2$. The situation becomes dramatically complex for a molecule containing two iodine nuclei, with each rotational level being split into 36 components, separated at the same order of magnitude as the rotational levels. Indeed, only one molecule, CH_2I_2, with two iodine atoms has been reported so far [91,92]. The spectrum exhibits a continuous distribution of lines, but it was nevertheless possible to identify some transitions, like the one shown in Figure 4 with all components falling in a range of a few hundred MHz; an initial assignment of the hyperfine structure was made on the basis of well-isolated splitting patterns observable in a supersonic jet and utilizing the intensity patterns as predicted from nuclear spin statistics. Subsequently, it was possible to assign thousands of lines resulting in the determination of χ_{aa}, χ_{bb}, and χ_{ab}.

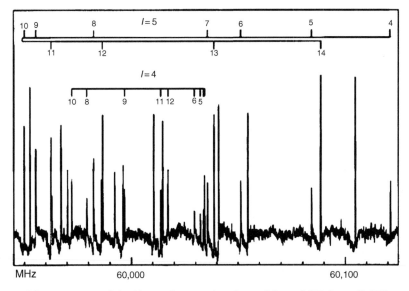

Figure 4   The spectrum of the $10_{2,9} \leftarrow 9_{1,8}$ rotational transition of CD_2I_2 at 60 GHz recorded with a free-jet spectrometer. The transition is expected to consist of 36 ($\Delta I = 0$, $\Delta F = 1$) hyperfine components. These are all visible in the spectrum that consists of 32 component lines because four lines are blends of two hyperfine components. The F quantum number assignment for the two highest values of I is indicated. Intensities of the $I = 5$ components are enhanced over those of the $I = 4$ components by a statistical weight ratio of 126:45. (Reprinted with permission from Ref. [92].)

2.4.2. A molecule with three chlorine atoms

Dore and Kisiel [93] succeeded in analyzing the hyperfine structure in the $J = 1 \leftarrow 0$ and $J = 2 \leftarrow 1$ transitions of $^{35}\text{Cl}_3\text{CCH}_3$ and $^{35}\text{Cl}_2{}^{37}\text{ClCCH}_3$ by utilizing supersonic-jet FTMW spectroscopy. The complete inertia principal axes system coupling tensor elements and principal quadrupole coupling tensor elements of the three chlorine nuclei were determined. The symmetric top treatment for $^{35}\text{Cl}_3\text{CCH}_3$ and the asymmetric top treatment for $^{35}\text{Cl}_2{}^{37}\text{ClCCH}_3$ are found to yield identical results for the principal tensor components of the ^{35}Cl nuclei. The quadrupole asymmetry parameter η for the chlorine nuclei in 1,1,1-trichloroethane is small, indicating near-cylindrical symmetry of the field gradient. Nevertheless, there is evidence for some deviation of orientation of the z symmetry axis of the field gradient from the direction of the C—Cl bond.

3. ISOLATED MOLECULES

Rotational spectroscopy was traditionally devoted to the investigation of the structure and shape of isolated molecules in the gas phase. Small stable polar molecules with a reasonable vapor pressure were suitable to be investigated with MW spectroscopic techniques. The rotational spectra of molecules with two large amplitude motions or with severe multiple quadrupole couplings could only be treated in some cases because of the complexity of the spectra. Recently, several techniques have been combined with MW spectroscopy, to facilitate the assignment of complicated spectra, to observe the spectra of essentially vapor-less compounds, and to easily prepare new and unstable species. The majority of these techniques rely on a jet expansion immediately after some special means of preparation: vaporization by laser ablation, electrical or MW discharge prior to complete expansion, pyrolysis, and photochemical reaction. In addition, the pure rotational spectra of several nonpolar molecules have been reported.

3.1. Chemistry and MW spectroscopy: new molecular species and in situ preparation

Pyrolysis and electrical and MW discharge, often combined with supersonic expansions, represent the most important tools to achieve new chemical species within MW spectroscopy. In some cases, species vaporized by laser ablation have been used to obtain new molecules by mixing them with reactive compounds in the pre-expansion channel. Rarely, photochemical preparations have been reported. The most common schemes that are used to date are (i) mm/sub-mm wave spectroscopy in an argon plasma produced by an abnormal electric discharge; (ii) mm/sub-mm wave spectroscopy in a RG plasma produced by a electric direct current (DC) discharge; (iii) supersonic jet FTMW spectroscopy employing a RG plasma produced by an electric DC discharge; (iv) mm/sub-mm wave direct absorption techniques under DC discharge conditions applied to a precursors flown through a furnace to vaporize metals; (v) supersonic jet FTMW spectroscopy employing a metal plasma prepared by laser ablation for reaction with a precursor. Many of these efforts are particularly directed to species of astrophysical importance. We can only give some examples of the many new species that have been studied in this course.

3.1.1. Neutral molecules

Neutral molecules are the easiest species to be obtained under the experimental conditions described above, and for this reason, the studies of neutral species are quite numerous. We try to divide them in meaningful groups.

3.1.1.1. Allotropic forms, alloys, metal chalcogenides, metal halides, pseudo-halides, and sulfides

Allotropic forms: While the structure of the second form of oxygen, ozone (O_3), has been known for more than 50 years [94], very little was known, up to a few years ago, about the geometries of the allotropic forms of other elements. Recently, Thaddeus et al. assigned the rotational spectra of S_3, [95] S_4, [96], and Si_3 [97]. The rotational spectra of S_3 and S_4 have been observed in an electrical discharge through sulfur vapor, while Si_3 was created in the throat of a small supersonic nozzle by applying a low-current DC discharge to a short gas pulse of silane (0.1%) in a neon or argon. The spectra were observed by supersonic-jet FTMW and, in the case of S_4, also with long-path mm wave absorption spectroscopy. Two of the molecules have C_{2v} symmetry; in the case of S_4, a small tunneling splitting at 14.1 kHz in its ground vibrational state suggests a low-lying transition state of D_{2h} symmetry, yielding interchange.

Alloys: Two rhomboidal isomers of SiC_3 have also been detected by supersonic-jet FTMW spectroscopy [98–100]. They are shown in Figure 5. While both isomers have C_{2v} symmetry, rhomboid II, characterized by a C—Si transannular bond, is calculated to lie $\sim$21 kJ/mol above the ground state rhomboid I. The strongest lines of SiC_3 were observed when expanding a mixture of 0.1% silane and 0.2% of diacetylene in Ne and then applying a discharge voltage of 1100 V across some portion of a short channel past the electromagnetic valve.

The pure rotational spectra of the metal carbide NiC has been measured in the laboratory using mm/sub-mm wave direct absorption methods [101]. The molecule was created by reacting metal vapor with CH_4 in a DC discharge.

Chalcogenides: FTMW spectra binary compounds of tin and lead with selenium and tellurium have been reported [102,103]. Gaseous samples of these chalcogenides (SnSe, SnTe, PbSe, and PbTe) were prepared by laser ablation of suitable target rods and were stabilized in supersonic jets of Ar. All together, 68 isotopic species have been analyzed, and all of them have a closed-shell $X^1\Sigma^+$ electronic ground state. Global multi-isotopologue analyses of all available high-resolution data produced spectroscopic Dunham parameters Y_{01}, Y_{11}, Y_{21}, Y_{31}, Y_{02}, and Y_{12} for all species, as well as Born–Oppenheimer breakdown coefficients δ_{01} for Sn, Pb, Se, and Te. Magnetic hyperfine interaction produced by the dipolar nuclei ^{119}Sn, ^{117}Sn, ^{77}Se,

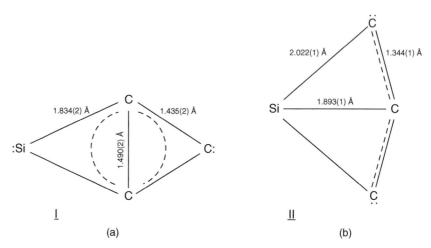

Figure 5 Two rhomboidal isomers of SiC$_3$: Form **II** is calculated to lie at about 21 kJ/mol higher in energy than form **I**. (Reused with permission from Ref. [98]. Copyright 1999, American Institute of Physics.)

^{125}Te, and ^{207}Pb was observed, yielding first determinations of the corresponding spin–rotation coupling constants.

Halides and pseudohalides: Gold(I)fluoride, AuF, historically one of the most elusive of all metal halides, has been prepared by laser ablation of Au metal in the presence of an F precursor and observed using a supersonic-jet FTMW spectrometer [104]. Both SF$_6$ and CF$_3$I were used as the F precursor, with SF$_6$ giving the stronger signals.

The FTMW spectra of CuI and AgI have been recently reported [105]. Gaseous samples of the diatomic metal iodides have been obtained by laser ablation of target rods made of suitable solid precursors. Improved rotational parameters, nuclear quadrupolar coupling constants, and spin–rotation interaction constants have been obtained, along with the first reported spin–spin coupling parameters for copper iodide. Information about the anisotropy of nuclear magnetic shielding and the indirect spin–spin coupling in the absence of intermolecular effects has also been obtained.

The MW absorption spectrum is reported also for a mixed alkali halide dimer, LiNaF$_2$ [106]. The sample was brought to different low rotational temperatures by an adiabatic expansion in a beam and by collisional cooling in a cold absorption cell. Precise rotational constants, quartic centrifugal constants, quadrupole coupling constants, and the electric dipole moment were determined from least-squares fits of the experimental transition frequencies. The dimer does have a planar rhomboidal shape and a C_{2v} symmetry, with Na and Li along the C_2 axis. The halogen–metal bonds appear to have a 100% ionic character.

Ziurys and collaborators prepared copper cyanide, CuCN, and aluminum isocyanide, AlNC, by gas-phase reactions between copper vapor and cyanogen, and aluminum vapor and trimethylsilyl cyanide, respectively, and subsequently measured their mm/sub-mm wave spectra [107,108]. Both molecules are linear. For CuCN, four isotopologues, $^{63}Cu^{12}C^{14}N$, $^{65}Cu^{12}C^{14}N$, $^{63}Cu^{13}C^{14}N$, and $^{63}Cu^{12}C^{15}N$, were measured and the r_0, r_s, and r_m structures determined. For AlNC, only the main and the ^{13}C species have been measured.

Sulfides: The rotational spectra of some earth alkaline sulfides are available. Magnesium sulfide, MgS, has been measured with supersonic-jet FTMW spectroscopy [109]. MgS was produced by reacting ablated magnesium metal with OCS (0.2–0.5%) in argon (400–600 kPa). The $J = 1 \leftarrow 0$ transition near 16 GHz has been measured for several isotopomers with the ^{25}Mg nuclear quadrupole hyperfine structure being observed and the nuclear quadrupole coupling constant, $eQq(^{25}Mg)$, being determined. Strontium sulfide, SrS, was characterized by mm/sub-mm wave spectroscopy [110], with SrS created in a reaction of strontium vapor with H_2S or CS_2 in the presence of a DC discharge. The Sr—S bond length is 244.1 pm.

3.1.1.2. Hydrides, hydroxides, metalamides, and metalmercaptanes

Hydrides: An interesting series of silicon hydrides has been directly prepared and observed by rotational spectroscopy. First Bogey and collaborators could assign the mm/sub-mm wave spectra of two forms of disilyne, Si_2H_2. The first observed species, form I, is characterized by a double hydrogen bridge between the two silicon atoms and possesses C_{2v} symmetry [111]. The second species, form II, is planar and has a single hydrogen bridge, with the second hydrogen attached only to one Si atom [112]. The molecule was obtained in a glow discharge on an Ar/SiH_4 mixture. Further work on the isotopic species, combined with theoretical calculations, indicate form I to be more stable than form II by 33–38 kJ/mol [113,114]. A monobridged isomer, $H_2Si(H)SiH$, has also been detected for Si_2H_4 by supersonic-jet FTMW spectroscopy, produced from discharge products of silane [115]. It is calculated to lie ~39 kJ/mol above disilene, H_2SiSiH_2, the most stable isomeric arrangement of Si_2H_4. Its rotational spectrum exhibits closely spaced doublets, characteristic of a molecule undergoing high-frequency inversion.

Hydroxides: The mm wave absorption spectra of AgOH and CuOH show the two molecules to be bent [116]. Both species were produced by sputtering from copper or silver cathode electrodes in the presence of a DC discharge of a 3:1 He/O_2 (2.6 Pa) gas mixture.

Metalamides: The pure rotational spectra of the monomeric forms of $LiNH_2$ and $NaNH_2$ have been recorded using mm/sub-mm wave direct absorption techniques [117,118]. These studies are the first gas-phase detections of the two molecules in the laboratory, and, for $LiNH_2$, the first preparation of the isolated molecule. The two metallo amines were generated by the reaction of the metal vapor with ammonia diluted in a RG, and in the presence of a DC discharge. Both molecules have a planar structure with alkali–nitrogen distances of $r_0(Li—N) = 173.6(3)$ and $r_0(Na—N) = 205.1$ pm, respectively.

The rotational spectrum of a metallo-imide, BaNH [119], has been measured with the same technique and the molecule produced in the same way, by the reaction of ammonia and barium vapor in the presence of a DC discharge. Transitions were measured in the ground vibrational state and the excited vibrational bending (0110) and heavy atom stretching (100) modes were measured. The reported r_s parameters are $r_{BaN} = 207.6$ and $r_{NH} = 101.3$ pm, respectively.

Metalmercaptanes: The rotational spectra of some sulfydrides, LiSH [120] and CuSH [121], have been investigated by using mm/sub-mm wave direct absorption techniques. These species were created by the reaction of H_2S and the corresponding metal vapor in a DC discharge. They are bent molecules with an M—S—H (M = Li, Cu) valence angle of 93°, very similar to the value in H_2S.

3.1.1.3. Organometallics

Archetypal organoalkalis, such as CH_3M (M = metal atom), have extensively been studied by Ziurys and collaborators using mm/sub-mm wave direct absorption techniques, because they may be present in astronomical sources. These species were synthesized in a gas-phase reaction employing a DC discharge on metal vapor, produced in a Broida-type oven, and $(CH_3)_4Sn$, $(CH_3)_2Hg$, or CH_3I. The CH_3M molecules are C_{3v} symmetric tops with a ^1A electronic ground state if M is an alkali metal and in some other cases, presented below. The first study reported the pure rotational spectrum of $AlCH_3$ [122] and supplied a precise r_{Al-C} distance of 199.4 pm.

With the same technique, structures of monomeric methyl alkalis, CH_3Li, CH_3Na, and CH_3K, in the absence of other ligands, were characterized [123,124]. The metal–carbon bond length decreases in ascending the periodic table, from 263.3 over 229.9–196.1 pm, while the H—C—H bond angles only change slightly, all being close to 107°.

For methylcopper, CH_3Cu, the mm wave spectra have been measured for five isotopologues, $^{63}Cu^{12}CH_3$, $^{65}Cu^{12}CH_3$, $^{63}Cu^{12}CD_3$, $^{63}Cu^{13}CH_3$, and $^{65}Cu^{13}CH_3$ [125]. From this data set, accurate spectroscopic constants and a structure for CH_3Cu were derived. The r_m geometry shows the shortest alkyl carbon–copper(I) bond length known, $r_m(C—Cu) = 188.09$ pm, and a rather large H—C—H valence angle of 109.9° within the methyl group.

The pure rotational spectra of two linear metal acetylides, LiCCH [126] and NaCCH [127], in their $^1\Sigma^+$ ground states have been reported. The species were created by the reaction of metal vapor and HCCH seeded in argon (5.3 Pa) and applying a DC discharge. The metal vapor was generated by a Broida-type oven attached to the bottom of the free-space cell. The spectra were recorded using mm/sub-mm wave direct absorption spectroscopy in the range of 105–538 GHz. Rotational, l-type doubling and vibration–rotation constants were determined for LiCCH, as well as r_s and r_0 structures.

3.1.1.4. Molecules containing a rare gas atom forming covalent bonds

Numerous van der Waals complexes have been observed by supersonic-jet FTMW spectroscopy (see Section 4.1). They are characterized by large centrifugal distortion effects, making it possible to size the dissociation energy of the complex. Gerry

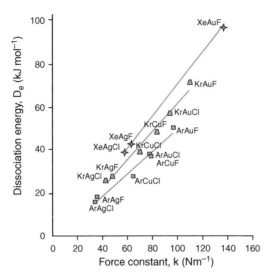

Figure 6 Plots of the ab initio dissociation energies versus experimental stretching force constants for RG—M bonds in RG—MX complexes. (Reprinted with permission from Ref. [131]. Copyright 2004 American Chemical Society.)

observed for complexes of the type RG—MX (RG = Xe, Kr, Ar; M = Ag, Au, Cu; X = F, Cl, Br) much smaller centrifugal distortion constants D_J than for other van der Waals complexes, corresponding to binding energies in the order of covalent bonds. Such effect was first noted within the series Ar—AgX (X = F, Cl, Br) [128], with van der Waals binding energies much larger than in normal van der Waals complexes, i.e., up to 23 kJ/mol. Later on, in complexes involving heavier metal halides (such as AuX and CuX) and heavier RG atoms, it was found that the dissociation energy could reach values typical of regular covalent bonds, i.e., up to 100 kJ/mol [129–132].

All these species have been generated by laser ablation of the metal in the presence of SF_6, Cl_2, or Br_2 for the fluorine, chlorine, or bromine derivatives, respectively, and the appropriated RG seeded in argon. In Figure 6, the force constants and dissociation energies for the various species are displayed while Figure 7 draws contour diagrams of two occupied valence molecular orbitals of XeAuF.

3.1.1.5. Carbon-based chain and ring molecules

Most of these species have been investigated by Thaddeus and collaborators, mainly by using supersonic-jet FTMW spectroscopy. Part of their work is compiled in two review articles, concerning 27 carbon-chain molecules prepared and detected in 1996–1997 [133], and 11 polyyne carbon chains [134]. Generally, these molecules were prepared by an electric DC discharge in the throat of the supersonic nozzle applied to precursors seeded in RGs. All of these molecules are of astronomical interest, and some of them have already been detected in the interstellar medium.

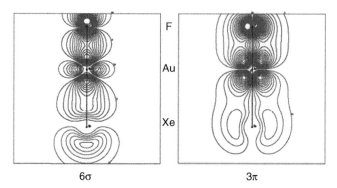

Figure 7   MOLDEN contour diagrams of two occupied valence molecular orbitals of XeAuF; in each case the value of the contours is $0.02n$, with $n = 1\text{--}25$. The different colors indicate opposite signs of the wavefunctions. (Reprinted with permission from Ref. [131]. Copyright 2004 American Chemical Society.)

These species can be divided in *hydrocarbons, cyano or isocyano* derivatives, *silicon-*containing chains, *sulfur* derivatives, *phosphorus* derivatives, and *fluoro*-substituted molecules.

Hydrocarbons: The hydrocarbons studied most by supersonic-jet FTMW spectroscopy share the sum formula C_nH_2. However, this group is comprised of a number of different electronic structures and shapes. Several cumulenes, characterized by linear chains of $C=C$ double bonds, from H_2C_5 to H_2C_{10}, were detected by Thaddeus and coworkers [135,136]. Like the shorter cumulenes in this series, all four were found to have singlet electronic ground states and linear carbon chain backbones. The strongest lines of the cumulene carbenes were observed in a DC discharge through a mixture of diacetylene (0.5%) in neon.

Species with C_nH_2 sum formula have also been found for electronic and conformational arrangements different from the linear cumulene chains. In the laboratory, for example, four C_5H_2 isomers have been detected and their rotational spectra been characterized to high accuracy by supersonic-jet FTMW spectroscopy [135,137,138]. All of them, shown in Figure 8, are good candidates for radioastronomical detection. The four species were obtained from a DC discharge in the throat of a supersonic nozzle applied to gas pulses of dilute precursors in Ne (100–200 kPa). During these experiments, optimized intensities of the various species were obtained by varying the precursor: diacetylene, allene, acetylene, or their mixtures.

Also "ring-chain" forms, with a terminal three-membered ring as in C_5H_2 isomer I of Figure 8, are observed. The corresponding form was also identified for C_7H_2 and C_9H_2 [139]. The strongest lines of C_9H_2, like those of the smaller ring chains were obtained from a diacetylene–neon mixture: C_9H_2 was first observed from diacetylene (1%) in neon subjected to a discharge in the throat of the supersonic nozzle. A bent singlet isomer of the C_6H_2 carbene, similar to species III of Figure 8, has been observed in the laboratory [140]. The three

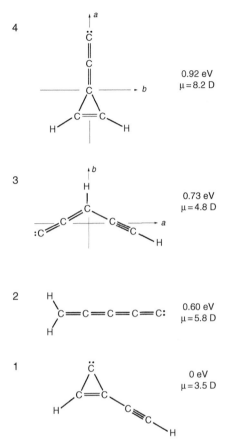

Figure 8 Four isomers of C_5H_2 identified with FTMW spectroscopy. Double dots indicate the characteristic carbene nonbonding electrons. The listed dipole moments and energies with respect to isomer 1 are from ab initio calculations. (Reprinted with the permission from Ref. [138].)

rotational constants and the leading centrifugal distortion constants were determined to high accuracy, allowing the entire rotational spectrum of this molecule below 120 GHz to be calculated to a fraction of 1 km/s equivalent radio velocity.

Supersonic-jet FTMW spectra of vinyldiacetylene and vinyltriacetylene are also reported [141]. They are planar, with a linear polyacetylenic chain and a vinyl bent. Finally, the rotational spectra of four symmetric top methylpolyynes, $CH_3(C{\equiv}C)_4H$, $CH_3(C{\equiv}C)_5H$, $CH_3(C{\equiv}C)_6H$, and $CH_3(C{\equiv}C)_7H$, have been reported [134]. All species were prepared in applying a DC discharge to a dilute mixture of methyldiacetylene in Ne.

Cyano and isocyano derivatives: Linear cyanopolyynes $HC_{2n+1}N$, with odd values of n, are well-studied molecules in the laboratory and in the space. They behave similarly to linear hydrocarbon carbenes C_nH_2. $HC_{15}N$ and $HC_{17}N$ [134,142] are the last investigated members of the family. With a molecular weight of 219 u and a rotational constant of slightly more than 50 MHz, $HC_{17}N$ is the longest of the carbon chains identified by high-resolution spectroscopy. Also highly polar "ring–chain" carbenes, HC_4N and HC_6N, formed by substituting either CN or CCCN for a hydrogen atom in cyclopropenylidene (c-C_3H_2), were detected [143]. Like the hydrocarbon carbenes C_5H_2, C_7H_2, and C_9H_2, the two molecules have planar ring–chain structures and singlet electronic ground states. The strongest lines were observed in a discharge through a mixture of cyanoacetylene (0.5%) in neon.

The supersonic-jet FTMW spectra of vinylcyanodiacetylene, $H_2C{=}CH{-}(C{\equiv}C)_2{-}C{\equiv}N$, having a plane of symmetry [141], and some symmetric top methylcyanopolyynes $CH_3(C{\equiv}C)_3CN$, $CH_3(C{\equiv}C)_4CN$, and $CH_3(C{\equiv}C)_5CN$ [134], all of them of astronomical interest, have also been reported.

The rotational spectra of two linear isocyanides, HC_4NC and HC_6NC, which are of key interest to interstellar cloud chemistry [144], are also available.

Silicon-containing chains: The rotational spectra of several highly polar linear silicon–carbon chains SiC_n, $n = 3$–8, were detected within the products of a DC discharge through a silane–diacetylene mixture by supersonic-jet FTMW. SiC_4, SiC_6, and SiC_8 were found to have a $X^1\Sigma^+$ ground state, while those with odd n have a $X^3\Sigma^-$ ground state [145,146]. We will consider the radicals more detailed in the next section. Similar to the closed-shell chains, accurate equilibrium structures were derived for the simplest molecules by converting the experimental rotational constants to equilibrium constants in using the vibration–rotation coupling constants obtained from coupled-cluster calculations, including connected triple substitutions. On the basis of the calculated vibration–rotation and l-type doubling constants, weak rotational satellites from a low-lying vibrational state of SiC_4 were assigned to ν_6, a bending mode calculated to lie $\sim$205 cm^{-1} above the ground state.

Sulfur derivatives: The linear carbon–chain radicals C_6S, C_7S, C_8S, and C_9S have been detected by supersonic-jet FTMW spectroscopy, and measurements of the previously studied chains C_4S and C_5S have been extended [147]. Oppositely to the silicon-containing chains, the electronic ground states are found to be a triplet for those with an even number of carbon atoms and a singlet for those with an odd number. Strong lines were observed using a low-current DC discharge through a precursor gas mixture of diacetylene, HCCCCH (0.5%), and carbon disulfide, CS_2 (0.3%), in Ne or Ar at a pressure of 330 kPa. The study of the MW spectra of the series H_2C_2S, H_2C_3S, H_2C_4S, H_2C_5S, H_2C_6S, and H_2C_7S identified all of them to be asymmetric top chains with C_{2v} symmetry [148].

Phosphourus derivatives: The linear, unstable HC_5P molecule has been first detected in the pyrolysis products of phosphorus trichloride and toluene mixtures. The rotational spectra of the ground and $\nu_{11} = 1$ excited state were investigated in the mm wave region from 78 to 195 GHz for the normal and deuterated species [149], obtaining accurate values of rotational, centrifugal distortion and q_{11} l-type doubling constants.

Fluoro-substituted molecules: Novick and collaborators reported the supersonic-jet FTMW spectra of four fluoromethyl polyynes, 5,5,5-trifluoro-1,3-pentadiyne, CF_3—C≡C—C≡C—H, 7,7,7-trifluoro-1,3,5-heptatriyne, CF_3—C≡C—C≡C—C≡C—H, 1,5,5,5-tetrafluoro-1,3-pentadiyne, CF_3—C≡C—C≡C—F, and 4,4,4-trifluoro-1-nitrile-2-butyne, CF_3—C≡C—C≡N [150]. The molecules were produced by pulsed high-voltage DC discharges of dilute mixtures of precursor gases such as trifluoropropyne in argon. The ^{13}C and deuterium-substituted isotopologous of trifluoropentadiyne were studied, and the molecular structure was determined.

Alonso and coworkers detected the supersonic-jet FTMW spectrum of the linear fluorotetraacetylene, F—(C≡C)$_4$—H [151], produced by pulsed high-voltage DC discharge on a mixture of 3,3,3-trifluoropropyne and acetylene highly diluted in neon.

Mixed hetero atom chains: The supersonic-jet FTMW spectra of the linear silicon- and sulfur-containing carbon chains $SiC_{2n}S$ ($n = 1$–3) in their singlet electronic ground state have been detected and characterized [152]. The best signal-to-noise ratio was achieved with mixtures of diacetylene (0.1%), silane (0.1%), and carbonyl disulfide (0.1%) in neon (3.3 kPa).

The closed-shell linear chain SiCCO was first generated by an electric DC discharge of silane, acetylene, and carbon monoxide highly diluted in neon and probed by supersonic-jet FTMW spectroscopy [153]. The precise spectroscopic constants allow predicting the radio spectrum with accuracy better than 1 km/s.

3.1.1.6. *Other closed-shell molecules*

Gerry and collaborators reported the supersonic-jet FTMW spectrum of sulfur chloride fluoride, ClSF, prepared by using a pulsed electric DC discharge on a mixture of SF_6 and SCl_2 in neon [154] along with the structure, harmonic force field, and Cl nuclear quadrupole hyperfine constants. An analogous investigation was performed for palladium monocarbonyl, PdCO [109]. The molecule was prepared by laser ablation of Pd in the presence of CO diluted in argon. The rotational constants from spectra of 15 isotopomers were used to determine the molecular geometry; the nuclear quadrupole coupling constant and nuclear spin–rotation constant have been determined for ^{105}Pd. Nuclear shielding parameters have been evaluated from the measured spin–rotation constant.

The mm wave spectrum of the transient phosphine oxide, H_3PO, a symmetric top, was first detected in the gas phase using a source-modulated spectrometer [155]. The H_3PO molecule was generated in a free space cell by a DC glow discharge on gas mixture of PH_3, CO_2, and H_2.

The supersonic-jet FTMW spectrum of cyanophosphine, H_2PCN, has been measured [156] and represents the first spectroscopic study of a phosphanitrile, H_2P—C≡N. The compound was obtained by expanding a gas mixture of PH_3 (0.3%) and CH_3CN (0.3%) in argon (100 kPa) through a pulsed DC discharge. The molecule has C_s symmetry, with the symmetry plane containing the three heavy atoms.

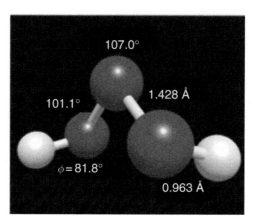

Figure 9 The molecular structure of HOOOH, a starting point species to investigate the ability of oxygen atoms to form chains. (Reprinted with permission from Ref. [157]. Copyright 2005, American Chemical Society.)

Most interesting is the preparation and detection of the rotational spectrum of H_2O_3, a molecule with an oxygen chain (see Figure 9). FTMW spectroscopy and FTMW/mm wave double-resonance and triple-resonance spectroscopy [157] were employed. For this study, H_2O_3 was produced in a pulsed nozzle applying a DC discharge to a gas mixture of O_2 (10%) in argon passed through a reservoir filled with a hydrogen peroxide solution (30%).

3.1.2. Radicals
The rotational spectra of many radicals are accessible by MW spectroscopy, employing mainly two techniques: mm wave absorption spectroscopy and supersonic-jet FTMW spectroscopy.

3.1.2.1. Alloys, oxides, and salts
Recently, the rotational spectra of a radical alloy, TeSe, has been investigated by FTMW spectroscopy combined with laser ablation [158,159]. Employing a global multi-isotopologue analysis to transitions of 43 isotopologues of TeSe in 7 vibrational states spectroscopic Dunham parameters $Y_{l,m}$, Born–Oppenheimer breakdown coefficients $\delta_{l,m}$, the equilibrium bond lengths r_e, and the vibration parameters ω_e and $\omega_e x_e$ were obtained for all analyzed isotopologues. For low vibrational states, the Morse-potential function describes the TeSe-potential very well and provides an estimate of the maximum dissociation energy for this semi-metal compound. In addition, the isotopologue-independent molecular constants $U_{l,m}$, and the corresponding Born–Oppenheimer breakdown coefficients $\Delta_{l,m}$ were determined. The large $\Delta_{0,1}$ coefficients were necessary for Watson's reference isotopologue-independent analysis for Te and Se, according to the interaction between the two sublevels of the electronic $^3\Sigma$-state.

Carbides: Several metallo-carbides have been investigated by Ziurys and collaborators, mainly with the intent to find their signals in the interstellar space. The rotational spectra of the NaC and KC radicals have been recorded by mm wave spectroscopy in their $X^4\Sigma^-$ ground states [160,161]. Both molecules were generated at DC discharge conditions from the reaction of the metal vapor with methane; the vapor was generated in a Broida oven. The quartet hyperfine structures were partially resolved for each transition. The rotational transition frequencies were modeled with an effective Hamiltonian. An ambiguity in the fitting procedure produced two alternative values of the spin–spin coupling constant, differing substantially in magnitude. A simple theoretical estimate of this parameter, based on atomic orbitals on the C^- atom, suggests the smaller value to be correct. The calcium carbide radical, CaC, was observed with the same techniques [162]. Here, each transition was found to be a triplet from the fine-structure interactions of a $X^3\Sigma^-$ ground state. The data were analyzed in a Hund's case (b) basis, and rotational, spin–spin, and spin–rotation constants have been accurately determined.

The mm wave spectra of three transition metal carbides have been reported. All compounds were obtained through the reaction of the metal vapor, produced in a high-temperature Broida-type oven, with methane gas at DC discharge conditions. However, the three spectra correspond to three different electronic states. NiC, as discussed in the previous section, is a closed-shell molecule [101] while the CoC [101] and FeC [163] radicals have $X^2\Sigma^+$ and $X^3\Delta_i$ ground states, respectively. For CoC, the study includes a complex hyperfine analysis of the magnetic, electric quadrupole, and nuclear spin–rotation terms. For FeC, transitions in the $\Omega = 2$ and 3 ladders were measured, allowing the determination of rotational and certain spin–orbit parameters.

Oxides: Most of the rotational spectra of oxides directly formed in the measurement cell have been reported by Saito and collaborators using mm wave spectroscopy. The pure rotational transitions of TiO in the 220–460 GHz spectral range indicated that the ground state of the radical is $X^3\Delta_r$ [164]. The radical was generated in the gas phase by placing a quartz boat filled with $Ti(C_5H_5)_2Cl_2$ powder inside of a stainless steel cathode generating a DC discharge in the presence of a He/O_2 mixture. By flowing liquid N_2 through a copper tube attached to a copper sheet, the absorption cell was cooled to a temperature of $-170°C$.

Using a DC sputtering absorption cell, the mm wave spectra of CuO [165] and AgO [166] were obtained and found to be compatible with $X^2\Pi_i$ ground states. The magnetic hyperfine parameters a, b_F, c, and d, have been determined for the $^{65}Cu(I = 3/2)$, $^{107}Ag(I = 1/2)$, and $^{109}Ag(I = 1/2)$. They are interpreted in terms of the $X^2\Pi_i$ electronic configuration. In addition, for ^{65}CuO, the χ_0 quadrupole coupling constant is given.

The MW spectra of the aluminum monoxide radical, exhibiting an $X^2\Sigma^+$ electronic ground state, for the vibrational states $v = 1$, 2 have been observed by using a source-modulated spectrometer combined with a free-space absorption cell [167]. The Al atom was generated by discharge in pure N_2O or argon with an aluminum hollow-cathode containing several pieces of alumina. The rotational, centrifugal distortion, spin–rotation coupling, and hyperfine coupling constants for the Al nucleus were precisely determined for the $v = 1$ and 2 states.

CoO in its $X^4\Delta_i$ ground state was observed using a source-modulated sub-mm wave spectrometer [168]. A DC sputtering method using cobalt powder placed in the hollow cathode electrode was used to generate the CoO radical in an atmosphere of oxygen and helium. Each rotational transition consists of eight components due to the ^{59}Co ($I=7/2$) hyperfine interaction. Rotational and hyperfine coupling constants, a, b_F, c, and eQq, centrifugal distortion corrections to rotation and spin–orbit interaction, and the less well known higher-order spin–orbit distortion term to the Fermi contact interaction, b_S, have been determined. A recent paper, reporting measurements in the $\Omega = 1/2$, 3/2, 5/2, and 7/2 spin levels, interprets the spectra in applying a slightly different Hamiltonian, which leads, however, to a large difference in the quadrupole coupling constant [169] ($\chi_0 = -37.9 \pm 1.5$ and -52.4 ± 7.4 MHz, respectively).

Salts: Most of the the radical salts presented below were investigated by Ziurys and collaborators by mm/sub-mm wave direct absorption techniques. Among them are several halogenides of transition metals, which exist in electronic ground states very different from each other.

In the pure rotational spectrum of TiF, four spin–orbit components were observed, proving the $X^4\Phi_r$ nature of its ground state [170]. Additional small splittings were resolved in several of the spin components in lower J transitions, which appear to arise from magnetic hyperfine interactions of the ^{19}F nucleus. The data indicate strong first-order spin–orbit coupling and minimal second-order effects, also evident from a small λ, the spin–spin parameter. Moreover, only a single higher-order term, the spin–orbit/spin–spin interaction parameter η, was needed in the analysis, also suggesting only limited perturbations in the ground state. The relative values of the a, b, and c hyperfine constants indicate that the three unpaired electrons of the radical occupy orbitals primarily located at the titanium atom and support the molecular orbital picture of TiF with a $\sigma^1\delta^1\pi^1$ single electron configuration. The Ti—F bond length is determined as 183.42 pm. To create TiF, titanium vapor was first produced in an oven from heating a solid metal rod. It was then reacted with SF_6 (0.4–0.7 Pa), which was introduced into the reaction chamber from underneath the oven. Neither a carrier gas nor a DC discharge was necessary for molecular synthesis.

The ZnF radical has an $X^2\Sigma^+$ electronic ground state, as revealed by the mm wave spectra of five isotopologues (^{64}ZnF, ^{66}ZnF, ^{67}ZnF, ^{68}ZnF, and ^{70}ZnF) taken for the ground and several vibrational excited vibrational ($v=1$, 2, and 3) states [171]. Each transition consists of spin–rotation doublets with a splitting of 150 MHz. Fluorine hyperfine splitting was observed in three isotopologues (^{64}ZnF, ^{66}ZnF, and ^{67}ZnF), and nuclear quadrupole hyperfine structure from the ^{67}Zn($I=5/2$) nucleus was additionally resolved in ^{67}ZnF. Rotational, fine-structure, and ^{19}F and ^{67}Zn hyperfine constants were determined for ZnF, as well as equilibrium structure parameters. The bond length of the main isotopologue ^{64}ZnF was calculated to be $r_e = 176.77$ pm. Evaluation of the hyperfine constants indicates that the σ orbital occupied by the unpaired electron is ~80% $4s$(Zn) in character with ~10% contributions from each of the $2p$(F) and $4p$(Zn) orbitals.

A third transition metal fluoride, MnF, has a very high spin ground electronic state, $X^7\Sigma^+$ [172]. MnF was created from SF_6 and manganese vapor, produced in a

Broida-type oven. The species exhibit a complex pattern in which the fine and [55]Mn and ^{19}F hyperfine structures are intermixed. Rotational, spin–rotation, spin–spin, and hyperfine parameters have been determined for MnF, with the latter three constants being interpreted in terms of bonding and electronic structure in metal fluorides.

Pure rotational spectra are reported for three transition metal mono-chlorides, MnCl [173], FeCl [174,175], and CoCl [176]; their ground states are $X^7\Sigma^+$, $X^6\Delta_i$, and $X^3\Phi_i$, respectively. All their spectra are complicated by the presence of [35]Cl and [37]Cl isotopologues, and from the chlorine $I = 3/2$ nuclear spins. The spectra are also affected by spin–rotation and spin–spin couplings, and lambda-doubling, all being different for the three salts. With the corresponding parameters being determined, information on the electronic structure was obtained. The spin–orbit pattern exhibited by CoCl is unusual, with the $\Omega = 3$ component significantly shifted compared with the other spin components. In addition, the regular octet hyperfine splittings become distorted above a certain J value for the $\Omega = 3$ transitions only. These effects suggest that the molecule is highly perturbed in its ground state, most likely a result of second-order spin–orbit mixing with a nearby isoconfigurational $^1\Phi_3$ state. The three radical salts were synthesized *in loco* by the reaction of metal vapor, produced in a Broida-type oven, with Cl_2.

Alkali sulfide radicals, LiS [177], NaS [178], and KS [179], have been characterized by mm/sub-mm wave direct absorption spectroscopy. All of them have an $X^2\Pi_i$ ground state, with rotational transitions in both the $\Omega = 1/2$ and 3/2 sublevels. Rotational, spin–orbit, and lambda-doubling parameters were determined. Splittings arising from Λ-doubling interactions were resolved in both spin–orbit components and were particularly large ($\sim$2 GHz) in the $\Omega = 1/2$ substate, indicating the presence of a near-by state. The species were created by the reaction of alkali vapor and CS_2 at DC discharge conditions. Both radicals may be detectable in the late-type star IRC 110216, given the observation of CS, SiS, NaCl, and NaCN in this astronomical object.

With a $X^2\Pi_i$ ground state, the rotational spectrum of CuS does have the same characteristics as the spectra of the alkali sulfides [180]. Spin multiplicity is much higher for the ground states of the two transition metal sulfides, MnS and CoS, $X^6\Sigma^+$ and $X^4\Delta_i$, respectively. Consequently, their mm/sub-mm wave pure rotational spectra are very complicated. For MnS, each rotational transition consists of six fine-structure components [181]. Additionally, the lower rotational lines show a hyperfine structure, arising from the [55]Mn($I = 5/2$) nuclear spin, which was also resolved in each spin component. These data were analyzed using a Hund's case (b) Hamiltonian, and rotational, fine-structure, and hyperfine parameters were determined. Third-order correction to the spin–rotation interaction, γ_S, and the fourth-order spin–spin coupling parameter, θ, were found necessary for a fit to near experimental accuracy. For CoS, four spin components were identified in the spectra, one of them exhibiting Λ-doubling, identifying the ground state as $^4\Delta_i$ [182]. The spectra were readily identified because each spin component exhibited an octet pattern arising from the [59]Co($I = 7/2$) nuclear spin. Using a Hund's case (a) Hamiltonian, the rotational, fine-structure, hyperfine, and Λ-doubling constants were determined. While MnS was synthesized in the gas phase by the reaction of manganese vapor and CS_2 in a high-temperature Broida-type oven, CoS was created by reacting cobalt vapor with H_2S.

The mm wave spectra of two isocyanides, CaNC [183] and MgNC [184], produced in a flowing metal reaction chamber in the presence of $(CN)_2$, reveal a $X^2\Sigma^+$ doublet ground state and a linear geometry. Also several cyanides were investigated by mm wave spectroscopy. Among them, ZnCN behaves like the two isocyanides in exhibiting a $X^2\Sigma^+$ ground state; it is even prepared in the same way [185]. The spectroscopic analysis established the rotational, spin-rotation, and l-type doubling parameters; r_0, r_s, and r_m structures were proposed. On the contrary, from mm/sub-mm wave direct absorption spectra, the ground state of the NiCN radical was found to be $X^2\Delta_i$ [186]. In the vibrational ground state, transitions from both spin–orbit components, $\Omega = 5/2$ and $3/2$, were identified; in the $\Omega = 3/2$ ladder, significant lambda-doubling was observed. Multiple vibronic components were found for each recorded bending quantum, a result of Renner–Teller interactions. These components were only observed in the $\Omega = 5/2$ lower spin–orbit ladder, however, suggesting that spin–orbit coupling dominates the vibronic effects. The ground-state data were analyzed with a Hund's case (a) Hamiltonian, giving rotational, spin–orbit, and Λ-doubling constants for ^{58}NiCN and ^{60}NiCN and r_0, r_s, and r_m structures were derived from the ground-state rotational constants. The NiCN radical was created by the reaction of nickel vapor, produced in a high-temperature Broida-type oven, with cyanogen gas.

Based on the measurement of three observed spin components, $\Omega = 4, 3, 2$, in the mm wave spectrum, the electronic configuration of the CoCN radical ground state was determined to be $^3\Phi_i$ [187]. Hyperfine splittings resulting from the ^{59}Co$(I = 7/2)$ nuclear spin were observed in every transition, each exhibiting an octet pattern. The ground state measurements of CoCN were analyzed with a case a_β Hamiltonian, establishing rotational, fine-structure, and hyperfine parameters. CoCN exhibits the linear cyanide structure, along with the Zn, Cu, and Ni analogs. The preference for this geometry, as opposed to the isocyanide form, may indicate a greater degree of covalent bonding in these species.

3.1.2.2. Hydrides, hydroxydes, amides

Hydrides: The sub-mm wave direct absorption spectrum of CrH has been recently reported for the $X^6\Sigma^+$ ground electronic state [188]. CrH was created in a DC discharge by the reaction of chromium metal, vaporized in a Broida-type oven, with H_2. The five unpaired electrons, as well as the proton nuclear spin, generate complicated fine and hyperfine structures for the $N = 1 \leftarrow 0$ transition. The five strongest observed transitions are spread over a 60 GHz region because of spin–spin and spin–rotation interactions.

Hydroxides: MgOH exhibits an $X^2\Sigma^+$ ground electronic state. Its mm wave spectrum has been recorded including the v_2 bending vibration satellites [189]. Multiple rotational transitions arise from the $v^l = 1^1$, 2^2, 2^0, 3^1, 3^3, 4^2, and 4^4 substates. Both the spin–rotation and l-type doubling interactions are resolved in the spectra and the complete data sets for MgOH and MgOD were analyzed using a linear model for the Hamiltonian taking higher order ($l = \pm 4$) l-type interactions into account. The r_0, r_s, and r_e structures determined from the data differ substantially, suggesting that MgOH is quasi-linear with competing ionic and covalent bonding.

Amides: The rotational spectra of the $MgNH_2$ and $CaNH_2$ radicals, created using Broida oven methods, were detected in their X^2A_1 ground state by mm wave spectroscopy [190,191]. Spin–rotation splittings were resolved in all components. The observed spectrum was modeled with an S-reduced Hamiltonian and very accurate rotational, centrifugal distortion, and spin–rotation parameters were determined for both radicals. The calculated r_0 structures strongly suggest a planar radical with C_{2v} symmetry, although small barriers to planarity cannot be ruled out.

Hydrosulfides: BaSH [192] and SrSH [110] in their X^2A' ground state have been characterized by mm wave absorption spectroscopy. These radicals were synthesized in a DC discharge by the reaction of metal vapor, produced in a Broida-type oven, and H_2S. Fine-structure splittings were observed in all transitions. The spectra of the two species are consistent with a bent molecule of C_s symmetry, following the trend established in the lighter earth alkaline hydrosulfides. From these measurements, rotational and fine-structure parameters were established for BaSH and SrSH. An r_0 analysis indicates that the M—S—H angles are 88° and 91.5° for M = Ca, Sr, respectively. These geometries indicate a covalently bonded MSH molecule, opposite to the linear (and ionic) MOH. The r_{M-S} distances are 280.7 and 270.5 pm, respectively.

3.1.2.3. Metallo organics

The pure rotational spectra of the free radicals $MgCH_3$, $CaCH_3$, $BaCH_3$, and $SrCH_3$ in their ground state, $X\,^2A_1$, have been measured by mm/sub-mm wave direct absorption spectroscopy [193–196]. The molecules were made in the presence of a DC discharge by the reaction of metal vapor and dimethyl mercury or tetramethyl tin. All these radicals are symmetric top molecules with net spin angular momentum originating in the unpaired electron. Consequently, both K structure and spin–rotation interactions were observed in every rotational transition. The data were analyzed with a $^2A_1$ Hamiltonian, and rotational, centrifugal distortion, and fine-structure parameters were accurately determined. For the ^{137}Ba isotopomer, hyperfine interactions were also resolved, arising from the spin of the barium nucleus. The fine and hyperfine structure constants established from this series suggest a predominantly ionic bond, but with a considerable covalent contribution.

The mm/sub-mm wave spectra have been reported for the NaCH and KCH alkali methylidyne radicals [197]. The molecules were created in the presence of a DC discharge by the reaction of metal vapor and CH_4. The data indicate that KCH and NaCH are linear molecules with $^3\Sigma^-$ electronic ground states arising from a π^2 configuration. Spectroscopic constants for KCH and NaCH have been determined from the data, including rotational, spin–spin, and spin–rotation parameters, as well as bond lengths. In comparison with other alkali and transition metal-bearing molecules, these results suggest some degree of covalent bonding in the alkali methylidynes, with the carbon atom undergoing sp hybridization.

Acetylide radicals MgCCH and SrCCH exist, according to their mm/sub-mm wave spectra [127,198], in the $X^2\Sigma^+$ electronic ground state. They were synthesized in a DC discharge by the reaction of metal vapor, produced in a Broida-type oven, and acetylene. Their fine-structure splittings have been resolved, and rotational and spin–rotation parameters were determined. They are linear molecules with an ionic character.

3.1.2.4. Radical carbon-based chain and ring molecules

We will order these radicals similarly to the corresponding section for closed-shell species, starting with hydrocarbons and followed by substituted chains:

Hydrocarbons: Rotational spectra for the series of carbon chain radicals C_nH are observed for $n = 1$–14. Its members from C_5H to $C_{14}H$ have been observed recently by supersonic-jet FTMW spectroscopy [199–202]. The radicals were produced in a DC discharge through a dilute diacetylene/neon mixture in the throat of a supersonic nozzle. All are linear with $^2\Pi$ electronic ground states, and all except $C_{14}H$ have resolved Λ-type doubling. Hyperfine structure in the rotational transitions of the lowest-energy fine-structure component ($^2\Pi_{1/2}$ for $n = $ odd, and $^2\Pi_{3/2}$ for $n = $ even) of each species was measured between 6 and 22 GHz, and precise rotational, centrifugal distortion, Λ-doubling, and ^{13}C hyperfine coupling constants were determined. However, for one of these radicals, C_5H, a second isomer, with a cyclic tri-membered ring, formed by substituting an ethynyl group CCH for the hydrogen atom in the astronomical molecule c-C_3H, has been detected in the laboratory by supersonic-jet FTMW spectroscopy [203]. Contrary to theoretical prediction, c-C_5H is apparently asymmetric, plausibly the result of pseudo–Jahn–Teller deformation.

The phenyl radical, C_6H_5, derived from benzene by the removal of one hydrogen atom, was detected by either supersonic-jet FTMW and mm wave spectroscopies, combined with pulsed DC- or low-pressure glow-discharge, respectively [204]. Phenyl is a prime candidate for astronomical detection, because it is the prototypical aromatic hydrocarbon radical and a possible progenitor of other aromatic species.

Cyano derivatives: The supersonic-jet FTMW spectrum of the cyanopropynyl radical, H_2CCCCN, produced by a DC discharge on a mixture of CH_3CCCN (1%) in argon, exhibits a $^2\Sigma$ electronic ground state [205]. The spectrum was analyzed with an effective linear Hamiltonian for a $^2\Sigma$ state, and the spectroscopic constants of H_2CCCCN, including the fine and hyperfine coupling constants, were determined to very high accuracy.

As seen above, odd-numbered cyanopolyynes, $HC_{2n+1}N$, are well-studied closed-shell molecules. Much less is known, however, on the even-numbered chains, $HC_{2n}N$, having a $^3\Sigma$ electronic ground state. The supersonic-jet FTMW spectra have recently been reported for two relatively long chains of this series, HC_6N [206] and HC_8N [207]. The radicals were prepared by applying a DC discharge in the throat of the supersonic nozzle to a gas mixture of cyanoacetylene, $HCCCN$ (0.5%), in neon. For HC_8N, eight spectroscopic constants, including six fine and hyperfine coupling constants, were determined to high accuracy, despite a complex spectrum of many closely spaced lines.

Another family of carbon chain radicals terminated with a nitrile group are those with a sum formula C_nN. The supersonic-jet FTMW spectra were studied for two members of this series, C_4N and C_6N [208]. Both of them are linear chains with $^2\Pi$ electronic ground states, and have resolvable hyperfine structure and Λ-type doubling in their lowest rotational levels. Several transitions were measured in the lowest-energy fine-structure component ($^2\Pi_{1/2}$), and precise sets of rotational, centrifugal distortion, spin–rotation, and hyperfine coupling constants were determined.

Silicon-containing chains: The linear silicon–carbon chains SiC_n have a $X\,^3\Sigma^-$ ground state for $n =$ odd. Rotational and centrifugal distortion constants and the spin–spin and spin–rotation coupling constants for these triplet states have been determined for SiC_3, SiC_5, and SiC_7 [146]. A steep increase in the magnitude of the spin–spin constant with chain length was observed in the rotational spectra of the three silicon carbides.

Oxygen- or sulfur-containing chains: The supersonic-jet FTMW spectra of the linear carbon-chain radicals HC_nO ($n = 5$–7) [209] and HC_nS ($n = 5$–8) [148] have been reported recently. All have linear heavy-atom backbones and $^2\Pi$ electronic ground states. For both HC_nO and HC_nS families of radicals, the ground states alternate with odd and even numbers of carbon atoms: those with $n =$ odd are $^2\Pi_{1/2}$, while those with $n =$ even are $^2\Pi_{3/2}$. The rotational, the centrifugal distortion, the Λ-type doubling, and the magnetic hyperfine constants were determined to high precision. For the production of these radicals a DC discharge was applied to gas mixtures of diacetylene, HCCCH, and CO (or CS_2), heavily diluted in neon.

The linear carbon-chain radicals C_6S, C_7S, C_8S, and C_9S, all investigated by supersonic-jet FTMW spectroscopy [147], are found to have a triplet $^3\Sigma$ electronic ground states for an even number of carbon atoms. Hund's case *a* is appropriate, because the spin–spin constant λ is positive and much larger than both B and the spin–rotation constant γ.

Two hetero atoms carbon chains: Seven carbon chain radicals NC_nS, where $n = 1$–7 have been detected by supersonic-jet FTMW spectroscopy [210]. Although NCCS is found to have a bent structure and an asymmetric top spectrum, the five longer chains have linear heavy-atom backbones, and like the isovalent HC_nS chains, the electronic ground states alternate with even and odd numbers of carbon atoms: NC_4S and NC_6S have $^2\Pi_{1/2}$ ground states, while NC_3S, NC_5S, and NC_7S have $^2\Pi_{3/2}$ ground states. In addition, the lowest-rotational transitions of the NCS radical have been detected in both fine-structure levels, allowing a precise determination of the ground state hyperfine coupling constants. The cm wave spectra of NC_3S and the four longer chains have been fully characterized and spectroscopic constants, including those that describe the Λ-type doubling and nuclear quadrupole hyperfine structure from the nitrogen nucleus, have been determined to high precision.

Halogen-substituted carbon chains: Pure rotational transitions of one halogen-substituted carbon-chain radical, CCCCCl, have been observed by FTMW spectroscopy [211]. The radical was produced in a supersonic-jet expansion by a pulsed discharge in a mixture of acetylene (0.2%) and carbon tetrachloride (0.2%) diluted in neon. Transitions with a $^2\Sigma$ type spectral pattern were observed in the regions from 11.6 GHz for $N = 4$–3 to 23.3 GHz for $N = 8$–7. The nuclear quadrupole hyperfine structures due to the Cl nucleus were clearly resolved, and the rotational, spin–rotation, and hyperfine coupling constants were determined.

3.1.2.5. Oxygen chains

Endo and collaborators have obtained interesting results on radicals containing two and three oxygen atom chains. Pure rotational spectra of the ClOO radical for the ^{35}Cl and ^{37}Cl isotopomers have been observed using supersonic-jet FTMW spectroscopy and supersonic-jet FTMW–mm wave double-resonance spectroscopy [212]. The rotational, centrifugal, spin–rotation coupling, and hyperfine coupling constants were determined, indicating that the electronic ground state is $^2A''$. The radical is bent, with an r_0 Cl—O—O angle of 116.4°. A similar investigation has been performed for the ^{79}Br and ^{81}Br isotopologues of the homologue BrOO radical with supersonic-jet FTMW spectroscopy [213]. ClOO and BrOO radicals were produced prior to the supersonic-jet expansion by discharging a gas mixture containing chlorine (or bromine) and oxygen diluted in argon. All transitions show spin doublets and hyperfine splittings due to the nuclear spin of the chlorine (or bromine) atom. The Cl—O and Br—O bonds are anomalously long, 207.5 and 240 pm, respectively, and correspond to a very weak X—O (X = halogen) bond, with a dissociation energy estimated to be ~10 kJ/mol in the case of BrOO.

With the same technique, Endo and collaborators could detect the rotational spectra of a radical, HOOO, with a three-oxygen-atom chain [214]. The radical has a *trans*-planar molecular structure, in contrast to the *cis*-planar structure predicted by most ab initio calculations. The bond linking the HO and O_2 moieties is fairly long (168.8 pm) and comparable with the F—O bond in the isoelectronic FOO radical. In contrast to catenated carbon, our knowledge of molecules containing extended oxygen chains is quite limited; even simple species such as H_2O_3 and H_2O_4 have not yet been well characterized. The HO_3 radical, which contains a chain consisting of three O atoms, can be regarded as adduct of OH and O_2, both being important open-shell species in the atmosphere. The existence of HO_3 has been considered for the reaction systems $H + O_3 \rightarrow OH + O_2$ and $O + HO_2 \rightarrow OH + O_2$, where the HO—$O_2$ adduct is proposed as a reaction intermediate [215]. Furthermore, if the reaction $OH + O_2 \rightarrow HO_3$ is exothermic and has a very small barrier, then a certain amount of HO_3 would exist in Earth's atmosphere, and the latter reaction would play a crucial role in reactions involving the OH radical [216,217].

3.1.2.6. Other radicals

The rotational spectrum of CHF_2 in the $^2A'$ ground electronic state was reported by Saito and collaborators [218]. The radical was obtained from the reaction of CH_2F_2 with the products of a MW discharge on CF_4.

Another important contribution describes the generation and FTMW investigation of transient species by laser photolysis (using a pulsed excimer laser), i.e., the SO radical in the vibrational ground state and the first two vibrationally excited states [219]. SO in its $X^3\Sigma$ electronic ground states has been produced by 193 nm photodissociation of SO_2 in a pulsed jet in Ar, according to the photochemical reaction: $SO + h\nu(193\,\text{nm}) \rightarrow SO(X^3\Sigma) + O(^3P)$.

Finally, we mention here the observation of three successive R-branch transitions of the H_2O—HO radical complex by FTMW spectroscopy [220]. The OH radical was generated by a DC discharge of H_2O diluted in Ar or Ne, and the radical was subsequently cooled to form the hydrated complex in a supersonic expansion. The spectrum exhibits extremely large spin doubling between the $J = N + 1/2$ and $N - 1/2$ components, suggesting a deviation from the case (b)-limit representation and approach of the case (a)-limit. Magnetic hyperfine structure of the rotational transitions is characteristic of a radical with two nonzero nuclear spins with $I_1 = 1/2$ and $I_2 = 1$, causing two different splittings. Pauli's exclusion principle restricts the complex structure to be of twofold symmetry around the a principal axis and the vibronic wavefunction to be anti-symmetric to the C_2 rotation, compatible with an effective planar C_{2v} structure.

3.1.3. Ions

Historically, the observation of rotational spectra of molecular ions has been a boon and bane of laboratory MW spectroscopy, now nearly 30 years ago. Indeed, the $J = 1 \leftarrow 0$ transition of HCO^+ near 89.2 GHz was observed in the interstellar medium [220] 5 years before the frequency was measured in the laboratory [221]. The detection of charged species is limited by their typically low achievable concentration. Ions were observed mainly with two techniques: mm/sub-mm wave absorption and supersonic-jet FTMW spectroscopy. The former technique was used since the early spectroscopic works, combined with glow discharge [222], and it has been improved by introducing longitudinal magnetic fields in the discharge cell [223] as well as velocity modulation techniques [224]. Supersonic-jet FTMW spectroscopy has only recently been devoted to the study of molecular ions, by Endo [225] and Thaddeus [226].

Molecular ions play important roles in physical chemistry and physics, but probably the main impetus for their studies comes from their importance in the chemical reactions assumed to take place in many astronomical sources.

3.1.3.1. Positive ions

Among the few new molecular ions investigated by mm/sub-mm wave absorption spectroscopy during the last 15 years, $C_2H_3^+$ was observed in a negative glow discharge, extended by an axial magnetic field, in a mixture of C_2H_2, H_2, and Ar cooled at liquid N_2 temperature [227]. No tunneling splitting was observed, and the rotational lines follow a conventional S-reduced Watson Hamiltonian. Line frequencies of astrophysical interest were measured or predicted with accuracy sufficient for radioastronomical searches.

The mm/sub-mm wave spectra of the protonated hydrogen cyanide ion, $HCNH^+$, and its isotopic species, $HCND^+$ and $DCND^+$, were measured in the laboratory [228], after its first observation in the interstellar space [229]. The ions were generated in a cell with a magnetically confined glow-discharge of HCN and/or DCN. Rotational and centrifugal distortion constants were precisely determined. The study of the deuterated species aims at the spectroscopic determination of the relative concentration of H and D in astronomical sources.

A strong impulse to the study of positive molecular ions has been given very recently by Ziurys and coworkers, using their velocity modulation mm/sub-mm wave absorption spectrometer [224]. The rotational spectra of four (radical) molecular ions were reported.

For SH^+, the $N = 1 \leftarrow 0$ transition of its $^3\Sigma^-$ electronic ground state was observed [230]. The molecular ion was created in an alternating current (AC) discharge of H_2S or CH_3SH in argon. Four hyperfine components of the transition were measured: three near 526 GHz and one near 346 GHz, arising from the $J = 2 \leftarrow 1$ and $J = 1 \leftarrow 0$ spin components. The study resulted in very accurate rest frequencies for SH^+, which is predicted to be an abundant species in photon-dominated regions.

The pure rotational spectra of TiF^+ [231] and $TiCl^+$ [232] radical ions have been measured with the same technique. Both ions have a $^3\Phi_r$ electronic ground state. The ions were created in an AC discharge of $TiCl_4$ and argon; for the generation of TiF^+, F_2 in helium was added. The determined hyperfine parameters are consistent with a $\delta^1\pi^1$ electron configuration with the electrons primarily located at the titanium nucleus. The nuclear spin–orbit constant a indicates that the unpaired electrons are closer to the fluorine nucleus in TiF^+ than in TiF, consistent with a decrease in bond length expected for the ion: the shorter bond distance is rationalized from the increased charge on the titanium nucleus as a result of a $Ti^{2+}F^-$ configuration. A similar decrease in bond length was found for $TiCl^+$ with respect to TiCl.

The mm/sub-mm wave spectrum of the molecular ion $FeCO^+$ in its $X^4\Sigma^-$ electronic ground state has been recorded using velocity modulation spectroscopy [233]. The molecular ion was created in an AC discharge on $Fe(CO)_5$ in argon. Employing a Hund's case (b) Hamiltonian, rotational, spin–rotation, and spin–spin constants were determined. Because of the presence of higher order spin–orbit interactions, probably caused in part by a nearby $^4\Pi$ excited state, numerous centrifugal distortion terms were needed for the spectral analysis. The value of γ_s, the third-order spin–rotation constant, was found remarkably large at -72.4 MHz. Rest frequencies for $FeCO^+$ are now available for interstellar and circumstellar searches. This species might exist in CO-abundant molecular clouds in the presence of gas-phase iron ions Fe^+. Molecular ions such as $FeCO^+$ could then be the hidden carriers of metallic elements in such clouds.

Supersonic-jet FTMW spectra of the molecular ions $HOCO^+$ and $HOCS^+$, and the ion complexes, D_3^+-Ar and sym-D_2H^+-Ar, were observed employing a pulsed DC discharge nozzle [227]. The $HOCO^+$ and $HOCS^+$ ions were produced by a DC discharge of a sample of H_2 ($\sim 50\%$) and Ar ($\sim 50\%$), premixed with CO_2 or OCS (0.1–0.2%), respectively. For the D_3^+—Ar complex, D_2 (2–10%) mixed with argon was used. Ion-formation efficiency for $HOCS^+$ compared with the parent molecule under these conditions was estimated to be 10^{-4}. Tunneling splitting in the lowest rotational transition of D_3^+–Ar was not resolved within the experimental linewidth of ~100 kHz.

The rotational spectra of the $HCCCNH^+$, $NCCNH^+$, and CH_3CNH^+ ions have been observed by supersonic-jet FTMW spectroscopy [228]. The rotational and centrifugal distortion constants were determined for all three ions and the nitrogen quadrupole hyperfine coupling constants for $HCCCNH^+$ and $NCCNH^+$. From the respective Doppler shifts, it is found that the velocities of the ions are 3% larger than those of the unprotonated parent molecules, and the linewidths were increased by ~50%. The concentration of the ions near the nozzle is approximately $10^{11}\,cm^{-3}$, sufficiently high for detection in the visible and the IR regions using present laser techniques. The ions were produced in the throat of a supersonic discharge nozzle by DC discharge of $HCCCN$, $NCCN$, or CH_3CN diluted in H_2.

3.1.3.2. Negative ions

Pure rotational spectra of anions are of rising interest because some of them have recently been identified in astronomical sources. Only for a few species, the spectra are known yet; this is expected to change rapidly however.

Bogey and collaborators measured a sub-mm wave transition of SH^-, created in the positive column of an electric discharge set-up employing a mixture of H_2S and argon [234]. The study of the Doppler shift – caused by the motion of charged species in the electric field of the discharge – allowed to distinguish between positively and negatively charged ions and neutrals.

Thaddeus and collaborators reported the rotational spectra of several negative closed-shell linear carbon chain ions, ranging from CCH^- to C_8H^-. The simplest of them is the acetylide anion, CCH^- [235], which was produced in DC discharge applied to a mixture of acetylene, $HCCH$ (85%), and argon (15%) flowing through a free-space mm wave spectrometer. The rotational spectra of the butadiyne anion C_4H^-, the hexatriene anion C_6H^-, and the octatetrayne anion C_8H^- have been detected in the laboratory [236,237], by using either mm wave absorption or supersonic-jet FTMW spectroscopy. Precise spectroscopic constants were obtained for all anions, and accurate transition rest frequencies can now be calculated, making the highly stable and fairly polar C_nH^- anions likely candidates for radioastronomical detection, given that their neutral analogues being among the most abundant molecules for a wide variety of astrophysical sources.

3.2. Transition metal complexes

The supersonic-jet FTMW spectra of a considerable number of transition metal complexes have been investigated by Kukolich and collaborators. Many transition metal complexes function as catalysts in reactions important in industry and biology. The information obtained from the rotational spectra is used to determine the structures of these complexes, often important in understanding the reactions. The electronic structure and electronic charge distribution can be related to the reactivity and relative stabilities of various complexes.

Complexes with 15 different transition metals were investigated, with the atomic number ranging from 22 (titanium) to 76 (osmium, mass of the most abundant isotope: 191.96 u). Most of these metals have a multiplicity of isotopes and nuclear spins $I > 1/2$, thereby significantly complicating the spectra. At the unrivaled precision typical for FTMW spectroscopy, several structural, conformational, and electronic properties of these complexes were determined for the first time.

Despite their complexity, several of these species were investigated. The first to be studied were symmetric tops, $Co(CO)_3NO$ [238] and $MnCp(CO)_3$, where Cp denotes cyclopentadienyl [239]. Apart the rotational and centrifugal distortion constants, for the first time the quadrupole coupling constants of transition metal nuclei, $\chi_{cc}(^{59}Co) = 35.14(30)$ and $\chi_{aa}(^{55}Mn) = 68.00(4)$ MHz, respectively, were obtained. Several other symmetric tops have been investigated: (i) cyclobutadiene iron tricarbonyl, where the vibrationally averaged structure of the cyclobutadiene ring is square and perpendicular to the a-molecular axis [240]. (ii) Manganese pentacarbonylhydride, an oblate top with C_{4v} symmetry, where $\chi_{cc}(^{55}Mn) = 45.24(4)$ MHz and $r_0(Mn—H) = 164(4)$ pm [241]. (iii) (Benzene) chromium tricarbonyl, also a symmetric top [242], but the MW rotational spectrum of ([1,2-D_2]benzene)chromium tricarbonyl contains transitions due to two different structural isomers of this complex [243]. These results are rationalized by a reduction of the benzene symmetry to C_{3v}, due to interactions with the $Cr(CO)_3$ moiety. One structural isomer (isomer E) occurs when the deuterium atoms are at the ends of a "long" C—C bond; the other (isomer S) occurs when the deuterium atoms are at the ends of a "short" C—C bond. The data indicate a difference of 1.6 pm in adjacent benzene C—C bond lengths in this complex. (iv) Cyclopentadienyl vanadium tetracarbonyl exhibits a MW spectrum that was interpreted in terms of a symmetric top, but one-half of the measured transition frequencies did not fit the pattern expected for a symmetric top [244] – probably originating from low-frequency CO bending modes that result in deviations from the C_{4v} symmetry of the $V(CO)_4$ group. (v) Cobalt tetracarbonyl hydride allowed for the determination of a full molecular structure; the quadrupole coupling constant was found to be $\chi_{cc}(^{59}Co) = 116.62(3)$ MHz [245]. (vi) Methyl rhenium trioxide is probably the first molecule containing a rhenium atom for which the rotational spectrum has been measured. The experimental quadrupole coupling constants for two isotopologues are $\chi_{aa}(^{187}Re) = 716.55(2)$ and $\chi_{aa}(^{185}Re) = 757.19(3)$ MHz, respectively [246]. (vii) The half-sandwich C_{5v} structures of cyclopentadienyl thallium and cyclopentadienyl indium also were proven by the supersonic-jet FTMW spectra of several isotopologues. The quadrupole coupling constants for the indium compounds are $\chi_{aa}(^{113}In) = -118.40(7)$ and $\chi_{aa}(^{113}In) = -119.98(3)$ MHz [247]. (viii) MW measurements have been performed for two isotopologues (^{187}Re and ^{185}Re) of cyclopentadienyl rhenium tricarbonyl [248]. The quadrupole coupling constants $\chi_{aa}(^{187}Re) = 614.464(12)$ and $\chi_{aa}(^{185}Re) = 649.273(14)$ MHz are smaller than those of methyl rhenium trioxide [246], suggesting a smaller electron density at the Re nucleus. (ix) The first measurement of a MW spectrum of niobium compounds was achieved through the study of a prolate top, the organometallic complex cyclopentadienyl niobium tetracarbonyl [249]. The quadrupole coupling constant of $^{93}Nb(I = 9/2)$, 100% natural abundance, was determined as

$\chi_{aa}(^{93}Nb) = -1.8(6)$ MHz. (x) The study of $MnRe(CO)_{10}$ represents the first rotational spectrum assigned for a dinuclear complex [250]. Interestingly, the rotational constants $B = 200.36871(18)$ and $200.5561(10)$ MHz for the ^{187}Re and ^{185}Re isotopologues, respectively, are smaller than the corresponding quadrupole coupling constants $\chi_{aa}(^{187}Re) = 370.4$ (4) and $\chi_{aa}(^{185}Re) = 390.9(6)$ MHz, respectively. $\chi_{aa}(^{55}Mn)$ has been determined to be $-16.52(5)$ MHz. The gas-phase Mn—Re bond distance is approximately 299 pm. The C_{4v} symmetry of the complex, in accounting for the zero nuclear spin of both ^{12}C and ^{16}O, allows only $K = 4n$ transitions to be observed. (xi) In the cyclopentadienyl cycloheptatrienyl titanium complex, $C_5H_5TiC_7H_7$, the titanium atom is located between two different symmetric rings, with the cycloheptadienyl ring being somewhat closer. The calculated Ti—C bond lengths are indeed shorter for the C_7H_7 ligand ($r(Ti—C) = 221$ pm) than for the C_5H_5 ligand ($r(Ti—C) = 234$ pm). $\chi_{aa}(^{47}Ti)$ was found to be 8.193(40) MHz [251].

Many of the asymmetric top metal complexes measured to date involve iron as central atom. The investigations of asymmetrically substituted ferrocenes, such as chloroferrocene [252], bromoferrocene [253], methylferrocene [254], 1,1′-dimethylferrocene [255], ethynylferrocene [256], and ferrocene-carboxaldehyde [257], were devoted to the determination of the structure. It was generally observed that the substituted cyclopentadienyl group is distorted. For 1,1′-dimethylferrocene, the spectra of two different eclipsed conformers, syn–periplanar and syn–clinal (see Figure 10) were measured. The investigation of iron tricarbonyl complexes with the unsaturated hydrocarbons butadiene [258], 1,3-cyclohexadiene [259], and cyclooctatetraene [260] are of considerable interest, because the structures of olefin–metal complexes are related to the heterogeneous catalysis of technically important alkene reactions on metal surfaces. Butadiene adopts a distorted cis-conformation in the complex with iron tricarbonyl, while the $trans$-configuration of the isolated molecule is ∼8 kJ/mol more stable. In the corresponding complexes, 1,3-cyclohexadiene and cyclooctatetraene are linked to

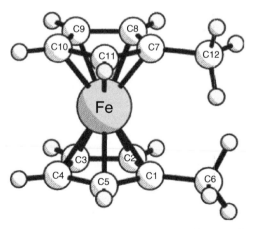

Figure 10 Structure of the eclipsed, synperiplanar (E0) conformer of 1,1′-dimethylferrocene, indicating the numbering scheme for the carbon atoms. (Reprinted with permission from Ref. [255]. Copyright 2004, American Chemical Society.)

iron tricarbonyl through an interaction that involves two alternate $C{=}C$ double bonds with well-defined relative orientations of the rings with respect to the carbonyl ligands.

The spectra of two metal hydrides complexes, $C_5H_5Mo(CO)_3H$ and $C_5H_5W(CO)_3H$ [261,262], and three metal dihydrides, $H_2Os(CO)_4$ [263], $H_2Rh(CO)_4$ [264], and $(C_5H_5)_2WH_2$ [265], are also known. The structural results give $r(H{-}H)$ distances between the two hydrogens attached to the metal atoms of 240, 236, and 213 pm, respectively, showing that these complexes are classical "dihydrides" rather than "dihydrogen" complexes.

3.3. Biomolecules

A small number of different building-block molecules, which are the same in all living organisms, combine to an immensely large number of different proteins and nucleic acids. They are sometimes called primordial biomolecules and can be grouped in (i) 20 proteogenic amino acids; (ii) $\sim$150 amino acids that occur in living cells, but not in proteins; (iii) three pyrimidinic and two purinic bases; (iv) several sugars; (v) some individual molecules that are a sugar alcohol (glycerol), a nitrogenous alcohol (coline), and a fatty acid (palmitic acid). A rapidly increasing number of these species and other biomolecules, such as drugs and neurotransmitters, are investigated by MW spectroscopy.

MW studies on biomolecules began in 1978, when the mm wave spectrum of the simplest amino acid, glycine (NH_2CH_2COOH), was first recorded independently by Brown et al. [266] as well as by Suenram and Lovas [267]. The main difficulties in obtaining the rotational spectra of biomolecules arise from their low volatility and their partial decomposition upon heating. Despite these problems, the rotational spectra of several biomolecules have been observed while heating the samples. Brown et al. combined a mm wave absorption spectrometer with a supersonic expansion [268] using it systematically for the investigation of several biomolecules, including amino acids, biological bases, and neurotransmitters. Biomolecular studies performed by Suenram and Lovas employing supersonic-jet FTMW spectroscopy were mainly directed on amino acids and sugars, both being of astrochemical interest. Among them, the first investigation of amino acids by employing laser ablation, rather than thermal heating, has been reported [269]. Some biomolecules, vaporized by heating, have been studied by the mm wave free-jet technique by Caminati and coworkers [270]. During the last few years, Alonso and coworkers systematically applied a combination of laser ablation and supersonic-jet FTMW techniques to the investigation of amino acids and biological basis [271]. In this way, important chemical information on the biomolecules, such as conformational and tautomeric preferences, H—bonds, and nonbonding interactions have been obtained. Here, we will review the most important results obtained on biomolecules using rotational spectroscopy.

3.3.1. Amino acids

Life is based on 21 coded amino acids. Probably their success as building blocks of proteins relies – apart chirality – on their high torsional flexibility that results in a large number of low-energy conformers. Investigation of the conformational behavior of amino acids and the forces that model their shape-free from solvent interactions – are accomplished in isolation; rotational spectroscopy utilizing the collision-free environment of a supersonic-jet expansion is the ad hoc technique for this purpose. Today, conformational preferences, structures, and other properties of several amino acids have been precisely determined by supersonic-jet FTMW spectroscopy. The studies include some nonnatural amino acids, which are nevertheless considered in bioengineering.

3.3.1.1. Glycine

Ab initio calculations indicate that the three more stable conformers of glycine are stabilized by NH_2—$O_{C=O}$, or OH—N, or NH_2—O_{OH} intramolecular H-bonds, They are labeled as I, II/III, and IV/V in Figure 11, where III and V are slightly distorted forms of II and IV, respectively. The relative energies are given as $E_{IV/V} > E_{II/III} > E_I$. However, due to a favorable dipole moment [272], the rotational spectrum of species II exhibits the strongest rotational spectrum and was observed first [256,257]. However, the rotational spectrum of species I was reported only shortly later [272], and it was shown that conformer I is more stable than II, by 6 kJ/mol. In 1995 Lovas et al. [269] carried out new laboratory studies of conformers I and II in the centimeter wave range, combining FTMW spectroscopy with laser ablation, to

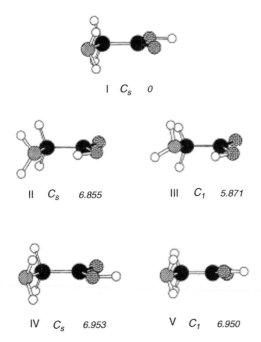

I C_s 0

II C_s 6.855   III C_1 5.871

IV C_s 6.953   V C_1 6.950

Figure 11   The five structures that embody the three conformers of glycine which are predicted from ab initio calculations to be of lowest energy. The Cs structures labeled II and IV are actually energy saddle-points between the two enantiometric forms of the adjacent C1 structures III and V, repectively. The numerical quantities shown are the relative ab initio MP2/6–31G(d,p) energies (kJ/mol). (Reprinted with permission from Ref. [274].)

investigate the ^{14}N nuclear quadrupole hyperfine structure and provide a more accurate determination of the electric dipole moment components for both conformers. In subsequent investigations of several isotopologues [273,274], it was shown that both observed conformers assume C_s symmetry. Though its calculated energy is very similar to that of conformer II, the conformer III was not observed. Godfrey et al. [275] suggest that in a jet-expansion experiment, glycine III will relax to glycine I. This is supported by a calculated low potential energy barrier to interconformational transformation. However, Ilyushin et al. investigated the mm wave spectrum of glycine [276] – for astronomical purposes – and, even in a stationary gas cell, glycine III was not detected.

3.3.1.2.   α- and β-Alanine

The two simplest amino acids following glycine, i.e., α- and β-alanine, have been investigated first with mm wave free-jet spectroscopy [277,278], and, more recently, by supersonic-jet FTMW spectroscopy combined with laser ablation [279,280]. Both investigations on α-alanine individuated the rotational spectra of the two main conformers, very similar to the two conformers of glycine. The conformer with a NH_2—$O_{C=O}$ H-bond is the global minimum, while the one with an internal OH—N H-bridge follows in the stability ordering. However, the FTMW investigation concerned many isotopologues and supplied more information, such as the ^{14}N quadrupole coupling constants and a complete structure of the two conformers.

β-Alanine is a natural amino acid not found in proteins. It is more flexible than α-alanine and their MW investigations supplied interesting chemical data. Both free-jet mm wave and FTMW studies revealed two conformers stabilized by a NH—$O_{C=O}$ H-bond. In addition, the FTMW investigation identified a conformer with an internal OH_2—N link and a fourth form displaying an $n-\pi^*$ interaction between the nonbonding electron pair of the nitrogen and the π^* orbital at the carboxylic carbonyl group. This kind of hyperconjugative delocalization is some times called Bürgi–Dunitz interaction [281], and its energy is equivalent to that of the H-bond mentioned above. The conformational assignment is reliably based on the ^{14}N hyperfine structures that, for the same rotational transitions, are completely different for the various conformers.

3.3.1.3.   Other proteogenic α- amino (or imino) acids

While a considerable spectroscopic attention has been devoted to glycine and alanine, the remaining amino acids remained practically ignored by MW spectroscopy for a number of years. The investigations were revived by Alonso and coworkers in employing the laser ablation supersonic-jet FTMW technique, which revealed itself as very powerful for the detection of a large number of these biological key species. In a short time, the rotational spectra of proline [282] (two conformers), valine [283] (two conformers), isoleucine [284] (two conformers), and leucine [285] (two conformers) were reported. The basic structures found for these amino acids correspond to the conformations I, II, and III of glycine (see Figure 11), but the complexity of the lateral chains increases the number of possible conformers.

3.3.1.4. Other natural nonproteogenic amino acids and nonnatural amino acids

Several nonproteogenic amino acids are important precursors or intermediates in metabolism, while also nonnatural amino acids are of considerable and increasing interest. They can be incorporated as new building blocks into the genetic codes of both prokaryotic and eukaryotic organisms to facilitate studies of protein structure and function [286]. Both kinds of noncoded amino acids have molecular flexibilities similar to those of coded amino acids. Alonso and coworkers characterized a number of them with laser ablation supersonic-jet FTMW spectroscopy. All studied species are α-amino acids: 4-hydroxyproline [287] (four conformers), N,N-dimethylglycine [288] (two conformers), α-aminobutyric acid [289] (three conformers), phenylglycine [290] (two conformers), and sarcosine [291] (two conformers).

3.3.2. Pyrimidine and purine nucleic acid basis

The three pyrimidine bases were first investigated by Brown and collaborators at the end of the 1980s [292–294] by free jet mm wave absorption spectroscopy. Only the rotational spectra of the diketo tautomers were observed for uracil [292] and thymine [293], revealing this tautomer to be considerably more stable than the monoketo forms. These results were also confirmed recently. Vice versa, cytosine indeed is a multiform molecular system and three species, keto-amino, keto-imino, and hydroxy-amino were detected [294]. The two former species are almost isoenergetic, and by 6 kJ/mol more stable than the latter form. Very recently, two reports on the FTMW spectrum of uracil (vaporized by laser ablation) with precisely determined ^{14}N quadrupole coupling constants have been published. Brünken et al. were mainly interested in improved spectroscopic data for a radioastronomical search [295], while Vaquero et al. [296] supplied information on the structure through the studies of several isotopologues. In both cases, only the diketo form was observed.

A laser ablation supersonic-jet FTMW study was also done for thymine [297], leading to the determination of the structure, ^{14}N nuclear quadrupole coupling constants, and V_3 barrier to internal rotation of the methyl group. The rotational spectrum of adenine (6-amino-purine) has been measured in a seeded supersonic beam by continuous wave MW spectroscopy, assigning the transitions to the N(9)H tautomer [298].

3.3.3. Sugars and glycerol

Very little has been reported on rotational studies of sugars. Practically only the MW spectra of glycolaldehyde was available [299–301] until a few years ago, when also the rotational spectra of some "C3 sugars," glyceraldehyde, 1,3-dihydroxy-2-propanone, and 2-hydroxy-2-propen-1-al, were reported [302].

Glycerol, one of the "primordial building blocks," has been first investigated by mm wave free-jet spectroscopy and two rotamers were assigned [303]. However, the three conceivable values of the OC—CO and HO—CC dihedral angles can generate many more stable conformers; a very recent FTMW investigation lead indeed to the observation of some more conformers [304]. One of these species undergoes a tunneling between two equivalent configurations, which involves a near 180° concerted internal rotation of the three OH groups.

3.3.4. Neurotransmitters and other biomolecules

Most of the MW investigations for this class of biomolecules have been performed by Brown and coworkers. They reported the rotational spectra of two neurotransmitters, 2-phenylethylamine [305] and amphetamine [306]. Five conformers of 2-phenylethylamine and nine conformers of amphetamine were predicted to be energetic minima by ab initio calculations. Two conformers of each molecule have been observed and characterized by mm wave free-jet spectroscopy.

Brown and collaborators reported also the rotational spectrum of a vitamin, nicotinamide [307]. They identified two conformers, E and Z (according to the *entgegen* or *zusammen* position of the amide oxygen with respect to the pyridine nitrogen). Both are nonplanar, and two states with lines of the same intensity, related to a tunneling motion, were observed for the Z-conformer.

Four different species (due to tautomeric and conformational equilibria) of histamine [308,309] were observed, showing an apparent discrepancy between experimental and calculated relative stabilities of the various forms, which was interpreted in terms of conformational relaxation of some species upon supersonic expansion [310].

The spectrum of glycine methyl ester was observed by conventional MW spectroscopy; it displays a bifurcated NH_2—$O_{C=O}$ H-bond [311]. The spectrum of purine, for the N(9)H tautomer, has been detected by mm wave free-jet spectroscopy [312].

3.4. Nonpolar Molecules

Rotational transitions of a rigid rotor only occur if the species possesses a permanent dipole moment. However, with real molecules there are three effects that allow the observation of the rotational spectra of nonpolar molecules. Watson pointed out [313] that centrifugal distortion of rotating molecules of certain symmetries can produce a nonzero transition moment. This mechanism is effective also in the ground vibrational state. Vibrational effects also can generate small dipole moments in nonpolar molecules. For example, the asymmetric substitution of hydrogen with deuterium atoms in benzene can generate a small dipole moment due to the different zero-point location of the heavier vibrating deuterium. Mizushima and Venkateswarlu [314] and Mills et al. [315] have shown that highly symmetric molecules in some degenerate vibrationally excited states gain a small permanent electric dipole moment induced by vibration. Investigations on such molecular systems have mainly been performed by Bauder and coworkers. Because of the very small induced dipoles, the sample was polarized with MW pulses of up to 40 W peak power. Typically, millions or several millions of pulsed MW excitation experiments were accumulated in a 512-channel analyzer.

3.4.1. Rotational spectra induced by centrifugal distortion

An interesting application of centrifugal distortion is the study of the ground-state rotational spectra of some spherical tops, CH_4 [315], SiH_4 [316], and GeH_4

[317], which have been measured by stationary gas FTMW spectroscopy. For tetrahedral XY_4 molecules, the magnitude of the distortion dipole moment θ_z^{xy} is only of the order of 10^{-5} D [318]. However, several tens of transitions have been measured for CH_4, SiH_4, and GeH_4, respectively. It has been possible to observe all three types of perturbation allowed Q-branch transitions, A_1–A_2, E–E, and F_1–F_2. The spectra were used for a new determination of the tensor centrifugal distortion constants D_t, H_{4t}, H_{6t}, L_{4t}, L_{6t}, and L_8 because it was expected that the larger and better balanced data set of all symmetry species would yield refined constants.

Other interesting examples are the rotational spectra obtained for D_{3h} symmetry nonpolar molecules BF_3 [319] and cyclopropane [320]. In this case the rotational spectrum in the vibrational ground state is characterized by $\Delta J = 0, \pm 1, \Delta k = \pm 3$ selection rules for the overall rotational angular momentum and for its projection along the symmetry axis of the molecule. The distortion dipole moments μ_D are expected to be only of the order of μD, i.e., only very weak spectra are expected. However, by accumulation of millions of measurements, quite good spectra were obtained, as shown in Figure 12 for a Q-branch of $^{11}BF_3$. Unlike conventional symmetric top spectra, these spectra allow for the determination of all sets of rotational constants. The symmetric top PF_5 also belongs to the D_{3h} symmetry group but, in contrast to BF_3, possesses two inequivalent sets of fluorine atoms. Berry postulated an intramolecular rearrangement process, called Berry pseudorotation [321], exchanging the inequivalent F atoms. Styger and Bauder reported the pure rotational spectrum of PF_5 [322], but they did not observe any tunneling related to this pseudorotation process.

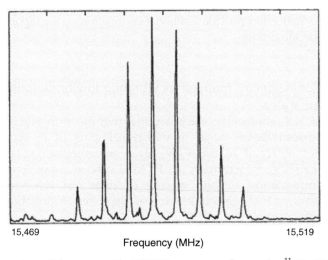

15,469   15,519
Frequency (MHz)

Figure 12 A section of the waveguide FTMW spectrum of nonpolar $^{11}BF_3$. The spectrum is due to a distortion induced dipole moment. The intensity distribution in the band is mainly due to the finite excitation bandwidth. (Reused with permission from Ref. [319]. Copyright 1987, American Institute of Physics.)

3.4.2. Rotational spectra induced by asymmetric $H \rightarrow D$ isotopic substitution

The dipole moment induced by asymmetric isotopic substitution was observed for the $H \rightarrow D$ case. Bauder reported the first of such a spectrum in 1983 for monodeuterated benzene [323] and investigated some more isotopic species later on [324]. This was a considerable step forward for the structure determination of nonpolar molecules. It is interesting to note that the spectrum of C_6H_5-D is a μ_a-type spectrum, which turns to a μ_b-type spectrum when passing to 1,3-dideuterobenzene.

Another interesting nonpolar molecule where the rotational spectrum was studied by $H \rightarrow D$ substitution is cyclobutane. Monodeutero-cyclobutane displays the *axial* and *equatorial* species, with well-defined and different spectra [76], while in the case of 1,1-dideutero-cyclobutane [77] each transition is split into a doublet, due to ring-puckering tunneling. From the corresponding vibrational splitting, combined with the structural relaxations deduced from all isotopic species, it has been possible to reliably size the PES.

Finally, let us consider glyoxal and butadiene, two molecules that are characterized by a conformational equilibrium neatly shifted toward the nonpolar *trans* form. The spectrum of *trans* glyoxal-d_1 has been measured and could be compared with that of the only very little abundant but strongly polar *cis* form [325]. In the case of butadiene, only the spectrum of the *trans* form, through the spectrum of s-*trans*-1,3-butadiene-1,1-d_2 has been observed [326], while that of the *cis* (or *cisoyd*) form is still a puzzle.

The supersonic-jet FTMW rotational spectrum has also been obtained for – nonpolar – cyclohexane, the prototype molecule for chair–boat configurations, relying on the small dipoles of its 1,1-d_2, ^{13}C-1,1-d_2, d_1 (equatorial and axial), and 1,1,2,2,3,3-d_6 isotopologues [86]. The rotational constants and the quartic centrifugal distortion constants have been determined, allowing for a complete substitution structure. Measurements of the Stark effect provide additional information on the dipole moment that is induced by the substitution.

3.4.3. Rotational spectra induced by vibration in vibrationally excited states

For tetrahedral XY_4 molecules, the rotational transitions in vibrational excited states, as mentioned above, occur in vibrational states of F_2 symmetry. Such transitions were observed in the v_2/v_4 dyad of methane [327], methane-d_4 [328], and silane [329], by stationary gas FTMW spectroscopy. From the spectra, very precise information on the spacing of rotational levels of the $v_2 = 1$ (E species) and $v_4 = 1$ (F_2 species) states of these tetrahedral molecules were obtained.

3.5. Polcyclic aromatic hydrocarbons and heterocycles

Polycyclic aromatic hydrocarbons (PAHs) have long been proposed as the source of the unidentified infrared bands between 3 and 16 mm, but so far, not a single

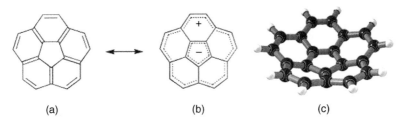

(a)   (b)   (c)

Figure 13 A polycyclic aromatic hydrocarbon (PAH) molecule: corannulene, $C_{20}H_{10}$. (a) Line-bond structural representation. (b) Polar resonance structure. (c) Three-dimensional geometry determined by gas-phase electron diffraction. (Reprinted with permission from Ref. [330]. Copyright 2005, American Chemical Society.)

PAH was identified in space, partly because PAHs generally have weak or non-existent radio spectra.

However, very recently, the rotational spectra of some polar PAHs have been reported. The supersonic-jet FTMW spectrum of the symmetric top nonplanar corannulene ($C_{20}H_{10}$, see Figure 13) has been measured [330] facilitated by its high dipole moment of 2.07 D.

The pure rotational spectrum of phenanthridine ($C_{13}H_9N$), a small polycyclic aromatic nitrogen heterocycle (PANH) with three fused aromatic rings, has been measured with free-jet mm wave absorption spectroscopy [331]. The negative value of its inertial defect $\Delta = -0.469\,u\,\text{Å}^2$ suggests a planar frame with low frequency out of plane vibrational modes. McNaughton et al. measured the free jet mm wave and FTMW spectra of acridine and 1,10-phenanthroline and found behaviors similar to that of phenanthridine [332].

A near equilibrium structure is reported for azulene (C_{2v} symmetry), whose FTMW spectrum has been measured for the parent and all possible singly substituted species [333].

3.6. Molecules of Astrophysical Interests

The pioneering observation with radio telescopes of the hydroxyl radical [334], ammonia [335] and water [336] in the 1960s, opened a new field of astrophysics, which lead to the identification of more than 146 molecules (July 2007) in the interstellar gas and circumstellar shells [337]. They are shown in Figure 14. For many of these molecules, the spectra have also been observed in the laboratory and are already discussed above. The largest among the molecules detected in space is a carbon-chain-based cyanopolyine with 13 atoms and molecular weight of 147.

Glycine itself has been claimed to be observed [338], but a recent analysis, based on new laboratory measurements [276], concludes that a consistent set of key lines necessary for a positive interstellar glycine identification has not yet been found [339].

2	3	4	5	6	7	8	9	10	11	12	13
H_2	H_2O	NH_3	SiH_4	CH_3OH	CH_3COH	CH_3CO_2H	CH_3CH_2OH	CH_3COCH_3			
OH	H_2S	H_2O^+	CH_4	NH_2CHO	CH_3NH_2	HCO_2CH_3	$(CH_3)_2O$	$CH_3(CC)_2CN$			
SO	SO_2	H_2CO	$CHOOH$	CH_3CN	$CH_3(CC)H$	$CH_3(CC)CN$	CH_3CH_2CN	$(CH_2OH)_2$			
SO^+	HN_2^+	H_2CS	$H(CC)CN$	CH_3NC	CH_3CHCN	C_7H	$H(CC)_2CN$	CH_3CH_2CHO			
SiO	HNO	$HNCO$	CH_2NH	CH_2NH	$H(CC)_2CN$	H_2C_6	$H(CC)_2CH_3$		**11**		
SiS	NNO	$HNCS$	NH_2CN	C_5H	C_6H	CH_2OHCHO	C_8H				
FeO	NH_2	CH_3	CH_2CN	$HCCCCN$	$I\text{-}H_2C_2HOH$	$H(CC)_3H$	$CH_2=CHCH_3$		$H(CC)_4CN$		
NO	H_3^+	$CCCN$	H_2CCO	$H(CC)CHO$	$c\text{-}CH_2OCH_2$	$CH_2=CHCHO$	$H(CC)_4^-$		$CH_3(CC)_3H$		
NS	$SiNC$	HCO_2^+	C_4H	$CH_2=CH_2$	$H(CC)_3^-$	$CH_2=CCHCN$					
HCl	HCO	$CCCH$	$c\text{-}C_3H_2$	H_2CCCC		$(CH_3)_2$				**12**	
$NaCl$	HCO^+	$c\text{-}C_3H$	H_2CCC	HC_3NH^+							
KCl	OCS	$CCCO$	C_5	C_5N						C_6H_6	
$AlCl$	CCH	$CCCS$	SiC_4	H_2CCNH							
AlF	HCS^+	$H(CC)H$	H_3CO^+	$c\text{-}C_3H_2O$							
PN	$c\text{-}SiCC$	$HCNH^+$	$H(CC)NC$	$H(CC)_2H$							**13**
SiN	CCO	$HCCN$	$HNCCC$								
NH	CO_2	H_2CN	$H(CC)_2^-$								$H(CC)_5CN$
SH	$AlNC$	$c\text{-}SiC_3$									
HF	$SiCN$										
N_2	CCS										
LiH	C_3										
PO	$MgNC$										
CO	$NaCN$										
CS	CH_2										
C_2	$MgCN$										
SiC	HOC^+										
CP	HCN										
CO^+	HNC										
CH	HCP										

Figure 14   Molecules identified in the interstellar gas and in circumstellar shells, ranked by number of atoms. (From Ref. [337].) (See color plate 27).

4.   MOLECULAR ADDUCTS

The first observation of a molecular adduct by MW spectroscopy in 1961 [340] consisted of a series of broad bands of $CF_3COOH—HCOOH$. A systematic investigation of molecular complexes took place, however, only after that the supersonic expansion of seeded RGs was combined with rotational spectroscopy. First Klemperer and coworkers observed in 1972 the rotational spectrum of HF dimer [341] by using molecular beam electric resonance (MBER) spectroscopy. A few years later, in 1979, the spectrum of Kr—HCl was observed with supersonic-jet FTMW spectroscopy [342]. The rotational spectra of numerous molecular complexes were obtained with these two techniques, the latter of the two, supersonic-jet FTMW, being adopted by most MW laboratories. About 10 years ago, in 1996, also a direct absorption free-jet technique was successfully introduced in the investigation of molecular complexes [343].

The methods provide access to very interesting chemical systems, difficult to achieve with other techniques. The results range from the nature and details of intermolecular interactions to, e.g., the electronic density at a specific nucleus. Several reviews on molecular complexes are available [34–37,344–346], as well as a website with the bibliography of rotational spectra of weakly bound complexes [38].

4.1. Van der Waals Complexes

The first van der Waals complex observed by rotational spectroscopy was a complex of an Ar atom interacting with an HCl molecule. Its MBER spectrum, for $K = 0$, showed the characteristics of a linear molecule [347]. Larger complexes of RGs with organic cyclic molecules were later observed by supersonic-jet FTMW spectroscopy, with the rotational spectrum of furan—Ar being the one reported first [348]. Since then, a considerable number of van der Waals complexes have been investigated by MW spectroscopy in supersonic expansions. Generally, their structures, energetics, and internal dynamics have been determined in detail.

4.1.1. Complexes between rare gas atoms

The supersonic-jet FTMW rotational spectra of all 1:1 RG hetero complexes, except those involving He, that is Ne—Ar [349], Ne—Kr and Ar—Kr [350], Ne—Xe, Ar—Xe, and Kr—Xe [351], have been investigated. The small induced electric dipole moments of the heteronuclear RG dimers give rise to pure rotational spectra. The stability of this kind of complexes is due to dispersion forces, which can be sized, at first approximation, from the centrifugal distortion constants [352–354]. Their basic parameters are shown in Table 2.

   Some detailed rotational studies of mixed noble gas trimers and tetramers containing Ne and Ar atoms (Ne_2Ar, $NeAr_2$, Ne_3Ar, Ne_2Kr, and Ne_2Xe) have been obtained with the same technique [355,356]. Accurate values for rotational constants and centrifugal distortion constants were obtained from the spectral analyses. Isosceles triangular geometries of the trimer systems and distorted tetrahedral arrangements for the tetramers have been determined. Further evidence for the geometries was found in the effects of spin statistics in the spectra of isotopomers with C_{2v} symmetry and of those with C_{3v} symmetry. The simplicity of these fundamental systems makes them well suited for the determination and characterization of many-body nonadditive interactions. The magnitudes of the induced dipole moments of these trimers have been estimated from the "$\pi/2$ condition" of the pulsed MW excitation. The nuclear quadrupole hyperfine patterns due to ^{21}Ne, ^{83}Kr, and ^{131}Xe have been resolved, and the corresponding quadrupole coupling constants were obtained.

Table 2 Rotational and centrifugal distortion constants (for the most abundant isotopic species), equilibrium bond lengths, and dissociation energies of the mixed rare gas dimers

	B (MHz)	D (kHz)	Re (A°)	E_D (kJ/mol)
Ne—Ar [349]	2914.9286(5)	231.01(13)	3.481	0.38
Ne—Kr [350]	2215.36414(12)	115.453(13)	3.648	0.44
Ne—Xe [351]	1824.81506(10)	72.791(7)	3.887	0.48
Ar—Kr [350]	1198.6914(2)	11.991(7)	3.894	1.29
Ar—Xe [351]	961.29377(4)	6.6466(8)	4.092	1.51
Kr—Xe [351]	549.68960(5)	1.6936(6)	4.201	1.95

4.1.2.   Complexes of rare gases with linear molecules

Following the first investigation of the rotational spectrum of a complex of a linear molecule with a RG atom, Ar—HCl [347], by MBER spectroscopy, and the first investigation of a similar complex, Kr—HCl, with supersonic-jet FTMW spectroscopy [349], several more complexes of RGs with linear molecules have been investigated with the two original techniques, mostly using supersonic-jet FTMW, but interesting results were also obtained by mm wave absorption spectroscopy, combined with supersonic expansions.

4.1.2.1.   *Complexes of rare gases with hydrohalogenic acids and HCN*

For the two complexes mentioned above, the linear molecules are hydrohalogenic acids. Today, all RG—HX complexes (with RG = Ne, Ar, Kr, Xe, and X = F, Cl, Br) are investigated, mostly by supersonic-jet FTMW spectroscopy, and to a lesser extend by the two complementary jet techniques. All of them are effective linear tops, with their spectra characterized by the rotational constant B and a limited set of centrifugal distortion constants. The hydrogen is located between the RG and halogen atoms; however, the two connecting lines RG—Hal and H—Hal intersect at an effective angle of 30–40°. In the case of Kr—HF [357], Kr—HCl [358], Xe—HCl [359], and Xe—HBr [360], it was possible to determine the corresponding ^{83}Kr or ^{131}Xe nuclear quadrupole coupling constants, which gave evidence for charge transfer effects. More investigations on these complexes have been performed recently. For Ar—HCl, the ^{36}Ar isotopologue has been observed in natural abundance (0.34%) [361]. For Ar—HBr, coaxial pulsed-jet sub-mm wave absorption spectra are reported for Ar—H^{79}Br and Ar—H^{81}Br [362]. The transitions, measured with sub-kHz precision, are rovibrational transitions between the ground state and the low frequency Σ bending mode. The two isotopic band origins are 329611.4284(10) and 329225.6778(10) MHz for the ^{79}Br and ^{81}Br species, respectively. The rotational constants B are much larger in the Σ bending mode than in the ground state (1236.4134 and 1106.6712, respectively, for the ^{79}Br species), outlining a considerable geometry change toward a T-shape configuration.

This trend is better described for the case of the Ar—HCN complex. Its rotational spectrum was first observed by Leopold et al. [363] using the MBER and by Klots et al. [364] utilizing the supersonic-jet FTMW techniques. The anomalous ratio between rotational constants B of the Ar—DCN isotopologue and that of the normal species as well as large isotopic effects on centrifugal distortion constants [363] suggested an unusually strong angular-radial coupling in the intermolecular PES of the adduct. This effect was analyzed and satisfactorily described in several subsequent investigations. Drucker and collaborators [365] measured indeed the lowest excited bending states, Σ_1 and Π_1 by mm wave electric resonance optothermal spectroscopy (EROS). They determined the origins of the rovibronic bands, $\nu_0 = 164890.790(12)$ and 181984.4126(47) MHz, respectively. In addition, they obtained a full set of molecular constants, including the Σ_1–Π_1 coupling constants and the transition dipole moment μ_b. The rotational constants suggest a nearly T-shaped average geometry for each state, which is in agreement with the analysis of the dipole moments and quadrupole coupling constants. Agreement between experimental and prior theoretical work supports an attribution of the anomalous distortion and isotope effects in the ground state to extreme angular-radial coupling. Assuming Laguerre angular distributions, the Σ_1 and Π_1 wavefunctions have a maximum at 108° and 80° with respect to linearity, respectively.

Subsequently, theoretical aspects of this provocative "rod and ball" system have been analyzed by Leopold et al. [346]. Finally, Tanaka and collaborators observed many more rovibrational bands by mm wave absorption spectroscopy combined with a pulsed-jet expansion technique [366,367] to measure the internal rotation $j = 2$–1 hot band in the frequency region of 147–287 GHz and assigned the Σ_2–Σ_1, Σ_2–Π_1, Π_2–Σ_1, Π_2–Π_1, Δ_2–Σ_1, and Δ_2–Π_1 subbands. For the Σ_2, Π_2, and Δ_2 internal rotation substates, the subband origins, rotational constants, nuclear quadrupole coupling constants, and Coriolis interaction constants were determined.

4.1.2.2. Complexes between rare gases and CO

Considerable attention has been paid to the rotational spectra of complexes of RGs with CO. He—CO has been investigated by supersonic-jet FTMW spectroscopy, a MW-THz double-resonance spectrometer [368], and an intracavity spectrometer based on a mm wave generator known as OROTRON [369]. In total, eight rotational transitions have been measured for the most abundant species, and transitions were reported also for exotic isotopologues such as ^{3}He—CO and HeC^{17}O, but it has not been possible, due to large amplitude motions of the complex, to determine any of the usual spectroscopic constants, except for HeC^{17}O. In this case, from the value of the ^{17}O quadrupole coupling constants, $\chi_{aa} = 0.0316(96)$ MHz; it was possible to estimate the average value of β, the angle between the CO bond and the a-inertial axis in the Eckart axis system, such that $<\cos^2\beta> = 0.34$ [368].

Two MW reports are available for Ne—CO, based on supersonic-jet FTMW [370], and on mm wave OROTRON spectroscopy [371], respectively. The former investigation supplies an interpretation of the spectrum in terms of "traditional" spectroscopic constants (three rotational and six distortional parameters), although the rotational constant A was held fixed at the value obtained from IR data. The latter reports seven b-type mm wave transitions.

Several publications are devoted to the rotational spectrum of Ar—CO [372–376]. The first three investigations, based on supersonic-jet FTMW spectroscopy [372], on double-resonance technique being capable of extending the accessible frequency range of a typical resonator FTMW spectrometer beyond the cm to the mm wave range [373], and on mm wave absorption spectroscopy utilizing supersonic-jet expansions [374], were directed to the analysis of the vibrational ground state and allowed for a precise determination of all rotational constants (including A) and eight centrifugal distortion constants, consistent with a T-shaped structure of the complex. The two latter studies, based on sub-mm wave absorption spectroscopy in a pulsed supersonic jet, aim at the rovibronic spectra of the van der Waals bending [375] and stretching [376] vibrations, and supply precise values of the origins, being v_0 (bending) $= 360179.215$ (25) and v_0 (stretching) $= 542791.26(26)$ MHz, respectively.

The rotational spectra of the vibrational ground states have also been measured for the two heaviest members of the series, Kr—CO and Xe—CO, using supersonic-jet FTMW [370] as well as pulsed supersonic jet mm wave [377] spectroscopies. In the first investigation, only K_{-1} transitions have been measured for the Xe—CO, such that only $(B + C)/2$ was determined; for Kr—CO, also K_{-1} transitions were available, such that all rotational constants could be estimated. In the second study, several b-type transitions have been measured, such that – in principle – all rotational constants should be determined. However, a different set of spectroscopic constants has been used, which lead to the determination of $(B + C)/2$, $(B - C)/2$ (there called b) and σ, a parameter which gives the energy of the $K_{-1} = 1$ states with respect to the $K_{-1} = 0$ states.

4.1.2.3. Complexes of rare gases with other linear molecules

The rotational spectra of about five complexes of He-linear molecule species have been recently reported [378]. The very shallow van der Waals PES, with several slightly spaced minima, generates complex spectra, some times corresponding to considerably different configurations. The most detailed investigation concerns He—ClF, where pure rotational transitions are observed for the lowest linear Σ_0 state and for an excited T-shaped $K = 0$ Σ_1 state. The observed energy difference between the $J = 0$ level of the linear state and the $J = 0$ level of the T-shaped state is $2.320\,\mathrm{cm}^{-1}$. In addition, transitions into the two $J = 1$ levels and one $J = 2$ level of the $K = 1$ T-shaped state, Π_1, have been observed for the most abundant isotopomer, He^{35}ClF. A wealth of spectroscopic constants (rotation, centrifugal distortion, quadrupole coupling, dipole moment, interaction) for He^{35}ClF has been obtained. The values of the rotational constant B are 5586.8312(34), 7056.161(17), and 7430.338(32) MHz for the Σ_0 (ground), Σ_1, and Π_1 states, respectively, consistant with a linear structure for the former and a T-shape for the latter states. The origins of the two vibrational excited states (or conformational energy differences) are $\nu_0 = 69565.023(35)$ and 100302.239(46) for the Σ_1 and Π_1 states, respectively.

For four more He complexes, only the rotational spectrum of the ground state has been detected. In the case of He—OCS, the theoretical potential is found to have three minima, with the three lowest calculated bound states loosely localized in each of the minima and the ground state being T-shaped. Ten rotational transitions of the ground state are observed [379] in the frequency region 1.5–45.0 GHz. For He—CO$_2$, the 1_{01}–0_{00} transition has been observed for several isotopologues, including He—OC^{17}O, which exhibits quadrupole coupling constants that support an almost linear arrangement of the ground state [380]. The supersonic jet FTMW spectra of He—N$_2$O and of three minor isotopologues (He—^{14}N^{15}NO, He—^{15}N^{14}NO, and He—^{15}N^{15}NO) have been measured [381], and in the case of ^{14}N containing isotopologues, nuclear quadrupole hyperfine structure of the rotational transitions was observed and analyzed. The resulting spectroscopic parameters were used to determine geometrical and dynamical information on the complex. He—HCCCN is, according to the results obtained by supersonic-jet FTMW spectroscopy, an asymmetric rotor. The precise values of the rotational and ^{14}N and D nuclear quadrupole coupling constants agree with a near T-shaped complex in its ground state [382]. However, the unusual positive value of the inertial defect, $\Delta_c = 14.4$ uÅ^2, reflects the overall large amplitude motions of the complex.

The MW spectra of complexes with four linear molecules, N_2, CO_2, OCS, and N_2O have been studied for the full series of the heavier RGs. In each case, only the ground vibrational state has been observed, with spectra typical of asymmetric tops. Within the series RG—N_2 [383–386], the supersonic-jet FTMW spectra have been observed for several isotopologues of each adduct. They have T-shaped conformations of C_{2v} symmetry, which are confirmed by the values of the ^{14}N quadrupole coupling constants, and by the ^{83}Kr ($I = 9/2$) or ^{131}Xe ($I = 3/2$) nuclear quadrupole hyperfine constants of the respective complexes.

For the RG—CO_2 complexes, still of T-shape with C_{2v} symmetry (only $K_{-1} = even$ transitions have been detected) are found [387–389]. Endo and collaborators [388] report the dipole moments of the four complexes, which are given in Table 3. These dipole moments, in the order of 10^{-2}–10^{-1} D, can be regarded as a probe to investigate the charge redistribution by complex formation, which serves as important experimental sources for detailed understandings of the intermolecular interactions.

The first study on RG—OCS complexes by Klemperer and collaborators was realized with MBER on Ar—OCS [390], proving its T-shape. More recently, several isotopologues have been studied by supersonic-jet FTMW spectroscopy [391], and the new data, including ^{17}O ($I = 5/2$) quadrupole coupling constants in Ar—^{17}OCS, confirm the results of the early analysis. According to a supersonic-jet FTMW study, Ne—OCS does have a similar behavior [392]. For this complex, besides the quadrupole hyperfine structure (hfs) of Ne—^{17}OCS, also those of Ne—OC^{33}S and ^{21}Ne—OCS have been measured. The measurements provide an opportunity to estimate the changes in the electronic structures of OCS and Ne upon formation of the complex, since ^{21}Ne does have a nuclear spin quantum number $I = 3/2$. This study reports just the second measurements of Ne quadrupole coupling effects in a molecule, sometime after the hfs of the ^{21}Ne—Ar complex was observed and analyzed [349]. T-shaped Kr—OCS has been investigated by supersonic-jet FTMW spectroscopy, and its dipole moment has been measured, together with those of Ne—OCS and Ar—OCS [393].

The MW spectra of complexes of RGs with linear N_2O are available for Ne—N_2O and Ar—N_2O. A supersonic-jet FTMW investigation of Ne—N_2O reports the spectra of ^{20}Ne—^{14}N^{14}NO, ^{22}Ne—^{14}N^{14}NO, ^{20}Ne—^{14}N^{15}NO, ^{22}Ne—^{14}N^{15}NO, ^{20}Ne—^{15}N^{14}NO, and ^{22}Ne—^{15}N^{14}NO [394]. The experimental structures indicate that in the T-shaped complex, the Ne atom is – on average – closer to the O atom than to the terminal N atom. Ar—N_2O has been first investigated by the MBER technique [395], and then by supersonic-jet FTMW spectroscopy [396,397]. With the MBER investigation, Klemperer and collaborators determined the T-shaped configuration of the complex, and measured its dipole moment. Two studies followed, observing

Table 3 Dipole Moments (D) of Ne—, Ar—, Kr—, and Xe—CO_2 [388]

Ne—CO_2	0.0244 (13)
Ar—CO_2	0.0667 (32)
Kr—CO_2	0.0832 (30)
Xe—CO_2	0.1027 (42)

various isotopic species. Besides giving precise nuclear quadrupole coupling constants for each ^{14}N nucleus, a partial substitution r_s structure was obtained, indicating that the Ar atom is tilted towards the O atom.

The supersonic-jet FTMW spectra of complexes with Ne and Ar have been reported for HCCH. Eight isotopologues were analyzed for Ne—HCCH. For Ne—$H^{12}C^{12}CH$ and Ne—$H^{13}C^{13}CH$, the Σ_0 ground state and the Σ_1 first excited van der Waals bending state were measured. Also for the Ne—DCCD isotopologue, Σ_0 and Π_1 transitions were recorded, indicating a reversal of the energy level ordering upon deuterium substitution. The observed transitions exhibit a Coriolis perturbation of the Σ_1 and Π_1 energy levels. The Σ_1/Π_1 state is no longer metastable for the Ne—DCCH isotopomers, so only ground state transitions were recorded [398]. For Ar—HCCH, unusually large standard deviations of the spectroscopic fits are indicators of large-amplitude internal motions of the acetylene subunit. Separate fits of the individual K-stacks yielded lower standard deviations, and their results were used to interpret some of the unusual spectroscopic observations [399].

The MW spectra reported for Ar—ClF [400], Ar—Cl_2 [401], and Ar—ICl [402] show that these adducts have a near-linear shape in the case of hetero-halogen molecules, and a T-shape in the case of a homo-halogen molecule. For the adducts of Ar with hetero molecules, the Ar atom is on the side of the heavier atom, that is close to Cl for ClF and to I for ICl. The interpretation of the complicated hyperfine structures of ICl uses the changes in the nuclear quadrupole coupling constants $\chi_{aa}(I)$ and $\chi_{aa}(Cl)$ of Ar—ICl with respect to those of the free molecule, $\chi_0(I)$ and $\chi_0(Cl)$, respectively, to show that the electric charge redistribution within ICl, when Ar—ICl is formed, is equivalent to the transfer of 5.4×10^{-3} e from I to Cl.

4.1.3. Complexes of rare gases with cyclic (aromatic) molecules

After the first investigation of the molecule furan—Ar [348], several complexes of ring molecules with all of the RGs have been reported. The complexes with benzene (Bz) can be considered as the prototype system for this family of adducts. The rotational spectra of Bz—Ne [403,404], Bz—Ar [404,405], Bz—Kr [406], and Bz—Xe [404] have been observed by FTMW spectroscopy. Only the rotational spectrum of the adduct Bz—He has not yet been reported, due to its very low stability. However, the MW spectrum of the complex of pyridine (Py) with He has recently been reported [407]. Helium pressures in between 800 and 4000 kPa were needed to generate the Py—He complex in the supersonic expansion. Doubling of some transitions, presumably due to He tunneling, has been observed. Apart Py—He, the series of Py—RG complexes has been investigated for Py—Ne [408], Py—Ar [343,409,410], and Py—Kr [409], but it is not complete because the Py—Xe complex has not yet been investigated.

Some complexes with Ar and derivatives of benzene [411–413] and with diazines have also been investigated [414–417]. In all cases, except Py—He, the RG atom was quite firmly on one side of the ring, i.e., exhibiting no tunneling splittings. Some models have been developed to obtain approximate information on the van der Waals PESs. Makarewicz et al. [418] proposed a model to fit potential energy parameters to the centrifugal distortion constants, or to directly fit those parameters to the spectral frequencies. Caminati et al. [417] proposed to fit the mass distribution and Coriolis coupling terms to changes of the planar moments of inertia by using a 2D flexible model [30].

4.1.4. Complexes of rare gases with small asymmetric rotors

Although for the previous class of van der Waals complexes equivalent minima were characterizing the van der Waals PES, long pathways combined with large reduced masses did not allow for the observation of tunneling effects. However, when the RG atom is involved in noncovalent interactions with smaller molecules, splittings of the rotational transitions related to tunneling effects have easily been observed. Sometimes, for complexes with the lighter RGs, wide and irregular splittings of the rotational transitions into multiplets make the assignment of the spectra very complicated if not impossible. For example, the complexes of Ne with oxirane [419], difluoromethane [420], and dimethylether (DME) are characterized by splittings of several GHz, and so far only the spectrum of DME—Ne [421,422] was satisfactorily assigned. In these cases, the barriers connecting equivalent minima and the reduced masses of the motions are quite small. Very little information is available for such complexes with He, which has difficulties in possessing bound states. The three heaviest RG atoms Ar, Kr, and Xe behave in a regular way. For example, splittings due to the tunneling motions in DME—RG drop considerably in going from RG = Ne to RG = Ar, Kr, Xe [421–425] as shown in Table 4. Occasionally, additional splittings due to other motions have been observed, such as in the case of acetaldehyde-Ar [426,427] and acetaldehyde-Kr [428]. Here, the methyl group internal rotation is coupled to the Ar tunneling between two equivalent conformations, generating a quartet of lines. Rigorous treatments of the measured transitions were presented by Hougen et al. [426], but satisfactory fits can also be achieved with a general program like Pickett's suite CALPGM [9,10], simultaneously fitting the lines of the four states [427,428].

Table 4 Trend of the potential energy parameters of van der Waals motions in going from dimethyl ether—Ne to f dimethyl ether—Xe

ΔE_{0-0}^{+-} (MHz)[a]	Ne [421]	Ar [423]	Kr [424]	Xe [425]
	807.2(9)	0.9980(6)	0.26(5)	0.105(1)
k_S (Nm^{-1})[b]	1.0	2.3	2.6	3.0
v_s (cm^{-1})[c]	36	42	38	39
E_B (kJ mol^{-1})[d]	1.0	2.5	2.9	3.7
V_{inv}(kJ mol^{-1})[e]	0.19	0.69	1.33	1.40

[a] Tunneling splitting.
[b] Stretching force constant from pseudodiatomic approximation.
[c] Stretching frequency.
[d] Dissociation energy.
[e] Barrier to inversion.

4.1.5. Complexes of rare gases with radicals and ions

Rotational spectra have also been reported for a few complexes of RGs with radicals and ions, in the majority of cases by Endo and collaborators.

4.1.5.1. Complexes of rare gases with radicals

The most investigated radical is SH, whose complexes have been studied with Ne, Ar, and Kr. The RG—SH species were generated in a pulsed electric discharge of H_2S (0.3%) and suitable combinations of RGs. Ar—SH was the first to be investigated by supersonic-jet FTMW [429], observing several R-branch transitions in the lower spin component ($\Omega = 3/2$) for the linear $^2\Pi_i$ radical (both Ar—SH and Ar—SD). The analysis supplied effective rotational and centrifugal distortion and the hyperfine constants. The negative value of the centrifugal distortion constant D obtained in the effective Hamiltonian fit was successfully reproduced with a potential exhibiting a minimum at Ar—SH being lower in energy than at Ar—HS. Then, by MW–mm wave double-resonance spectroscopy, rotation–vibration transitions of a van der Waals bending vibration, $P = 1/2 \leftarrow 3/2$, have been observed [430], clarifying the rotational energy-level structure for the two isotopologues, with hyperfine structure due to the hydrogen or deuterium nuclei and parity doublings in the $P = 1/2$ state.

R-branch transitions in the lower-spin component ($\Omega = 3/2$), corresponding to a linear $^2\Pi_i$ radical, were observed also for Ne—SH and Kr—SH [431]. The spectral pattern of Kr—SH is relatively regular, while that of Ne—SH is irregular with the J dependence of the parity doublings quite different from other RG—SH complexes. Two-dimensional intermolecular potential energy surfaces (IPSs) for both of the species have been determined from the least-squares fits to the observed rotational transitions.

The Ar—OH open-shell complex is one of the spectroscopically most investigated radical complexes. Its rotational spectrum was observed for the first time by Endo and collaborators [432] using supersonic-jet FTMW spectroscopy in combination with the production of the OH radical by a pulsed electric discharge on a water/Ar mixture in a channel extending the pulsed valve. The B rotational constant suggests an angle of 59° between the OH intermolecular axis and the axis connecting Ar with the OH center of mass. Subsequent supersonic-jet FTMW investigations of Ar—OH [433,434] provided additional data that, in combination with other spectroscopic information, allowed the construction of a three-dimensional PES.

Some open-shell complexes of RGs with triatomic radicals are reported. Howard and collaborators investigated the supersonic-jet FTMW spectra of Ar—NO_2 [435], Kr—NO_2 [436], and Xe—NO_2 [437]. The MW spectra consisted solely of the a–type transitions involving the $K_a = 0$ and $K_a = 2$ states. These transitions showed structure due to fine, magnetic hyperfine and electric quadrupole interactions. The derived parameters are analyzed in terms of those of the free NO_2 radical. The absence of $K_a = $ odd states was rationalized in terms of a high-frequency tunneling motion of the NO_2 within the complex. For ^{131}Xe—NO_2 intermolecular super-hyperfine interactions were observed between the Xe nuclear magnetic moment and the unpaired electron on the NO_2 radical. The measured transition frequencies could be fitted by the parameters of a semi-rigid molecule Hamiltonian, but the spectra show significant evidence of large internal motion, in particular a tunneling motion between the two equivalent nonplanar structures of the complex.

Ar—ClO$_2$ displays a- and c-type spectra, the latter of which is shifted by internal motion of the ClO$_2$ unit [438]. Changes of the electronic structure of ClO$_2$ upon complexation are shown to be very small. The observed internal motion effect was analyzed with a model of an internal rotation of the ClO$_2$ subunit around it's a axis. The rotational (N), electron spin ($S = 1/2$), and nuclear spin ($I = 3/2$) angular momenta are coupled in the following way: $N + S = J$, $J + I = F$. Therefore, all rotational levels with $N > 1$ are split into eight hyperfine components, with the strongest electric dipole transitions obeying the selection rule $\Delta F = \Delta J = \Delta N$.

Finally, the rotational spectrum of Ar—HO$_2$ has been observed by supersonic-jet FTMW and MW/mm wave double-resonance [439] exhibiting a- and b-type transitions, which provide precise molecular constants. The structure was determined to be planar and almost T-shaped, with the Ar atom slightly shifted toward the hydrogen atom of HO$_2$. The effects on the unpaired electron distribution upon complex formation were found to be fairly small, because the fine and hyperfine constants of Ar—HO$_2$ are well explained by those of the HO$_2$ monomer.

4.1.5.2. Complexes of rare gases with ions

The rotational spectra of the complexes of two molecular ions with RGs are available: Bogey and collaborators reported the first rotational spectrum of a weakly bound cluster ion, Ar—H$_3^+$ [440,441]. Also the isotopologue ArD$_3^+$ was observed. The ionic clusters were produced from Ar/H$_2$/D$_2$ mixtures inside a negative glow extended by a magnetic field; the spectra were measured by mm/sub-mm wave spectroscopy. Most of the observed rotational lines were split by internal motion; spin statistical weights as well as intensity ratios for the components were determined from symmetry considerations. The splittings were interpreted in terms of internal motion of the H$_3^+$ triangle. Within a third study [442], the measurements were extended to higher frequencies and the new data were analyzed in using an IAM-like approach, which accounts for the large amplitude internal rotation motion displayed by both species and gives insight into the geometry of the intermediate configuration for the large amplitude motion of H$_3^+$. The ion complexes, Ar—H$_3^+$, Ar—D$_3^+$, and sym Ar—D$_2$H$^+$ have also been observed by supersonic-jet FTMW spectroscopy combined with a pulsed DC discharge nozzle [225].

Endo and collaborators could assign the supersonic-jet FTMW spectra of two more ionic clusters, Ar—HCO$^+$ [443] and Kr—HCO$^+$ [444]. The studies of several isotopologues proved that both ionic complexes have a linear proton-bound form. The RG...H interaction is much stronger than in the corresponding neutral complexes. The RG...H distances, 213.4 and 222.2 pm for the Ar and Kr adducts, respectively, are, indeed, much shorter (by 60–70 pm) than those of the RG-halogen, consistent with intermolecular stretching force constants that are determined to be one order of magnitude larger.

4.2. Hydrogen-bonded Complexes

Hydrogen bonding involves many research areas and is often invoked to explain the energetic and structural features of inorganic, organic, and biological chemical systems. It is the most important noncovalent interaction, which can be rationalized in terms of electrostatic and charge transfer effects. Because of its paramount importance, it deserves special attention as a class by its own. Several text books are available on this topic, the three most recent being written by Jeffrey [445], Scheiner [446], and Desiraju and Steiner [447]. Many detailed information on the hydrogen bond (HB) have been the results of rotational investigations of hydrogen-bonded molecular complexes, as described below.

4.2.1. Molecular complexes involving water

The most important molecule involved in hydrogen bonding is water. Probably, the most investigated molecular complex of this kind is the water dimer. It is an extremely flexible system, which, taking into account the feasible internal motions, is characterized by the G16 MS group derived from PI group theory [448]. The rotational spectrum is very complicated, with the observed rotational transitions involving several of the tunneling ground state sublevels. However, supersonic-jet FTMW and sub-mm wave rotation-tunneling spectra of the normal and of several mixed deuterated–protonated isotopologues of the water dimer have been fitted to experimental uncertainties, obtaining information on the major tunneling processes, such as the tunneling interchange of the two protons on the proton–acceptor subunit and the tunneling interchange of the two protons on the proton–donor subunit [449]. Recently, Coudert et al. measured and analyzed the hyperfine patterns of several transitions of $(H_2O)_2$ and accounted for the magnetic spin–rotation and spin–spin hyperfine couplings [450]. Symmetry-adapted nuclear spin wavefunctions were built to account for the interaction of the hyperfine coupling with the large amplitude motions displayed by the water dimer and to build total rotational-tunneling-hyperfine wavefunctions obeying the Pauli exclusion principle.

4.2.1.1. Complexes of water with small molecules

The small complexes with water are characterized by pronounced internal dynamics and tunneling effects. Among them, all the 1:1 adducts of water with hydrohalogenic acids have been investigated by MW spectroscopy. Probably in all these complexes the acid molecule is delocalized between the two electronic lone pairs of water, but this effect has been clearly observed only in the case of H_2O—HF, with a tunneling splitting due to the HF transfer of $\sim 70\,cm^{-1}$ [451], which spectrum has been observed in standard MW absorption cells. In these conditions, it was possible to simultaneously observe H_2O, HF, and H_2O—HF in thermodynamical equilibrium and determine the zero-point and equilibrium dissociation energies as $D_0 = 34.3(3)\,kJ/mol$ and $D_e = 42.9\ (8)\,kJ/mol$, respectively, from absolute intensities of the rotational transitions [452]. The rotational spectra of the remaining complexes of water with hydrohalogenic acid, H_2O—HCl [453], H_2O—HBr [454], H_2O—HI [455], and H_2O—HCN [456], have been obtained by supersonic-jet FTMW spectroscopy. In the jet conditions, it has been possible to measure only rotational transitions of the 0^+

component of the two tunneling states, because the 0^- sub-state is too high in energy to be reasonably populated. However, centrifugal distortion constants and some effects on the hyperfine structure due to the nuclear quadrupole moment of the N, Cl, Br, and I nuclei indicate a Coriolis interaction with a "blind" state.

The supersonic-jet FTMW spectra of adducts with water have been investigated for two more inorganic acids, HNO_3 and H_2SO_4. HNO_3—H_2O is characterized by a near-linear, $r = 178$ pm, HB to the oxygen of the water and a second HB formed between a water hydrogen and one of the HNO_3 oxygens [457]. The resulting cyclic structure adopts a planar configuration except for the non-hydrogen-bonded proton of the H_2O. The a- and b-type transitions of the H_2O- and D_2O-containing species exhibit a doubling that disappears in the DOH complex, providing direct evidence for the existence of a proton interchange motion in the system. Moreover, c-type rotational transitions do not appear at their predicted rigid rotor positions, even for the DOH species, providing indirect evidence for a second motion that is interpreted as a large amplitude wagging of the non-hydrogen-bonded proton of the water. H_2SO_4—H_2O forms a distorted six-membered ring with the water unit acting as both a HB donor and a HB acceptor toward the sulfuric acid [458]. Extensive isotopic substitution has also permitted a re-determination of the structure of the H_2SO_4 unit within the complex, showing that the O—H bond of the hydrogen-bonding OH of the sulfuric acid elongates by 7(2) pm compared with the O—H bond length in free H_2SO_4, and the S—O bond involved in the secondary interaction stretches by 4(1) pm.

The MBER spectra of H_2O—NH_3 [459] and the MW and FIR spectra of H_3N—HOH [460] measured within a planar supersonic jet/tunable laser sidebands spectrometer have been reported. The latter paper gives details on the internal dynamics of the complex. Splitting into two sets arises from water tunneling, while the overall band structure is due to internal rotation of the ammonia top. Analysis based on an internal rotor Hamiltonian provided rotational constants and an estimate of $V_3 = 10.5(50)$ cm^{-1} for the barrier height to internal rotation for the NH_3 monomer. An estimated barrier of $V = \sim 700$ cm^{-1} is derived for the water tunneling motion about its c axis.

The rotational spectra of several 1:1 adducts of water with oxides, anhydrides, and sulfides were also reported. The supersonic-jet FTMW spectrum of H_2O—CO (and five isotopologues) reveals a tunneling motion that exchanges the equivalent hydrogens originating two states in the H_2O and D_2O complexes [461]. The water is hydrogen bonded to the carbon of CO; however, the bond is nonlinear. At equilibrium, the O—H bond of water is oriented at an angle $\phi = 11.5°$ with respect to the a axis of the complex; the C_{2v} axis of water deviates by $\phi = 64°$ from the a axis of the complex. The HB length is about $r = 241$ pm. From the difference in the dipole moment between the symmetric and antisymmetric states, the barrier to exchange of the bound with the free hydrogens is determined as 210(20) cm^{-1}.

The series H_2O—CO_2, H_2O—COS, and H_2O—CS_2 of water complexes was investigated by supersonic-jet FTMW spectroscopy. H_2O—CO_2 is T-shaped and shows a splitting of the two energy levels from internal rotation between the water and carbon dioxide molecules around the van der Waals bond of the complex [462]. As a consequence of the ^{16}O spin statistics in carbon dioxide, only half of the energy levels are populated except for the H_2O—$C^{16}O^{18}O$ species. Hyperfine splittings of the rotational transitions due to the hydrogen and deuterium nuclei have been resolved. Rotational and centrifugal distortion constants were fitted to the transition frequencies for the different internal rotation states. Quadrupole coupling constants were determined from the deuterium hyperfine splittings. The hydrogen hyperfine splittings for the parent species have been attributed to the spin–spin interaction of the hydrogen nuclei. A barrier equivalent to $285.6(14)\,cm^{-1}$ has been adjusted with the help of a flexible model for the internal rotation in the complex of water about its C_{2v} symmetry axis. Vice versa, both H_2O—COS [463] and H_2O—CS_2 [464] have an almost linear arrangement of the heavy atoms, with the water oxygen in contact with a sulfur atom. The most probable structures of H_2O—SCO and H_2O—SCS are near C_{2v}, with oxygen-sulfur van der Waals bond lengths of $r = 314$ and $320\,pm$, respectively.

N_2O is isoelectronic to CO_2, and its adduct with water, H_2O—N_2O, behaves like H_2O—CO_2. According to its MBER spectrum [465], it has, indeed, a planar T-shaped arrangement of the heavy atoms, with the distance $r = 297(2)\,pm$ from the H_2O oxygen to the central nitrogen of N_2O. In addition, the internal rotation of the two moieties produces two tunneling states.

The supersonic-jet FTMW spectra of the complexes of water with the two sulfur anhydrides, SO_2 and SO_3, are also available. H_2O—SO_2 presents two states with a- and c-type spectra which are split by internal rotation of the water unit [466]. The geometry obtained from a fit to the derived moments of inertia has the planes of the two monomer units tilted approximately 45° from the parallel orientation with the oxygen atom of the water closest to the S atom of SO_2, giving an S—O distance of $r = 282.4(16)\,pm$ and a center-of-mass distance $R_{c.m.} = 296.2(5)\,pm$.

H_2O—SO_3, which is considered an important precursor to the formation of H_2SO_4 in the atmosphere, assumes a structure in which the oxygen of the water approaches the sulfur of the SO_3 above its plane, reminiscent of a donor–acceptor complex [467]. The intermolecular S—O bond length is $r = 243.2\,pm$.

Complexes of H_2O with homo or hetero di-halogen molecules have been investigated by Legon and collaborators using supersonic-jet FTMW with a modified combined (fast-mixing) nozzle. By isotopic substitution it was shown for H_2O—F_2 [468], H_2O—ClF [469], and H_2O—ICl [470] that in each complex the H_2O molecule acts as the electron donor and either F_2 or ClF or ICl acts as the electron acceptor, with nuclei arranged as $H_2O\ldots FF$, $H_2O\ldots ClF$, and $H_2O\ldots ICl$. For $H_2O\ldots ClF$, the angle between the bisector of the HOH angle and the $O\ldots Cl$ internuclear line is $\phi = 59°$, while the distance $r(O\ldots Cl) = 260.8\,pm$. The corresponding quantities for H_2O—F_2 are $\phi = 48°$ and $r(O\ldots F_i) = 274.8\,pm$, where F_i indicates the inner F atom.

Finally, the supersonic-jet FTMW spectrum of the radical $OH\ldots H_2O$, prepared by electrical discharge, was reported independently by Oshima et al. [217] and Brauer et al. [471]. Both studies agree on a $^2A'$ electronic state and a $OH\ldots OH_2$ configuration with partial quenching of the OH orbital angular momentum, which dramatically affects the rotational spectra. However, there are some discrepancies in the $(B + C)$ parameter of the two analyses, with the calculated FIR transition frequencies exhibiting deviations beyond the experimental accuracy. In Ref. [471], the $^2A'$–$^2A''$ energy separation ρ is determined to be $-146.50744(42)\,cm^{-1}$.

4.2.1.2. Structural and energetic features of bonds between water and organic or biomolecules

Organic and biological molecules are often large systems and can present several different sites to interact with a water molecule. The structural and energetic factors which lead to the preferred conformation of adducts are precisely investigated by high-resolution spectroscopy of the hydrogen-bonded complexes. Bonds such as O—H...O, O—H...N, O—H...S, N—H...O, and O—H...X (X = halogen) are commonly considered as classical HBs. These interactions were characterized in detail through rotational spectroscopy studies of molecular complexes of water with alcohols and phenols, organic acids, amides, ethers, aldehydes, ketones, amines, aminoacids, diazines, imino groups contained in aromatic rings, thioethers, and halogenated hydrocarbons. All of them are relatively strong HBs, with energies in the range 15–25 kJ/mol.

Very few adducts of water with alcohols and phenols have been investigated by MW spectroscopy. The supersonic-jet FTMW spectrum of $CH_3OH...H_2O$ is very complicated by tunneling effects due to the methyl group and water moiety internal rotations [472]. However, it has been possible to determine spectroscopic constants precise enough to demonstrate unambiguously that the lower-energy conformation formed by supersonic cooling corresponds to a water–donor, methanol–acceptor complex. The supersonic-jet FTMW spectrum of the 3-hydroxytetrahydrofuran–water has been assigned to the networked structure of the complex, with intermolecular HBs from the hydroxyl to the water oxygen and from water to the furanose-ring oxygen [473], corresponding to the lowest-energy ring-puckering conformation of the 3-hydroxytetrahydrofuran monomer.

The rotational spectrum of phenol–water has been investigated by supersonic-jet FTMW and free-jet mm wave absorption techniques [474,475]. The water molecule acts as a proton acceptor, forming a nearly linear $O_{ph}H_{ph}...O_w$ HB. Tunneling splittings, due to combined internal rotation and inversion motions of water, were observed. The corresponding 2D PES has been modeled to reproduce several experimental data [475].

Formic acid–water ($HCOOH—H_2O$) has been investigated by supersonic-jet FTMW spectroscopy, and it was found to adopt a planar structure [476]. Water and formic acid form a six-membered ring with two HBs, which see water acting as proton donor to the C=O group and as acceptor from the OH group.

Three conformers – I, II, and III in order of stability – of monohydrated formamide ($HCONH_2—H_2O$) have been observed by supersonic-jet FTMW [477]. Conformer I adopts a closed planar ring structure stabilized by two intermolecular HBs [N—H... O(H)—H...O=C]. Conformer II is stabilized by an O—H...O=C and a weak C—H...O HB, while, in the less stable form III, water accepts an HB from the *anti* hydrogen of the amino group.

The rotational spectra for several ether–water complexes, either utilizing free-jet mm wave absorption or supersonic-jet FTMW techniques, are reported. The complexes with saturated cyclic ethers, such as oxirane–water [478], oxetane–water [479], 1,4-dioxane–water [480], and tetrahydropyran–water [481], are characterized by relatively simple spectra, where a well-defined conformation is found. For example, for 1,4-dioxane–water and tetrahydropyran–water, the water molecule is linked in an *axial* arrangement, with the "free" hydrogen *entgegen* with respect to the ring, and with O—H . . . O HBs far away from linearity. The complexes of water with non-cyclic ethers turned out to be quite complicated or to present unusual effects. An interesting feature was observed, indeed, in the case of anisole–water [482], where the deuteration of the water moiety produces a conformational change in the complex: the secondary interaction O . . . H$_{Me}$ is replaced by the O . . . H$_{\underline{Ph}}$ interaction (see Figure 15). This phenomenon is likely due to the Ubbelohde effect [483], associated with a very flat potential energy function of the in plane bending of the water molecule. The rotational spectrum of DME–water shows a doubling of the rotational transitions, clearly due to the tunneling of water between the two equivalent lone-pairs of the ether oxygen [484]. A similar doubling was found for furane–water [485]. In the case of dimethoxymethane–water, the water molecule forms a five-membered ring with a part of the substrate molecule, stabilized by a O—H . . . O and a weak C—H . . . O HB with one of the two methyl groups, making them no longer equivalent [486]. Each rotational transition is then split into quintuplets, with the spacings allowing for the determination of the two V_3 barriers to internal rotation of the methyl groups. From the difference between the two V_3 values, one can estimate the strength of the weak C—H . . . O interaction.

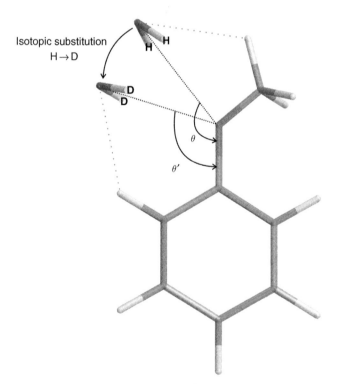

Figure 15 The deuteration of water produces a conformational change in the anisole. . . water complex, as shown by the scheme. The value of the θ angle decreases from 138 to 128°, while the secondary interaction O . . . H$_{Me}$ is replaced by the O . . . H$_{\underline{Ph}}$ interaction. (Reproduced with permission from Ref. [482]. Copyright Wiley-VCH Verlag GmbH & Co. KgaA.)

Two water complexes with aldehydes or ketones were studied by supersonic-jet FTMW spectroscopy: formaldehyde–water (CH_2O—H_2O) is a planar and quite strongly bound adduct [487], with a harmonic pseudodiatomic stretching force constant, $k_s = 8.93$ N/m, and HB lengths of approximately $r = 268$ pm between the water oxygen atom and a CH_2 hydrogen atom, and $r = 203$ pm between a water hydrogen atom and the oxygen atom of H_2CO. With respect to the HB features, cyclobutanone (CBU)–water [488] is quite similar to formaldehyde–water. The complex is effectively planar and the water molecule is linked to CBU through a strong O—H...O HB, with an O...H distance of $r = 195$ pm, and two C—H...O interactions of the water oxygen with the two hydrogens of one of the two CH_2 groups adjacent to the carbonyl group.

Several MW reports are available on amines–water complexes. With the symmetric tops trimethylamine and quinuclidine, water acts as a proton donor and originates adducts characterized by effective symmetric top spectra [489,490]. In the case of trimethylamine, a near-free internal rotation of the water about the symmetry axis of trimethylamine is invoked [489]. Also in the case of dimethylamine–water, the rotational constants of four isotopologues are consistent with a structure in which the water acts as an HB donor to the dimethylamine [491].

The supersonic-jet FTMW spectrum of aniline–water leads to a configuration of the complex with a N...H—O HB and the water molecule located in the symmetry plane of aniline, orienting the free water proton toward the aniline ring. This corresponds to water being a double proton donor, towards the N lone pair and toward the ring π-system [492]. A similar N...H—O H-bond is also formed in the morpholine–water complex [493].

Pyrrolidine (PRL)-water has been investigated by free-jet mm wave absorption spectroscopy. PRL itself exists as two, axial and equatorial conformers [494], but only the adduct with water acting as proton donor to the equatorial-PRL was detected to date [495]. The water molecule lies in the plane of symmetry of PRL; the water hydrogen involved in the HB is *axial* with respect to the ring, while the "free" hydrogen is *entgegen* to the ring. The three atoms involved in the hydrogen bond adopt a bent arrangement with an N_{ring}...H distance of about $r = 189$ pm and $\angle(N_{ring}...H—O) \cong 163°$.

Very recently, the supersonic jet FTMW spectrum of laser-ablated glycine and water was reported [496]. In the observed conformer, water interacts through the carboxyl group, simultaneously acting as a proton donor to the carbonyl oxygen atom and accepting a proton from the hydroxyl group. This configuration benefits from two intermolecular HBs, with water bridging the carboxylic acid group in a cyclic planar structure. The water hydrogen atom not participating in the intermolecular HB is predicted to lie out of the plane of the glycine skeleton. The distance of the O_G—H_G...O_W interaction $r = 180.6$ pm is significantly shorter than that of the O_W—H_W...O_G with the value $r = 207.3$ pm. In addition, the nonlinearity of the O_W—H_W...O_G bond at an angle of $\phi = 126.5°$ is much more pronounced than that of the O_G—H_G...O_W bond at $\phi = 161.9°$. A similar interaction of water with the carboxyl group is found for the lowest energy conformation of the complex alaninamide–water, with similar values for the two intermolecular HB lengths [497].

The rotational spectra of the 1:1 adducts of the three diazines and water are also available. Pyrazine is a highly symmetric molecule (D_{2h}), and part of this symmetry pertains to the complex, which has been found to be planar, with a σ-type O—H...N H-bond and does have an effective C_{2v} symmetry [498]. As a result, the rotational lines are split into two components, due to the W-shaped potential of the internal rotation of water with respect to the H...N H-bond.

Also pyrimidine–water and pyridazine–water are characterized by a σ-type O—H...N H-bond, but, due to their reduced symmetry [C_s], they look like rigid tops without tunneling effects [499,500]. The HB features are similar for the three complexes, with H...N bond lengths of about $r = 200$ pm and deviations of about 20° of the O—H...N angle from linearity.

A σ-type O—H...N H-bond is presumably also found in isoxazole–water, being indicated by the ^{14}N quadrupole coupling constants resulting from the supersonic-jet FTMW spectrum [501]. The rotational constants alone cannot discriminate between possible alternative shapes of the water complexes with an O—H...O H-bond.

For complexes of water with imino groups inserted in an aromatic ring, such as pyrrole–water, indole–water, and 2-pyridinone–water, the imino group has always been found to act as the proton donor. In pyrrole–water [502] and indole–water [503] – being investigated by supersonic-jet FTMW spectroscopy – the water molecule undergoes an internal rotation resulting in splittings of the rotational transitions, which have been used to obtain information on the corresponding PESs. For 2-pyridinone–water, the rotational spectrum has been observed by free-jet mm wave absorption spectroscopy [504]. The water molecule is strongly linked to the partner group through a double-HB, N—H...O$_w$ and O$_w$—H$_w$...O$_{py}$, such that its internal rotation is practically inhibited.

H_2O—CH_3NO_2 has been investigated by supersonic-jet FTMW spectroscopy [505]. It was found that the complex has a plane of symmetry and that the water molecule is firmly linked to nitromethane through a O$_w$—H$_w$...O$_{NO2}$ and a C—H...O$_w$ interaction.

The tetrahydrothiophene–water complex, C_4H_8S—H_2O, was studied by free-jet mm wave absorption and supersonic.jet FTMW spectroscopies [506]. The rotational parameters were interpreted in terms of a geometry where the water molecule acts as proton donor lying close to the plane bisector to the CSC angle of tetrahydrothiophene. The "free" hydrogen is *entgegen* to the ring. The parameters characterizing the HB are the distance between the sulfur and hydrogen atoms, $r(S...H) = 237(4)$ pm, and the angle between the line bisecting the CSC angle of tetrahydrothiophene and the S...H bond $\phi = 85.(3)°$.

The O—H...F linkage has also been investigated by the analysis of the rotational spectrum of the complex of difluoromethane (Freon-32) with water [507]. The dissociation energy of this complex, ~ 8 kJ/mol, is considerably lower than those of the above mentioned systems, and it includes contribution of secondary C—H...O interactions. The investigation of the complex of chlorofluoromethane (Freon-31) suggests that a water molecule prefers to form a hydrogen bond with a Cl rather than with an F atom, as indicated by the OH...Cl linkage which

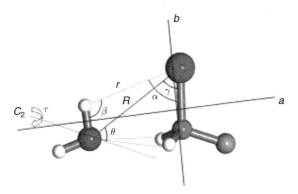

Figure 16   Water prefers an O—H ... Cl to a O—H ... F interaction. The observed conformer of chlorofluoromethane-H_2O displays one O—H ... Cl and two weak C—H ... O H-bonds. Some structural parameters are shown. (Reproduced with permission from Ref. [508]. Copyright 2006, Wiley-VCH Verlag GmbH & Co. KGaA.)

has been found [508] (see Figure 16). Finally, H_2O ... CF_4 [509] does have O ... F contacts without HB, outlining the little propensity of the F atom in being a proton acceptor.

4.2.2. Complexes with hydrohalogenic acids

The rotational spectrum of the HF dimer [341] was the first spectrum of a molecular complex to be observed with modern – high sensitivity, high resolution – MW techniques, by Klemperer and coworkers in 1972, using MBER. In the following years, several small clusters, with hydrohalogenic acids involved in internal HBs, were characterized by MW spectroscopy.

More recently, the adducts of hydrohalogenic acids with organic molecules were targeted. For example, in saturated cyclic ethers, such as tetrahydropyran, two different lone pairs (axial and equatorial) are available at the ether oxygen to accept a proton. Alonso and collaborators could assign the supersonic-jet FTMW spectra of both axial and equatorial complexes in the case of tetrahydropyran-HCl [510], tetrahydropyran-HF [511], and pentamethylene sulfide-HF [512], characterizing relative energies and structural differences.

The rotational spectra have been reported for several other complexes of hydrohalogenic acids and ethers. The supersonic-jet FTMW spectra are available for cyclohexene oxide–HCl, for which only the equatorial adduct, probably the far more stable one, was observed [513], for oxetane–HF and 2,5-dihydrofuran–HCl [514], which have C_s symmetries with HX (X = halogen) lying in the symmetry plane bisecting the COC ring angle [515], and for tetrahydrofuran–HF [516]. The rotational transitions of the latter adduct exhibit small tunneling splittings, which were not observed for C_4D_8O—HF and the four single ^{13}C isotopologues. An analysis of these observations in terms of symmetry considerations attributes the splittings to the pseudorotation of the tetrahydrofuran subunit in the complex.

The rotational spectra of the dimethyl ether (DME) complexes DME–HF [517] and DME—HCl [518] were assigned by free-jet mm wave absorption spectroscopy (in the case of DME—HF, some fine details of the spectrum were later resolved by supersonic-jet FTMW). Either HF or HCl group act as proton donors and tunnels at rates of 44178.2(7) and 8182(7) MHz between the two oxygen lone pairs, respectively. This tunneling splittings correspond to an inversion barrier of 0.7–0.8 kJ/mol for both adducts.

The spectroscopic constants of the adducts of an aldehyde and of a ketone with HCl, H_2CO—HCl [519], and CBU—HCl [520] have been interpreted in terms of coplanarity of HCl with the H_2CO plane and with the "effective" plane of the CBU heavy atoms, respectively.

The structural features of the S—H...F HB have been established from the supersonic-jet FTMW spectra of tetrahydrothiophene–HF [521] and thiophene–HF [522]. The $r(S...H)$ bond lengths are very different in the two complexes, being $r = 215(4)$ pm and $r = 298(5)$ pm, respectively. The long HB in thiophene–HF is presumably much weaker.

Quite interestingly, the molecular adducts of HBr with CO_2 and OCS do have completely different shapes: the supersonic-jet FTMW spectra show that HBr—CO_2 has a T-like shape [523], with an Br...C interaction, while HBr—OCS is nearly linear with an Br—H...OCS arrangement [524]. For HBr—CO_2, the $\angle$C...Br—H angle was determined as 110.6° (from the Br nuclear quadrupole coupling constants), which constitutes a double minimum potential for the hydrogen atom having equivalent positions towards both oxygens. The corresponding tunneling splitting has been measured for several isotopomers, e.g., 1187.26 ± 0.05 MHz for the $H^{79}Br$—CO_2 species, leading to a tunneling barrier of $V = 250$ cm^{-1}. It is worthwhile pointing out that, due to the Bose–Einstein statistics of the spin-zero oxygen nuclei, for the ground vibrational state only symmetric tunneling states are allowed for even K_a and only antisymmetric tunneling states are allowed for odd K_a.

Finally, the complexes of SO_3 with HF, HCl, and HBr were studied by supersonic-jet FTMW spectroscopy. In all three systems, the halogen atom approaches the SO_3 on or near its C_3 axis, and the vibrationally averaged structure is that of a symmetric top, with S...X (X = halogen) contacts [525]. The S...X bond lengths are 266 pm, 313 pm, and 323 pm for the HF, HCl, and HBr complexes, respectively. The nuclear quadrupole hyperfine structures – the DF isotopologue for HF—SO_3 – can be used to probe how much the hydrogen points away from the averaged C_3 axis of the complexes: 47.7°, 72.8°, and 73.0° for the HF, HCl, and HBr complexes, respectively.

4.2.3. Strong hydrogen bonding

Strong hydrogen bonding, within neutral species, characterizes complexes with a double-HB, such as the carboxylic acid dimers observed with low-resolution MW spectroscopic methods by Costain in 1961 [340] and Bellot and Wilson in 1975 [526]. Detailed supersonic-jet FTMW analyses are now available for some carboxylic acid bimolecules. According to the resonance-assisted hydrogen bond model suggested by Gilli [527], the monomers are held together by more than 60 kJ/mol. By Bauder and coworkers, the structures of $CF_3COOH...HCOOH$ and $CF_3COOH...CH_3COOH$ have been determined through the analyses of the rotational spectra of several isotopologues [528]. For the latter complex, also the V_3

barrier to internal rotation of the methyl group was determined. Antolinez et al. reported the MW spectrum of the trifluoroacetic acid–cyclopropanecarboxylic acid bimolecule [529]. In none of these cases, doubling of the rotational transitions attributable to a double proton transfer tunneling is observed.

4.2.4. Weak hydrogen bonding

Weak molecular interactions not only as C—H...O, C—H...F, C—H...Cl, C—H...N, C—H...S, and C—H...π, but also O—H...π, N—H...π an Hal—H...π, are often encountered in nature. They are commonly termed weak hydrogen bonds (WHB) and are a central topic in HB research [447]. These interactions exhibit energies within a few kJ/mol, which approach those of van der Waals forces, but maintain the directional properties of "classical" HBs [530,531]. Detailed structural and energetic information on these WHBs is obtained from investigations of the rotational spectra of hydrogen-bonded molecular complexes generated in supersonic jets. Two of the most common or most investigated WHB linkages are the C—H...O and C—H...F. Among the first complexes to be investigated with this technique was the difluoromethane dimer, which is characterized by three of these interactions (C—H...F in this case) [532]. The dimer of DME exhibits three C—H...O contacts [533]. Following these two prototype complexes, several more were investigated by MW spectroscopy, many of them with several mixed C—H...F or C—H...O interaction [534–539]. All of them have three WHBs and exhibit dissociation energies of about 5–8 kJ/mol. However, in the case of CHF_3—CH_3F [537], the dipole–dipole interaction energy was found to be an important factor in determining the global minimum configuration. This complex displays considerable internal dynamics, with both subunits, as shown in Figure 17, approximately rotating around their individual symmetries axes at

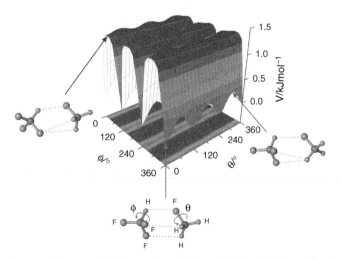

Figure 17 Shape and internal dynamics of the CHF_3—CH_3F molecular complex. (Reproduced with permission from Ref. [537]. Copyright 2005, Wiley-VCH Verlag GmbH & Co. KGaA.) (See color plate 28).

low barriers. The C—H...N interaction appears to be the dominating factor in establishing the conformation of Pyridine—CHF_3 [540], with the CHF_3 proton linked to the sp lone pair of pyridine. A CH...S contact was observed in the complex thiirane–trifluoromethane (C_2H_4S—HCF_3) [541]. Its equilibrium C_s symmetry is stabilized by one C—H...S and two C—H...F—C WHBs.

Any X—H...π contact investigated so far, although of a significant directionality, is characterized by a low interaction energy. Examples are found in the supersonic-jet FTMW spectra of the complexes of water, ammonia, HBr, and CHF_3 with benzene, which allow for a study of the O—H...π, N—H...π, Halogen-H...π, and C—H...π WHBs, respectively.

In benzene–water, a π O—H...π H-bond is formed between the water hydrogens and the benzene π system on one side of the ring [542]. In its spectrum, the complex appears as a pseudo-symmetric top, rationalized by a water molecule undergoing nearly free internal rotation with both hydrogen atoms pointing toward the π cloud, exchanging their role in a "one-leg" HB. Benzene–NH_3 behaves in a similar way, exhibiting an $m = 0$ torsional ground state, which – interpreted as a symmetric top state [543] – suggests a low barrier to internal rotation of the ammonia unit. The value of χ_{aa} (^{14}N) indicates that the C_3 axis of NH_3 forms an angle of $\phi = 57.6°$ with the benzene plane. Also benzene–HBr in the ground state has an effective C_{6v} symmetry [544]. The complex Bz—CHF_3, however, is a real symmetric top. Its rotational spectrum has been measured up to $J = 8$, $K = 8$, and $m = 2$ [545], and the dissociation energy was estimated to be 8.4 kJ/mol, which is in line with the values for the other three complexes.

An interesting and puzzling result has been obtained from the supersonic-jet FTMW spectrum of the benzene dimer, where – in a T-shaped arrangement – the first unit of benzene acts as proton donor, and the second as proton acceptor [546]. An approximate symmetric top model was adopted to interpret the central frequencies of quadruplet lines, arising from the internal motions of the complex. There is more work in progress on this system and a recent communication assigns the observed quadruplet splitting to the internal rotation of the "stem" benzene ring (H-donator), which is the motion exhibiting the higher barrier [547]. The splitting from the lower barrier motion, i.e., the "top" benzene ring (H-acceptor), is not assigned. The new interpretation is supported by the splitting pattern predicted by the "high-barrier tunneling matrix formalism for the "stem" rotation. There are unresolved discrepancies with the spin statistics though.

4.3. Charge Transfer Complexes

Two types of molecular complexes can be distinguished: those of the outer type B...XY (weak, no significant charge transfer) and those of the inner type $[BX]^+...Y^-$ (strong, substantial charge transfer). This represents the definition of "charge transfer complexes", although, Mulliken, who developed a detailed theory of such interactions [548], preferred to use the term "electron donor–acceptor complexes". In the realm of MW spectroscopy, Legon demonstrated the isolation of prereactive adducts of both the Mulliken outer and inner types in the gas phase by utilizing a supersonic coexpansion of the components, and their precise characterization by observation of their rotational spectra [35]. There, Legon has shown how the values of nuclear quadrupole coupling can be utilized as a probe of electric charge redistribution in a homonuclear dihalogen X_2 in $B \ldots X_2$ ($X = $ halogen). Within the series of complexes $(CH_3)_3N \ldots HX$, the estimated contributions of the ionic form $[(CH_3)NH]^+ \ldots X^-$ increases from $X = F$ to $X = I$; namely it is 5%, 62%, 80%, and 93% for $X = F$, Cl, Br, and I, respectively [549–553]. For $H_3N \ldots HCl$, $H_3N \ldots HBr$, and $H_3N \ldots HI$, the ionic character is zero, but for $CH_3NH_2 \ldots HCl$ it is found to be 23% [554].

Leopold and collaborators investigated several donor–acceptor complexes and wrote a review on partially bonded molecules [555], putting emphasis on interactions that lie in the intermediate regime *between* bonding and nonbonding, and on their implications on the atmospheric chemistry. Mostly complexes involving the BF_3 and SO_3 as Lewis' acids were investigated. Molecules having a terminal nitrogen atom with a lone pair were used as Lewis' bases. It was found that complexes between molecules with a terminal cyano group and SO_3, such as $HCN—SO_3$ and $CH_3CN—SO_3$ [556], have a moderate charge transfer and a moderate binding energy. These values increase considerably when the complexes are formed between SO_3 and molecules with aminic or iminic nitrogens, such as $H_3N—SO_3$ [557], $(CH_3)_3N—SO_3$ [558], and pyridine-SO_3 [559]. The corresponding binding energies and quotes of electron transfer are given in Table 5. The complexes with BF_3, $HCN—BF_3$ [560] and $CH_3CN—BF_3$ [561], follow the same trend and have moderate charge transfers and binding energies, while the corresponding values are higher for $H_3N—BF_3$ [562]. The effect of the charge transfer in complexes is also reflected in substantial deviations of dipole moment with respect to vectorial additivity of the monomer values [563]. The rotational spectrum of the trimer $H_3N—SO_3—H_2O$ [564] shows that the binding with a water molecule further increases the stability of a donor–acceptor complex.

Another very interesting system displaying charge transfer effects is solvated NaCl. Endo and collaborators investigated the supersonic-jet FTMW spectra of $NaCl—(H_2O)_n$, ($n = 1–3$) [565]. The clusters were formed in an adiabatic expansion of laser-vaporized NaCl with an Ar stream containing a trace amount of water. The observed lines have been assigned to two asymmetric tops ($n = 1$ and 2) and one symmetric top ($n = 3$). The $n = 2$ cluster was confirmed to have C_2 symmetry on the basis of the observed intensity alternation and lack of a-type transitions. Analyses with a rotational Hamiltonian including centrifugal distortion terms and hyperfine interactions yield rotational constants, which indicate ring forms of all three adducts

Table 5 Binding energies (kJ/mol) of some donor–acceptor complexes (From Ref. 555).

$NCCN—BF_3$	15
$HCN—BF_3$	19
$CH_3CN—BF_3$	24
$(CH_3)_2HN—SO_2$	43
$(CH_3)_3N—SO_2$	56
$H_3N—SO_3$	80
$H_3N—BF_3$	80
$(CH_3)_3N—BF_3$	138

NaCl-H₂O

NaCl-(H₂O)₂

NaCl-(H₂O)₃

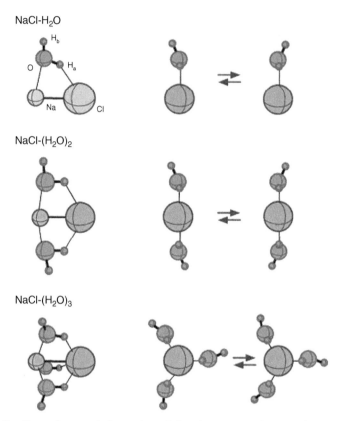

Figure 18 Configurations and dynamics of the charge transfer complexes NaCl-(H$_2$O)$_n$ ($n=1$–3), all observed by FTMW spectroscopy. Because of this dynamical removal of chirality, the vibrationally averaged structures belong to C_s, C_{2v}, and C_{3v} symmetries for $n=1$, 2, and 3, respectively. (Reprinted with permission from Ref. [565]. Copyright 2003, American Chemical Society.)

(Figure 18). A substantial increase in the r(Na—Cl) distance with the successive addition of water was observed. The distance in the $n=1$ cluster is longer by 6(2) pm with respect to the free NaCl. The increments per water unit addition for $n=2$ and 3 becomes much larger, yielding a 48(6) pm increase for $n=3$ compared to free NaCl. Both Na and Cl nuclei have 3/2 spins, causing nuclear quadrupole coupling observed in the spectra. The corresponding χ components along the Na—Cl axis have small negative values for both the Na and the Cl nuclei in free NaCl, representing the small covalent-bond character in the ionic molecule. They gradually increase with the number of attached water units, finally becoming positive for $n=3$. The changes indicate that the principal directions of the electron distribution in the atoms are tilted away from those in the free molecule because of polarization and/or charge transfer by the directly contacting O and H atoms of the water molecules. Such a charge rearrangement around Na and Cl is the initial step toward the complete charge separation in the ion pair during progressing hydration.

4.4. Molecular Recognition

Molecular recognition is the specific interaction between two or more molecules, which exhibit molecular complementarity, through noncovalent bonding such as hydrogen bonding, metal coordination, hydrophobic forces, van der Waals forces, π–π interactions, and/or electrostatic effects. Such a phenomenon is of crucial importance in the areas of molecular biology, crystal engineering, and supramolecular assembly [566]. Recently, high-resolution techniques have been dedicated to the study of model systems to facilitate a detailed understanding of the "lock and key" principle, enunciated by Fisher in 1894 [567], to explain the specificity of enzyme reactions.

Lately, molecular recognition gains importance in spectroscopic investigations [568]. Very recently, a special *issue of Physical Chemistry Chemical Physics* has been recently dedicated to "Spectroscopic probes of molecular recognition." In particular, MW spectroscopy, being very sensitive to the shape of a complex . . ., is significantly contributing to the spectroscopic efforts, e.g., in probing the conformational equilibria within chiral molecules or in studying the induced chirality upon formation of a complex from two non-chiral molecules. Two conformers, stabilized by a O—H . . . π interaction, have been identified by free-jet mm wave absorption spectroscopy for 1-phenyl-1-propanol [569], a chromophore chiral molecule often used for electronic spectroscopy investigations, while – with respect to a previous free-jet mm wave absorption study [305] – two additional conformers have been observed by supersonic-jet FTMW spectroscopy for the simplest amine neurotransmitter 2-phenylethylamine [570]. The ^{14}N nuclear quadrupole hyperfine structure, resolved for all four conformers, provides a conclusive test for their identification.

Supersonic-jet FTMW spectroscopy has been used to elucidate the diastereomeric interactions in 1:1 complexes of ethanol, a transient chiral alcohol, hydrogen-bonded to the achiral oxirane or *trans*-2,3-dimethyloxirane (DMO) with 2 stereocenters [571]. Two conformers of oxirane—ethanol and three conformers of DMO—ethanol have been identified, and their structures as well as their stability ordering were determined.

Techniques providing rotational resolution are the key to access precise information on the shape of homo-hetero chiral adducts. Indeed, a number of important results have been obtained by supersonic-jet FTMW spectroscopy. The rotational spectrum of a dimeric complex of a chiral molecule has been first reported for butan-2-ol [572]. Later on, also induced chiral dimers were observed for ethanol [573].

The rotational spectra of three homochiral and of three heterochiral dimers of propylene oxide (PO) have been observed [574]. Here, the different stereospecificity of the R and S forms lead to the formation of different RR (or SS) and RS (or SR) adducts, as shown for PO dimers in Figure 19. The rotational spectra of all six possible hydrogen-bonded PO—CH_3CH_2OH conformers are detected and analyzed [575]. The chiral discriminating forces at play were examined with complementary high-level ab initio calculations and a clear stability ordering of the conformers was extracted from the experiment.

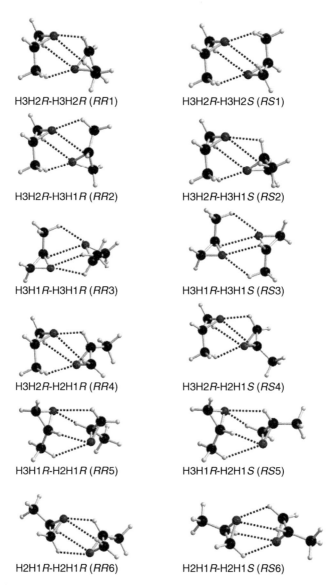

H3H2R-H3H2R (RR1)

H3H2R-H3H2S (RS1)

H3H2R-H3H1R (RR2)

H3H2R-H3H1S (RS2)

H3H1R-H3H1R (RR3)

H3H1R-H3H1S (RS3)

H3H2R-H2H1R (RR4)

H3H2R-H2H1S (RS4)

H3H1R-H2H1R (RR5)

H3H1R-H2H1S (RS5)

H2H1R-H2H1R (RR6)

H2H1R-H2H1S (RS6)

Figure 19 Ab initio optimized geometries of six homochiral (*RR*) and six heterochiral (*RS*) conformers of propylene oxide dimer. Six species (RR2 RR4 RR5 and RS2 RS4 RS5) have been identified in the FTMW spectrum. (Reprinted with permission from Ref. [574]. Copyright 2006, American Chemical Society.)

4.5. Molecular Aggregation

One of the fundamental goals of cluster research is the interpretation of the properties of condensed phases in terms of those of their constituent atoms and molecules [576]. However, the clusters that can be investigated in detail by MW spectroscopy are limited in size, i.e., the number n of constituent molecules. To date, the largest systems are those with $n = 3$ for oligomers and $n = 4$ for hetero-mers. Because of their special properties, we will discuss clusters formed prevalently by RG atoms in a separate section. Many of the molecular aggregates have been described in a recent review, "Microwave spectroscopy of ternary and quaternary van der Waals clusters", by Xu et al. [577]. For these studies, generally supersonic-jet FTMW spectroscopy was used, unless otherwise specified.

4.5.1. Oligomers
The results obtained by MW spectroscopy for some homo-dimers and homo-trimers are presented as a separate class of molecular adducts, because − in the limit of a very high number n of constituent molecules − a pure substance is formed. It is an interesting question, which value of n can be associated with a transition between microscopic and macroscopic behavior.

4.5.1.1. Dimers
Apart the water dimer and the hydrohalogenic acid dimers discussed above, much attention was dedicated to the dimers of some small molecules. Considerable effort was also devoted to the investigation of the dimers of some aromatic molecules.

The dimer of ammonia was first investigated by Klemperer and collaborators using MBER [578,579], determining $(B + C)$, $\chi_{aa}(N1)$, $\chi_{aa}(N2)$, and μ_a for several isotopologues of two nearby vibrational states, α and β. It was shown that the dimer does not have a "classical" N—H—N HB, rather the two subunits seek to have anti-parallel dipole moments, thus reducing the dipole–dipole interaction energy. Later on, Heineking et al. [580] assigned the α and β tunneling states as G states in the symmetry group G_{36}, which can be viewed either as states with one unit of angular momentum associated with one internal rotor (the *para* monomer) and zero internal angular momentum associated with the other rotor (the *ortho* mono-mer) or as two interchange tunneling partners. They derived expressions for the nuclear quadrupole splittings in the E_3 and E_4 (*para–para*) states of $(NH_3)_2$ and showed that these can be matched with the standard expressions for rigid rotors with two identical quadrupolar nuclei. In this way, χ_{aa} and $(\chi_{bb}-\chi_{cc})$ for the E_3 and E_4 were determined. Finally, Saykally and collaborators measured the mm/sub-mm wave spectrum of $(ND_3)_2$ using the EROS technique [581]. Similar to $(NH_3)_2$, the spectrum is complicated by the threefold internal rotation of the ND_3 subunits, the interchange tunneling of the two subunits, and the inversion of the subunits through their respective centers of masses. The transitions were assigned to rotation-tunneling states correlating with $A–A$ (*ortho–ortho*) combina-tions of the ND_3 monomer states, where A designates the rovibronic symmetries of the ND_3 subunits.

The MBER spectrum of $(NO)_2$ allowed for the determination of precise values of rotational, nuclear quadrupole coupling and spin–rotation constants of the ground electronic state treated as a singlet state [582]. From these data, $(NO)_2$ is interpreted to have a symmetric *cis*-planar structure. The FTMW spectra for $^{14}NO—^{14}NO$, $^{14}NO—^{15}NO$, and $^{15}NO—^{15}NO$ species allowed an unambiguous structure determination [583]. More recently, the measurements of the rotational spectrum of $(NO)_2$ have been extended to the mm wave region [584].

Also the CO dimer has been extensively investigated by rotational spectroscopy. The more recent investigation studies the $(^{12}C^{18}O)_2$ isotopologue utilizing an Orotron jet spectrometer [585]. The assigned six mm wave transitions belong to two branches which connect four lower levels of A^+ symmetry to three previously unknown levels of A^- symmetry. The discovery of the lowest state of A^- symmetry, which corresponds to the projection $K = 0$ of the total angular momentum J onto the intermolecular axis, identifies the geared bending mode of the $^{12}C^{18}O$ dimer at $3.607 \, cm^{-1}$.

The FTMW spectrum of the polar isomer of the OCS dimer, at higher energy than the nonpolar isomer of $(OCS)_2$ well known from IR spectroscopy, has been recently observed by Minei and Novick [586] exhibiting C_s symmetry. This polar isomer of $(OCS)_2$ has been produced by high pressure expansion of dilute OCS in helium and it has a surprisingly strong MW spectrum.

The FTMW spectrum of $(SO_2)_2$ indicates that the SO_2 dimer undergoes a high-barrier tunneling motion [587], identified as a geared interconversion motion similar to that displayed by $(H_2O)_2$. From the analysis of the moments of inertia of the various isotopic species, an ac plane of symmetry is established for the dimer and the tilt angles of the C_2 axes of each subunit with respect to the line joining their centers of mass were determined.

The formaldehyde dimer behaves similarly to the water dimer: its FTMW spectrum is split by large amplitude motions of the monomers, which interchange their roles of donor–acceptor bonding [588]. The planes of the two monomer units are perpendicular to each other, with the two CO groups in a near antiparallel orientation.

Three FTMW investigations of $(CH_3OH)_2$ were performed [589–591]. The data from the first two investigations confirm the theoretical prevision that each rotational level is split into 16 components, according to the 16 states generated by 25 distinct tunneling motions. From the internal-rotation analysis of the two inequivalent methyl groups, it was also found that the effective barrier to internal rotation for the donor and acceptor methyl groups, $V_3 = 183.0$ and $120 \, cm^{-1}$, respectively, are much lower than that of the methanol monomer ($370 \, cm^{-1}$). In the third study – employing PI group theory – the dimer is considered in terms of the G_{36} MS group. In addition, the lone-pair exchange tunneling splitting were determined experimentally.

Vice versa, no splittings have been observed in the FTMW spectrum of the dimethylamine dimer [592]. A cyclic structure with C_s symmetry was found to reproduce the inertial data most satisfactorily.

The dimers of two large aromatic molecules – apart from the dimer of benzene discussed above – were characterized by FTMW spectroscopy. In a study of the pyrrole dimer, the rotational constants are consistent with an essentially T-shaped structure [593]. The planes of the two pyrrole monomers form an angle of $55.4(4)°$ with the nitrogen side of one ring directed to the π electron system of the other ring, establishing a WHB. Only c-type transitions were found for the 1,2-difluorobenzene dimer, all of them split into two tunneling components separated by

110 kHz [594]. The ring planes are assumed to be parallel at a distance of 345 pm. Both rings are rotated by an angle of 130.3° against each other.

4.5.1.2. Trimers

To date, pure rotational spectra of homo-trimers are available only for two linear molecules, $(HCN)_3$ [595] and $(OCS)_3$ [596,597], and for one asymmetric top, $(CH_2F_2)_3$. [598].

The three trimers have completely different assembling motifs. $(HCN)_3$ has a rod-like structure – linear or very linear – with the three subunits held together by two strong HBs. The centers of mass of the two outer HCNs are far away by 879 pm. $(OCS)_3$ has the shape of a corral, with the three units acting as its poles. One of the three units is upside down with respect to the other two. Small splittings, arising from the inversion of one of the two equivalent "poles," are observed.

$(CH_2F_2)_3$ is the first trimer of an asymmetric rotor, where the pure rotational spectrum was reported. It has a cage structure, with the three monomers held together by nine C—H—F contacts [598], as shown in Figure 20. It exhibits nonzero coordinates for all three C-atoms, and all dipole moment components differ from zero. Therefore, it is a chiral complex, with the chirality induced by the formation of the adduct.

4.5.2. Mixed trimers and tetramers

Pioneering results on the investigation of molecular complexes larger than dimers have been obtained by Gutowsky and coworkers. Many of the studied trimers contain the linear dimer of HCN, which generally acts a proton donor to a third group. These trimers can be represented as X—HCN—HCN, where X = HF, HCl, HCF_3, CO_2, CO, N_2, NH_3, and H_2O [599,600]. Linear structures are observed for X = HF, HCl, CO, and N_2, a symmetric top structure was found for X = HCF_3 and NH_3, while HCN—HCN—CO_2 is T-shaped and also HCN—HCN—H_2O is an asymmetric top. Further contributions facilitated an understanding of the shape of

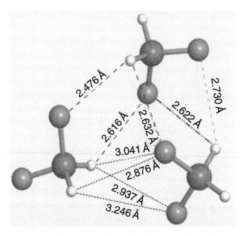

Figure 20 Nine CH...F weak hydrogen bonds characterize the structure of the $(CH_2F_2)_3$ trimer, as deduced from the rotational spectra of several isotopologues. (Reprinted with permission from Ref. [598]. Copyright 2007, American Chemical Society.)

H_2O—$(CO_2)_2$, where the two CO_2 subunits are not co-planar [601]. The first "all-hetero"-trimer studied was H_3N—HCN—HF [602].

The observation of the tetramer, HCN—$(CO_2)_3$, showed a spectrum with the features of a symmetric top [603]. Here, the $(CO_2)_3$ group has a pinwheel configuration, with each individual CO_2 tilted by ~37° with respect to the C—C—C plane.

Peterson et al. assigned the rotational spectra of the complexes $(H_2O)_2$—CO_2 [604] and H_2O—$(CO_2)_2$ [605]. Several trimers constituted of CO_2, OCS, N_2O, HCCH, and SO_2 were investigated by Kuczkowski and collaborators, i.e., $(CO_2)_2$—OCS [606], CO_2—$(OCS)_2$ [607], $(CO_2)_2$—N_2O [608], $(OCS)_2$—HCCH [609], and SO_2—$(N_2O)_2$ [610].

Leopold and coworkers found a strong dipole moment enhancement with respect to isolated HCN and HCN—SO_3 in forming HCN—HCN—SO_3 [611]. Interestingly, in the trimer, the N—S bond distance is 10 pm shorter than that in HCN—SO_3. The study of $(HF)_2$—NH_3 [612], which forms a six-membered HN-HF-HF ring, showed that both the linear hydrogen bond in the H_3N—HF moiety and the F—H—F angle of $(HF)_2$ are perturbed with respect to those in the corresponding dimers. The N—F and F—F distances in the trimer are 245.09(74) pm and 265.1(11) pm, respectively. In the complex, the NH_3 unit undergoes internal rotation, with a V_3 barrier of 118(2) cm^{-1}.

Kisiel and coworkers studied the complexes of the water dimer with hydrohalogenic acids, HCl—$(H_2O)_2$ [613] and HBr—$(H_2O)_2$ [614]. The two trimers are nearly planar, as evidenced by inertial defect, and the constituent units are bound in a triangular arrangement through O . . . HO and O . . . HX (X = halogen) HBs and a primarily dispersive OH . . . X bond; all atoms, with the exception of two nonbonded water hydrogens, lie near a common plane. All rotational transitions, in addition to the quadrupolar hyperfine structure from the halogen nuclei, are split into components belonging to four low-lying vibration-rotation-tunneling substates.

Bauder and collaborators observed the supersonic-jet FTMW spectra of several hydrogen-bonded complexes between formic acid and water [476]. Among them, the mixed trimers formic acid–(water)₂, $(H_2O)_2$—HCOOH, and (formic acid)₂–water, H_2O—$(HCOOH)_2$, were identified and characterized to have the shapes shown in Figure 21.

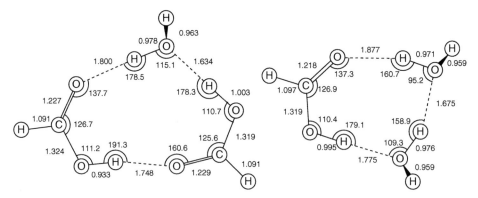

Figure 21 Molecular structure of the two observed trimers involving formic acid and water. Bond lengths are given in angstroms and angles in degrees [466]. (Reused with permission from [476]. Copyright 2000, American Institute of Physics.)

4.6. Quantum solvation

Because of the rather unusual observations in the course of solvation of a substrate molecule with RG atoms, the clusters formed prevalently by RG atoms are rather special. The effects encountered in systematic studies of such systems are described as quantum solvation.

4.6.1. Complexes involving two or more Ne and/or Ar rare gas atoms

Molecular complexes involving a number $n > 1$ of RG atoms can be viewed as model systems for solvation. The complexes with $RG = He$, since connected to superfluidity properties, are of particular interest and will be discussed in the following subsection.

4.6.1.1. Complexes RG_n with linear molecules

Complexes of two or more RG atoms with small linear molecules are the ones studied most. Several of them are formed from the same two RG atoms with a hydrohalogenic acid, such as Ar_2—HX (X = halogen or pseudo-halogen): Ar_2—HF [615], Ar_2—HCl [616], Ar_2—HBr [617], and Ar_2—HCN [618] trimers are found to be planar T-shaped asymmetric rotors, with the equilibrium position of the HX located on the C_{2v} axis and the H atom pointing toward the two Ar atoms. For Ar_2—HCl, a ~7° average displacement with respect to the T-shape is found. In Ar_2—HCN, the HCN axis rotates about the C_2 axis maintaining an angle of ~40° between the two axes for the $m = 0$ internal rotation state. Such internal rotation accounts at least qualitatively for the otherwise anomalous rotational behavior of the Ar_2—HCN cluster. Three complexes with a higher number of Ar atoms have also been studied: Ar_3—HF [619], Ar_3—HCN [620], and Ar_4—HF [621]. All of them have a K-fine structure limited to $K = 0$, ± 3, and ± 6, indicative of symmetric tops with the threefold axis of symmetry generated by $I = 0$ nuclei, e.g., ^{40}Ar. The former are oblate and the latter prolate rotors. The HF or HCN unit lies along the threefold axis of the Ar_3 group, with the H end being closest to Ar_3. Ar_3—HCN displays a long list of "abnormalities," including a J-dependent nuclear quadrupole coupling constant $\chi(^{14}N)$; a very large projection angle θ; large centrifugal distortion including higher-order terms in H_J and H_{JK}; splitting of the $K = 3$ transitions into J-dependent doublets; and the ready observation of an excited vibrational state.

Ar_4—HF is the largest investigated complex of this type. The cluster consists of the trigonal Ar_3—HF tetramer discussed above, with the fourth argon on the back side of the Ar_3 group to form a tetrahedral or near tetrahedral Ar_4 subunit.

Several MW spectra of complexes of Ne_2 and Ar_2 with triatomic linear molecules such as OCS, CO_2, and N_2O have been reported. The ternary clusters Ne_2—OCS [622] and Ar_2—OCS [623] have asymmetric top rotational spectra. The rotational constants of several isotopologues are consistent with distorted tetrahedral shapes, placing the O and S atoms outside the RG—RG—C plane. Ne_2—N_2O and Ar_2—N_2O behave similarly [624]. Both trimers were found to have distorted tetrahedral structures with the RG position tilted toward the O atom of the N_2O subunit. Not surprisingly, the MW spectrum of Ar_2—CO_2 indicates a C_{2v} distorted tetrahedral shape [625].

Three ternary clusters formed by one Ne, one Ar, and one linear molecule units were investigated by supersonic-jet FTMW spectroscopy. Their studies are of considerable importance for the understanding of the correction terms proposed in investigations of nonadditive contributions in weakly bound molecular systems. One of the studied trimers is NeArHCl [626], a symmetric top where the hydrogen atom is located on the triangle being defined by the three heavy atoms. The nuclear quadrupole hyperfine structures due to the ^{35}Cl, ^{37}Cl, and D nuclei were observed and assigned. The resulting nuclear quadrupole coupling constants provide information on the angular anisotropy of the NeArHCl PES. NeArN$_2$O [627] and NeArCO$_2$ [628] are the two other "all-hetero" trimers investigated by MW spectroscopy. Similarly to the cases of RG$_2$N$_2$O and RG$_2$CO$_2$, the linear molecules are almost perpendicular to the triangle containing the two RG atoms and the central atom of the linear molecule.

4.6.1.2. Complexes RG$_n$ with small nonlinear molecules.

The supersonic-jet FTMW spectra of complexes of two or three RG atoms with the symmetric top NH$_3$ and with two small asymmetric tops, H$_2$O and H$_2$S, have been reported. The Ne$_2$—NH$_3$ [629] and Ar$_2$—NH$_3$ [630] van der Waals trimers and the Ne$_3$—NH$_3$ [631] and Ar$_3$—NH$_3$ [632] van der Waals tetramers were studied by van Wijngaarden and Jäger. For all of them, the ^{14}N nuclear quadrupole coupling constants was found to be sensitive to the orientation and the large amplitude internal rotation of the NH$_3$ moiety. The trimers are asymmetric tops, with the minimum energy orientation of NH$_3$ corresponding to a geometry in which the C$_3$-axis of NH$_3$ is aligned perpendicular to the Ne—Ne or Ar—Ar axis with two of the hydrogen atoms pointing toward the RG atoms. The normal tetramers are symmetric tops; the ^{20}Ne$_2$^{22}Ne- and ^{20}Ne ^{22}Ne$_2$-isotopologues, however, are asymmetric tops. All deuterated and partially deuterated species exhibit splittings due to the inversion of ammonia, while this tunneling splitting was not observed for the NH$_3$ and ^{15}NH$_3$ isotopologues because of their nuclear spin statistics.

The MW spectrum of the Ar$_2$—H$_2$O trimer agrees with a planar T-shaped structure with C$_{2v}$ symmetry and the bidentate protons point at the argons [633]. Two sets of asymmetric top transitions were found, corresponding to two internal rotor states of the water. The Ar$_3$—H$_2$O and Ar$_3$—H$_2$S tetramers were both found to be symmetric tops, the former being oblate and the latter being prolate [634]. The angle between the C$_2$ axis of the H$_2$X unit and the C$_3$ axis of the heavy atoms is estimated to be 74° for H$_2$O and 13° for the H$_2$S complex, respectively.

4.6.1.3. Complexes of Ar$_2$ with large aromatic molecules

The complexes Ar$_2$–furan and Ar$_2$–pyridine have been investigated by supersonic-jet FTMW spectroscopy [410,635]. Both conformers have C$_{2v}$ symmetry placing the argon atoms on both sides of the molecular plane while the C$_2$ axes coincide with the C$_2$ axes of the isolated molecule.

4.6.2. Complexes with He and He nanodroplets

The complexes with He are of particular interest because helium can exhibit superfluidity properties. The binding energies of He atoms in a complex are very small, some tens of a wavenumber. This makes the MW observation of its complexes very difficult: the jet experiments require strong cooling conditions as well as the spectra turn out to be rather complicated from extremely large amplitude motions within multi-minima PESs.

In the area of superfluidity, the recent development of He nanodroplet isolation spectroscopy [636] helps to close the gap between cluster and bulk studies. It offers possibilities for synthesizing, stabilizing, and characterizing novel chemical species [637] and also constitutes an important step toward detailed microscopic understanding of superfluidity, a collective bulk property. In an elegant nanomatrix study, the microscopic Andronikashvili experiment, Grebenev and coworkers [638] used the appearance of sharp infrared (IR) spectral features of dopant molecules in He nanodroplets (consisting of several thousand He atoms) as an indicator of the onset of the superfluidity.

Impressive MW studies of molecules solvated by He were performed by Jäger and collaborators, mostly using carbonyl sulfide, OCS, as the dopant molecule. The 1:1 complex He—OCS was investigated by Higgins and Klemperer [379], while Xu and Jäger studied the rotational spectrum of He_2—OCS [639]. Still Xu and Jäger [640] and Tang et al. [641] reported the rotational spectra for He_n—OCS, with n from 2 to 8. In Figure 22 the shape of the complex with eight He atoms is shown. Finally, the measurements of rotational spectra have been extended up to

Figure 22 Shape of the cluster OCS—He_8 . (Reprinted with permission from Ref. [641]. Copyright 2003, by Science.) (See color plate 29).

$n = 39$ [642]. The observed rotational constants B, being reciprocal to the moment of inertia, pass through a minimum at $n = 9$, then rises due to onset of superfluid effects, and exhibits broad oscillations with maxima at $n = 24$, 47 and minima at 36, 62. These unexpected oscillations are interpreted as a manifestation of the *aufbau* of a nonclassical helium solvation shell structure. These results bridge an important part of the gap between individual molecules and bulk matter with atom by atom resolution, providing new insight into microscopic superfluidity and a critical challenge for theory. The experimental rotational constants B are plotted in Figure 23 as a function of cluster size n. The dashed horizontal line indicates the limiting value for larger He nanodroplets [639].

Another important doping molecule is N_2O. First, the rotational spectra of He_n—N_2O, with n from 3 to 12 [643] and of the complex He—N_2O [381] have been reported. In a following paper, extending the observed species He_n—N_2O to $n = 19$, experimental evidence of recurrences in the rotational dynamics of doped helium clusters have been presented by Xu et al. [644]. The interpretation of the evolution of B based on the superfluid response suggests that a molecular dopant can be used as a quantitative experimental probe of superfluidity at the microscopic level.

Finally, also of importance in the understanding superfluidity, studies on the $(orthoH_2)_n$—HCCCN and the $(paraH_2)_n$—HCCCN van der Waals clusters are interesting. $ParaH_2$ is suspected, indeed, to exhibit the bulk properties of superfluidity, analogously to He [645].

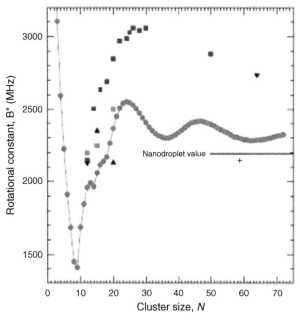

Figure 23 Variation of the rotational constant B with the size of clusters OCS—He_n. (Reprinted figure with permission from Ref. [642]. Copyright 2006, by the American Physical Society.)

ACKNOWLEDGMENTS

All microwavers are acknowledged for providing (p)reprints of their work and thus made this task simpler. W.C. is indebted to his former collaborators and teachers P. G. Favero, E. B. Wilson, A. Bauder, R. Meyer, and to the members of his group, S. Melandri, A. Maris, B. Velino, L. B. Favero, P. Ottaviani, and B. M. Giuliano. The help of Z. Kisiel, J. L. Alonso, J. C. Lopez, and A. Lesarri is gratefully acknowledged. J.-U.G. is thankful to G. T. Fraser, J. T. Hougen, W. J. Lafferty, F. J. Lovas, A. S. Pine, R. D. Suenram, A. R. Hight-Walker, and S. E. Novick.

REFERENCES

[1] C. H. Townes and A. L. Schawlow, "Microwave Spectroscopy", McGraw-Hill, New York, 1955.

[2] W. Gordy and L. R. Cook, "Microwave Molecular Spectra", 3rd Edition, in Weissberger, A. (ED.), Techniques of Chemistry, vol. XVIII, John Wiley & Sons Inc., New York, 1984.

[3] T. M. Sugden and C. N. Kenney, "Microwave Spectroscopy of gases", Van Nostrand, London, 1965.

[4] H. W. Kroto, "Molecular Rotation Spectra", John Wiley & Sons Inc., New York, 1975.

[5] J. M. Brown and A. Carrington, "Rotational Spectroscopy of Diatomic Molecules", Cambridge University Press, Cambridge, 2003.

[6] J. E. Wollrab, "Rotational Spectra and Molecular Structure", Academic press, New York, 1967.

[7] D. G. Lister, J. N. Macdonald and N. L. Owen, "Internal Rotation and Inversion: An Introduction to Large Amplitude Motions in Molecules", Academic Press, New York, 1978.

[8] P. R. Bunker and P. Jensen, "Molecular Symmetry and Spectroscopy", NRC Research Press, Ottawa, 1998.

[9] H. M. Pickett, J. Mol. Spectrosc., 148: 371–377, 1991.

[10] http://spec.jpl.nasa.gov/

[11] http://info.ifpan.edu.pl/~kisiel/prospe.htm

[12] http://www.ph1.uni-koeln.de/vorhersagen/

[13] N. Ohashi and J. T. Hougen, J. Mol. Spectrosc., 203: 170–174, 2000.

[14] J. T. Hougen, I. Kleiner and M. Gotefroid, J. Mol. Spectrosc., 163: 559–586, 1994.

[15] P. Groner, J. Chem. Phys., 107: 4483–4498, 1997.

[16] D. F. Plusquellic, R. D. Suenram, B. Mate, J. O. Jensen and A. C. Samuels, J. Chem. Phys., 115: 3057–3067, 2001.

[17] http://physics.nist.gov/Divisions/Div844/facilities/uvs/jb95userguide.htm

[18] Program IAMCALC, Available at http://spec.jpl.nasa.gov/

[19] N. Ohashi and J. T. Hougen, J. Mol. Spectrosc., 170: 493–505, 1995.

[20] D. Christen, L. H. Coudert, R. D. Suenram and F. J. Lovas, J. Mol. Spectrosc., 172: 57–77, 1995.

[21] Gaussian 03, Revision B.02, M. J. Frisch, G. W. Trucks, H. B. Schlegel, G. E. Scuseria, M. A. Robb, J. R. Cheeseman, J. A. Montgomery, Jr., T. Vreven, K. N. Kudin, J. C. Burant, J. M. Millam, S. S. Iyengar, J. Tomasi, V. Barone, B. Mennucci, M. Cossi, G. Scalmani, N. Rega, G. A. Petersson, H. Nakatsuji, M. Hada, M. Ehara, K. Toyota, R. Fukuda, J. Hasegawa, M. Ishida, T. Nakajima, Y. Honda, O. Kitao, H. Nakai, M. Klene, X. Li, J. E. Knox, H. P. Hratchian, J. B. Cross, C. Adamo, J. Jaramillo, R. Gomperts, R. E. Stratmann, O. Yazyev, A. J. Austin, R. Cammi, C. Pomelli, J. W. Ochterski, P. Y. Ayala, K. Morokuma, G. A. Voth, P. Salvador, J. J. Dannenberg, V. G. Zakrzewski, S. Dapprich, A. D. Daniels, M. C. Strain, O. Farkas,

D. K. Malick, A. D. Rabuck, K. Raghavachari, J. B. Foresman, J. V. Ortiz, Q. Cui, A. G. Baboul, S. Clifford, J. Cioslowski, B. B. Stefanov, G. Liu, A. Liashenko, P. Piskorz, I. Komaromi, R. L. Martin, D. J. Fox, T. Keith, M. A. Al-Laham, C. Y. Peng, A. Nanayakkara, M. Challacombe, P. M. W. Gill, B. Johnson, W. Chen, M. W. Wong, C. Gonzalez and J. A. Pople, Gaussian, Inc., Pittsburgh PA, 2003.

[22] M. W. Schmidt, K. K. Baldridge, J. A. Boatz, S. T. Elbert, M. S. Gordon, J. J. Jensen, S. Koseki, N. Matsunaga, K. A. Nguyen, S. Su, T. L. Windus, M. Dupuis, J. A. Montgomery, J. Comput. Chem., 14: 1347–1363, 1993; M. S. Gordon and M. W. Schmidt, "Advances in Electronic Structure Theory: GAMESS a Decade Later", in "Theory and Applications of Computational Chemistry, the first forty years", Dykstra, C. E., Frenking, G., Kim, K. S., Scuseria G. E. (Eds), Elsevier, Amsterdam, 2005, pp. 1167–1189.

[23] A. J. Stone, J. Chem Theory Comput., 1: 1128–1132, 2005.

[24] A. J. Stone, A. Dullweber, O. Engkvist, E. Fraschini, M. P. Hodges, A. W. Meredith, D. R. Nutt, P. L. A. Popelier and D. J. Wales, Orient: a program for studying interactions between molecules, version 4.5, University of Cambridge, 2002. Enquiries to A. J. Stone, ajs1@cam.ac.uk.

[25] http://www-stone.ch.cam.ac.uk/documentation/

[26] Z. Kisiel, J. Phys. Chem., 95: 7605–7612, 1991.

[27] Z. Kisiel, P. W. Fowler and A. C. Legon, J. Chem. Phys., 95: 2283–2291, 1991.

[28] A. D. Buckingham and P. W. Fowler, Can. J. Chem. 63: 2018–2025, 1985.

[29] Z. Kisiel, P. W. Fowler and A. C. Legon, J. Chem. Phys., 93: 3054–3062, 1990; J. Chem. Phys., 93: 6249–6255, 1990; J. Chem. Phys., 101: 4635–4643, 1994.

[30] R. Meyer, J. Mol. Spectrosc., 76: 266–300, 1979.

[31] R. Meyer and W. Caminati, J. Mol. Spectrosc., 150: 229–237, 1991.

[32] M. C. McCarthy and P. Thaddeus, Chem. Soc. Rev., 30: 177–185, 2001.

[33] F. J. Lovas, J. Phys. Chem. Ref. Data, 33: 137–355, 2004.

[34] A. Bauder, J. Mol. Struct., 408/409: 33–37, 1997.

[35] A. C. Legon, Angew. Chem. Int. Ed. Engl., 38: 2686–2714, 1999.

[36] Y. Xu, J. van Wijngaarden and W. Jaeger, Int. Rev. Phys. Chem., 24: 301, 2005.

[37] E. Arunan, S. Dev and P. K. Mandal, Appl. Spectrosc. Rev., 39: 131–181, 2004.

[38] http://www.wesleyan.edu/chem/faculty/novick/vdw.html

[39] E. Hirota, J. Chem. Phys., 37: 283–291, 1962.

[40] W. Caminati, J. Mol. Spectrosc., 92: 101–116, 1982.

[41] S. Melandri, P. G. Favero, D. Damiani, W. Caminati and L. B. Favero, J. Chem. Soc. Faraday Trans. 90: 2183–2188, 1994.

[42] G. T. Fraser, R. D. Suenram and C. L. Lugez, J. Phys. Chem. A, 104: 1141–1146, 2000.

[43] J. M. Fisher, L.-H. Xu, R.D Suenram, B. Pate and K. Douglass, J. Mol. Struct., 795: 143–154, 2006.

[44] G. T. Fraser, R. D. Suenram and C. L. Lugez, J. Phys. Chem. A, 105: 9859–9864, 2001.

[45] S. Blanco, J. C. López, A. Lesarri, W. Caminati and J. L. Alonso, Mol. Phys., 103: 1473–1479, 2005.

[46] Z. Kisiel, O. Dorosh, A. Maeda, F. C. De Lucia and E. Herbst, 61st Ohio State University International Conference on Molecular Spectroscopy, Columbus, Ohio, June 2006, Comm. RI14 and 62nd Ohio State University International Conference on Molecular Spectroscopy, Columbus, Ohio, June 2007, Comm. WG06.

[47] S. L. Baughcum and Z. Smith, E. B. Wilson and R. W. Duerst, J. Am. Chem. Soc., 106: 2260–2265, 1984, and references therein.

[48] W. Caminati and J.-U. Grabow, J. Am. Chem. Soc., 128: 854–858, 2006.

[49] L. D. Hatherley, R. D. Brown, P. D. Godfrey, A. P. Pierlot, W. Caminati, D. Damiani, S. Melandri and L. B. Favero, J. Phys. Chem., 97: 46–51, 1993.

[50] R. Sanchez, B. M. Giuliano, S. Melandri, L. B. Favero and W. Caminati, J. Am. Chem. Soc., 129: 6287–6290, 2007.

[51] C. E. Cleeton and N. H. Williams, Phys. Rev., 45: 234–237, 1934.

[52] W. Caminati, G. Cazzoli and D. Troiano, Chem. Phys. Lett., 43: 65–68, 1976.

[53] W. Caminati and F. Scappini, J. Mol. Spectrosc., 117: 184–194, 1986.

[54] L. H. Xu, R. M. Lees and J. T. Hougen, J. Chem. Phys., 110: 3835–3841, 1999.

[55] I. Kleiner, J. T. Hougen, J.-U. Grabow, S. P. Belov, M. Y. Tretyakov and J. Cosléou, J. Mol. Spectrosc., 179: 41–60, 1996.

[56] L. H. Xu and J. T. Hougen, J. Mol. Spectrosc., 173: 540–551, 1995.

[57] P. Groner, E. Herbst, F. C. De Lucia, B. J. Drouin and H. Mäder, J. Mol. Struct., 795: 173–178, 2006.

[58] N. Ohashi, J. T. Hougen, R. D. Suenram, F. J. Lovas, Y. Kawashima, M. Fujitake and J. Pyka, J. Mol. Spectrosc., 227: 28–42, 2004.

[59] M. Fujitake, Y. Kubota and N. Ohashi, J. Mol. Spectrosc., 236: 97–109, 2006.

[60] M. Schnell and J.-U. Grabow, Angew. Chem. Int. Ed. Engl., 45: 3465–3470, 2006.

[61] M. Schnell and J.-U. Grabow, Phys. Chem. Chem. Phys., 8: 2225–2231, 2006.

[62] M. Schnell and J.-U. Grabow, Chem. Phys., 343: 121–128, 2008.

[63] M. Schnell, J. T. Hougen and J.-U. Grabow, J. Mol. Spectrosc., Available online 17 January 2008.

[64] Y.-C. Chou and J. T. Hougen, J. Chem. Phys., 124: 074319, 2006.

[65] N. D. Sanders, J. Mol. Spectrosc., 86: 27–42, 1981.

[66] W. Caminati and R. Meyer, J. Mol. Spectrosc., 90: 303–314, 1981.

[67] A. C. Fantoni, W. Caminati and R. Meyer, Chem. Phys. Lett., 133: 27–33, 1987.

[68] E. Mathier, A. Bauder and Hs. H. Günthard, J. Mol. Spetrosc., 37: 63–76, 1971.

[69] R. Sanchez, B. M. Giuliano, S. Melandri and W. Caminati, Chem. Phys. Lett., 425: 6–9, 2006.

[70] J. J. Oh, B. J. Drouin and E. A. Cohen, J. Mol. Spetrosc., 234: 10–24, 2005.

[71] M. Quack, Angew. Chem. Int. Ed. Engl., 44: 3623–3626, 2005.

[72] J. C. Pearson, K. V. L. N. Sastry, E. Herbst and F. C. De Lucia, J. Mol. Spetrosc., 175: 246–261, 1996.

[73] S. Melandri, P. G. Favero and W. Caminati, Chem. Phys. Lett., 223: 541–545, 1994.

[74] E. Hirota and Y. Kawashima, J. Mol. Spetrosc., 207: 243–253, 2001.

[75] J.-U. Grabow, A. M. Andrews, G. T. Fraser, K. K. Irikura, R. D. Suenram, F. J. Lovas, W. J. Lafferty and J. L. Domenech, J. Chem. Phys., 105: 7249–7262, 1996.

[76] B. Vogelsanger, W. Caminati and A. Bauder, Chem. Phys. Lett., 141: 245–250, 1987.

[77] W. Caminati, B. Vogelsanger, R. Meyer, G. Grassi and A. Bauder, J. Mol. Spectrosc., 131: 172–184, 1988.

[78] J. C. López, J. L. Alonso, M. E. Charro, G. Wlodarczak and J. Demaison, J. Mol. Spetrosc., 155: 143–157, 1992.

[79] T. L. Smithson, J. A. Duckett and H. Wieser, J. Phys. Chem., 88: 1102–1109, 1984; K. H. Hassan and J. M. Hollas, J. Mol. Spectrosc., 147: 100–113, 1991.

[80] W. Caminati, D. Damiani, G. Corbelli and L. B. Favero, Mol. Phys., 75: 857–865, 1992.

[81] W. Caminati, L. B. Favero, B. Velino and F. Zerbetto, Mol. Phys., 78: 1561–1574, 1993.

[82] W. Caminati, D. Damiani and L. B. Favero, Mol. Phys., 76: 699–708, 1993.

[83] W. Caminati, S. Melandri, G. Corbelli, L. B. Favero and R. Meyer, Mol. Phys., 80: 1297–1315, 1993.

[84] P. Ottaviani and W. Caminati, Chem. Phys. Lett., 405: 68–72, 2005.

[85] Z. Kisiel, L. Pszczółkowski, G. Pietraperzia, M. Becucci, W. Caminati and R. Meyer, Phys. Chem. Chem. Phys., 6: 5469–5475, 2004.

[86] J. Dommen, Th. Brupbacher, G. Grassi and A. Bauder, J. Am. Chem. Soc., 112: 953–957, 1990.

[87] R. Meyer, J. C. López, J. L. Alonso, S. Melandri, P. G. Favero and W. Caminati, J. Chem. Phys., 111: 7871–7880, 1999.

[88] D. G. Melnik, S. Gopalakrishnan, T. A. Miller and F. C. De Lucia, J. Chem. Phys., 118: 3589–3599, 2003.

[89] D. G. Melnik, T. A. Miller and F. C. De Lucia, J. Mol. Spectrosc., 221: 227–238, 2003.

[90] W. Caminati, S. Melandri, A. Maris, J. C. López, J. L. Alonso and R. Meyer, 60th Ohio State University International Conference on Molecular Spectroscopy, Columbus, Ohio, June 2005, Comm. RC09.

[91] Z. Kisiel, L. Pszczólkowski, W. Caminati and P. G. Favero, J. Chem. Phys., 105: 1778–1785, 1996.
[92] Z. Kisiel, L. Pszczólkowski, L. B. Favero and W. Caminati, J. Mol. Spectrosc., 189: 283–290, 1998.
[93] L. Dore and Z. Kisiel, J. Mol. Spectrosc., 189: 228–234, 1998.
[94] R. Trambarulo, S. N. Ghosh, C. A. Burrus, Jr. and W. Gordy, J. Chem. Phys., 21: 851–855, 1953.
[95] M. C. McCarthy, S. Thorwirth, C. A. Gottlieb and P. Thaddeus, J. Am. Chem. Soc., 126: 4096–4097, 2004.
[96] M. C. McCarthy, S. Thorwirth, C. A. Gottlieb and P. Thaddeus, J. Chem. Phys., 121: 632–635, 2004.
[97] M. C. McCarthy and P. Thaddeus, Phys. Rev. Lett., 90: 2130031997.
[98] M. C. McCarthy, A. J. Apponi and P. Thaddeus, J. Chem. Phys., 110: 10645–10648, 1999.
[99] A. J. Apponi, M. C. McCarthy, C. A. Gottlieb and P. Thaddeus, J. Chem. Phys., 111: 3911–3918, 1999.
[100] M. C. McCarthy, A. J. Apponi and P. Thaddeus, J. Chem. Phys., 111: 7175–7178, 1999.
[101] M. A. Brewster and L. M. Ziurys, Astrophys. J., 559: L163–L166, 2001.
[102] L. Bizzocchi, B. M. Giuliano, M. Hess and Jens-Uwe Grabow, J. Chem. Phys., 126: 114305, 2007.
[103] B. M. Giuliano, L. Bizzocchi, S. Cooke, D. Banser, M. Hess, J. Fritzsche and Jens-Uwe Grabow, Phys. Chem. Chem. Phys., 10: 2078–2088, 2008.
[104] C. J. Evans and M. C. L. Gerry, J. Am. Chem. Soc., 122: 1560–1561, 2000.
[105] L. Bizzocchi, B. M. Giuliano and J.-U. Grabow, J. Mol. Struct., 833: 175–183, 2007, and references therein.
[106] S. Biermann, J. Hoeft, T. Törring, R. Mawhorter, F. J. Lovas, R. D. Suenram, Y. Kawashima and E. Hirota, J. Chem. Phys., 105: 9754–9761, 1996.
[107] D. B. Grotjahn, M. A. Brewster and L. M. Ziurys, J. Am. Chem. Soc., 124: 5895–5901, 2002.
[108] J. S. Robinson, A. J. Apponi and L. M. Ziurys, Chem. Phys. Letters, 278: 1–8, 1997.
[109] K. A. Walker and M. C. L. Gerry, J. Mol. Spectrosc., 182: 178–183, 1997.
[110] D. T. Halfen, A. J. Apponi, J. M. Thompsen and L. M. Ziurys, J. Chem. Phys., 115: 1131–1138, 2001.
[111] M. Bogey, H. Bolvin, C. Demuynck and J.-L. Destombes, Phys. Rev. Lett., 66: 413–416, 1991.
[112] M. Cordonnier, M. Bogey, C. Demuynck and J.-L. Destombes, J. Chem. Phys., 97: 7984–7989, 1992.
[113] M. Bogey, H. Bolvin, M. Cordonnier, C. Demuynck, J. L. Destombes and A. G. Császár, J. Chem. Phys., 100: 8614–8624, 1994.
[114] M. C. McCarthy and P. Thaddeus, J. Mol. Spectrosc., 222: 248–254, 2003.
[115] M. C. McCarthy, Z. Yu, L. Sarib, H. F. Schaefer, III and P. Thaddeus, J. Chem. Phys., 124: 074303, 2006.
[116] C. J. Whitham, H. Ozeki and S. Saito, J. Chem. Phys., 112: 641–646, 2000.
[117] J. Xin, M. A. Brewster and L. M. Ziurys, Astrophys. J., 530: 323–328, 2000.
[118] D. B. Grotjahn, P. M. Sheridan, I. Al Jihad and L. M. Ziurys, J. Am. Chem. Soc., 123: 5489–5494, 2001.
[119] A. Janczyk, D. L. Lichtenberger and L. M. Ziurys, J. Am. Chem. Soc., 128: 1109–1118, 2006.
[120] A. Janczyk and L. M. Ziurys, Chem. Phys. Lett., 365: 514–524, 2002.
[121] A. Janczyk, S. K. Walter and L. M. Ziurys, Chem. Phys. Lett., 401: 211–216, 2005.
[122] J. S. Robinson and L. M. Ziurys, Astrophys. J., 472: L131–L134, 1996.
[123] D. B. Grotjahn, T. C. Pesch, J. Xin and L. M. Ziurys, J. Am. Chem. Soc., 119: 12368–12369, 1997.
[124] D. B. Grotjahn, T. C. Pesch, M. A. Brewster and L. M. Ziurys, J. Am. Chem. Soc., 122: 4735–4741, 2000.
[125] D. B. Grotjahn, D. T. Halfen, L. M. Ziurys and A. L. Cooksy, J. Am. Chem. Soc., 126: 12621–12627, 2004.
[126] A. J. Apponi, M. A. Brewster and L. M. Ziurys, Chem. Phys. Lett., 298: 161–169, 1998.

[127] M. A. Brewster, A. J. Apponi, J. Xin and L. M. Ziurys, Chem. Phys. Lett., 310: 411–422, 1999.
[128] N. R. Walker, J. K.-H. Hui and M. C. L. Gerry, J. Phys. Chem. A, 106: 5803–5808, 2002.
[129] C. J. Evans and M. C. L. Gerry, J. Chem. Phys., 112: 1321–1329, 2000.
[130] C. J. Evans, A. Lesarri and M. C. L. Gerry, J. Am. Chem. Soc., 122: 6100–6105, 2000.
[131] S. A. Cooke and M. C. L. Gerry, J. Am. Chem. Soc., 126: 17000–17008, 2004.
[132] J. M. Michaud and M. C. L. Gerry, J. Am. Chem. Soc., 128: 7613–7621, 2006.
[133] P. Thaddeus, M. C. McCarthy, M. J. Travers, C. A. Gottlieb and W. Chen, Faraday Discuss., 109: 121–135, 1998.
[134] M. C. McCarthy, W. Chen, M. J. Travers and P. Thaddeus, Astrophys. J. Suppl. Ser., 129: 611–623, 2000.
[135] M. C. McCarthy, M. J. Travers, A. Kovács, W. Chen, S. E. Novick, C. A. Gottlieb and P. Thaddeus, Science, 275: 518–520, 1997.
[136] A. J. Apponi, M. C. McCarthy, C. A. Gottlieb and P. Thaddeus, Astrophys. J., 530: 357–361, 2000.
[137] M. J. Travers, M. C. McCarthy, C. A. Gottlieb and P. Thaddeus, Astrophys. J., 483: L135–L138, 1997.
[138] C. A. Gottlieb, M. C. McCarthy, V. D. Gordon, J. M. Chakan, A. J. Apponi and P. Thaddeus, Astrophys. J., 509: L141–L144, 1998.
[139] M. C. McCarthy, M. J. Travers, W. Chen, C. A. Gottlieb and P. Thaddeus, Astrophys. J., 498: L89–L92, 1998.
[140] M. C. McCarthy and P. Thaddeus, Astrophys. J., 569: L55–L58, 2002.
[141] S. Thorwirth, M. C. McCarthy, J. B. Dudek and P. Thaddeus, J. Chem. Phys., 122: 184308, 2005.
[142] M. C. McCarthy, J.-U. Grabow, M. J. Travers, W. Chen, C. A. Gottlieb and P. Thaddeus, Astrophys. J., 494: L231–L234, 1998.
[143] M. C. McCarthy, J.-U. Grabow, M. J. Travers, W. Chen, C. A. Gottlieb and P. Thaddeus, Astrophys. J., 513: 305–310, 1999.
[144] P. Botschwina, Ä Heyl, W. Chen, M. C. McCarthy, J.-U. Grabow, M. J. Travers and P. Thaddeus, J. Chem. Phys., 109: 3108–3115, 1998.
[145] V. D. Gordon, E. S. Nathan, A. J. Apponi, M. C. McCarthy and P. Thaddeus, J. Chem. Phys., 113: 5311–5320, 2000.
[146] M. C. McCarthy, A. J. Apponi, C. A. Gottlieb and P. Thaddeus, Astrophys. J., 538: 766–772, 2000.
[147] V. D. Gordon, M. C. McCarthy, A. J. Apponi and P. Thaddeus, Astrophys. J. Suppl. Ser., 134: 311–317, 2001.
[148] V. D. Gordon, M. C. McCarthy, A. J. Apponi and P. Thaddeus, Astrophys. J. Suppl. Ser., 138: 297–303, 2002.
[149] L. Bizzocchi, C. Degli Esposti and P. Botschwina, J. Chem. Phys., 119: 170–175, 2003.
[150] L. Kang and S. E. Novick, J. Phys. Chem. A, 106: 3749–3753, 2002.
[151] S. Blanco, J. C. López, M. E. Sanz, A. Lesarri, H. Dreizler and J. L. Alonso, J. Mol. Spectrosc., 227: 202–205, 2004.
[152] P. Botschwina, M. E. Sanz, M. C. McCarthy and P. Thaddeus, J. Chem. Phys., 116: 10719–10729, 2002.
[153] M. E. Sanz, J. L. Alonso, S. Blanco, A. Lesarri and J. C. López, Astrophys. J., 621: L157–L159, 2005.
[154] J. Preusser and M. C. L. Gerry, J. Chem. Phys., 106: 1037–1047, 1997.
[155] I. K. Ahmad, H. Ozeki and Shuji Saito, J. Chem. Phys., 110: 912–917, 1999.
[156] L. Kang and S. E. Novick, J. Mol. Spectrosc., 225: 66–72, 2004.
[157] K. Suma, Y. Sumiyoshi and Y. Endo, J. Am. Chem. Soc., 127: 14998–14999, 2005.
[158] D. Banser, M. Schnell, J.-U. Grabow, E. J. Cocinero, A. Lesarri and J. L. Alonso, Angew. Chem. Int. Ed. Engl., 44: 6311–6315, 2005.
[159] D. Banser, M. Schnell, J.-U. Grabow, E. J. Cocinero, A. Lesarri and J. L. Alonso, J. Mol. Struct., 795: 163–172, 2006.
[160] P. M. Sheridan, J. Xin, L. M. Ziurys, S. A. Beaton, S. M. Kermode and J. M. Brown, J. Chem. Phys., 116: 5544–5550, 2002.

[161] J. Xin and L. M. Ziurys, J. Chem. Phys., 110: 4797–4802, 1999.

[162] D. T. Halfen, A. J. Apponi and L. M. Ziurys, Astrophys. J., 577: L67–L70, 2002.

[163] M. D. Allen, T. C. Pesch and L. M. Ziurys, Astrophys. J., 472: L57–L60, 1996.

[164] K. Namiki, S. Saito, J. S. Robinson and T. C. Steimle, J. Mol. Spectrosc., 191: 176–182, 1998.

[165] T. Steimle, K. Namiki and S. Saito, J. Chem. Phys., 107: 6109–6113, 1997.

[166] T. Steimle, M. Tanimoto, K. Namikia and S. Saito, J. Chem. Phys., 108: 7616–7622, 1998.

[167] M. Goto, S. Takano, S. Yamamoto, H. Ito and S. Saito, Chem. Phys. Lett., 227: 287–292, 1994.

[168] K. C. Namiki and S. Saito, J. Chem. Phys., 114: 9390–9394, 2001.

[169] S. K. McLamarrah, P. M. Sheridan and L. M. Ziurys, Chem. Phys. Lett., 414: 301–306, 2005.

[170] P. M. Sheridan, S. K. McLamarrah and L. M. Ziurys, J. Chem. Phys., 119: 9496–9503, 2003.

[171] M. A. Flory, S. K. McLamarrah and L. M. Ziurys, J. Chem. Phys., 125: 194304, 2006.

[172] P. M. Sheridan and L. M. Ziurys, Chem. Phys. Lett., 380: 632–646, 2003.

[173] D. T. Halfen and L. M. Ziurys, J. Chem. Phys., 122: 0543092005.

[174] M. D. Allen, B. Z. Li and L. M. Ziurys, Chem. Phys. Lett., 270: 517–526, 1997.

[175] M. Tanimoto, S. Saito and T. Okabayashi, Chem. Phys. Lett., 242: 153–156, 1995.

[176] M. A. Flory, D. T. Halfen and L. M. Ziurys, J. Chem. Phys., 121: 8385–8392, 2004.

[177] M. A. Brewster and L. M. Ziurys, Chem. Phys. Lett., 349: 249–256, 2001.

[178] B.-Z. Li and L. M. Ziurys, Astrophys. J., 488: L137–L140, 1997.

[179] J. Xin and L. M. Ziurys, Astrophys. J., 495: L119–L122, 1998.

[180] J. M. Thompsen and L. M. Ziurys, Chem. Phys. Lett., 344: 75–84, 2001.

[181] J. M. Thompsen, M. A. Brewster and L. M. Ziurys, J. Chem. Phys., 116: 10212–10219, 2002.

[182] M. A. Flory, S. K. McLamarrah and L. M. Ziurys, J. Chem. Phys., 123: 164312, 2005.

[183] T. C. Steimle, S. Saito and S. Takano, Astrophys. J., 410: L49–L51, 1993.

[184] T. C. Steimle, S. Saito and S. Takano, Astrophys. J., 410: L39–L42, 1993.

[185] M. A. Brewster and L. M. Ziurys, J. Chem. Phys., 117: 4853–4860, 2002.

[186] P. M. Sheridan and L. M. Ziurys, J. Chem. Phys., 118: 6370–6379, 2003.

[187] P. M. Sheridan, M. A. Flory and L. M. Ziurys, J. Chem. Phys., 121: 8360–8368, 2004.

[188] D. T. Halfen and L. M. Ziurys, Astrophys. J., 611: L65–L68, 2004.

[189] A. J. Apponi, M. A. Anderson and L. M. Ziurys, J. Chem. Phys., 111: 10919–10925, 1999.

[190] P. M. Sheridan and L. M. Ziurys, Astrophys. J., 540: L61–L64, 2000.

[191] M. A. Brewster and L. M. Ziurys, J. Chem. Phys., 113: 3141–3149, 2000.

[192] A. Janczyk and L. M. Ziurys, J. Chem. Phys., 119: 10702–10712, 2003.

[193] M. A. Anderson and L. M. Ziurys, Astrophys. J., 452: L157–L160, 1995.

[194] M. A. Anderson and L. M. Ziurys, Astrophys. J., 460: L77–L80, 1996.

[195] J. Xin, J. S. Robinson, A. J. Apponi and L. M. Ziurys, J. Chem. Phys., 108: 2703–2711, 1998.

[196] M. A. Anderson, J. S. Robinson and L. M. Ziurys, Chem. Phys. Lett., 257: 471–480, 1996.

[197] J. Xin and L. M. Ziurys, J. Chem. Phys., 110: 3360–3367, 1999.

[198] B. P. Nuccio, A. J. Apponi and L. M. Ziurys, Chem. Phys. Lett., 247: 283–288, 1995.

[199] M. C. McCarthy, M. J. Travers, P. Kalmus, C. A. Gottlieb and P. Thaddeus, Chem. Phys. Lett., 264: 252–256, 1997.

[200] C. A. Gottlieb, M. C. McCarthy, M. J. Travers, J.-U. Grabow and P. Thaddeus, J. Chem. Phys., 109: 5433–5438, 1998.

[201] M. C. McCarthy and P. Thaddeus, J. Chem. Phys., 122: 174308, 2005.

[202] M. C. McCarthy, W. Chen, A. J. Apponi, C. A. Gottlieb and P. Thaddeus, Astrophys. J., 520: 158–161, 1999.

[203] A. J. Apponi, M. E. Sanz, C. A. Gottlieb, M. C. McCarthy, and P. Thaddeus, Astrophys. J., 547: L65–L68, 2001.

[204] R. J. McMahon, M. C. McCarthy, C. A. Gottlieb, J. B. Dudek, J. F. Stanton and P. Thaddeus, Astrophys. J., 590: L61–L64, 2003.

[205] W. Chen, M. C. McCarthy, M. J. Travers, E. W. Gottlieb, M. R. Munrow, S. E. Novick, C. A. Gottlieb and P. Thaddeus, Astrophys. J., 492: 849–853, 1998.

[206] V. D. Gordon, M. C. McCarthy, A. J. Apponi and P. Thaddeus, Astrophys. J., 540: 286–291, 2000.

[207] M. C. McCarthy and P. Thaddeus, J. Mol. Spectrosc., 232: 351–357, 2005.

[208] M. C. McCarthy, G. W. Fuchs, J. Kucera, G. Winnewisser and P. Thaddeus, J. Chem. Phys., 118: 3549–3557, 2003.

[209] S. Mohamed, M. C. McCarthy, A. L. Cooksy, C. Hinton and P. Thaddeus, J. Chem. Phys., 123: 2343012005.

[210] M. C. McCarthy, A. L. Cooksy, S. Mohamed, V. D. Gordon and P. Thaddeus, Astrophys. J. Suppl. Ser., 144: 287–297, 2003.

[211] Y. Sumiyoshi, K. Katoh and Y. Endo, Chem. Phys. Lett., 414: 82–86, 2005.

[212] K. Suma, Y. Sumiyoshi and Y. Endo, J. Chem. Phys., 121: 8351–8359, 2004.

[213] K. Suma, Y. Sumiyoshi and Y. Endo, J. Chem. Phys., 123: 024312, 2005.

[214] K. Suma, Y. Sumiyoshi and Y. Endo, Science, 308: 1885–1886, 2005.

[215] D. M. Silveira, P. J. S. B. Caridade and A. J. C. Varandas, J. Phys. Chem. A, 108: 8721–8730, 2004.

[216] F. Cacace, G. de Petris, F. Pepi and A. Troiani, Science, 285: 81–82, 1999.

[217] Y. Ohshima, K. Sato, Y. Sumiyoshi and Y. Endo, J. Am. Chem. Soc., 127: 1108–1109, 2005.

[218] N. Inada, K. Saito, M. Hayashi, H. Ozeki and S. Saito, Chem. Phys. Lett., 284: 142–146, 1998.

[219] N. Hansen, U. Andresen, H. Dreizler, J.-U. Grabow, H. Mäder and F. Temps, Chem. Phys. Lett., 289: 311–318, 1998.

[220] D. Buhl and L. E. Snyder, Nature, London, 228: 267–269, 1970.

[221] R. C. Woods, T. A. Dixon, R. J. Saykally and P. G. Szanto, Phys. Rev. Lett., 35: 1269–1272, 1975.

[222] R. J. Saykally and R. C. Woods, Annu. Phys. Rev. Chem., 32: 403–431, 1981.

[223] F. C. De Lucia, E. Herbst and G. M. Plummer, J. Chem. Phys., 85: 2312–2316, 1983.

[224] C. Savage and L. M. Ziurys, Rev. Sci. Instrum., 76: 043106, 2005.

[225] Y. Ohshima and Y. Endo, Chem. Phys. Lett., 256: 635–640, 1996.

[226] C. A. Gottlieb, A. J. Apponi, M. C. McCarthy, P. Thaddeus and H. Linnartz, J. Chem. Phys., 113: 1910–1915, 2000.

[227] M. Bogey, M. Cordonnier, C. Demuinck and J. L. Destombes, Astrophys. J., 399: L103–L105, 1992.

[228] M. Araki, H. Ozeki and S. Saito, Astrophys. J., 496: L53–L55, 1998.

[229] L. M. Ziurys and B. E. Turner, Astrophys. J., 302: L19–L23, 1986.

[230] C. Savage, A. J. Apponi and L. M. Ziurys, Astrophys. J., 608: L73–L76, 2004.

[231] D. T. Halfen and L. M. Ziurys, J. Mol. Spectrosc., 240: 58–63, 2006.

[232] D. T. Halfen and L. M. Ziurys, J. Mol. Spectrosc., 234: 34–40, 2005.

[233] D. T. Halfen and L. M. Ziurys, Astrophys. J., 657: L61–L64, 2007.

[234] S. Civiš, A. Walters, M. Yu Tretyakov, S. Bailleux and M. Bogey, J. Chem. Phys., 108: 8370–8373, 1998.

[235] S. Brünken, C. A. Gottlieb, H. Gupta, M. C. McCarthy and P. Thaddeus, Astron. AstroPhys., 464: L33–L36, 2007.

[236] M. C. McCarthy, C. A. Gottlieb, H. Gupta and P. Thaddeus, Astrophys. J., 652: L141–L144, 2006.

[237] H. Gupta, S. Brünken, F. Tamassia, C. A. Gottlieb, M. C. McCarthy and P. Thaddeus, Astrophys. J., 655: L57–L60, 2007.

[238] S. G. Kukolich, M. A. Roehrig, S. T. Haubrich and J. A. Shea, J. Chem. Phys., 94: 191–194, 1991.

[239] M. A. Roehrig, D. W. Wallace and S. G. Kukolich, J. Chem. Phys., 94: 2449–2452, 1992.

[240] M. A. Roehrig, P. Wikrent, S. R. Huber, D. E. Wigley and S. G. Kukolich, J. Mol. Spectrosc., 154: 355–360, 1992.

[241] S.G Kukolich and S. M. Sickafoose, Chem. Phys. Lett., 215: 168–172, 1993.

[242] S. G. Kukolich, S. M. Sickafoose, L. D. Flores and S. M. Breckenridge, J. Chem. Phys., 100: 6125–6128, 1994.

[243] S. G. Kukolich, J. Am. Chem. Soc., 117: 5512–5514, 1995.

[244] R. T. Mckay, J. L. Hubbard and S. G. Kukolich, J. Mol. Spectrosc., 152: 378–383, 1995.

[245] S. G. Kukolich and S. M. Sickafoose, J. Chem. Phys., 105: 3466–3471, 1996.

[246] S. M. Sickafoose, P. Wikrent, B. J. Drouin and S. G. Kukolich, Chem. Phys. Lett., 263: 191–196, 1996.

[247] B. J. Drouin, P. A. Cassak, P. M. Briggs and S. G. Kukolich, J. Chem. Phys., 107: 3766–3773, 1997.

[248] B. J. Drouin, P. A. Cassak and S. G. Kukolich, J. Chem. Phys., 108: 8878–8883, 1998.

[249] J. C. Earp, D. S. Margolis, C. Tanjaroon, T. E. Bitterwolf and S. G. Kukolich, J. Mol. Spectrosc., 211: 82–85, 2002.

[250] C. Tanjaroon, K. S. Keck, S. G. Kukolich, M. H. Palmer and M. F. Guest, J. Chem. Phys., 120: 4715–4725, 2004.

[251] K. S. Keck, C. Tanjaroon and S. G. Kukolich, J. Mol. Spectrosc., 232: 55–60, 2005.

[252] B. J. Drouin, J. Dannemiller and S. G. Kukolich, J. Chem. Phys., 112: 747–751, 2000.

[253] B. J. Drouin, T. G. Lavaty, P. A. Cassak and S. G. Kukolich, J. Chem. Phys., 107: 6541–6548, 1997.

[254] D. S. Margolis, C. Tanjaroon and S. G. Kukolich, J. Chem. Phys., 117: 3741–3747, 2002.

[255] C. Tanjaroon, K. S. Keck and S. G. Kukolich, J. Am. Chem. Soc., 126: 844–850, 2004.

[256] R. Subramanian, C. Karunatilaka, K. S. Keck and Stephen G. Kukolich, Inorg. Chem., 44: 3137–3145, 2005.

[257] R. Subramanian, C. Karunatilaka, R. O. Schock, B. J. Drouin, P. A. Cassak and S. G. Kukolich, J. Chem. Phys., 123: 054317, 2005.

[258] S. G. Kukolich, M. A. Roebrig, D. W. Wallace and G. L. Henderson, J. Am. Chem. Soc., 115: 2021–2027, 1993.

[259] G. L. Henderson, M. A. Roehrig, P. Wikrent and S. G. Kukolich, J. Phys. Chem., 96: 8303–8306, 1992.

[260] S. G. Kukolich, S. M. Breckenridge-Estes and S. M. Sickafoose, Inorg. Chem., 36: 4916–4918, 1997.

[261] C. Tanjaroon, K. S. Keck, M. M. Sebonia, C. Karunatilaka and S. G. Kukolich, J. Chem. Phys., 121: 1449–1453, 2004.

[262] C. Tanjaroon, C. Karunatilaka, K. S. Keck and Stephen G. Kukolich, Organometallics, 24: 2848–2853, 2005.

[263] S. G. Kukolich, S. M. Sickafoose and S. M. Breckenridge, J. Am. Chem. Soc., 118: 205–208, 1996.

[264] T. G. Lavaty, P. Wikrent, B. J. Drouin and S. G. Kukolich, J. Chem. Phys., 109: 9473–9478, 1998.

[265] B. S. Tackett, C. Karunatilaka, A. M. Daly and S. G. Kukolich, Organometallics, 26: 2070–2076, 2007.

[266] R. D. Brown, P. D. Godfrey, J. W. V. Storey, M. P. Bassez, J. Chem. Soc. Chem. Commun.: 547–548, 1978.

[267] R. D. Suenram and F. J. Lovas, J. Mol. Spectrosc., 72: 372–382, 1978.

[268] R. D. Brown, J. G. Crofts, P. D. Godfrey, D. McNaughton and A. P. Pierlot, J. Mol. Struct., 190: 185–193, 1988.

[269] F. J. Lovas, Y. Kawashima, J.-U. Grabow, R. D. Suenram, G. T. Fraser and E. Hirota, Astrophys. J., 455: L201–L204, 1995.

[270] S. Melandri, W. Caminati, L. B. Favero, A. Millemaggi and P. G. Favero, J. Mol. Struct., 352/353: 253–258, 1995.

[271] A. Lesarri, S. Mata, J. C. López and J. L. Alonso, Rev. Sci. Instrum., 74: 4799–4804, 2003.

[272] R. D. Suenram and F. J. Lovas, J. Am. Chem. Soc., 102: 7180–7184, 1980.

[273] P. D. Godfrey and R. D. Brown, J. Am. Chem. Soc., 117: 2019–2023, 1995.

[274] S. J. McGlone, P. S. Elmes, R. D. Brown and P. D. Godfrey, J. Mol. Struct., 485/486: 225–238, 1999.

[275] P. D. Godfrey, R. D. Brown and F. M. Rodgers, J. Mol. Struct., 376: 65–81, 1996.

[276] V. V. Ilyushin, E. A. Alekseev, S. F. Dyubko, R. A. Motiyenko and F. J. Lovas, J. Mol. Spectrosc., 231: 15–22, 2005.

[277] P. D. Godfrey, S. Firth, L. D. Hatherley, R. D. Brown and A. P. Pierlot, J. Am. Chem. Soc., 115: 9687–9691, 1993.

[278] S. J. McGlone and P. D. Godfrey, J. Am. Chem. Soc., 117: 1043–1048, 1995.

[279] S. Blanco, A. Lesarri, J. C. López and J. L. Alonso, J. Am. Chem. Soc., 126: 11675–11683, 2004.

[280] M. E. Sanz, A. Lesarri, M. I. Peña, V. Vaquero, V. Cortijo, J. C. López and J. L. Alonso, J. Am. Chem. Soc., 128: 3812–3817, 2006.

[281] H. B. Bürgi and J. D. Dunitz, Acc. Chem. Res., 16: 153–161, 1983.

[282] A. Lesarri, S. Mata, E. J. Cocinero, S. Blanco, J. C. López and J. L. Alonso, Angew. Chem. Int. Ed., 41: 4673–4676, 2002.

[283] A. Lesarri, E. J. Cocinero, J. C. López and J. L. Alonso, Angew. Chem. Int. Ed. Engl., 43: 605–610, 2004.

[284] A. Lesarri, R. Sánchez, E. J. Cocinero, J. C. López and J. L. Alonso, J. Am. Chem. Soc., 127: 12952–12956, 2005.

[285] E. J. Cocinero, A. Lesarri, J.-U. Grabow, J. C. López and J. L. Alonso, ChemPhysChem., 8: 599–604, 2007.

[286] L. Wang and P. G. Schultz, Angew. Chem. Int. Ed., 44: 34–66, 2005.

[287] A. Lesarri, E. J. Cocinero, J. C. López and J. L. Alonso, J. Am. Chem. Soc., 127: 2572–2579, 2005.

[288] A. Lesarri, E. J. Cocinero, J. C. López and J. L. Alonso, ChemPhysChem., 6: 1559–1566, 2005.

[289] E. J. Cocinero, A. Lesarri, M. E. Sanz, J. C. López and J. L. Alonso, ChemPhysChem., 7: 1481–1487, 2006.

[290] M. E. Sanz, V. Cortijo, W. Caminati, J. C. López and J. L. Alonso, Chem. Eur. J., 12: 2564–2570, 2006.

[291] E. J. Cocinero, P. Villanueva, A. Lesarri, M. E. Sanz, S. Blanco, S. Mata, J. C. López and J. L. Alonso, Chem. Phys. Lett., 435: 336–341, 2007.

[292] R. D. Brown, P. D. Godfrey, D. McNaughton and A. P. Pierlot, J. Am. Chem. Soc., 110: 2329–2330, 1988.

[293] R. D. Brown, P. D. Godfrey, D. McNaughton and A. P. Pierlot, J. Chem. Soc. Chem. Commun.: 37–38, 1989.

[294] R. D. Brown, P. D. Godfrey, D. McNaughton and A. P. Pierlot, J. Am. Chem. Soc., 111: 2308–2310, 1989.

[295] S. Brünken, M. C. McCarthy, P. Thaddeus, P. D. Godfrey and R. D. Brown, Astron. Astrophys., 459: 317–320, 2006.

[296] V. Vaquero, M. E. Sanz, J. C. López and J. L. Alonso, J. Phys. Chem. A, 111: 3443–3445, 2007.

[297] J. C. López, M. I. Peña, M. E. Sanz and J. L. Alonso, J. Chem. Phys., 126: 191103, 2007.

[298] R. D. Brown, P. D. Godfrey, D. McNaughton and A. P. Pierlot, Chem. Phys. Lett., 156: 61–63, 1989.

[299] K. M. Marstokk and H. Møllendal, J. Mol. Struct., 5: 205–213, 1970.

[300] K. M. Marstokk and H. Møllendal, J. Mol. Struct., 16: 259–270, 1973.

[301] R. A. H. Butler, F. C. De Lucia, D. T. Petkie, H. Møllendal, A. Horn and E. Herbst, Astrophys. J. Suppl. Ser., 134: 319–321, 2001.

[302] F. J. Lovas, R. D. Suenram, D. Plusquellic and H. Møllendal, J. Mol. Spectrosc., 222: 263–272, 2003.

[303] G. Maccaferri, W. Caminati and P. G. Favero, J. Chem. Soc. Faraday Trans., 93: 4115–4117, 1997.

[304] V. V. Ilyushin, R. A. Motiyenko, F. J. Lovas, D. F. Plusquellic, 2008, in press, Available online 10 March 2008.

[305] P. D. Godfrey, L. D. Hatherley and R. D. Brown, J. Am. Chem. Soc., 117: 8204–8210, 1995.

[306] P. D. Godfrey, S. J. McGlone and R. D. Brown, J. Mol. Struct., 599: 139–152, 2001.

[307] B. Vogelsanger, R. D. Brown, P. D. Godfrey and A. P. Pierlot, J. Mol. Spectrosc., 145: 1–11, 1991.

[308] B. Vogelsanger, P. D. Godfrey and R. D. Brown, J. Am. Chem. Soc., 113: 7864–7869, 1991.

[309] P. D. Godfrey and R. D. Brown, J. Am. Chem. Soc., 120: 10724–10732, 1998.

[310] W. Caminati and R. Cervellati, J. Am. Chem. Soc., 104: 4748–4752, 1982.

[311] W. Caminati, G. Maccaferri, P. G. Favero and L. B. Favero, Chem. Phys. Lett., 251: 189–192, 1996.

[312] J. K. G. Watson, J. Mol. Spectrosc., 40: 536–544, 1971.

[313] M. Mizushima and P. Venkateswarlu, J. Chem. Phys., 21: 705–709, 1953.
[314] I. M. Mills, J. K. G. Watson and W. L. Smith, Mol. Phys., 16: 329–344, 1969.
[315] M. Oldani, M. Andrist and A. Bauder, J. Mol. Spectrosc., 110: 93–105, 1985.
[316] M. Oldani, A. Bauder and A. G. Robiette, J. Mol. Spectrosc., 117: 60–68, 1986.
[317] W. Stahl, H. Dreizler, L. Jörissen and W. A. Kreiner, Z. Naturforsch., 41a: 747–751, 1986.
[318] W. A. Kreiner, H. D. Rudolph and A. G. Robiette, J. Mol. Spectrosc., 91: 499–502, 1982.
[319] M. Oldani and A. Bauder, J. Chem. Phys., 86: 624–628, 1987.
[320] Th. Brupbacher, C. Styger, B. Vogelsanger, I. Ozier and A. Bauder, J. Mol. Spectrosc., 138: 197–203, 1989.
[321] R. S. Berry, J. Chem. Phys., 32: 933–938, 1960.
[322] C. Styger and A. Bauder, J. Mol. Spectrosc., 148: 479–493, 1991.
[323] M. Oldani and A. Bauder, Chem. Phys. Lett., 108: 7–10, 1984.
[324] M. Oldani, R. Widmer, G. Grassi and A. Bauder, J. Mol. Struct., 190: 31–40, 1988.
[325] M. Rodler, M. Oldani, G. Grassi and A. Bauder, J. Chem. Phys., 86: 5365–5369, 1988.
[326] W. Caminati, G. Grassi and A. Bauder, Chem. Phys. Lett., 148: 13–16, 1988.
[327] M. Oldani, A. Bauder and J. C. Hilico, unpublished results, 1985.
[328] M. Oldani, A. Bauder, M. Loëte, J. P. Champion, G. Pierre, J. C. Hilico and A. G. Robiette, J. Mol. Spectrosc., 113: 229–242, 1985.
[329] M. Oldani, A. Bauder and G. Pierre, J. Mol. Spectrosc., 117: 435–443, 1986.
[330] F. J. Lovas, R. J. McMahon, J.-U. Grabow, M. Schnell, J. Mack, L. T. Scott and R. L. Kuczkowski, J. Am. Chem. Soc., 127: 4345–4349, 2005.
[331] D. McNaughton, P. D. Godfrey, R. D. Brown and S. Thorwirth, Phys. Chem. Chem. Phys., 9: 591–595, 2007.
[332] D. McNaughton, P. D. Godfrey, R. D. Brown, S. Thorwirth and J. -U. Grabow, Astrophys. J., 678: 309–315, 2008.
[333] S. Huber, G. Grassi and A. Bauder, Mol. Phys., 103: 1395–1409, 2005.
[334] S. Weinreb, A. H. Barrett, M. L. Meeks and J. C. Henry, Nature, 200: 829–831, 1963..
[335] A. C. Cheung, D. M. Rank, C. H. Townes, D. D. Thornton and W. J. Welch, Phys. Rev. Lett., 21: 1701–1705, 1968.
[336] A. C. Cheung, D. M. Rank, C. H. Townes and W. J. Welch, Nature, 221: 626–628, 1969.
[337] Dr. Lovas web-site (NIST). http://physics.nist.gov/PhysRefData/Micro/Html/contents.html
[338] Y.-J. Kuan, S. B. Charnley, H.-C. Huang, W.-L. Tseng and Z. Kisiel, Astrophys. J., 593: 848–867, 2003.
[339] L. E. Snyder, F. J. Lovas, J. M. Hollis, D. N. Friedel, P. R. Jewell, A. Remijan, V. V. Ilyushin, E. A. Alekseev and S. F. Dyubko, Astrophys. J., 619: 914–930, 2005.
[340] C. C. Costain and G. P. Srivastava, J. Chem. Phys., 35: 1903–1904, 1961.
[341] T. R. Dyke, B. J. Howard and W. Klemperer, J. Chem. Phys., 56: 2442–2454, 1972.
[342] T. J. Balle, E. J. Campbell, M. R. Keenan and W. H. Flygare, J. Chem. Phys., 71: 2723–2724, 1979.
[343] S. Melandri, G. Maccaferri, A. Maris, A. Millemaggi, W. Caminati and P. G. Favero, Chem. Phys. Lett., 261: 267–271, 1996.
[344] S. E. Novick, K. R. Leopold and W. Klemperer. The Structure of Weakly Bound Complexes as Elucidated by Microwave and Infrared Spectroscopy, in Atomic and Molecular Clusters, Bernstein, E. R. (Ed.), p. 359 (Elsevier, Amsterdam, 1990).
[345] A. C. Legon and D. J. Millen, Chem. Soc. Rev., 21: 71–78, 1992.
[346] K. R. Leopold, G. T. Fraser, S. E. Novick and W. Klemperer, Chem. Rev., 94: 1807–1827, 1994.
[347] S. E. Novick, P. Davies, S. J. Harris and W. Klemperer, J. Chem. Phys., 59: 2273–2275, 1973.
[348] S. G. Kucolich and J. A. Shea, J. Chem. Phys., 77: 5242–5243, 1982.
[349] J.-U. Grabow, A. S. Pine, G. T. Fraser, F. J. Lovas, R. D. Suenram, T. Emilsson, E. Arunan and H. S. Gutowsky, J. Chem. Phys., 102: 1181–1187, 1995.
[350] Y. Xu, W. Jäger, J. Djauhari and M. C. L. Gerry, J. Chem. Phys., 103: 2827–2833, 1995.
[351] W. Jäger, Y. Xu and M. C. L. Gerry, J. Chem.Phys., 99: 919–927, 1993.
[352] D. J. Millen, Can. J. Chem., 63: 1477–1479, 1985.
[353] W. G. Read, E. J. Campbell and J. Hederson, J. Chem. Phys., 78: 3501–3508, 1983.

[354] R. P. A. Bettens, R. M. Spycher, A. Bauder, Mol. Phys., 86: 487–511, 1995.

[355] Y. Xu, W. Jäger and M. C. L. Gerry, J. Chem. Phys., 100: 4171–4180, 1994.

[356] Y. Xu and W. Jäger, J. Chem. Phys. 107: 4788–4796, 1997.

[357] E. J. Campbell, M. R. Keenan, L. W. Buxton, T. J. Balle, P. O. Soper, A. C. Legon W. H. Flygare, Chem. Phys. Lett., 70: 420–424, 1980.

[358] E. J. Campbell, L. W. Buxton, M. R. Keenan and W. H. Flygare, Phys. Rev. A, 24: 812–821, 1981.

[359] M. R. Keenan, L. W. Buxton, E. J. Campbell, T. J. Balle and W. H. Flygare, J. Chem. Phys., 73: 3523–3529, 1980.

[360] S. G. Kukolich and E. J. Campbell, Chem. Phys. Lett., 94: 73–76, 1983.

[361] Z. Kisiel and L. Pszczółkowski, Chem. Phys. Lett., 291: 190–196, 1998.

[362] B. A. McElmurry, R. R. Lucchese, J. W. Bevan, I. I. Leonov, S. P. Belov and A. C. Legon, J. Chem. Phys., 119: 10688–10695, 2003.

[363] K. R. Leopold, G. T. Fraser, F. J. Lin, D. D. Nelson Jr. and W. Klemperer, J. Chem. Phys., 81: 4922–4931, 1984.

[364] T. D. Klots, C. E. Dykstra and H. S. Gutowsky, J. Chem. Phys., 90: 30–38, 1989.

[365] S. Drucker, A. L. Cooksy and W. Klemperer, J. Chem. Phys., 98: 5158–5183, 1993.

[366] K. Tanaka, S. Bailleux, A. Mizoguchi, K. Harada, T. Baba, I. Ogawa and M. Shirasaka, J. Chem. Phys., 113: 1524–1534, 2000.

[367] A. Mizoguchi, K. Harada, M. Shirasaka and K. Tanaka, J. Mol. Spectrosc., 222: 74–85, 2003.

[368] A. R. W. McKellar, Y. Xu, W. Jäger and C. Bissonnette, J. Chem. Phys., 110: 10766–10773, 1999.

[369] L. A. Surin, D. A. Roth, I. Pak, B. S. Dumesh, F. Lewen and G. Winnewisser, J. Chem. Phys., 112: 4064–4068, 2000.

[370] K. A. Walker, T. Ogata, W. Jäger, M. C. L. Gerry and I. Ozier, J. Chem. Phys., 106: 7519–7530, 1997.

[371] G. Winnewisser, B. S. Dumesh, I. Pak, L. A. Surin, J. Mol. Spectrosc., 192: 243–246, 1998.

[372] T. Ogata, W. Jäger, I. Ozier and M. C. L. Gerry, J. Chem. Phys., 98: 9399–9404, 1993.

[373] W. Jäger and M. C. L. Gerry, J. Chem. Phys., 102: 3587–3592, 1995.

[374] M. Hepp, W. Jäger, I. Pak and G. Winnewisser, J. Mol. Spectrosc., 176: 58–63, 1996.

[375] M. Hepp, R. Gendriesch, I. Pak, F. Lewen and G. Winnewisser, J. Mol. Spectrosc., 183: 295–299, 1997.

[376] R. Gendriesch, I. Pak, F. Lewen, L. Surin, D. A. Roth and G. Winnewisser, J. Mol. Spectrosc., 196: 139–145, 1999.

[377] K. A. Walker and A. R. W. McKellar, J. Mol. Spectrosc., 205: 331–337, 2001.

[378] K. Higgins, F.-M. Tao and W. Klemperer, J. Chem. Phys., 109: 3048–3061, 1998.

[379] K. Higgins and W. Klemperer, J. Chem. Phys., 110: 1383–1388, 1999.

[380] Y. Xu and W. Jäger, J. Mol. Struct., 599: 211–217, 2001.

[381] X. G. Song, Y. Xu, P.-N. Roy and W. Jäger, J. Chem. Phys., 121: 12308–12314, 2004.

[382] C. Topic and W. Jäger, J. Chem. Phys., 123: 064303, 2005.

[383] W. Jäger, Y. Xu, G. Armstrong, M. C. L. Gerry, F. Y. Naumkin, F. Wang and F. R. W. McCourt, J. Chem. Phys., 109: 5420–5432, 1998.

[384] W. Jäger and M. C. L. Gerry, Chem. Phys. Lett., 196: 274–279, 1992.

[385] W. Jäger, Y. Xu, N. Heineking and M. C. L. Gerry, J. Chem. Phys., 99: 7510–7520, 1993.

[386] Q. Wen and W. Jäger, J. Chem. Phys., 122: 214310, 2005.

[387] J. M. Steed, T. A. Dixon and W. Klemperer, J. Chem. Phys., 70: 4095–4100, 1979.

[388] M. Iida, Y. Ohsbima and Y. Endo, J. Phys. Chem., 97: 357–362, 1993.

[389] Y. Xu and W. Jäger, J. Mol. Spectrosc., 192: 435–440, 1998.

[390] S. J. Harris, K. C. Janda, S. E. Novick and W. Klemperer, J. Chem. Phys., 63: 881–884, 1975.

[391] Y. Xu, W. Jäger and M. C. L. Gerry, J. Mol. Spectrosc., 151: 206–216, 1992.

[392] Y. Xu and M. C. L. Gerry, J. Mol. Spectrosc., 169: 542–554, 1995.

[393] A. M. Andrews, L. Nemes, S. L. Maruca, K. W. Hillig II, R. L. Kuczowski and J. S. Muenter, J. Mol. Spectrosc., 160: 422–433, 1993.

[394] M. S. Ngari and and W. Jäger, J. Mol. Spectrosc., 192: 320–330, 1998.

[395] C. H. Joyner, T. A. Dixon, F. A. Baiocchi and W. Klemperer, J. Chem. Phys., 75: 5285–5290, 1981.
[396] H. O. Leung, D. Gangwani and J.-U. Grabow, J. Mol. Spectrosc., 184: 106–112, 1997.
[397] M. S. Ngari and W. Jäger, J. Mol. Spectrosc., 192: 452–454, 1998.
[398] Y. Liuy and W. Jäger, Phys. Chem. Chem. Phys., 5: 1744–1751, 2003.
[399] Y. Liu and W. Jäger, J. Mol. Spectrosc., 205: 177–182, 2001.
[400] S. J. Harris, S. E. Novick, W. Klemperer and W. E. Falconer, J. Chem. Phys., 61: 193–197, 1974.
[401] Y. Xu, W. Jäger, I. Ozier and M. C. L. Gerry, J. Chem. Phys., 98: 3726–3731, 1993.
[402] J. B. Davey, A. C. Legon and E. R. Waclawik, Chem. Phys. Lett., 306: 133–144, 1999.
[403] E. Arunan, T. Emilsson, and H. S. Gutowsky, J. Chem. Phys., 101: 861–868, 1994.
[404] Th. Brupbacher, J. Makarewicz and A. Bauder, J. Chem. Phys., 101: 9736–9746, 1994.
[405] Th. Brupbacher and A. Bauder, Chem. Phys. Letters, 173: 435–438, 1990.
[406] T. D. Klots, T. Emilsson, and H. S. Gutowsky, J. Chem. Phys., 97: 5335–5340, 1992.
[407] C. Tanjaroon and W. Jäger, J. Chem.Phys., 127: 034302 2007.
[408] A. Maris, W. Caminati and P. G. Favero, Chem. Commun., 2625–2626, 1998.
[409] T. D. Klots, T. Emilsson, R. S. Ruoff and H. S. Gutowsky, J. Phys. Chem., 93: 1255–1261, 1989.
[410] R. M. Spycher, D. Petitprez, F. L. Bettens and A. Bauder, J. Phys. Chem., 98: 11863–11869, 1994.
[411] W. Stahl and J.-U. Grabow, Z. Naturforsch., 47a: 681–684, 1992.
[412] E. Jochims, J.-U. Grabow and W. Stahl, J. Mol. Spectrosc, 158: 278–286, 1992.
[413] E. Jochims, J.-H. Mader and W. Stahl, J. Mol. Spectrosc, 180: 116–120, 1996.
[414] W. Caminati and P. G. Favero, Chem. Eur. J., 5: 811–814, 1999.
[415] W. Caminati, S. Melandri, A. Dell'Erba and P. G. Favero, Phys.Chem.Comm, 1:2000.
[416] W. Caminati, A. Millemaggi, P. G. Favero and J. Makarewicz, J. Phys. Chem., 101: 9272–9275, 1997.
[417] W. Caminati, P. G. Favero, S. Melandri and R. Meyer, Chem. Phys. Lett., 268: 393–400, 1997.
[418] J. Makarewicz and A. Bauder, Mol. Phys., 84: 853–878, 1995.
[419] W. Caminati, D. Petitprez, L. H. Coudert et al., unpublished results.
[420] W. Caminati, Sixteenth Colloquium on High Resolution Molecular Spectroscopy, Dijon, 6–10 September 1999, Communication P2.
[421] A. Maris and W. Caminati, J. Chem. Phys., 118: 1649–1652, 2003.
[422] Y. Morita, N. Ohashi, Y. Kawashima and E. Hirota, J. Chem. Phys., 124: 094301, 2006.
[423] P. Ottaviani, A. Maris, W. Caminati, Y. Tatamitani, Y. Suzuki, T. Ogata and J. L. Alonso, Chem. Phys. Lett., 361: 341–348, 2002.
[424] B. Velino, S. Melandri and W. Caminati, J.Phys.Chem. A, 108: 4224–4227, 2004.
[425] W. Caminati, A. Millemaggi, J. L. Alonso, A. Lesarri, J. C. Lopez and S. Mata, Chem. Phys. Lett., 392: 1–6, 2004.
[426] I. I. Ioannu, R. L. Kuczkowski and J. T. Hougen, J. Mol. Spectrosc., 171: 265–286, 1995.
[427] S. Melandri, A. Dell'Erba, P. G. Favero and W. Caminati, J. Mol. Spectrosc., 222: 121–128, 2003.
[428] S. Melandri, P. G. Favero, W. Caminati and B. Velino, J. Chem. Phys., 122: 134310, 2005.
[429] Y. Sumiyoshi, Y. Endo and Y. Ohshima, J. Chem. Phys., 113: 10121–10129, 2000.
[430] Y. Sumiyoshi, H. Katsunuma, K. Suma and Y. Endo, J. Chem. Phys., 123: 054324, 2005.
[431] K. Suma, Y. Sumiyoshi and Y. Endo, J. Chem. Phys., 120: 6935–6943, 2004.
[432] Y. Ohshima, M. Iida and Y. Endo, J. Chem. Phys., 95: 7001–7003, 1991.
[433] Y. Endo, H. Kohguchi and Y. Ohshima, Faraday Discuss. 97: 341–350, 1994.
[434] Y. Sumiyoshi, I. Funahara, K. Sato, Y. Ohshima and Y. Endo, J. Chem. Phys., 125: 124307, 2006.
[435] R. J. Low, M. D. Brookes, C. J. Whitham and B. J. Howard, J. Chem. Phys., 105: 6756–6770, 1996.
[436] S. Blanco, C. J. Whitham, H. B. Qian and B. J. Howard, Phys. Chem. Chem. Phys., 3: 3895–3900, 2001.
[437] J. Whitham, R. J. Low, and B. J. Howard, Chem. Phys. Lett., 286: 408–414, 1998.

[438] M. Schäfer, T. K. Ha and A. Bauder, J. Chem. Phys., 119: 8404–8416, 2004.

[439] K. Suma, W. Funato, Y. Sumiyoshi and Y. Endo, J. Chem. Phys., 122: 184302, 2005.

[440] M. Bogey, H. Bolvin, C. Demuynck and J. L. Destombes, Phys. Rev. Lett., 58: 988–991, 1987.

[441] M. Bogey, H. Bolvin, C. Demuynck, J. L. Destombes and B. P. Van Eijck, J. Chem. Phys., 88: 4120–4126, 1988.

[442] S. Bailleux, M. Bogey, H. Bolvin, S. Civiš, M. Cordonnier, A. F. Krupnov, M.Yu Tretyakov, A. Walters and L. H. Coudert, J. Mol. Spectrosc., 190: 130–139, 1998.

[443] Y. Ohshima, Y. Sumiyoshi and Y. Endo, J. Chem. Phys., 106: 2977–2979, 1997.

[444] K. Seki, Y. Sumiyoshi and Y. Endo, Chem. Phys. Lett., 331: 184–188, 2000.

[445] G. A. Jeffrey, An Introduction to Hydrogen Bonding, (Oxford University Press, New York, 1997).

[446] S. Scheiner, Hydrogen Bonding, (Oxford University Press, New York, 1997).

[447] G. R. Desiraju and T. Steiner (Eds), The weak hydrogen bond in structural chemistry and biology IUCr Monographs on crystallography, Vol. IX, (Oxford University Press, Oxford, 2001.

[448] T. R. Dyke, J. Chem. Phys., 66: 492–497, 1977.

[449] G. T. Fraser, F. J. Lovas, R. D. Suenram, E. N. Karyakin, A. Grushow, W. A. Burns and K. R. Leopold, J. Mol. Spectrosc., 181: 229–245, 1997, and references therein.

[450] L. H. Coudert, W. Caminati, M. Schnell and J.-U. Grabow, J. Mol. Spectrosc., 242: 118–128, 2007.

[451] G. Cazzoli, P. G. Favero, D. G. Lister, A. C. Legon, D. J. Millen and Z. Kisiel, Chem. Phys. Lett., 117: 543–549, 1985.

[452] A. C. Legon, D. J. Millen and H. M. North, Chem. Phys. Lett., 135: 303–306, 1987.

[453] Z. Kisiel, B. A. Pietrewicz, P. W. Fowler, A. C. Legon and E. Steiner, J. Phys. Chem. A, 2000: 6970–6978, 104.

[454] A. C. Legon and A. P. Suckley, Chem. Phys. Lett., 150:6970–6970, 104.

[455] A. McIntosh, T. Walther, R. R. Lucchese, J. W. Bevan, R. D. Suenram and A. C. Legon, Chem. Phys. Lett., 314: 57–64, 1999.

[456] H. S. Gutowsky, T. C. Germann, J. D. Augspurger and C. E. Dykstra, J. Chem. Phys., 96: 5808–5816, 1992.

[457] M. Canagaratna, J. A. Phillips, M. E. Ott and K. R. Leopold, J. Phys. Chem. A, 102: 1489–1497, 1998.

[458] D. L. Fiacco, S. W. Hunt and K. R. Leopold, J. Am. Chem. Soc., 124: 4504–4511, 2002.

[459] P. Herbine and T. R. Dyke, J. Chem. Phys., 83: 3768–3774, 1985.

[460] P. A. Stockman, R. E. Bumgarner, S. Suzuki and G. A. Blake, J. Chem. Phys., 96: 2496–2510, 1992.

[461] D. Yaron, K. I. Peterson, D. Zolandz, W. Klemperer, F. J. Lovas and R. D. Suenram, J. Chem. Phys., 92: 7095–7109, 1990.

[462] G. Columberg, A. Bauder, N. Heineking, W. Stahl and J. Makarewicz, Mol. Phys., 93: 215–228, 1998.

[463] Y. Tatamitani and T. Ogata, J. Chem. Phys., 121: 9885–9890, 2004.

[464] T. Ogata and F. J. Lovas, J. Mol. Spectrosc., 162: 505–512, 1993.

[465] D. Zolandz, D. Yaron, K. I. Peterson and W. Klemperer, J. Chem. Phys., 97: 2861–2868, 1992.

[466] K. Matsumura, F. J. Lovas and R. D. Suenram, J. Chem. Phys., 91: 5887–5894, 1989.

[467] J. A. Phillips, M. Canagaratna, H. Goodfriend and K. R. Leopold, J. Phys. Chem., 99: 501–504, 1995.

[468] S. A. Cooke, G. Cotti, C. M. Evans, J. H. Holloway, Z. Kisiel, A. C. Legon and J. M. A. Thumwood, Chem. Eur. J., 7: 2295–2305, 2001.

[469] S. A. Cooke, G. Cotti, J. H. Holloway and A. C. Legon, Angew. Chem. Int. Ed. Engl., 36: 129–130, 1997.

[470] J. P. Davey, A. C. Legon and E. R. Waclawik, Phys. Chem. Chem. Phys., 2: 1659–1665, 2000.

[471] C. S. Brauer, G. Sedo, E. M. Grumstrup, K. R. Leopold, M. D. Marshall and H. O. Leung, Chem. Phys. Lett., 401: 420–425, 2005.

[472] P. A. Stockman, G. A. Blake, F. J. Lovas and R. D. Suenram, J. Chem. Phys., 107: 3782–3790, 1997.
[473] D. Richard, J. Lavrich, C. R. Torok and M. J. Tubergen, J. Phys. Chem. A, 105: 8317–8322, 2001.
[474] M. Gerhards, M. Schmitt, K. Kleinermanns and W. Stahl, J. Chem. Phys. 104: 967–971, 1996.
[475] S. Melandri, A. Maris, P. G. Favero and W. Caminati, Chem. Phys., 283: 185–192, 2002.
[476] D. Priem, T.-K. Ha and A. Bauder, J. Chem. Phys., 113: 169–175, 2000.
[477] S. Blanco, J. C. López, A. Lesarri and J. L. Alonso, J. Am. Chem. Soc., 128: 12111–12121, 2006, and references therein.
[478] W. Caminati, P. Moreschini, I. Rossi and P. G. Favero, J. Am. Chem. Soc., 120: 11144–11148, 1998.
[479] P. Ottaviani, M. Giuliano, B. Velino and W. Caminati, Chem. Eur. J., 10: 538–543, 2004.
[480] W. Caminati, A. Dell'Erba, S. Melandri and P. G. Favero, J. Am. Chem. Soc., 120: 5555–5558, 1998.
[481] U. Spoerel, W. Stahl, W. Caminati and P. G. Favero, Chem. Eur. J., 4: 1974–1981, 1998.
[482] B. M. Giuliano and W. Caminati, Angew. Chem. Int. Ed., 44: 603–605, 2005.
[483] A. R. Ubbelhode and K. J. Gallagher, Acta Crystallogr., 8: 71–83, 1955.
[484] W. Caminati, P. G. Favero, A. Maris, S. Melandri and B. Velino, XIV Conference – Workshop "Horizons in Hydrogen Bond Research", Sept. 3–7 2001, Torino, Italy, Communication MS3-P14.
[485] F. L. Bettens, R. P. A. Bettens, A. Bauder, 49th International Symposium on High Resolution Spectroscopy, Columbus OHIO USA, June 1994, Lecture TC05.
[486] L. B. Favero, B. M. Giuliano, S. Melandri, A. Maris and W. Caminati, Chem. Eur. J., 13: 5833–5837, 2007.
[487] F. J. Lovas and C. L. Lugez, J. Mol. Spectrosc., 179: 320–323, 1996.
[488] S. Melandri, A. Maris, B. M. Giuliano and W. Caminati, J. Chem. Phys., 123: 164304, 2005.
[489] M. J. Tubergen and R. L. Kuczkowski, J. Am. Chem. Soc., 115: 9263–9266, 1993.
[490] D. Consalvo and W. Stahl, J. Mol. Spectrosc., 174: 520–535, 1995.
[491] M. J. Tubergen, R. L. Kuczkowski, J. Mol. Struct., 352/353: 335–344, 1995.
[492] U. Spoerel and W. Stahl, J. Mol. Spectrosc., 190: 278–289, 1998.
[493] O. Indris, W. Stahl and U. Kretschmer, J. Mol. Spectrosc., 190: 372–378, 1998.
[494] W. Caminati, A. Dell'Erba, G. Maccaferri and P. G. Favero, J. Mol. Spectrosc., 191: 45–48, 1998.
[495] W. Caminati, A. Dell'Erba, G. Maccaferri and P. G. Favero, J. Am. Chem. Soc., 120: 2616–2621, 1998.
[496] J. L. Alonso, E. J. Cocinero, A. Lesarri, M. E. Sanz and J. C. López, Angew. Chem. Int. Ed., 45: 3471–3474, 2006.
[497] R. J. Lavrich and M. J. Tubergen, J. Am. Chem. Soc., 122: 2938–2943, 2000.
[498] W. Caminati, L. B. Favero, P. G. Favero, A. Maris and S. Melandri, Angew. Chem. Int. Ed., 37: 792–795, 1998.
[499] S. Melandri, M. E. Sanz, W. Caminati, P. G. Favero and Z. Kisiel, J. Am. Chem. Soc., 120: 11504–11509, 1998.
[500] W. Caminati, P. Moreschini and P. G. Favero, J. Phys. Chem. A, 102: 8097–8100, 1998.
[501] S. McGlone, P. Moreschini, T.-K. Ha and A. Bauder, Mol. Phys., 99: 1353–1364, 2001.
[502] M. J. Tubergen, A. M. Andrews and R. L. Kuczkowski, J. Phys. Chem., 97: 7451–7457, 1993.
[503] S. Blanco, J. C. López, J. L. Alonso, P. Ottaviani and W. Caminati, J. Chem. Phys., 119: 880–886, 2003.
[504] A. Maris, P. Ottaviani and W. Caminati, Chem. Phys. Lett., 360: 155–160, 2002.
[505] F. J. Lovas, N. Zobov, G. T. Fraser and R. D. Suenram, J. Mol. Spectrosc., 171: 189–199, 1995.
[506] M. E. Sanz, J. C. López, J. L. Alonso, A. Maris, P. G. Favero and W. Caminati, J. Phys. Chem. A, 103: 5285–5290, 1999.
[507] W. Caminati, S. Melandri, I. Rossi and P. G. Favero, J. Am. Chem. Soc., 121: 10098–10101, 1999.
[508] W. Caminati, S. Melandri, A. Maris and P. Ottaviani, Angew. Chem. Int. Ed., 45: 2438–2442, 2006.

[509] W. Caminati, A. Maris, A. Dell'Erba and P. G. Favero, Angew. Chem. Int. Ed., 45: 6711–6714, 2006.

[510] S. Antolínez, J. C. López, J. L. Alonso, Angew. Chem. Int. Ed., 38: 1772–1774, 1999.

[511] S. Antolínez, J. C. López, J. L. Alonso, Chem. Phys. Chem., 2: 114–117, 2001.

[512] M. E. Sanz, J. C. López and J. L. Alonso, Chem. Eur. J., 5: 3293–3298, 1999.

[513] R. Sánchez, S. Blanco, J. C. López and J. L. Alonso, J. Mol. Struct., 780/781: 57–64, 2006.

[514] A. C. Legon and J. C. Thorn, Chem. Phys. Lett., 227: 472–479, 1994.

[515] M. E. Sanz, V. M. Sanz, J. C. López and J. L. Alonso, Chem. Phys. Lett., 342: 31–38, 2001.

[516] J. L. Alonso, J. C. López, S. Blanco, A. Lesarri and F. J. Lorenzo, J. Chem. Phys., 113: 2760–2767, 2000.

[517] P. Ottaviani, W. Caminati, B. Velino, S. Blanco, A. Lesarri, J. C. López and J. L. Alonso, Chem. Phys. Chem., 5: 336–341, 2004.

[518] P. Ottaviani, W. Caminati, B. Velino and J. C. López, Chem. Phys. Lett., 394: 262–265, 2004.

[519] A. C. Legon, J. Chem. Soc. Faraday Trans., 92: 2677–2679, 1996.

[520] S. Antolínez, J. C. López, J. L. Alonso, Chem. Phys. Lett., 323: 130–136, 2000.

[521] M. E. Sanz, J. C. López, J. L. Alonso, Chem. Phys. Lett., 288: 760–766, 1998.

[522] S. A. Cooke, G. K. Corlett, A. C. Legon, Chem. Phys. Lett., 291: 269–276, 1998.

[523] J. K. Rice, F. J. Lovas, G. T. Fraser and R. D. Suenram, J. Chem. Phys., 103: 3877–3884, 1995.

[524] A. R. Hight Walker, W. Chen, S. E. Novick, B. D. Bean and M. D. Marshall, J. Chem. Phys., 102: 7298–7305, 1995.

[525] M. Canagaratna, J. A. Phillips, H. Goodfriend, D. L. Fiacco, M. E. Ott, B. Harms and K. R. Leopold, J. Mol. Spectrosc., 192: 338–347, 1998.

[526] E. M. Bellott, Jr. and E. B. Wilson, Tetrahedron, 31: 2896–2898, 1975.

[527] P. Gilli, V. Bertolasi, V. Ferretti and G. Gilli, J. Am. Chem. Soc., 122: 10405–10417, 2000.

[528] L. Martinache, W. Kresa, M. Wegener, U. Vonmont and A. Bauder, Chem. Phys., 148: 129–140, 1990.

[529] S. Antolinez, H. Dreizler, V. Storm, D. H. Sutter and J. L. Alonso, Z. Naturforsch., 52a: 803–806, 1997.

[530] P. L. A. Popelier, J. Phys. Chem. A, 102: 1873–1878, 1998.

[531] P. Hobza and Z. Havlas, Chem. Rev., 100: 4253–4264, 2000, and references therein.

[532] W. Caminati, S. Melandri, P. Moreschini and P. G. Favero, Angew. Chem. Int. Ed., 38: 2924–2925, 1999.

[533] Y. Tatamitani, B. Liu, J. Shimada, T. Ogata, P. Ottaviani, A. Maris, W. Caminati and J. L. Alonso, J. Am. Chem. Soc., 124: 2739–2743, 2002.

[534] J. L. Alonso, S. Antolínez, S. Blanco, A. Lesarri, J. C. López and W. Caminati, J. Am. Chem. Soc., 126: 3244–3249, 2004.

[535] S. Blanco, J. C. López, A. Lesarri, W. Caminati and J. L. Alonso, Chem. Phys. Chem., 5: 1779–1782, 2004.

[536] Y. Tatamitani, K. Yamanou, H. Kanno and T. Ogata, J. Mol. Spectrosc., 242: 150–155, 2007.

[537] W. Caminati, J.-U. Grabow, J. C. López and J. L. Alonso, Angew. Chem. Int. Ed. Engl., 44: 3840–3844, 2005.

[538] L. B. Favero, B. M. Giuliano, S. Melandri, A. Maris, P. Ottaviani, B. Velino and W. Caminati, J. Phys. Chem. A, 109: 7402–7404, 2005.

[539] P. Ottaviani, W. Caminati, L. B. Favero, S. Blanco, J. C. López and J. L. Alonso, Chem. Eur. J., 12: 915–920, 2006.

[540] B. M. Giuliano, A. Maris, S. Melandri, P. Ottaviani, W. Caminati, L. B. Favero and B. Velino, 62th International Symposium on High Resolution Spectroscopy, Columbus OHIO USA, June 18–22, 2007, Lecture RG07.

[541] E. J. Cocinero, R. Sánchez, S. Blanco, A. Lesarri, J. C. López and J. L. Alonso, Chem. Phys. Lett., 402: 4–10, 2005.

[542] S. Suzuki, P. G. Green, R. E. Bumgarner, S. Dasgupta, W. A. Goddard III and G. A. Blake, Science, 257: 942–945, 1992.

[543] D. A. Rodham, S. Suzuki, R. D. Suenram, F. J. Lovas, S. Dasgupta, W. A. Goddard III and G. A. Blake, Nature; 362: 6422–6424, 1993.

[544] S. A. Cooke, G. K. Corlett, C. M. Evans and A. C. Legon, Chem. Phys. Lett., 272: 61–68, 1997.

[545] J. C. López, J. L. Alonso and W. Caminati, Angew. Chem. Int. Ed., 45: 290–293, 2006.
[546] E. Arunan and H. S. Gutowsky, J. Chem. Phys., 98: 4294–4296, 1993.
[547] U. Erlekam, G. Meijer, M. Schnell, G. Von Helden, J.-U. Grabow and P. R. Bunker, 62th International Symposium on High Resolution Spectroscopy, Columbus OHIO USA, June 18–22, 2007, Lecture MH13.
[548] R. S. Mulliken, W. B. Person, Molecular Complexes: A Lecture and Reprint Volume, Wiley-Interscience, New York, 1969, and references therein.
[549] A. C. Legon, Chem. Soc. Rev., 22: 153–163, 1993.
[550] A. C. Legon and C. A. Rego, Chem. Phys. Lett., 154: 468–472, 1989.
[551] A. C. Legon and C. A. Rego, J. Chem. Phys., 90: 6867–6876, 1989.
[552] A. C. Legon, A. L. Wallwork and C. A. Rego, J. Chem. Phys., 92: 6397–6407, 1990.
[553] A. C. Legon and C. A. Rego, J. Chem. Phys., 99: 1463–1468, 1993.
[554] A. C. Legon and C. A. Rego, J. Chem. Soc. Faraday Trans., 89: 1173–1178, 1993.
[555] K. R. Leopold, M. Canagaratna and J. A. Phillips, Acc. Chem. Res., 30: 57–64, 1997.
[556] W. A. Burns, J. A. Phillips, M. Canagaratna, H. Goodfriend and K. R. Leopold, J. Phys. Chem. A, 103: 7445–7453, 1999.
[557] M. Canagaratna, J. A. Phillips, H. Goodfriend and K. R. Leopold, J. Am. Chem. Soc., 118: 5290–5295, 1996.
[558] D. L. Fiacco, A. Toro and K. R. Leopold, Inorg. Chem., 39: 37–43, 2000.
[559] S. W. Hunt and K. R. Leopold, J. Phys. Chem. A, 105: 5498–5506, 2001.
[560] S. M. A. Dvorak, R. S. Ford, F. J. Lovas, R. D. Suenram and K. R. Leopold, J. Am. Chem. Soc., 114: 108–115, 1992.
[561] S. W. Reevet, W. A. Burns, F. J. Lovas, R. D. Suenram and K. R. Leopold, J. Phys. Chem., 97: 10630–10637, 1993.
[562] D. J. Hankinson, J. Almlöf and K. R. Leopold, J. Phys. Chem., 100: 6904–6909, 1996.
[563] D. L. Fiacco, Y. Mo, S. W. Hunt, M. E. Ott, A. Roberts and K. R. Leopold, J. Phys. Chem. A, 105: 484–493, 2001.
[564] S. W. Hunt, C. S. Brauer, M. B. Craddock, K. J. Higgins, A. M. Nienow and K. R. Leopold, Chem. Phys., 305: 155–164, 2004.
[565] A. Mizoguchi, Y. Ohshima and Y. Endo, J. Am. Chem. Soc., 125: 1716–1717, 2003.
[566] See, for example, Chem. Rev., 97(5), 1997, fully dedicated to "molecular recognition".
[567] E. Fischer, Ber.Dtsch. Chem. Ges., 27: 2985–2993, 1894.
[568] A. D. Buckingham, A. C. Legon, and S. M. Roberts (Eds), Principles of Molecular Recognition, Springer, Heidelberg, 1993.
[569] B. M. Giuliano, P. Ottaviani, L. B. Favero, W. Caminati, J.-U. Grabow, A. Giardini and M. Satta, Phys. Chem. Chem. Phys., 9: 4460–4464, 2007.
[570] J. C. López, V. Cortijo, S. Blanco and J. L. Alonso, Phys. Chem. Chem. Phys., 9: 4521–4527, 2007.
[571] N. Borho and Y. Xu, Phys. Chem. Chem. Phys., 9: 4514–4520, 2007.
[572] A. K. King and B. J. Howard, Chem. Phys. Lett., 348: 343–349, 2001.
[573] J. P. I. Hearn, R. V. Cobley and B. J. Howard, J. Chem. Phys., 123: 134324, 2005.
[574] Z. Su, N. Borho and Y. Xu, J. Am. Chem. Soc., 128: 17126–17131, 2006.
[575] N. Borho and Y. Xu, Angew. Chem. Int. Ed. Engl., 46: 2276–2279, 2007.
[576] A. Stace, Science, 294: 1292–1293, 2001.
[577] Y. Xu, J. van Wijngaarden and W. Jäger, Int. Rev. Phys. Chem., 24: 301–338, 2005.
[578] D. D. Nelson, Jr., G. T. Fraser and W. Klemperer, J. Chem. Phys., 83: 6201–6208, 1985.
[579] D. D. Nelson, W. Klemperer, G. T. Fraser, F. J. Lovas and R. D. Suenram, J. Chem. Phys., 87: 6364–6372, 1987.
[580] N. Heineking, W. Stahl, E. H. T. Olthof, P. E. S. Wormer, A. van der Avoird and M. Havenith, J. Chem. Phys., 102: 8693–8703, 1995.
[581] E. N. Karyakin, G. T. Fraser, J. G. Loeser and R. J. Saykally, J. Chem. Phys., 110: 9555–9559, 1999.
[582] C. M. Western, P. R. R. Langridge-Smith, B. J. Howard and S. E. Novick, Mol. Phys., 44: 145–160, 1981.
[583] S. G. Kukolich, J. Mol. Spectrosc., 98: 80–86, 1983.

[584] M. D. Brookes, A. R. W. McKellar and T. Amano, J. Mol. Spectrosc., 185: 153–157, 1997.

[585] L. A. Surin, D. N. Fourzikov, T. F. Giesen, S. Schlemmer, G. Winnewisser, V. A. Panfilov, B. S. Dumesh, G. W. M. Vissers and A. van der Avoird, J. Chem. Phys., 125: 094304, 2006, and refs therein.

[586] A. J. Minei and S. E. Novick, J. Chem. Phys., 126: 101101, 2007.

[587] A. Taleb-Bendiab, K. W. Hillig II and R. L. Kuczkowski, J. Chem. Phys., 94: 6956–6963, 1991, and references therein.

[588] F. J. Lovas, R. D. Suenram, L. H. Coudert, T. A. Blake, K. J. Grant and S. E. Novick, J. Chem. Phys., 92: 891–898, 1990.

[589] F. J. Lovas, S. P. Belov, M. Y. Tretyakov, W. Stahl and R. D. Suenram, J. Mol. Spectrosc., 170: 478–492, 1995.

[590] F. J. Lovas and H. Hartwig, J. Mol. Spectrosc., 185: 98–109, 1997.

[591] C. L. Lugez, F. J. Lovas, J. T. Hougen and N. Ohashi, J. Mol. Spectrosc., 194: 95–112, 1999.

[592] M. J. Tubergen and R. L. Kuczkowski, J. Chem. Phys., 100: 3377–3383, 1994.

[593] G. Columberg and A. Bauder, J. Chem. Phys., 106: 504–510, 1997.

[594] T. Goly, U. Spoerel and W. Stahl, J. Chem. Phys., 283: 289–296, 2002.

[595] R. S. Ruoff, T. Emilsson, T. D. Klots, C. Chuang and H. S. Gutowsky, J. Chem. Phys., 89: 138–148, 1988.

[596] J. P. Connelly, A. Bauder, A. Chisholm and B. J. Howard, Mol. Phys., 88: 915–929, 1996.

[597] R. A. Peebles and R. L. Kuczkowski, J. Phys. Chem. A, 103: 6344–6350, 1999.

[598] S. Blanco, S. Melandri, P. Ottaviani and W. Caminati, J. Am. Chem. Soc., 129: 2700–2703, 2007.

[599] R. S. Ruoff, T. Emilsson, C. Chuang, T. D. Klots and H. S. Gutowsky, J. Chem. Phys., 90: 4069–4078, 1989.

[600] R. S. Ruoff, T. Emilsson, C. Chuang, T. D. Klots and H. S. Gutowsky, J. Chem. Phys., 93: 6363–6370, 1990.

[601] H. S. Gutowsky and C. Chuang, J. Chem. Phys., 93: 894–896, 1990.

[602] T. Emilsson, T. D. Klots, R. S. Ruoff and H. S. Gutowsky, J. Chem. Phys., 93: 6371–6376, 1990.

[603] H. S. Gutowsky, P. J. Hajduk, C. Chuang and R. S. Ruoff, J. Chem. Phys., 92: 862–869, 1990.

[604] K. I. Peterson, R. D. Suenram and F. J. Lovas, J. Chem. Phys., 94: 106–117, 1991.

[605] K. I. Peterson, R. D. Suenram and F. J. Lovas, J. Chem. Phys., 90: 5964–5970, 1989.

[606] S. A. Peebles and R. L. Kuczkowski, J. Chem. Phys., 109: 5276–5282, 1998.

[607] S. A. Peebles and R. L. Kuczkowski, J. Phys. Chem. A 102, 8091–8096, 1998.

[608] R. A. Peebles, S. A. Peebles and R. L. Kuczkowski, Mol. Phys., 96: 1355–1365, 1999.

[609] S. A. Peebles and R. L. Kuczkowski, J. Chem. Phys., 111: 10511–10519, 1999.

[610] R. A. Peebles and R. L. Kuczkowski, J. Chem. Phys., 112: 8839–8846, 2000.

[611] D. L. Fiacco, S. W. Hunt and K. R. Leopold, J. Phys. Chem. A, 104: 8323–8327, 2000.

[612] S. W. Hunt, K. J. Higgins, M. B. Craddock, C. S. Brauer and K. R. Leopold, J. Am. Chem. Soc., 125: 13850–13860, 2003.

[613] Z. Kisiel, E. Bialkowska-Jaworska, L. Pszczółkowski, A. Milet, C. Struniewicz, R. Moszynski and J. Sadlej, J. Chem. Phys., 112: 5767–5776, 2000.

[614] Z. Kisiel, B. A. Pietrewicz, O. Desyatnyk, L. Pszczółkowski, I. Struniewicz and J. Sadlej, J. Chem. Phys., 119: 5907–5917, 2003.

[615] H. S. Gutowsky, T. D. Klots, C. Chuang, C. A. Schmuttenmaer and T. Emilsson, J. Chem. Phys., 86: 569–576, 1987.

[616] T. D. Klots, C. Chuang, R. S. Ruoff, T. Emilsson and H. S. Gutowsky, J. Chem. Phys., 86: 5315–5322, 1987.

[617] Z. Kisiel, B. A. Pietrewicz and L. Pszczółkowski, J. Chem. Phys., 117: 8248–8255, 2002.

[618] H. S. Gutowsky, T. D. Klots, and C. E. Dykstra, J. Chem. Phys., 93: 6216–6225, 1990.

[619] H. S. Gutowsky, T. D. Klots, C. Chuang, J. D. Keen, C. A. Schmuttenmaer and T. Emilsson, J. Am. Chem. Soc., 109: 5633–5638, 1987.

[620] H. S. Gutowsky, E. Arunan, T. Emilsson, S. L. Tschopp and C. E. Dykstra, J. Chem. Phys., 103: 3917–3926, 1995.

[621] H. S. Gutowsky, C. Chuang, T. D. Klots, T. Emilsson, R. S. Ruoff and K. R. Krause, J. Chem. Phys., 88: 2919–2924, 1988.
[622] Y. Xu and W. Jäger, Phys. Chem. Chem. Phys., 2: 3549–3553, 2000.
[623] Y. Xu, M. C. L. Gerry, J. P. Connelly and B. J. Howard, J. Chem. Phys., 98: 2735–2741, 1993.
[624] M. S. Ngari and W. Jäger, J. Chem. Phys., 111: 3919–3928, 1999.
[625] Y. Xu, W. Jäger and M. C. L. Gerry, J. Mol. Spectrosc., 157: 132–140, 1993.
[626] Y. Xu, G. S. Armstrong, and W. Jäger, J. Chem. Phys., 110: 4354–4362, 1998.
[627] M. S. Ngari and W. Jäger, Mol. Phys., 99: 13–24, 2001.
[628] Y. Xu and W. Jäger, Mol. Phys., 93: 727–737, 1998.
[629] J. van Wijngaarden and W. Jäger, J. Chem. Phys., 283: 29–46, 2002.
[630] J. van Wijngaarden and W. Jäger, Phys. Chem. Chem. Phys., 4: 4883–4889, 2002.
[631] J. van Wijngaarden and W. Jäger, J. Am. Chem. Soc., 125: 14631–14641, 2003.
[632] J. van Wijngaarden and W. Jäger, J. Chem. Phys., 116: 2379–2387, 2002.
[633] E. Arunan, C. E. Dykstra, T. Emilsson and H. S. Gutowsky, J. Chem. Phys., 105: 8495–8501, 1996.
[634] E. Arunan, T. Emilsson, H. S. Gutowsky and C. E. Dykstra, J. Chem. Phys., 114: 1242–1248, 2001.
[635] R. M. Spycher, P. M. King and A. Bauder, Chem. Phys. Lett., 191: 102–106, 1992.
[636] C. Callegari, K. K. Lehmann, R. Schmied and G. Scoles, J. Chem. Phys., 115: 10090–10110, 2001.
[637] K. Nauta, D. T. Moore, P. L. Stiles, R. E. Miller, Science, 292: 481–484, 2001.
[638] S. Grebenev and J. P. Toennies, A. F. Vilesov, Science, 279: 2083–2086, 1998.
[639] Y. Xu and W. Jäger, Chem. Phys. Lett., 350: 417–422, 2001.
[640] Y. Xu and W. Jäger, J. Chem. Phys., 119: 5457–5466, 1999.
[641] J. Tang, Y. Xu, A. R. W. McKellar and W. Jäger, Science, 297: 2030–2033, 2002.
[642] A. R. W. McKellar, Y. Xu and W. Jäger, Phys. Rev. Lett., 97: 183401, 2006.
[643] Y. Xu, W. Jäger, J. Tang and A. R. W. McKellar, Phys. Rev. Lett., 91: 163401, 2003.
[644] Y. Xu, N. Blinov, W. Jäger and P.-N. Roy, J. Chem. Phys., 124: 081101, 2006.
[645] J. M. Michaud and W. Jäger, 62th International Symposium on High Resolution Spectroscopy, Columbus OHIO USA, June 18–22 2007 Lecture FB05.

RAMAN SPECTROSCOPY OF VIRUSES AND VIRAL PROTEINS

Daniel Němeček *and* George J. Thomas, Jr.

Contents

Abstract

In this chapter, we consider the implementation of methods of Raman spectroscopy, ultraviolet resonance Raman (UVRR) spectroscopy, and polarized Raman spectroscopy as structural probes of native viruses, precursor viral assemblies, and their constituent proteins and nucleic acids. Focus is placed on recent applications from the author's laboratory to icosahedral (P22 and HK97) and filamentous (fd, Pf1, Pf3, and PH75) bacteriophages. These studies provide novel insights into the many macromolecular structural transformations — including conformational rearrangements of viral proteins and nucleic acids and site-specific interactions of key molecular subgroups — that define the viral morphogenetic pathway. Emphasis is given to structural, thermodynamic, and kinetic information that is well suited to investigation by Raman methods and cannot be addressed by alternative biophysical probes. Experimental requirements appropriate to each Raman application are considered. We also review the application of the singular value decomposition (SVD) algorithm for quantitative treatment of Raman spectroscopic data collected from viral proteins and related biological molecules.

Keywords: Raman; Spectroscopy; Virus; Protein; DNA; RNA; Structure

1.   INTRODUCTION

    The vibrational Raman effect, which was predicted on the basis of theoretical considerations by Smekal in 1923 [1] and confirmed experimentally by Raman in 1928 [2], was first applied to model peptides in the 1950s by Edsall and collaborators [3–7]. Raman spectroscopy was subsequently demonstrated as an effective experimental probe of nucleic acids and native proteins by Lord and collaborators [8–11] in the late 1960s and early 1970s. With the advent of the continuous wave laser as an effective excitation source for biomolecular applications, the Raman method soon gained use as a structural probe of supramolecular biological assemblies, including viruses [12–21]. Biological applications of Raman spectroscopy are greatly advantaged over infrared spectroscopy by both the relatively low interference of liquid water in the Raman effect and the diversity of sample states amenable to Raman analysis (solutions, suspensions, gels, precipitates, fibers, single crystals, amorphous solids, etc.). These considerations facilitate the correlation of biomolecular solution structures with corresponding high-resolution structures determined by X-ray crystallography [22–26]. More recently, the development of improved statistical approaches has allowed exploitation of Raman spectroscopy for detailed quantitative analyses of thermodynamic and kinetic properties of complex biomolecular systems [27–30]. Concurrent developments in Raman instrumentation [charge-coupled device (CCD) detectors, efficient solid-state lasers, and Raman microscopes] have similarly advanced biochemical, biophysical, pharmaceutical, and commercial applications of the related techniques of polarized Raman spectroscopy, Raman optical activity (ROA), ultraviolet resonance Raman (UVRR) spectroscopy, and surface-enhanced Raman spectroscopy (SERS) [31–37].

    The objective of the present chapter is to review applications of Raman spectroscopy – including UVRR and polarized Raman spectroscopy – that address fundamental questions of virus structure and assembly. The focus is on bacterial viruses (bacteriophages) that comprise only protein and nucleic acid (either DNA or RNA) constituents. Typical questions addressed in these studies are as follows: What are the molecular mechanisms of viral shell assembly, shell maturation, and genome packaging? What protein and/or nucleic acid residues interact with one another during these processes? How are molecular conformations and hydrogen-bonding states altered in the course of assembly, maturation, and packaging processes? What are the thermodynamic parameters governing key intermolecular interactions and structural transformations? Often, the structural, mechanistic, and thermodynamic information obtainable from Raman spectra can be augmented by time-resolved data, such as that collected in the course of hydrogen/deuterium (H/D) exchanges of labile protein and nucleic acid sites. UVRR approaches offer advantages of selectivity and sensitivity over conventional (nonresonance) Raman spectroscopy, whereas polarized Raman spectroscopy is particularly well suited to determining orientations of specific moieties in viral assemblies.

2. PATHWAYS OF VIRUS ASSEMBLY

The maturation pathways of highly organized biomolecular assemblies, including viruses, typically proceed in several discrete steps. Figure 1 shows this stepwise process for a representative double-stranded (ds) DNA virus, the bacteriophage P22, which infects strains of the bacterium *Salmonella typhimurium* and is considered a paradigm for the assembly of many icosahedral dsDNA viruses [38]. As

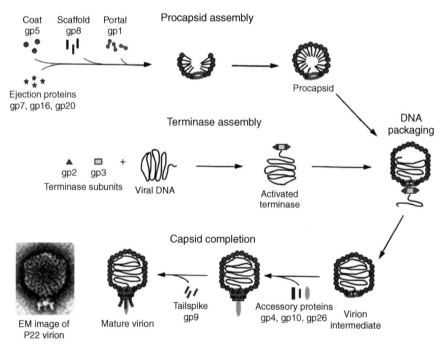

Figure 1 The assembly pathway of the *Salmonella* bacteriophage P22, which serves as a model for the assembly of many icosahedral dsDNA viruses [38]. The initial step (upper left) is the coassembly of 420 copies of the coat protein (gp5), 12 copies of the portal protein (gp1), approximately 300 copies of the scaffolding protein (gp8), and a few copies of minor ejection proteins (gp7, gp16, and gp20) to form a precursor shell (procapsid) incorporating a unique portal vertex. Subsequently, DNA packaging through the portal is fueled by ATP hydrolysis under the control of the phage terminase (subunits gp2 and gp3) and is accompanied by release of scaffolding and shell maturation. The final steps involve closure of the portal channel by accessory proteins (gp4, gp10, and gp26) and attachment of six trimeric tailspike proteins (gp9). The inset at the lower left shows an electron micrograph of the mature P22 virion. (See color plate 30).

shown in Figure 1, the pathway begins with the coassembly of 420 copies of the coat protein subunit (product of viral gene 5 or simply gp5) plus 12 copies of the portal protein subunit (gp1) and approximately 300 copies of the scaffolding protein subunit (gp8) to form a *procapsid*. The procapsid exterior consists of the gp5 shell incorporating the portal channel at a unique vertex, whereas the gp8 molecules are enclosed within. The procapsid also contains a few copies of the minor proteins gp7, gp16, and gp20. At the time of DNA packaging, the procapsid shell matures into an expanded *capsid* shell and ejects its scaffolding component. DNA translocation, which proceeds through the portal channel, is fueled by the hydrolysis of ATP through the action of a virally encoded hetero-oligomeric enzyme called terminase (proteins gp2 and gp3). The last steps in the assembly/maturation pathway involve the closure of the portal channel by accessory proteins (gp4, gp10, and gp26) and the attachment of six trimeric tailspike proteins (gp9), which serve as the host cell adhesin.

The filamentous bacterial viruses exhibit assembly pathways distinct from those of icosahedral viruses. The best studied example is that of the filamentous virion Ff (strains fd, f1, and M13) (Figure 2), which infects *Escherichia coli*. The viral particle recognizes the conjugative pilus of the host cell and exploits the host transmembrane network of proteins (Tol–Pal system) for penetration. When the viral adhesion protein (pIII) binds to the Tol–Pal complex, the viral capsid is compromised and its single-stranded (ss) DNA genome is released. The major

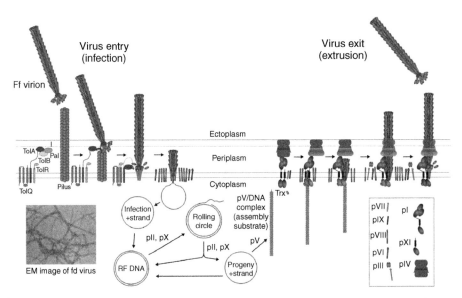

Figure 2 The life cycle and assembly pathway of the filamentous bacteriophage Ff, which infects *Escherichia coli* [39]. Essential features of membrane recognition and penetration are discussed in the text. The mature viral particle consists of a ssDNA genome in the form of a covalently closed circle that extends from one end of the capsid sheath to the other. The sheath comprises thousands of copies of the major coat protein (pVIII) plus a few copies of minor proteins at each filament end. The inset at the lower left shows an electron micrograph of the mature Ff virion. (See color plate 31).

(pVIII) and minor (pIII, pVI, pVII, and pIX) proteins of the capsid are disassembled and retained in the host inner membrane for subsequent use in the assembly of progeny virions. The ssDNA genome is converted into a replicative form for expression of viral structural and nonstructural proteins. Ultimately, the progeny virions exit the bacterial cell through a complex transmembrane channel formed from multiple nonstructural viral proteins [39]. The mature viral particle consists of the ssDNA genome in the form of a covalently closed circle extending from one end of the filament to the other and sheathed by thousands of copies of the major coat protein plus a few copies of each of two minor proteins at each of the filament ends.

3. ADVANTAGES OF RAMAN SPECTROSCOPY

Raman spectroscopy, like infrared spectroscopy, is a method for probing structures and interactions of molecules by measuring the frequencies of molecular vibrations. Although the two methods differ fundamentally in the mechanism of interaction between radiation and matter, one obtains in both cases a vibrational spectrum consisting of a number of discrete bands, the frequencies and intensities of which are determined by the nuclear masses in motion, the equilibrium molecular geometry, and the molecular force field. An important advantage of Raman over IR for applications to viral proteins, nucleic acids, and their complexes is the virtual transparency of liquid water (both H_2O and D_2O) in the Raman effect. This greatly simplifies the analysis of solution spectra and also facilitates the investigation of hydrogen-isotope-exchange phenomena. Other notable advantages of Raman spectroscopy are ease of implementation for samples in different physical states (solutions, suspensions, gels, precipitates, fibers, single crystals, amorphous solids, etc.), applicability to large supramolecular assemblies, a spectroscopic time scale ($\sim10^{-14}$ s) that is fast in comparison to both biomolecular structure transformations and protium/deuterium (H/D) exchanges, nondestructiveness of data collection protocols, relatively low mass (<1 mg) and volume ($\sim1\,\mu$L) requirements for most sampling protocols, and freedom from the need for chemical labels or probes. Additionally, a large database of Raman spectra of model compounds has been established for applicability to viral proteins and nucleic acids.

The technique of Raman difference spectroscopy provides a complementary and powerful experimental approach for applications to viruses. For example, digital difference methods can be exploited to (i) facilitate definitive Raman band assignments by detection of small frequency shifts and/or intensity changes accompanying residue-specific isotopic substitutions or mutations [40–42], (ii) assess qualitatively and quantitatively the structural changes induced in proteins and nucleic acids when solution conditions, such as temperature, pH, ionic strength, or ligand concentration, are altered [43–45], (iii) detect localized interactions and conformational changes in viruses [46–49], and (iv) probe changes in solvent-accessible surfaces along viral assembly pathways [50,51].

Methods of polarized Raman spectroscopy have also been successfully applied to viruses. The data obtained from polarized Raman measurements are particularly informative when the virus specimen can be precisely oriented with respect to the directions of polarization of incident and scattered laser light. This is readily achieved for the examination of oriented filamentous viruses in the Raman microscope [52]. A comprehensive review of applications has been given recently [37].

4.   EXPERIMENTAL RAMAN METHODS

Instrumentation for Raman spectroscopy of viruses is generally similar to that employed for investigating other macromolecules in condensed states (solutions, crystals, fibers, etc.) [31]. Diode-pumped solid-state lasers are available to provide stable excitations throughout the visible and near-infrared wavelength regions. Excitation in the 750–800 nm interval, such as that provided at 785 nm by the Invictus laser (Kaiser Optical Systems, Inc., Ann Arbor, MI, USA), offers the additional advantage of freedom from potentially interfering fluorescence transitions. Spectrometers for biomolecular applications ordinarily incorporate the holographic notch filter for rejection of elastically scattered light, a single axial transmissive grating to maximize photon throughput, and a liquid-nitrogen-cooled CCD detector to optimize spectrophotometric sensitivity and selectivity. Recent sampling innovations, which are particularly useful for studying time-dependent processes in viruses, include solution flow dialysis [53,54] and crystal diffusion [25,55]. In principle, time-resolved Raman spectroscopy can span a wide temporal domain, from tens of femtoseconds (using pulse lasers) to indefinitely long periods. The slower structural transformations that characterize viral assembly and maturation processes, including genome packaging enzyme activities, are relatively easily addressed [54,56,57].

Solutions of viruses are generally sealed in thin glass capillaries and thermostated at the appropriate temperature [58] for collection of Raman spectra. The exciting laser beam is focused on the sample volume (1–10 μL), and the light scattered at 90° is directed to the entrance slit of the spectrograph, where it is analyzed and directed to the CCD detector as shown in Figure 3 [59]. Solid samples, including oriented fibers and crystals, are most effectively probed by means of a Raman microspectrophotometer, which consists of a confocal microscope coupled to an efficient single-grating spectrometer. The microscope functions as the sample illumination chamber and typically is coupled to a video camera and monitor to simplify optical alignment and focusing of the laser beam with respect to the sample. The incident laser beam is directed onto the oriented specimen through the microscope objective; the Raman scattering, preferably at 180°, is collected with the same objective, then directed to a beam splitter, polarization analyzer, optical scrambler, monochromator, and CCD detector as shown in Figure 4 [37,60].

We note that near-infrared (NIR) lasers and NIR-optimized detectors are finding increasing use in commercial Raman spectrometers. Often, it is desirable to compare the spectra collected using such instrumentation with spectra reported

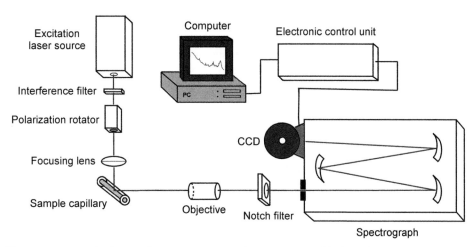

Figure 3 Optical scheme of a Raman spectrophotometer suitable for collecting spectra from solutions of viral proteins, viral nucleic acids, and their macroassemblies [59]. Typically, a 1–10 μL aliquot of dilute protein, nucleic acid, or virus solution (1–10 μg/μL concentration) is illuminated by the laser (∼100 mW power at the sample), and the Raman photons scattered at 90° are collected and directed to the monochromator and charge-coupled device detector. Rayleigh scattering is eliminated by an efficient notch filter.

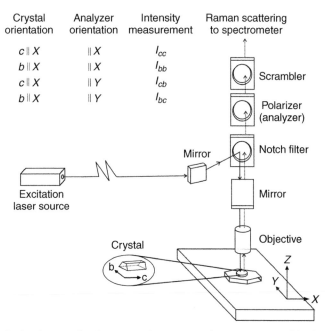

Figure 4 Optical scheme of a Raman microspectrophotometer suitable for spectral data collections from oriented fibers and crystals of viruses or their molecular constituents. The microscope, which functions as a sample illumination chamber, is coupled to a video camera and monitor to simplify laser focusing and optical alignment. The incident laser beam is directed through the microscope objective, and the Raman scattering, preferably at 180°, is collected with the same objective. Raman photons are then directed to a beam splitter, polarization analyzer, optical scrambler, monochromator, and detector [37,60].

in the literature that were collected using excitation wavelengths in the visible region. It should be kept in mind that the relative intensities of bands in NIR–excited Raman spectra of proteins or nucleic acids may differ significantly from those observed in the corresponding spectra obtained using visible excitation wavelengths [61]. This is exemplified for the case of nucleic acids by the data of Table 1.

Table 1   Raman band wavenumbers and intensities of nucleosomal calf-thymus DNA obtained with near-infrared (752 nm) and visible (532 nm) excitations [151]

752 nm		532 nm		Assignment[a]
Wavenumber	Intensity	Wavenumber	Intensity	
415	0.8	412	0.9	
432	0.7	437	0.6	
452	0.6			
472	0.7			
490s	1.0			
501	2.8	497	2.0	G, T
534	0.7	532	0.6	A
567	0.6	562	0.5	T
583	0.5	578	0.6	
596	0.8	594	1.0	C, G
626	0.4			
646	0.8	642	0.6	A, C
670	2.4	670	2.5	T
682	3.3	681	3.4	G
729	3.8	727	3.5	A
750	2.6	748	2.3	T
784	10.0	785	10.0	C
788s	9.9			bk [σ(OPO)]
835	2.9	834	2.9	bk [σ(OPO)]
894	1.3	894	1.6	bk
917	1.1	919	1.2	bk
925	1.0			bk
970	0.9	971	0.5	bk
		994	1.2	bk
1011	2.7	1013	2.2	T
1053	3.0	1052	2.2	bk [σ(CO)]
1067	2.6	1070	2.1	bk [σ(CO)]
1092	8.2	1092	8.2	bk [$\sigma(PO_2^-)$]
1107s	3.2			bk
1142	1.2	1142	0.9	T
1175	2.0	1177	1.8	T
1190	2.0	1191	1.5	bk
1212	3.5	1214	3.2	T, A
1237	3.8	1237	1.1	T
1254	5.7	1255	5.1	C, A, T
1292s	3.8	1292s	4.1	C
1301	4.8	1302	5.2	A, T
1318	3.8	1315	3.8	G

Table 1   (Continued)

752 nm		532 nm		Assignment[a]
Wavenumber	Intensity	Wavenumber	Intensity	
1336	6.7	1337	7.1	A, G
1375	8.2	1374	8.8	T, G, A
1385s	3.5			
1420	4.1	1420	3.1	A, bk, δ(C5$'$H$_2$)
1443	2.1	1443	1.1	bk, δ(C5$'$H$_2$)
1460	3.4	1461	2.1	bk, δ(C2$'$H$_2$)
1488	6.6	1488	8.8	G, A, T
1510	2.7	1510	1.9	A, C
1532	0.7	1528	0.4	C, G
1577	7.5	1577	7.6	G, A
		1600	1.6	C
1647	3.8	1650	3.3	T, σ(C4=O/ C5=C6)
1670	5.8	1668	4.6	T, σ(C4=O/ C5=C6)
1686s	4.7	1693	2.9	T, σ(C2=O)
1716s	2.4	1715	1.7	G, σ(C6=O)

σ, stretch; δ, deformation.

Wavenumbers (cm^{-1} units) are accurate to ± 1 cm^{-1}. Intensities are on an arbitrary 0–10 scale, with 10 assigned to the most intense band (784 cm^{-1}) in each spectrum; s indicates a partially resolved shoulder.

[a] A, T, G, C, and bk indicate bands assigned to base ring vibrations and the DNA backbone. More specific bond vibrational assignments, when known, are given in parentheses. See also references cited in [151].

5.   ANALYSIS OF SPECTRAL DATA

5.1.   General considerations

Raman spectra of viral proteins and nucleic acids are usually collected repetitively, and the individual data collections are averaged to achieve an acceptable signal-to-noise ratio ($\sim$10:1 or greater). The length of each accumulation (exposure time) is ordinarily dictated by the stability of the sample at the experimental conditions employed and the objective of the experiment. The total accumulation time may vary from seconds to hours, depending on the sample concentration and the intrinsic Raman band intensities. Figure 5 shows a typical viral protein Raman spectrum, which was obtained from a 2% weight/volume solution (20 µg/µL) of the bacteriophage P22 terminase small subunit (Figure 1) dissolved in 10 mM Tris buffer at pH 7.8. This result was generated by averaging 40 spectral accumulations, each of 15 s exposure. The raw data collection was also subjected to the following refinement steps to standardize the spectral format and simplify its subsequent interpretation and comparison to Raman spectra of other proteins: (i) precise wavenumber calibration versus known standards (e.g., inert gas emission spectra and/or Raman spectra of organic liquids), (ii) removal of transient drift in the

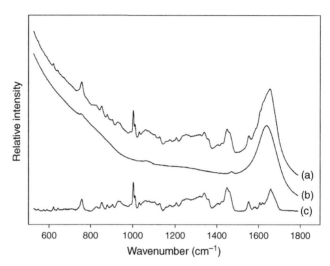

Figure 5 (a) Raman spectrum of a typical viral protein at approximately 20 µg/µL in buffer solution. The spectrum was collected in 10 min with laser power of approximately 120 mW at the sample. (b) Raman spectrum of the buffer acquired at the same conditions as (a). (c) Raman signature of the protein obtained by digital subtraction of (b) from (a).

spectral baseline by polynomial fitting, (iii) elimination of scattering contributions from the silica glass capillary by spectral subtraction, (iv) elimination of detector artifacts caused by cosmic rays or spurious light reaching the CCD, and (v) normalization of the intensities of all exposures in the accumulation to the same ordinate scale. Artificial refinements of the data, such as the use of computational algorithms for smoothing of natural spectral noise, narrowing of natural spectral bandwidths, or reduction of spectral band overlap, should be scrupulously avoided. Whenever a series of spectra is processed, particular care must be taken to avoid introduction of any systematic errors [62,63].

Quantitative analysis of the ordered secondary structures present in native proteins – specifically α-helix and β-strand conformations of the polypeptide chain – can be accomplished by decomposition of the prominent amide I (1640–1680 cm^{-1}) or amide III (1230–1310 cm^{-1}) bands of the Raman spectrum. Several approaches have been described [64–67], including decomposition of the amide I profile in terms of a relatively simple basis set [68]. Although the method of Berjot et al. is constrained by the limited reference data set upon which it is based (obtained mainly from spectra of small globular proteins), it is particularly useful for viral proteins and their assemblies, especially when quantifying a change in structure. Fourier deconvolution of complex bands can also prove helpful in certain cases although such methods should be used with caution and the results interpreted accordingly [69].

Analysis of Raman spectral changes that reflect structural transformations or interactions of viral proteins and nucleic acids can be exploited to elucidate kinetic and thermodynamic parameters. Relevant applications have been described recently [51,70]. More powerful approaches – based upon factor analysis [71] – offer greater sensitivity and selectivity and improved insight into the structural basis of biomolecular interaction. Several equivalent algorithms are available to statistically treat large sets of Raman spectra, including principal component analysis (PCA) and singular value decomposition (SVD). While the former facilitates qualitative evaluations, the latter enables testing of specific theoretical models and provides quantitative parameters [29,72–74]. Physicochemical properties and structural information are derived simultaneously and can provide a comprehensive characterization of the system under investigation.

5.2. The singular-value decomposition approach

Using the SVD algorithm, a series of N Raman spectra can be represented as $Y_i(\nu)$, where $i = 1, \ldots, N$ and ν is the number of data points in each spectrum. The set $Y_i(\nu)$ is transformed into a set of N orthonormal components (singular vectors) given as $S_j(\nu)$, where $j = 1, \ldots, N$. The vectors S_j decrease in statistical weight with increasing j. We express the statistical weights as W_j (singular values); the relative contribution of the j-th component to the i-th Raman spectrum is expressed in terms of the unitary matrix V_{ij} by Equation (1):

$$Y_i(\nu) = \sum_{j=1}^{N} W_j V_{ij} S_j(\nu) \qquad (1)$$

The series of N Raman spectra $[Y_i(\nu)]$ can be approximated within the experimental error(s) by a subset of S_j ($j = 1, 2, \ldots, M$) components, where $M < N$. This value of M also represents the number of independent spectral species in the original series and constitutes the dimension of the factor analysis [71].

The SVD approach is illustrated in Figure 6 for a simple series of spectra, viz. 28 spectra ($N = 28$) of Na_2SO_4 at two different concentrations (4 and 20 mM) in H_2O and as a function of temperature over the interval $11 < t < 70°C$. Here we have monitored the spectral interval $800–1200 \text{ cm}^{-1}$ ($n = 400$), which includes the prominent 980 cm^{-1} band of the sulfate ion (symmetrical S—O stretching mode). The SVD analysis of Figure 6 shows that all of the spectral information is expressed by three independent spectral components $S_1(\nu)$, $S_2(\nu)$, and $S_3(\nu)$. Thus, in this simple example, the number of independent species is identified as three ($M = 3$). In biomolecular systems, the number of independent species in usually not known and the determination of factor dimension by SVD can be crucial. In such cases, the corresponding SVD analysis can also provide information of direct physicochemical significance, as will be illustrated by an example below (see Section 7.8.2).

An important and convenient criterion for determination of the factor dimension in the SVD analysis is the residual error between the parent spectra and those generated from the M subspectra. As long as an increase of M results in a better than linear decrease of the residual error, the additional subspectra describe significant and relevant spectral changes [71]. Auxiliary criteria are as follows: (i) randomness of the relative contributions to the original Raman spectra (rows of the matrix V_{ij}),

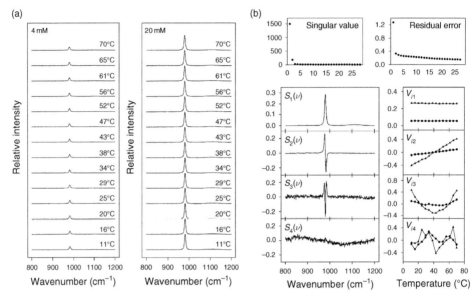

Figure 6 (a) A series of Raman spectra showing the symmetrical stretching band of the sulfate ion (SO_4^{-2}) near $980\,cm^{-1}$. Data were obtained from solutions of Na_2SO_4 (4 and 20 mM) over the temperature range $11 < t < 70°C$, as labeled. (b) Factor analysis treatment of the spectral series using the SVD algorithm and showing singular value (upper left), singular vectors with the highest singular value $S_1(\nu)$, $S_2(\nu)$, $S_3(\nu)$, $S_4(\nu)$ (lower left panel), parameter-dependent distribution of the singular vectors V_{i1}, V_{i2}, V_{i3}, V_{i4} (lower right), and residual error (upper right). The vector $S_1(\nu)$ represents a weighted average spectrum of the series, where contributions V_{i1} reflect the higher intensity of the $980\,cm^{-1}$ band in the more concentrated solution (▲). The singular vectors $S_2(\nu)$, $S_3(\nu)$, and $S_4(\nu)$ represent linearly independent spectral changes of respectively decreasing statistical importance. $S_2(\nu)$ shows the shift of the $980\,cm^{-1}$ band to lower wavenumber with increasing t; $S_3(\nu)$ shows the corresponding broadening; and S_4 (and other vectors not shown) represent random noise in the data. Relative contributions of the significant singular vectors to the original Raman spectra are expressed in the V_{i1}, V_{i2}, and V_{i3} scores, respectively.

(ii) signal-to-noise ratio in the component $S_j(\nu)$, and (iii) singular value W_j. The factor dimension can, however, exceed the number of genuine spectral components owing to the presence of various artifacts in the Raman spectra. Typical artifacts may be (i) nonuniform background correction within the series, (ii) wavenumber scale drift, or (iii) traces of parasitic signals, such as the contributions from the glass capillary cell. Exceptional care must therefore be taken to eliminate all artifacts during the processing of the spectral series.

Overall, SVD accomplishes a substantial reduction in the size of the experimental data set, generally from many experimentally obtained spectra (N) to a few computed subspectra (M). Fitting of a theoretical model leads to additional data reduction (elimination of the spectral components). In effect, the model is applied only to the relevant columns of the matrix V_{ij} multiplied by their corresponding singular values W_j [29,72–74].

6. GENERAL CHARACTERISTICS OF VIRAL PROTEIN AND NUCLEIC ACID RAMAN SPECTRA

The structural interpretation of Raman and UVRR spectra of viral proteins and nucleic acids requires definitive spectral band assignments, often referred to as group frequencies. Considerable effort has been devoted to this objective by numerous investigators, and the subject has been adequately reviewed elsewhere [37,44,45,61,75–77]. In brief, the required definitive assignments can be achieved by diverse approaches, including (i) comparison of Raman wavenumbers and intensities with corresponding infrared spectral data; (ii) determination of spectral wavenumber shifts accompanying stable isotope substitutions, usually $^1H \rightarrow$ ^2H (D), $^{12}C \rightarrow {}^{13}C$, $^{16}O \rightarrow {}^{18}O$, or $^{14}N \rightarrow {}^{15}N$; (iii) effects of pH, temperature, and/or environmental parameters on the Raman and infrared spectra; (iv) detailed and well-analyzed Raman and infrared spectra of simpler molecules, which are structurally related to the viral protein or nucleic acid; (v) Raman analyses of single crystals of related molecules, the structures of which have been authenticated by X-ray diffraction analysis, or analysis of corresponding solution data obtained by high-resolution NMR methods; (vi) *ab initio* calculations of intramolecular force fields and normal coordinate calculations based upon them; (vii) depolarization ratios of the Raman bands for randomly oriented molecules; (viii) polarized Raman measurements on oriented fibers or single crystals, and (ix) Raman excitation profiles, which correlate Raman intensities with laser excitation wavelengths.

Generally, Raman spectra of viral proteins are dominated by localized vibrations of multiply-bonded atoms, such as the amide vibrational modes (amide I and amide III) associated with the peptide group (C^α—NH—C=O) and aromatic ring modes associated with the amino acid side chains of tyrosine (Tyr), tryptophan (Trp), and phenylalanine (Phe). The sulfhydryl (S—H) stretching mode of cysteine (Cys) is also a significant contributor. Conversely, contributions from vibrational modes of nonaromatic side chains are generally weak. However, when the numbers of such side chains are large, their cumulative contributions can be significant. Additionally, chemical groups that are common to many side chains, such as the $C^\beta H_2$ moiety, can contribute importantly to the protein Raman spectrum[41]. Prominent bands represented in Raman spectra of H_2O and D_2O solutions of most viral proteins and their corresponding group assignments are summarized in Table 2.

Raman spectra of viral nucleic acids (DNA and RNA) are similarly dominated by spectral bands attributable to ring vibrations involving multiply-bonded atoms of the purine and pyrimidine nucleotide bases (A, T or U, G, C). Symmetrical stretching modes of the phosphodiester (C—O—P—O—C) and phosphodioxy (O—P=O) linkages in the sugar–phosphate backbone are also major contributors [78]. By comparison, the vibrational modes of the furanosyl methylenes in the deoxyribosyl (DNA) or ribosyl (RNA) sugars lead to feeble Raman scattering. However, as noted above for aliphatic groups of nonaromatic amino acid side chains of proteins, the intrinsically weak Raman scattering of the furanosyl groups can achieve prominence in the spectra of DNA or RNA because of the relatively large numbers of such groups represented in the nucleic acid chain. Prominent

Table 2 Raman wavenumbers and assignments for the prominent bands in spectra of H_2O and D_2O solutions of proteins [41,86,130]

Wavenumber $(cm^{-1})^a$		Assignment
H_2O	D_2O	
621	621	Phe
643	642	Tyr
699	–	Amide V
722	722	Met, Tyr, $\delta(C—C—C)$
758	754	Trp, Ala
825	825	Tyr (Fermi doublet)
850	850	Tyr (Fermi doublet)
877	–	Trp
897	887	Lys, Arg, Ala
–	910	Amide III′, side chains
–	927	Amide III′ (β-strand)
934	–	$\nu(C^\alpha—C^\beta)$ (α-helix)
–	950	Amide III (α-helix)
957	–	Val. Leu, $\delta(CH_3)$
–	963	Amide III′ (coil)
–	992	Amide III′ (extended conformation)
1003	1003	Phe
1012	–	Trp
1032	1032	Phe
1060	–	Lys, Arg, Ala, $\nu(C—C, C—O)$
1081	–	Lys, $\nu(C—N)$
1101	–	Pro, Ala, $\nu(C^\alpha—N)$, $\nu(C—C)$
1127	–	Trp, $\nu(C—C)$, $\nu(C—N)$
1156	–	$\delta(CH_3)$
1174	1174	Tyr, $\delta(CH_3)$
1206	1206	Tyr, Phe, Trp
1230–1240	–	Amide III (β-strand)
1245–1255	–	Amide III (coils, turns)
1264–1272	–	Amide III (α-helix)
1264	1260	Tyr
1300	–	Amide III (α-helix)
1318	1318	$\delta(CH_2)$
1340	–	Trp (Fermi doublet), $\delta(C—H)$
1358	–	Trp (Fermi doublet), $\delta(C—H)$
1406	1404	$\nu(CO_2)$
1450	1450	$\delta(CH_2)$, $\delta(CH_3)$
–	1458	Amide II′ (α-helix)
–	1470	Amide II′ (extended conformation)
1552	1552	Trp
1605	1608	Phe, Tyr
1618	1614	Tyr, Trp
1645–1654	1633–1640	AmideI/I′ (α-helix)
1665–1674	1655	Amide I/I′ (β-strand)
1660–1665	–	Amide I/I′ (coil)
1684	1669	Amide I/I′ (extended conformation)

ν, stretching vibration; δ, deformation vibration.
^{a} Wavenumber values are accurate to $\pm 1\,cm^{-1}$.

bands in the Raman spectrum of a chromosomal DNA fiber and the corresponding assignments are summarized in Table 1 [43–45]. The same Raman spectral signature is obtained from B DNA in H_2O solution.

6.1. The protein Raman spectrum

6.1.1. Amide I and amide III modes: markers of main chain conformation

The Raman spectrum of a specific viral protein, the terminase small subunit (gp3) of bacteriophage P22, is shown in Figure 7. Assignments for the most prominent spectral bands of gp3 correspond to the listing given in Table 2. The dominating amide I and amide III markers, which occur respectively in the intervals 1640–1680 cm^{-1} and 1225–1320 cm^{-1} [65,76,79–82], are particularly useful as quantitative indicators of protein secondary structure [64,68]. Deuteration effects on these amide markers, which shifts them respectively to approximately 1630–1670 cm^{-1} (designated amide I′) and approximately 900–1000 cm^{-1} (amide III′), have also been profitably exploited [17,83,84].

In addition to the amide I/I′ and amide III/III′ markers, other conformation-sensitive bands have been identified in Raman spectra of peptide model compounds and proteins (Table 3). For example, vibrations localized in the peptide main chain near 950 and 1330 cm^{-1}, although not considered amide modes, are sensitive quantifiers of α-helix content [85,86]. The amide II′ mode ($\sim$1400–1500 cm^{-1}),

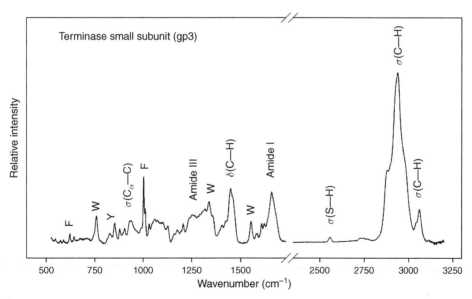

Figure 7 Raman signature of the terminase small subunit (gp3) of bacteriophage P22. The spectrum is divided into (i) the region 520–1750 cm^{-1}, where the often utilized amide modes of the polypeptide main chain and skeletal stretching and bending modes of the side chains predominate, and (ii) the often neglected region 2300–3200 cm^{-1}, where hydrogenic stretching modes of cysteinyl SH and aliphatic and aromatic CH groups occur. Assignments for many prominent peaks are shown. Labels indicate stretching (σ) and bending (δ) modes of specific bonds or assigned side chain markers (1-letter amino acid abbreviations). See also Table 2.

Table 3 Raman amide I and amide III bands of representative polypeptide and protein secondary structures [43–45]

Molecule	Amide I	Amide III	Secondary structure
α-poly-L-alanine	1655	1265–1348	α-helix
α-poly-L-glutamate	1652	1290	α-helix
α-poly-L-lysine	1645	1295–1311	α-helix
β-poly-L-alanine	1669	1226–1243	β-strand
β-poly-L-glutamate	1672	1236	β-strand
β-poly-L-lysine	1670	1240	β-strand
poly-L-lysine, pH 4	1665	1243–1248	Irregular
poly-L-glutamate, pH 11	1656	1249	Irregular
fd subunit	1650	1270–1300	α-helix
Pf1 subunit	1650	1270–1300	α-helix
cI repressor (1–102)	1650	1286–1297	α-helix
	1675	1245	Turns/irregular
P22 subunit	1653	1273	α-helix
	1655	1235	β-strand/turns
P22 tailspike	1668	1231	β-strand
	>1668	1249	Turns/irregular

which involves peptide C—N stretching and achieves prominence only in spectra of deuterated proteins, can also be exploited for conformational analyses [51,87].

Because all Raman amide modes typically undergo significant and well-characterized frequency shifts with peptide group deuteration (NH → ND), they are conveniently exploited as probes of main chain dynamics as well as conformation [17,51].

6.1.2. Tryptophan markers

Raman bands of tryptophan have been investigated extensively, and many useful structural correlations have been developed for tryptophan side chains in proteins [88–93]. The intense and sharp tryptophan marker in the interval 1540–1560 cm^{-1} is indicative of the magnitude of the indolyl side chain torsion angle $\chi^{2,1}$ ($C^{\delta 1}$— C^{γ}—C^{β}—C^{α}). The tryptophan Fermi doublet with components at 1340 and 1360 cm^{-1} reflects the local environment of the indolyl ring, such that the intensity ratio I_{1360}/I_{1340} increases with increasing hydrophobicity. The tryptophan marker near 875 cm^{-1} is somewhat sensitive to indolyl NH hydrogen bonding. Strong N—H$\cdots$X interactions lead to a band center near 871 cm^{-1}, whereas a hydrophobic environment shifts the band center to 883 cm^{-1}. The breathing vibration of the indolyl ring generates a tryptophan marker near 755 cm^{-1}, the intensity of which has been considered inversely related to the hydrophobicity of the local ring environment. Typically, the band intensity increases with decreasing hydrophobicity of the indolyl ring environment. Assignments, deuteration effects, and structural interpretations for tryptophan markers have been discussed in greater detail by Takeuchi [92]. Recently, the Raman markers associated with a nonhydrogen-bonding environment for the tryptophan side chain were inferred from a comprehensive vibrational analysis of the indolyl moiety in the gas phase [93].

6.1.3. Tyrosine markers

It has long been recognized that the para-substituted phenyl ring of tyrosine can generate a pair of relatively intense Raman bands near 830 and $850\,\mathrm{cm}^{-1}$ that corresponds to a Fermi interaction doublet [10,94]. The correlation between the ratio of intensities of the doublet components (I_{850}/I_{830}) and the hydrogen-bonding state of the phenolic OH group, which was initially quantified by Siamwiza et al. [94] and later extended to the nonhydrogen-bonded phenoxyl state by Arp et al. [95], is quite useful in viral protein applications [40,47,52,96,97]. The doublet intensity distribution is governed by Fermi coupling between the phenolic ring-stretching fundamental expected near $840\,\mathrm{cm}^{-1}$ and the first overtone of the phenolic ring deformation mode occurring near $420\,\mathrm{cm}^{-1}$, such that the following intensity distribution applies: (i) if the tyrosine phenoxyl group functions as a strong hydrogen-bond donor only, then $I_{850}/I_{830} = 0.30$, (ii) if the phenoxyl group functions as both a donor and an acceptor, then $I_{850}/I_{830} = 1.25$, (iii) if the phenoxyl group functions as a strong hydrogen-bond acceptor only, then $I_{850}/I_{830} = 2.5$, and (iv) if the phenoxyl group is not hydrogen bonded, then $I_{850}/I_{830} = 6.7$. For a protein containing a single tyrosine, the I_{850}/I_{830} ratio is unambiguously interpretable. Conversely, if the protein contains multiple tyrosines, then the observed I_{850}/I_{830} reflects the average hydrogen-bonding state of all tyrosine phenoxyls. In the latter eventuality, site-directed isotopic substitutions or mutagenesis can be exploited to deduce the hydrogen-bonding states of individual tyrosyl residues [40]. The nonhydrogen-bonded phenoxyl state, although unlikely to occur for tyrosines either within folded globular proteins or at solvent-exposed protein surfaces, is often encountered in the tightly packed α-helical subunits of filamentous virus capsids [40,95,96].

6.1.4. Cysteine and cystine markers

The Raman band resulting from the cysteine sulfhydryl bond (S—H) stretching vibration is a unique probe of the local S—H structure and dynamics of the cysteine side chain [59,98,99]. The sulfhydryl Raman marker occurs in a region of the spectrum (2500–$2600\,\mathrm{cm}^{-1}$) that is virtually devoid of interference from any other Raman band of a protein or its aqueous solvent. The relatively large Raman cross section associated with the S—H stretching mode and the sensitivity of the bond stretching vibrational frequency to hydrogen-bonding interactions and side chain conformation are well documented [99]. The stoichiometric dependence of the S—H Raman band on thiol concentration also confers a unique spectrophotometric value on the S—H marker for assessing the pK_a values of thiol/thiolate equilibria [100,101]. Because the S—H marker requires the reduced state of cysteine, it can also prove useful for distinguishing this state from its oxidized counterpart, cystine, when either is possible in a protein. Correlations established between sulfhydryl hydrogen bonding and the wavenumber of the S—H bond stretching vibration are summarized in Table 4.

Raman bands originating from bond stretching vibrations of the disulfide network of cystine (C^{α}—C^{β}—S—S—C^{β}—C^{α}) have been considered in detail [10,11,102–105]. Most useful is the disulfide (S—S) stretching marker, which

Table 4 Dependence of the Raman S—H stretching band on the hydrogen-bonding state of the cysteinyl sulfhydryl group [98,99]

Hydrogen-bonding state	Wavenumber[a] (cm^{-1})	Bandwidth[b] (cm^{-1})
No hydrogen bond	2581–2589	12–17
S acceptor	2590–2595	2–17
Weak S—H donor	2575–2580	20–25
Moderate S—H donor	2560–2575	25–30
Strong S—H donor	2525–2560	35–60
S—H donor and S acceptor	2565–2575	30–40

[a] Range of values observed for the peak of the Raman S—H stretching band.
[b] Full width of the Raman S—H stretching band at half maximum.

ordinarily occurs within the 150–550 cm^{-1} interval and is sensitive in frequency to the C^β—S—S—C^β dihedral angle [106].

6.1.5. Histidine markers

Raman bands of histidine and the histidinium ion are relatively weak compared with those of the aromatic amino acids (tyrosine, tryptophan, and phenylalanine). However, when the protein does not contain many aromatic residues, some histidine bands can be identified. At neutral pH, the Raman spectrum of the imidazolium ring of histidine exhibits several weak bands, whereas at acidic pH, their Raman intensities are redistributed to fewer bands of intermediate intensities. In its deuterio form, the imidizolium ring exhibits a relatively strong band at 1408 cm^{-1}, which has been assigned to the N_1—C_2—N_3 symmetric stretch [107]. The band is particularly useful for monitoring histidine H/D exchange dynamics as well as histidine–histidinium equilibria in D_2O [108,109]. A linear correlation between the frequency of the 1408 cm^{-1} Raman band and the histidine side chain conformation has been proposed [92,110]. The histidine markers are also sensitive to the binding of transition metals at imidazole sites, and the Raman spectrum has been extensively exploited for monitoring such interactions [92,111].

6.2. The nucleic acid Raman spectrum

The concept of localized group frequencies, as referred to above, may break down in biological macromolecules as a consequence of mechanical coupling between vibrational modes of covalently linked groups. Although this phenomenon is rarely encountered in Raman spectra of proteins, it occurs often in Raman spectra of nucleic acids – in the form of coupling between modes of the nucleotide bases and those of the sugar–phosphate moiety. The resulting delocalized vibrational modes may then generate Raman bands that combine the relatively high intensity typical of localized purine and pyrimidine (base) vibrational modes with the conformational dependency expected of localized backbone (sugar–phosphate) modes. Several such Raman bands have been identified for both DNA and RNA [112,113]. These Raman bands manifest dependence of frequency and/or intensity on the nucleic acid macromolecular conformation (secondary structure) and are often referred to as Raman conformation markers.

Two categories of Raman conformation markers have been delineated [114]: those relating to the overall conformation of the sugar–phosphate backbone and those relating to the conformation of the constituent nucleosides. The former are sensitive primarily to changes in the torsions of the phosphodiester bond network of the backbone (—P—O5′—C5′—C4′—C3′—O3′—P—) but do not depend strongly on the identity of the base attached at C1′. These are listed in Table 5. Thus, the occurrence of unique backbone conformation markers for the three known secondary structures of DNA, Z DNA ($745 \pm 3 \, \text{cm}^{-1}$), A DNA ($807 \pm 3 \, \text{cm}^{-1}$) and B DNA ($835 \pm 7 \, \text{cm}^{-1}$), facilitates the use of Raman spectroscopy to distinguish the A, B, and Z conformations from one another. The relatively large range quoted in Table 5 for the B marker ($\pm 7 \, \text{cm}^{-1}$) reflects the structural complexity of B DNA in comparison to A and Z DNAs. In fact, the diversity of Raman markers observed for B DNAs of different base composition suggests a family of structures, the precise conformations of which are dependent on the AT and GC base pair content and sequence. Thus, regions of B DNA containing predominantly GC and CG base pairs exhibit the B marker near $830 \, \text{cm}^{-1}$, whereas those containing predominantly AT and TA pairs exhibit the marker near $840 \, \text{cm}^{-1}$. This distinction is due to the slightly different phosphodiester torsions associated with the groove dimensions of GC/CG-rich (narrow minor groove) and AT/TA-rich (wide minor groove) sequences. Consistent with this finding is a resultant broadband centered near $835 \, \text{cm}^{-1}$ in genomic DNAs containing comparable percentages of GC/CG and AT/TA pairs.

Nucleoside conformation markers, on the contrary, originate primarily from certain base ring stretching vibrations that are weakly coupled via the glycosyl bond to motions of the furanose ring atoms. These are listed in Table 6. For each nucleoside listed in Table 6, the markers are sensitive to either furanose pucker, as measured by torsion angle δ (defined by atoms C5′—C4′—C3′—O3′ of the sugar), or glycosyl orientation, as measured by the torsion angle χ (defined by atoms O4′—C1′—N1—C2 or O4′—C1′—N9—C4 for pyrimidine or purine nucleosides, respectively), or both.

Table 5 Raman bands diagnostic of the A, B, and Z backbone conformations of double-stranded DNA

Mode type	A DNA	B DNA	Z DNA
σ(C—O)	706 ± 5		
σ_{sym}(O—P—O)	807 ± 3^a	790 ± 5	745 ± 3
σ_{sym}(O—P—O)		835 ± 7^b	
σ_{sym}(PO$_2^-$)	1099 ± 1	1092 ± 1	1095 ± 2
δ(C2′H$_2$)	1418 ± 2	1420 ± 2	1425 ± 2

σ, stretching; sym, symmetric; δ, deformation.
Wavenumbers (cm^{-1} units) are determined from Raman spectra of DNA crystals and fibers of known structure.
[a] The corresponding band in A RNA is observed at $813 \pm 2 \, \text{cm}^{-1}$.
[b] In B DNA containing both GC and AT pairs, the band is broad and centered close to $835 \, \text{cm}^{-1}$. In B DNA containing only GC or only AT pairs, the band is sharp and centered, respectively, close to 830 or $840 \, \text{cm}^{-1}$.

Table 6   Raman bands diagnostic of the conformations of deoxynucleosides and nucleosides of nucleic acids

Residue	C3' endo/anti	C2' endo/anti	C3' endo/syn	C2' endo/syn
G	664 ± 2^a	682 ± 2	625 ± 3	671 ± 2
	1318 ± 2	1333 ± 3^b	1316 ± 2	1324 ± 2
A	644 ± 4	663 ± 2	624 ± 3	
	1335 ± 2	1339 ± 2	1310 ± 5	
C	780 ± 2	782 ± 2	784 ± 2	
	1252 ± 2	1255 ± 5	1265 ± 2	
T	642 ± 2	665 ± 2		
	745 ± 2	748 ± 2		
	777 ± 2	790 ± 3		
	1239 ± 2	1208 ± 2		

Wavenumbers (cm^{-1} units) are determined from Raman spectra of DNA crystals and fibers of known structure.
[a] The corresponding band in structures containing rG is usually observed at $668 \pm 1\,cm^{-1}$.
[b] A weak companion band near $1316\,cm^{-1}$ is also observed in *B* DNA.

These considerations apply also to RNA, with the important distinctions that the base uracil (U) replaces thymine (T) and the sugar ribose replaces deoxyribose. Importantly, only one type of secondary structure occurs for dsRNA, namely *A* RNA, which exhibits many architectural and spectroscopic similarities to *A* DNA.

The following additional generalizations can be made regarding relative Raman band intensities of the nucleic acid constituents: (i) in-plane stretching vibrations of conjugated CC, CN, and CO double bonds of the bases yield Raman bands of the highest intensity. Conversely, out-of-plane ring vibrations yield relatively weak bands [115]. (ii) Symmetrical stretching vibrations of the diesterified orthophosphate moiety, namely phosphodioxy (PO_2^-) and phosphodiester (C—O—P—O—C) group vibrations, generate intense Raman bands. On the contrary, antisymmetrical and asymmetrical modes of the orthophosphate moiety are usually too weak to be observed [116]. (iii) Vibrations localized in the sugar moieties of DNA (deoxyribose) or RNA (ribose) exhibit very low Raman intensities [113,115–117]. (iv) On the basis of Raman excitation profiles reported by several groups [118–122], UVRR intensities of the nucleic acid bases are generally greatest when excitation wavelengths in the vicinity of 260 nm are employed. Sugar and phosphate group vibrations are absent from UVRR spectra excited at wavelengths in the range 200–300 nm [123].

7.   APPLICATIONS IN STRUCTURAL STUDIES OF VIRUSES

7.1.   Assembly and maturation of the bacteriophage P22 capsid

The shell of bacteriophage P22 is constructed from 420 copies of a 47-kDa coat protein subunit (gp5) (Figure 1). Raman spectroscopy has provided details of the self-assembly of gp5 into the viral shell precursor (procapsid) and the subsequent rearrangement of the subunits of the procapsid shell to form the architecture of the mature capsid lattice [17,51,124,125]. Structural characterization of the initial assembly step, that is the formation of the procapsid shell from gp5 monomers, was made possible by improvements in the sensitivity of Raman instrumentation that allow acquisition of spectra of the monomer at very low concentration [125]. Although Raman signatures of gp5 in the monomeric and procapsid shell states share many features, it is clear that procapsid assembly significantly perturbs the structure of the unassembled monomer (Figure 8). The computed difference spectrum, as shown in the bottom trace of Figure 8, highlights the spectral changes accompanying monomer-to-procapsid assembly. Importantly, the center of the Raman amide I marker is shifted from 1667 cm^{-1} in the monomer to 1663 cm^{-1} in the procapsid shell. This shift gives rise to the Raman difference peaks observed at 1655 cm^{-1} (assigned to α-helix) and 1673 cm^{-1} (β-strand) and corresponding troughs at lower and higher wavenumber values. The difference band pattern reflects a more ordered secondary structure for the subunit in the viral shell than in the monomer at the expense of disordered chain segments. The more highly ordered secondary structure of the shell subunit is confirmed in the Figure 8 difference spectrum by the diminished intensity of the Raman amide III band of the monomer (1251 cm^{-1} band center), which leads to a difference trough at

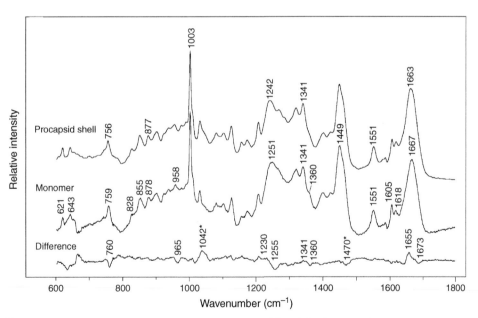

Figure 8 Raman characterization of in vitro assembly of the bacteriophage P22 procapsid. Data were obtained with 514 nm excitation at low gp5 concentration (~15 µg/µL) and temperature (~10°C) [125]. The spectral difference profile (bottom trace) between gp5 monomer (middle trace) and shell (top trace) is modest. The amide band peaks and troughs reflect only minor reordering of the polypeptide backbone. Other differences are relatively small but significant, consistent with coupling between subunit folding and assembly. The monomer fold is, nevertheless, largely conserved with shell assembly.

$1255\,\mathrm{cm}^{-1}$ (diminished irregular chain conformation) and difference peak at $1230\,\mathrm{cm}^{-1}$ (enhanced β-strand). These results show that the polypeptide backbone of the gp5 monomer undergoes refolding during procapsid assembly. Overall, about $4 \pm 1\%$ of the backbone, or 17 of 428 peptide bonds, are estimated to undergo refolding. Thus, assembly of the P22 procapsid leads to local reordering of a small but significant percentage of the peptide residues and demonstrates limited coupling of subunit folding with the shell assembly process. The native fold of the gp5 subunit is thus largely conserved with procapsid assembly.

In addition to the changes in subunit secondary structure noted above, the Raman spectrum also provides information on the effect of procapsid assembly on subunit tertiary structure. For example, the shift of the tryptophan (Trp) marker band at $760\,\mathrm{cm}^{-1}$ in the monomer to $756\,\mathrm{cm}^{-1}$ in the procapsid shell with a concomitant large decrease in intensity signifies a change in the amphipathic environment of the average Trp side chain. Broadening of the $1360\,\mathrm{cm}^{-1}$ component of the Trp Fermi doublet indicates that this change is due to a more hydrophilic indolyl environment in the procapsid shell subunit. Overall, the change in average Trp environment is small. Additional Raman markers show that the cysteine (Cys), sulfhydryl, and tyrosine (Tyr) phenoxyl hydrogen-bonding environments are essentially invariant to the assembly process. Thus, despite measurable refolding of the coat protein backbone with procapsid assembly, the tertiary structure of the subunit is not greatly affected [125].

The folding state of gp5 in the monomeric and procapsid shell states has also been characterized by deuterium exchange of peptide NH groups [125]. Figure 9a compares Raman profiles of the exchanged ND groups (amide III′ region, 900–$1000\,\mathrm{cm}^{-1}$) of gp5 in monomeric and procapsid shell states, following exposure of each to D_2O (pD 7.8) under native conditions (2°C for monomer and 10°C for procapsid shell). Also shown is the amide III′ profile of a gp5 control in which all NH groups were exchanged by deuterium under denaturing conditions prior to refolding of the fully exchanged protein in native D_2O buffer. These data show that NH → ND exchange is more extensive in the monomer than in the procapsid shell subunits. In particular, greater amide III′ intensity is observed at 930–945 cm^{-1} assignable to deuterated α-helix [79,126], at $960\,\mathrm{cm}^{-1}$ assignable to deuterated disordered peptides [51,127,128], and at 980–990 cm^{-1} assignable to deuterated β-strand [51,79,127,128]. The amide III′ exchange profile of the monomer indicates that the native fold of the monomer comprises both unprotected and protected residues in regions of α-helix, β-strand, and disordered secondary structures. Figure 9a also shows extensive exchange protection. Neither the monomer nor procapsid shell subunit undergoes complete NH → ND exchange of α-helical, β-stranded, or disordered regions. The data suggest that the most protected residues of the monomer are those located in regions of β-strand structure.

Raman-monitored H/D exchange of selected side chains can be used to identify residues of the protein folding core [43]. For example, the degree of protection of the S—H group of the single cysteine (Cys 405) in the coat protein is a measure of the accessibility of the C-terminal region to H/D exchange. Figure 9b depicts Raman spectra in the region 1840–1890 cm^{-1}, where the stretching vibration of the deuterated sulfhydryl group (S—D) is expected [125]. The absence of an S—D

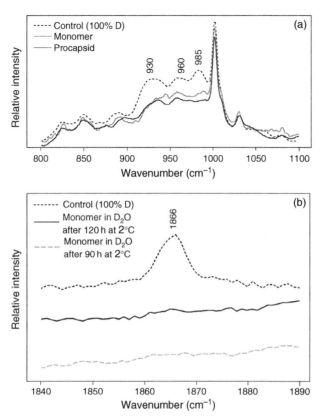

Figure 9 Normalized Raman signatures (514 nm excitation) depicting native-state peptide NH → ND exchanges of P22 gp5 in the monomeric state (2°C) and procapsid shell state (10°C), each following 12 h exposure to D₂O buffer. (a) Amide III′ profiles of the P22 procapsid, native gp5 monomer and fully exchanged gp5 control. (b) Deuteriocysteinyl S—D stretching bands of the native gp5 monomer, following different conditions of exposure to D₂O [125].

band in the spectrum of the native gp5 monomer shows that the Cys 405 sulfhydryl is fully protected against H/D exchange. Protection persists for samples maintained at native conditions (pD 7.8, 2°C) for up to 6 days. The S—H group can be exchanged only at conditions that favor global unfolding (40°C). Thus, the Cys 405 sulfhydryl is buried within the hydrophobic folding core of the gp5 monomer. A similar degree of sulfhydryl exchange protection is observed when gp5 is assembled into the procapsid shell, but not upon maturation to the capsid shell [51]. This shows that the procapsid-to-capsid lattice transformation is accompanied by sequestering of the subunit C-terminal region from H/D exchange.

In summary, during P22 shell assembly, about 20 peptide linkages of the gp5 monomer undergo local refolding without greatly affecting the global structure of the monomer. Compared to monomeric gp5, the procapsid subunits achieve enhanced protection against peptide H/D exchange. On the contrary, the monomer possesses a robust exchange-protected folding core that encompasses the Cys 405 sulfhydryl. Both the monomeric and procapsid states of the subunit constitute a distinct and compact late-folding intermediate, which ultimately matures to the native state of the capsid. Details of the additional changes in gp5 subunit structure and exchange protection that occur with the final procapsid → capsid lattice transformation (Figure 1) have been quantified by Raman methods and are discussed in detail elsewhere [51].

7.2. The scaffolding protein of bacteriophage P22

Efficient assembly of the properly dimensioned procapsid of bacteriophage P22 requires the participation of the virally encoded scaffolding protein (gp8, 34 kDa). The scaffolding protein, which is not found in the mature virus particle, is ejected from the procapsid at the time of DNA packaging and recycled for use in subsequent rounds of de novo procapsid assembly. On average, approximately 300 copies of gp8 per procapsid participate in the in vivo P22 assembly process (Figure 1) The molecular mechanisms of interaction of gp8 with other proteins of the procapsid shell, such as the coat (gp5), portal (gp1), and minor proteins (gp7, gp16, and gp20), are not known. The scaffolding protein is highly soluble and has provided an excellent target for Raman spectroscopic analysis [17,129,130]. The Raman results show that (i) the native gp8 secondary structure is highly α-helical, (ii) the α-helices melt noncooperatively, and (iii) the protein lacks an exchange-protected folding core.

7.3. The portal of bacteriophage P22

The portal of bacteriophage P22 is self-assembled from 12 copies of an 84-kDa subunit (gp1; Figure 1). The assembly process is accompanied by little change in the amide I and amide III markers of the protein, but significant changes in Raman markers sensitive to side chain local environments [18]. Thus, the gp1 tertiary structure, but not secondary structure, is altered by dodecamer formation (Figure 10). For example, intensity changes in Raman markers of the gp1 tyrosines, most of which (20 of 26) occur within the central 50% of the sequence, implicate these residues in the subunit–subunit interface. Assembly-related perturbations to Raman markers of tryptophan, the majority of which are located within the N-terminal segment of gp1, implicate the N-terminal domain in the subunit interface. Detailed analysis of the amide I and amide III markers (Figure 11) demonstrates the marginal changes occurring in gp1 secondary structure with portal assembly.

Figure 11c shows the assembly-related Raman spectral changes that occur in the region of the S—H stretching vibrations ($2500–2600 \, \mathrm{cm}^{-1}$) of gp1 because of changes in the local environments of the four cysteines per subunit (Cys 153, Cys 173, Cys 283, and Cys 516). The spectral profile of the monomer is relatively broad with the central peak at $2562 \, \mathrm{cm}^{-1}$ and with a prominent shoulder of nearly equal intensity at $2557 \, \mathrm{cm}^{-1}$. This suggests that all four cysteines of the subunit are engaged in relatively strong S—H $\cdots$ X hydrogen bonds. The sulfhydryl bands are collectively upshifted in the spectrum of the dodecamer, which is indicative of general weakening of the S—H $\cdots$ X hydrogen bonds. The major peak appears at $2566 \, \mathrm{cm}^{-1}$ and a less intense though broader peak occurs at $2587 \, \mathrm{cm}^{-1}$, indicative

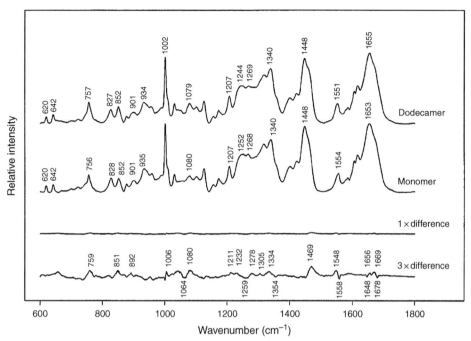

Figure 10 Raman signatures (532 nm excitation) of the P22 portal protein monomer (middle trace) and dodecamer (top trace) and the digital difference spectrum (second trace from the bottom) obtained by subtracting the spectrum of the monomer from that of the ring. Also shown (bottom trace) is an approximate three-fold amplification of the difference spectrum. Proteins were dissolved to a final concentration of approximately 40 mg/mL in 20 mM Tris (pH 7.5), 100 mM NaCl, and 2 mM EDTA [18].

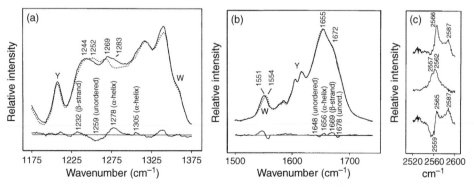

Figure 11 Raman profiles in the amide III (panel a) and amide I (panel b) regions of the P22 portal protein, gp1, for monomeric (···) and dodecameric (—) states. Raman SH bands of dodecameric (top trace) and monomeric (middle trace) gp1 are shown in panel c. In each panel, the digital difference spectrum between monomer and decamer is included as the bottom trace [18].

of moderately strong and weak hydrogen bonds, respectively. Consequently, portal assembly likely disrupts two relatively strong intrasubunit S—H $\cdots$ X linkages and results in exposure of Cys sites, enabling weaker S—H $\cdots$ OH$_2$ hydrogen bonds. The absence of any assembly-related changes in the 500–550 cm^{-1} interval of the Raman spectrum confirms that no disulfide bridges are formed [18].

The roles and interactions of the cysteine residues in the portal subunit were further investigated by difference Raman spectroscopy of single-site Cys → Ser mutants of the portal subunit [131,132]. Raman spectra of the wild-type (WT) and mutant subunits are shown in Figure 12. The Raman signature of each Cys site was obtained by spectral subtraction of the WT and mutant spectra. Table 7 summarizes the classifications of the S—H $\cdots$ X hydrogen bonds in both the unassembled and the assembled subunits of these portal variants. The results show that the composite S—H Raman band envelope of the WT monomer comprises moderate S—H $\cdots$ X interactions from Cys 173 and Cys 283 and strong S—H $\cdots$ X interactions from Cys 153 and Cys 516. An analogous analysis of the assembled portal spectra suggests that each S—H side chain of the WT portal protein participates in two types of hydrogen-bonding interactions and that the population distribution is specific to each cysteine site. The population distributions, estimated on the basis of integrated band intensities [42,98], are listed in Table 7. The rank order of increasing average hydrogen bond strength in the WT portal is Cys 173 < Cys 516 < Cys 283 < Cys 153. The Raman spectra of the monomeric and dodecameric WT subunits correspond to the sum of the spectra of the single-site mutants. This excellent agreement between the experimental and the summation spectra shows that the hydrogen-bonding properties of all cysteine sites in the portal subunit are mutually independent. Importantly, the in vitro assembly

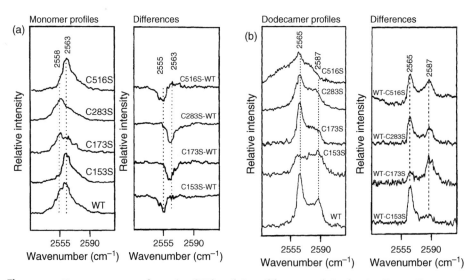

Figure 12 Raman spectra of cysteine SH bands in wild-type and single-site Cys → Ser mutant variants of gp1. (a) Monomeric gp1 profiles (left) and derived SH signatures (right). (b) Dodecameric gp1 profiles (left) and derived SH signatures [131,132].

Table 7 Cysteine S—H Raman markers and S—H···X hydrogen-bond strengths of the P22 portal protein in monomeric and dodecameric forms [131,132]

Cysteine site	Monomer state[a]		Dodecamer state[b]	
	cm^{-1}	H-bond	cm^{-1} (abundance)	H-bond
Cys 153	2558	Strong	2565 (80%)	Moderate
			2587 (20%)	Very weak
Cys 173	2563	Moderate	2565 (24%)	Moderate
			2587 (76%)	Very weak
Cys 283	2563	Moderate	2565 (60%)	Moderate
			2587 (40%)	Very weak
Cys 516	2558	Strong	2565 (50%)	Moderate
			2587 (50%)	Very weak

[a] Data for the monomeric portal subunit in 100 mM NaCl solution.
[b] Data for the dodecameric portal ring in 400 mM NaCl solution. Numbers in parentheses are the percentages of the total SH Raman band intensity contributed at the indicated wavenumber value.

properties of each of the single-site mutants of gp1 have been quantified and serious assembly defects identified only for the Cys 516 and Cys 283 mutants. Additionally, the in vitro assembly defects correspond precisely with in vivo defects of P22 virions incorporating the mutant portals [132].

7.4. The terminase of bacteriophage P22

Recent studies of the terminase large (gp2, 58 kDa) and small (gp3, 19 kDa) subunits of bacteriophage P22 using spectroscopic, hydrodynamic, and imaging techniques have provided the first insights into the structure and assembly of the P22 genome packaging machine [20]. Improvements in the sensitivity of Raman instrumentation and in the precision of spectral data analyses have yielded excellent vibrational profiles of the two terminase subunits. Figure 13a shows the Raman spectrum of gp2 obtained from a highly dilute solution (0.7 mg/mL). The amide I (1664 cm^{-1} peak and broad shoulder at lower wavenumber) and amide III (1241 cm^{-1} peak and shoulder at higher wavenumber) bands identify an α/β-fold. Quantitative analysis of the amide I profile [68] indicates 36% α-helix and 36% β-strand in the native structure. This α/β-fold is typical of ATPase-activated DNA-translocating motors [133], whereas the significant amount (28%) of unordered structure in gp2 may reflect a flexible linker between the ATPase domain and one or more additional modules (endonuclease domain) of the protein. More complete folding of the protein may also be dependent on assembly with gp3 and/or DNA to form an activated terminase complex.

The terminase small subunit (gp3) self-assembles into a highly stable ring that is readily visualized by electron microscopy [20]. Figure 13b shows the Raman signatures of the distinctive rings formed by the WT gp3 protein (nonamer) and the single-site gp3 mutant, Ala 112 → Thr (decamer). Despite the dramatic difference in quaternary structure, the Raman spectra are very similar, indicative

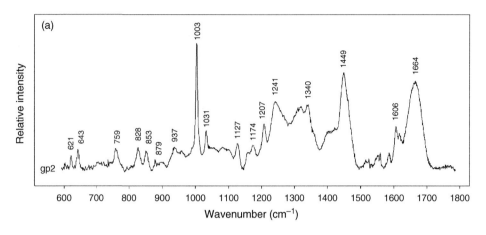

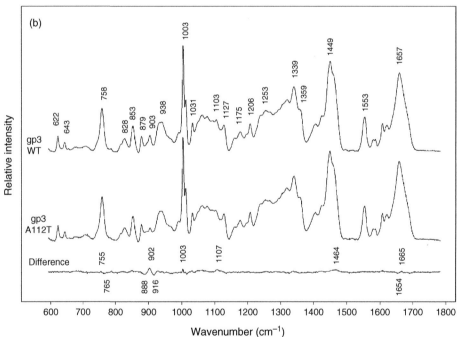

Figure 13 (a) Raman signature of the P22 terminase large subunit (gp2) at 0.7 mg/mL. (b) Raman signatures of the P22 terminase small subunit (gp3) at 20 mg/mL, showing the wild-type protein (WT, top trace), the single-site Ala 112 → Thr mutant (A112T, middle trace), and their spectral difference (bottom trace). All prominent difference bands in (b) can be attributed to the substitution of Thr for Ala at position 112, indicating that the subunit fold (secondary and tertiary structures) is not significantly perturbed by the mutation [20].

of a common subunit fold. The major Raman difference bands between the WT and mutant (A112T) rings (bottom trace of Figure 13b) can be assigned to the different amino acid side chain at position 112. Indeed, the amide I ($1657 \, \text{cm}^{-1}$) and amide III ($1253 \, \text{cm}^{-1}$) bands of both WT and A112T gp3 signify a highly α-helical subunit fold. Decomposition of the amide I band [68] of WT gp3 gives 40% α-helix, 22% β-strand, and 38% irregular structure. The computed difference spectrum between WT and A112T suggests at most a marginal reduction (~1.2% or two peptide bonds) in α-helical secondary structure with the mutation.

Sulfhydryl Raman markers of the two cysteine side chains (Cys 32 and Cys 33) of gp3 generate a single overlapping band centered at $2557 \, \text{cm}^{-1}$ (data not shown). Both cysteines therefore participate in strong S—H···X hydrogen bonds, which are unaffected by the A112T mutation. Because the A112T mutation perturbs the protein quaternary structure (from nonamer to decamer), but not the Cys Raman signatures, it is likely that Ala 112, but neither Cys 32 nor Cys 33, is located at the intersubunit interface. The presence of strong and assembly-independent S—H···X bonding further suggests that Cys 32 and Cys 33 may be located within the subunit folding core.

7.5. The tailspike of bacteriophage P22

The molecular structure of the trimeric tailspike protein (gp9, 71 kDa) of bacteriophage P22 has been solved at high resolution by methods of X-ray crystallography [134–136]. While the eight cysteines of the gp9 main chain (Cys 169, Cys 267, Cys 287, Cys 290, Cys 458, Cys 496, Cys 613, and Cys 635) are identified in the native fold (Figure 14), the locations of the S—H protons and their possible

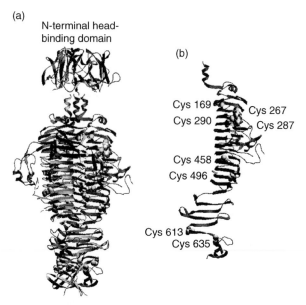

Figure 14 (a) Ribbon diagram of the X-ray crystal structure of the trimeric P22 tailspike protein (gp9), showing from top-to-bottom the N-terminal head-binding domain, the central parallel β-helix (or β-coil) domain, and the C-terminal β-prism domain [134–136]. (b) The β-helix and β-prism domains of a single gp9 subunit, showing the locations of the eight cysteines with space-filling atoms [42]. (See color plate 32).

S—H···X hydrogen-bonding states remain unresolved in the crystal structure. Cysteine sulfhydryl interactions of gp9 are considered important because selective and transient oxidation of a subset of S—H groups to the disulfide state may be involved in the trimer assembly pathway [137].

The eight cysteines of the trimeric tailspike subunit generate a prolific S—H stretching band in the 2500–2600 cm^{-1} region of the Raman spectrum [42,138,139]. The complexity of the band in the spectrum of the WT tailspike can be deconstructed by the exploitation of single-site Cys → Ser mutations. Thus, the unique Raman S—H signatures of each of the eight cysteines have been resolved and interpreted to identify putative hydrogen-bonding interactions relevant to the trimer assembly pathway [42]. Relevant results are shown in Figure 15, which reveals a diverse and unexpected range of S—H···X interactions in the native structure. On the basis of previously established correlations between the state of sulfhydryl hydrogen bonding and the Raman S—H frequency [98,99], the tailspike cysteines can be ranked according to the strength of S—H···X hydrogen bonding

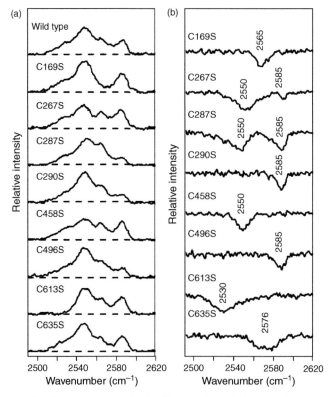

Figure 15 Raman SH signatures of P22 tailspike cysteines. (a) Raman S—H profiles observed for the wild-type and single-site Cys → Ser mutant variants of gp9, as labeled. (b) Raman difference spectra (computed as mutant minus wild-type) for each tailspike variant revealing the specific SH signature of the indicated cysteine [42].

Table 8 Cysteine S—H Raman markers and S—H···X hydrogen-bond strengths of the P22 tailspike protein [42]

Cysteine site	Wavenumber (cm^{-1})^a	H-bond strength
Cys 169	2565	Moderate
Cys 267	2550 (90%)	Strong
	2585 (10%)	Very weak
Cys 287	2550 (63%)	Strong
	2585 (37%)	Very weak
Cys 290	2585	Very weak
Cys 458	2550	Strong
Cys 496	2585	Very weak
Cys 613	2530	Very strong
Cys 635	2576	Weak

^{a} Numbers in parentheses are the percentages of the total SH Raman band intensity contributed at the indicated wavenumber value.

as follows: Cys 613 > Cys 458 > Cys 267 > Cys 287 > Cys 169 > Cys 635 > Cys 290 = Cys 496 (Table 8). The sulfhydryl of Cys 613 forms the strongest S—H···X hydrogen bond so far encountered in a protein. Interestingly, this cysteine is also one of the candidates for assembly-dependent transient disulfide bond formation [137].

The Raman data of Figure 15 also indicate that the Cys signatures are not coupled to one another, which is similar to the situation observed for cysteines of the P22 portal protein. These studies demonstrate a general approach for probing cysteine environments and local SH interactions in viral proteins and their assemblies.

7.6. Accessory proteins of the bacteriophage P22 portal/tail vertex

Raman spectra of the portal accessory proteins of P22 (gp4, 18 kDa; gp10, 52 kDa; and gp26, 25 kDa) [140] have been obtained at low protein concentrations, as shown in Figure 16 (unpublished results of S. A. Overman and G. J. Thomas, Jr.). At the conditions of these experiments, gp4 and gp10 remain monomeric, whereas gp26 forms a highly thermostable trimeric coil of three largely α-helical chains [141]. The Raman spectrum of gp26 exhibits the amide I (1653 cm^{-1}) and amide III (1270 cm^{-1}) markers expected for a largely α-helical structure. The Raman spectrum of monomeric gp4 similarly shows Raman amide markers (1653 and 1270 cm^{-1}) diagnostic of a highly α-helical secondary structure, whereas gp10 exhibits Raman markers (1671 and 1242 cm^{-1}) indicative of high β-strand content.

7.7. Filamentous bacteriophages

The capsid of a filamentous bacterial virus, such as that of the coliphage Ff, consists of several thousand copies of a major coat protein (pVIII) and a few copies of minor proteins at the filament ends (Figure 2). Fiber X-ray diffraction from highly oriented specimens of Ff provides useful information about the symmetry of the

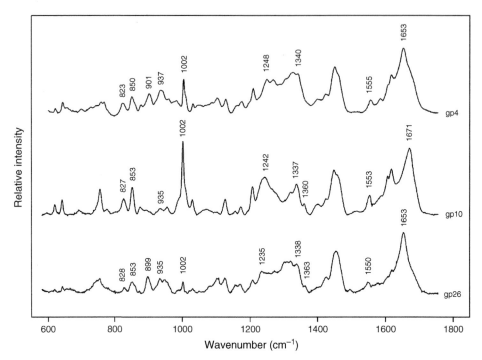

Figure 16 Raman signatures of the portal completion proteins of bacteriophage P22, gp4 (top), gp10 (middle), and gp26 (bottom). Proteins were dissolved to a final concentration of ~1 mg/mL in 10 mM Tris (pH 7.8), 300 mM NaCl [39].

capsid subunit arrangement. Also, the electron density deduced from the fiber diffraction patterns is consistent with a pVIII capsid subunit containing a highly α-helical secondary structure [142]. Raman spectroscopy has been used extensively over the past several decades to probe further details of the molecular architectures of Ff and other filamentous bacteriophages of similar (class I) and distinctive (class II) capsid symmetries. Methods of off-resonance Raman, UV-resonance Raman, and polarized Raman have all been exploited effectively for this purpose. Details of pVIII side chain orientations and hydrogen-bonding interactions, as well as information about the conformation and organization of the packaged ssDNA within the viral capsids, have been obtained. Comprehensive reviews of these applications have been given recently [37,61]. Here, we consider representative applications.

Knowledge of the viral capsid symmetry and appropriate Raman tensors (which are independently determined from polarized Raman studies of model compounds of known structure) permits assessment of viral subgroup orientations with respect to the filament axis. For example, knowledge of the amide I Raman tensor [143] enables the polarized Raman spectrum of the class I Ff phage (strain fd) to be interpreted in terms of the angle of orientation of the subunit α-helix axis with respect to the virion axis (16 ± 4°) [144]. Similar protocols were applied to the polarized Raman spectra of the class II bacteriophages Pf1, Pf3, and PH75

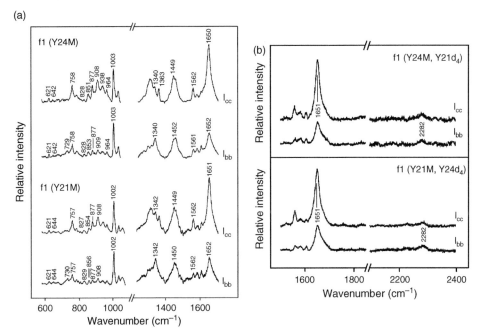

Figure 17 (a) Polarized Raman spectra (I_{cc} and I_{bb}) in the spectral regions 600–1100 and 1200–1700 cm^{-1} of oriented fibers of two variants of the Ff virus (strain f1), incorporating the single-site mutations Tyr 24 → Met [f1(Y24M), top two traces] and Tyr 21 δ Met [f1(Y21M), bottom two traces] in their respective capsid subunits. Raman intensities are normalized to the intensity of the isotropic marker band at 1002 cm^{-1} (Phe). (b) Polarized Raman spectra (I_{cc} and I_{bb}) in the spectral regions 1600–1800 and 2200–2400 cm^{-1} of oriented fibers of the same Ff variants in (a), which additionally carry tetradeuterio isotopomers of Tyr 21 [f1(Y24M,Y21d$_4$)] and Tyr 24 [f1(Y21M,Y24d$_4$)] in their respective capsid subunits. Accordingly, the Raman bands in the C–D stretching region are markers of the individual tyrosines [97].

[19,52,85,145]. Polarized Raman methods were also used to determine orientations of key tryptophan and tyrosine side chains in the subunits of class I and class II filamentous viruses [19,97,145–147]. Representative data are shown in Figure 17.

Polarized Raman bands of the deoxynucleoside residues of the packaged ssDNA genomes in filamentous viruses have also been structurally interpreted. Thus, bands at 680 (dG), 720 (dA), 750 (dT), and 782 cm^{-1} (dC) were exploited to infer the organization of the packaged Pf1 DNA genome [52]. The results show that (i) the packaged ssDNA genome is ordered specifically with respect to the capsid super-helix and the bases are nonrandomly arranged along the capsid interior, (ii) deoxynucleosides are uniformly in the C2′-endo/anti conformation, and (iii) the average DNA phosphodioxy group (PO$_2^-$) is oriented so that the line connecting the oxygen atoms (O···O) forms an angle of 71 ± 5° with the virion axis.

Analysis of the polarized Raman amide I and amide III signatures of the filamentous virus PH75 reveals that while the capsid subunit is predominantly α-helical (87%), it also contains a significant number of residues (6 ± 1 or 13 ± 3%) differing from the canonical α-helix. This minor structural component is not apparent in the capsid subunits of other filamentous phages (fd, Pf1, and Pf3). The Raman spectrum also manifests local environments for alanine, tryptophan, and tyrosine side chains that are distinct from those of fd, Pf1, or Pf3. These structural anomalies were proposed to contribute to the high thermostability of the PH75 virion [145]. Interestingly, the polarized Raman spectra of PH75 reveal that (i) the average peptide group of the PH75 capsid subunit is oriented with greater angular displacement from the virion axis (25°) than peptide groups of fd, Pf1, or Pf3 subunits (16°), (ii) the indolyl ring of Trp 37 in the PH75 capsid subunit projects nearly equatorially from the α-helix axis, in contrast to the more axial orientations that occur for Trp 26 of fd and Trp 38 of Pf3, and (iii) the phenolic rings of Tyr 15 and Tyr 39 of PH75 project along the subunit α-helix axis, and one phenoxyl engages in hydrogen-bonding interaction that has no counterpart in tyrosines of either fd or Pf1.

7.8. Assembly and maturation of the bacteriophage HK97 head

7.8.1. Structural considerations

The assembly pathway of the *E. coli* bacteriophage HK97, which differs fundamentally from that of P22, is shown schematically in Figure 18. Icosahedral structures of precursor (*prohead*) and mature (*head*) states of the bacteriophage HK97 shell have been determined by application of electron cryomicroscopy and X-ray crystallography to shells constructed in vitro from recombinantly expressed subunits (gp5, 42 kDa) [148,149]. Mechanistic steps involving the initial assembly of the prohead (Prohead I), its transformation to Prohead II, and eventually its maturation to the Head are not well understood. Of particular interest is the structure and role of the gp5 N-terminal segment (Δ-domain, residues 1–102), which is cleaved from the Prohead I subunits by a virally encoded protease. This cleavage results in the reorganization of the shell lattice and the Prohead II architecture [150].

Raman spectra of Prohead I and Prohead II particles identify a conserved subunit domain with an α/β-fold, as well as a highly α-helical (74%) Δ-domain specific only to the Prohead I subunit (Figure 19a). The diagnostic tryptophan (1339 and 1551 cm^{-1}) and tyrosine (825 and 850 cm^{-1}) Raman markers are identical in

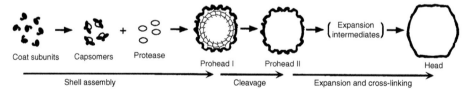

Coat subunits   Capsomers   Protease

Prohead I   Prohead II   Head

Shell assembly   Cleavage   Expansion and cross-linking

Figure 18 Schematic representation of the in vitro assembly pathway of the icosahedral capsid of *Escherichia coli* bacteriophage HK97. This pathway differs fundamentally from that of P22 (Figure 1) in that the precursor shell (Prohead I) is assembled from the coat of the protein monomer (gp5) via an oligomeric intermediate (capsomer) and without the participation of scaffolding protein subunits. Prohead I subsequently submits to a virally encoded protease that cleaves from gp5 the N-terminal domain (Δ-domain, residues 1–102), yielding the Prohead II shell. Subsequent shell expansion and subunit cross linking generate the mature Head architecture [86,150].

subunits of both shells and indicate minimal changes in the tertiary structure of the C-terminal segment of gp5 (residues 104–385). However, a shift of the sulfhydryl Raman marker of Cys 362 from 2579 to 2573 cm^{-1} indicates a strengthening of the cysteinyl S—H $\cdots$ X bond with the prohead shell transformation [86].

The Raman signature of the Δ-domain was calculated by subtracting the Prohead II spectrum from the Prohead I spectrum (Figure 19b) [86]. Interestingly, this difference spectrum exhibits a close correspondence to the Raman spectrum of the P22 scaffolding protein [17] and supports the notion that the Δ-domain fulfills a role

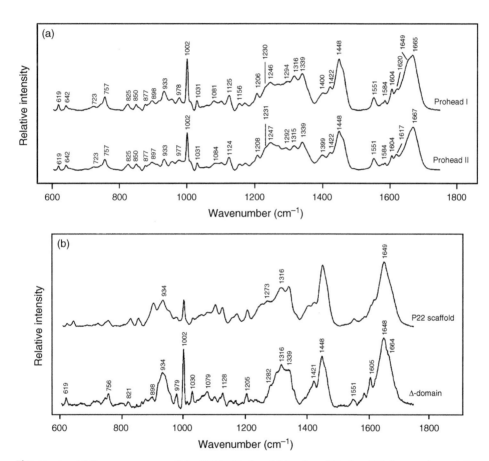

Figure 19 (a) Raman spectra of the HK97 Prohead I (top) and Prohead II (bottom) particles. The two spectra have been scaled to match the intensities of environment-invariant tyrosine Raman markers at 642 cm^{-1}, as both particles contain the same number of tyrosine residues. (b) The Raman spectrum of the Δ-domain (bottom) obtained by subtracting the Raman spectrum of Prohead II from that of Prohead I. The intensity scale is expanded three-fold over that shown in a. The Raman signature of the P22 scaffolding protein (top), which is shown for comparison with that of the Δ-domain, reveals the fundamental similarities between the two protein structures. (c) Raman spectra of the HK97 mature Head (top) and Prohead II precursor (second from top), each normalized to the intensity of the 1449 cm^{-1} peak. The computed difference spectrum (bottom) is also shown [86].

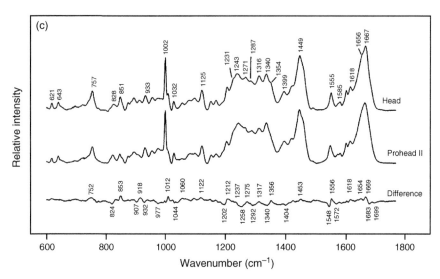

Figure 19 (Continued)

in HK97 assembly similar to that of the scaffolding protein in P22 assembly. In contrast, the maturation of Prohead II to the Head involves a significant reordering of the secondary structure of gp5 (residues 104–385), as well as extensive changes in the subunit tertiary structure. The difference spectrum between the Prohead II and the Head of HK97 (Figure 19c) demonstrates (i) an increase in ordered secondary structure, (ii) altered interactions of tyrosine and tryptophan side chains, and (iii) additional strengthening of the S—H $\cdots$ X bond of the Cys 362 side chain.

7.8.2. Thermodynamic properties of the Δ-domain

Recently, Raman spectra of exceptional signal-to-noise quality were obtained from highly purified samples of Prohead I and Prohead II as a function of solution temperature in the interval 10–92°C (Figure 20a and b). Quantitative analysis of the data allows elucidation of the thermostability of the Δ-domain within the framework of the prohead shell assembly pathway (D. Němeček and G. J. Thomas, Jr., unpublished results). For both Prohead I and Prohead II, major spectral changes are observed between 85 and 92°C, which can be attributed to thermal denaturation of the subunits that are released upon shell disassembly. Spectral changes between 10 and 85°C, on the contrary, reflect initial partial unfolding events in the shell subunits followed by the onset of shell disassembly. In the case of Prohead I, the observed changes include the effects of unfolding of the Δ-domain. The latter were deciphered and quantified by factor analysis of the Prohead I and Prohead II spectral series using the SVD algorithm. Figure 20c and d shows the factor analysis results for the amide I bands of Prohead I and Prohead II, inclusive of the third SVD component (i.e., V_{i3} terms). The contribution of the second component (V_{i2} term) to the respective parent spectrum reveals two sigmoidal transitions for Prohead I, but only a single transition for Prohead II. The first transition for Prohead I can be

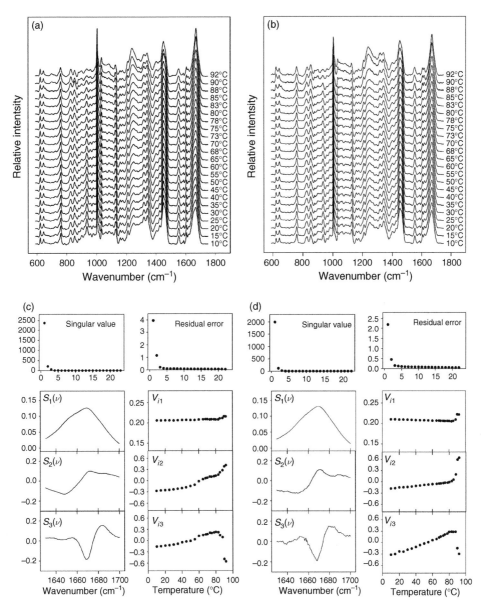

Figure 20 Raman spectra of the bacteriophage HK97 Prohead I (a) and Prohead II (b) shells were collected over the temperature interval $10 < t < 92°C$. Changes of secondary structure were analyzed by factor analysis (SVD algorithm) of the amide I conformational marker (1630–1700 cm^{-1}) of Prohead I (c) and Prohead II (d). Each spectral set can be described within the experimental noise level by three independent components. The scores V_{i2} of the second component manifest two distinct structural transitions in Prohead I, whereas Prohead II exhibits only one transition. For both Prohead I and Prohead II, the high-temperature transition (~89°C) corresponds to melting (thermal denaturation of the subunit). The lower temperature transition in Prohead I (~60°C) represents unfolding of the Δ-domain (D. Němeček and G. J. Thomas, Jr., manuscript in preparation).

attributed to unfolding of the Δ-domain, whereas the second likely corresponds to Prohead I disassembly. Only the second transition is observed for Prohead II, consistent with the absence of the Δ-domain from the Prohead II subunit. Both parent spectral series can be approximated within the precision of the observed noise level by three spectral components [$S_1(\nu)$, $S_2(\nu)$, and $S_3(\nu)$]. Consequently, each spectral series is said to contain three independent spectral species. For Prohead I, these are (i) the natively folded, (ii) the Δ-domain unfolded, and (iii) the globally unfolded shell subunit (residues 1–385). For Prohead II, these are (i) the natively folded, (ii) the locally unfolded, and (iii) the globally unfolded shell subunit (residues 104–385). Species (ii) of the Prohead II spectral series represents only a small structural change in comparison to that of species (ii) of Prohead I. Thermodynamic parameters that describe the stability of the Δ-domain in Prohead I were obtained by fitting V_{ij} scores of the significant components. A specific thermodynamic model for Δ-domain unfolding has been developed from these results [30].

Raman spectra of the independent spectral species considered in the SVD analysis were also derived from the fitted parameters [27]. It can be concluded from these studies that the Prohead I assembly confers structural thermostability on the Δ-domain and enhances the cooperativity of Δ-domain unfolding. This is accomplished by a secondary structural fold – specifically α-helix – that is converted to a β-strand conformation during Δ-domain unfolding. In a forthcoming publication, these results will be further discussed [30].

8. OVERVIEW AND CONCLUSIONS

Raman, UV-resonance Raman, and polarized Raman spectroscopy are valuable methods for investigating structures of viral proteins and nucleic acids. The methods are also well suited to probing the assembly pathways and architectures of native viruses. The sensitivity and selectivity shared by these spectroscopic approaches are well exemplified by the applications reviewed here to icosahedral and filamentous viruses and their constituent molecules. Significant developments in Raman instrumentation and data processing, coupled with the diversity of available sample-handling procedures and molecular biological techniques, have vastly improved the scope of Raman methods for studies not only of viruses but also of other biological macromolecular assemblies.

In the applications considered above, it must be stressed that accurate Raman vibrational band assignments are essential. Only with reliable assignments can the Raman spectrum be effectively exploited to investigate molecular structural details, including sites of intermolecular recognition and local and global changes in structure attendant with biomolecular recognition events. We have shown how the use of isotope-editing procedures and recombinantly engineered mutational substitutions can be used to advance Raman vibrational assignments that lead to definitive interpretations of the spectroscopic data. Statistical treatment of the spectra is also shown to facilitate quantitative analysis of physicochemical properties and provide improved understanding of the biological significance of the results.

The Raman structural applications considered here exemplify the important roles of local hydrogen-bonding interactions and global conformations in directing viral protein recognition and assembly events. Additionally, we have shown how Raman methods can be implemented to assess kinetic and thermodynamic parameters governing structural transitions of viral proteins and transformations of virion architecture.

Finally, we note that the Raman methods discussed here are often applicable to problems in virus structure analysis that are not amenable to investigation by alternative biophysical probes.

ACKNOWLEDGMENTS

We thank Dr. Stacy A. Overman for assistance with the preparation of illustrations and proofreading. Research support from the National Institutes of Health (Grant GM50776 to GJT) is gratefully acknowledged.

REFERENCES

[1] A. G. Smekal, Die Naturwissenschaften, 43: 873–875, 1923.
[2] C. V. Raman and K. S. Krishnan, Indian J. Phys., 2: 419, 1928.
[3] J. T. Edsall, J. W. Otvos and A. Rich, J. Am. Chem. Soc., 72: 474–477, 1950.
[4] D. Garfinkel and J. T. Edsall, J. Am. Chem. Soc., 80: 3823–3826, 1958.
[5] D. Garfinkel, J. Am. Chem. Soc., 80: 3827–3831, 1958.
[6] D. Garfinkel and J. T. Edsall, J. Am. Chem. Soc., 80: 3818–3823, 1958.
[7] D. Garfinkel and J. T. Edsall, J. Am. Chem. Soc., 80: 3807–3812, 1958.
[8] R. C. Lord and G. J. Thomas, Jr., Biochim. Biophys. Acta, 142: 1–11, 1967.
[9] R. C. Lord and G. J. Thomas, Jr., Dev. Appl. Spectrosc., 6: 179–199, 1968.
[10] R. C. Lord and N. T. Yu, J. Mol. Biol., 50: 509–524, 1970.
[11] R. C. Lord and N. T. Yu, J. Mol. Biol., 51: 203–213, 1970.
[12] K. A. Hartman, N. Clayton and G. J. Thomas, Jr., Biochem. Biophys. Res. Commun., 50: 942–949, 1973.
[13] G. J. Thomas, Jr. and P. Murphy, Science, 188: 1205–1207, 1975.
[14] G. J. Thomas, Jr., B. Prescott, P. E. McDonald-Ordzie and K. A. Hartman, J. Mol. Biol., 102: 103–124, 1976.
[15] G. J. Thomas, Jr., B. Prescott and D. E. Olins, Science, 197: 385–388, 1977.
[16] T. Miura and G. J. Thomas, Jr., Biochemistry, 34: 9645–9654, 1995.
[17] R. Tuma and G. J. Thomas, Jr., Biophys. Chem., 68: 17–31, 1997.
[18] A. Rodriguez-Casado, S. D. Moore, P. E. Prevelige, Jr. and G. J. Thomas, Jr., Biochemistry, 40: 13583–13591, 2001.
[19] M. Tsuboi, S. A. Overman, K. Nakamura, A. Rodriguez-Casado and G. J. Thomas, Jr., Biophys. J., 84: 1969–1976, 2003.
[20] D. Němeček, E. B. Gilcrease, S. Kang, P. E. Prevelige, Jr., S. Casjens and G. J. Thomas, Jr., J. Mol. Biol., 374: 817–836, 2007.
[21] M. Tsuboi, J. M. Benevides and G. J. Thomas, Jr., Biophys. J., 92: 928–934, 2007.
[22] T. J. Thamann, R. C. Lord, A. H. J. Wang and A. Rich, Nucleic Acids Res., 9: 5443–5457, 1981.
[23] T. Li, Z. Chen, J. E. Johnson and G. J. Thomas, Jr., Biochemistry, 31: 6673–6682, 1992.

[24] T. Miura and G. J. Thomas, Jr., Biochemistry, 33: 7848–7856, 1994.

[25] M. S. Helfand, M. A. Taracila, M. A. Totir, R. A. Bonomo, J. D. Buynak, van den Akker F. and P. R. Carey, Biochemistry, 46: 8689–8699, 2007.

[26] M. A. Totir, M. S. Helfand, M. P. Carey, A. Sheri, J. D. Buynak, R. A. Bonomo and P. R. Carey, Biochemistry, 46: 8980–8987, 2007.

[27] J. Hanuš, J. Štěpánek, P.-Y. Turpin and J. Bok, J. Mol. Struct., 480–481: 437–442, 1999.

[28] J. Hanuš, I. Barvík, K. Ruszová-Chmelová, J. Štěpánek, P. Y. Turpin, J. Bok, I. Rosenberg and M. Petrová-Endová, Nucleic Acids Res., 29: 5182–5194, 2001.

[29] M. Procházka, J. Štěpánek, P. Y. Turpin and J. Bok, J. Phys. Chem. B, 106: 1543–1549, 2002.

[30] D. Němeček, S.A. Overman, R.W. Hendrix and G.J. Thomas, Jr., Biochemistry (submitted, 2008).

[31] D. Pappas, B. W. Smith and J. D. Winefordner, Talanta, 51: 131–144, 2000.

[32] I. R. Lewis and H. G. M. Edwards (Eds.), Handbook of Raman Spectroscopy: From the Research Laboratory to the Process Line, Marcel Dekker, New York, 2001.

[33] E. W. Blanch, L. Hecht, C. D. Syme, V. Volpetti, G. P. Lomonossoff, K. Nielsen and L. D. Barron, J. Gen. Virol., 83: 2593–2600, 2002.

[34] A. V. Mikhonin and S. A. Asher, J. Phys. Chem. B, 109: 3047–3052, 2005.

[35] W. R. Premasiri, D. T. Moir, M. S. Klempner, N. Krieger, G. Jones and L. D. Ziegler, J. Phys. Chem. B, 109: 312–320, 2005.

[36] A. V. Mikhonin, S. V. Bykov, N. S. Myshakina and S. A. Asher, J. Phys. Chem. B, 110: 1928–1943, 2006.

[37] M. Tsuboi and G. J. Thomas, Jr., Polarized Raman and polarized FTIR spectroscopy, in Methods in Protein Structure and Stability Analysis Vibrational Spectroscopy (V.N. Uversky and E.A. Permyakov, eds.), pp. 153–194, Nova, Hauppage, New York, 2007.

[38] S. Casjens and P. Weigele, DNA packaging by bacteriophage P22, in Viral Genome Packaging Machines: Genetics, Structure, and Mechanism (C. E. Catalano, ed.), pp. 80–88, New York, Plenum, 2005.

[39] S. A. Overman and G. J. Thomas, Jr., Filamentous ssDNA bacterial viruses, in Encyclopedia of Virology (B. Mahy, B. and M. van Regenmortel, eds.), pp. xxx–xxx, Elsevier, Oxford, UK, in press.

[40] S. A. Overman, K. L. Aubrey, N. S. Vispo, G. Cesareni and G. J. Thomas, Jr., Biochemistry, 33: 1037–1042, 1994.

[41] S. A. Overman and G. J. Thomas, Jr., Biochemistry, 38: 4018–4027, 1999.

[42] S. W. Raso, P. L. Clark, C. Haase-Pettingell, J. King and G. J. Thomas, Jr., J. Mol. Biol., 307: 899–911, 2001.

[43] G. J. Thomas, Jr., Annu. Rev. Biophys. Biomol. Struct., 28: 1–27, 1999.

[44] R. Tuma and G. J. Thomas, Jr., Raman spectroscopy of viruses, in Handbook of Vibrational Spectroscopy, Volume 5, Applications in Life, Pharmaceutical and Natural Sciences (J. M. Chalmes and P. R. Griffiths, eds.), pp. 3519–3535, Wiley, Chichester, U.K., 2002.

[45] J. M. Benevides, S. A. Overman and G. J. Thomas, Jr., J. Raman Spectrosc., 36: 279–299, 2005.

[46] K. L. Aubrey, S. R. Casjens and G. J. Thomas, Jr., Biochemistry, 31: 11835–11842, 1992.

[47] Z. Q. Wen, S. A. Overman and G. J. Thomas, Jr., Biochemistry, 36: 7810–7820, 1997.

[48] S. A. Overman, K. L. Aubrey, K. E. Reilly, O. Osman, S. J. Hayes, P. Serwer and G. J. Thomas, Jr., Biospectroscopy., 4: S47–S56, 1998.

[49] Z. Q. Wen, A. Armstrong and G. J. Thomas, Jr., Biochemistry, 38: 3148–3156, 1999.

[50] K. E. Reilly and G. J. Thomas, Jr., J. Mol. Biol., 241: 68–82, 1994.

[51] R. Tuma, P. E. Prevelige, Jr. and G. J. Thomas, Jr., Proc. Natl. Acad. Sci. U.S.A., 95: 9885–9890, 1998.

[52] M. Tsuboi, Y. Kubo, T. Ikeda, S. A. Overman, O. Osman and G. J. Thomas, Jr., Biochemistry, 42: 940–950, 2003.

[53] J. P. Caille, M. Pigeon-Gosselin and M. Pezolet, Biochim. Biophys. Acta, 758: 121–127, 1983.

[54] R. Tuma and G. J. Thomas, Jr., Biophys. J., 71: 3454–3466, 1996.

[55] P. R. Carey, Annu. Rev. Phys. Chem., 57: 527–554, 2006.

[56] T. Li, J. E. Johnson and G. J. Thomas, Jr., Biophys. J., 65: 1963–1972, 1993.

[57] R. H. Jenkins, R. Tuma, J. T. Juuti, D. H. Bamford and G. J. Thomas, Jr., Biospectroscopy., 5: 3–8, 1999.

[58] G. J. Thomas, Jr. and J. Barylski, Appl. Spectrosc., 24: 463–464, 1970.

[59] R. Tuma, S. Vohník, H. Li and G. J. Thomas, Jr., Biophys. J., 65: 1066–1072, 1993.

[60] J. M. Benevides, M. Tsuboi, A. H. J. Wang and G. J. Thomas, Jr., J. Am. Chem. Soc., 115: 5351–5359, 1993.

[61] J. M. Benevides, S. A. Overman and G. J. Thomas, Jr., Raman spectroscopy of proteins, in Current Protocols in Protein Science (J. E. Coligan, B. M. Dunn, H. L. Ploegh, D. W. Speicher and P. T. Wingsfield, eds.), pp. 17.8.1–17.8.35, Wiley, New York, 2005.

[62] J. M. Benevides, J. T. Juuti, R. Tuma, D. H. Bamford and G. J. Thomas, Jr., Biochemistry, 41: 11946–11953, 2002.

[63] L. Movileanu, J. M. Benevides and G. J. Thomas, Jr., Biopolymers, 63: 181–194, 2002.

[64] R. W. Williams, J. Mol. Biol., 166: 581–603, 1983.

[65] J. Bandekar, Biochim. Biophys. Acta, 1120: 123–143, 1992.

[66] S. U. Sane, S. M. Cramer and T. M. Przybycien, Anal. Biochem., 269: 255–272, 1999.

[67] S. A. Asher, A. Ianoul, G. Mix, M. N. Boyden, A. Karnoup, M. Diem and R. Schweitzer-Stenner, J. Am. Chem. Soc., 123: 11775–11781, 2001.

[68] M. Berjot, J. Marx and A. J. P. Alix, J. Raman Spectrosc., 18: 289–300, 1987.

[69] G. J. Thomas, Jr. and D. A. Agard, Biophys J., 46: 763–768, 1984.

[70] L. Movileanu, J. M. Benevides and G. J. Thomas, Jr., Nucleic Acids Res., 30: 3767–3777, 2002.

[71] E. R. Malinowski (ed.), Factor Analysis in Chemistry, Third Edition, Wiley, New York, 2002.

[72] J. Hanus, D. Němeček, J. Stepanek and P.-Y. Turpin, J. Raman Spectrosc., 35: 418–425, 2004.

[73] I. Notingher, G. Jell, P. L. Notingher, I. Bisson, O. Tsigkou, J. M. Polak, M. M. Stevens and L. L. Hech, J. Mol. Struct., 744–747: 179–185, 2005.

[74] C. Owen, I. Notingher, R. Hill, S. Molly and L. Hench, J. Mater. Sci., 17: 1019–1023(5), 2006.

[75] J. C. Austin, T. Jordan and T. G. Spiro. Ultraviolet resonance Raman studies of proteins and related compounds, in Biomolecular Spectroscopy, Part A (R. J. H Clark and R. E. Hester, eds.), pp. 55–127, Wiley, New York, 1993.

[76] T. Miura and G. J. Thomas, Jr., Subcell. Biochem., 24: 55–99, 1995.

[77] R. Tuma, J. Raman Spectrosc., 36: 307–319, 2005.

[78] G. J. Thomas, Jr. and M. Tsuboi, Raman spectroscopy of nucleic acids and their complexes, in Advances in Biophysical Chemistry (C. A. Bush, ed.), pp, 1–70, JAI Press, Greenwich, CN, 1993.

[79] T. J. Yu, J. L. Lippert and W. L. Peticolas, Biopolymers, 12: 2161–2175, 1973.

[80] M. C. Chen and R. C. Lord, J. Am. Chem. Soc., 96: 4750–4752, 1974.

[81] J. L. Lippert, D. Tyminski and P. J. Desmeules, J. Am. Chem. Soc., 98: 7075–7080, 1976.

[82] G. J. Thomas, Jr., B. Prescott and D. W. Urry, Biopolymers, 26: 921–934, 1987.

[83] R. W. Williams, T. Cutrera, A. K. Dunker and W. L. Peticolas, FEBS Lett., 115: 306–308, 1980.

[84] G. J. Thomas, Jr and L. A. Day, Proc. Natl. Acad. Sci. U.S.A., 78: 2962–2966, 1981.

[85] M. Tsuboi, M. Suzuki, S. A. Overman and G. J. Thomas, Jr., Biochemistry, 39: 2677–2684, 2000.

[86] J. M. Benevides, P. Bondre, R. L. Duda, R. W. Hendrix and G. J. Thomas, Jr., Biochemistry, 43: 5428–5436, 2004.

[87] R. Tuma, J. K. H. Bamford, D. H. Bamford, M. P. Russell and G. J. Thomas, Jr., J. Mol. Biol., 257: 87–101, 1996.

[88] H. Takeuchi and H. Harada, Spectrochim. Acta, 42A: 1069–1078, 1986.

[89] I. Harada, T. Miura and H. Takeuchi, Spectrochim. Acta, 42A: 307–312, 1986.

[90] T. Miura, H. Takeuchi and I. Harada, Biochemistry, 27: 88–94, 1988.

[91] T. Miura, H. Takeuchi and I. Harada, J. Raman Spectrosc., 20: 667–671, 1989.

[92] H. Takeuchi, Biopolymers, 72: 305–317, 2003.

[93] A. Combs, K. McCann, D. Autrey, J. Laane, S. A. Overman and G. J. Thomas, Jr., J. Mol. Struct., 735–736: 271–278, 2005.

[94] M. N. Siamwiza, R. C. Lord, M. C. Chen, T. Takamatsu, I. Harada, H. Matsuura and T. Shimanouchi, Biochemistry, 14: 4870–4876, 1975.

[95] Z. Arp, D. Autrey, J. Laane, S. A. Overman and G. J. Thomas, Jr., Biochemistry, 40: 2522–2529, 2001.

[96] S. A. Overman and G. J. Thomas, Jr., Biochemistry, 34: 5440–5451, 1995.

[97] M. Tsuboi, K. Ushizawa, K. Nakamura, J. M. Benevides, S. A. Overman and G. J. Thomas, Jr., Biochemistry, 40: 1238–1247, 2001.

[98] H. Li and G. J. Thomas, Jr., J. Am. Chem. Soc., 113: 456–462, 1991.

[99] H. Li, C. J. Wurrey and G. J. Thomas, Jr., J. Am. Chem. Soc., 114: 7463–7469, 1992.

[100] H. Li, C. Hanson, J. A. Fuchs, C. Woodward and G. J. Thomas, Jr., Biochemistry, 32: 5800–5808, 1993.

[101] S. Vohník, C. Hanson, R. Tuma, J. A. Fuchs, C. Woodward and G. J. Thomas, Jr., Protein Sci., 7: 193–200, 1998.

[102] H. E. Van Wart and H. A. Scheraga, Proc. Natl. Acad. Sci. U.S.A., 83: 3064–3067, 1986.

[103] H. E. Van Wart, A. Lewis, H. A. Scheraga and F. D. Saeva, Proc. Natl. Acad. Sci. U.S.A., 70: 2619–2623, 1973.

[104] H. Sugeta, Spectrochim. Acta, 31A: 1729–1737, 1975.

[105] H. E. Van Wart and H. A. Scheraga, Proc. Natl. Acad. Sci. U.S.A., 74: 13–17, 1977.

[106] H. E. Van Wart and H. A. Scheraga, Proc. Natl. Acad. Sci. U.S.A., 83: 3064–3067, 1986.

[107] M. Tasumi, I. Harada, T. Takamatsu and S. Takahashi, J. Raman Spectrosc., 12: 149–151, 1982.

[108] I. Harada, T. Takamatsu, M. Tasumi and R. C. Lord, Biochemistry, 21: 3674–3677, 1982.

[109] A. Okada, T. Miura and H. Takeuchi, Biochemistry, 42: 1978–1984, 2003.

[110] H. Takeuchi, Y. Kimura, I. Koitabashi and I. Harada, J. Raman Spectrosc., 22: 233–236, 1991.

[111] T. Miura, T. Satoh, A. Hori-i and H. Takeuchi, J. Raman Spectrosc., 29: 41–47, 1998.

[112] G. J. Thomas, G. C. Medeiros and K. A. Hartman, Biochem. Biophys. Res. Commun., 44: 587–592, 1971.

[113] S. C. Erfurth, E. J. Kiser and W. L. Peticolas, Proc. Natl. Acad. Sci. U.S.A., 69: 938–941, 1972.

[114] G. J. Thomas, Jr. and A. H. J. Wang, Nucleic Acids Mol. Biol., 2: 1–30, 1988.

[115] R. C. Lord and G. J. Thomas, Jr., Spectrochim. Acta, 23A: 2551–2591, 1967.

[116] T. Shimanouchi, M. Tsuboi and Y. Kyogoku, Adv. Chem. Phys., 7: 435–498, 1964.

[117] G. J. Thomas, Jr. and K. A. Hartman, Biochim. Biophys. Acta, 312: 311–322, 1973.

[118] D. C. Blazej and W. L. Peticolas, Proc. Natl. Acad. Sci. U.S.A., 74: 2639–2643, 1977.

[119] L. D. Ziegler, B. Hudson, D. P. Strommen and W. L. Peticolas, Biopolymers, 23: 2067–2081, 1984.

[120] W. L. Kubasek, B. Hudson and W. L. Peticolas, Proc. Natl. Acad. Sci. U.S.A., 82: 2369–2373, 1985.

[121] M. Tsuboi, Y. Nishimura, A. Y. Hirakawa and W. L. Peticolas, Resonance Raman spectroscopy and normal modes of the nucleic acid bases, in Biological Applications of Raman Spectroscopy, Volume 2, (T. G. Spiro, ed.), pp. 109–179, Wiley, New York, 1987.

[122] J. C. Austin, T. Jordan and T. G. Spiro, Ultraviolet resonance Raman studies of proteins and related compounds, in Biomolecular Spectroscopy, Part A (R. J. H Clark and R. E. Hester, eds.), pp. 55–127, New York, 1993.

[123] Z. Q. Wen and G. J. Thomas, Jr., Biopolymers, 45: 247–256, 1998.

[124] P. E. Prevelige, Jr., D. Thomas, J. King, S. A. Towse and G. J. Thomas, Jr., Biochemistry, 32: 537–543, 1993.

[125] R. Tuma, H. Tsuruta, J. M. Benevides, P. E. Prevelige, Jr. and G. J. Thomas, Jr., Biochemistry, 40: 665–674, 2001.

[126] S. A. Overman and G. J. Thomas, Jr., Biochemistry, 37: 5654–5665, 1998.

[127] M. C. Chen and R. C. Lord, J. Am. Chem. Soc., 96: 4750–4752, 1974.

[128] M. C. Chen, R. C. Lord and R. Mendelsohn, J. Am. Chem. Soc., 96: 3038–3042, 1974.

[129] R. Tuma, P. E. Prevelige, Jr. and G. J. Thomas, Jr., Biochemistry, 35: 4619–4627, 1996.

[130] R. Tuma, M. H. Parker, P. Weigele, L. Sampson, Y. Sun, N. R. Krishna, S. Casjens, G. J. Thomas, Jr. and P. E. Prevelige, Jr., J. Mol. Biol., 281: 81–94, 1998.

[131] A. Rodriguez-Casado and G. J. Thomas, Jr., Biochemistry, 42: 3437–3445, 2003.

[132] Y. Sun, S. A. Overman and G. J. Thomas, Jr., Virology, 365: 336–345, 2007.

[133] C. E. Catalano (ed.), Viral Genome Packaging Machines: Genetics, Structure, and Mechanism, Plenum, New York, 2005.

[134] S. Steinbacher, R. Seckler, S. Miller, B. Steipe, R. Huber and P. Reinemer, Science, 265: 383–386, 1994.

[135] S. Steinbacher, S. Miller, U. Baxa, N. Budisa, A. Weintraub, R. Seckler and R. Huber, J. Mol. Biol., 267: 865–880, 1997.

[136] S. Steinbacher, S. Miller, U. Baxa, A. Weintraub and R. Seckler, Biol. Chem., 378: 337–343, 1997.

[137] A. S. Robinson and J. King, Nat. Struct. Biol., 4: 450–455, 1997.

[138] D. Sargent, J. M. Benevides, M. H. Yu, J. King and G. J. Thomas, Jr., J. Mol. Biol, 199: 491–502, 1988.

[139] G. J. Thomas, Jr., R. Becka, D. Sargent, M.-H. Yu and J. King, Biochemistry, 29: 4181–4187, 1990.

[140] H. Strauss and J. King, J. Mol. Biol., 172: 523–543, 1984.

[141] L. Tang, W. R. Marion, G. Cingolani, P. E. Prevelige and J. E. Johnson, EMBO J., 24: 2087–2095, 2005.

[142] D. A. Marvin, L. C. Welsh, M. F. Symmons, W. R. Scott and S. K. Straus, J. Mol. Biol, 355: 294–309, 2006.

[143] M. Tsuboi, T. Ikeda and T. Ueda, J. Raman Spectrosc., 22: 619–626, 1991.

[144] S. A. Overman, M. Tsuboi and G. J. Thomas, Jr., J. Mol. Biol., 259: 331–336, 1996.

[145] M. Tsuboi, J. M. Benevides, P. Bondre and G. J. Thomas, Jr., Biochemistry, 44: 4861–4869, 2005.

[146] M. Tsuboi, T. Ueda, K. Ushizawa, Y. Ezaki, S. A. Overman and G. J. Thomas, Jr., J. Mol. Struct., 379: 43–50, 1996.

[147] M. Tsuboi, S. A. Overman and G. J. Thomas, Jr., Biochemistry, 35: 10403–10410, 1996.

[148] W. R. Wikoff, L. Liljas, R. L. Duda, H. Tsuruta, R. W. Hendrix and J. E. Johnson, Science, 289: 2129–2133, 2000.

[149] W. R. Wikoff, J. F. Conway, J. Tang, K. K. Lee, L. Gan, N. Cheng, R. L. Duda, R. W. Hendrix, A. C. Steven and J. E. Johnson, J. Struct. Biol., 153: 300–306, 2006.

[150] R. W. Hendrix and R. L. Duda, Adv. Virus Res., 50: 235–288, 1998.

[151] J. M. Benevides and G. J. Thomas, Jr., Biochemistry, 44: 2993–2999, 2005.

VIBRATIONAL SPECTROSCOPY VIA INELASTIC NEUTRON SCATTERING

Bruce S. Hudson

Contents

Abstract

The use of vibrationally inelastic neutron scattering (INS) as a technique in molecular spectroscopy is described and illustrated. INS is particularly sensitive to the motions of hydrogen atoms relative to that of most other atoms including deuterium. The use of H/D isotopic partitioning in INS is described. When there are multiple exchangeable sites and a fixed H/D ratio, the arrangement that lowers the zero-point level is favored at low temperature. Overtone and combination transitions are allowed in INS even in the harmonic approximation. Measurement of intensity as a function of momentum transfer can distinguish the order of the transition as fundamental, first overtone, etc. and differs for harmonic and quartic oscillators. INS spectra have no symmetry selection rules and thus transitions that are otherwise forbidden can be observed. Neutrons interact with the motion of nuclei rather than to the response of the electrons to the motions of nuclei. INS intensities can be computed with good reliability using known atomic cross sections. Modern quantum chemical methods provide accurate intensities and so spectral assignments are very reliable. Applications of these principles to several hydrocarbons and to the short strong symmetric hydrogen bond of the $H_5O_2^+$ Zundel cation are presented. In this case, the coupling of motions in neighboring unit cells is very strong as revealed in spectral changes seen with isotopic dilution. For hydrocarbons where intermolecular interactions are very weak, INS can be used to establish molecular conformations in crystals.

Keywords: Inelastic neutron scattering; spallation; dodecahedrane; $C_{20}H_{20}$; $DMol^3$; aCLIMAX; conformational analysis; azulene; short; strong; symmetrical hydrogen bonds; Zundel cation; CASTEP

1. INTRODUCTION

Inelastic neutron scattering (INS) is a well-established technique for vibrational spectroscopy [1–3] and as a method for testing *ab initio* calculations [2–28]. For reasons of spectral resolution, this method is usually applied to the solid state at low temperature. INS has several unique aspects in comparison with conventional "optical" methods of vibrational spectroscopy. These are that (i) there are no symmetry selection rules for INS, (ii) hydrogen atoms dominate the scattering relative to all other atoms including deuterium, and (iii) INS spectra can be easily computed in intensity as well as spectral position given a description of the nuclear dynamics without any description of the response of the electrons to the nuclear motions or the radiation field.

The capability for performing neutron scattering experiments of all kinds will be greatly enhanced in the next few years with the completion of the Spallation Neutron Source (SNS) now in operation at the Department of Energy Oak Ridge National Laboratory [29]. This enhancement in overall capability includes vibrational spectroscopy studies of the type described here, which were carried out at the ISIS facility of the Rutherford Appleton Laboratory in the United Kingdom [30] using the spectrometer TOSCA [31,32]. The neutron flux at SNS will be an at least an order of magnitude higher than at ISIS. This factor plus design changes that are being planned for the new VISION spectrometer at SNS (http://materials.chem.utk.edu/visionintro.html) will result in an increase in the rate of data collection by roughly two orders of magnitude relative to that of TOSCA at ISIS. Experiments that are performed on 1 g samples at ISIS in 6 h will be performed with 100 mg samples at SNS in 30 min.

2. BASIC SCATTERING THEORY

The general theory of neutron scattering is described in the text by Squires [33]. It is useful to begin with the definition of the neutron scattering wavevector $\mathbf{Q} = \mathbf{k}_i - \mathbf{k}_f$. The momentum transferred to the sample in a scattering event is $\hbar\mathbf{Q}$. The effective distance range of a scattering event is $2\pi/Q$, which may range from 0.01 to 100 Å or even greater in small angle experiments. $\mathbf{Q}$ depends on the incident and final neutron energies and on the scattering angle. For a given energy change, $\mathbf{Q}$ is made smallest by using a high initial energy and forward scattering. The power of neutron scattering stems from the fact that inelastic events may be measured with a range of values of $\mathbf{Q}$. This means that dynamics may be measured as a function of both frequency and distance making neutron scattering a two-dimensional method. The variables of the method are usually designated as Q and ω. The corresponding transform variables are distance and time. This aspect of neutron scattering is essential in the determination of dispersion curves for external (phonon) motions of solids. In studies of internal molecular vibrations, the Q dependence can be used as a method for separation of fundamental and overtone transitions and as an aid in maximizing signal intensity in various spectral regions. Molecular vibrations do not usually have significant dispersion and can thus be treated as localized to the molecule. However, in some of the cases to be considered here, some of the internal degrees of freedom result in large charge redistribution and this, in turn, results in significant dispersion.

The peak of the thermal energy distribution for room temperature corresponds to an optical (far infrared) photon with a wavevector magnitude of $200 \, \text{cm}^{-1}$. For a neutron with the same energy, the wavevector is $\sim 3 \, \text{Å}^{-1}$ and the deBroglie wavelength is 1.8 Å. Vibrational transitions typically result in very large fractional changes in neutron energy and also large Q. This can be used to perform experiments that probe vibrational wavefunctions with resolution of 0.01 Å providing a direct picture of a nuclear distribution given by the square of the wavefunction.

The quantity that describes the scattering of a nucleus is a scattering length, b, with dimensions of distance. The magnitude if b is isotope-dependent with typical magnitudes of a few fermis where a fermi is 10^{-15} m. The sign of b may be positive or negative depending on whether the interaction with a thermal neutron is attractive or repulsive. Diffraction is based on a coherent superposition of signals from chemically identical sites in the sample (identical atoms in neighboring or distant unit cells, for example). Disorder results in loss of this coherence. For X-ray diffraction, this disorder is usually associated with positional disorder. An extra contribution to disorder in the case of neutron scattering comes from variation in the value of b from one otherwise identical position to another. Variation in b can arise from either random isotopic variation or random spin orientation relative to the neutron spin orientation (which is usually also random). This "incoherent" scattering is proportional to the ensemble average quantity $<b^2> - <b>^2$, which is zero if b is the same at all equivalent positions. Hydrogen has a particularly large incoherent scattering cross section due to random spin orientation relative to the neutron spin. Incoherent scattering is a source of background noise in a diffraction experiment. This is particular problem in practice for powder studies and so it is useful to replace hydrogen by deuterium for such studies. Some values of the scattering length b, the coherent cross section $(4 \pi b^2)$, the incoherent cross section, and the absorption cross section for several atoms are given in Table 1. The usual unit of cross section is the barn or 10^{-28} m^2.

The origin of the contrast in X-ray scattering is the difference in the number of electrons for an atom. The intensity of scattering is proportional to Z^2. This makes it difficult to locate a hydrogen atom in the presence of other atoms except with very good data usually collected from a low temperature sample. However, as shown in Table 1, the cross section for scattering of neutrons by deuterium is about the same as that for iodine. The scattering length depends on nuclear properties and varies in a complex way with Z although at large Z a $Z^{1/3}$ trend is seen. The scattering lengths of H and D are opposite in sign. This permits the location of these species as if they were distinct elements.

We are concerned here with vibrational INS spectroscopy, analogous to infrared and Raman spectroscopy. The location of the inelastic processes on an energy scale is

Table 1 Neutron scattering properties of some elements

	Coherent σ (barns)	Scattering length, b (fm)	Incoherent σ (barns)	Absorption cross section (barns)
Hydrogen	1.8	−3.74	80.3	0.33
Deuterium	5.6	6.67	2.0	0.005
Boron	3.5	5.30–0.213i	1.7	767
(natural)				
^{10}B	0.14	−0.1–1.066i	3	3835
^{11}B	5.6	6.65	0.21	0.006
Carbon	5.6	6.64	0.001	0.005
Nitrogen	11.0	9.36	0.5	1.9
Oxygen	4.2	5.80	0.001	0.0002
Fluorine	4.0	5.65	0.001	0.01
Chlorine	11.5	9.56	5.3	33.5
Bromine	5.8	6.8	0.1	6.9
Iodine	3.5	5.28	0.31	6.15
Aluminum	1.5	3.5	0.26	0.23
Vanadium	0.018	−0.38	5.08	5.08

determined in the usual way by the interatomic forces of the material. However, the intensity of a particular inelastic event is determined by considerations that are very different from those of optical spectroscopies. In those cases, the radiation has a wavelength that is very much larger than the distances between atoms and a multi-polar expansion is appropriate. The dipole approximation is nearly universally applic-able. In the neutron case, however, the interaction of a neutron is with a nucleus. This interaction is very short range on the molecular distance scale. Because of this, it is valid to use an isotropic delta function for the interaction. The strength of the interaction is related to the scattering length, which depends on the nuclide involved and the relative spin orientations. Furthermore, the effective wavelength of a thermal neutron is in the order of or smaller than the interatomic separation of the atoms in a molecule, and a multipole expansion is not appropriate.

The starting point for the probability of incoherent scattering of a neutron is derived from Fermi's Golden Rule:

$$P(I \rightarrow F) = |<F| \exp(ik_f r)\Sigma_n V(r-R_n)\exp(-ik_i r)|I>|^2$$

where I and F are the initial and final states of the material system, r is the position of the neutron, R_n the position of nucleus n, the exponential factors represent the initial and final plane-wave states of the neutron, and $V(r-R_n)$ is the interaction of the neutron with nucleus n. Use of the "Fermi pseudopotential" delta function for $V(r-R_n)$.

$$V(r-R_n) = (2\pi\hbar^2/m)\, b\, \delta(r-R_n)$$

restricts the evaluation of the neutron plane waves to the positions of the nuclei and so

$$P(I \rightarrow F) = |\Sigma_n b_n <F| e^{iQ\cdot r}| I>|^2$$

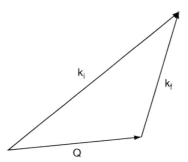

Figure 1 A vector diagram showing the relationship between the incident neutron wavevector k_i, the final neutron wavector k_f, and the momentum transfer vector Q. The momentum transferred from the neutron to the sample is $\hbar Q$.

Thus the intensity of a transition induced by interaction with a neutron is the spatial Fourier transform of the overlap of the initial and final states of the system. The transform variable is the momentum transfer, $Q = k_i - k_f$ (Figure 1), which is subject to experimental variation at fixed $\hbar\omega$. If the initial and final states of the matter are the same, $I = F$, then this is diffraction, the spatial Fourier transform of the nuclear density distribution. If $I \neq F$ due to inelasticity, then the expansion of the summation and the subsequent absolute square leads to terms with $n \neq n'$ as well as $n = n'$. The implicit ensemble average results in the multiplication of the $n \neq n'$ term by $\sigma_{coh} = 4\pi <b>^2$, while the $n = n'$ term is multiplied by $\sigma_{inc} = 4\pi(<b^2> - <b>^2)$ For hydrogen $\sigma_{coh} = 1.8$ barns while $\sigma_{inc} = 80.2$ barns. Thus the scattering from individual atoms, "self scattering," is much larger than that from the cross terms, "coherent scattering," For deuterium atoms, the value of $\sigma_{coh} = 5.6$ barns while $\sigma_{inc} = 2.0$ barns. These values are typical of the rest of the periodic table (except that for some cases σ_{inc} vanishes) and the increased atomic mass as one descends the periodic table decreases the mean square amplitude and thus the scattering as discussed below.

The incoherent scattering term now has the form

$$P(I \rightarrow F) = \Sigma_n \sigma_{n,inc} |<F([R_n])|e^{iQ \cdot r}|I([R_n])>|^2$$

The initial and final states, $I([R_n])$ and $F([R_n])$, are functions of the positions of all the nuclei, $[R_n]$. If these functions can be approximated by products of collective harmonic functions ("normal modes"), then the mode or modes that undergo excitation in an $I \rightarrow F$ transition can be distinguished from those that do not. This results in the following simple, exact (within the harmonic approximation) expressions:

$$I(\text{mode } i) = \Sigma_n \exp\left(-U_n^2 Q^2\right)\sigma_n (Q \cdot U_{ni})^2 \text{ fundamental}$$

$$I(\text{mode } i+j) = \Sigma_n \exp\left(-U_n^2 Q^2\right)\sigma_n (Q \cdot U_{ni})^2 (Q \cdot U_{nj})^2 \text{ } i+j \text{ combination}$$

plus similar terms for higher order overtones and combinations. U_{ni} is the motion of atom n in the ith normal mode eigenvector and Q is the momentum transfer vector with magnitude Q. The factor $\exp(-U_n^2 Q^2)$ is the Debye–Waller factor. Here U_n is the amplitude of motion of atom n summed over all normal modes. Additional contributions to the Debye–Waller factor come from translation and rotation or, in a crystal, the corresponding libration and translation phonons. The amplitude of motion of the nth atom in the ith normal mode U_{ni} is given by the coefficient of that atom in the normal mode eigenvector for that mode, C_{ni}, multiplied by the mean square amplitude of motion of atom n when normal mode i is excited, $U_{ni} = C_{ni}/(m_n\omega_i/\hbar)^{1/2}$. The vibrational eigenvectors are normalized $\Sigma_i (C_{ni})^2 = 1$.

Overtones and combinations are allowed in INS in the harmonic approximation. If one were to measure the intensity of a fundamental scattering process as a function of Q, then the variation would be of the form $I(Q) = U^2 Q^2 \exp(-U^2 Q^2)$. This has a maximum when $U^2 Q^2 = 1$. A binary overtone $(i = j)$ will have $I(Q) = U^4 Q^4 \exp(-U^2 Q^2)$. This has a maximum when $U^2 Q^2 = 2$. Thus a spectrometer that permits Q resolution permits separation of the observed spectra into their harmonic order.

It is straightforward to evaluate the intensity for scattering of nuclei in anharmonic potentials by simply evaluating the off-diagonal Fourier transform matrix element for the corresponding initial and final states (or, equivalently, the spatial Fourier transform of the overlap of the initial and final wavefunctions). Figure 2 illustrates the harmonic behavior indicated above and compares the result with that for a quartic oscillator, $V(x) = qx^4$, where q has been chosen so that the frequency of the $0\rightarrow 1$ transition of a hydrogen atom in the quartic oscillator is the same as that of

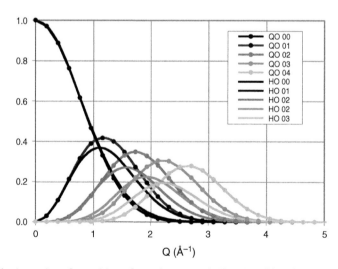

Figure 2 The intensity of transitions from the ground vibrational level to excited vibrational levels with 0–4 quanta of vibrational excitation for a harmonic oscillator and for a quartic oscillator. The quartic oscillator is chosen so that its fundamental transition frequency is the same as that for the harmonic oscillator. The elastic scattering (black) is nearly identical for the two cases. This calculation is done for a single oscillator so that only one degree of freedom contributes to the Debye–Waller factor.

the harmonic oscillator used for this simulation. This particular nonharmonic oscillator is chosen because of its relevance to cases discussed below. The variation of I with Q for the quartic oscillator is similar to that of the harmonic case but sufficiently different that the two cases could be distinguished in certain cases.

3. EXPERIMENTAL NEUTRON SCATTERING

3.1. Neutron sources

The major research neutron sources are based on either nuclear reactors or proton accelerators that produce neutrons using the spallation process. In a reactor-based facility, the neutrons are moderated with a surrounding layer of water (or D_2O) or a cryogen such as liquid hydrogen. The distribution of neutron energies from a moderator is a Maxwell–Boltzmann distribution characterized by the temperature of the moderator. Neutrons with energies in excess of $1000\,cm^{-1}$ are a small fraction of the total neutron flux in such sources. Examples of reactor-based research facilities are the NIST Center for Neutron Research [34], the HFIR facility at Oak Ridge National Laboratory [35], and the Institute Laue Langevin (ILL) in Grenoble, France [36]. For vibrational spectroscopy at energies above $800\,cm^{-1}$, the higher content of high-energy neutrons and the pulsed nature of the neutron beam provided by a spallation source make such sources preferable. The highest power spallation source in current operation as a user facility is the ISIS Facility of the Rutherford Appleton Laboratory in the UK. The first spallation source, Intense Pulsed Neutron Source (IPNS), is a user facility at Argonne National Laboratory [37], and LANSCE includes a user facility at Los Alamos National Laboratory [38]. The new SNS at the Oak Ridge National Laboratory began production of neutrons on April 29, 2006. On August 30, 2007, the beam on target power exceeded that of ISIS. SNS is expected to produce usable neutrons at a rate that will be about 10 times that of ISIS.

In a spallation neutron facility, protons with an energy of 0.8 GeV (ISIS) to 2 or even 3 GeV (SNS) are incident on a heavy metal target of tantalum (ISIS) or mercury (SNS). Each incident proton results in 20–50 (increasing with energy) high-energy neutrons. Because the proton accelerator is pulsed, the neutrons are produced in a pulse. These neutrons strike a moderator where they undergo neutron thermalization. The result is a distribution that looks like a Boltzmann distribution at low energy but which has a tail to very high energy. The details of the neutron energy distribution and the shape of the resulting pulse of neutrons depend on the moderator material, size, and temperature and the presence or absence of neutron-absorbing materials. For vibrational spectroscopy, the epithermal distribution from a water moderator is usually used.

3.2. Types of neutron scattering

Several types of neutron scattering experiments are listed in Table 2. This list is not exhaustive but is meant to provide a framework for distinguishing between elastic and inelastic scattering, on the one hand, and coherent and incoherent scattering on

Table 2 Varieties of neutron scattering experiments

Name	Energy change	Basic process	Information
Diffraction	Elastic	Coherent	Structure
SANS	Elastic	Coherent	Big picture
Reflectometry	Elastic	Coherent	Depth structure
QENS	Quasielastic	Incoherent	Diffusion, Dynamics
NSE (spin echo)	Quasielastic	Incoherent	NMR time scale
Backscattering	Inelastic	Coherent/ Incoherent	
Vibrational	Inelastic	Coherent/ Incoherent	
DINS (Neutron compton)	Inelastic		H/D Wavefunctions

the other. Elastic scattering involves no change in energy of the neutron and results in one form or another of diffraction providing structural information. Inelastic scattering involves energy changes ranging from energies that are more commonly associated with nuclear magnetic resonance (NMR) up to multiple electron volts. The varieties of vibrational spectroscopy are subdivided in Table 3.

The first three entries in this table are varieties of diffraction with distance scales ranging from interatomic spacing to macromolecular association arrangements and surface layer structure at short distances. Quasielastic neutron scattering is the premier method for the determination of the time scale of diffusive degrees of freedom as a function of the distance scale of that motion. Rotational and translational diffusion can be distinguished on the basis of their different Q dependence. Neutron spin echo spectroscopy is based on using spin polarized neutron precession as a clock to determine very small changes in neutron velocity. Backscattering spectroscopy takes advantage of the diffraction properties of three-dimensional crystal gratings in this particular geometry. The grating is physically displaced to produce a Doppler scanned spectrum. The resolution of the backscattering

Table 3 Varieties of vibrational inelastic neutron scattering spectrometers

Method	Example(s)	Range (cm^{-1})	Resolution (cm^{-1})	Observation
Back scattering	HFBS (NIST)	0–0.003	0.00001	Tunneling
Inverse Scanning	FANS (NIST)	0–1200		Vibrational
Time-focusing inverse geometry TOF	TOSCA (ISIS) VISION (SNS)	0–4000	1% $\Delta E/E$ (0–1000)	Full vibrational spectrum
Direct chopper TOF	HRMECS (ANL) HET, MARI, MAPS (ISIS)	0–8000	2% $\Delta E/E$	Q-Resolution

spectrometer at NIST is 1 at $36\,\mu eV$ (8 at $290\,\mu\,cm^{-1}$). A spectrum of 2,6-dimethylpyridine (lutidine) is given at http://www.ncnr.nist.gov/instruments/hfbs/HFBS_perfNew.html. The methyl groups exhibit multiple resolved tunneling features in the $0-300\,\mu\,cm^{-1}=0.0003\,cm^{-1}$ region. High-resolution neutron spectroscopy is relatively easy for low-energy transfer. Deep INS (DINS) is neutron Compton scattering. DINS is a scattering process that uses neutrons of such high energy that their scattering can be interpreted in the impulse approximation. The resulting information can be used to determine the momentum distribution function for H and D atoms. In favorable cases, this information can be used to obtain the wavefunction for H atoms. The relevant instrument for such studies is VESUVIO at ISIS (http://www.fisica.uniroma2.it/~vesuvio/dins/index.html).

3.3. Vibrational inelastic scattering

The INS spectrometer used in the present work is the inverse geometry, time-focusing, crystal analyzer TOSCA [31,32] spectrometer of the ISIS facility of the Rutherford Appleton Laboratory. The principles of this inverse geometry spectrometer are shown in Figure 3. Details are given in the figure legend. The major point is that, in this inverse geometry design, the final energy is specified to be approximately $1.25\,meV = 10\,cm^{-1}$. Such low-energy, low-velocity neutrons have a very high counting efficiency. All of the neutrons from the moderator are incident on the sample. The neutrons that excite a vibrational transition of, say, $1000\,cm^{-1}$ have must have an initial energy of $1010\,cm^{-1}$. The time of arrival of the neutron at the detector determines the incident neutron velocity and thus its energy. In a spectrometer like TOSCA, the value of Q is determined primarily by the vibrational energy, which, in turn, determines the initial energy. This leads to the approximate expression $Q^2 = (const)\,\omega$. For the range of $0-3000\,cm^{-1}$, the value of Q ranges from 2 to $13\,\mathring{A}^{-1}$ so $2\pi/Q$ is in the range from 0.48 to $3.1\,\mathring{A}$, comparable to intra-atomic distances.

The relationship $Q^2 = (const)\,\omega$ means that the intensity of a fundamental transition $I(\text{mode }i) = (const)\Sigma_n\,exp(-U_n^2Q^2)\,(\sigma_n/m_n)\,(C_{ni})^2$. The Q^2 dependence has been cancelled by the inverse ω dependence of the mean square amplitude when it is replaced by the normal mode eigenvector. The inverse atomic mass dependence of the mean square amplitude can be combined with the atomic cross section. This further enhances the dominance of H atom scattering. The fact that H atoms dominate means that if it can be assumed that all H atoms have the same overall mean square amplitude U_n^2, then $I(\text{mode }i) = (const)\,exp(-U_H^2Q^2)\,\Sigma_H(C_{Hi})^2$, the sum over all H atoms of their motion in a particular normal mode. Methyl rotations with the property $\Sigma_H(C_{Hi})^2 = 1$ are thus the strongest transitions in INS spectra. Methyl torsions are usually weak or absent in IR and Raman spectra. The increase in Q with frequency for TOSCA means that the Debye–Waller factor results in a decrease in intensity with ω.

Vibrational neutron spectrometers that resolve Q are direct geometry, chopper spectrometers. Examples are HRMECS at Argonne National Laboratory (http:// www.pns.anl.gov/instruments/hrmecs/), MAPS at ISIS (http:// www. isis.rl.ac.uk/excitations/maps/), and ARCS to be installed at SNS

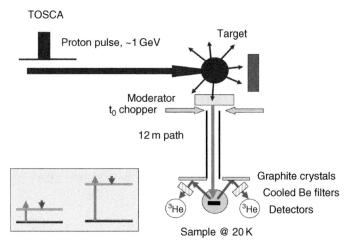

TOSCA

Figure 3 Schematic illustration of an inverse geometry spectrometer such as TFXA or TOSCA at ISIS or VISION under development for SNS. A pulse of high-energy protons (blue arrrow) strikes a tantalum target producing a burst of high-energy neutrons. These are moderated by collisions in a water moderator (yellow). The partially thermalized neutrons (green) pass through a t_0 chopper, which opens slightly after the neutron burst to block all very high energy neutrons. After a 12 m flight path, this full distribution of neutrons strikes the sample (black) in a cryostat at 20 K (blue gray). The neutrons that are scattered backward at 135° (red) are diffracted by graphite crystals and passed through cold Be filters. Only neutrons with energy of ∼1.25 meV are passed to the detectors. Two backscattering crystal and filter detectors are shown in this diagram as was the configuration for TFXA. TOSCA now has 10 backscattering detectors arranged on a circle. Another set of 10 detectors utilizes neutrons scattered in the forward direction at 45°. The detectors contain ³He at an elevated pressure such that >90% of the neutrons are absorbed. The subsequent relaxation of the excited ⁴He★ nucleus produces ionizing radiation that is detected by a breakdown production of photons and a photomultiplier. The energy-level diagram in the yellow box shows how low (high) energy incident neutrons can excite low (high) frequency vibrations to produce the same final energy neutrons (red). Time of flight analysis of the arrival time of the detector events determines the initial energy and thus the vibrational energy as the final energy is fixed. (See color plate 33).

(http://www.cacr.caltech.edu/ projects/ARCS/). These devices specify the incident neutron energy using a chopper that is phased with respect to the proton pulse and thus selects an energy slice from the burst of neutrons from the moderator. Time of flight in the secondary flight path then determines the final energy for a specified initial energy. A large number of detectors at different scattering angles, in combination with variation of the initial energy, permits Q resolution. Such devices are also used for magnetic scattering. When a high-frequency motion (e.g., a hydrogen stretch) is of interest, the use of high incident energy and low scattering angle permits relatively low Q (compared to TOSCA) even at high ω, thus diminishing the damping effect of the Debye–Waller factor.

A summary of spectrometers used for vibrational spectroscopy is given in Table 3. The FANS spectrometer at NCNR NIST is based on a fixed final energy specified by a large bank of cooled Be filters. Monochromatic incident neutrons are provided by a crystal diffraction monochromator that will be fitted with curved crystals providing a focused neutron beam. This device should be capable of studies of small samples.

4. ADVANTAGES OF INS FOR VIBRATIONAL SPECTROSCOPY

The advantages and disadvantages (or differences) of INS compared with IR and Raman methods for applications in vibrational spectroscopy have been described elsewhere [1–3] and are summarized in Table 4. The absence of selection rules is an advantage with high-symmetry systems but a disadvantage in many more common systems except if the analysis of the spectra is augmented by computational methods. The point here is that when lines overlap in an INS spectrum, a simulation will result in enhanced intensity so that such occurrences can be detected. There are, of course, cases where a vibration, such as methyl torsion mode, is nominally allowed by symmetry but has no intensity in either IR or Raman spectroscopy. This tends often to be the case at low-frequency modes and so INS spectra tend to be relatively more complex than IR or Raman spectra at low frequency. The TOSCA spectrometer extends to $20\,\text{cm}^{-1}$. Spectra through the elastic region are possible with other spectrometers as discussed above.

The unique aspect and major advantage of INS is the differential sensitivity to H/D substitution. This permits effective masking of components in multi-component systems. This is the aspect of INS that most clearly distinguishes it from other methods. Studies of mixed crystal systems, co-crystals, and ionic species with hydrogen in both ions can be incisively investigated using deuterium substitution of one component but leaving hydrogen in the other. Furthermore, as described for one case below, it is possible to use the natural partitioning of hydrogen and deuterium in cases where there is free exchange of these species but unequal zero-point levels as a combined spectroscopic and potential energy probe of a complex system.

The major practical advantage of neutron radiation in general is its penetrating nature. This permits the use of aluminum for the fabrication of sample containers (except for diffraction where vanadium is used because its scattering is almost entirely incoherent). This makes "special" sample containers relatively inexpensive to fabricate. It also permits experiments to be performed at low temperature on a routine basis as well as permitting experiments at high pressure.

Table 4 Advantages and disadvantages of INS vibrational spectroscopy

Advantages of INS	Disadvantages of INS
Absence of selection rules; all vibrations observed	Large samples needed
Sensitivity to isotopic substitution, especially H/D	Lower resolution
Ease of calculation of intensities	Longer experiment time
Penetrating radiation	Low temperature required

The major disadvantages of INS in comparison with IR and Raman methods are the larger sample needed and the lower resolution. The need for low temperature and long experiment times may also be considered a disadvantage. To some extent, these disadvantages of INS will be reduced with the SNS facility. For example, the 1 g samples currently used routinely at ISIS with a run time of about 6 h will give comparable results in a few minutes at SNS.

Resolution is a more complex issue. The fact that the samples are held at low temperature and the limitation of the resolution for cold samples are both related to the fact that neutrons scatter from translational excitations. This permits the use of neutron scattering to measure diffusion coefficients. The TOSCA spectrometer has a resolution that is limited by timing error at high frequency (where the neutrons that caused the excitation have high velocity). The resulting timing error is partly due to the finite pulse duration. This contribution to the resolution can be reduced by the use of a longer primary flight path – in exchange for lower flux. At low frequency the resolution is determined by the crystal monochromators. It is found that high molecular weight samples result in sharper spectral features. This is a demonstration of the expectation that some of the observed spectral width is due to residual translational recoil broadening.

Recent developments in computational chemistry are well known. Density Functional Theory (DFT) has become very popular as a cost-effective correlation treatment. A parallel development has occurred in what might be called "computational material science." This has resulted in a proliferation of periodic DFT methods that impose the periodicity of a crystal lattice on the wavefunction. These methods permit computation of the atomic dynamics of a periodic lattice from first principles (given a choice of functional). This allows computation of the molecular structure, the vibrational dynamics, and an INS spectrum without additional assumptions regarding intermolecular forces. The comparison of the results of such a computation with the experimental data provides a detailed test of the validity of the method employed.

5.   EXAMPLES OF INS SPECTRA AND THEIR UTILITY

5.1.   Hydrocarbons

5.1.1.   Dodecahedrane, $C_{20}H_{20}$

Several aspects of INS spectroscopy are dramatically illustrated by the case of dodecahedrane, $C_{20}H_{20}$. This molecule has I_h symmetry in its isolated state. Because of this high symmetry, the vibrations of this molecule are highly degenerate (up to fivefold) and most of the vibrational transitions are not allowed in either infrared or Raman spectroscopy. Specifically, of the 30 distinct vibrations, only 11 are permitted in total in IR (3 T_{1u} modes) plus Raman (2 A_g and 6 H_g modes). Figure 4 shows infrared (top), Raman (middle), and INS spectra of crystalline $C_{20}H_{20}$ [17]. The face centered cubic ($Z=1$) crystal structure is shown in Figure 5. In this crystal the site symmetry is T_h.

The spectra of Figure 4 show the expected Raman and IR bands, but there are clearly some additional features in the IR spectrum and some splitting of the

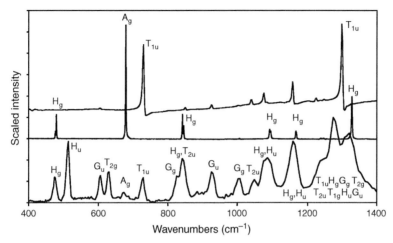

Figure 4 The infrared, Raman, and inelastic neutron scattering (INS) spectra of crystalline dodecahdedrane, $C_{20}H_{20}$. Adapted from Ref. [17].

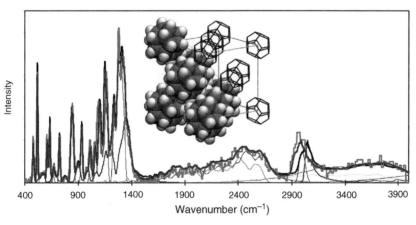

Figure 5 The inelastic neutron scattering (INS) spectrum of crystalline dodecahedrane at 20 K compared with the results of a periodic Density Functional Theory (DFT) treatment. The face centered cubic crystal structure is also shown. The experimental results are shown in red. The heavy black curve is the summation of the calculated intensities of all fundamental and overtone contributions including the phonon wings using the program aCLIMAX [40]. The experimental phonon part of the spectrum has been used to generate the phonon wings. The heavy blue curve is the same except that the calculated frequency of all of the CH stretching modes have been uniformly scaled down by a factor of 0.98 as determined from the Raman spectrum in order to correct for anharmonicity. In the region from 1500 to 2800 cm^{-1}, the intensity is due entirely to binary combinations and overtones and phonon-wings added to these. From 3300 cm^{-1} to higher frequency, the intensity is due to ternary overtones and combinations. Adapted from Ref. [17]. (See color plate 34).

Raman features. These effects reflect the deformation of the I_h molecular frame in the T_h environment [17]. This deformation splits the five- and four-fold degenerate modes, makes all g-modes Raman active, and induces infrared activity in all u-symmetry modes. These details are reflected in the solid-state periodic DFT calculations, but the agreement at the resolution of the Raman spectra is not quantitative [17].

Figure 5 shows the degree of agreement obtained when this spectrum is simulated using the program DMol^3 [39]. This program uses an atom-centered basis set to perform periodic DFT calculations. In this calculation, the positions of the contents of the unit cell are optimized to minimize the energy and the second derivatives are determined using numerical second derivatives. The resulting normal mode eigenvalues and eigenvectors are used as input for the program aCLI-MAX [40], which computes the INS spectrum including the contributions of overtones and combinations and phonon wings. Dodecahedrane is a model case for testing these methods. Figure 5 shows the experimental spectrum compared with the results of two simulations, each including all overtones and combinations. The lighter traces are individual components of the overall intensity from binary and ternary overtones. The two simulations differ in whether the well-known anharmonicity of the CH stretch is ignored or is incorporated. Incorporation of CH stretch anharmonicity is performed by scaling the fundamental frequencies above $2500\,cm^{-1}$ (CH stretch vibrations) by a factor of 0.98. The degree of agreement observed in this case is nearly exact. This is accomplished in the high-frequency region with this empirical scaling procedure to correct for failure of the harmonic approximation. The agreement between theory and experiment in the low-frequency region shows that the harmonic approximation is valid for the C—C stretching modes. The well-known correlation between force constant and bond length, combined with the good agreement obtained for the C—C stretching region, further suggests that the computed value of the C—C bond length is also likely to be close to the correct value. The average C—C bond length is computed to be 1.558 Å. The value deduced from X-ray diffraction data is 1.544 Å. We conclude [17] that the computed value is closer to reality and that the diffraction value is shortened by disorder [41].

A final "practical" point should be noted in the context of this INS experiment. In the performance of these experiments, the sample of dodecahedrane used consisted of 260 mg and was the world's supply. The point is that, because of the nature of INS experiment, it is possible to borrow or even rent [18,20] materials that would be prohibitively expensive in the amounts needed, for subsequent return.

5.1.2. Conformational analysis: Cycloalkanes

In some cases, it is reasonable to expect that the vibrations of the crystal may be approximated by a treatment of the isolated molecule. This is the case, for example, for dodecahedrane [7], porphine and its zinc complex [18], the cis- and trans-isomers of stilbene [26], and conformationally flexible methoxybenzene derivatives [42]. There are differences in detail between the computed and the observed spectra, but these are often in the nature of an overall scale factor. This suggests that we can combine computational approaches with experimental INS spectra using the comparison to determine what conformation is present in a crystal when this is otherwise unknown. This will provide clear-cut results when the molecule in question has several thermally accessible conformations that differ significantly in their spectra. The spectra for different conformations are most likely to differ in the low frequency part of the spectrum, which is where INS excels. In this application the ability to compute INS spectral intensities greatly aids in the evaluation of the agreement of the observed spectrum with several possible computed spectra.

Figure 6 shows the INS spectrum obtained with TOSCA for cyclooctane compared with the spectra anticipated for each of three conformations [20]. The boat-chair conformation is calculated to be lowest in energy with the twist boat-chair and twist chair-chair being higher in energy by less than 8 kJ/mol. It is clear that the computed spectra differ from each other and that the boat-chair conformation is the only species found in appreciable amounts in the low-temperature crystal. The computed boat-chair spectrum agrees with the observed INS outside the $0–250 \, \text{cm}^{-1}$ region, which is dominated by phonons. The crystal structure of cyclooctane is unknown. When this procedure is applied to several other cycloalkane crystals with known structure, the conformation deduced from the spectral comparison agrees with that deduced from the diffraction data. In each case the species present in the crystal is the one computed

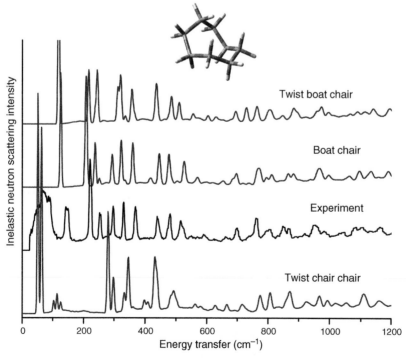

Figure 6 Experimental and computed inelastic neutron scattering (INS) spectra for cyclooctane, C_8H_{16}. The calculations are for three of the low-energy conformations of the isolated species. Adapted from Ref. [20].

to have lowest energy as an isolated species. This is somewhat surprising and is not expected to be the case in general. [43] A dramatic example of polymorphism in which one crystal contains a conformation that is not in a minimum energy conformation for the isolated species has recently been described [44]. This method can be applied to assist or confirm the results of analysis of powder diffraction data where current methods can be complicated by conformational flexibility [42].

5.1.3. The crystal and molecular structure of azulene

The crystal structure of the aromatic hydrocarbon azulene is unknown because of crystal disorder. The structure of the isolated molecule is known from a detailed microwave study [45]. This is an unusual situation for such a large molecule in which the structure of the isolated species is known with greater precision than that of the molecular structure determined from diffraction data. Figure 7 compares the observed INS spectrum (b) with a DFT treatment for an isolated molecule (c) and a DMol^3 treatment of a hypothetical crystal structure based on the known unit cell parameters and a hypothetical anti-parallel arrangement of the two azulene molecules in the unit cell [28]. The DMol^3 treatment of the model crystal reproduces even small features of the INS spectrum. The discrepancies at higher energy are probably due to the harmonic approximation being not strictly valid or due to changes in the crystal lattice parameters and resulting shifting of vibrations for the

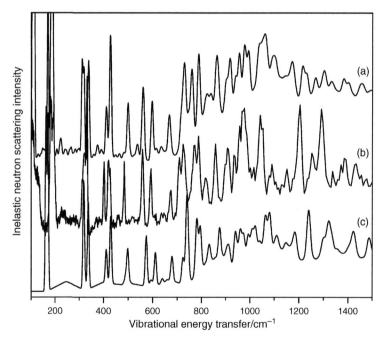

Figure 7 Calculated spectra using (a) periodic Density Functional Theory (DFT) with two azulene molecules arranged antiparallel to one another, and (c) an isolated molecule DFT calculation compared with (b) the observed inelastic neutron-scattering spectrum. Adapted from Ref. [28].

low temperature of the experimental spectrum. The isolated molecule calculation provides a reasonable treatment as expected because intermolecular interactions are expected to be weak for this hydrocarbon. The major differences between the isolated molecule result and the DMol^3 periodic calculations and the experimental results are overall shifts of the spectral features as seen, for example, in the 450–650 cm^{-1} region. There remain some discrepancies between the calculated structure for the isolated molecule and that determined from the microwave spectra. The origin of these differences is not known. The objective of such studies is the development of a method for permitting comparisons of isolated molecule calculations (which can be performed at high levels of theory) with experimental results where crystal packing forces may result in small conformational changes.

5.2. Short, strong, symmetric hydrogen bonds

5.2.1. General considerations and the hydrogen bisacetate anion

Hydrogen bonds sometimes have very short heavy atom distances. When two electronegative atoms (O, N, F) are close together, they can share a proton equally. If the two heavy atoms are of the same type and have the same environment, then the hydrogen bond will be symmetric. In that case the potential energy for motion of the central proton along the line between the heavy atoms will be as depicted in Figure 8. This particular set of potentials has been calculated for the case of the bisacetate anion, $D_3CCOO—H—OOCCD_3$ [22]. Here it is indicated that the acetate methyl groups are deuterated. This material, as well as the fully protonated

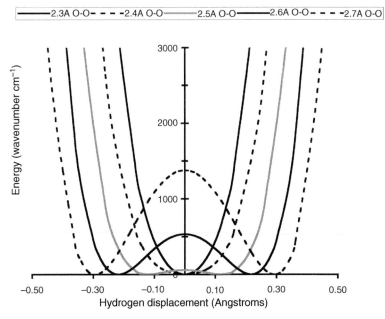

Figure 8 The potential energy for motion of a proton between anionic oxygen atoms for several values of the O—O separation. The calculations (B3LYP/631G★★) are for two acetate anions.

and the $H_3CCOO-D-OOCCH_3$ species, has been studied as the sodium salts. In the methyl-deuterated case, the central H atom is the only H atom present resulting in a relatively simple spectrum [3,22].

Calculations performed for an isolated bisacetate anion illustrate the general behavior for short, strong, symmetric hydrogen-bonded systems. When the O—O (or other heavy atom pair) distance is large, there is a barrier to the transfer of the proton from one side to the other. In the limit of very large separation, this barrier is the bond dissociation energy to form the ionic products. The proton dynamics in this large R_{OO} limit can be described in terms of two (equivalent) localized minima with an OH stretch in the 3000 cm^{-1} region. In this high-barrier limit, the tunneling splitting will be very small. When the O—O distance is reduced, the bond to the adjacent acetate forms before the acetic acid bond breaks and the barrier decreases. The cases we are interested in have O—O distances in the 2.4–2.6 Å range. In this case, the barrier is expected to be well below the zero-point level. In the region of experimental importance, the computed potentials are well described as a positive quartic oscillator with a small harmonic term that may have either a positive or a negative sign. As the separation increases, the negative harmonic term becomes larger and contributes a barrier.

These types of hydrogen-bonded materials, often called Speakman–Hadzi complexes exhibit a variety of interesting features including heavy atom distances that are different for the H and D forms because of the zero-point level effects (the Ubbelohde effect) and infrared spectra with strong, very broad features and sometimes sharp negative features (Evans' holes). They are common features of materials that are ferroelectric, nonlinear optical materials and protonic conductors.

Figure 9 shows the lowest two vibrational frequencies for the one-dimensional (1D) motion of a proton between two acetate ions computed as a function of distance between the O—O atoms. For each value of R_{OO}, there is a hydrogen atom displacement potential curve of the type shown in Figure 8 which can be treated using the FGH method of Balint-Kurti and co-workers [46]. For large O—O separation, the lowest transition is the tunneling splitting transition with near-zero frequency and the higher transition is the usual OH stretch. As R_{OO} decreases, the OH stretch frequency first decreases and then increases as the walls of

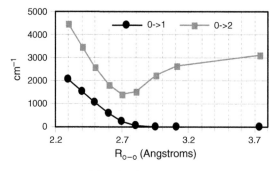

Figure 9   The frequencies computed for the first two transitions for the potentials shown in Figure 5 plus others of intermediate O—O separation. The FGH [46] method was used to determine the eigenvalues. Adapted from Ref. [22].

the well begin to crowd the particle into a smaller box. At this same value of R_{OO}, the tunneling splitting begins to rise and moves into the usual vibrational range. The mixed harmonic quartic oscillator is barrier-less for $R_{OO} < 2.7$ Å for this case.

Figure 10 shows the experimental spectrum of Na $[D_3COO—H—OOCD_3]$ [3,22]. Adiabatic motion of the H atom changes the C—O bond lengths and the O—C—O bond angles. For this and other reasons, the axial motion of the central H atom may not be a major contributor to any of the normal modes of the molecule. Possibly by coincidence, the large feature observed at $1275\,cm^{-1}$ is in good agreement with the 1D calculation of Figure 9 for the R_{OO} separation of 2.46 Å found in the X-ray diffraction study. It would appear that this is a relatively simple system but this may be deceptive. A more thorough solid state treatment is made difficult by the very large size of the unit cell for this material. However, if the computational method used takes symmetry into account the cubic nature of this unit cell makes the problem tractable.

5.2.2. Perchloric acid dihydrate
The second example of a short strong symmetric hydrogen bond is the aquonium or Zundel cation, $[H_2O—H—OH_2]^+$. We have investigated this species as the perchlorate salt where the cation is at a crystallographic center of symmetry. The O—O distance in the $H_2O—H—OH_2$ cation is 2.42 Å. At this distance it is expected that the motion of the proton will have a single minimum with the potential being dominated by the quartic oscillator term. It should be noted, of

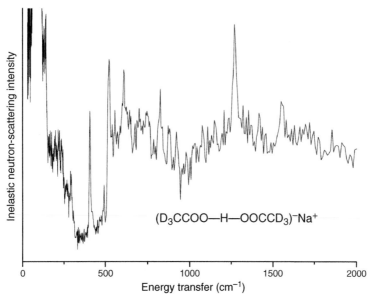

Figure 10 The INS spectrum of Na $[D_3COO—H—OOCD_3]$. The central H atom is the only hydrogen in the sample. The unit cell consists of 12 formula units in a cubic arrangement. Adapted from Ref. [22].

course, that this conclusion is based on a single ion behavior and coupling of the motions in multiple ions may result in distinctly different collective behavior.

Figure 11 shows the INS spectrum of $H_5O_2ClO_4$ as the middle black trace. The lower trace is a harmonic spectrum computed for this crystalline material using the periodic DFT method DMol^3 [39]. In this calculation, all of the vibrational displacements in all unit cells are in-phase (Γ-point). However, for this $Z=4$ unit cell, the motions of the atoms in each cation adopt all of the relative phases according to the unit cell symmetry with a resulting factor group splitting. The DMol^3 harmonic treatment is not expected to provide an adequate description for the axial H atom motion of the central proton in $[H_2O{-}H{-}OH_2]^+$, but it should provide a reasonable description for the four "peripheral" hydrogen atoms. This appears to be the case in the region below $250\ cm^{-1}$ and the region from 500 to $\sim1000\ cm^{-1}$. The region near $450\ cm^{-1}$ appears to have the greatest discrepancy. (The broad experimental feature near $300\ cm^{-1}$ is an artifact as discussed below.)

The factor group splitting of the axial OHO motion of the four molecules in the unit cell computed by this DMol^3 harmonic model is shown by the arrows in the

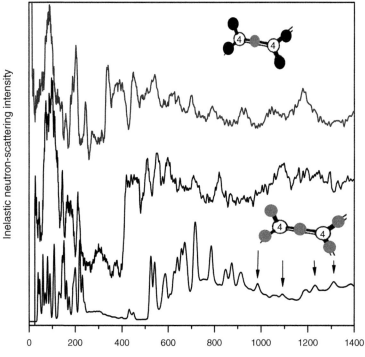

Figure 11   The INS spectra of $HD_4O_2\ ClO_4$ (top, blue) and $H_5O_2\ ClO_4$ (middle, black) and a computed spectrum for the $H_5O_2\ ClO_4$ case (bottom, black). The red balls represent H atoms, and the black balls represent deuterium atoms in the Zundel cation. The calculation at the bottom was performed for the unit cell ($Z=4$) at the Γ point using DMol^3 [15] and the INS plotting program aCLIMAX [16]. The vertical arrows indicate the factor group split axial motions of the central H atom in this harmonic treatment of the $Z=4$ unit cell. Adapted from Ref. [22].

bottom of Figure 11. This splitting of approximately $350 \, \text{cm}^{-1}$ is ascribed to very large charge redistribution associated with motion of the central H. This will also lead to very strong dispersion and hence broad lines in an INS spectrum. A 1D anharmonic treatment for an isolated $[H_5O_2]^+$ cation of the kind discussed above for the bisacetate anion predicts the axial OHO motion for the Zundel cation to be near $1440 \, \text{cm}^{-1}$ for $R_{OO} = 2.42 \, \text{Å}$. There is no feature in this spectrum in the $1000–1500 \, \text{cm}^{-1}$ region. This might be because this degree of freedom exhibits very large dispersion.

A treatment that takes dispersion into account is provided by the plane-wave basis method CASTEP [47] using a perturbative technique to establish the inter-cell interactions [48] has been carried out for crystalline $H_5O_2ClO_4$ by Anibal Ramirez-Cuesta of ISIS. The resulting Γ-point and dispersive CASTEP treatments of the spectrum of $H_5O_2{}^+ClO_4{}^-$ are shown in Figure 12 [49]. In this treatment, 12 points are taken along three directions in k-space. The overall shape of the observed spectrum is reproduced with this $12 \times 12 \times 12$ calculation. As the calculated and observed spectra begin to converge, it is important to note that at least some of the intensity of the large feature near $100 \, \text{cm}^{-1}$ and the smaller feature near $300 \, \text{cm}^{-1}$ are artifacts caused by thermalization of the incident neutron beam in the beryllium filters. This has been eliminated by installation of neutron-absorbing sheets in the filters parallel to the beam direction. This

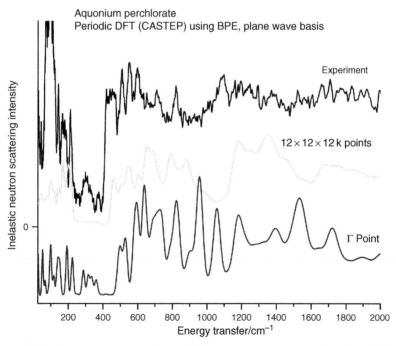

Figure 12 Inelastic neutron scattering (INS) spectrum of $H_5O_2ClO_4$ (top black trace) compared with the results of CASTEP simulations at the Γ-point (bottom blue trace) and with inclusion of dispersion sampled at 12 points along each of three directions is k-space [49].

CASTEP simulation of the INS spectrum of $H_5O_2ClO_4$ includes dispersion but is still based on the harmonic approximation. This will result in incorrect placement of the overtone and combination transitions. The higher overtone and combinations, such as those shown in Figure 5 for dodecahedrane, have the appearance of a nearly unstructured "background" underlying the fundamental and low-order overtone transitions. The simulations of Figure 2 indicate that this "background" should be higher for a quartic oscillator in comparison with the harmonic case. Further, the highly anharmonic axial O—H—O degree of freedom, to the extent that it corresponds to an isolated mode of motion, will be incorrectly located in this calculation.

In order to simplify this spectrum of $H_5O_2ClO_4$, the isotopic species containing a 4:1 ratio of D:H has been prepared [3,22,50]. It is anticipated that there will be preferential partitioning of the H atom into the central O—H—O position with the deuterium locating primarily in the peripheral D_2O positions. Comparison of the isotopomerically isomeric species $[D_2O—H—OD_2]^+$ with the $[D_2O—D—ODH]^+$ form shows that the centered H atom is favored by ~ 200–$400\,cm^{-1}$ because of the difference in the zero-point level. This is because the peripheral OD (or OH) groups make a larger contribution to the zero-point level than does the central O—H—O (or O—D—O) group. If the H atom occupies the central position, then the peripheral positions can be occupied by OD with a low overall zero-point level, but if one of the D nuclei is in the central position, then an H has to be in a peripheral position where it makes a very high contribution to the zero-point level. Given the four-fold degeneracy of the $[D_2O—D—ODH]^+$ case, this zero-point level difference results in ~ 50:50 ratio of the central H to peripheral H at room temperature. At $20\,K$ as used for the INS spectrum, the 200–$400\,cm^{-1}$ zero-point energy difference is enough to tip the balance in favor of the centered proton by 1000:1. The spectrum of the $D_4O_2ClO_4$ is shown as the upper blue trace in Figure 11 [3,22,50]. This spectrum was obtained after the retrofit of baffles in the beryllium filters of TOSCA so that the thermalization artifacts have been removed. There is a general downward shift of the spectrum of $HD_4O_2ClO_4$ in comparison with that of $H_5O_2ClO_4$ and emergence of a feature near $1200\,cm^{-1}$.

The interpretation of the vibrational spectra of the isotopically mixed species is dependent on attainment of isotopic equilibrium. The sample of $HD_4O_2ClO_4$ used for the INS spectrum has been investigated by Gerry Harbison of the University of Nebraska Lincoln using low-temperature solid-state deuterium NMR [50]. The ^{2}H-NMR spectrum at $190\,K$ is shown in Figure 13. The spectrum is complex but contains two discernable components: the outer "water-like" deuterons have a large quadrupole splitting, indicating weak hydrogen bonding, contributes the wings that extent to $\sim 85\,kHz$, and a central peak that arises from the residual O—D—O deuterons. The ^{2}H with large splitting are rapidly relaxed by hopping motion about the local two fold axis. The sharp central feature has a very small quadrupole coupling, characteristic of a deuteron in a very strong hydrogen bond. This feature is chemically shifted from the water deuterons leading to a slight asymmetry of the spectrum. It also has a much longer T_1 relaxation time than the rapidly flipping outer deuterons, allowing it to be distinguished from them. The

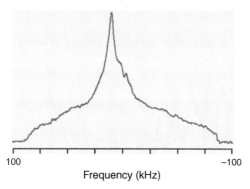

Figure 13 Solid–state ^{2}H-NMR spectrum of HD$_4$O$_2{}^+$ClO$_4{}^-$, obtained at 190 K and a field of 14 T[50].

center peak has an intensity that is roughly 40% of that predicted for random deuteration. This supports the hypothesis that the protons are preferentially partitioning into the center position consistent with attainment of equilibrium at this temperature with a zero-point energy difference of 530 cm^{-1}, which is consistent with the computations. Studies of the analogous sulfate salt that are a bit more advanced are consistent with a zero point energy difference of 325 cm^{-1}.

This spectrum of HD$_4$O$_2$ClO$_4$ in the top trace of Figure 11 still includes the effects of factor group splitting and dispersion. In order to begin to see what effect these factors have on the spectra complexity, a study of a solution of HD$_4$O$_2$ClO$_4$ diluted in 3.4 parts of D$_5$O$_2$ClO$_4$ was performed [50]. The resulting INS spectrum is compared to with of the neat HD$_4$O$_2$ClO$_4$ material in Figure 14. Several features seen in the neat HD$_4$O$_2$ClO$_4$ material disappear with this modest dilution. The feature near 1200 cm^{-1} becomes more clearly discernible. It is anticipated that further dilution will reveal a spectrum that can be more directly attributable to that of an uncoupled [D$_2$O—H—OD$_2$]$^+$ species.

Despite the preliminary nature of the analysis of the spectra of the bisacetate and Zundel cation salts presented here, it is hoped that these studies illustrate the utility of INS spectroscopy and especially the power of isotopic substitution, preferential isotopic partitioning, and isotopic dilution in INS spectroscopy. Isotopic substitution, i.e., taking advantage of exchangeable and nonexchangeable groups, as illustrated with the methyl deuteration of the bisacetate anion, has been widely applied in previous INS studies. Isotopic dilution is widely used in optical studies of the solid state but is rare in neutron scattering studies. The general subject of isotope effects has recently been reviewed in detail [51] including a historical survey by J. Bigeleisen, the pioneer in this field [52]. Isotopic partitioning makes use of the preferential location of H versus D at equilibrium. It is a well-established method for the study of low-barrier hydrogen bonds. We believe that the studies presented in the work reviewed here are the first case of the use of preferential isotopic partitioning in INS studies. This spectrum even at moderate dilution is of interest in respect to the clear changes that are observed including emergence of the 1200 cm^{-1} feature and enhanced resolution in the low-frequency region.

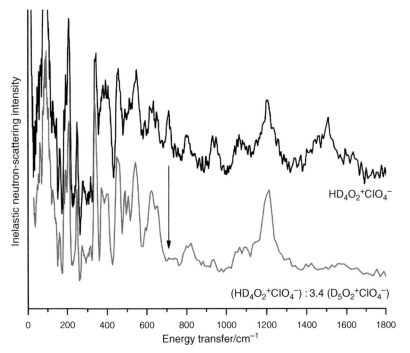

Figure 14   INS spectra of $HD_4O_2ClO_4$ neat (top) and diluted at 1:4.3 in $D_5O_2ClO_4$ (bottom). The loss of bands such as the one at 700 cm^{-1} is attributed to removal of factor group splitting or to the removal of Zundel cations with two or more H atoms [50].

ACKNOWLEDGMENTS

The author thanks the Rutherford Appleton Laboratory for neutron beam access at the ISIS Facility, where the TOSCA spectrometer was used, and the TOSCA staff, Stewart F. Parker and Timmy Ramirez-Cuesta, for their assistance. This work was supported by US National Science Foundation grant CHE 0240104 and by the US Department of Energy grant DE-FG02-01ER14245.

Gerry Harbison of the University of Nebraska is thanked for the ^{2}H-NMR results of Figure 13[50]; Anibal Ramirez-Cuesta of ISIS for the CASTEP simulations of Figure 12[49]; and Nina Verdal of National Institute of Science and Technology for Figures 10, 11, and 14[3,22,49,50].

REFERENCES

[1] P. C. H. Mitchell, S. F. Parker, A. J. Ramirez-Cuesta and J. Tomkinson, Vibrational Spectroscopy with Neutrons, World Scientific, Singapore, 2005. ISBN 981-256-013-0.
[2] B. S. Hudson, J. Phys. Chem. A, 105: 3949–3960, 2001.
[3] B. S. Hudson, Vib. Spectrosc., 42: 25–42, 2006.

[4] S. F. Parker, D. A. Braden, J. Tomkinson and B. S. Hudson, J. Phys. Chem. B, 102: 5955–5956, 1998.

[5] D. A. Braden, S. F. Parker, J. Tomkinson and B. S. Hudson, J. Chem. Phys, 111: 429–437, 1999.

[6] S. F. Parker, J. Tomkinson, D. A. Braden and B. S. Hudson, Chem. Commun., 2000: 165–166.

[7] B. S. Hudson, D. A. Braden, S. F. Parker and H. Prinzbach, Angew. Chem. Int. Ed., 39: 514–516, 2000. [Angew. Chem. 112: 524–526, 2000].

[8] D. A. Braden and B. S. Hudson, J. Phys. Chem., 104: 982–989, 2000.

[9] B. S. Hudson, J. Tse, M. Z. Zgierski, S. F. Parker, D. A. Braden and C. Middleton, Chem. Phys., 261: 249–260, 2000.

[10] B. S. Hudson, Mol. Cryst. Liq. Cryst., 356: 423–432, 2001.

[11] J. Tomkinson, S. F. Parker, D. A. Braden and B. S. Hudson, Phys. Chem. Chem. Phys., 4: 716–721, 2002.

[12] M. R. Johnson, K. Parlinski, I. Natkaniec and B. S. Hudson, Chem. Phys., 291: 53–60, 2003.

[13] D. G. Allis and B. S. Hudson, Chem. Phys. Lett., 385: 166–172, 2004.

[14] D. G. Allis and B. S. Hudson, Chem. Phys. Lett., 386: 356–363, 2004.

[15] D. G. Allis, M. Kosmowski and B. S. Hudson, J. Am. Chem. Soc., 126: 7756–7757, 2004.

[16] B. S. Hudson, D. A. Braden, D. G. Allis, T. Jenkins, S. Baronov, C. T. Middleton, R. Withnall and C. M. Brown, J. Phys. Chem. A, 108: 7356–7363, 2004.

[17] B. S. Hudson, D. G. Allis, S. F. Parker, A. J. Ramirez-Cuesta, H. Herman and H. Prinzbach, J. Phys. Chem. A, 109: 3418–3424, 2005.

[18] N. Verdal, P. M. Kozlowski, B. S. Hudson, J. Phys. Chem. A, 109: 5724–5733, 2005.

[19] J. A. Ciezak and B. S. Hudson, THEOCHEM J. Mol. Struct., 755: 195–202, 2005.

[20] N. Verdal, J. J. Wilke and B. S. Hudson, J. Phys. Chem. A, 110: 2639–2646, 2006.

[21] D. G. Allis and B. S. Hudson, J. Phys. Chem. A, 110: 3744–3749, 2006.

[22] B. S. Hudson and N. Verdal, Physica B, 385–386: 212–215, 2006.

[23] J. A. Ciezak, J. B. Leao, and B. S. Hudson, THEOCHEM, 767: 23–28, 2006.

[24] S. F. Parker, K. Refson, K. P. J. Williams, D. A. Braden, B. S. Hudson and K. Yvon, Inorg. Chem., 45: 10951–10957, 2006.

[25] N. Verdal and B. S. Hudson, Chem. Phys. Lett., 434: 241–244, 2007.

[26] N. Verdal and B. S. Hudson, J. Neutron Res., 15: 69–74, 2007.

[27] R. W. Williams, S. Schlücker and B. S. Hudson, Chem. Phys., 343: 1–18, 2008.

[28] N. Verdal, S. A. Rivera and B. S. Hudson, Chem. Phys. Lett., 437: 38–44, 2007.

[29] http://neutrons.ornl.gov/facilities/facilities_sns.shtml

[30] http://www.isis.rl.ac.uk/

[31] http://www.isis.rl.ac.uk/molecularSpectroscopy/tosca/index.htm A library of INS spectra can be found at this site.

[32] D. Colognesi, M. Celli, F. Cilloco, R. J. Newport, S. F. Parker, V. Rossi-Alertini, F. Sacchetti, J. Tomkinson and M. Zoppi, Appl. Phys., A74: S64–S66, 2002.

[33] G. L. Squires, Introduction to the Theory of Thermal Neutron Scattering, Dover Publications, Mineola, NY, 1996. ISBN 0-486-69447-X

[34] http://www.ncnr.nist.gov/ [An excellent site for neutron information in general].

[35] http://neutrons.ornl.gov/

[36] http://www.ill.fr/

[37] http://www.pns.anl.gov/ The Intense Pulsed Neutron Source (IPNS) was closed for budgetary reasons on 25 January 2008. The Spallation Neutron Source (SNS; http://neutrons.ornl.gov/aboutsns/aboutsns.shtml) is now accepting user applications. The Neutron Spallation Source (NSS) of J-PARC (Japan Proton Accelerator Research Center), http://j-parc.jp/MatLife/en/index.html, is scheduled for first neutrons in May 2008.

[38] http://lansce.lanl.gov/

[39] B. Delley, J. Chem. Phys., 113: 7756–7764, 2000.

[40] A. J. Ramirez-Cuesta, Comput. Phys. Commun., 157: 226–238, 2004.

[41] J. Dillen, J. Phys. Chem. A, 104: 7734–7737, 2000; J. L. M. Dillen, S. Afr. J. Chem., 44: 62–63, 2000.

[42] N. Verdal, Z. Pan, K. D. M. Harris and B. S. Hudson, in preparation.

[43] J. Bernstein and A. T. Hagler, J. Am. Chem. Soc., 100: 673–681, 1978;
     J. Bernstein, Polymorphism in Molecular Crystals, Clarendon Press, Oxford, 2002. .
[44] P. U. Biedermann, J. J. Stezowski and I. Agranat, Chem. Eur. J., 12: 3345–3354, 2006.
[45] S. Huber, G. Grassi and A. Bauder, Mol. Phys., 103: 1395–1409, 2005.
[46] C. C. Martson and G. G. Balint-Kurti, J. Chem. Phys., 91: 3571–3576, 1989.
[47] S. J. Clark, M. D. Segall, C. J. Pickard, P. J. Hasnip, M. J. Probert, K. Refson, and M. C. Payne
     Z. Kristallographie, 220: 567–570, 2005; M. D. Segall, P. J. D. Lindan, M. J. Probert, C. J. Pickard,
     P. J. Hasnip, S. J. Clark and M. C. Payne, J. Phys.: Condens. Matter, 14: 2717–2744, 2002.
[48] K. Refson, P. R. Tulip and S. J. Clark, Phys. Rev. B 73: 155114, 2006.
[49] A. J. Ramirez-Cuesta, N. Verdal and B. S. Hudson, in preparation.
[50] N. Verdal, A. Huq, G. S. Harbison and B. S. Hudson, in preparation.
[51] A. Kohen and H. -K. Limbach (eds), Isotope Effects in Chemistry and Biology, Taylor &
     Francis, Boca Raton, 2006.
[52] J. Bigeleisen and M. G. Mayer, J. Chem. Phys., 15: 261–267, 1947.

CHAPTER 18

OPTIMAL SIGNAL PROCESSING IN CAVITY RING-DOWN SPECTROSCOPY

Kevin K. Lehmann *and* Haifeng Huang

Contents

Abstract

In this chapter, the authors present a systematic statistical analysis of cavity ring-down signal extraction. The traditional uncorrelated least squares fit can be generalized to the situation with data correlation (e.g. caused by data filtering, which is essential to minimize noise). If the data is sufficiently highly sampled, the effect of the data correlation can be included by introducing an effective variance of the data. This correction has substantial influence on the estimation of the standard error of the fit parameters for correlated data, especially for the fitted decay rate k', because this determines the final sensitivity. For both the white noise dominated and the shot noise dominated situations, the sensitivity limit is given. The authors found that the bias of k in the white noise situation is normally very small and can be neglected. The authors also compared several commonly used alternative algorithms in cavity ring-down community. These mathods include linearized weighted least squares fit, determining the decay rate by Fourier transform, corrected successive integration (CSI) method, and several analog methods in extracting k from decay signal. The bias, dispersion prediction (including optimum results), and speed of these methods are discussed in detail. Among all these methods, the least square fit gives the smallest dispersion estimation of fit parameters. Considering the special properties of exponential function, the authors found that the least squares fit has lower computation cost than the Fourier transform and CSI methods. Lastly, the effect of residual mode beating on the extracted k is also analyzed.

Keywords: cavity ring-down spectroscopy; signal processing; data correlation; optimal signal processing; cavity-enhanced spectroscopy; generalized least squares fit; white noise; fourier transform method; successive integration method; analog detected cavity ring-down; shot noise; residual mode beating

1. INTRODUCTION

Since the seminal work of O'Keefe and Deacon [1], cavity ring-down spectroscopy (CRDS) has become an important method for obtaining highly sensitive absolute absorption spectra of weak transitions or rarefied species. Twenty years after its introduction, the popularity of this and related methods that use high-finesse optical cavities continues to grow. Our bibliographic database now contains over 900 entries, the majority of entries from the last 4 years alone. The work in the field up to 1999 was well represented in a collective volume edited by Busch and Busch [2]. In 2008, another collective volume is due to be published, edited by Berden and Engel [3]. There have been a number of excellent reviews published over the years, including Refs [4,5] and most recently Ref. [6].

In standard CRDS, a spectrum is obtained by observing changes in the optical decay rate of a high-finesse optical cavity as a function of excitation frequency. It therefore follows that the sensitivity of CRDS is directly proportional to the stability and signal-to-noise ratio of the cavity decay rate, which can be extracted in a number of different ways from the ring-down transient of cavity output power as a function of time. Many different methods have been described for determining the cavity decay rate, which is not surprising given the rich variety of lasers that have been used and samples that have been studied.

Starting with Romanini and Lehmann [7], several papers have reported the analysis of the expected signal-to-noise ratio of the cavity decay rate with different assumptions about the data analysis method and the statistical character of the decay. However, this chapter presents what the authors believe to be the most systematic investigation to date into the statistical analysis of cavity ring-down signal extraction. We review and derive sensitivity limits for all the methods commonly used to extract cavity decay rates in the CRDS community. In Section 2, we lay out the mathematical model of the data, including the data correlation that arises from a low pass filtering of the data. Such filtering is required to obtained an optimized signal-to-noise ratio and thus minimize statistical fluctuations in the extracted cavity decay rate, which determines the sensitivity of a CRDS experiment. Such data correlations are ignored in standard least squares fit treatments. Section 3 sets out the form of the least squares fit problem with data correlation [8]. Section 4 derives the tridiagonal fit weight matrix that arises for exponential data correlation as is created by a low-pass filter. For CRDS data under cases where the data are sufficiently highly sampled, the least squares fit equations only change by the introduction of an effective variance of the data. This correction of the data variance is essential to correctly predict the standard error of the fit parameters, including the cavity decay rate. Section 5 presents an analysis of least squares fitting of CRDS data in the case where detector and other forms of intensity independent noise dominate, including derivation of the sensitivity limit that can be expected. Section 6 examines the weighted least squares fit that results from linearization of the problem by calculation of the log of the data points. In high signal-to-noise situations, this gives the same sensitivity as the direct least squares fit to the data, but the non-linear transformation introduces some bias that become important at modest signal-to-noise ratios. At low signal to noise, the method introduces substantial bias [9]. Sections 7 and 8 examine the noise and bias expected when the cavity decay rate is determined using the Fourier transform [10] and the successive integration (SI) [11] methods, respectively. These are two noniterative methods that do not require prior knowledge of the baseline and are being used in the CRDS community as alternatives to the least squares fitting of the data. In both cases, the predicted noise in the decay rate is moderately higher than for the least squares fit to the data. We do not find that these methods offer computational advantages over an optimized least squares fitting program. Section 9 examines three different analog processing approaches for the determination of the ring-down rate. The first is the phase shift method [12,13], which uses lock-in demodulation of the cavity transmission; the second calculates the decay rate from the ratio of signal sampled by two gated integrator windows [7]; and the last uses a log-amplifier followed by a differentiator [14]. These methods are useful for cases of very high cavity decay event rates, where all the individual decays can no longer be digitized and fit, though the last appears to have a poorer sensitivity limit. Section 10 presents an analysis of the standard least squares fit in the case where shot noise in the ring-down signal dominates. This proves to be a particularly simple case where the solution of the least squares fit equations proves to be equivalent to taking the normalized first moment of the signal transient. Section 11 considers the effect of residual mode beating in the ring-down decay upon the expected sensitivity. Such beating is usually unavoidable when using a pulsed laser to excite the ring-down cavity. Section 12 presents some closing remarks.

 ## 2. THE MODEL

We will begin with a model for the signal observed in a cavity ring-down experiment. Let $y(t)$ be the time dependent optical power on the detector. We expect the signal, expressed in watts, to be of the following form:

$$y(t) = F(t) + \epsilon(t) = Ae^{-kt} + B + \epsilon(t), \tag{1}$$

where $\epsilon(t)$ is the noise term. We will start by assuming that we have white noise with spectral density $\sigma'(t)$ (units of watts per $\sqrt{Hz}$), which implies the following statistical properties:

$$<\epsilon(t)>=0$$
$$<\epsilon(t) \cdot \epsilon(t')>=\sigma'^2(t)\delta(t-t'), \tag{2}$$

where $<>$ denotes an ensemble averaging over a large number of macroscopically identical experimental runs. A is the amplitude of the ring-down signal. k is the ring-down rate and is typically in the range $(0.001 - 1) \cdot 10^6 \, s^{-1}$ in most CRDS experiments. $\tau = k^{-1}$ is the decay time constant for the cavity. B represents a baseline signal after the light inside the cavity has decayed to a negligible value. In cases where B is stable with time, it will not be treated as a fit parameter, and its value can be subtracted from the data prior to fitting. In any real experiment, we will not have an infinite bandwidth for our detector. Furthermore, to optimize our signal-to-noise ratio, we may want to further limit the detection bandwidth. We assume that we have a single pole low-pass filter with a time constant $t_f = 1/k_f$. This leads to the following redefinition of the time dependent signal:

$$\bar{y}(t) = \int_0^\infty k_f \, y(t-\tau) \exp(-k_f \tau)d\tau = \bar{F}(t) + \bar{\epsilon}(t) \tag{3}$$

with analogous definitions for $\bar{F}(t)$ and $\bar{\epsilon}(t)$. If we further assume that $\sigma'(t)$ is effectively constant over time intervals $\sim t_f$, then it can be shown from these definitions that our averaged noise has the following correlations:

$$<\bar{\epsilon}(t) \cdot \bar{\epsilon}(t')>=\frac{k_f}{2}\sigma'^2\left(\frac{t+t'}{2}\right) \cdot \exp(-k_f|t-t'|). \tag{4}$$

Thus, the time constant leads to a "memory" of t_f for the noise. The form for $\bar{F}(t)$ depends on what we assume for the signal for negative time. Two choices are $F(t<0) = B$, which would be a model for excitation of the ring-down cavity by a short pulse at $t=0$, and $F(t<0) = A+B$, which is a model for quasi-cw excitation of the ring-down cavity. In these two cases we have:

$$\bar{F}(t>0) = \begin{cases} \dfrac{k_f}{k_f - k}A\left(e^{-kt} - e^{-k_f t}\right) + B & \text{if } F(t<0) = B \\[3mm] \dfrac{k_f}{k_f - k}A\left(e^{-kt} - \dfrac{k}{k_f}e^{-k_f t}\right) + B & \text{if } F(t<0) = A+B. \end{cases} \tag{5}$$

In most CRDS experiments, the very beginning of the decay is not fit because it is distorted by various effects, including the finite response time of the detector and electronics. Regardless, as long as $k_f \gg k$ (i.e. $\tau \gg t_f$), which will typically be the case, then the extra term in $\bar{F}(t)$ will quickly decay (compared to the ring-down time), and thus we can fit most of the decay without needing to include the corrections to a single exponential decay. For pulsed excitation in CRDS, one has

100% amplitude modulation of the signal on the 1–10 nsec time scale and the use of a higher order filter to drastically reduce this modulation without cutting into the detection bandwidth is advised. However, as long as t_f is the shortest filter time constant, the present treatment should give a good approximation.

We will now consider that the ring-down transient is digitized by an analog-to-digitial (A/D) convertor with $\bar{y}(t)$ sampled at equally spaced times $t_i = i \cdot \Delta t$ with $i = 0 \ldots N-1$. Spence et al. [14] discussed some of the effects that A/D nonidealities can introduce, but we will not consider them further as our experience is that these are not important if the root-mean-squared (RMS) noise input of the A/D is greater than the least significant bit. To most faithfully reproduce the ring-down signal, we want $k\Delta t \ll 1$. Δt is usually determined by the hardware limitations. An important practical trade-off is between the dynamic range (how many bits of resolution) versus speed of the digitizer. Even without the hardware constraints, the memory and computational cost to determine the ring-down rate from a decay scales inversely with Δt, and thus oversampling may limit the number of decay per second that can be processed. Below, we will examine the sensitivity of the determination of the ring-down rate to the value of $k\Delta t$. We will denote the set of measured values as y_i, or collectively as $\vec{y}$. Likewise, we define $F_i = \bar{F}(t_i)$ and $\epsilon_i = \bar{\epsilon}(t_i)$, with $\vec{F}$ and $\vec{\epsilon}$ denoting the vector of values.

We also need a model for the noise density. The two intrinsic noise sources in CRDS are (i) shot noise and (ii) detector noise. If $\bar{F}(t)$ is expressed in watts, we have light frequency of ν, our detector has quantum efficiency Q, and noise equivalent power P_N, then we can write for the noise density:

$$\sigma'^2(t) = \frac{h\nu}{Q} A e^{-kt} + P_N^2, \tag{6}$$

where h is Planck's constant. P_N has contributions from the shot noise of the detector dark current and also Johnston noise in the resistor that is used to convert the photocurrent into a voltage. Thus, P_N is minimized by having that resistance large enough that the dark current noise dominates, but this will often restrict the detector bandwidth.

The measurements will have variance $\sigma_i^2 = (k_f/2)\sigma'^2(t_i)$. Under the assumption of slowly changing $\sigma'^2(t)$, we can write the covariance matrix, σ^2, for the data as

$$(\sigma^2)_{i,j} = <\epsilon_i \epsilon_j> = \frac{\sigma_i^2 + \sigma_j^2}{2} e^{-|i-j|k_f \Delta t}. \tag{7}$$

In this chapter, we will consider the two limiting cases, namely shot noise limited CRDS, where the first term dominates and the baseline offset, B, from detector dark current can be neglected and detector noise limited CRDS, where the second term dominates. Of course, because of the exponential decay, the second term will dominate the first sufficiently far into the decay, but the tail of the decay makes only a modest contribution to the fitted parameters, so no instability will be presented by the assumption of shot noise limited decay. As will be shown below, the assumption of shot noise limited decay leads to simpler expressions, including the fact that the exact least squares solutions can be written down directly.

We will now consider the statistical properties of a fit of the ring-down curve. This fit will give parameters A', k', and possibly B', which are estimates for unknown ring-down amplitude, rate, and signal baseline, of which k' is the one used in extracting the spectrum or determining analyte concentration. The obvious goal is to use estimates that are unbiased (i.e. on average will give the correct values) and have the smallest possible dispersion. If we assume that the noise has a Gaussian distribution, then these criteria are met by a properly weighted least squares fit of the data [8]. An important point, however, is that the assumption that is typically made, that the experimental data points have uncorrelated errors, is not valid unless $k_f \Delta t \gg 1$. However, it is possible to properly account for correlation by a minor modification of the least squares equations.

3. GENERALIZED LEAST SQUARES FIT WITH CORRELATED DATA

The mathematical treatments of linear and linearized nonlinear least squares fits to uncorrelated experimental data are well known and described in detail in many texts [15,16]. However, the treatment of the more general problem with correlated data is much less well known, at least in the spectroscopy community, though the proof of the general solution of the linear model has been presented by Albritton et al. [8]. In this section, the solutions for this more general least squares problem will be given in the context of fitting CRDS data, which is a nonlinear model.

If we assume that the individual ϵ_i are Gaussian distributed, then by going to the basis set that diagonalizes the $\tilde{\sigma}^2$ matrix [8] (in this basis set the errors are uncorrelated), it is easily shown that we can write the probability density for any given set of experimental errors as follows:

$$P\left(\vec{\epsilon}\right) = \frac{1}{\sqrt{(2\pi)^N det\left(\tilde{\sigma}^2\right)}} \exp\left(-\frac{1}{2}\vec{\epsilon}\cdot\left(\tilde{\sigma}^2\right)^{-1}\cdot\vec{\epsilon}\right). \qquad (8)$$

We will denote the fitted function by $F'(t) = A'e^{-k't} + B'$, and its value at the measured times as F'_i. The principle of maximum likelihood states that the fit parameters $\{A'\ B'\ k'\}$ that best estimate the unknown decay parameters $\{A\ B\ k\}$ are those that maximize the probability that the observed set of data will occur.

For any given parameters, we can write $\vec{\epsilon} = \vec{y} - \vec{F}'$. We also introduce a generalized weight matrix:

$$\widetilde{W} = \left(\widetilde{\sigma^2} \right)^{-1}. \tag{9}$$

Thus, we have a generalized χ^2 function:

$$\chi^2(A', B', k') = \left(\vec{y} - \vec{F}' \right) \cdot \widetilde{W} \cdot \left(\vec{y} - \vec{F}' \right), \tag{10}$$

for which we will find the minimum. Note that under the above assumption of Gaussian errors, the $\chi^2(A, B, k)$ should follow a chi-squared distribution with N degrees of freedom, whereas $\chi^2(A', B', k')$ should follow a chi-squared distribution with $N - 3$ degrees of freedom. The generalized least squares equations, found by minimizing χ^2 with respect to the parameters $p_j = \{A', B', k'\}$, are as follows:

$$\left(\vec{y} - \vec{F}'(A', B', k') \right)^{\dagger} \cdot \widetilde{W} \cdot \widetilde{\nabla F} = \vec{0} \tag{11}$$

$$\nabla F_{ij} = \left(\partial \frac{F'_i}{\partial p_j} \right). \tag{12}$$

Note that we do not have derivatives of the $\widetilde{W}$ matrix with respect to fit parameters even if, because of shot noise, the noise of the signal is intensity dependent. In the limit of a diagonal $\widetilde{W}$, this reduces to the standard least squares equations found in numerous texts [15,16]. Given the solution to the above equations and assuming linear behavior near the solution (i.e. $\frac{\partial \vec{F}'}{\partial p_j}$ are constant), we can write the covariance matrix, $\widetilde{c}$, for the fit parameters in terms of the curvature matrix, $\widetilde{\alpha}$, by

$$\widetilde{\alpha} = \widetilde{\nabla F}^{\dagger} \cdot \widetilde{W} \cdot \widetilde{\nabla F} \tag{13}$$

$$\widetilde{c} = \widetilde{\alpha}^{-1}. \tag{14}$$

The diagonal elements of $\widetilde{c}$ give the variances of the fit parameters and the off-diagonal elements of the covariances $c_{ij}/\sqrt{c_{ii}c_{jj}}$ (which is between -1 and 1) gives the correlation coefficient between the ith and j^{th} parameters. The least squares equations can be solved iteratively, given an initial estimate for the parameters and by calculating the change that will solve the equations under the assumption of linearity:

$$\vec{\delta p} = \widetilde{c} \cdot \vec{\beta} \quad \text{with} \quad \vec{\beta} = \widetilde{\nabla F}^{\dagger} \cdot \widetilde{W} \cdot \left(\vec{y} - \vec{F}' \right). \tag{15}$$

This equation can also be used, with $\vec{y} - \vec{F}'$ replaced by $\vec{\epsilon}$, to give the sensitivity of the parameters to a perturbation of the input data.

Because of parameter correlation, the standard error of k will be higher for a fit that adjusts the baseline parameter, B. Often, the detector and electronics offsets are more stable with time than the accuracy to which B can be determined in a fit to a single decay. In these cases, lower fluctuations in the fitted k' values are expected if only the two parameters A', k' are floated in the fit. This is done simply by subtracting the fixed value B from the experimental data and then removing the B row and column from the curvature matrix, $\widetilde{\alpha}$.

If we have an incorrect model for the variances and covariances of the noise, the predicted weight matrix, $\widetilde{W}'$, will be in error. How do we estimate the effect that this will have on the fit results? The mean value of the fit parameters will still be the true parameters, but the covariance matrix calculated from Equation (14) using $\widetilde{W}'$ (let us call it $\widetilde{c}'$) will not give correct estimates for the true uncertainties in the fit parameters. If we neglect data correlation (set off-diagonal elements of the weight matrix to zero), the resulting $\widetilde{c}'$ gives smaller estimates for the parameter standard errors than predicted by $\widetilde{c}$. However, in general, we expect the true covariance matrix for an improperly weighted fit, which we will denote by $\widetilde{c}''$, to give larger uncertainties than those of the appropriately weighted fit, $\widetilde{c}$, as it is known that the properly weighted least squares fit is the lowest dispersion linear estimator of the unknown parameters [8]. By using the standard linearized equation for error propagation,

$$c''_{ab} = \langle \delta p_a \delta p_b \rangle = \sum_{ij} \left(\frac{\partial p_a}{\partial y_i} \right) \sigma^2_{ij} \left(\frac{\partial p_b}{\partial y_j} \right) \tag{16}$$

combined with the modified version of Equation (15) ($\widetilde{W}$ replaced by $\widetilde{W}'$), we have

$$\widetilde{c}'' = \widetilde{U}^\dagger \cdot \widetilde{\sigma^2} \cdot \widetilde{U}$$
$$\widetilde{U} = \widetilde{W}' \cdot \widetilde{\nabla F} \cdot \widetilde{c}' \tag{17}$$
$$\widetilde{c}' = \left(\widetilde{\nabla F}^\dagger \cdot \widetilde{W}' \cdot \widetilde{\nabla F} \right)^{-1}.$$

The matrix $\widetilde{c}'$ is the covariance matrix we would calculate assuming $\widetilde{W}'$ is the weight matrix. It can easily be shown that $c'' \to \widetilde{c}$ if $\widetilde{W}' = $ constant $\times \widetilde{W}$. Equation (17) will be used below to calculate the statistical dispersion of the fit parameters for approximate fit models that neglect parameter correlation.

4. WEIGHT MATRIX FOR MODEL OF CAVITY RING-DOWN DATA

Returning to our specific problem, we need to evaluate the various terms in the above generalized least squares expressions. For the equally time spaced points and again assuming slowly changing noise variance, we can analytically invert the covariance matrix to give the weight matrix. Let us define $a_f = \exp(-k_f \Delta t)$, then we have

$$W_{i,j} = \frac{2}{\left(\sigma_i^2 + \sigma_j^2\right)} \times \begin{cases} \dfrac{1 + a_f^2}{1 - a_f^2} & \text{if } i=j, i \neq 0, N-1 \\[2ex] \dfrac{1}{1 - a_f^2} & \text{if } i=j, i=0, N-1 \\[2ex] -\dfrac{a_f}{1 - a_f^2} & \text{if } i=j \pm 1 \\[2ex] 0 & \text{otherwise.} \end{cases} \tag{18}$$

If we have a function, $f(t)$, that changes slowly over time interval Δt, we can write

$$\left(\widetilde{W} \vec{f}\right)_i \approx \frac{1}{\sigma_i^2}\left[\left(\frac{1 - a_f}{1 + a_f}\right)f(t_i) - \left(\frac{a_f \Delta t^2}{1 - a_f^2}\right)\left(\frac{d^2 f}{dt^2}(t_i)\right)\right]. \tag{19}$$

The definition of $\vec{f}$ is the same as $\vec{y}$. Thus, we can see that in the limit of sufficiently small sampling time,

$$|f(t)| \gg \frac{a_f}{\left(1 - a_f\right)^2}\left|\left(\frac{d^2 f}{dt^2}(t_i)\right)\right|\Delta t^2, \tag{20}$$

the effect of the data correlation can be approximated by multiplying a diagonal weight matrix by a scale factor of $\left(\frac{1-a_f}{1+a_f}\right)$, or equivalently, the variances of the data points should be multiplied by a uniform factor of

$$G = \left(\frac{1 + \exp\left(-k_f \Delta t\right)}{1 - \exp\left(-k_f \Delta t\right)}\right), \tag{21}$$

$G > 1$. This same factor will need to be applied to the covariance matrix calculated for a least squares fit that ignores data correlation. For CRDS data, $\left(\frac{d^2 f}{dt^2}(t)\right)\Delta t^2 = (k\Delta t)^2 f(t)$ (the baseline can be neglected for this purpose) and thus the Equation (20) holds to high accuracy for well-sampled CRDS data, for which $k\Delta t \sim 0.01$ is typical.

An obvious question is how does the variance in our parameters changes with k_f? We will first consider the simpler problem of averaging N data points, each separated by Δt for a constant signal. In this case, we can treat the noise density $\sigma'(t)$ as constant. The best variance we can expect for the average over this interval is $\sigma'^2/(N\Delta t)$. It is the authors' experience that many people believe, at least intuitively, that the signal to noise will improve as the time constant, t_f, is increased to values much beyond Δt. However, by using the above generalized least squares equations, it can be found that the variance of the mean, μ, is given by

$$\sigma^2(\mu) = \xi^2 \frac{\sigma'^2}{N\Delta t} = \left(\frac{k_f}{2N}\right)\sigma'^2\left(\frac{1 + \exp\left(-k_f \Delta t\right)}{1 - \exp\left(-k_f \Delta t\right)}\right), \tag{22}$$

ξ is the fractional "excess" noise in the mean over the ideal limit. In the limit that $k_f \Delta t \ll 1$, $\xi \to 1$, which means that we have sufficiently sampled the signal to extract the maximum information. Figure 1 shows a plot of ξ as a function of $k_f \Delta t$. Specific values are $\xi = 1.04$, 1.15, and 1.44 for $k_f \Delta t = 1$, 2, and 4. One way to view the increase in noise for larger values of $k_f \Delta t$ is that the digitization process will alias the noise at all frequency components that are harmonics of $1/\Delta t$ to zero frequency, where they will survive the averaging. Thus, we want to pick the 3–dB roll off frequency of the low–pass filter (which equals $k_f/(2\pi)$) low enough that the noise contributions from these harmonics are small compared to the near DC noise density. Thus, there is simply no signal-to-noise advantage to making the time constant of the filter much longer than sampling time interval. Longer time constants will reduce the RMS noise in the data but will make the data highly correlated and thus largely redundant. The gain in the reduced noise will be almost exactly compensated for by a reduction in the effective number of independent data points. Fits that ignore the correlation of the data will predict a variance of the average that will underestimate the true variance by a factor of $\approx k_f \Delta t/2$ when $k_f \Delta t \ll 1$. From the above considerations, it should be clear that for optimal signal processing in CRDS, one wants to select values of k_f and Δt such that $k_f \Delta t \sim 1$ and $k \Delta t \ll 1$.

The authors have constructed an ensemble of 1000 fits to decays with parameters $\sigma' = 1$, $A = 100$, $\Delta t = 1$, $k = 0.05$, and $k_f = 1$, with the data fit with and without including the correlation of the data. It was found that statistical properties of the two sets of decay rates agreed well within their uncertainty. Further, the standard deviation of the difference in decay rate for the two fits was an order of magnitude below the standard deviation of either set of fits. Thus, in this case, there is no practical difference

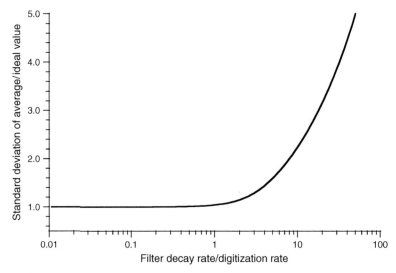

Figure 1 Plot of excess noise created when digitizing and then averaging a DC signal with white noise that has been passed through a low–pass filter with a filter time constant $1/k_f$. The ordinate is equal to $k_f \Delta t$ in the notation used in the text.

in the two set of fits. However, the variance in the ensemble of decay rates is only correctly predicted by the correlated least squares fit calculation. The uncorrelated fit underestimates the ensemble variance by a factor of ≈ 2.2. For an ensemble with the same parameters except $k_f = 0.2$, the variance of the ring-down decay rates for the uncorrelated fits was found to be $\approx 16\%$ higher than for the correlated fits. In this case, the variance predicted by the uncorrelated fits is a factor of 10 below that found for the ensemble. These simulation results were confirmed by the general calculations of the expected variance of both correlated and uncorrelated fits, which will be given below.

5. DETECTOR NOISE LIMITED CAVITY RING-DOWN DATA

We will now look at the predicted dispersion of the ring-down rate for different assumptions about the fit. First, we consider a fit with σ' constant. Ideally, this noise density arises from the intrinsic detector noise, P_N, though excess noise from the amplifier and/or digitizer (including bit quantization noise) would likely be of this form.

As demonstrated above, as long as $k_f \gg k$, we can replace the general least squares problem with a nondiagonal weight matrix, $\widehat{W}$ (Equation (18)), with a simple diagonal matrix, but with the variance scaled by G (Equation (21)). This gives for the effective variance per point $\sigma^2 = \xi^2 P_N^2 / \Delta t$, where ξ is as defined in Equation (22) and is plotted in Figure 1. It is nearly unity for $k_f \Delta t \leq 1$.

In this case, we can use geometric series expressions $\sum_{i=0}^{N-1} i^n a^i = [a(d/da)]^n [(1 - a^N)/(1 - a)]$ to write explicit closed form expressions for the elements of the curvature matrix, $\widetilde{\alpha}$. In terms of $a = \exp(-k\Delta t)$, the matrix elements of the symmetric curvature matrix, $\widetilde{\alpha}$, and vector β are

$$\alpha_{B,B} = \frac{N}{\sigma^2}$$

$$\alpha_{B,A} = \left(\frac{1 - a^N}{1 - a} \right) \frac{1}{\sigma^2}$$

$$\alpha_{B,k} = -A\Delta t \left(\frac{a(1 - a^N)}{(1 - a)^2} - \frac{Na^N}{1 - a} \right) \frac{1}{\sigma^2}$$

$$\alpha_{A,A} = \left(\frac{1 - a^{2N}}{1 - a^2} \right) \frac{1}{\sigma^2} \tag{23}$$

$$\alpha_{A,k} = -A\Delta t \left(\frac{a^2(1 - a^{2N})}{(1 - a^2)^2} - \frac{Na^{2N}}{1 - a^2} \right) \frac{1}{\sigma^2}$$

$$\alpha_{k,k} = (A\Delta t)^2 \left(\frac{2a^4(1 - a^{2N})}{(1 - a^2)^3} + \frac{a^2 - (2N + 1)a^{2N+2}}{(1 - a^2)^2} - \frac{N^2 a^{2N}}{1 - a^2} \right) \frac{1}{\sigma^2}$$

$$\beta_B = \left(\sum_i \gamma_i - A \left(\frac{1 - a^N}{1 - a} \right) - BN \right) \frac{1}{\sigma^2}$$

$$\beta_A = \left(\sum_i \gamma_i a^i - A \left(\frac{1 - a^{2N}}{1 - a^2} \right) - B \left(\frac{1 - a^N}{1 - a} \right) \right) \frac{1}{\sigma^2} \qquad (24)$$

$$\beta_k = -A \Delta t \left(\sum_i \gamma_i i a^i - A \left(\frac{a^2 (1 - a^{2N})}{(1 - a^2)^2} - \frac{N a^{2N}}{1 - a^2} \right) - B \left(\frac{a(1 - a^N)}{(1 - a)^2} - \frac{N a^N}{1 - a} \right) \right) \frac{1}{\sigma^2}.$$

Note that for each cycle of the least squares fit, we have to do only three sums over the data points that can be calculated with only three floating point multiplications and three additions per data point. We need to invert $\tilde{\alpha}$ to determine $\tilde{c}$. For the 2 × 2 case (fixed baseline), $c_{ii} = \alpha_{jj}/\det(\alpha)$ and $c_{ij} = -\alpha_{ij}/\det(\alpha)$, where $\det(\alpha)$ is the determinant of α. For the 3 × 3 case, $c_{ii} = \left(\alpha_{jj} \alpha_{kk} - \alpha_{jk}^2 \right)/\det(\alpha)$ and $c_{ij} = (\alpha_{ik} \alpha_{jk} - \alpha_{ij} \alpha_{kk})/\det(\alpha)$, where ijk is a cyclic permutation of A, B, k. The diagonal elements of $\tilde{c}$ give the variances (standard errors) in the corresponding parameters and the off-diagonal elements the corresponding covariances.

If we take N larger enough that the terms a^N and a^{2N} can be neglected ($Nk\Delta t \gg 1$), $\tilde{\alpha}$ is simple enough that one can write a compact, closed form expression for the covariance, $\tilde{c}$. This gives for the kk element

$$\sigma^2(k') = c_{kk} = \left(\frac{\xi P_N}{A} \right)^2 \frac{(1 - a^2)^3}{\Delta t^3 a^2} \left[\frac{N(1 - a) - (1 + a)}{N(1 - a) - 2(1 + a)} \right]. \qquad (25)$$

In the limit that $k\Delta t \ll 1$ and that $\xi \to 1$ ($k_f \Delta t < 1$), we approach the "ideal" or minimum value for the variance in the fitted value of the decay rate, k',

$$\sigma^2(k')_{\text{ideal}} = 8k^3 \left(\frac{P_N}{A} \right)^2. \qquad (26)$$

Below, we will examine the convergence of $\sigma^2(k')$ for finite experimental parameters. We will denote by the reduced variance the ratio $\sigma^2(k')/\sigma^2(k')_{\text{ideal}}$. The reduced dispersion is just the square root of the reduced variance.

We will first assume that we are fitting well into the tail of the ring-down transient, so that $Nk\Delta t \gg 1$, but we have a finite value for $k\Delta t$. In this case, we will get the same variance for k' regardless of whether we fit a baseline in our model, because the baseline will be essentially determined by the tail of the transient. From the above curvature matrix, it can be shown that the correlation coefficient between k and B is $\sqrt{2/(Nk\Delta t + 2)}$ and between k and A is $1/\sqrt{2}$ in the $k\Delta t \ll 1$ and $Nk\Delta t \gg 1$ limit. Figure 2 shows the predicted reduced dispersion as a function of $k\Delta t$ for both correlated and uncorrelated fits. It is important to note that here and for the rest of this chapter, these are variances calculated using Equation (17) and thus reflect influence of the data correlation, even though the fit itself is assumed to neglect data correlation. As discussed above, if we had

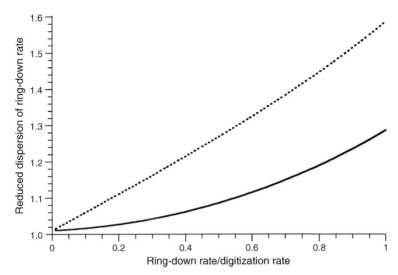

Figure 2   Plot of the reduced dispersion (fractional increase in the standard deviation over the "ideal" case) of the ring-down rate as a function of the $k\Delta t$, which is the inverse of the number of data points per ring-down time constant. A constant value of $k_f\Delta t = 0.5$ was used to construct this plot, and data were taken for at least eight ring-down time constants. The upper curve is for ring-down rates extracted from a fit that ignores data correlation, the lower for fits that properly account for data correlation. The ring down is assumed to be detector noise limited, that is, has constant noise.

calculated the variance using Equations (13) and (14), but with a diagonal weight matrix, we would predict variances that are too small by a factor in the order of $1/G$, at least for $k \ll k_f$. The abscissa is the inverse of the number of data points per ring-down time constant. A value of $k_f\Delta t = 0.5$ was used to make this plot. We see from the figure that in the limit that $k\Delta t \to 0$, both correlated and uncorrelated fits are predicted to have dispersions close to ideal. For the correlated fit, the dispersion in the ring-down rate increases rather slowly with $k\Delta t$, whereas for the uncorrelated fit, the "cost" of under sampling the decay is much higher. This is in large part because of the need, for both fits, to decrease the bandwidth of the filtering to keep from under sampling the noise.

Figure 3 shows the reduced dispersion as a function of $k_f\Delta t$ for both correlated and uncorrelated fits. Values of $k\Delta t = 0.01$ and $Nk\Delta t \gg 1$ are used in this calculation. This plot, which can be compared with Figure 1, demonstrates that for white noise, there is no significant improvement in the predicted sensitivity of CRDS if $k_f\Delta t < 0.5$. For the correlated fit, the dispersion continues to decrease, though very slowly, as $k_f\Delta t$ decreases below 0.5. For the uncorrelated fit, however, the dispersion slowly rises. This is because of the increasing importance of data correlations for smaller $k_f\Delta t$, which is of course neglected in such fits. It is interesting to note that if $F(t)$ instead of $\overline{F}(t)$ is used in the least squares equations, then predicted dispersion continues to get smaller with decreasing $k_f\Delta t$ and dips below the "ideal" values for fixed A. This is an artifact of the fact that to get the single exponential

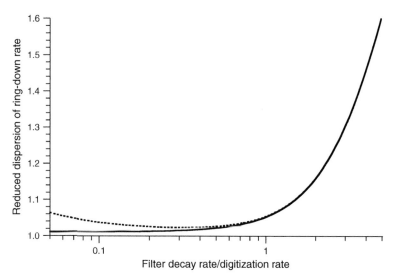

Figure 3 Plot of the reduced dispersion of the ring-down decay rate as a function of $k_f \Delta t$, which equals the number of low-pass filter time constants between each data point. A value of $k\Delta t = 0.01$ was used in calculating these curves. Again, the upper curve is for fits that ignore data correlation, whereas the lower is for fits that properly account for it. The ring down is assumed to be detector noise limited.

decay, one must delay the fit until $k_f t \gg 1$, which means that the appropriate amplitude of the ring down should be decreased.

We will now examine the convergence of the reduced dispersion as a function of the fraction of the ring-down transient that is sampled and fit. For a finite interval, we will get different results depending upon whether we include a variable or fixed baseline in the fit. Figure 4 shows the reduced dispersion as a function of $Nk\Delta t$ calculated by the correlated fit with or without a variable baseline. Values of $k\Delta t = 0.01$ and $k_f \Delta t = 0.5$ were used in this calculation. Fits that assume uncorrelated data give almost indistinguishable results for this plot with these parameters. It can be seen that fits with fixed baseline converge quickly, and essentially no improvement in the dispersion of the ring-down rate is expected once three time constants of data are included. In contrast, fits including an adjustable baseline parameter converge rather slowly with increasing size of the data set. When fitting up to three time constants, the dispersion in k for variable baseline fits is more than twice as large as for fits that fix the baseline. Basically, this reflects the effect of correlation between the baseline and the decay rate parameters. This correlation goes to zero if an infinite number of data points are sampled, but slowly.

Because the least squares fit model is nonlinear in k, we will in general expect some bias, that is, the ensemble average of the cavity decay rate, $\langle k' \rangle$ will differ from k by an amount that is proportional to σ^2_- for an "unbiased" method. Starting with the least squares fit solution equations $\left(\beta = 0 \right)$, we have derived

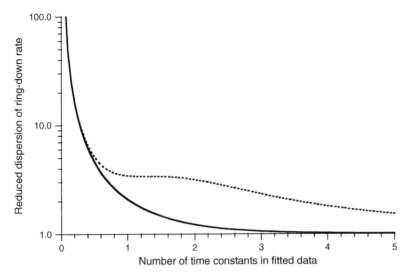

Figure 4 Plot of the reduced dispersion of the ring-down decay rate as a function of $Nk\Delta t$, which equals the number of cavity decay times sampled by the data. The lower curve is for fits that treat the baseline as a fixed number, whereas the upper curve is for fits that include a baseline offset as a fit parameter. Values of $k\Delta t = 0.01$ and $k_f\Delta t = 0.5$ were used. These curves were calculated including data correlation, but corresponding curves for fits ignoring data correlation are almost identical for these parameters.

$$\langle k' \rangle - k = \frac{1}{2}\sum_i \left(\frac{\partial^2 k'}{\mathrm{d}y_i^2}\right)\sigma^2(y_i) \xrightarrow[\substack{k\Delta t <<1 \\ Nk\Delta t >>1}]{} 4k^2\left(\frac{P_N}{A}\right)^2. \tag{27}$$

Note that the bias in k' scales as the inverse square of the signal-to-noise ratio unlike the standard error of k' which scales as the inverse of the signal to noise. The bias scales as k^2 compared to $k^{3/2}$ for the standard error. For typical CRDS experiments, the bias in k' is so much smaller than its fluctuations that it can be neglected.

Since the least squares solution is in general iterative, it is useful to characterize the rate of convergence. We ran numerical calculations for an ensemble of 1000 decays with $k\Delta t = 0.01$, $A = 100\sigma$, and $N = 2000$ and iterated to convergence with different initial values for the decay rate. It was found numerically that the mean error in each fit (compared to the converged value for a given simulated decay) is approximately quadratic in the initial error in k', with a mean and standard deviation of the error of $-0.91(6)\%$ for a 10% initial error. This can be compared to a 0.30% fluctuation of k' for the ensemble after full convergence of the fits. A second iteration reduced the convergence error by more than two orders of magnitude. Thus, even with a crude, $\sim10\%$, initial estimate of the decay rate, two cycles of the least squares fit will converge to well below the noise. With a moderately accurate initial guess, $\sim1\%$, a single cycle of the fit should produce convergence well below the noise.

5.1. To average and then fit, or fit each decay and average the fit results?

If one truly has Gaussian distributed noise, there should be no difference in the standard error of the determined cavity decay rate whether one sums the transients and fits the resulting curve or if one fits each individual curve and calculates the weighted mean (based on predicted fit standard error) of the individual decay rates. Given the computational cost of fitting up to several thousand data points, one may expect that it is better to average and then fit.

It is our experience that even in a well running CRDS instrument, there is a small fraction of "bad decays" that give cavity decay rates that are outliers from the Gaussian distribution of k' values that is expected if the detector has Gaussian noise characteristics. Recently, we published a paper that presents an analysis of one source of such outliers – resonances with high order transverse modes that are coupled to the TEM_{00} mode by mirror scattering [17]. But even with such resonances killed by control of the cavity Fresnel number [17], occasional outliers continue to exist. One likely source of such outliers is the drifting of a small particle through the mode of the cavity. Given a typical TEM_{00} mode diameter of 1 mm, a particle with scattering cross section in the order of $1\,\mu m^2$ will introduce ~ 1 ppm per pass loss, which is orders of magnitude above the noise level we achieve. Luckily, usually the χ^2 of the fit to the cavity decay is typically significantly higher than expected for our observed bad shots. If one examines the residuals of the fit, a pronounced "ringing" is often evident. The modulation in the residuals suggests that part of the loss is modulated during the ring-down transient. Motion of a particle along the optic axis of the cavity could produce such modulation since one expects the scattering loss to be larger for the center of the particle at the antinodes of the standing wave of the excited cavity mode than at the nodes. Miller and Orr-Ewing [18] recently demonstrated that significant spatial loss modulation is expected even for particles whose diameters are many times the wavelength of the scattered light. We note that for $\lambda = 1\,\mu m$, a particle velocity of 1 cm/s will produce a loss modulation of 20 kHz, which will strongly distort a ring-down transient with $k \sim 10^4\,s^{-1}$ as is typical in our experiments. When processing real CRDS data, we have the software reject data points that have excessive χ^2 values. Such filtering to reject outliers is not possible if a large number of transients are averaged and then fit. In that case, a single bad decay can cause significant deviation of the cavity decay rate. It is our suggestion, whenever possible without significantly reducing the rate of cavity decays observed, that each transient be fit and a χ^2 test be performed on the fit. In cases of very high data rates that often occurs when using mirrors of only modest reflectivity, fitting each decay will no longer be possible, but we recommend that the smallest practical sets of averaged data be fit.

6. LINEARIZATION OF THE FIT IN CAVITY RING-DOWN SPECTROSCOPY

In many applications of CRDS, requiring rapid fitting of a large number of observed ring-down transients, the ring-down signal is converted $y_i \rightarrow \ln(y_i)$, which converts $F(t) \rightarrow \ln(A) - kt$, that is, we now have a linear least squares fit problem. The advantage of a linear least squares fit is that it can be solved exactly in a single step, instead of requiring an iterative solution, as most nonlinear least squares fits require. In the limit that the fractional error in the data is small over the entire region fit, this "linearization" will not change the least square solution or variances, provided that the uncertainties of each data point are appropriately transformed as well, $\sigma_i \rightarrow \sigma_i / F(t_i)$. A problem with this approach is that errors in the wing of the decay are not properly estimated. In fact, one will attempt to calculate the log of a negative number should the noise ever dip below the zero value, which it clearly must at long time. As a result, if this procedure is used, one must be careful to restrict the fit to only the high signal-to-noise portion of the decay, say requiring that $y_i \geq m\sigma_i$, where m is some predetermined multiplier. If this point occurs sufficiently far out in the decay, then this truncation results in a negligible decrease in precision. Another issue that should be kept in mind is that the nonlinear transformation of the experimental data will generate a bias in the distribution of errors

$$\ln\left(y(t)\right) = \ln\left(F(t)\right) + \frac{\epsilon(t)}{F(t)} - \frac{1}{2}\frac{\epsilon(t)^2}{F(t)^2} + \cdots \tag{28}$$

with the ensemble average of $\epsilon(t)^2 = \sigma_i^2$. For the case of detector noise limited CRDS (i.e. constant σ^2), $k\Delta t \ll 1$ but $\exp(-Nk\Delta t) \ll 1$, it can be shown by propagation of error that the second-order term in the above Taylor expansion for the transformation introduces a bias in the fitted decay rates of

$$<k'> -k = 2k\left(\frac{\sigma}{A}\right)^2 (Nk\Delta t)^2 \left(1 - (2k\Delta tN)^{-1}\right). \tag{29}$$

Another source of bias occurs if one uses $\sigma_i \rightarrow \sigma_i / y_i$ for the uncertainty and thus weight the data proportional to y_i^2 instead of F_i^2, which are not known. This has the effect of giving more weight to points with positive noise and less to those with negative noise. Numerical simulations indicate that this bias is of the opposite sign as the bias due to the nonlinear transformation and about twice as large (at least for the parameter range that was explored). Figure 5 displays two scattering plots of fitted τ (equals $1/k'$) versus signal-to-noise ratio of decay transients. For the upper part, this linearization process is used in the fit, without corrections to log transformation and with the weight proportional to y_i^2. One can see clearly from it that the fit is biased and the bias increases with the decrease of the signal-to-noise ratio. For an initial signal-to-noise ratio of 100:1, the noise in k was close to that predicted by Equation (26), but the mean value of k was shifted by -0.5% from the value used to generate the data ($k\Delta t = 0.01$, $m = 1.5$), which was close to twice the ensemble standard deviation of k. With an initial signal-to-noise ratio of 10:1, the standard deviation of k was nearly twice the prediction of Equation (26), and the bias was -8.5%. Even with a signal-to-noise ratio of 1000:1, the bias (-0.015%) is still 58% of the ensemble fluctuations in k and thus will be significant with only modest signal averaging to the decay rates. The bias in the log-transformed fit at low signal-to-noise ratios was previously discussed by von Lerber and Sigrist [9].

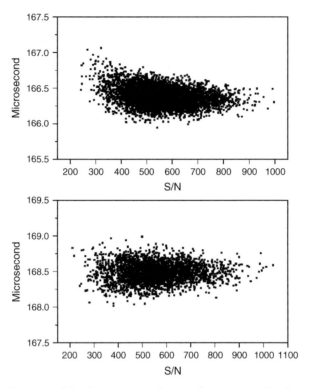

Figure 5 Scattering plots of the fitted τ versus the signal-to-noise ratio of experimental decay transients. For the upper part, linearized least squares fit is used, without corrections to the log transformation and with the weight proportional to y_i^2. The fit is significantly biased for transients with signal-to-noise ratio less than 500. For the lower part, a single pass of the nonlinear least squares fit process is added to the fitting program and it is unbiased.

Numerical simulations have demonstrated that the effects of the bias can be greatly reduced by two simple changes in the routine. To correct for the leading order bias term, $\ln y_i + (1/2)\sigma_i^2/y_i^2$ is used for the "data" in the least squares expressions. Also, the weight is calculated from an estimate of the value of k (the assumed values of A does not effect the results of the least fit results). Using an incorrect value (even with 20% deviation from k) for the decay rate to calculate the weights, k_w, will not introduce any significant bias, but the lowest fluctuations in the fitted decay rate, k', will be obtained if $k_w = k$. However, the dependence is relatively weak. The variance of k' increases from the optimal value by a factor of $\kappa^4(2\kappa^2 - 2\kappa + 1)/(2\kappa - 1)^3$, where $\kappa = k_w/k$. The variance increases by only 10% for $\kappa = 0.87$ or 1.18. Thus, one could use a rough estimate for k for calculation of the weight with little penalty in precision. With this change in the fitting procedure ($k\Delta t = 0.01$, $m = 1.5$), the bias in k' with signal to noise of 100:1 is reduced to -0.027%. With a signal to noise of 10:1, the bias is reduced to -2.1%, but the fluctuations in k' were still about 66% higher than predicted. For such low signal to noise, the nonlinear least squares fit to the data is needed to reduce the standard error in the fitted decay rate close to the theoretical limit.

Our suggestion for an optimal strategy is a two-step fit. First, linearization is used to provide a good estimate of the parameters A' and k'. The detector value corresponding to zero light (B') is either assumed (best if it is stable with time) or else estimated from the tail of the decay. Speed can be improved by storing a table of possible $\ln(y_i)$ values rather than recomputing these for each data point. Then a single step of refinement to the solution of the original least squares fit problem be carried out. As discussed earlier, for an initial estimate of k that is in error by 1%, this single step will reduce the convergence error to the order of 0.01%. Only two or three sums over the data points (depending upon whether the baseline is constrained for fit) need to be calculated, so the single round of the nonlinear fit in fact requires less floating point operations than the linearized log fit. The lower part of the Figure 5 shows that the fitted τ is unbiased after the nonlinear fit process is added to the fitting program. An alternative strategy is to use k' from the previous decay as the initial value for the least squares fit and dispense with the log transformation entirely. This should be an ideal approach when decays are being observed very rapidly, because in that case the cavity loss (and thus k) cannot change much decay to decay.

7. DETERMINATION OF RING-DOWN RATE BY FOURIER TRANSFORM METHOD

One method widely used to extract one or more decay rates from a transient is based on taking a Fourier transform of the decay [19], which for points evenly spaced in time can be done with the fast Fourier transform method [16]. This has been proposed for the rapid analysis of CRDS data by Mazurenka et al. [10].

Let f_j be the $\omega_j = 2\pi j/N\Delta t$ frequency component of the discrete Fourier transform of the time series $F(t_i) = A\ \exp(-k\Delta t i) + B$ $(i = 0\ldots N-1)$. This is easily calculated to be

$$f_j = \left(A\frac{1 - \exp(-k\Delta t N)}{1 - \exp\left(-\left(k + i\omega_j\right)\Delta t\right)} + BN\delta_{j0} \right)\Delta t. \qquad (30)$$

δ_{j0} is 1 if $j = 0$ and 0 if $j \neq 0$. For $\omega_j \neq 0$, we can take the ratio

$$\omega_j\frac{\Re\left(f_j\right)}{Im\left(f_j\right)} = \omega_j\frac{1 - e^{-k\Delta t}\cos\left(\omega_j\Delta t\right)}{e^{-k\Delta t}\sin\left(\omega_j\Delta t\right)} \to k, \qquad (31)$$

where the right arrow applies if both $k,\omega_j \ll (\Delta t)^{-1}$. This is the relationship we get from a continuous Fourier transform. This suggests that we can evaluate k

from any frequency, except the zero frequency for which the baseline makes a contribution. It has been traditional to use the first nonzero frequency point in the computation of k' from the FFT. For finite values of k and ω_j, bias is introduced by the above procedure. However, the first half of Equation (31) can be solved for $\exp(-k\Delta t)$, and this was used to evaluate k' from the ratio without bias (without noise)

$$k' = \frac{1}{\Delta t} \ln \left[\sin(\omega_j \Delta t) \frac{\Re e(f_j)}{Im(f_j)} + \cos(\omega_j \Delta t) \right]. \tag{32}$$

Error propagation for $k, \omega_j \ll (\Delta t)^{-1}$ gives the following expression for the variance in k' extracted from the ω_j frequency component of the FFT

$$\sigma^2(k') = \frac{\Delta t^2 \left(k^2 + \omega_j^2 \right)^3 N}{2\omega_j^2} \left(\frac{\sigma}{A} \right)^2. \tag{33}$$

Note that the variance grows with N, so one does not want to take too much baseline. This is particularly true if one takes the first nonzero frequency, as $\omega_1 = 2\pi/(\Delta t N)$ and thus the variance grows as N^3 because typically $\omega_1 \ll k$. Treating ω_j as a continuous variable, it is easily shown that the noise will be minimized by taking $\omega_j = k/\sqrt{2}$, for which $\sigma^2(k') \rightarrow (3/2)^3 k^4 \Delta t^2 N \left(\frac{\sigma'}{A} \right)^2$. Because of the nonlinear relationship between k' and f_j (which is linearly related to the data), noise will create a bias because of the second derivatives of the k' with respect to the data points used to calculate it, as we had for the linearized least squares fit model above. For $k, \omega_j \ll (\Delta t)^{-1}$ and $Nk\Delta t \gg 1$, this bias can be calculated analytically, giving

$$\langle k' \rangle - k = -\frac{kN}{2} \left(\frac{\Delta t \left(k^2 + \omega_j^2 \right)}{\omega_j} \frac{\sigma}{A} \right)^2. \tag{34}$$

The optimal estimate of k, at least in cases of small noise where the first order error propagation is valid, would be to average the value of k' extracted from each frequency component, but with a weight inversely proportional to the variance, as given by Equation (33). Because the number of values of ω_j that will contribute significantly to the sum will be proportional to N, this extra step should produce a value for k' with a variance that converges for large N, like for the least squares solution. However, if N is too large, the noise in the values of f_j becomes comparable to their magnitude, and the first-order treatment of the noise will no longer be accurate. Using only the first point, we have found by numerical simulation with parameters $k\Delta t = 0.01$ and $N = 512$ that the extracted values of k' showed fluctuations that are about 84% larger than for the ideal least squares limit. Because the number of numerical operations to evaluate an FFT scales as $N \ln N$, it appears that this method should be more computationally expensive than a direct least squares fit to the data.

8. THE SUCCESSIVE INTEGRATION METHOD FOR EXPONENTIAL FITTING

The SI method for exponential fitting was introduced by Matheson [20]. Using the fact that the integration of an exponential function is still an exponential function, the exponential fitting problem is transformed into a linear regression problem, which is a noniterative method. Halmer et al. [11] found the method can be improved by introducing a correction factor to account for errors introduced by the use of Simpson's rule to evaluate the integration from the discrete experimental points and called the improved estimate the corrected successive integration (CSI) method. Both the SI and the CSI methods give nearly the same results as the least squares algorithm. However, the dispersion estimation for fitted parameters given in both papers is incorrect. The authors utilized the curvature matrix as in the normal least squares fit. However, in the SI and CSI methods, the form of χ^2 has changed. The new curvature matrix is a function of the dependent data, y_i, not just the independent data, t_i, and parameters. As a consequence, the curvature matrix in the SI and CSI methods contains fluctuations caused by the noise in the data. These fluctuations contribute to the variance of fitted parameters. Below, we present an evaluation of standard error of the fit parameters that properly accounts for this effect.

The fitting model can be changed into a different form after one integrates the model equation [11].

$$y(t) = Ae^{-kt} + B \tag{35}$$

$$\int_0^{i\Delta t} y(t')dt' = \frac{A+B}{k} - \frac{y(i)}{k} + Bi\Delta t \tag{36}$$

$$y(i) = A + B - k\int_0^{i\Delta t} y(t')dt' + kBi\Delta t. \tag{37}$$

The correction factor $CT(k)$ was introduced by the Halmer et al. [11].

$$\int_{n\Delta t}^{(n+1)\Delta t} y(t')dt' = B\Delta t + CT(k)\left(\frac{y(n) + y(n+1)}{2} - B\right)\Delta t \tag{38}$$

$$CT(k) = \frac{2}{k\Delta t} \cdot \frac{1 - e^{-k\Delta t}}{1 + e^{-k\Delta t}}. \tag{39}$$

With this correction, one can calculate the integral in Equation (37).

$$\int_0^{i\Delta t} y(t')dt' = CT(k)X_i\Delta t + B[1 - CT(k)]i\Delta t, \tag{40}$$

with X_i as defined by Halmer et al. [11]. $X_0 = 0$ and for $i > 0$,

$$X_i = \sum_{j=0}^{i-1} \frac{y_j + y_{j+1}}{2}. \tag{41}$$

The last expression on the right results from using Simpson's rule to evaluate the integral in terms of the discrete data points, y_i. Substituting Equation (40) into Equation (37), we have a new equation:

$$y_i = A + B - k\Delta t C T(k) X_i + k\Delta t B C T(k) i. \tag{42}$$

Following normal least squares fit procedures [11], one can have the matrix equation for the best fit parameters A, B, and k.

$$M \cdot \beta = v$$

$$M = \begin{pmatrix} N & SY & t \\ SY & SY.SY & t.SY \\ t & t.SY & t.t \end{pmatrix}, \quad \beta = \begin{pmatrix} A+B \\ -k\Delta t C T(k) \\ k\Delta t C T(k) B \end{pmatrix}, \quad v = \begin{pmatrix} Y \\ Y.SY \\ Y.t \end{pmatrix} \tag{43}$$

with definitions

$$Y = \sum_{i=0}^{N-1} y_i \qquad\qquad SY = \sum_{i=0}^{N-1} X_i \qquad\qquad Y.SY = \sum_{i=0}^{N-1} y_i X_i$$

$$SY.SY = \sum_{i=0}^{N-1} X_i^2 \qquad\qquad Y.t = \sum_{i=0}^{N-1} i y_i \qquad\qquad t.SY = \sum_{i=0}^{N-1} i X_i$$

$$t = N(N-1)/2 \quad t.t = N(N-1)(2N-1)/6$$

The definitions of the elements in the matrix and v vector are the same as those in the Ref. [11] though we have used sums $i = [0 \ldots N-1]$ to be consistent with the notation of this chapter. Using Equation (39), k' is easily calculated in terms of β_1. These elements are functions of data points y_i and contain the noise fluctuations. This makes it much more complex to estimate the variance of parameters in the fitting than the normal curvature situation. Derivation of a fully analytic expression for the standard errors of the parameters is quite involved, and so we used a numerical method. We start with the standard expression for propagation of errors, neglecting correlation of the data points:

$$\delta\beta_i^2 = \sigma^2 \sum_{j=0}^{N-1} \left(\frac{\partial \beta_i}{\partial y_j} \right)^2 \quad i \in 0, 1, 2, \tag{44}$$

where we have assumed that detector noise dominates with a constant noise per data point. As above, data correlation can be included by taking for $\sigma^2 = \xi^2 P_N^2/\Delta t$, which is a factor of G larger than the variance of the data points without correlation. Thus, we need to evaluate the partial derivatives of the fit parameters in terms of the input data. Differentiation of Equation (43) gives

$$\left(\frac{\partial \beta}{\partial y_k} \right) = -M^{-1} \cdot \left(\frac{\partial M}{\partial y_k} \right) \cdot \beta + M^{-1} \left(\frac{\partial v}{\partial y_k} \right). \tag{45}$$

By defining the matrix J,

$$J = \begin{pmatrix} 1 & 0 & 0 \\ 1 & X_1 & 1 \\ \vdots & \vdots & \vdots \\ 1 & X_{N-1} & N-1 \end{pmatrix}, \tag{46}$$

we can express matrix $M = J^\dagger J$ and vector $v = J^\dagger \vec{y}$ and thus express their derivatives in terms of the derivatives of J.

$$\left(\frac{\partial M}{\partial \gamma_k} \right) = \left(\frac{\partial J}{\partial \gamma_k} \right)^\dagger \cdot J + J^\dagger \cdot \left(\frac{\partial J}{\partial \gamma_k} \right) \tag{47}$$

$$\left(\frac{\partial v}{\partial \gamma_k} \right) = \left(\frac{\partial J}{\partial \gamma_k} \right)^\dagger \cdot \vec{y} + J^\dagger \left(\frac{\partial \vec{y}}{\partial \gamma_k} \right) \tag{48}$$

$\left(\frac{\partial \vec{y}}{\partial \gamma_k} \right) = \delta_{ki}$. Using the definition of X_i (Equation (41)), we can write the nonzero elements of $\left(\frac{\partial J}{\partial \gamma_k} \right)$ as follows (the column and row numbering start at zero): $\left(\frac{\partial J_{i1}}{\partial \gamma_0} \right) = 1/2 \ (i > 0)$; $\left(\frac{\partial J_{i1}}{\partial \gamma_k} \right) = 1 \ (0 < k < i)$; $\left(\frac{\partial J_{k1}}{\partial \gamma_k} \right) = 1/2 \ (k > 0)$.

With the same values of parameters as in Ref. [11], that is, $A = 1200$, $k = 0.02$, $\Delta t = 1$, $B = -13$, and $\sigma = 20$, we generated 5000 decay transients and fit them with different methods. We found with the usual nonlinear fit method, the averages of fitted parameters and their standard deviations, $A' = 1200.08 \pm 5.54$, $k' = 0.02000 \pm 0.00014$, and $B' = -13.00 \pm 0.55$. The predicted standard deviations are 5.55, 0.00014, and 0.55, respectively, by using the curvature matrix. With the CSI method fitting the same data, the averages of fitted parameters and their standard deviations are $A' = 1199.36 \pm 7.21$, $k' = 0.01999 \pm 0.00021$, and $B' = -13.01 \pm 0.59$. The predicted standard deviations by Equation (44) are 7.22, 0.00021, and 0.54, respectively, with Equation (44), that is, the standard deviation of k' in the CSI fit is about 50% higher than for a direct least squares fit to the same data. We note that Table II of the Ref. [11] gives parameter standard deviations of 2.3, 0.00048, and 0.98 matching neither of our results. Although the SI and CSI methods are noniterative, one disadvantage is that the M matrix is badly conditioned, increasingly so with increasing N, which can generate numerical unstability in the calculation of its inverse.

The principal advantage that Halmer et al. [11] ascribe to the CSI method is its computational speed; their Table II shows fitting times two orders of magnitude faster than for the least squares fit using the Levenberg – Marquardt algorithm. This appears, however, to be an artifact of their using a general purpose fitting package for the least squares fit. The SI and CSI fits each requires the calculation of seven sums over data points, requiring eight floating point additions and five floating point multiplications. This can be compared to 3, 3, and 3 for the same quantities in the least squares fit using the expressions given above. Thus, one cycle of the least squares fit should take about half the number of floating point operations if an efficient code is used that exploits the closed form expressions for the curvature matrix. Because the least squares fit converges very rapidly and does not require more than two cycles, we see no computational speed advantage to the use of the CSI method. Certainly, the substantial advantage reported in Ref. [11] is not correct.

9. ANALOG-DETECTED CAVITY RING-DOWN

9.1. Phase shift method

There are several analog electronic methods to determine the ring-down decay rate of an optical cavity. The oldest is the phase shift CRDS method (ps-CRDS) [12,13]. In this method, the light injected into an optical cavity is modulated at an angular frequency Ω, and the light transmitted by the cavity is demodulated by a vector lock-in amplifier that determines both the in-phase (S_c) and out-of-phase (S_s) components of the signal and the phase shift of the signal determined by the equation $\tan(\theta) = S_s/S_c$. The cavity decay time can be determined by $\tan(\theta) = \Omega\tau$. We will now examine the expected noise in this signal extraction method.

Let us begin by assuming that the cavity is excited by a pulsed laser with pulse length much shorter than the cavity decay time, such that one can treat the excitation as impulses at times $t_n = n2\pi/\Omega$. In each time interval (t_n, t_{n+1}), the detector signal will be $\overline{F}(t) = A_n\exp(-kt)$. Demodulation gives the average of the following two signals:

$$F_c = \frac{\Omega}{2\pi} \int_{t_n}^{t_{n+1}} \overline{F}(t')\cos(t')dt' = A_n \frac{\Omega k\left[1 - e^{-k2\pi/\Omega}\right]}{2\pi\left(k^2 + \Omega^2\right)} \qquad (49)$$

$$F_s = \frac{\Omega}{2\pi} \int_{t_n}^{t_{n+1}} \overline{F}(t')\sin(t')dt' = A_n \frac{\Omega^2\left[1 - e^{-k2\pi/\Omega}\right]}{2\pi\left(k^2 + \Omega^2\right)} \qquad (50)$$

$$k' = \frac{\Omega}{\tan(\theta)} = \Omega\frac{F_c}{F_s}. \qquad (51)$$

If we have detector-limited CRDS, then the time averaged of both F_c and F_s will have noise with variance given by $\sigma^2 = P_N^2 BW/2$, where BW is the detection bandwidth on the lock-in amplifier and (as before) P_N is the noise-equivalent power of the detector. Adding these noise terms to the above ratio and doing standard error propagation, we find that the variance in the calculated cavity decay rate is given by

$$\sigma^2(k') = \frac{4\pi^2\left(k^2 + \Omega^2\right)^3}{\Omega^4\left(1 - e^{-k2\pi/\Omega}\right)^2}\left(\frac{\sigma}{\langle A_n\rangle}\right)^2. \qquad (52)$$

If we assume that the light field of different laser pulses adds incoherently, then it can be shown that $\langle A_n\rangle = J(\Omega)T^2t_r^{-1}\left(1 - e^{-k2\pi/\Omega}\right)^{-1}$, where $J(\Omega)$ is the energy per pulse (which is a function of the laser repetition rate $= \Omega/2\pi$), T is the power transmission of the mirrors (assumed to be the same), and t_r is the cavity round trip time. The last term corrects for the energy left in the cavity from previous pulses. This gives for the noise in the estimated decay rate

$$\sigma^2(k') = \frac{4\pi^2(k^2 + \Omega^2)^3}{\Omega^4} \left(\frac{\sigma t_r}{J(\Omega)T^2}\right)^2. \tag{53}$$

We consider two limiting cases for the laser pulse energy $J(\Omega)$. One is that it is a constant, J, independent of repetition rate, such as when a high repetition rate laser is pulse picked or a continuous wave optical source is chopped. In that case, the lowest variance in k' occurs for $\Omega = \sqrt{2}k$, and for this modulation frequency, $\sigma^2(k') = 125\pi^2 k^2 \left(\frac{\sigma t_r}{JT^2}\right)^2$. The other limit is where the average power of the pulsed laser is fixed (such as for many Q-switched lasers at high repetition rate), in which case $J(\Omega) = 2\pi I_{\text{avg}}/\Omega$. In this case, the optimal modulation angular frequency is $\Omega = k/\sqrt{2}$, and at this modulation frequency, $\sigma^2(k') = (27/4)k^4 \left(\frac{\sigma t_r}{I_{\text{avg}}T^2}\right)^2$. It is worth noting that shot-to-shot fluctuations in the amplitude of the light will cause correlated noise in F_s and F_c and will not degrade the ability to extract the cavity decay rate using the phase shift method.

Because one must take the ratio of the in-phase to out-of-phase lock-in outputs to compute the decay rate, noise will generate a bias in k' equal to $(1/2)(d^2k'/dF_s^2)\sigma^2(F_s)$. Evaluation using the above expressions gives the following prediction for the bias

$$\langle k' \rangle - k = \frac{4\pi^2 k(k^2 + \Omega)^2}{\Omega^4} \left(\frac{\sigma t_r}{J(\Omega)T^2}\right)^2 \tag{54}$$

For the two limiting cases of $J(\Omega)$ described above, we find for the optimal modulation frequency $\langle k' \rangle - k = 9\pi^2 k \left(\frac{\sigma t_r}{JT^2}\right)^2$ for the constant pulse energy case and $\langle k' \rangle - k = (9/2)k \left(\frac{\sigma t_r}{I_{\text{avg}}T^2}\right)^2$ for the constant average power case.

If one is exciting the cavity with a continuous wave source that is chopped, one will want to leave the excitation source for a significant fraction of the modulation cycle. We will now consider the case of 50% duty cycle excitation of the cavity. We will also assume that the coherence time of the excitation source is much less than the cavity decay time, so that we will have a rate equation for the cavity intensity (instead of the cavity field). We will define the modulation cycle such that the input light intensity drops to zero at the beginning of the cycle and turns on to a constant value for the second half of the cycle. In that case, the mean output intensity of the cavity will be $F(t) = A \exp(-kt)$ for $0 < t < \pi/\Omega$ and $F(t) = A[1 + \exp(-k\pi/\Omega) - \exp(t - \pi/\Omega)]$ for $\pi/\Omega < t < 2\pi/\Omega$. If we did not modulate the input light, we would have a mean cavity transmission of $I_{\text{avg}} = A[1 + \exp(-k\pi/\Omega)]$. If the bandwidth of the incident radiation is greater than the free spectral range of the cavity [21] or one is exciting many transverse modes to "fill in" the spectrum of the cavity [22], then $I_{\text{avg}} = kt_r T^2 I_{\text{inc}}$, where I_{inc} is the intensity incident on the cavity. Using the above defined definitions of F_c and F_s, we find for this $F(t)$ that

$$F_c = \frac{\Omega k}{\pi\left(k^2 + \Omega^2\right)} I_{\text{avg}} \tag{55}$$

$$F_s = \frac{k^2}{\pi\left(k^2 + \Omega^2\right)} I_{\text{avg}} \tag{56}$$

$$k' = -\Omega \tan(\theta) = \Omega \frac{F_s}{F_c}. \tag{57}$$

As above, if we have detector-limited CRDS, then the time averaged of both F_c and F_s will have noise with variance given by $\sigma^2 = P_N^2 \text{BW}/2$. Error propagation gives for the variance of the fitted cavity decay rate

$$\sigma^2(k') = \frac{\pi^2\left(k^2 + \Omega^2\right)^3}{k^2 \Omega^2} \left(\frac{\sigma}{I_{\text{avg}}}\right)^2. \tag{58}$$

The optimal modulation angular frequency is given by $\Omega = k/\sqrt{2}$, which gives $\sigma^2(k') = (27/4)\pi^2 k^2 (\sigma/I_{\text{avg}})^2$. As above, we can use the second derivative of k' with respect to F_c to predict the bias

$$\langle k' \rangle - k = -\frac{\pi^2\left(k^2 + \Omega^2\right)^2}{\Omega^2 k} \left(\frac{\sigma}{I_{\text{avg}}}\right)^2 = -\frac{9}{2}\pi^2 k \left(\frac{\sigma}{I_{\text{avg}}}\right)^2 \left(\text{For } \Omega = k/\sqrt{2}\right), \tag{59}$$

where the second equality holds for optimal modulation.

Unlike the case of pulsed excitation, fluctuations in the cavity excitation intensity will lead to noise in the values of F_c and F_s that are not perfectly correlated, that is, this form of CRDS is not immune to source noise. Most important, the field build up of each mode will, when the source coherence time is short compared to the build up, suffer "temporal speckle" with the light amplitude adding incoherently at different times [23,24]. As previously discussed, this leads to the intracavity intensity for each mode to fluctuate with a χ^2 in two degree of freedom, $P(I) = \exp(-I/\langle I \rangle)/\langle I \rangle$, where $\langle I \rangle$ is the mean intensity predicted from the intensity rate equation and $P(I)$ is the probability that the intensity will be I. In the case of one or few mode excitation of the cavity, this noise will likely be dominant over the detector noise, the effects of which we have calculated above. With many mode excitation, either because the input light is broad band and excites many longitudinal modes of the cavity or because light is injected off-axis to excite many transverse modes, the fluctuations will be greatly decreased. This type of excitation is widely used in cavity-enhanced spectroscopy [25] (also called cw integrated cavity output spectroscopy [26]) where one uses changes in the time averaged transmission of the cavity to determine absorption. In this type of spectroscopy, one must determine the empty cavity loss to convert the observed change in transmission into absolute loss [25]. In this case, the use of phase shift detected CRDS offers an attractive approach, particularly if one's excitation source is a broad band-width source, such as an incoherent light source like a lamp or light-emitting diode.

9.2. Gated integrator method

An alternative analog detection method was introduced by Romanini and Lehmann [7] and used a pair of integrator gates to extract the cavity decay rate. After the excitation pulse of the cavity has ended, the output light intensity will decay as $A \exp(-kt)$. We will assume that any offset in the detector has been subtracted off. Let gated integrator 1 integrate the signal from time $t = [0, \Delta t]$ and integrator 2 from $t = [\delta t, \delta t + \Delta t]$. (Note: the meaning of Δt has changed from previous sections.) As shown previously, the cavity decay rate can be evaluated as $k' = -\ln (F_2/F_1)/\delta t$, where $F_{1,2}$ is the output of gated integrator 1 or 2. Romanini and Lehmann [7] presented an analysis of the expected shot noise limited sensitivity. We will now present the analysis of the predicted sensitivity and bias for the case where detector noise dominates.

Integration of the signals over the detection windows gives the following values for $F_{1,2}$:

$$F_1 = \frac{A}{k} \left(1 - e^{-k\Delta t}\right) \tag{60}$$

$$F_2 = \frac{A}{k} e^{-k\delta t} \left(1 - e^{-k\Delta t}\right). \tag{61}$$

Error propagation easily demonstrates that $\sigma^2\left(F_{1,2}\right) = \Delta t P_N^2$.

$$\sigma^2(k') = \frac{1}{\delta t^2} \left(\frac{\sigma^2(F_2)}{F_2^2} + \frac{\sigma^2(F_1)}{F_1^2} \right)$$

$$\sigma^2(k') = \frac{k^2}{\delta t^2} \left(1 + e^{2k\delta t}\right) \left(1 - e^{-k\Delta t}\right)^{-2} \Delta t \left(\frac{P_N}{A}\right)^2 \tag{62}$$

Minimization of $\sigma^2(k')$ gives optimal values $\delta t = 1.11 k^{-1}$ and $\Delta t = 1.255 k^{-1}$ and leads to a predicted decay rate variance $\sigma^2(k') = 20.3 k^3 (P_N/A)^2$, which is only about 2.5 times larger than the ideal weighted least squares fit prediction given above. The above analysis ignores the correlation of the noise between F_1 and F_2, but this will be small as the optimized sample windows hardly overlap. In making comparison with the phase shift method, note that the present variance is computed for a single transient, and the variance will be reduced by the $\Omega/2\pi$, which is the repetition rate that ring-down transients are detected. If there are no constraints on the repetition rate of the light source, then the optimal gates will be reduced somewhat to allow for a higher value of Ω, which must be less than $2\pi(\delta t + \Delta t)^{-1}$.

We can also calculate the expected bias from the second derivatives of k' with respect to F_1 and F_2. This leads to the result

$$<k'> -k = -\frac{1}{2} k^2 \left(\Delta t k \left(2 - e^{-k\Delta t}\right)^{-2}\right) \left(\frac{e^{2k\delta t} - 1}{k\delta t}\right) \left(\frac{P_N}{A}\right)^2. \tag{63}$$

Using the optimized values for Δt and δt given above, we find predicted bias $<k'> -k = -9.06k^2 \left(\frac{P_N}{A}\right)^2$.

We have assumed that the detector baseline is known and subtracted from the signal before integration. Alternatively, one could use a third integration gate placed after the ring-down signal has decayed to negligible level and the average over this time interval could be used to subtract the baseline from signals $F_{1,2}$. Ideally, the time interval over which the baseline should be integrated should significantly exceed Δt so that the noise in this baseline estimate is small compared to the noise in the signals $F_{1,2}$, but that may limit the repetition rate of cavity decays that could be sampled.

9.3. Logarithm-differentiator method

Spence et al. [14] presented an analysis and experimental results on a third analog detection method. Here, one passes the detector output through a log amplifier and then a differentiator. During the ring-down, the voltage output of the differentiator will be proportional to the cavity decay rate. If we assume that this analog signal is averaged over the time interval $[0, \Delta t]$, then the extracted cavity decay rate, k', is given by

$$k' = -\frac{1}{\Delta t} \int_0^{\Delta t} \frac{d \ln(y(t'))}{dt'} dt' = -\frac{1}{\Delta t} \left(\ln(y(\Delta t)) - \ln(y(0)) \right). \quad (64)$$

If we assume that the sample interval is long enough that we can treat the noise as uncorrelated at $t = 0$ and $t = \Delta t$, and taking the first derivative terms, it is easy to show that the variance in the extracted decay rate is

$$\sigma^2(k') = \frac{1}{\Delta t^2} \left(\frac{\sigma}{A}\right)^2 \left[1 + e^{2k\Delta t}\right]. \quad (65)$$

This is optimized with $\Delta t = 1.109k^{-1}$ for which, including that $\sigma^2 = (k_f/2)P_N^2$,

$$\sigma^2(k')_{\text{optimized}} = 4.14k^2 k_f \left(\frac{P_N}{A}\right)^2. \quad (66)$$

Comparing with the least squares fit results, we see that even with an optimal sample window, the use of a log amplifier and differentiator produces an estimate for the cavity decay rate that has a variance $\sim k_f/2k$ times larger than for optimal fitting. Note that this result is derived under the assumption that $k_f \gg k$ (since we started the fit at zero time instead of waiting for the initial decay transient due to detector time constant $(1/k_f)$ to die away). For typical CRDS conditions $k_f \sim 10^2 k$ or greater to avoid significant distortion of the cavity decay. This increased noise is a natural consequence of the differentiation of the signal, as this enhances the high frequency noise. Gaussian noise in the detector will produce a bias due to the log transformation, and this is easily seen to be

$$<k'> -k = \frac{k_f}{4\Delta t} \left(\frac{P_N}{A}\right)^2 \left[1 + e^{2k\Delta t}\right] = 2.29 k k_f \left(\frac{P_N}{A}\right)^2, \quad (67)$$

where the second expression is for the case that the lowest noise time window is used. This bias is in addition to any that arises from nonideality in the log and differential amplifiers.

The advantage of using analog over digitization methods is that one does not have to process and fit large quantities of data. A negative feature is that one extracts a cavity decay rate but no direct information on how well the cavity decay transients fit to a single exponential decay. As discussed earlier, CRDS experiments, at least in our experience, have outliers, a small fraction of individual decays that are far from the mean value. In the phase shift detection method, it does not appear that one eliminates these outliers. Even if the analog-determined decay rate is determined after each decay transient, which is possible with the dual gate and log amplifier approaches, one looses the ability to selectively reject decays that do not quantitatively fit a single exponential decay, unless further analog detection channels are added.

10. SHOT NOISE LIMITED CAVITY RING-DOWN DATA

Let us now consider the least squares fit in the limit that shot noise dominates over detector noise and that we have a negligible background term B (i.e., dark current). This should be applicable if the ring-down decay is observed using a multichannel scaler to count photon in time bins after the excitation pulse as was recently reported by Thompson and Myers [27], as long as dark counts (which determine B) are sufficiently rare. As before, we can derive a lowest possible variance in k' as

$$\sigma^2 (k')_{\text{SNL}} = k^3 \left(\frac{h\nu}{QA} \right), \tag{68}$$

SNL refers to the "shot noise limit" and represents the lowest possible noise in the ring-down rate (and thus highest possible sensitivity for CRDS) for a given mirror reflectivity and transmitted optical power on the detector. This result was previously derived by Romanini and Lehmann [7].

As a point in passing, we note that we are neglecting the possibility of filling the ring-down cavity with amplitude "squeezed" light, and thereby reducing the noise in the ring-down below the shot noise limit. Even if this were done, linear loss destroys squeezing of the light [28], and thus we would expect the cavity output to rapidly evolve shot noise even if it was not present at the beginning of the decay.

Comparison of Equations (26) and (68) shows that for identical initial signal to noise at the beginning of the ring down, the limiting variance in the case of detector noise will be eight times larger than for the shot noise-limit. This is a consequence of the fact that for the shot noise limited case, the signal to noise of the data points falls only half as fast as for the detector noise limit and thus it is as if the k value was half as large. Thus, the factor of 8 is a direct consequence of the k^3 dependence of the variance.

Figure 6 shows the reduced dispersion of the ring-down rate (compared to the shot noise limiting value) as a function of $k\Delta t$ for both correlated and uncorrelated fits

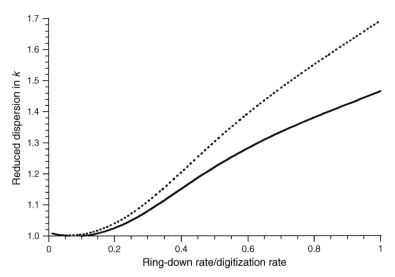

Figure 6   Plot of the reduced dispersion (fractional increase in the standard deviation over the "ideal" case) of the ring-down rate as a function of the $k\Delta t$. Calculations are identical as given in Figure 2 except that the ring down is assumed to be shot noise limited, that is, the noise variance at any time is proportional to the expected light intensity.

to shot noise limited ring-down decays. The decays were fit for in excess of 12 time constants, with $k_f\Delta t = 0.5$. Compared to the data presented in Figure 2, it is seen that for modest values of $k\Delta t \sim 0.1$, the shot noise-limited case is much closer to the limiting value, particularly for the uncorrelated fit to the data. Figure 7 shows the

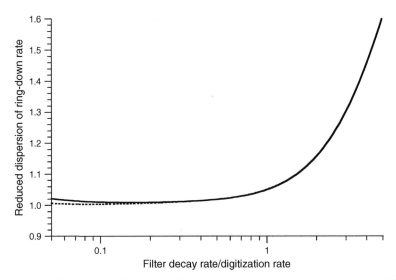

Figure 7   Plot of the reduced dispersion of the ring-down decay rate as a function of $k_f\Delta t$ for shot noise-limited ring down. Otherwise, calculations are the same as in Figure 4.

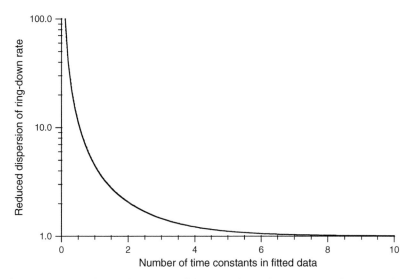

Figure 8   Plot of the reduced dispersion of the ring-down decay rate as a function of $Nk\Delta t$ for shot noise-limited cavity ring down. Only fits with a fixed baseline are presented for reasons given in text.

reduced dispersion as a function of $k_f \Delta t$, calculated with $k\Delta t = 0.01$ and $Nk\Delta t = 8$. These curves look qualitatively similar to the equivalent curves for the detector noise limited case (Figure 3), but again the shot noise-limited case is more forgiving, particularly with regard to using uncorrelated fits. Figure 8 shows the reduced dispersion as a function of $Nk\Delta t$, calculated with $k\Delta t = 0.01$ and $k_f \Delta t = 0.5$. Both correlated and uncorrelated fits are plotted but are essentially indistinguishable for these parameters. Here we see that the shot noise limiting case requires that we go about twice as far into the tail of the decay to recover the limiting dispersion. This is as expected given that the signal-to-noise decays half as fast.

If we do a fit to parameters A', k', and also assume $Nk\Delta t \gg 1$, we can analytically solve the uncorrelated least squares equations (or the correlated equations neglecting the second derivative term in Equation (19)) to give

$$k' = \frac{1}{\Delta t} \ln \left(1 + \frac{\sum y_i}{\sum i y_i} \right) \tag{69}$$

$$A' = [1 - \exp(-k'\Delta t)] \sum y_i = \frac{\left(\sum y_i\right)^2}{\sum (i+1) y_i}. \tag{70}$$

This is just the finite step size analog of the well-known integral for an exponential

$$k = \frac{\displaystyle\int_0^\infty A\exp(-kt)\,dt}{\displaystyle\int_0^\infty tA\exp(-kt)\,dt},$$

which is applicable in the continuum limit, $k\Delta t \ll 1$. Romanini and Lehmann [7] had previously derived that the predicted noise of the integral estimate was the same as for a weighted least squares fit but failed to observe that it is the same estimate. Well into the tail of the exponential decay, the number of photons detected will be small and the fluctuations should follow a Poisson, not Gaussian distribution. However, making the assumption of Poisson distribution for number of detected phonons and selecting A', k' by the maximum likelihood principle leads to precisely Equations (69) and (70).

The covariance matrix is also easily calculated analytically and is given by

$$\tilde{c} = \frac{Gk_f h\nu}{2QA} \begin{pmatrix} (1-a^2)A^2 & \dfrac{(1-a)^2 A}{\Delta t} \\[2ex] \dfrac{(1-a)^2 A}{\Delta t} & \dfrac{(1-a)^3}{\Delta t^2 a} \end{pmatrix} \tag{71}$$

$$\xrightarrow{k\Delta t \ll 1} \frac{Gk_f h\nu}{2QA}(k\Delta t)\begin{pmatrix} 2A^2 & 1Ak \\ 1Ak & 1k^2 \end{pmatrix}.$$

In the shot noise limited case, because of the simple analytical solution, there is no reason to introduce the linearization of the decay. Romanini and Lehmann [7] presented the expected noise for the gated integrator approach for the case of shot noise limited detection.

For a finite value of $Nk\Delta t$, the least squares equations have terms proportional to $\exp(-Nk\Delta t)$ and cannot be solved in closed form. If the fit includes a variable baseline correction, one must be careful to specify how the shot noise is estimated. If it is assumed that the baseline error arises from an offset in the electronics, then $\sigma'^2(t)$ is proportional to $A\exp(-kt)$. In this case, the least squares fit equations and the curvature matrix become singular for large $Nk\Delta t$, because the model predicts that the signal to noise of the fit to the baseline becomes infinite at large time. However, if the baseline correction, B, represents dark counts on the photodetector, then $\sigma'^2(t)$ is proportional to $A\exp(-kt) + B$. In this case, the least squares fit equations and the curvature matrix remain stable even as $Nk\Delta t \to \infty$.

Realistic modeling of real data will often require including both shot and detector noise or dark count contributions to $\sigma'^2(t)$ (see Equation (6)). In this case, one cannot write down closed form expressions for the curvature matrix, because the sums are no longer simple geometric series. Converting the sums to integrals over t (which should be accurate for $k\Delta t \ll 1$ and the errors of which can be corrected by using a Euler – MacLaurin summation formula), closed form expressions for the α_{AA}, α_{AB}, and α_{BB} can be expressed in terms of ln and exp functions. α_{Ak} and α_{Bk} can be expressed in terms of the dilogarithm function while α_{kk} can be expressed in terms of the trilogarithm function [29, 30]. Alternatively, numerical evaluation of the sums is possible and can be used to calculate the covariance matrix for particular model though at an increased computational cost. We can expect the results to vary smoothly, with covariances to increase monotonically as the contribution of detector noise is introduced into the model and is increased to the point that it dominates the shot noise throughout the decay, which takes us between the two limiting cases considered explicitly above. One can do a two or three parameter least square fit that treats the noise as constant during the decay, using the closed form expressions for $\tilde{\alpha}$ and $\tilde{\beta}$ given in Section 5. In the standard limit ($k\Delta t \ll 1$, $Nk\Delta t \gg 1$), Equation (17) predicts that, for $B = 0$, the variance of the determined k' value will be (80/27), approximately three times, that given by Equation (68), which applies the shot noise weight to each point. For $|B| \gg |A|$, the noise changes little between points, so use of constant weights in the fit should be close to optimal and will never be worse than for the $B = 0$ case just discussed.

11. EFFECT OF RESIDUAL MODE BEATING IN THE RING-DOWN DECAY

The model used so far in this chapter has assumed a single exponential decay of the ring-down transient. In many CRDS experiments, one has multimode excitation of the ring-down cavity, and thus the decay contains residual beating terms that arise from the interference of different longitudinal and transverse modes excited in the cavity. The longitudinal mode beating is relatively easy to filter from the detection bandwidth, because it typically occurs on a time scale three to five orders of magnitude shorter than the ring-down time. The effects of excitation of transverse modes can be more problematic, because for a general cavity, the transverse and longitudinal mode spacings are irrationally related [31]; this means that at least some of the potential beating terms will be within the detection bandwidth of the transient that is fit in the experiment. These effects can be mitigated by careful control of the optical alignment and mode matching of the excitation source into the cavity, by an intracavity aperture, by insuring that there are no external limiting apertures, and that the entire transverse profile of the output radiation is focused onto a detector with uniform quantum efficiency [17]. Further, the cavity length can be adjusted to ensure that at least no low-order transverse modes are nearly degenerate with the TEM_{00} longitudinal modes of the cavity and also that the digitization of the decay transient does not "alias" one of the stronger beating frequencies so that it appears close to DC. One of the authors has observed, using a digital oscilloscope, a case where the primary longitudinal mode beating near 150 MHz was "aliased" down to ~100 kHz, giving the appearance of a strongly modulated ring down curve!

Even with careful attention to these factors, residual mode beating in the detection bandwidth can often not be completely eliminated, particularly if using an excitation source of poor spatial and temporal coherence, as is produced by many multimode pulsed dye lasers. For this reason, it is useful to have an estimate of how much the residual beating will affect the fitted ring down rate. We will model the beating by assuming a ring down of the following form:

$$F(t) = A(1 + c_{mb}\cos(\omega t) + s_{mb}\sin(\omega t))\exp(-kt). \tag{72}$$

If we go the continuum limit, $k\Delta t$, $\omega\Delta t \ll 1$, the formulae for error propagation (Equation (15)) can be solved in closed form for the detector and shot noise limiting cases. For the case of detector noise limited decay, we have the fractional shift in the fitted ring–down rate induced by the mode beating is given by

$$\frac{\delta k'}{k} = \left(\frac{16k^2\omega^2}{(4k^2 + \omega^2)^2} \right) c_{mb} + \left(\frac{4k\omega(\omega^2 - 4k^2)}{(4k^2 + \omega^2)^2} \right) s_{mb}$$

$$\xrightarrow{\omega \gg k} \left(\frac{16k^2}{\omega^2} \right) c_{mb} + \left(\frac{4k}{\omega} \right) s_{mb}.$$

(73)

While in the shot noise limited case we have

$$\frac{\delta k'}{k} = \left(\frac{2k^2\omega^2}{(k^2 + \omega^2)^2} \right) c_{mb} + \left(\frac{k\omega(\omega^2 - k^2)}{(k^2 + \omega^2)^2} \right) s_{mb}$$

$$\xrightarrow{\omega \gg k} \left(\frac{2k^2}{\omega^2} \right) c_{mb} + \left(\frac{k}{\omega} \right) s_{mb}.$$

(74)

If we had several different frequency beating terms in the decay, then under the linear correction terms used here, the total shift in k' would just have additive contributions. From these expressions, it is clear that for the same fractional amplitude of mode beating in the decay, the shot noise weighted fit is considerably less perturbed. These results were derived for the case of uncorrelated fits to the data (i.e. diagonal weight matrix the correction factor G). For the typical case, where $\omega \gg k$, including the data correlation, which can be approximated by adding corrections proportional to the second derivative with respect to time of $\bar{F}(t)$, will tend to enhance the effect of the mode beating! This is an example where the ideal procedure for the case of random statistical error is not the optimal procedure to use in the case of nonstatistical error.

It is clear from Equations (73) and (74) that the effect of the mode beating will be second order if we start the data collection at a time corresponding to an antinode of the mode beating. This suggests one strategy to minimize the effect of mode beating (assuming we have a single dominant beating frequency) by picking the starting point of the fit to minimize the χ^2 of the fit. The phase of all the excited modes of the cavity is fixed at $t=0$ for the case of impulsive excitation of the cavity, and the beat frequencies fixed by the cavity spectrum. Thus, this "best" time should be constant as the excitation spectrum is scanned, and the starting point can be fixed after a series of preliminary measurements. The "optimal" method of data analysis is to include the amplitude of the known mode beating terms in the least squares fit of the data. Including the extra parameter will raise the predicted dispersion of k due to parameter correlation, but simple calculations indicate that this will be by a minor amount, 1% if $\omega = 10(2\pi k)$.

12. CONCLUSIONS

This work has presented a statistical analysis of data processing in CRDS. In particular, it is shown that to approach the "ideal" limits in the standard deviation or dispersion of the ring-down decay rate, one must filter the input signals to ensure that the noise is sufficiently well sampled, that is, the noise density aliased to near-zero frequency should be negligible. This filtering, however, introduces correlation of the data. As long as the filter time constant is much less than the ring-down time, the data correlation is found to have only a modest effect on the ring-down time extracted by a least squares fit. However, a least squares fit that ignores the data correlation will predict parameter dispersions substantially below the correct statistical value. The effect of "linearization" of the fit is considered, and it is found to have minimal effect of the least squares fit, provided the signal to noise of the decay is initially sufficiently large. Other methods used to determine cavity decay rates from CRDS data and their noise and bias have been analyzed. As could have been anticipated, all give higher dispersion in the extracted decay rate, which translates into decreased sensitivity in CRDS experiments.

A natural extension of this chapter would be a similar analysis of related cavity-enhanced spectroscopic methods, particularly the cavity-enhanced absorption method [25,26] which is gaining popularity. These alternative methods are not immune to source amplitude and frequency (or phase) noise. The last is most important as one has interference between light stored in the cavity and the instantaneous excitation field. Noise analysis will depend on the details of the temporal coherence of the laser used to excite the cavity.

ACKNOWLEDGMENTS

The authors' work in CRDS has been supported by the National Science Foundation, the New Jersey Commission on Science and Technology, and the University of Virginia.

REFERENCES

[1] A. O'Keefe and D. A. G. Deacon, Review of Scientific Instrument **59**(12), 2544 (1988).

[2] K. Busch and M. Busch, eds., *Cavity Ringdown Spectroscopy – An Ultratrace-Absorption Measurement Technique*, vol. 720 of *ACS Symposium Series*, (American Chemical Society, Washington, D.C., 1999).

[3] G. Berden and G. Engel, *Cavity Ring-down Spectroscopy: Techniques and Applications* (Blackwell Publishing, Oxford, 2008).

[4] J. J. Scherer, J. B. Paul, A. O'Keefe, and R. J. Saykally, Chemical Reviews **97**(1), 25 (1997).

[5] G. Berden, R. Peeters, and G. Meijer, International Reviews in Physical Chemistry **19**(4), 565 (2000).

[6] C. Vallance, New Journal of Chemistry **29**(7), 867 (2005).

[7] D. Romanini and K. K. Lehmann, Journal of Chemical Physics **99**(9), 6287 (1993).

[8] D. L. Albritton, A. L. Schmeltekopf, and R. N. Zare, chap. 1 in Rao, ed., *Molecular Spectroscopy: Modern Research*, vol. II, (Academic Press, New York, 1974).

[9] T. von Lerber and M. W. Sigrist, Chemical Physics Letters **353**(1–2), 131 (2002).

[10] M. Mazurenka, R. Wada, A. J. L. Shillings, T. J. A. Butler, J. M. Beames, and A. J. Orr-Ewing, Applied Physics B **81**(1), 135 (2005).

[11] D. Halmer, G. Basum, P. Hering, and M. Murtz, Review of Scientific Instruments **75**(6), 2187 (2004).

[12] J. M. Herbelin, J. A. McKay, M. A. Kwok, R. H. Ueunten, D. S. Urevig, D. J. Spencer, and D. J. Benard, Applied Optics **19**(1), 144 (1980).

[13] R. Engeln, G. vonHelden, G. Berden, and G. Meijer, Chemical Physics Letters **262**(1–2), 105 (1996).

[14] T. G. Spence, C. C. Harb, B. A. Paldus, R. N. Zare, B. Willke, and R. L. Byer, Review of Scientific Instruments **71**(2), 347 (2000).

[15] P. R. Bevington and D. K. Robinson, *Data Reduction and Error Analysis for the Physical Sciences* 2nd Edition. (McGraw-Hill Inc., New York, 1992).

[16] W. H. Press, B. P. Flannery, S. A. Teukolsky, and W. Vetterling, *Numerical Recipes* (Cambridge University Press, Cambridge, 1986).

[17] H. Huang and K. K. Lehmann, Optics Express **15**(14), 8745 (2007).

[18] J. L. Miller and A. J. Orr-Ewing, Journal of Chemical Physics **126**(17), 174303 (2007).

[19] A. A. Istratov and O. F. Vyvenko, Review of Scientific Instruments **70**(2), 1233 (1999).

[20] I. B. C. Matheson, Analytical Instrumentation **16**(3), 345 (1987).

[21] K. K. Lehmann and D. Romanini, Journal of Chemical Physics **105**(23), 10263 (1996).

[22] G. Meijer, M. G. H. Boogaarts, R. T. Jongma, D. H. Parker, and A. M. Wodtke, Chemical Physics Letters **217**(1–2), 112 (1994).

[23] J. Morville, D. Romanini, M. Chenevier, and A. Kachanov, Applied Optics **41**(33), 6980 (2002).

[24] J. B. Dudek, P. B. Tarsa, A. Velasquez, M. Wladyslawski, P. Rabinowitz, and K. K. Lehmann, Analytical Chemistry **75**(17), 4599 (2003).

[25] R. Engeln, G. Berden, R. Peeters, and G. Meijer, Review of Scientific Instruments **69**(11), 3763 (1998).

[26] A. O'Keefe, J. J. Scherer, and J. B. Paul, Chemical Physics Letters **307**(5–6), 343 (1999).

[27] J. E. Thompson and K. Myers, Measurement Science and Technology **18**(1), 147 (2007).

[28] D. Walls and G. Milburn, *Quantum Optics* (Springer, 1995).

[29] L. Lewin, *Polylogarithms and Associated Functions* (North Holland, New York, 1981).

[30] L. Vepstas, *An Efficient Algorithm for Accelerating the Convergence of Oscillatory Series, Useful for Computing the Polylogarithm and Hurwitz Zeta Functions* (2007), http://arxiv.org/abs/math/0702243.

[31] A. E. Siegman, *Lasers* (University Science Books, Mill Valley, California, 1986).

Spectroscopy and Dynamics of Neutrals and Ions by High-Resolution Infrared–Vacuum Ultraviolet Photoionization and Photoelectron Methods

Cheuk-Yiu Ng

Contents

Abstract

By using a broadly tunable infrared (IR) optical parametric oscillator laser and a comprehensive tunable vacuum ultraviolet (VUV) laser, together with the supersonic molecular beam and pulsed-field ionization (PFI) detection techniques, we have demonstrated an array of novel two-color IR–VUV and VUV–IR photoion–photoelectron methods. The isomeric and conformation sensitivity of IR excitation and detection sensitivity of VUV photoionization detection, together with the long lifetimes of IR excited rovibrational states and VUV excited high-n Rydberg states, make the combination of IR and VUV excitations ideal for two-color spectroscopic and photoionization dynamics probes of polyatomic species and their ions. Selected experiments are presented to illustrate the principles, information contents, and unique capabilities of these IR–VUV and VUV–IR photoionization methods.

Keywords: photoionization; photoelectron; Rydberg states; vacuum ultraviolet; pulsed-field ionization; photo-induced Rydberg ionization

1. INTRODUCTION

Vacuum ultraviolet (VUV) photoion–photoelectron spectroscopy represents a major technique for the study of photoionization dynamics and obtaining energetic and spectroscopic information of neutrals and ions [1–3]. Because of the space charge effect, ions cannot be prepared in high concentration. Consequently, conventional spectroscopic techniques applicable to neutral molecules cannot be generally employed for ion species. The lion share of spectroscopic and energetic data for molecular ions in the literature has been provided by photoionization and photoelectron measurements. The development of VUV lasers [4] and the introduction of the high-resolution pulsed-field ionization–photoelectron (PFI–PE) technique [5,6] have had a profound impact on the field of VUV photoionization spectroscopy and thermochemistry. The VUV–PFI–PE spectra of many diatomic, triatomic, and simple hydride molecules can now be measured to the rotational-resolved level [3,4]. The analyses of the rotationally resolved spectra have provided definitive ionization energies (IEs) for these molecular species. It has been well-known that the cooling of a gas sample to a lower temperature by supersonic expansion can reduce the thermal rotational and vibrational populations, and thus, lead to better resolved spectral transitions [7]. By using the VUV–PFI–PE scheme, combined with the supersonic expansion technique, vibrationally resolved photoelectron spectra for many simple polyatomic molecules have been measured [8–16]. The VUV–PFI–PE vibrational bands observed for polyatomic species generally exhibit a full width at half maximum (FWHM) of $10–40\,cm^{-1}$, which arises from overlapping photoionization rotational branches. The typically achieved PFI–PE resolutions of $1–1.5\,cm^{-1}$ (FWHM) are still insufficient for the resolution of individual rotational transitions in VUV–PFI–PE measurements of polyatomic molecules.

In order to overcome this spectral congestion problem, we have recently introduced the two-color infrared (IR)–VUV laser photoionization efficiency (PIE) and PFI–PE methods, which have made possible state-to-state photoionization measurements of several polyatomic molecules [17–30]. These methods involve the setting of the IR laser at a selected rovibrational transition of the neutral molecule prior to scanning the VUV laser frequency for PIE or PFI–PE detection. We have shown that by IR laser excitation, a molecule in single rovibrational states can be selected for PIE and PFI–PE studies [20,21,23–30]. As a spectroscopic technique, the preparation of the molecule in a single rovibrational state by IR excitation is superior to cooling the molecule to 0 K such that only the lowest rotational level is populated because many rovibrational states can be accessible by IR excitation. From a technical point of view, the long lifetimes of IR excited rovibrational states, the structural and conformational sensitivity of IR excitation, and the detection sensitivity of VUV photoionization are expected to make IR–VUV photoionization an ideal double-resonance technique for spectroscopic and dynamics probes of polyatomic molecules and their ions.

In addition to IR–VUV–PFI–PE measurements, the combination of IR and VUV laser excitations has given rise to a series of new photoionization methods

[24,31–34]. We have also illustrated that the IR–VUV–photoion (IR–VUV–PI) detection scheme is a highly sensitive method for vibrational spectroscopy studies of molecular neutrals [25–31]. In this spectroscopic scheme, the VUV laser frequency is set below the IE of the molecule of interest, whereas IR laser frequency is scanned to excite the rovibrational transitions of the molecule. Because the ion signal is observed only when the IR laser frequency matches an IR transition, the IR–VUV–PI spectrum thus recorded provides a good measure of the IR absorption spectrum for the neutral molecule. In IR–VUV–PI measurements of many small molecules, the delay between the IR and VUV lasers can be set in the range from a few nanoseconds to a few microseconds. Excellent signal-to-noise ratios can be achieved in IR–VUV–PI measurements because of the negligible ion background produced by the VUV laser. The IR–VUV–PI method has the ability to identity the mass of the neutral IR absorber, and thus, is suitable for IR spectroscopy measurements of reactive and cluster species, which usually coexist with other impure species. This represents a major advantage over the traditional Fourier-transform IR method [35], which has no capability to identify the mass of the neutral IR absorber.

The IR–VUV–PIE depletion [24] and IR–VUV–PFI–PE depletion [19,32–34] methods are variants of the IR–VUV–PIE, IR–VUV–PI, and IR–VUV–PIE detection schemes. The IR–VUV–PIE depletion method concerns with the depletion measurement of the VUV–PIE signal as the IR laser frequency is fixed to excite molecules out of a selected rotational level. Considering that setting the IR laser at a selected rotational transition depletes the population of a specific rotational level in the ground vibrational state, the fixing of the IR frequency at a specific rotational transition and scanning the VUV laser energy can give rise to a negative or depletion VUV–PIE signal of an autoionizing Rydberg peak if it is formed by excitation from the same neutral rotational level of the neutral. The IR–VUV–PIE depletion method provides information about the rotational character of Rydberg transitions, and thus, is valuable for the rotational assignment of Rydberg series. The IR–VUV–PFI–PE depletion measurement involves the detection of VUV–PFI–PE depletion by scanning the IR laser frequency as the VUV laser frequency is fixed to a given VUV–PFI–PE photoionization rotational transition. Although both the IR–VUV–PFI–PE depletion and IR–VUV–PI methods provide information about IR rovibrational transitions of neutral molecules, the IR–VUV–PFI–PE method is different from the IR–VUV–PI measurement [34].

The major progress in photoion–photoelectron spectroscopy in the past decade is based on the exploitation of the properties of high-n Rydberg states [3,5,36], such as in high-resolution measurements using the PFI techniques. Because high-n Rydberg states converging to a specific ionic state have an ion core structure similar to the converging ionic state along with a weakly interacting Rydberg electron, the measurement of the Rydberg electron ejected or the bare ion produced by PFI of the Rydberg species conveys spectroscopic information about the ion. The high detection efficiency for charged particles has made the spectroscopic techniques based on PFI of high-n Rydberg states highly sensitive, eliminating the need of forming ions in high concentration for conventional spectroscopic studies. Instead of ejecting the Rydberg electron by a Stark field, the ionization of high-n Rydberg species can also be achieved by other mechanisms such as photo-induced autoionization. If the excitation begins with a high-n Rydberg level lying below the IE of the neutral species, the autoionization process can be induced by photoexcitation of the Rydberg state to a state above the ionization onset. Because the electron associated with a high-n Rydberg state is similar to free electrons and does not couple well with the optical radiation field, the photoabsorption should occur with the ion core [36]. The idea that autoionization can be induced by core excitations is rooted in atomic spectroscopy and has been applied for lifetime measurements of excited atomic Rydberg states [37]. In the case of excitation involving high-n molecular Rydberg states, the light absorption is expected to populate discrete rovibronic levels of the molecular ion core. If the resulting system has a total energy greater than the IE of the molecule, finite couplings between the Rydberg electron and the rovibronically excited ion core can induce the ejection of the Rydberg electron. The detection of the electron or ion produced in such a photo-induced Rydberg ionization (PIRI) process would make possible the spectroscopic study of the molecular ion. Because the production of PIRI ions does not require a Stark electric field, which has the effect of broadening the energy resolution, the achievable resolution for PIRI detection is expected to be higher than that for PFI measurements.

While the PFI techniques are highly mature and have been widely used in the past decade, the PIRI detection scheme for spectroscopy studies of molecular ions is still in the developing stage. The early PIRI experiments concern autoionization induced by electronic excitation using a visible laser [36,38,39]. Variations of the PIRI experiment by using a tunable IR laser to induce autoionization have also been developed [40–42]. In all previous IR–PIRI experiments, high-n Rydberg states were prepared by a UV 1+1 resonance-enhanced mechanism, which requires the existence of a long-lived, bound intermediate electronic state. However, because many molecular species do not have such excited states, the UV–IR–PIRI scheme can only be applied to selected molecular species. The VUV–IR–PIRI measurement requires the reversed order for applying the IR and VUV lasers. That is, the VUV laser is used first to excite the molecule to a high-n Rydberg state lying below its IE followed by the scanning of the IR laser frequency to induce autoionization of the Rydberg species. The fact that the VUV–IR–PIRI scheme does not require an UV intermediate state of the neutral species makes it a general scheme, applicable to all molecular species. As described below, we have demonstrated that the VUV–IR–PIRI method can be used to probe the IR absorption of molecules in excited high-n Rydberg states.

The high-resolution IR–VUV photoion–photoelectron methods have been successfully employed for the study of a series of polyatomic molecules and their cations [18–33]. Thus, we can conclude that these methods are generally applicable for spectroscopic studies of neutrals and cations. This chapter presents a review of the recent high-resolution IR–VUV and VUV–IR photoionization and photo-electron experiments of selected molecules, including ammonia (NH_3) [23,24], methyl bromide (CH_3Br) [30], methyl iodide (CH_3I) [29], ethylene (C_2H_4) [20,25,34], and propyne (C_3H_4) [28], to illustrate the merit of these two-color photoionization methods. The prospect for future applications of these methods is also assessed.

2. IR–VUV LASER PHOTOION–PHOTOELECTRON APPARATUS

The VUV, IR–VUV, and VUV–IR photoion–photoelectron measurements for photoionization dynamics and spectroscopy studies of neutrals and their cations are made using the IR–VUV photoion–photoelectron apparatus [19–34], which is equipped with a comprehensive tunable VUV laser and a high-resolution, broadly tunable single-mode IR-optical parametric oscillator (IR-OPO) laser, and a photo-ion–photoelectron apparatus. The detailed experimental arrangement of this appa-ratus is described below.

Figure 1 shows the schematic diagram of the IR–VUV laser photoion–photoelectron apparatus used in this study. Because this apparatus has not been reported in detail previously, a short description is given below. The VUV laser system [11–13,20,21] consists of an injection seeded Nd:YAG pumped laser (1) (repetition rate $= 30$ Hz, pulsed energies $= 1.6$ J at 1064 nm, 550 mJ at 532 nm, and 325 mJ at 355 nm), two dye lasers [both dye laser 1 (2) and dye laser 2 (3) have a specified optical bandwidth of ≈ 0.03 cm^{-1} (FWHM)], a nonlinear mixing chamber (9) for frequency mixings, and a windowless VUV monochro-mator (11). Dye lasers 1 and 2 can be pumped by either the 532 or 355 nm outputs of the Nd:YAG laser. The rare gases (Ar, Kr, and Xe) or Hg vapor are used as the nonlinear media for frequency mixings, giving a tunable range from the UV to

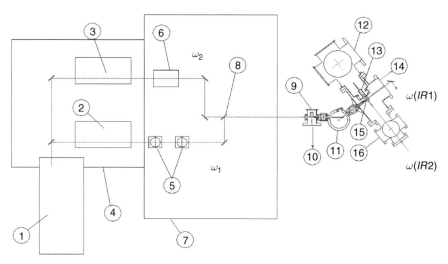

Figure 1 Schematic diagram showing the basic components of the vacuum ultraviolet (VUV)-laser photoion–photoelectron apparatus. (1) Injection seeded Q-switched Nd:YAG laser (550 mJ at 532 nm and 355 mJ at 355 nm, pulse duration of 5 ns); (2) dye laser 1; (3) dye laser 2; (4) $4' \times 4'$ optical table; (5) BBO crystals (2 units); (6) optional frequency doubler; (7) $6' \times 8'$ optical table; (8) dichroic mirror; (9) nonlinear frequency mixing chamber equipped with pulsed supersonic jet; (10) to turbomolecular pump; (11) VUV monochromator; (12) beam source chamber; (13) pulsed valve; (14) photoionization chamber; (15) photoionization region. The IR laser can enter the photoionization region in either counter propagating the VUV laser beam via ω (IR1) or counter propagating the molecular beam via ω(IR2) [22]. (See color plate 35).

19.5 eV. The output of dye laser 1 after appropriate frequency doubling and/or mixing by BBO crystals is fixed at ω_1 to match the two-photon UV-resonance of a nonlinear gas. The visible output of dye laser 2 (ω_2) is tuned to generate the VUV laser radiation of interest. The laser beams ω_1 and ω_2 are combined with a dichroic mirror (8) and focused into the nonlinear gas (either in the form of a free jet or a gas cell) in the nonlinear mixing chamber. The resulting VUV sum frequencies ($2\omega_1 + \omega_2$) or difference frequencies ($2\omega_1 - \omega_2$) are selected by the windowless VUV monochromator before intersecting the gas sample beam at 90° in the photoionization region (15) of the photoionization chamber (14). The electrons and ions thus produced at the photoionization region are detected by the electron and ion microchannel plate (MCP) detectors (not shown in Figure 1) situated below and above the photoionization region, respectively. The nature of the electrons and the ions detected can be distinguished by time-of-flight (TOF) analyses. After passing through the photoionization region, the intensity of the VUV beam is measured by a copper photoelectric detector, such that the photoion and PFI–PE intensities can be normalized by the VUV intensities. The output of ω_2 is continuously monitored by a wave meter for frequency calibration during the experiment. The VUV laser system is equipped with a Raman shifter in H_2 (D_2, or CH_4) to extend the tunable range of ω_2, such that the full VUV range of 7–19.5 eV can be generated without energy gaps. The optical bandwidth of the VUV radiation thus generated has been measured to be $0.12\,cm^{-1}$ (FWHM) [11–13].

Two IR-OPO laser systems have been used for the IR–VUV measurements presented here. In the earlier experiments, an IR-OPO laser with an optical bandwidth of $0.25\,cm^{-1}$ (FWHM) was used, which will be referred to as the low resolution IR-OPO laser below. A single-mode IR-OPO laser has been employed in the recent IR–VUV study. Both IR-OPO systems were pumped by the 1064 nm output of an injection seeded Nd:YAG laser (pulsed energy = 550 mJ) operated at 15 Hz. The stabilization of the single-mode operation is maintained by a piezo located on the rear mirror of the oscillator. The voltage on the piezo can be adjusted either manually or automatically according to the feedback from the etalons rings displayed on a monitor. The IR-OPO laser system is tunable in the range of 1.35–5.0 μm (7407–2000 cm^{-1}) with typical pulse energies in the range of 5–12 mJ at full power and a specified optical bandwidth of $0.007\,cm^{-1}$ (FWHM). For most experiments presented here, the IR pulse energies used were in the range of 1–4 mJ at ≈3 μm. A spherical lens of 1-m focal length is used to slightly focus the IR laser beam, resulting in a spot size of ≈2 mm × 4mm at the photoionization region.

The photoion–photoelectron apparatus is partitioned into five differentially pumped chambers. The nonlinear mixing chamber (9), the monochromator (11), the photoionization chamber (14), and the ion TOF chamber (not shown in Figure 1) are separately evacuated by a 500, a 500, a 1000, and a 300 L/s turbomolecular pump (TMP), respectively. The beam source chamber (12) is pumped by a 10″ water-cooled diffusion pump (DP) with a nominal pumping speed of 5000 L/s.

The gaseous sample is generally premixed with a rare gas (He or Ar) before introducing into the photoionization region as a pulsed supersonic molecular beam. The pulsed beam is formed through a pulsed valve (General valve, nozzle diameter = 0.5 mm, repetition rate $= 30\,Hz$) and is skimmed by a conical skimmer prior to intersecting the VUV laser beam perpendicularly at the photoionization region. During the experiment, the photoionization chamber and the ion TOF mass spectrometer are maintained at pressures of $\approx 10^{-7}\,Torr$.

The timing sequence for opening of the pulsed valve, firing of the IR and VUV lasers, and applying the PFI field is controlled by two digital delay units (Stanford Research DG535). The photoion and photoelectron signals from the MCP detectors are pre-amplified and integrated by independent boxcar integrators before processing by a personal computer.

3. EXPERIMENTAL MEASUREMENTS

3.1. VUV–PIE and VUV–PFI–PE measurements

The reliable VUV–PIE and VUV–PFI–PE measurements are fundamental for the success of photoion–photoelectron measurements using the IR–VUV excitation schemes. The procedures for VUV–PIE and VUV–PFI–PE measurements have been described in detailed previously, and thus, will not be substantiated here [11–13]. Briefly, for VUV laser PFI–PE measurements, a DC field of 0.1 V/cm is applied to disperse the prompt background photoelectrons. After a delay of 1 μs with respect to the VUV laser pulse, a pulsed field of 0.3 V/cm is applied to field ionize the high-n Rydberg molecules and to extract the PFI–PEs toward the electron MCP detector. The PFI–PE resolution achieved in this study is ≈ 1.2– $1.5\,cm^{-1}$ (FWHM) [11,12].

3.1.1. VUV–PIE and VUV–PFI–PE measurements of C_3H_4

As an example, we show in Figure 2a the VUV–PIE (upper curve) and VUV–PFI– PE (lower curve) spectra of C_3H_4 obtained near its ionization threshold [28]. The rise of the PIE onset spans $\approx 50\,cm^{-1}$, which is consistent with the FWHM of $\approx 40\,cm^{-1}$ observed for the VUV–PFI–PE origin band. Without a reliable simulation of the origin band, a precise $IE(C_3H_4)$ value cannot be obtained. A reliable simulation of the PFI–PE origin band for $C_3H_4^+$ would require knowledge about the rotational temperature of the neutral C_3H_4 sample and detailed rotationally selected and resolved state-to-state photoionization cross sections of C_3H_4 [28]. To determine these state-to-state photoionization cross sections and the rotational temperature of the gas sample is the main motivation for the development of the IR–VUV photoionization and photoelectron methods.

3.2. IR–VUV–PI measurements

As described above, the VUV laser radiation selected by the windowless VUV monochromator intersects with the sample molecular beam at 90° in the photoionization region. The molecular beam axis lies in the plane defined by the axes of the monochromator entrance and exit arms (or the Roland circle) and is

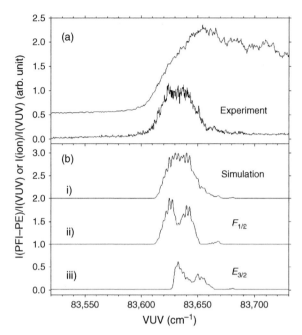

Figure 2 (a) VUV–PIE spectrum (upper curve) and VUV–PFI–PE (lower curve) of C_3H_4 near its ionization threshold in the energy range of 83,520–83,730 cm^{-1}. (b) The origin VUV–PFI–PE band for C_3H_4. (b) Curves (i), (ii), and (iii) are the overall fitted spectrum, the fit spectrum to the lower $^2E_{1/2}$, component, and the fit to the higher $^2E_{3/2}$ component, respectively [28].

perpendicular to the exit arm of the VUV monochromator. Two configurations have been used for the IR–VUV–PI measurements. Referring to Figure 1, the first configuration involves the introduction of the IR laser beam [ω(IR1)] into the photoionization region counter propagating with the VUV laser beam. When the optimal spatial overlap of the IR and VUV laser beams is achieved at the photo-ionization region, the IR laser beam can pass through the monochromator and exit its entrance arm. In the second configuration, the IR laser beam [ω(IR2)] enters the photoionization region along the molecular beam axis. In both configurations, the optimal overlap of the IR, VUV, and molecular beams can be monitored by the photoion signal. All IR–VUV–PI data presented here have been obtained using the first experimental configuration.

In IR–VUV–PI measurements, the VUV laser frequency is usually set at $\approx 100 \, cm^{-1}$ below the IE of the molecule such that the VUV alone is insufficient to ionize the molecule. However, a small background signal of the molecular ions can be produced due to thermal populations of the molecules in excited rovibra-tional states. The fact that the repetition rate (30 Hz) for the VUV laser is by design twice that (15 Hz) of the IR-OPO laser allows the shot-to-shot subtraction of this ion background. The application of the VUV laser pulse is delayed by ≈ 50–$100 \, ns$ with respect to the IR laser pulse. The IR–VUV–PI spectra are obtained by scanning the IR laser to cover the frequency range of the vibrational band of interest. Although the single-mode IR–OPO laser is specified to have an optical bandwidth of $0.007\,cm^{-1}$ (FWHM), the best measured spectral resolution is $\approx0.012\,cm^{-1}$ (FWHM), which is observed in a photoacoustic measurement of CH_4 in the mid-IR using a room temperature CH_4 gas cell. A small contribution to the IR absorption linewidth observed in the latter experiment due to the Doppler broadening effect is to be expected in the latter measurement. As shown below, the achieved resolution for IR–VUV–PI measurements presented in this study using the single-mode IR–OPO laser is $\approx0.02\,cm^{-1}$ (FWHM), suggesting that some broadening may have been arisen from VUV photoionization processes [26–31].

3.2.1. IR–VUV–PI measurements of CH_3X (X = Br and I)

The IR–VUV–PI spectrum for the ν_1 (C—H stretching) = 1 vibrational band of CH_3I obtained using the single-mode IR–OPO laser [29] is plotted in Figure 3a for comparison with that (Figure 3b) obtained previously [20] using the low-resolution IR–OPO laser. This is a parallel transition with the usual $\Delta J = 0, \pm1$ and $\Delta K = 0$ selection rules, resulting in the normal structure of the P-, Q-, and R-branches. The assignments of rotational transitions are marked on top of the spectra. The spectral simulation shown as a stick plot below the spectrum of Figure 3b is consistent with a rotational temperature of 8 K for the supersonically cooled CH_3I beam [20]. At this temperature, the CH_3I molecules are predominantly in the $K'' = 0$ or 1 rotational states with a very small portion at $K'' = 2$ or higher levels. The different K'-transitions are well resolved in the high-resolution scan of Figure 3a, while only individual J'-transitions are resolved in the low-resolution spectrum.

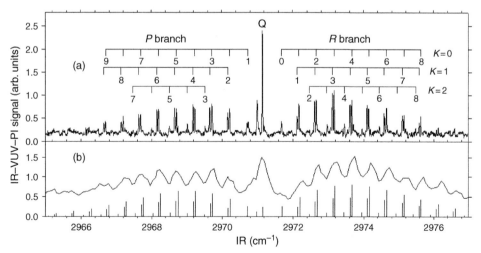

Figure 3 IR–VUV–PI spectra of CH_3I ($\nu_1 = 1$) C—H stretching band in the IR range of 2965–2977 cm^{-1} obtained with (a) a high-resolution single-mode IR–OPO laser (resolution = 0.02 cm^{-1}) and (b) a low-resolution IR–OPO laser (resolution = 0.25 cm^{-1}). The resolutions achieved in the high- and low-resolution measurements are 0.02 and 0.25 cm^{-1}, respectively [20,29].

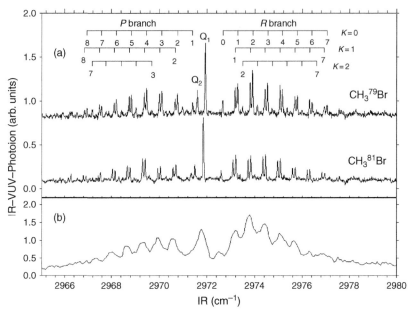

Figure 4 IR–VUV–PI spectra for the $\nu_1 = 1$ (C—H stretching) vibrational band of (a) CH_3Br^{79} (upper spectrum) and CH_3Br^{81} (lower spectrum) obtained using the high-resolution single-mode IR-OPO laser and (b) CH_3Br^{79} obtained using the lower resolution IR-OPO laser. Rotational assignments are marked on top of the spectra in (a). The resolutions achieved in the high- and low-resolution measurements are 0.02 and 0.25 cm^{-1}, respectively [30].

Similar to the CH_3I experiments, Figure 4a and b compares the IR–VUV–PI spectra of CH_3Br using the high- and low-resolution IR-OPO laser, respectively [31]. The two spectra shown in Figure 4a are the high-resolution (J', K')-resolved IR spectra of the ν_1 bands of CH_3Br^{79} and CH_3Br^{81}. A slight shift in frequency for these IR spectra of the two isotopes is observed. Because the low-resolution scan is unable to differentiate rotational transitions originating from CH_3Br^{79} and CH_3Br^{81}, only the IR–VUV–PI spectrum for CH_3Br^{79} is plotted in Figure 4b. The spectra for CH_3Br^{79} and CH_3Br^{81} illustrate the merit of the IR–VUV–PI technique for IR spectroscopic measurements. The fact that the mass of the IR absorber can be identified by photoionization mass spectrometry makes the IR–VUV–PI method applicable for IR spectroscopic measurements of isotopomers, reactive radicals, and clusters, which usually exist as impure samples.

3.2.2. IR–VUV–PI measurements of C_2H_4

We have measured the IR–VUV–PI spectra for the $\nu_{11}(b_{1u})$, $\nu_2 + \nu_{12}(b_{1u})$, and $\nu_9(b_{2u})$ vibrational bands of ethylene $[C_2H_4(\tilde{X}^1A_g)]$ as shown in Figures 5a and b, 6a and b, and 7a and b, respectively, where $\nu_2(a_g)$, $\nu_9(b_{2u})$, $\nu_{11}(b_{1u})$, and $\nu_{12}(b_{1u})$ represent the respective C—C stretching, CH_2 stretching, CH_2 stretching, and CH_2 bending modes of $C_2H_4(\tilde{X}^1A_g)$[21,25,31]. The IR–VUV–PI vibrational bands shown in Figures 5a, 6a, and 7a are high-resolution spectra recorded using the single-mode IR-OPO laser, whereas the IR–VUV–PI vibrational bands of Figures 5b, 6b, and 7b

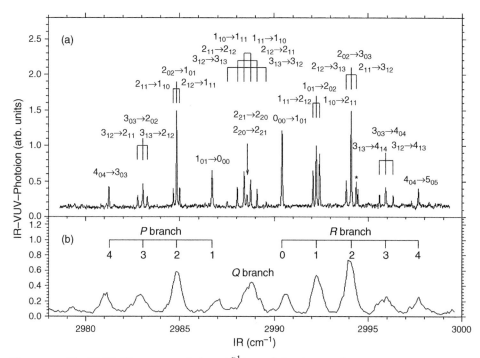

Figure 5 IR–VUV–PI spectra of $C_2H_4(\tilde{X}^1A_g; \nu_{11})$ in the IR frequency range of 2978–3000 cm^{-1} recorded by using (a) the single-mode IR-OPO laser [resolution achieved = 0.02 cm^{-1} (FWHM)] and (b) the lower resolution IR-OPO laser [resolution achieved = 0.25 cm^{-1} (FWHM)]. The assignment of rotational transitions $N''_{K''_a K''_c} \rightarrow N'_{K'_a K'_c}$ are marked on top of the spectrum plotted in (a), where (N'', K_a'', K_c'') and (N', K_a', K_c') are the respective rotational quantum numbers for the ground vibrational state and the $\nu_{11} = 1$ excited state of C_2H_4. The spectrum of (b) exhibits the P-, Q-, and R-branches. The assignment of the $P(N'')$, $Q(N'')$, and $R(N'')$ rotational lines are marked on top of (b) [21,25,27].

were recorded by using the low-resolution[8] IR-OPO laser. Compared with the observed rotational linewidths in the present and previous IR–VUV–PI measurement, we conclude that the resolution achieved in the single-mode IR–VUV–PI measurements of Figures 5a, 6a, and 7a is more than 12-fold higher than that observed in the low-resolution spectra of Figures 5b, 6b, and 7b.

The IR–VUV–PI band of $C_2H_4(\tilde{X}^1A_g; \nu_{11})$ observed in the range of 2978–3000 cm^{-1} (Figure 5a and b) is an A-type IR transition band, where the selection rules: $\Delta N' = N' - N'' = 0$, ±1, $\Delta K_a' = K_a' - K_a'' = 0$, and $\Delta K_c' = K_c' - K_a'' = 1$ are valid [25]. As marked on top of Figure 5b, the low-resolution spectrum exhibits the typical P-, Q-, and R-branches. The comparison of Figure 5a and b shows that the Q-branch and many of the $P(N'')$ and $R(N'')$ lines observed in Figure 5b consists of multiple unresolved transition lines. The assignments of the allowed $N''_{K''_a K''_c} \rightarrow N'_{K'_a K'_c}$ rotational transitions are marked on top of Figure 5a, where (N'', K_a'', K_c'') are rotational quantum numbers for the ground vibrational state and (N', K_a', K_c') are rotational quantum numbers for the excited $\nu_{11} = 1$ state of C_2H_4. After taking into account the nuclear

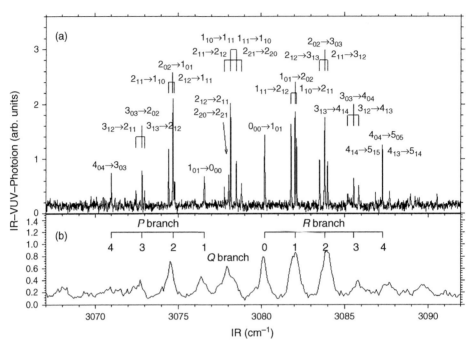

Figure 6 (a) IR–VUV–PI spectrum of $C_2H_4(\tilde{X}^1A; \nu_2+\nu_{12})$ in the IR frequency range of 3067–3092 cm^{-1} recorded by using (a) the single-mode IR-OPO laser [resolution achieved $= 0.02$ cm^{-1} (FWHM)] and (b) the low-resolution IR-OPO laser [resolution achieved $= 0.25$ cm^{-1} (FWHM)]. The assignment of the rotational transitions $N''_{K''_a K''_c} \rightarrow N'_{K'_a K'_c}$ are marked on top of the spectrum plotted in (a), where (N'', K_a'', K_c'') and (N', K'_a, K'_c) are the respective rotational quantum numbers for the ground vibrational state and the $\nu_2+\nu_{12}$ excited state of C_2H_4. The spectrum of (b) exhibits the P-, Q-, and R-branches. The assignment of the $P(N'')$, $Q(N'')$, and $R(N'')$ rotational lines are marked on top of (b) [21,25,30].

spin statistical weights of the rotational levels, the simulation of the spectrum of Figure 5a indicates that the rotational temperature of the C_2H_4 sample achieved in the supersonic expansion is ≈ 8–10 K [25]. As a result, more than 95 percent of the molecules are expected to populate the rotational levels $N'' < 5$ of the ground state. For ethylene in its vibronic ground state, the nuclear spin statistical weights are 7, 3, 3, and 3 for $K_a K_c =$ ee, eo, oe, and oo, respectively, where e and o stand for even and odd. The largest nuclear spin statistic weight for the $K_a K_c =$ ee levels contributes to the outstanding intensities for the transitions of $2_{02} \rightarrow 1_{01}$ in $P(2)$, $0_{00} \rightarrow 1_{01}$ in $R(0)$, and $2_{02} \rightarrow 3_{03}$ in $R(2)$ as shown in Figure 5a.

The IR–VUV–PI spectrum for the $\nu_2 + \nu_{12}$ vibrational band measured in range of 3067–3092 cm^{-1} (Figure 6a and b) is also an A-type band [25]. The rotational assignments are marked on top of Figure 6a and b. The fact that both the ν_{11} and $\nu_2+\nu_{12}$ bands are governed by the same rotational selection rules results in the nearly identical rotational structures for these two bands.

The IR–VUV–PI spectra of the $C_2H_4(\tilde{X}^1A_g; \nu_9)$ band (Figure 7a and b) are governed by the selection rules of the B-type IR transitions: $\Delta N' = 0, \pm 1, \Delta K'_a = \pm 1$,

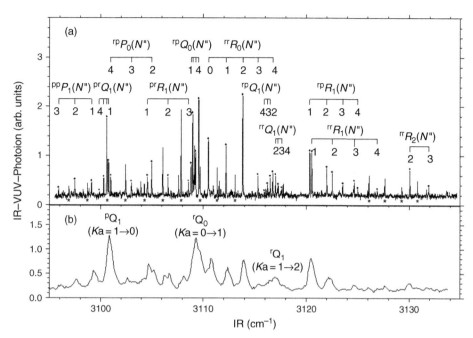

Figure 7   IR–VUV–PI spectrum of $C_2H_4(\tilde{X}^1A; \nu_9)$ in the IR frequency range of 3095–3135 cm^{-1} recorded by using (a) the single-mode IR-OPO laser [resolution achieved = 0.02 cm^{-1} (FWHM)] and (b) the low-resolution IR-OPO laser [resolution achieved = 0.25 cm^{-1} (FWHM)]. The rotational transitions are marked as $^{\Delta K_a' \Delta K_c'} \Delta N'_{K_a''}(N'')$ on top of the spectrum, where $\Delta K_a' = K_a'' - K_a'$ and $\Delta K_c' = K_c'' - K_c'$ with values of −1 and 1 are denoted as p and r, respectively. The assigned peaks are marked by solid dots on top of the peaks, while the extra peaks originated from other vibrational bands are marked by asterisks below the peaks [21,25,30].

and $\Delta K_c' = \pm1$ [8]. Thus, the rotational structures are quite different from those of the ν_{11} and $\nu_2 + \nu_{12}$ bands. The IR–VUV–PI band of ν_9 consists of three overlapping sub-bands, which are classified by $\Delta K_a' = 0 \leftarrow 1$, $\Delta K_a' = 1 \leftarrow 0$, and $\Delta K_a' = 2 \leftarrow 1$ transitions. Each of these sub-bands comprises of the P-, Q-, and R-branches. The comparison of Figure 7a and b reveals that many of the rotational peaks resolved in the Figure 7b correspond to single rotational transitions. This observation can be ascribed to the relatively large separations of the K_a' levels. The rotational assignments marked on top of Figure 7a are designated as $^{\Delta Ka\, \Delta Kc} \Delta N_{Ka''}(N'')$, where ΔK_a and ΔK_c with values of −1 and 1 denoted as p and r, respectively. All rotational lines that are assigned are marked by a black dot on top of the peaks. Some extra lines observed in Figure 7a that are not predicted by the spectral simulation of the ν_9 band are marked by asterisks under the peaks. These peaks are found to be equally spaced by ≈ 1.7 cm^{-1}, indicating the existence of other A-type vibrational bands in this energy region.

3.2.3.   IR–VUV–PI measurements of C_3H_4

We have also measured the IR–VUV–PI spectra for the C—H stretching vibration bands of C_3H_4, including ν_1 (acetylenic C—H stretch), ν_2 (symmetric CH$_3$ stretch), and ν_6 (anti-symmetric CH_3 stretch) [26,28,31]. The IR–VUV–PI spectra for the ν_1 and ν_6 bands obtained using the high-resolution IR-OPO laser are shown in Figures 8a and 9a, respectively, for comparison with the corresponding spectra depicted in Figures 8b and 9b observed using the low-resolution IR-OPO laser. The ν_1 band is a parallel transition band, the rotational assignments of which are marked on top of Figure 8a. The simulation of the IR–VUV–PI spectrum of the ν_1 band indicates a rotational temperature of ≈ 8 K for the supersonically cooled C_3H_4 sample. At such a rotational temperature, C_3H_4 molecules mostly populate in the lowest $K'' = 0$ and 1 states with nearly equal intensities. Since the rotational constant for the ground state is very close to that for the $\nu_1 = 1$ excited state, the $K' \leftarrow K'' = 0 \leftarrow 0$ and $1 \leftarrow 1$ transitions cannot be resolved in these spectra.

Figure 9a and b depicts the ν_6 band obtained using the single-mode IR-OPO and the low-resolution IR-OPO laser system, respectively. The ν_6 band is a perpendicular band governed by the $\Delta J = 0, \pm 1$ and $\Delta K = \pm 1$ selection rules, and is shown in the high-resolution spectrum of Figure 9a to consist of five sub-bands. Each sub-band consists of a prominent $Q(\Delta J = 0)$ and the $P(\Delta J = -1)$ and $R(\Delta J = 1)$ branches. In comparison, only three Q bands are discernible in the low-resolution spectrum of Figure 9b. Furthermore, the $P(\Delta J = -1)$ and $R(\Delta J = 1)$

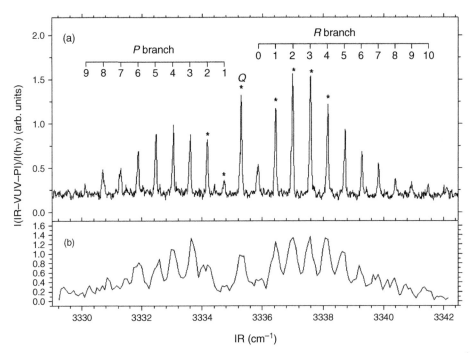

Figure 8 IR–VUV–PI spectra for the $C_3H_4(\nu_1 = 1)$ band in the IR region of 3329–3342 cm^{-1} obtained by using (a) the single-mode IR-OPO laser [resolution achieved $= 0.02$ cm^{-1} (FWHM)] and (b) the lower resolution IR-OPO laser [resolution achieved $= 0.25$ cm^{-1} (FWHM)] The transitions marked with (*) are selected for IR–VUV–PFI–PE measurements [28,30].

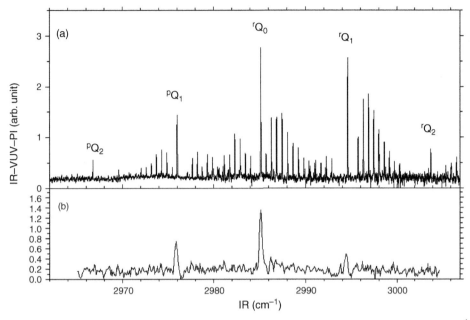

Figure 9 IR–VUV–PI spectra of $C_3H_4(v_6 = 1)$ band in the IR range of 2965–3005 cm^{-1} recorded by using (a) the single-mode IR-OPO laser [resolution achieved = 0.02 cm^{-1} (FWHM)] and (b) the low-resolution IR-OPO laser [resolution achieved = 0.25 cm^{-1} (FWHM)] [30].

branches for the sub-bands resolved in Figure 9a are indiscernible in Figure 9b. This observation again indicates that the increase in IR resolution by using the high-resolution IR-OPO laser actually enhances the sensitivity of IR–VUV–PI measurements.

3.3. IR–VUV–PFI–PE measurements

The IR–VUV–PI measurements presented above are required for IR–VUV–PFI–PE measurements to be presented below. The higher sensitivity in IR–VUV–PI measurements can translate directly into improved sensitivity for IR–VUV–PFI–PE studies.

3.3.1. IR–VUV–PFI–PE measurements of CH_3I

Using the low-resolution IR-OPO laser, we have recorded the IR–VUV–PFI–PE spectra for the $CH_3I(v_1 = 1)$ band with selected rotational levels, $J' = 1, 2, 5, 7$, and 10 [20]. In the latter experiment, because $K' = 0$ and 1 are not resolved in the IR–VUV–PI spectrum, individual J' transitions consist at least two transitions from the $K' = 0$ and 1 states. Using the high-resolution IR-OPO laser, we have obtained the IR–VUV–PFI–PE spectra for the $CH_3I(v_1 = 1)$ band with (J', K') prepared at (3, 0), (4, 0), (5, 0), (6, 0), (7, 0), and (8, 0) as shown in Figure 10a–f, respectively. The spectra for (4, 1), (5, 1), and (6, 1) have also been measured (not shown here) [29]. Almost identical rotational structures are observed for $K' = 0$ and $K' = 1$,

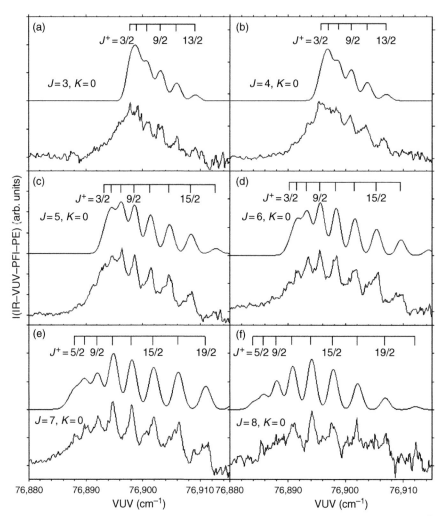

Figure 10 (J', K') selected IR–VUV–PFI–PE spectra (lower curves) for $CH_3I^+(\tilde{X}^2E_{3/2}, \nu_1^+ = 1, J^+, P^+) \leftarrow CH_3I(\tilde{X}^1A_1, \nu_1 = 1, J', K')$ obtained by setting the single-mode IR-OPO laser at the transitions: (a) $J' = 3$, $K' = 0$; (b) $J' = 4$, $K' = 0$; (c) $J' = 5$, $K' = 0$; (d) $J' = 6$, $K' = 0$; (e) $J' = 7$, $K' = 0$; (f) $J' = 8$, $K' = 0$ resolved in Figure 3(a). The simulated spectra are shown as the upper curves [28].

except for a minor difference in their intensities. The observation of identical rotational structures for $K' = 0$ and 1 is consistent with the fact that the photoionization transitions, $P^+ = \frac{1}{2} \leftarrow K' = 0$ and $P^+ = -1/2, 3/2 \leftarrow K' = 1$, overlap with each other in the spectra, where $P^+ = K^+ + 1/2$. The similar intensities observed for the $K' = 0$ and 1 transitions indicate that the photoionization cross sections are only weakly dependent on K' [29].

The simulation of the rovibrationally selected and resolved IR–VUV–PFI–PE spectra has provided a precise IE value of $76,896.9 \pm 0.2\ cm^{-1}$ for the photoionization transition, $CH_3I^+\left(\tilde{X}^2E_{3/2}\,;\nu_1^+\right)\leftarrow CH_3I(\tilde{X}^1A_1;\nu_1)$. Assuming that the IR–VUV–PI intensities observed are proportional to the IR absorption strengths, we have determined the relative state-to-state $(\nu_1^+,J^+)\leftarrow(\nu_1,J)$ photoionization cross sections (σ_{rel}) by normalizing the relative intensities for the J^+ levels of the ion with the corresponding populations of the J levels observed in the IR–VUV–PI spectrum. The σ_{rel} values thus determined were found to depend strongly on the $|\Delta J^+|=|J^+-J|$, but not on J and K. This trend can be clearly seen by plotting σ_{rel} as a function of ΔJ^+ in Figure 11, showing that for a given J, the σ_{rel} peaks at the smallest $|\Delta J^+|$ value and falls off rapidly as $|\Delta J^+|$ is increased. It is interesting that the maximum $|\Delta J^+|$ of up to 5.5 was observed. Furthermore, for transitions starting from different J values, the profiles for σ_{rel} versus ΔJ^+ are nearly superimposable after taking into account the experimental uncertainties. The σ_{rel} values of the negative ΔJ^+ branches are slightly higher than those of the positive branches. This observation is consistent with the mechanism of forced autoionization [43]. The selection rules for the photoionization transitions are not restricted by $\Delta J^+=0,\pm1$ as for the bound-to-bound transitions. Instead, the photoionization rules are expressed by $|\Delta J^+|\le l+s+1$, where l and s are the orbital and spin angular momenta of the Rydberg electron, respectively. For the photoionization transition of $CH_3I^+(\tilde{X}^2E_{3/2})\leftarrow CH_3I(\tilde{X}^1A_1)$, the selection rule can be written as $|\Delta J^+|\le l+3/2$ and $|\Delta K^+|=|P^+-K|\le|\Delta J^+|$, where $P^+=K^++1/2$. Because the highest occupied 2e orbital of CH_3I has mostly p character, the Rydberg states excited by a VUV photon are expected to compose predominantly of the ns ($l=0$) and nd ($l=2$) characters in accordance with atomic-like selection rules, which should give rise to the maximum $|\Delta J^+|=3.5$ according to the photoionization selection rules. The observation of larger changes in $|\Delta J^+|$ up to 5.5 for $J=7$ and 10

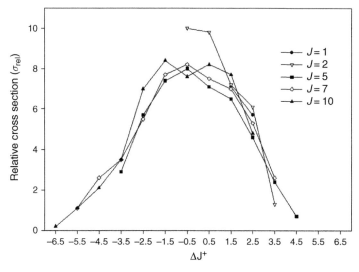

Figure 11 Relative cross sections (σ_{rel}) for the state-to-state photoionization transitions $CH_3I^+(\tilde{X}^2E_{3/2};\nu_1^+;J^+)\leftarrow CH_3I(\tilde{X}^1A_1;\nu_1;J)$ with selected $J=1,2,5,7,$ and 10 plotted as a function of $\Delta J^+=J^+-J$ [20,29].

signifies that the highest l value for the Rydberg electron produced is up to 4. The l-mixing mechanism for the Rydberg electrons induced by stray electric field may give rise to higher l values, which in turn will result in larger ΔJ^+ changes as observed in the IR–VUV–PFI–PE spectrum. The observed l-distribution for the photoelectron and the rotational distribution of the resulting CH_3I^+ cation contain interesting dynamical information on the photoionization process.

3.3.2. IR–VUV–PFI–PE measurements of C_2H_4

In the IR–VUV–PFI–PE study using the low-resolution IR-OPO laser, the $P(1)$, $P(2)$, $R(1)$, and $R(2)$ lines of the ν_{11} band [see Figure 5b] were excited for the preparation of the respective $N' = 0$, 1, 2, and 3 rotational levels of $C_2H_4(\nu_{11})$ [20,25,27]. The IR–VUV–PFI–PE spectra [shown as the lower curves in Panels (a1), (a2), and (a3) of Figure 12, respectively] for the $\nu_{11} \rightarrow \nu_2^+ + \nu_{12}^+$ band thus obtained consist of transitions from three rotational levels, i.e., the 1_{01} level via $2_{02} \rightarrow 1_{01}$ [$P_{101}(2)$], the 1_{10} level via $2_{12} \rightarrow 1_{10}$ [$P_{110}(2)$], and the 1_{11} level via $2_{12} \rightarrow 1_{11}$ [$P_{111}(2)$] for $P(2)$; the 2_{02} level via $1_{01} \rightarrow 2_{02}$ [$R_{202}(1)$], the 2_{11} level via $1_{10} \rightarrow 2_{11}$ [$R_{211}(1)$], and the 2_{12} level via $1_{11} \rightarrow 2_{12}$ [$R_{212}(1)$] for $R(1)$; and the 3_{03} via $2_{02} \rightarrow 3_{03}$ [$R_{303}(2)$], the 3_{12} level via $2_{11} \rightarrow 3_{12}$ [$R_{312}(2)$], and the 3_{13} level via $2_{12} \rightarrow 3_{13}$ [$R_{313}(2)$] for $R(2)$. Using the high-resolution single-mode IR-OPO laser, each of these rotational levels can be distinctly selected for VUV–PFI–PE measurements [27,31]. The IR–VUV–PFI–PE spectra obtained using the single-mode IR-OPO laser with $C_2H_4(\nu_{11})$ prepared in individual $N'_{Ka'Kc'} = (1_{01}, 1_{10}$, and $1_{11})$, $(2_{02}, 2_{11}$, and $2_{12})$, and $(3_{03}, 3_{12}$, and $3_{13})$ rotational levels are plotted as the lower curves in Panels [(b1), (c1), and (d1)], [(b2), (c2), and (d2)], and [(b3), [(c3), and (d3)] of Figure 12, respectively. The upper spectra marked as "sim" shown in Panels (b1)–(d1), (b2)–(d2), and (b3)–(d3) of Figure 12 are simulated spectra. The assignments of rotational transitions are marked on top of these figures. In order to show the relative intensities for the rotational transitions, the vertical scales of these spectra have the same I(PFI–PE)/I(hv) units, where I(PFI–PE) and I(hv) represent the PFI–PE intensity and VUV intensity, respectively. As expected, the structures of the lower resolution IR–VUV–PFI–PE spectra (lower curves) of Panels (a1), (a2), and (a3) are different from those of individual high-resolution spectra of Panels (b1)–(d1), (b2)–(d2), and (b3)–(d3), respectively, but have contributions from the corresponding high-resolution spectra. The upper curves of Panels (a1), (a2), and (a3) are synthesized spectra obtained by adding the $P_{101}(2)$, $P_{110}(2)$, and $P_{111}(2)$ spectra of Panels (b1)–(d1), the $R_{202}(1)$, $R_{211}(1)$, and $R_{212}(1)$ spectra of Panels (b2)–(d2), and the $R_{303}(2)$, $R_{312}(2)$, and $R_{313}(2)$ spectra of Panels (b3)–(d3), respectively. As shown in Panels (a1), (a2), and (a3), these synthesized spectra are in excellent agreement with the low-resolution [4,8] IR–VUV–PFI–PE spectra. Furthermore, the synthesized spectra are shown to have significantly better signal-to-noise ratios than the low-resolution [21,25] IR–VUV–PFI–PE spectra.

The IR–VUV–PFI–PE spectra for the ionization transitions $C_2H_4(\tilde{X}^1 A_g; \nu_{11} = 1, N'_{Ka'Kc'} = 3_{03}) \rightarrow C_2H_4^+(\tilde{X}^2 B_{3u}; \nu_i^+, N^+_{Ka^+Kc^+})$ have also been measured in a broad VUV frequency range [25,27]. The rovibrational level of $(\nu_{11}, N'_{Ka'Kc'} = 3_{03})$ was selected by parking the single-mode IR-OPO laser frequency at the transition

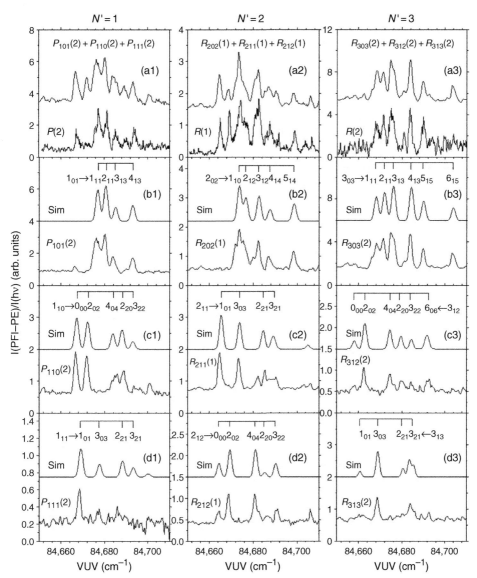

Figure 12 IR−VUV−PFI−PE spectra of the $C_2H_4(\tilde{X}^1A_g; \nu_{11}; N'_{Ka'Kc'}) \rightarrow C_2H_4^+(\tilde{X}^2B_{3u}; \nu_2^+ + \nu_{12}^+; N^+_{Ka+Kc+})$ transition in the VUV frequency range of 84,650–84,730 cm^{-1}. $N'_{Ka''Kc''} = 1_{01}, 1_{10}, 1_{11}, 2_{02}, 2_{11}, 2_{12}, 3_{03}, 3_{12}$, and 3_{13} are selected by fixing IR at different transitions resolved in the IR−VUV−PI spectrum of ν_{11}. The experimental spectra were shown as the lower curves in a1−d1, a2−d2, a3−d3, while the simulations (sim) and assignments are shown on top of these figures. The PFI−PE intensities [I(PFI−PE)] have been normalized by the VUV intensity [I(hν)] and have the same units for all figures. The upper curves of (a1), (a2), and (a3) are the synthesized spectra obtained by adding the $P_{101}(2)$, $P_{110}(2)$, and $P_{111}(2)$ spectra of (b1)–(d1), the $R_{202}(1)$, $R_{211}(1)$, and $R_{212}(1)$ spectra of (b2)–(d2), and the $R_{303}(2)$, $R_{312}(2)$], and $R_{313}(2)$ spectra of (b3)–(d3), respectively. The lower curves represent the IR−VUV−PFI−PE spectra obtained previously by using the low-resolution IR-OPO laser [27].

$2_{02} \rightarrow 3_{03}$ of $R(2)$ of the spectrum of Figure 5a. A total of 24 vibrational bands are identified in the IR–VUV–PFI–PE spectra thus obtained [27], whereas only the 7 strongest bands have been observed in the previous IR–VUV–PFI–PE study using the lower resolution IR-OPO laser [25], illustrating that the sensitivity of IR–VUV–PFI–PE measurements using the high-resolution single-mode IR-OPO laser is enhanced compared with that using the low-resolution IR-OPO laser. Rovibrational transitions of the $\nu_2 + \nu_{12}(b_{1u})$, and $\nu_9(b_{2u})$ bands of C_2H_4 have also been prepared by IR excitation for VUV—PFI–PE measurements (not shown here).

As the simplest hydrocarbon with an C=C double bond, C_2H_4 and its cation $C_2H_4^+$ represent a model system for detailed experimental and theoretical investigations. While all vibrational modes of the neutral have been accurately measured experimentally, only 4 $[\nu_2^+(a_g)$ (C—C stretching) $= 1487.7(4)$ cm^{-1}, $\nu_3^+(a_g)$ (CH$_2$ bending) $= 1258.7(4)$ cm^{-1}, $\nu_4^+(a_u)$ (C=C torsion) $= 83.7(4)$ cm^{-1}, and $\nu_7^+(b_{3u})$ (CH$_2$ wagging) $= 901.3(8)$ cm$^{-1}]$ of the 12 vibrational modes have been experimentally determined by the single-photon VUV–PFI–PE method [25,27,44]. We have performed *ab initio* anharmonic (AH) vibrational frequency calculations at different levels of theory [25]. Although nearly all the AH vibrational frequency predictions for $C_2H_4/C_2H_4^+$ calculated at the CCSD(T)/aug-cc-pVQZ level of theory are found to agree with known experimental vibrational frequencies to better than 1 percent, the predicted AH frequency for the ν_4^+ torsional mode is nearly four-fold greater than the experimental value. This indicates that it remains necessary for accurate determination of the fundamental frequencies for $C_2H_4^+$. Based on the IR–VUV–PFI–PE measurements of $C_2H_4(\nu_{11})$ and $C_2H_4(\nu_2 + \nu_{12})$, we have obtained the fundamental frequencies $\nu_8^+(b_{2g}) = 1103 \pm 10$ cm^{-1}, $\nu_{10}^+(b_{2u}) = 813 \pm 10$ cm^{-1}, $\nu_{11}^+(b_{1u}) = 2978.7(2)$ cm^{-1}, and $\nu_{12}^+(b_{1u}) = 1411.7(2)$ cm^{-1} [27]. Including the 4 fundamental frequencies determined previously by VUV–PFI–PE measurements, 8 of the 12 fundamental frequencies of $C_2H_4^+$ are now known.

3.3.3. IR–VUV–PFI–PE measurements of C_3H_4

The IR–VUV–PFI–PE spectra for the $C_3H_4(\nu_1, J_K) \rightarrow C_3H_4^+(\nu_1^+, J^+{}_{K+})$ photoionization band have been measured by setting the single-mode IR-OPO laser at individual rotational transitions $P(1)$, $P(2)$, $R(1)$, $R(2)$, $R(3)$, and $R(4)$ resolved in the IR–VUV–PI spectrum of Figure 8a, which correspond to the respective rotational levels $J' = 0$, 1, 2, 3, 4, and 5 of the $\nu_1 = 1$ state [28]. The IR–VUV–PFI–PE spectra thus obtained are depicted in Figures 13a–f (lower curves). As pointed out above, each J'-transition line has about equal contributions of transitions to the $K' = 0$ and 1 levels, except for $P(1)$, which can only have the transition to the $K' = 0$ level.

The spectrum of Figure 13a, which is obtained by selecting only one rotation level $J'_{K'} = 0_0$ of the C_3H_4 $(\nu_1 = 1)$ state, shows two prominent peaks followed by two small peaks, which can be accounted for by two sets of the $K^+ = 1$, 2, and 3 stacks associated with the $E_{1/2}$ and $E_{3/2}$ spin–orbit components of $C_3H_4^+$. That is, this spectrum allows the determination of the spin–orbit constant A $= -13.0 \pm 0.2$ cm^{-1} for the $C_3H_4^+$ cation [28]. Using this A value, we have successfully simulated the same band at different selected J' levels ($J' = 1 - 5$) as shown in Figure 13b–f [28]. In these cases, for each J' value, both $K' = 0$ and 1 can contribute. Although the contributions

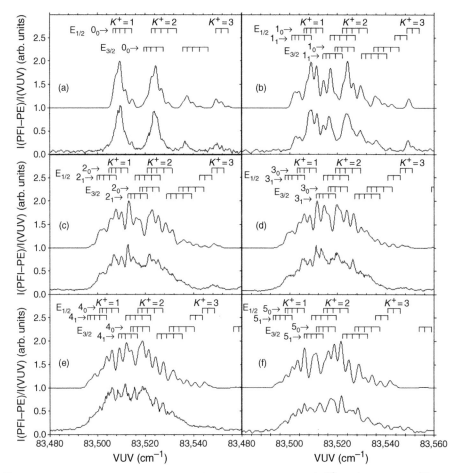

Figure 13 IR–VUV–PFI–PE spectra (lower curves) for $C_3H_4^+(\tilde{X}^2E; v_1^+) \leftarrow C_3H_4(\tilde{X}^1A_1; v_1)$ band with the IR laser frequency setting at (a) $P(1)$, (b) $P(2)$, (c) $R(1)$, (d) $R(2)$, (e) $R(3)$, and (f) $R(4)$ of the IR–VUV–PI spectrum of Figure 8(a). The upper curves represent the simulated spectra obtained using a spin–orbit constant $A = -13.0 \pm 0.2$ cm^{-1}. The rotational assignments $(J_K \rightarrow N^+{}_{K+})$ are marked on top of the spectra [28].

of these two rotational states have blurred the band heads of individual K^+ stacks, the spectra of Figure 13b–f can still be simulated satisfactorily with nearly all the individual rotational transitions marked on top of the spectra identified. All simulations of Figures 13a–f yield the same band origin of 83,501.0 ± 0.2 cm^{-1} [28], corresponding to the photoionization transition energy from $C_3H_4(\tilde{X}^1A_1; v_1 = 1)$ to $C_3H_4^+(\tilde{X}^2E_{1/2}; v_1^+ = 1)$. Knowing the relative intensities of $J' = 0$–5 levels formed by IR excitations as measured in Figure 8a, relative rotationally selected and resolved state-to-state cross sections for photoionization transitions originating from $J' = 0$–5 with $K' = 0$ and 1 can be determined from the simulations of the spectra of Figure 13a–f [20,28].

Assuming that the A constant and state-to-state photoionization cross sections determined from the rotational analysis of the $v_1^+ = 1 \leftarrow v_1 = 1$ transitions are identical to those of the origin band for $C_3H_4^+ (\tilde{X}^2 E_{3/2,1/2})$, we have simulated the VUV–PFI–PE origin band (lower curve) of Figure 2a[28]. We used all the relative state-to-state photoionization cross sections of $J' = 1$–5 and $K' = 0$ and 1, which were determined by fitting the spectra of Figure 13a–f. The overall simulated spectrum and its deconvoluted contributions from the $E_{1/2}$ and $E_{3/2}$ sub-bands for $C_3H_4^+$ are depicted as curves (i), (ii), and (iii), respectively, in Figure 2b. Although ignoring the contributions from high J'''s results in a narrower distribution of the overall simulated spectrum compared with the VUV–PFI–PE origin band, the excellent agreement observed between the simulated and experimental spectra is a compelling support of the simulating procedures used. The origins or IE values were determined to be $83,619.0 \pm 1.0$ and $83,632.0 \pm 1.0\,\mathrm{cm}^{-1}$ for the formation of the respective $E_{1/2}$ and $E_{3/2}$ spin–orbit states. Knowing the vibrational frequency of $C_3H_4(v_1)$ and the origin of the vibrational band $C_3H_4^+ (\tilde{X}^2 E_{1/2}; v_1^+ = 1) \leftarrow C_3H_4 (\tilde{X}^1 A_1; v_1 = 1)$, the vibrational frequency of $C_3H_4(v_1^+)$ is determined to be $v_1^+ = 3217.1 \pm 0.2\,\mathrm{cm}^{-1}$ [28]. Selected rotational levels of the $C_3H_4(v_2$ and $v_6)$ states have also been prepared by IR excitations for VUV–PFI–PE measurements.

3.4.   IR–VUV–PIE and IR–VUV–PIE depletion measurements

The IR–VUV–PIE and the IR–VUV–PIE depletion measurements at VUV energies above the IE of the molecule of interest can be made simultaneously because both schemes involve fixing the IR laser at a given rotational transition of a neutral vibrational band based on the IR–VUV–PI measurement, while scanning the VUV laser frequency and setting the boxcar to collect the ions produced at 30 Hz with the "toggle mode" [24]. Because the VUV beam is fired at 30 Hz and the IR is set at 15 Hz, the toggle mode allows shot-to-shot corrections of VUV-only photoion background by taking the differences of two adjacent ion signals as the net signals for IR–VUV–PIE and IR–VUV–PIE depletion measurements. The net signals for IR–VUV–PIE measurements are positive signals, whereas the IR–VUV–PIE depletion peaks have negative intensities.

3.4.1.   IR–VUV–PIE depletion measurements of NH_3

We have applied the IR–VUV–PIE depletion method on the Rydberg series of NH_3 [24]. In order to apply the IR–VUV–PIE and IR–VUV–PIE depletion methods, we first measured the IR–VUV–PI spectrum for the v_1 band of NH_3 in the IR range of 3300–3390 cm^{-1}, revealing seven rotational transitions from $NH_3(v_1 = 0; J'', K'')$ to $NH_3(v_1 = 1; J', K')$ vibrational state. On the basis of the assignment of this IR–VUV–PI spectrum, the IR-OPO laser can be used to prepare NH_3 in a known rotational level of $NH_3(v_1 = 1)$, i.e., $J' = 0$ via $P(1)$ $[0_0(a) \leftarrow 1_0(s)]$ at 3317.2 cm^{-1}, $J' = 1$ via $R(0)$ $[1_0(s) \leftarrow 0_0(a)]$ at 3355.0 cm^{-1}, or $J' = 2$ via both $R(1)$ $[2_0(a) \leftarrow 1_0(s)]$ and $R(1)$ $[2_1(a) \leftarrow 1_1(s)]$ at 3376.3 cm^{-1}, prior to VUV–PIE or VUV–PFI–PE measurements.

Figure 14 depicts the VUV–PIE spectrum (upper spectrum) of NH_3 in the VUV range of $82,100$–$83,150\,cm^{-1}$ obtained using a supersonically cooled NH_3 sample. This spectrum reveals complex autoionizing Rydberg structures and is consistent with the high-resolution VUV–PIE measurement reported previously by Merkt and co-workers [45]. The major autoionization features have been identified as members of two nd Rydberg series. The stronger Rydberg series, nd $(v_2^+ = 1,\ 1_0 \leftarrow 0_0)$ (quantum defect $\delta = 0.036$), arises from the rotational transition $1_0 \leftarrow 0_0$ and has a converging limit of NH_3^+ $(\tilde{X};\ v_2^+ = 1,\ 1_0)$ at $83,082.2\,cm^{-1}$. The weaker series, $nd(v_2^+ = 1,\ 2_1 \leftarrow 1_0)$ ($\delta = 0.055$), results from the rotational transition $2_1 \leftarrow 1_0$ and has a converging limit of NH_3^+ $(\tilde{X},\ v_2^+ = 1,\ 2_1)$ at $83,102.3\,cm^{-1}$.

Using the "toggle" mode for NH_3^+ ion detection and setting the IR laser at the $R(0)$ $[1_0(s) \leftarrow 0_0(a)]$ $(J' = 1)$ transition, we have obtained the IR–VUV–PIE and IR–VUV–PIE depletion spectrum for NH_3^+ in the VUV range of $82,050$–$83,250\,cm^{-1}$ (see lower spectrum of Figure 14). The positive IR–VUV–PIE signal corresponds to NH_3^+ formed from the intermediate state NH_3 $[\tilde{X};\ v_1 = 1,\ 1_0(s)]$ prepared by IR excitation, whereas the negative IR–VUV–PIE or IR–VUV–PIE depletion signal arises from depletion of the NH_3^+ ion intensity that result from photoionization of NH_3 $[\tilde{X};\ v_1 = 0,\ 0_0(a)]$. In other words, the observation of the IR–VUV–PIE depletion signal indicates that the autoionizing Rydberg resonance has a contribution from neutral $NH_3(\tilde{X};\ v_1 = 0)$ populating the $0_0(a)$ rotational level. As shown in Figure 14, the IR–VUV–PI depletion spectrum exhibits a series of negative peaks, the positions of which match members of the $nd(v_2^+ = 1,\ 1_0 \leftarrow 0_0)$ Rydberg series as marked in the upper VUV–PIE spectrum of NH_3. This observation indicates that this Rydberg series originates from the $0_0(a)$ rotational level of $NH_3(\tilde{X};\ v_1 = 0)$.

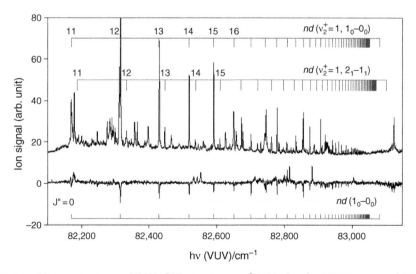

Figure 14 Upper spectrum: VUV–PIE spectrum of NH_3 in the VUV range of $82,100$–$83,150\,cm^{-1}$. Lower spectrum: IR–VUV–PIE and IR–VUV–PIE depletion spectra of NH_3^+ observed by setting the IR laser at $3355.0\,cm^{-1}$, which corresponds to the $R(0)$ $[1_0(s) \leftarrow 0_0(a)]$ $(J' = 1)$ transition [24].

As expected, the pattern of positive going PIE peaks resolved in the lower IR–VUV–PIE spectrum of Figure 14 is significantly simplified compared with that observed in the upper VUV–PIE spectrum. However, the nature of these IR–VUV–PIE peaks is not known. Considering that the energy range for the IR–VUV–PIE spectrum of Figure 14 corresponds to the total energy (the sum of IR, VUV, and rotational energies of the initial state of NH_3) range of 85,450–86,600 cm^{-1}, these IR–VUV–PIE peaks mostly belong to members of the nd Rydberg series converging to $\nu_2^+ = 4$ and 5 states of NH_3^+ because the ionization thresholds for these states are 85,965.6 and 86,979.5 cm^{-1}, respectively.

3.5. IR–VUV–PFI–PE depletion measurements

The IR–VUV–PFI–PE depletion measurement involves setting the VUV laser frequency at the IE of the molecule of interest (or the central peak of the VUV–PFI–PE origin band), while looking for depletion in the PFI–PE signal as the IR–OPO laser frequency is scanned [19,32–34]. With the VUV frequency set at the IE, the VUV–PFI–PE intensity is expected to depend on the population of the molecule in the ground state. Thus, the IR excitation of the molecule out of the ground state to an excited vibrational state would give rise to a dip in the VUV–PFI–PE signal.

Although the radiative lifetime for a vibrationally excited state is expected to be longer than μs, the lifetime for intramolecular vibrational redistribution (IVR) processes can be faster than ns [46], which may prevent the application of the IR–VUV–PI method. Because both the IR–VUV–PI and IR–VUV–PFI–PE depletion methods provide information about the IR absorption spectroscopy of neutral molecules, the IR–VUV–PFI–PE method can serve as an alterative of the IR–VUV–PI method, particularly in the case of fast IVRs.

3.5.1. IR–VUV–PFI–PE depletion measurements of trichloroethene

The application of the IR–VUV–PFI–PE depletion method was first demonstrated in the measurements of the ν_{12} and ν_{11} C—H stretching vibrational bands of trichloroethene [19] (CHCl=CCl$_2$) and cis-dichloroethene [33] (cis-CHCl=CHCl), respectively. In both experiments, the vibrational bands observed by the IR–VUV–PFI–PE depletion method are found to be essentially identical to the corresponding bands measured by the IR–VUV–PI detection scheme.

As an example, we compare the IR–VUV–PI spectrum (lower spectrum) and the IR–VUV–PFI–PE depletion spectrum (upper spectrum) for CHCl=CCl$_2$ in Figure 15 [19]. The IR–VUV–PI spectrum was obtained by measuring the CHCl=CCl$_2^+$ ion intensity by scanning the IR frequency while holding the VUV frequency fixed at 76,176 cm^{-1} [266 cm^{-1} below the IE(CHCl=CCl$_2$)]. This spectrum reveals a peak at 3,097.7 cm^{-1} with a FWHM of 5 cm^{-1}, which is consistent with the ν_{12} (C—H stretching) frequency for CHCl=CCl$_2$. The IR–VUV–PFI–PE-depletion spectrum was obtained by measuring the PFI–PE intensity as a function of the IR laser frequency with the VUV energy fixed at the IE(CHCl=CCl$_2$). Because the PFI–PE intensity depends on the population of CHCl=CCl$_2$ in the ground state, the excitation of CHCl=CCl$_2$ to $\nu_{12} = 1$

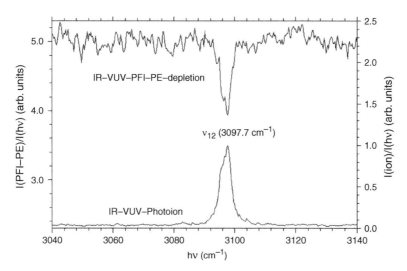

Figure 15 Comparison of the IR–VUV–PFI–PE-depletion spectrum of CHCl=CCl$_2$ with the IR–VUV–PI spectrum in the IR energy range of 3040–3140 cm^{-1} with the VUV frequency fixed at 76,176 cm^{-1}. The dip and peak of the respective depletion and photoion spectra at 3097.7 cm^{-1} is assigned as excitation to CHCl=CCl$_2$(ν_{12} = 1) [19].

manifests as a depletion at 3,097.7 cm^{-1} in the PFI–PE spectrum. As shown in Figure 15, the dip observed in the IR–VUV–PFI–PE depletion spectrum appears as the mirror image of the peak resolved in the IR–VUV–PI spectrum. The FWHM observed for the IR–VUV–PFI–PE spectrum is essentially the same as that for the IR–VUV–PI spectrum. In addition to identifying the ν_{12} frequency, the IR–VUV–PFI–PE-depletion spectrum shows that $\approx$25 percent of the CHCl=CCl$_2$ molecules in the supersonic beam are excited to ν_{12} = 1 [19]. Because the ion background due to VUV photoionization is negligible in the IR–VUV–PI detection, its sensitivity is higher than that in the IR–VUV–PFI–PE-depletion method. Similar double-resonance schemes for IR spectroscopic measurements using UV laser photoionization based on nonresonant two-photon ionization or resonant enhanced 1+1 ionization ion dip detection have been reported in the literature [47,48]. Compared with the resonant enhanced 1+1 UV–PI method, the single-photon VUV–PI method does not require the existence of a well-characterized, long-lived electronically excited intermediate state, and thus is a general method, applicable to all molecules. For experiments that concern the mechanisms for the formation of photoions in specific internal states as in this study, the IR–VUV and IR–UV photoionization schemes are certainly different because they are governed by the selection rules of two- and three-photon excitations, respectively.

3.5.2. IR–VUV–PFI–PE depletion measurements of C$_2$H$_4$

Because the VUV–PFI–PE origin bands and the IR–VUV–PI and IR–VUV–PFI–PE depletion spectra of trichloroethene and *cis*-dichloroethene observed in these experiments are not rotationally resolved [19,33], the difference between the

IR–VUV–PI and IR–VUV–PFI–PE depletion methods has not been revealed. We have performed a rotationally resolved IR–VUV–PFI–PE depletion study of the $\nu_9(b_{2u})$ and $\nu_{11}(b_{1u})$ CH$_2$ vibrational bands of C$_2$H$_4$ [34]. The comparison of the rotationally resolved VUV–PFI–PE, IR–VUV–PFI–PE depletion, and IR–VUV–PI spectra of C$_2$H$_4$ has revealed the different state-specific natures of IR–VUV–PFI–PE depletion and IR–VUV–PI measurements for IR spectroscopic studies of neutral species.

Figure 16a depicts the VUV–PFI–PE origin band for C$_2$H$_4^+$ in the region of 84,760–84,840 cm^{-1} obtained with a PFI–PE resolution of 1.5 cm^{-1} (FWHM) [21,25]. The rotational structures resolved in the VUV–PFI–PE origin band have been well analyzed, and their assignments are marked on top of the spectrum of Figure 16a. The intensity of the strongest peak of this spectrum is predominantly contributed by the photoionization rotational transition $0_{00} \rightarrow 3_{12}$. By setting the VUV laser frequency at this transition, we have obtained the IR–VUV–PFI–PE depletion spectrum [shown as the blue spectrum in Figure 16b] in the IR region of 2977–3000 cm^{-1}, which is found to exhibit a single sharp dip at 2990.5 cm^{-1} [34]. This IR region covers the IR–VUV–PI spectrum of the C$_2$H$_4(\nu_{11})$ band (bottom spectrum in Figure 16b) obtained with the low-resolution IR–OPO laser. As shown in Figure 16b, the IR–VUV–PFI–PE depletion spectrum is very different from and simpler than the IR–VUV–PI spectrum [25,34]. The dip of the IR–VUV–PFI–PE depletion spectrum is found to coincide with the $R(0)$ transition line of the ν_{11} band. This observation is consistent with the fact that the $R(0)$ line corresponds to the $0_{00} \rightarrow 1_{01}$ transition. As the molecules are excited out of the 0_{00} neutral ground level, the depletion of the $0_{00} \rightarrow 3_{12}$ photoionization transition is to be expected, and thus, results in a dip in the IR–VUV–PFI–PE depletion spectrum of C$_2$H$_4(\nu_{11})$ at the $R(0)$ or $0_{00} \rightarrow 1_{01}$ transition.

Figure 16c compares the IR–VUV–PFI–PE depletion spectrum (upper blue spectrum) [34] and the IR–VUV–PI spectrum (lower spectrum) [25] for the C$_2$H$_4(\nu_9)$ band in the region of 3093–3132 cm^{-1}. This IR–VUV–PFI–PE depletion spectrum is also measured by setting the VUV laser frequency at the strongest peak (or the $0_{00} \rightarrow 3_{12}$ photoionization rotational transition) of the VUV–PFI–PE spectrum shown in Figure 16a. Similar to the comparison of IR–VUV–PFI–PE depletion and IR–VUV–PI spectrum of the ν_{11} band in Figure 16b, the IR–VUV–PFI–PE depletion spectrum of the ν_9 band is also found to exhibit only one dip and is significantly simpler than the IR–VUV–PI spectrum of C$_2$H$_4(\nu_9)$. As expected, the dip observed in the depletion spectrum at 2992 cm^{-1} is found to coincide with the $0_{00} \rightarrow 1_{11}$ rotational transition peak (marked in Figure 16b) of the IR–VUV–PI spectrum of the ν_9 band.

To further test our understanding on the information content of rotationally resolved IR–VUV–PFI–PE depletion measurements, we have also obtained the IR–VUV–PFI–PE depletion spectrum for C$_2$H$_4(\nu_{11})$ in the region of 2988–3000 cm^{-1} (shown as the red spectrum in Figure 16b) by setting the VUV laser frequency at the second strongest peak of the VUV–PFI–PE spectrum. This VUV–PFI–PE peak has contributions from two photoionization rotational transitions, $1_{01} \rightarrow 1_{11}$ and $2_{02} \rightarrow 2_{12}$. Thus, we expect to observe depletion in VUV–PFI–PE signal at the $R(1)$ and $R(2)$ lines. This expectation is consistent with the

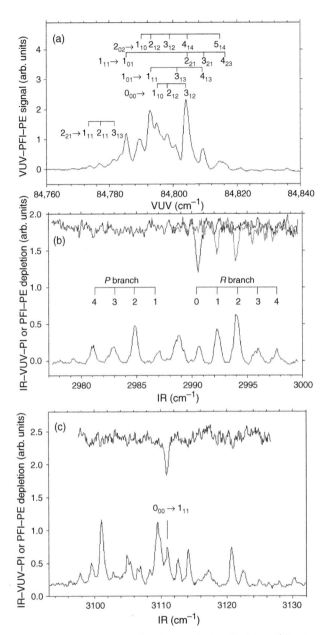

Figure 16 (a) VUV–PFI–PE spectrum for the origin band of $C_2H_4^+$ in the region of 94,760–84,840 cm^{-1}. (a) Comparison of the IR–VUV–PI spectrum for the ν_{11} band of C_2H_4 obtained using the low-resolution IR-OPO laser with the IR–VUV–PFI–PE depletion spectra for C_2H_4 in the region of 2977–3000 cm^{-1}. The IR–VUV–PFI–PE spectra marked in blue and red are obtained by setting the VUV laser at the strongest and the second strongest peaks of the VUV–PFI–PE spectrum of (a), which correspond to the photoionization transitions of $0_{00} \rightarrow 3_{12}$ and $1_{01} \rightarrow 1_{11}$ and $2_{02} \rightarrow 2_{12}$, respectively. (c) Comparison of the IR–VUV–PI spectrum for the ν_9 band of C_2H_4 obtained using the low-resolution IR-OPO laser with the IR-VUV-PFI-PE depletion spectrum (blue curve) for C_2H_4 in the region of 3096–3126 cm^{-1} measured by setting the VUV laser frequency at the strongest peak of the VUV–PFI–PE spectrum of (a), which corresponds to the photoionization transition $0_{00} \rightarrow 3_{12}$ [34]. (See color plate 36).

experimental IR–VUV–PFI–PE depletion spectrum, showing two dips at the $R(1)$ and $R(2)$ lines of the $C_2H_4(\nu_{11})$ band. Assuming that the rotational assignment of the VUV–PFI–PE peak selected for IR–VUV–PFI–PE depletion measurement is not known, the observed IR–VUV–PFI–PE depletion spectrum can also provide information about the rotational character of the photoionization transition of the selected VUV–PFI–PE peak.

3.6. VUV–IR–PIRI measurements

The concept of PIRI spectroscopy is similar to that of the ion spectroscopic technique based on photo-induced dissociation of cluster ions [32,33]. The latter technique begins with attaching a rare-gas atom or H_2 to the molecular ion of interest [49]. The loss of the weakly interacting rare-gas atom or H_2 due to photoexcitation of the ion is used to probe the energy levels of the molecular ion. In the VUV–IR–PIRI measurement, an IR photon is used to eject the Rydberg electron of a high-n Rydberg state prepared by VUV excitation in order to probe the spectroscopy of the ion core. The VUV laser is set at a frequency below the IE of the molecule of interest. The firing of the IR laser is delayed by 20–50 ns with respect to the VUV laser pulse. After a delay of $\approx$150 ns, a pulsed electric field of 6 V/cm (width = 6 μs) is switched on at the photoionization region to extract the PIRI ions toward the ion detector. The VUV–IR–PIRI spectrum is recorded by measuring the PIRI ion signal as a function of the IR frequency.

3.6.1. VUV–IR–PIRI measurements of *trans*-dichloroethene

We have successfully employed the VUV–IR–PIRI method for the measurements of the IR-active C—H stretching vibrational frequencies of several polyatomic ions [31–33]. As an example, we show the VUV–IR–PIRI spectrum of *trans*-ClCH=CHCl$^+$ in Figure 17 [31]. The *trans*-ClCH=CHCl$^+$ cation has two C—H stretching modes, $\nu_{11}{}^+(b_u)$ and $\nu_{12}{}^+(a_g)$. From the IR selection rule, only the $\nu_{11}{}^+(b_u)$ mode is IR-active. The $\nu_{12}{}^+(a_g)$ mode is allowed in VUV photoionization, but its intensity is extremely low because of the limited Franck–Condon factor (FCF). Based on our FCF calculations, the predicted intensity of the $\nu_{12}{}^+$ mode is negligible compared with that of the VUV–PFI–PE origin band. The IR–PIRI curves obtained with the VUV frequency fixed at 77,134.8 and 77,596.6 cm^{-1} are plotted as curves (i) and (ii) in Figure 17 respectively. Knowing the IE(*trans*-ClCH=CHCl) value from VUV–PFI–PE measurements, the corresponding effective n^*-values of the Rydberg states excited in curves (i) and (ii) are calculated as $n^* = 14$ and 36, respectively Although the n^*-value are different, the VUV–IR–PIRI peaks appear at the same position centered at 3068 cm^{-1}, corresponding to the $\nu_{11}{}^+(b_u)$ vibrational mode of *trans*-ClCH=CHCl$^+$. Similar phenomenon is also observed in the VUV–IR–PIRI spectra of other molecules [32–34], where the positions of the VUV–IR–PIRI bands are independent of n^*.

If autoionization follows vibrational autoionization and follows the $\Delta\nu = -1$ propensity rule in the VUV–IR–PIRI measurement of *trans*-dichloroethene, the resulting ion should be in the ground vibrational state. It should be interesting to determine the final ion vibrational state distribution by measurement of the kinetic

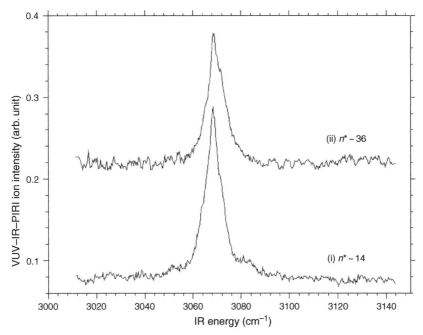

Figure 17 VUV–IR–PIRI spectra of *trans*-ClCH=CHCl in the region of 3010–3150 cm^{-1}. (i) VUV energy fixed at 77,134.8 cm^{-1}; (ii) VUV energy fixed at 77,596.6 cm^{-1} [31].

energies of the photoelectrons. The independence of n^* for the VUV–IR–PIRI peak frequency suggests that the Rydberg electron behaves as a spectator during the excitation process and has little effect on the vibrational structure of the ion core.

The VUV–IR–PIRI experiments show that better signal-to-noise ratios are observed for VUV–IR–PIRI spectra obtained via lower n^*-value ($n^* < 50$) [32,33]. These n^*-values involved are significantly lower than those ($n^* > 100$) involved in PFI–PE spectroscopy. In vibrational autoionization, the low-n Rydberg states are known to have higher autoionization cross-section because the Rydberg electron is better coupled to the ion core. On the contrary, there is a higher probability that the low-n Rydberg states decay before the autoionization. That it, the low-n Rydberg states have a shorter lifetime. In order for these short-lived low-n Rydberg states to be detected in the PIRI autoionization process, the orbital angular momentum quantum number l involved must be high. However, from the optical selection rules, the l-value of the low-n Rydberg states initially prepared by VUV excitation should be low. This suggests that subsequent l-mixing processes may occur to promote the l-value, which is probably induced by stray electric fields presented in the photoionization region [50].

3.6.2. IR–VUV–PIRI measurements of *cis*-dichloroethene

Employing the VUV–IR–PIRI method, we have measured the IR vibrational spectra for the *cis*-ClCH=CHCl(n^*) excited Rydberg states with the effective principal

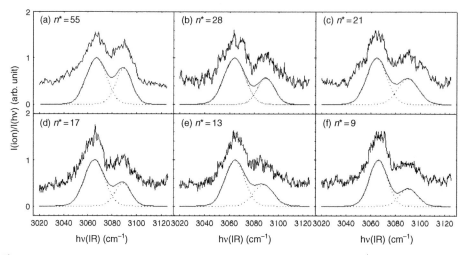

Figure 18 VUV–IR–PIRI spectra (upper traces) for *cis*-ClCH=CHCl$^+$ via (a) $n^* = 55$, (b) $n^* = 28$, (c) $n^* = 21$, (d) $n^* = 17$, (e) $n^* = 13$, and (f) $n^* = 9$ respectively, where n^* is the effective principal quantum number. The corresponding lower traces represent the Gaussian fits. The maximum intensities for all the experimental and simulated spectra have been normalized to unity [33].

quantum numbers $n^* = 9$, 13, 17, 21, 28, and 55, which are displayed as the upper spectra in Figures 18a–f, respectively [33]. These spectra reveal two overlapping IR absorption bands, the energies of which are expected to be similar to the IR absorption bands of the *cis*-ClCH=CHCl$^+$ cation. Guided by high-level *ab initio* calculations of the *cis*-ClCH=CHCl$^+$, we have assigned the bands centered at 3067 and 3090 cm^{-1} to be the respective ν_{11}^*(asymmetric C—H stretching) and ν_{12}^*(symmetric CH stretching) vibrational bands of *cis*-ClCH=CHCl(n^*). The lower spectra of Figure 18a–f represent the Gaussian fits to the two vibrational bands observed in the experimental spectra. The most interesting observation in this experiment is that although the IR frequencies observed for the ν_{11}^* and ν_{12}^* vibrational bands are essentially unchanged for different n^* states, suggesting that the IR absorption predominantly involves the ion core and that the Rydberg electron behaves as a spectator, the intensity ratio for the ν_{11}^* and ν_{12}^* bands $[R(\nu_{11}/\nu_{12})]$ is found to decrease as n^* is increased, from 2.7 at $n^* = 9$ to 1.3 at $n^* = 56$. This observation indicates that the IR absorption spectra of Rydberg states with different n^* are distinguishable. This trend is consistent with a model *ab initio* quantum calculation of $R(\nu_{11}^*/\nu_{12}^*)$ for excited *cis*-ClCH=CHCl in $n^* = 3$–18 states and the calculations of $R(\nu_{11}/\nu_{12}) = 57$ and $R(\nu_{11}^+/\nu_{12}^+) = 1.4$, where $R(\nu_{11}/\nu_{12})[R(\nu_{11}^+/\nu_{12}^+)]$ represents the intensity ratio of the ν_{11} (ν_{11}^+) asymmetric to the ν_{12} (ν_{12}^+) symmetric C—H stretching vibrational bands for *cis*-ClCH=CHCl (*cis*-ClCH=CHCl$^+$).

We have also measured the IR–VUV–PI and IR–VUV–PFI–PE-depletion spectra for *cis*-ClCH=CHCl [33]. These spectra are consistent with *ab initio* calculations, indicating that IR absorption cross section for the ν_{12} band is negligibly small compared with that for the ν_{11} band. While the VUV–IR–PIRI measurements have allowed the determination of $\nu_{11}^+ = 3067 \pm 2$ cm^{-1},

$v_{12}{}^{+} = 3090 \pm 2\ \mathrm{cm}^{-1}$, and $R(v_{11}{}^{+}/v_{12}{}^{+}) \approx 1.3$ for cis-ClCH$=$CHCl^{+}, the IR–VUV–PI and IR–VUV–PFI–PED measurements have provided the value $v_{11} = 3088.5 \pm 0.2\ \mathrm{cm}^{-1}$ for cis-ClCH$=$CHCl.

4. SUMMARY AND FUTURE PROSPECTS

Using a broadly tunable single-mode IR-OPO laser and a comprehensive tunable VUV laser, together with the molecular beam and the PFI detection techniques, we have demonstrated an array of new photoion–photoelectron methods. The IR–VUV–PI method is shown to be a generally applicable, highly sensitive method for IR spectroscopy studies of neutral molecules. The high-resolution and enhanced sensitivity of the IR–VUV–PI method achieved by using the single-mode IR-OPO laser translates directly into the high resolution and sensitivity for IR–VUV–PFI–PE measurements of cations. The rovibrationally selected and resolved state-to-state PFI–PE spectra thus obtained by the IR–VUV–PFI–PE method have made possible definitive assignments of photoelectron spectra of the test molecules. The sensitivity of the experiments demonstrated here shows that there are good prospects for the application of the IR–VUV–PI and IR–VUV–PFI–PE methods for spectroscopy studies of radicals and clusters and their cations.

In the IR–VUV–PI demonstration experiments presented here, we have focused on the IR measurements of the C—H stretching vibrational bands of several simple polyatomic molecules at around 3 μm. However, this method should not be limited to the study of the C—H vibrational modes. The current tunable range of the single-mode IR-OPO laser is $\approx$1.35–5.0 μm. A wavelength extension unit based on the silver–gallium–selenide crystal (AgGaSe2) can extend the IR wavelength down to 16 μm (625 cm^{-1}). The use of this extension unit would allow spectroscopy studies of chemically interesting vibrational modes such as C—C stretching modes, which are usually in the range of $\approx$900 cm^{-1}. The IR–VUV–PI experiments of this study indicate that the IR–VUV–PI method has the sensitivity for IR spectroscopy studies of radicals and clusters. The expansion of the tuning range of the IR-OPO laser is also expected to increase the ability of the IR–VUV–PFI–PE method to cover more vibrational bands of the cations.

While the IR–VUV–PI and IR–VUV–PFI–PE methods have been well proven to be generally useful high-resolution techniques for spectroscopic studies of neutrals and ions, high-resolution VUV–IR–PIRI measurements have not been made. Further studies to understand the fundamental IR–PIRI mechanism are required before the VUV–IR–PIRI method can be used for high-resolution spectroscopic studies.

ACKNOWLEDGMENTS

This work is supported by the NSF Grant No. CHE 0517871. The authors also acknowledge partial supports by the DOE Contract No. DE-FG02-02ER15306, the AFOSR Grant No. FA9550-06-1-0073, and the NASA Grant No.

07-PATM07-0012. The calculations were performed at the National Energy Research Scientific Computing Center and the Molecular Science Computing Facility supported by DOE.

REFERENCES

[1] J. Berkowitz, Photoabsorption, Photoionization, and Photoelectron Spectroscopy, Academic Press, New York, 1979.
[2] C. Y. Ng (Ed.), Vacuum Ultraviolet Photoionization and Photodissociation of Molecules and Clusters, World Scientific, Singapore, 1991.
[3] C.Y. Ng, Annu. Rev. Phys. Chem., 53: 101–140 ((2002)).
[4] I. Powis, T. Baer, and C. Y. Ng (Eds.), High Resolution Laser Photoioniztion and Photoelectron Studies, Wiley Series in Ion Chem. & Phys., Wiley, Chichester, 1995.
[5] E.W. Schlag, ZEKE Spectroscopy, Cambridge University Press, Cambridge, 1998.
[6] K. Müller-Dethlefs and E.W. Schlag, Annu. Rev. Phys. Chem., 42: 109–136 (1991).
[7] R.E. Smalley, B.L. Ramakrishna, D.H. Levy, and L. Wharton, J. Chem. Phys., 61: 4363 (1974).
[8] C.-W. Hsu and C.Y. Ng, J. Chem. Phys., 101: 5596 (1994).
[9] Y.-S. Cheung and C.Y. Ng, Int. J. Mass Spectrom Ion Processes., 185/186/187: 533 (1999).
[10] Y.-S. Cheung, J.-C. Huang, and C.Y. Ng, J. Chem. Phys., 109: 1781 (1998).
[11] H.K. Woo, J. Zhan, K.-C. Lau, C.Y. Ng, and Y.-S. Cheung, J. Chem. Phys., 116: 8803 (2002).
[12] H.K. Woo, K.-C. Lau, J.-P. Zhan, C.Y. Ng, Y.-S. Cheung, W.K. Li, and P.M. Johnson, J. Chem. Phys., 119: 7789 (2003).
[13] H.K. Woo, P. Wang, K.-C. Lau, X. Xing, and C.Y. Ng, J. Chem. Phys., 120: 9561 (2004).
[14] J. Yang, Y.X. Mo, K.C. Lau, Y. Song, X.M. Qian, and C.Y. Ng, Phys. Chem. Chem. Phys., 7: 1518 (2005).
[15] K.C. Lau, H.K. Woo, P. Wang, X. Xing, and C.Y. Ng, J. Chem. Phys., 124: 224311 (2006).
[16] X. Xing, B. Reed, K.-C. Lau, C.Y. Ng, X. Zhang, and G.B. Ellison, J. Chem. Phys., 126: 171101 (2007).
[17] X.M. Qian, T. Zhang, C.Y. Ng, A.H. Kung, and M. Ahmed, Rev. Sci. Instrum., 74: 2784–2890 (2003).
[18] X.-M. Qian, A.H. Kung, T. Zhang, K.C. Lau, and C.Y. Ng, Phys. Rev. Lett., 91: 233001 (2003).
[19] H.K. Woo, P. Wang, K.-C. Lau, X. Xing, C. Chang, and C.Y. Ng, J. Chem. Phys., 119: 9333 (2003).
[20] P. Wang, X. Xing, K.-C. Lau, H.K. Woo, and C.Y. Ng, J. Chem. Phys., 121: 7049 (2004).
[21] P. Wang, X. Xing, S.J. Baek, and C.Y. Ng, J. Phys. Chem., A, 108: 10035 (2004).
[22] C.Y. Ng, J. Electron Spectrosc. Relat. Phenom., 142: 179 (2005).
[23] M.-K. Bahng, X. Xing, S.J. Baek, and C.Y. Ng, J. Chem. Phys., 123: 084311 (2005).
[24] M.-K. Bahng, X. Xing, S.J. Baek, X.-M. Qian, and C.Y. Ng, J. Phys. Chem. A, 110: 8488 (2006).
[25] Xi Xing, M.-K. Bahng, P. Wang, K.C. Lau, S.-J. Baek, and C.Y. Ng, J. Chem. Phys., 125: 133304 (2006).
[26] X. Xing, B. Reed, K.-C. Lau, S.-J. Baek, M.-K. Bahng, and C.Y. Ng, J. Chem. Phys., 127: 044313 (2007).
[27] X. Xing, B. Reed, M.-K. Bahng, and C. Y. Ng, J. Phys. Chem. A, 112: 2572 (2008).
[28] X. Xing, M.-K. Bahng, B. Reed, C. S. Lam, K.-C. Lau, and C. Y. Ng, J. Chem. Phys., 128: 094311 (2008).
[29] X. Xing, B. Reed, P. Wang, M.-K. Bahng, S. J. Baek, P. Wang, and C. Y. Ng, J. Chem. Phys., 128: 104306 (2008).
[30] X. Xing, B. Reed, M.-K. Bahng, P. Wang, H.-K. Woo, S.-J. Baek, C. S. Lam, and C. Y. Ng, Chinese J. Chem. Phys., (in press).
[31] H.-K. Woo, P. Wang, K.c. Lau, and C.Y. Ng, J. Chem. Phys., 108: 9637 (2004).

[32] H.K. Woo, P. Wang, K.C. Lau, X. Xing, and C.Y. Ng, J. Chem. Phys., 120: 1756 (2004).
[33] P. Wang, H.K. Woo, K.C. Lau, X. Xing, C.Y. Ng, A.S. Zyubin, and A.M. Mebel, J. Chem. Phys., 124: 064310 (2006).
[34] X. Xing, B. Reed, M.-K. Bahng, S.-J. Baek, P. Wang, H.-K. Woo, C. S. Lam, and C. Y. Ng, Chem. Phys. Lett., 455: 321(2008).
[35] T.L. Tan, S.Y. Lau, P.P. Ong, K.L. Goh, and H.H. Teo, J. Mol. Spectrosc., 203: 310 (2000).
[36] P.M. Johnson, Adv. Ser. Phys. Chem., 10A: 296 (1999), and references therein.
[37] W. Cooke, T. Gallagher, S. Edelstein, and R. Hill, Phys. Rev. Lett., 40: 178 (1978).
[38] D.P. Talor, J.G. Goode, J.E. LeClaire, and P.M. Johnson, J. Chem. Phys., 103: 6293 (1995).
[39] J.D. Hofstein and P.M. Johnson, Chem. Phys. Lett., 316: 229 (2000).
[40] A. Fujii, A. Iwasaki, T. Ebata, and N. Mikami, J. Phys. Chem. A, 101: 5963 (1997).
[41] M. Honda, A. Fujii, E. Fujimaki, T. Ebata, and N. Mikami, J. Phys. Chem. A, 107: 3678 (2003).
[42] M. Gerhards, M. Schiwek, C. Unterberg, and K. Kleinermanns, Chem. Phys. Lett., 297: 515 (1998).
[43] W.R.S. Garton, W.H. Parkinson, and E.M. Reeves, Proc. Phys. Soc. A, 8: 860 (1962).
[44] S. Willitsch, U. Hollenstein, and F. Merkt, J. Chem. Phys., 120: 1761 (2004).
[45] R. Seiler, U. Hollenstein, T.P. Softley, and F. Merkt, J. Chem. Phys., 118: 10024 (2003).
[46] J. Keske, D.A. McWhorter, and B.H. Pate, Int. Rev. Phys. Chem., 19: 363 (2000).
[47] T. Ebata, A. Fujii, and N. Mikami, Int. Rev. Phys. Chem., 17: 331 (1998).
[48] M. Schmitt, Ch. Jacoby, M. Gerhards, C. Unterberg, W. Roth, and K. Kleinermanns, J. Chem. Phys., 113: 2995 (2000).
[49] J.M. Lisy, J. Chem. Phys., 125: 132302 (2006).
[50] W.A. Chupka, J. Chem. Phys., 98: 4520 (1993).

INDEX

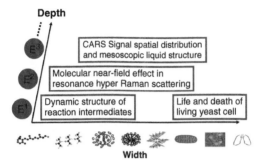

Plate 1 Width and depth of Raman spectroscopy and the four topics discussed this article. (See Figure 1 in p. 14 of Chapter 2).

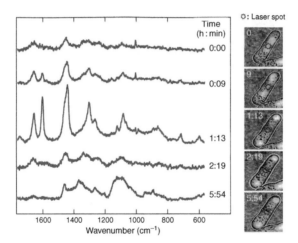

Plate 2 Time- and space-resolved Raman spectra of a dividing *Schizosaccharomyces pombe* cell. (See Figure 9 in p. 24 of Chapter 2).

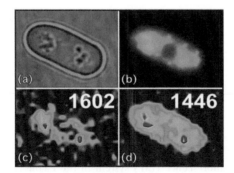

Plate 3 Optical image (a), GFP image, (b) and Raman images (c and d) of an *Schizosaccharomyces pombe* cell whose mitochondria are GFP labeled. (See Figure 10 in p. 25 of Chapter 2).

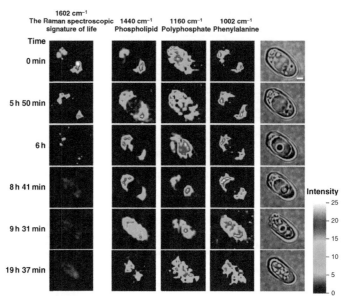

Plate 4 Time-resolved Raman mapping of a spontaneous cell death process of a staving *S. cerevisiae* cell. (See Figure 12 in p. 27 of Chapter 2).

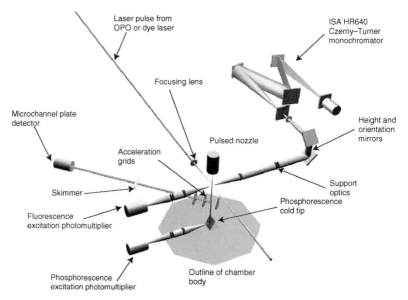

Plate 5 Laser-induced fluorescence (LIF) system in the Laane laboratories. (See Figure 8 in p. 74 of Chapter 4).

Plate 6 Jarrell Ash far-infrared spectrometer at MIT in the 1960s. (See Figure 9 in p. 75 of Chapter 4).

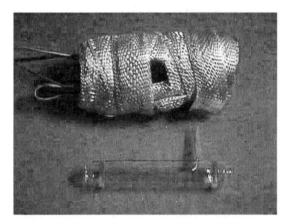

Plate 7 High-temperature Raman cell for vapors. (See Figure 11 in p. 76 of Chapter 4).

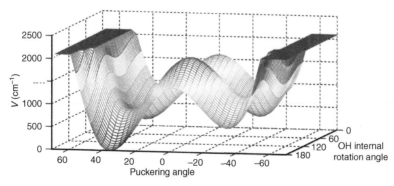

Plate 8 Calculated potential energy surface of 2-indanol in terms of its ring-puckering angle (degrees) on OH internal rotation (degrees relative to 180° at the A conformation). (See Figure 79 in p. 125 of Chapter 4).

Plate 9 Stinging insect inclusion in a Baltic amber matrix, ~30 My old. (See Figure 5 in p. 139 of Chapter 5).

(a)

(b)

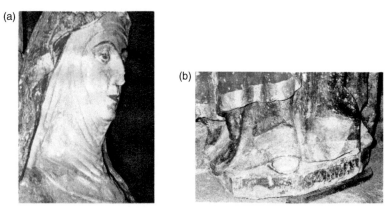

Plate 10 Statue of Santa Ana, monastery of Real Santo Domingo de Silos, Castille y Leon, Spain; pigments on marble, dating from the late 13th century. The Raman spectrum of the golden yellow hem of the cape of this statue indicated that it was not gold leaf as expected (which does not possess a Raman spectrum) but rather *mosaic gold*, a pigment that apparently did not exist in Europe in the 13th century, but was well known in China at that time. (See Figure 6 in p. 141 of Chapter 5).

(a) (b)

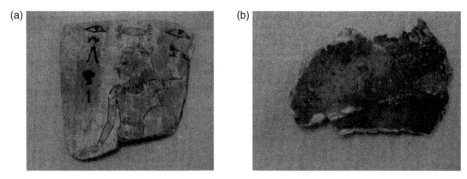

Plate 11 Fragment of Egyptian sarcophagus, Graeco-Roman period, showing several pigments and a human figure with hieroglyphics; the Raman spectrum of the yellow wash analysed as a degraded terpenoid resin of the *Pinus/Pistacia* variety, confirmed by GCMS analysis to be a degraded Pistacia resin, not native to Egypt, which has some implications of possible novel trade routes of interest to art historians. (See Figure 8 in p. 144 of Chapter 5).

Plate 12 PalaeoIndian rock art, Pecos culture, Seminole Canyon, Rio Grande–Devil River region, Texas/New Mexico; ca. 3000 BC. (See Figure 9 in p. 146 of Chapter 5).

Plate 13 The 14th century fresco of *Christ in Majesty* in central nave of the church of SS Damien and Cosmo at Basconcillos del Tozo, Castille y Leon, Spain. Source of religious hierarchical use of pigments detected by Raman spectroscopy. (See Figure 10 in p. 147 of Chapter 5).

Plate 14 Sarcophagi from the 12th dynasty *Tomb of the Two Brothers*, Der Rifeh, Egypt, ca. 2100 BC, excavated by Sir William Flinders Petrie in 1905; the tomb with the mummies of Nekht-ankh and Khnum-nakht was intact on discovery and contained items of funerary furniture, some of which are seen in the photograph. The sarcophagi and mummies are now in the Manchester Museum. (See Figure 18 in p. 154 of Chapter 5).

Plate 15 The *Mary Rose*, flagship of King Henry VIII's navy, sunk during an engagement with a French naval force in the English Channel in 1545. From the Antony Roll in the Pepys Library, Magdalene College, University of Cambridge; reproduced with the permission of the Master and Fellows of Magdalene College, Cambridge. (See Figure 24 in p. 158 of Chapter 5).

Plate 16 Painting by J.M.W. Turner of *HMS Victory* breaking the line and attacking the Franco-Spanish fleet at Trafalgar, 21 October 1805; commissioned by King George IV in 1825 for St. James' Palace and now in the National Maritime Museum at Greenwich. The foretopmast with foretopsail attached is shown being brought down early in the engagement by gunfire from the *Neptune* as the *Victory* is engaging the *Bucentaure* and *Redoubtable*. Reproduction by courtesy of the National Maritime Museum, Greenwich. (See Figure 28 in p. 162 of Chapter 5).

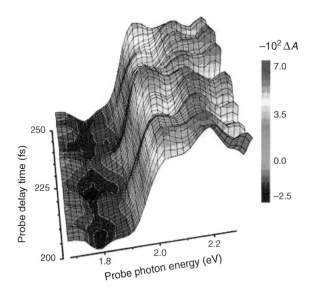

Plate 17 Time evolution (200–250 fs after excitation) of the sequential difference absorption spectra of polydiacetylene-3-butoxycarbonylmethylurethane (PDA-3BCMU) with resolution of ≤ 1 fs (vertical axis) and ≤ 10 meV (horizontal axis) obtained by the pump–probe measurement. Positive and negative difference absorbances are caused by the induced absorption from $2^1 A_g$ state and the absorption saturation in $1^1 A_g$ state, respectively. (See Figure 12 in p. 196 of Chapter 6).

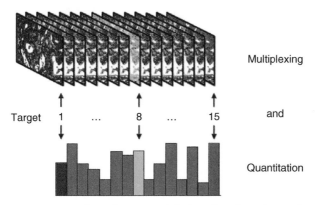

Histology	PSA	SERS spectrum
Epithelium	(+)	
Stroma	(−)	
Lumen	(−)	

Plate 18 Proof of principle for SERS microscopy [31], demonstrated by the localization of prostate-specific antigen (PSA) in biopsies from patients with prostate cancer. Prostate tissue (left) contains different histological classes in which PSA is either abundant (+) or not (−) (right). SERS-labelled anti-PSA antibodies are detected in the PSA-(+) epithelium (right), showing the characteristic Raman signals of the SERS label. Locations in the PSA-(−) stroma and lumen serve as negative controls at which no spectral contributions of the SERS-labelled antibody are detected (right). Tissue autofluorescence in the stroma is observed upon 632.8 nm excitation, while in the lumen only glass substrate contributions are detected (right). (See Figure 16 in p. 286 of Chapter 9).

Multiplexing

Target 1 ... 8 ... 15   and

Quantitation

Plate 19 SERS microscopy has the potential for the detection of multiple targets (multiplexing) and target quantitation, two important aspects for tissue diagnostics. (See Figure 17 in p. 286 of Chapter 9).

(a)   (b)

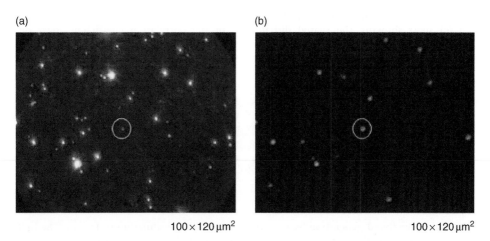

100 × 120 µm²   100 × 120 µm²

Plate 20 Plasmon resonance Rayleigh scattering (a) and surface-enhanced Raman scattering (SERS) spectroscopy (b) images of Ag nanoaggregates dispersed on a glass plate. White circles in (c) and (d) correspond to areas selected by a pinhole shown in Figure 1. (See Figure 2 in p. 294 of Chapter 10).

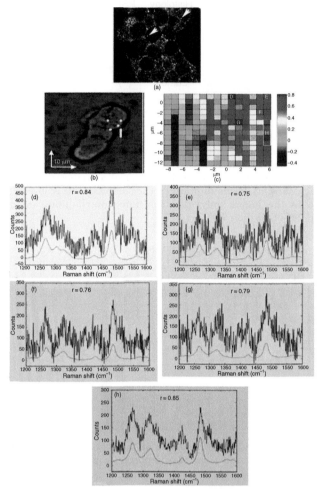

Plate 21 (a) Gold nanoparticle adsorbed by PC12 cell. (b) Cells showing the measured area and (c) spots showing the spatial distribution of the correlation coefficients over one particular cell (dashed rectangle). Corresponding surface-enhanced Raman scattering (SERS) spectra and their similarities with neurotransmitters is also given. [61]. (See Figure 11 in p. 306 of Chapter 10).

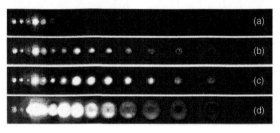

Plate 22 Broadband generation by vibrational molecular modulation in D_2 gas. The light beam emerging from the output of the molecular gas cell is dispersed by a prism and projected onto a screen. (a) P = 71 Torr and $\Delta\omega = -400\,$MHz, (b) P = 71 torr and $\Delta\omega = 100\,$MHz, (c) P = 71 torr and $\Delta\omega = 700\,$MHz, and (d) P = 350 torr and $\Delta\omega = 700\,$MHz. At low pressures, the Raman sidebands are emitted collinearly, with all sidebands having similar divergences. An increase in pressure results in a divergent structure for the higher order beams (reprinted from Ref. [57]). (See Figure 1 in p. 351 of Chapter 12).

Plate 23 Broadband generation by rotational molecular modulation in H_2 gas, driven by a pair of nanosecond lasers with frequencies equal to integer multiples of their frequency difference. The gas pressure is 270 torr and the temperature is 77 K. (See Figure 2 in p. 352 of Chapter 12).

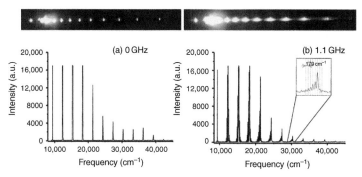

Plate 24 The spectrum generated in D_2 at two different vibrational Raman detunings. (a) A picture of a pure vibrational dispersed spectrum taken with a digital camera and the same spectrum recorded by the spectrometer at a zero detuning from vibrational Raman resonance (adapted from Refs [81] and [67]). (b) A ro-vibrational dispersed spectrum taken with a digital camera and the same spectrum recorded by the spectrometer at a 1.1 GHz detuning from vibrational Raman resonance. The 807–nm laser energy is 8 mJ/pulse, whereas the energy of the 1064–nm laser is 180 mJ/pulse, and the D_2 pressure is 300 Torr (adapted from Ref. [81]). (See Figure 7 in p. 357 of Chapter 12).

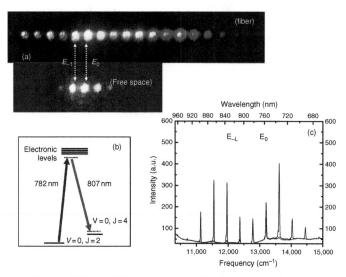

Plate 25 A pair of lasers (807 and 782 nm) drive the $v'' = 0, J'' = 2 \to v' = 0, J' = 4$ transition. The frequency difference of the driving lasers is slightly detuned from the Raman resonance to allow adiabatic preparation of molecules in a superposition state. (a) A comparison of the dispersed spectrum generated in a hollow fiber filled with D_2 gas and (b) in free space at a pressure of 704 torr. (c) Some of the generated sideband frequencies as measured by a spectrometer. The driving fields are indicated by E_{-1} and E_0. (See Figure 10 in p. 361 of Chapter 12).

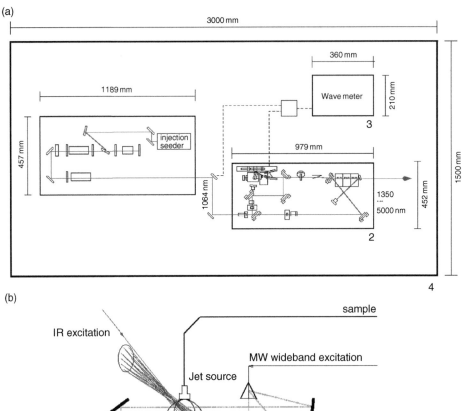

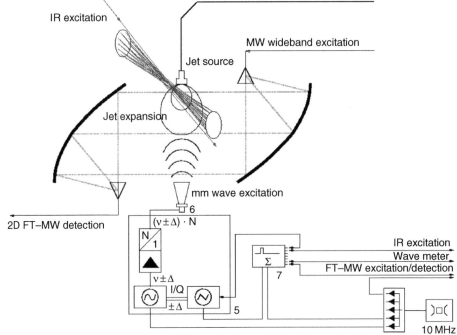

Plate 26 Supersonic-jet expansion FP 2D FTMW IR/mm wave-TR experiments: (a) tunable mid-IR-pulsed laser source: (1) Q-switched Nd : YAG laser, injection-seeded, 1064 nm, 1000 mJ, 20 Hz, (2) single-mode KTP/KTA OPO/OPA, 1350–5000 nm, ~10 mJ, 20 Hz, $\Delta\lambda$ <0.02 cm^{-1}, (3) pulsed wavemeter, $\Delta\lambda$ <0.02 cm^{-1}, (4) optical table; (b) supersonic-jet TR excitation cell; (5) mm wave source, 60–325 GHz, (6) mm wave horn antennae, 60–325 GHz, >24 dBi, (7) PXIe-based experiment control system. For components not numbered, refer to Figure 36 and respective caption. *Source*: Figure 6 of Ref. [103]. (See Figure 37 in p. 450 of Chapter 14).

2	3	4	5	6	7	8	9	10	11	12	13
H_2	H_2O	NH_3	SiH_4	CH_3OH	CH_3COH	CH_3CO_2H	CH_3CH_2OH	CH_3COCH_3			
OH	H_2S	H_3O^+	CH_4	NH_2CHO	CH_3NH_2	HCO_2CH_3	$(CH_3)_2O$	$CH_3(CC)_2CN$			
SO	SO_2	H_2CO	$CHOOH$	CH_3CN	$CH_3(CC)H$	$CH_3(CC)CN$	CH_3CH_2CN	$(CH_2OH)_2$			
SO^+	HN_2^+	H_2CS	$H(CC)CN$	CH_3NC	CH_2CHCN	C_7H	$H(CC)_3CN$	CH_3CH_2CHO			
SiO	HNO	$HNCO$	CH_2NH	CH_3SH	$H(CC)CN$	H_2C_6	$H(CC)_2CH_3$				
SiS	NNO	$HNCS$	NH_2CN	C_5H	C_6H	CH_2OHCHO	C_8H		**11**		
FeO	NH_2	CH_3	CH_2CN	$HCCCCN$	$l\text{-}H_2C_2HOH$	$H(CC)_3H$	$CH_2=CHCH_3$				
NO	H_3^+	$CCCN$	H_2CCO	$H(CC)CHO$	$c\text{-}CH_2OCH_2$	$CH_2=CHCHO$	$H(CC)_4^-$				
NS	$SiNC$	HCO_2^+	C_4H	$CH_2=CH_2$	$H(CC)_3^-$	$CH_2=CCHCN$			$H(CC)_4CN$		
HCl	HCO	$CCCH$	$c\text{-}C_3H_2$	H_2CCCC		$(CH_3)_2$			$CH_3(CC)_3H$		
$NaCl$	HCO^+	$c\text{-}C_3H$	C_5	HC_3NH^+							
KCl	OCS	$CCCO$	SiC_4	C_5N							
$AlCl$	CCH	$CCCS$	H_2CO^+	H_2CCNH							
AlF	HCS^+	$H(CC)H$	$H(CC)NC$	$c\text{-}C_3H_2O$						**12**	
PN	$c\text{-}SiCC$	$HCNH^+$	$HNCCC$	$H(CC)_2H$						C_6H_6	
SiN	CCO	$HCCN$	$H(CC)_2^-$								
NH	CO_2	H_2CN									
SH	$AlNC$	$c\text{-}SiC_3$									**13**
HF	$SiCN$										$H(CC)_5CN$
N_2	CCS										
LiH	C_3										
PO	$MgNC$										
CO	$NaCN$										
CS	CH_2										
C_2	$MgCN$										
SiC	HOC^+										
CP	HCN										
CO^+	HNC										
CH	HCP										

Plate 27 Molecules identified in the interstellar gas and in circumstellar shells, ranked by number of atoms. (From Ref. [337].) (See Figure 14 in p. 502 of Chapter 15).

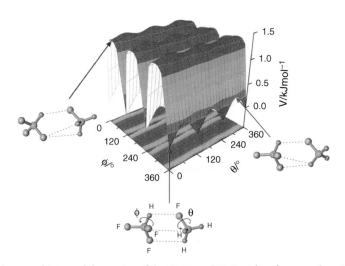

Plate 28 Shape and internal dynamics of the CHF_3—CH_3F molecular complex. (Reproduced with permission from Ref. [537]. Copyright 2005, Wiley-VCH Verlag GmbH & Co. KGaA.) (See Figure 17 in p. 521 of Chapter 15).

Plate 29   Shape of the cluster OCS—He$_8$ [613]. (Reprinted with permission from Ref. [641].) (See Figure 22 in p. 533 of Chapter 15).

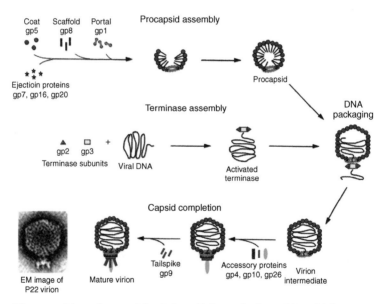

Plate 30   The assembly pathway of the *Salmonella* bacteriophage P22, which serves as a model for the assembly of many icosahedral dsDNA viruses [38]. The initial step (upper left) is the coassembly of 420 copies of the coat protein (gp5), 12 copies of the portal protein (gp1), approximately 300 copies of the scaffolding protein (gp8), and a few copies of minor ejection proteins (gp7, gp16, and gp20) to form a precursor shell (procapsid) incorporating a unique portal vertex. Subsequently, DNA packaging through the portal is fueled by ATP hydrolysis under the control of the phage terminase (subunits gp2 and gp3) and is accompanied by release of scaffolding and shell maturation. The final steps involve closure of the portal channel by accessory proteins (gp4, gp10, and gp26) and attachment of six trimeric tailspike proteins (gp9). The inset at the lower left shows an electron micrograph of the mature P22 virion. (See Figure 1 in p. 555 of Chapter 16).

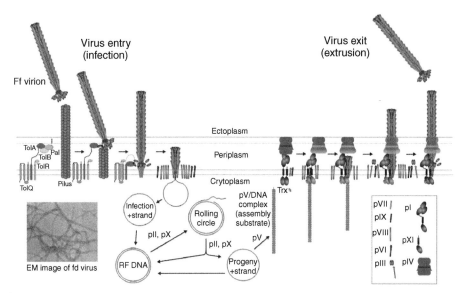

Plate 31 The life cycle and assembly pathway of the filamentous bacteriophage Ff, which infects *Escherichia coli* [39]. Essential features of membrane recognition and penetration are discussed in the text. The mature viral particle consists of a ssDNA genome in the form of a covalently closed circle that extends from one end of the capsid sheath to the other. The sheath comprises thousands of copies of the major coat protein (pVIII) plus a few copies of minor proteins at each filament end. The inset at the lower left shows an electron micrograph of the mature Ff virion. (See Figure 2 in p. 556 of Chapter 16).

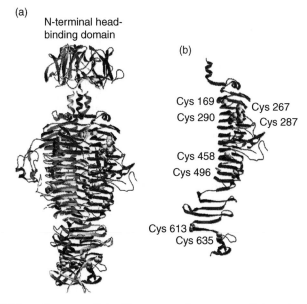

Plate 32 (a) Ribbon diagram of the X-ray crystal structure of the trimeric P22 tailspike protein (gp9), showing from top-to-bottom the N-terminal head-binding domain, the central parallel β-helix (or β-coil) domain, and the C-terminal β-prism domain [134–136]. (b) The β-helix and β-prism domains of a single gp9 subunit, showing the locations of the eight cysteines with space-filling atoms [42]. (See Figure 14 in p. 581 of chapter 16).

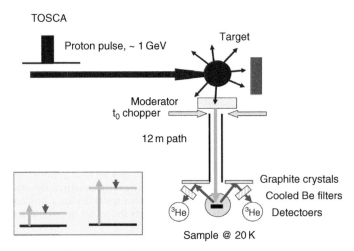

Plate 33 Schematic illustration of an inverse geometry spectrometer such as TFXA or TOSCA at ISIS or VISION under development for SNS. A pulse of high-energy protons (blue arrrow) strikes a tantalum target producing a burst of high-energy neutrons. These are moderated by collisions in a water moderator (yellow). The partially thermalized neutrons (green) pass through a t_0 chopper, which opens slightly after the neutron burst to block all very high energy neutrons. After a 12 m flight path, this full distribution of neutrons strikes the sample (black) in a cryostat at 20 K (blue gray). The neutrons that are scattered backward at 135° (red) are diffracted by graphite crystals and passed through cold Be filters. Only neutrons with energy of ~1.25 meV are passed to the detectors. Two backscattering crystal and filter detectors are shown in this diagram as was the configuration for TFXA. TOSCA now has 10 backscattering detectors arranged on a circle. Another set of 10 detectors utilizes neutrons scattered in the forward direction at 45°. The detectors contain ^{3}He at an elevated pressure such that >90% of the neutrons are absorbed. The subsequent relaxation of the excited ^{4}He* nucleus produces ionizing radiation that is detected by a breakdown production of photons and a photomultiplier. The energy-level diagram in the yellow box shows how low (high) energy incident neutrons can excite low (high) frequency vibrations to produce the same final energy neutrons (red). Time of flight analysis of the arrival time of the detector events determines the initial energy and thus the vibrational energy as the final energy is fixed. (See Figure 3 in p. 606 of Chapter 17).

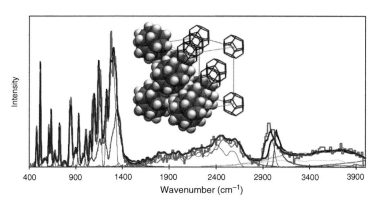

Plate 34 The inelastic neutron scattering (INS) spectrum of crystalline dodecahedrane at 20 K compared with the results of a periodic Density Functional Theory (DFT) treatment. The face centered cubic crystal structure is also shown. The experimental results are shown in red. The heavy black curve is the summation of the calculated intensities of all fundamental and overtone contributions including the phonon wings using the program aCLIMAX [40]. The experimental phonon part of the spectrum has been used to generate the phonon wings. The heavy blue curve is the same except that the calculated frequency of all of the CH stretching modes have been uniformly scaled down by a factor of 0.98 as determined from the Raman spectrum in order to correct for anharmonicity. In the region from 1500 to 2800 cm^{-1}, the intensity is due entirely to binary combinations and overtones and phonon-wings added to these. From 3300 cm^{-1} to higher frequency, the intensity is due to ternary overtones and combinations. Adapted from Ref. [17]. (See Figure 5 in p. 609 of Chapter 17).

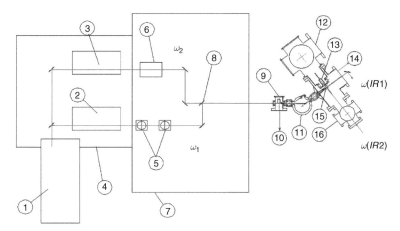

Plate 35 Schematic diagram showing the basic components of the vacuum ultraviolet (VUV)-laser photoion–photoelectron apparatus. (1) Injection seeded Q-switched Nd:YAG laser (550 mJ at 532 nm and 355 mJ at 355 nm, pulse duration of 5 ns); (2) dye laser 1; (3) dye laser 2; (4) 4′ × 4′ optical table; (5) BBO crystals (2 units); (6) optional frequency doubler; (7) 6′ × 8′ optical table; (8) dichroic mirror; (9) nonlinear frequency mixing chamber equipped with pulsed supersonic jet; (10) turbomolecular pump; (11) VUV monochromator; (12) beam source chamber; (13) pulsed valve; (14) photoionization chamber; (15) photoionization region. The IR laser can enter the photoionization region in either counter propagating the VUV laser beam via ω (IR1) or counter propagating the molecular beam via ω (IR2) [22]. (See Figure 1 in p. 663 of Chapter 19).

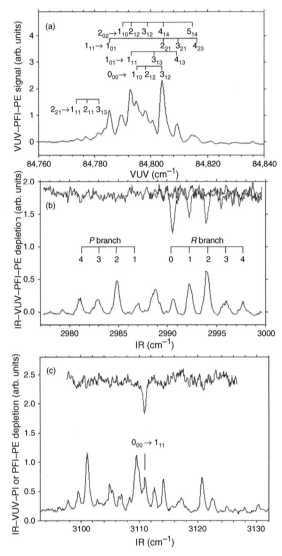

Plate 36 (a) VUV–PFI–PE spectrum for the origin band of $C_2H_4^+$ in the region of 94,760–84,840 cm^{-1}. (a) Comparison of the IR–VUV–PI spectrum for the ν_{11} band of C_2H_4 obtained using the low-resolution IR-OPO laser with the IR–VUV–PFI–PE depletion spectra for C_2H_4 in the region of 2977–3000 cm^{-1}. The IR–VUV–PFI–PE spectra marked in blue and red are obtained by setting the VUV laser at the strongest and the second strongest peaks of the VUV–PFI–PE spectrum of (a), which correspond to the photoionization transitions of $0_{00} \rightarrow 3_{12}$ and $1_{01} \rightarrow 1_{11}$ and $2_{02} \rightarrow 2_{12}$, respectively. (c) Comparison of the IR–VUV–PI spectrum for the ν_9 band of C_2H_4 obtained using the low-resolution IR-OPO laser with the IR-VUV-PFI-PE depletion spectrum (blue curve) for C_2H_4 in the region of 3096–3126 cm^{-1} measured by setting the VUV laser frequency at the strongest peak of the VUV–PFI–PE spectrum of (a), which corresponds to the photoionization transition $0_{00} \rightarrow 3_{12}$ [34]. (See Figure 16 in p. 685 of Chapter 19).

Printed and bound by CPI Group (UK) Ltd, Croydon, CR0 4YY

08/05/2025

01864814-0001